Nature-Inspired Computing
Physics- and Chemistry-Based Algorithms

Nazmul Siddique
Ulster University, UK

Hojjat Adeli
The Ohio State University, USA

CRC Press
Taylor & Francis Group
Boca Raton London New York

CRC Press is an imprint of the
Taylor & Francis Group, an **informa** business

CRC Press
Taylor & Francis Group
6000 Broken Sound Parkway NW, Suite 300
Boca Raton, FL 33487-2742

© 2017 by Taylor & Francis Group, LLC
CRC Press is an imprint of Taylor & Francis Group, an Informa business

No claim to original U.S. Government works

Printed on acid-free paper

International Standard Book Number-13: 978-1-4822-4482-3 (Hardback)

Library of Congress Cataloging-in-Publication Data

Names: Siddique, Nazmul, author. | Adeli, Hojjat, 1950- author.
Title: Nature-inspired computing : physics and chemistry-based algorithms /
Nazmul Siddique, Hojjat Adeli.
Description: Boca Raton : CRC Press, 2017. | Includes bibliographical
references and index.
Identifiers: LCCN 2016040264 | ISBN 9781482244823 (hardback : acid-free paper)
Subjects: LCSH: Biomathematics. | Computational intelligence.
Classification: LCC QH323.5 .S49 2017 | DDC 570.1/51--dc23
LC record available at https://lccn.loc.gov/2016040264

Visit the Taylor & Francis Web site at
http://www.taylorandfrancis.com

and the CRC Press Web site at
http://www.crcpress.com

To Kaniz, Oyndrilla, Opala, and Orla

–Nazmul

To Nahid, Amir, Anahita, Cyrus, and Mona

–Hojjat

Contents

Foreword

Some might argue that science began the first time someone started to ask why the world is the way it is, or perhaps pondered how to predict how the world might be in the future. If so, then science is many thousands of years old, at least as ancient as the twelfth century. Science is the tool that lifted mankind from complete ignorance—from witch doctors to sequencing entire genomes—from believing that dragons were eating the sun during an eclipse to understanding the physics needed to rocket to the moon and back. Science has helped us to understand our place in the universe. It has revealed how matter and energy relate, how space and time relate, and how order can come from chaos.

Basic science asks questions without regard for a specific application. It seeks knowledge for its own sake. But then comes applied science, which seeks to put what we know in practice and for practical benefit. In this book, Nazmul Siddique and Hojjat Adeli have looked to the principles found in our understanding of physics, chemistry, and biology and compiled a series of optimization algorithms that rely on these foundations. These algorithms can be used to find practical solutions to real-world engineering problems.

In the past 50 years, there has been increasing interest in modeling nature for problem solving. The term computational intelligence reflects a specific perspective that seeks inspiration from living systems to create algorithms or other tools to solve problems. Among these techniques are evolutionary algorithms, neural networks, fuzzy logic, swarm algorithms, and others. These are models based on our knowledge of biology. In parallel, there is interest in looking to physics and chemistry for other ideas. Siddique and Adeli offer numerous alternative approaches to optimization based on these ideas. In addition, they develop historical context for each algorithm and the reader will be able to determine if a particular approach may be of specific interest for his or her own problems.

The algorithms offered here certainly comprise a subset of all that may be eventually invented or discovered based on modeling nature or the physical world around us. The more science explains our world, the more opportunity we have for applying those explanations to other contexts. For now, I encourage you to study the opportunities that Siddique and Adeli present and consider how to expand upon our knowledge to inspire novel approaches in the future.

David B. Fogel
President, Natural Selection, Inc.
San Diego, California, USA

Preface

The very simple question and the most unresolved problem hitherto in the history of science is the origin of life. There are many hypotheses on the origin of life. Many philosophers believed that the origin of life was the result of a supernatural event. Many scientists believed that life arose spontaneously from nonliving matter in short period of time. The excellent scientific evidence behind the emergence of man was discerned by Charles Darwin in his groundbreaking work, the *Origin of Species*, in 1859. He established that all species of life have evolved or descended over time from common ancestors. A modern rephrasing of his theory would be that organisms slowly evolve toward greater complexity from simple organisms through evolution or adaptation. Adaptation is a process resulting from the dialectics of nature, where species compete among themselves for survival and evolve to complex organisms.

In a very similar fashion, the most popular belief, even among many brilliant philosophers, is that the universe was created by a supernatural event. The question of whether the universe had a beginning in time was posed by philosopher Immanuel Kant in 1781. Much later in the history of science, Edwin Hubble made a landmark discovery in 1929 that the distant galaxies are moving away. This discovery changed our conventional belief and made us think that the universe is rather expanding. The Russian physicist Alexander Friedmann made the exact prediction some years before, in 1922. Extrapolating the expansion of the universe backward in time using general theory of relativity yields an infinite density and temperature at a finite time in the past. This is known as singularity theory or big-bang singularity. The big-bang theory was finally proved by Roger Penrose and Stephen Hawking in 1970. This big-bang singularity theory helps break down the general relativity and all the laws of physics. In other words, the big-bang singularity refers to the birth of the universe and explains the subsequent evolution of the universe. The theory became very popular with the publication of the book *A Brief History of Time: From the Big Bang to Black Hole* by Stephen Hawking in 1987. The evolution of the universe resulted from the dialectics of the forces, in which heavenly bodies in the universe such as moons, planets, suns, stars, galaxies, and nebular and stellar materials attract each other and this sets them in motion.

Behind all of the changes, adaptation, evolution, and the ultimate motion of all the bodies in the universe, it is the dialectics of nature that puts them, be it living or nonliving, in a process of development. In dialectics, phenomena are characterized by their abilities to change. Changes are initiated by contradiction. Contradiction is the sole basis of dialectics. Dialectics has so far been used by two thinkers, Aristotle and Hegel. The dialectics for the present-day natural sciences was introduced by German philosopher Friedrich Hegel (1770–1831). Inspired by Hegel, a popular natural philosophical movement began in the early nineteenth century. The concept of dialectics of nature was first applied by Friedrich Engels in the 1870s as a method of explaining the evolutionary processes occurring in nature and as a tool for analyzing science, the science of the historical development of human thought, society, their interconnection in general, and for revealing the metaphysical conceptions in natural sciences and the complex relationships among them. The two events, the origin of life (i.e., origin of species) and the origin of the universe (i.e., big-bang singularity), and their evolution over time are related in a metaphorical way, and only lend themselves to be explained with the help of the dialectics of nature.

The precise motions of the planets along with their moons in their destined orbits, the innumerable suns and solar systems bounded by outermost stellar rings of the Milky Way, motion of the stars and galaxies, supernova phenomena, creation of new stars, gravitational collapse of stars and black holes in the deep space of the universe, the interplay of gravitational, magnetic, and electrical forces are among many examples of the dialectics of nature. There are many wonders of chemistry. The great strength of a diamond, the most expensive stone and the hardest material comes from the tetragonal spatial structure of carbon. Plant leaves are green because of chlorophyll, a key component in photosynthesis, which lets plants use energy from sunlight to convert water and carbon dioxide into sugars. Fire comprising carbon dioxide, water vapor, oxygen, and nitrogen provides us heat and light. All these are simple examples of chemical reactions found in nature. There are also numerous examples from our lonely planet earth and the biological world with immense diversity of species. The swarm systems in nature are perhaps among the most charming things to observe. A flock of birds twisting in the evening light, v-shaped structure of the migrating geese, winter birds hunting for food, dancing of starlings in the evening light, ants marching for foraging, synchronized flashing of fireflies, and mound building by termites are some of the fascinating examples of swarm systems. But how do they produce such collective behavior without any central coordinator or leader? How do they communicate with each other? How does a bee which has found food tell other bees about the location of the food? How do the flocks of migrating geese maintain a v-shaped structure? How do even the fireflies know when to glow? Is there a central control or coordinator for the collective behaviors? The philosophers, scientists, and biologists have been trying over centuries to explain these phenomena. The visible phenomena appear to be the result of thousands of invisible, known, and unknown causes.

In the process of evolution whether an entity is living or nonliving, it has to satisfy a set of criteria to survive or move to the next stage of development. In a similar fashion, designing systems requires satisfying specific criteria in many domains of science, and engineering appeared due to design, economic, or material constraints. The problem of optimality then became one of the key issues in designing systems. In fact, the problem of optimality is a central issue in science, engineering, economy, and everyday life. The criterion of optimality (i.e., the functional) should be explicitly known *a priori* with sufficient information. The conditions of optimality only define local extrema. If the number of such extrema is large, the problem of finding the global extremum becomes a complex problem. Various clever algorithms, mathematical approaches, and derivative-based optimization techniques have been developed over the past many decades. These algorithms seem to work out well up to a certain level of complexity of the problems. As the complexity of problems grows, the efficiency and appeal of these algorithms or approaches become unsatisfactory or they even become inapplicable. On the other hand, very often these methods cannot be applied to a wide range of problems due to the fact that the functionals (or the objective functions) are not analytically treatable or are not available or demand high computational cost. Furthermore, many real-life optimization problems have constraints that either cannot be defined mathematically or are highly nonlinear implicit and discontinuous functions of design variables. Therefore, the search for unconventional algorithms and the identification of problems for which each of the methods (deterministic or stochastic) is suitable have to be pursued, together with an understanding of why and how they work or not. There arose the need for nontraditional approaches that are inspired by nature such as physical, chemical, and biological phenomena and the evolution observed in nature. Most of these phenomena have been well studied under different disciplines of natural sciences and can be well explained with the notion of dialectics of nature. These facts lead many researchers to apply nature-inspired techniques to optimization problems in engineering, economics, and science and technology that also enable searching a high-dimensional space.

The research interest for nature-inspired computing has grown considerably, exploring different phenomena observed in the nature and basic principles of physics, chemistry, and biology. The discipline has reached a mature stage, and the field has been well established. The inspiration for writing this book on *Nature-Inspired Computing* in three volumes came from the book *Machine Learning – Neural Networks, Genetic Algorithms, and Fuzzy Systems* (John Wiley and Sons, 1995) by Hojjat Adeli and S.L. Hung, which was the first treatise that presented the three main fields of computational intelligence in a single book. That led the authors first to write the book *Computational Intelligence – Synergies of Fuzzy Logic, Neural Networks, and Evolutionary Computing* (John Wiley and Sons, 2013). This endeavor is another attempt at investigation into various computational schemes inspired from nature, which are presented in this book with the development of a suitable framework and industrial applications.

Designed for final-year undergraduate, postgraduate, research students, and professionals, this book is written at a comprehensible level for those who have some basic knowledge in calculus and differential equation, and some exposure to optimization theory. Due to the focus on search and optimization, this book should be appropriate for electrical, control, civil, industrial and manufacturing engineering, business, and economics students as well as those of computer and information science. With the mathematical and programming references and applications in each chapter, this book is self-contained. It should also serve already-practicing scientists and researchers who intend to study the field of system science, natural computing, and optimization.

The final goal of the authors is adroit integration of three mainstream disciplines of the natural sciences and problem-solving paradigms: physics, chemistry, and biology. This volume is organized in ten chapters comprising methodologies and algorithms based on principles of physics and chemistry. It includes an introductory chapter, Chapter 1, on the brief history of the development of natural sciences, their impact on the development of science, engineering, technology, and industry and wider applications to the economy and the society. The limitations of the traditional approaches to computing and optimization have led to alternative approaches. Methods and algorithms based on the principles of physics with a wide range of application domains are presented in Chapters 2 through 8. Chapter 9 presents algorithms based on the principles of chemical reactions. Chapter 10 presents some relatively new algorithms based on the principles of physics and chemistry.

There are many texts, research monographs, and edited volumes published since the 1990s. They are all referenced in each chapter of this volume, which a reader may find useful in further reading for their research. There is no single book that covers algorithms and methods based on the philosophies of physics, chemistry, and biology or their combinations. This volume provides a mixture of algorithms and methods based on the laws of physics and chemistry. The proposed book is not written for any specific course. However, it can be used for courses such as computational intelligence, evolutionary computing, and intelligent systems. As such, it is appropriate for beginners in the computational intelligence field and is also applicable as prescribed material for a final-year undergraduate course. This book is written based on many years of experience of following pedagogical features, with illustrations and step-by-step algorithms.

Nazmul Siddique
Londonderry, UK

Hojjat Adeli
Columbus, Ohio, USA

Acknowledgments

First of all, the authors would like to thank all the staff at the learning resource center of Magee Campus of Ulster University. The initial material was developed and used for different courses and invited talks. Useful feedback was received from many sources. This is a voluminous work in which many original works are explored and used. The authors acknowledge the immense contributions of numerous researchers whose works have been described and cited in this book. Special thanks are expressed to Martin Doherty who was a great help in advising on such matters.

The authors would like to thank many of their collaborators: Dr. Bala Amavasai, Professor Richard Mitchell, Dr. Michael O'Grady, Dr. Mourad Oussalah, Dr. Osman Tokhi, Dr. Takatoshi Okuno, Dr. David Fogel, Dr. Kathryne Merrick, Professor Mohammad Kaykobad, Professor Alamgir Hossain, Professor Ali Hessami, Professor Hydeuki Takagi, Professor Akira Ikuta, Professor Nikola Kazabov, Professor Matthias Riebisch, and Professor Bernard Widrow. The authors would like to thank all the staff of CRC Press associated with the publication of this book in three volumes, especially Randi Cohen and Jill Jurgensen for their support and help throughout the preparation of this book.

The first author would like to thank his wife Kaniz for her love and patience during the entire process and without which this book would have never been published, and to his daughters Oyndrilla, Opala, and Orla for their forbearance as well.

Authors

Nazmul Siddique is with the School of Computing and Intelligent Systems, Ulster University. He obtained Dipl.-Ing. degree in cybernetics from the Dresden University of Technology, Germany, MSc in computer science from Bangladesh University of Engineering and Technology and PhD in intelligent control from the Department of Automatic Control and Systems Engineering, University of Sheffield, England. His research interests include: cybernetics, computational intelligence, bio-inspired computing, stochastic systems, vehicular communication, and opportunistic networking. He has published over 150 research papers in the broad area of cybernetics, intelligent control, computational intelligence and robotics, including four books published by John Wiley, Springer, and Taylor & Francis (in press). He guest edited eight special issues of reputed journals and coedited seven conference proceedings on Cybernetic Intelligence, Computational Intelligence, Neural Networks, and Robotics. He is a Fellow of the Higher Education Academy, a senior member of IEEE, and member of different committees of IEEE SMC Society and UK-RI Chapter. He is on the editorial board of seven international journals.

Hojjat Adeli earned his PhD from Stanford University in 1976 at the age of 26. He is professor of civil, environmental, and geodetic engineering, and by courtesy a professor of biomedical engineering, biomedical informatics, electrical and computer engineering, neuroscience, and neurology at The Ohio State University. He has authored over 560 research and scientific publications in various fields of computer science, engineering, applied mathematics, and medicine including 16 books. In 1998 he received the Distinguished Scholar Award, from The Ohio State University "in recognition of extraordinary accomplishment in research and scholarship." He is the recipient of numerous other awards and honors such as The Ohio State University College of Engineering Lumley Outstanding Research Award (quadruple winner); Peter L. and Clara M. Scott Award for Excellence in Engineering Education, and Charles E. MacQuigg Outstanding Teaching Award, the 2012 IEEE-EMBS Outstanding Paper Award (IEEE Engineering in Medicine and Biology Society), a Special Medal from The Polish Neural Network Society "in Recognition of Outstanding Contribution to the Development of Computational Intelligence," the Eduardo Renato Caianiello Award for Excellence in Scientific Research from the Italian Society of Neural Networks, and an Honorary Doctorate from Vilnius Gediminas Technical University, Lithuania. In 2010 he was profiled as an Engineering Legend in the ASCE journal *Leadership and Management in Engineering* and Wiley-Blackwell, established the Hojjat Adeli Award for Innovation in Computing awarded annually. In 2011 World Scientific established the Hojjat Adeli Award for Outstanding Contribution in Neural Systems awarded annually. He is a Thompson Reuters Highly Cited Researcher (2014, 2015, and 2016) in two categories of computer science and engineering, a corresponding member of the Spanish Royal Engineering Society, a foreign member of Lithuanian Academy of Sciences, a Distinguished Member of ASCE, and a Fellow of AAAS, IEEE, AIMBE, and American Neurological Association.

1 Dialectics of Nature
Inspiration for Computing

Out of intense complexities intense simplicities emerge

Winston Churchill

1.1 INSPIRATION FROM NATURE

Nature has recourse at times to crucial measures and radical changes in its way forward, but never after any choice or intent of mankind (Le Bon, 1896). Nature does things in an amazing way and we cannot foresee the effects of their reciprocal influences. Behind the visible phenomena, there are innumerable invisible causes hidden at every time instant. Philosophers and scientists have been observing these phenomena in nature for centuries and trying to understand, explain, adapt, and replicate the artificial systems. There are innumerable agents and forces within the living and nonliving world, most of which are unknown and the underlying complexity of which is impossible to grasp as a whole. These agents are acting in parallel and very often against each other, giving forms, shapes, and features to nature, regulating the harmony, and balancing the ecosystem, beauty, and vigor of life. This is seen as the dialectics of nature, which lies in the concept of the evolution of the natural world and the emergence of new qualities with being at new stages of evolution. The evolution of complexity in nature follows a distinctive order. There is also information processing in nature performed in a distributed, self-organized, and optimal manner without any central control. This whole series of forms, mechanical, physical, chemical, biological, and social is distributed according to complexity from lower to higher. This sequence expresses their mutual dependence and relationship in terms of structure and history. The activities change due to changed circumstances. All these phenomena known or partially known so far are emerging as new fields of science, technology, and computing that study the problem-solving techniques inspired by nature as well as attempts to understand the underlying principles and mechanisms of natural, physical, chemical, or biological organisms that perform complex tasks in a befitting manner with limited resources and capability.

Science is the means of communication between scientists and nature (Prigogine, 1996), which itself has evolved over the centuries, getting enriched with new concepts, methods, and tools and developing into well-defined disciplines of scientific knowledge. Mankind has been trying to understand nature since time immemorial by developing new tools and techniques. The field of nature-inspired computing (NIC) is interdisciplinary in nature, combining computing science with knowledge from different branches of sciences, for example, physics, chemistry, biology, mathematics, and engineering, that allows development of new computational tools such as algorithms, hardware or wetware for problem solving, synthesis of patterns, behaviors, and organisms (Bounds, 1987; Brady, 1985; de Castro, 2007; Kari and Rozenberg, 2008; Packel and Traub, 1987). A very brief summary of these developments is presented in this chapter, which would be helpful in understanding different phenomena observed in nature, theories developed to explain those, and concepts behind nature-inspired approaches, methods, and algorithms presented in the rest of the chapters of the current volume.

1.2 BRIEF HISTORY OF NATURAL SCIENCES

The study of nature has its beginning in antiquity. The three mathematicians during the Hellenic age who became more well-known than most of their predecessors and successors of the time are Euclid of Alexandria, Archimedes of Syracuse, and Apollonius of Perga (Boyer, 1991). Antiquity had enlightened human knowledge and civilization with Euclidean geometry, Archimedes' lever and hydrostatic principles, Apollonius' conics, and the Ptolemaic solar system. In antiquity, the Arabs made some sporadic discoveries, which soon disappeared. Very little was achieved during the whole of the middle ages in Europe. The Arabs also made distinct contributions to decimal notation, beginning of algebra, modern numerals, and alchemy. The twelfth century Persian mathematician, astronomer, and geographer al-Khwarizmi (Latinized as Algoritmi) introduced the decimal point number system and first systematic solution of linear and quadratic equations and is considered the original inventor of algebra. The word algorithm originates from his Latin name, Algoritmi.

The most fundamental natural science, namely the mechanics of terrestrial and heavenly bodies, was developed in the fifteenth century in Europe. The basic features of many fundamental mathematical methods were developed in the sixteenth and seventeenth century such as logarithms by John Napier (1550–1617) (Hobson, 1914; see Napier, 1969), analytical geometry by Rene Descartes (1596–1650) (see Descartes, 1954; Forbes, 1977), and differential and integral calculus by Pierre de Fermat* (1601–1665) (Anderson, 1983), Sir Isaac Newton (1642–1727) (Cohen, 1971; see Newton, 1972), and Gottfried Wilhelm Leibnitz (1646–1716) (Child, 1920; see Leibnitz, 1975). In the eighteenth century, Europe saw a number of mathematical prodigies such as Leonhard Euler (1707–1783), who made significant contribution to the development of mathematics (Grattan-Guinness, 1971); Joseph-Louis Lagrange (1736–1813), who made tremendous contribution to the development of analytical mechanics, calculus of variations, and celestial mechanics (Fraser, 1983; see Lagrange, 1901); Pierre-Simon Laplace (1749–1827), who contributed to celestial mechanics (see Laplace, 1966); Legendre (1752–1833), who contributed to geometry, differential equations, and theory of function and numbers (Boyer, 1991); and Jean-Baptiste Joseph Fourier (1768–1830), who contributed to mathematical physics (Grattan-Guinness, 1972; Jourdian, 1912).

Nicholas Copernicus (1473–1543) established the first consistent theory of the heliocentric solar system, placing the sun in the center (Hallyn, 1993); Johannes Kepler (1571–1630) contributed to pure planetary kinematics and described the laws of planetary motions without reference to governing forces (Brasch, 1931; Hallyn, 1993). Galileo Galilei (1564–1642), well known for his observations, introduced the concepts of modern dynamics (Drake, 1973; see Galilei, 1960) and Isaac Newton formulated the general laws of motion and the relation between the forces acting on a body and the motion of the body, that is, the law of inertia, the law of force, and the law of action and reaction. Galileo also made major progress in understanding the properties of natural motion and simple accelerated motion of earthly bodies. Galileo showed through experimentation that the earth's gravitational acceleration is independent of the body mass. During the sixteenth and seventeenth century, several other scientific contributions to the problem of earthly and celestial motion set the stage for Newton's gravitational theory. In fact, Galileo Galilei and Isaac Newton created the system for modern science over the centuries. These achievements have been remarkable in this era with the postulation of the major theories and explanations to physical phenomena in terms of basic laws of motion and gravitation. This gave birth to some trends in physics, especially the field that was later named mechanics. Physics had its first beginning and mathematics, especially the concepts of calculus, played the most important role in the development of modern science.

---

* Pierre Laplace said about Fermat "Fermat—the true inventor of differential calculus."

1.2.1 LAWS OF MOTION

Isaac Newton (1643–1727) published his three laws of motion in his book *Mathematical Principles of Natural Philosophy* in 1687 (see Newton, 1999). The three physical laws laid the foundation of classical mechanics, which describe the relationship between a body and the forces acting upon it, and its motion in response to said forces. The laws cover the law of inertia, the law of force, and the law of action and reaction. The first law of inertia states that if a body is at rest or moving at a constant speed in a straight line, it will remain at rest or keep moving at constant speed in the same direction. The second law of force states that the acceleration a (also defined as the rate of change of velocity) is directly proportional to applied force F and inversely proportional to mass m of the body. The third law also referred to as the law of action and reaction states that every action has an equal and opposite reaction.

1.2.2 LAW OF GRAVITATION

Newton published the law of universal gravitation in the year 1687 in his famous book *Mathematical Principles of Natural Philosophy.* According to the law of universal gravitation, every body in the universe attracts every other body. The force of attraction between two bodies is directly proportional to the product of their masses and inversely proportional to the square of the distance between them. The modern quantitative science of gravitation began with Newton's law of universal gravitation. The law of gravitation has been expressed, explained, and applied by many researchers in many fields over the last three centuries until Einstein's work in the early twentieth century. Newton's theory is of sufficient accuracy for most applications.

But there remained unanswered questions of what gave rise to the orbits of the planets. Many scientists believed that was purposiveness of the arrangements of nature. Isaac Newton explained it as the divine first impulse imposed by God. This belief dominated the first half of the nineteenth century. It was the German philosopher Immanuel Kant (1724–1804) who, in his book published in 1755 on *"Allgemeine Naturgeschichte und Theorie des Himmels,"* laid out the nebular hypothesis (see Kant, 1781). The nebular hypothesis states that the solar system formed from a large cloud of gas called nebula. Kant's discovery was a point of departure from all previous theories. Years later, Pierre-Simon Laplace (1749–1827) provided a deeper foundation, which led to the discovery of glowing nebular masses and motion of the stars (Eisberg, 2003). His typical study was the conditions for equilibrium of a rotating fluid mass, a subject closely related to the nebular hypothesis of the origin of the solar system. According to Laplace's nebular theory, the solar system evolved from an incandescent gas rotating about an axis. The gas contracted as it cooled, causing rotation with higher speed according to the law of conservation of momentum. As the speed increased, the successive rings broke off from the outer edge to condense and form the planets. The rotating sun constitutes the remaining central force of the nebula. Laplace mathematically showed the development of a solar system from an individual nebular mass.

The suns and solar systems of the universe, bounded by the outermost stellar rings of the Milky Way, are developed by contraction and cooling from the spinning masses of vapor. The formation of suns and solar systems and the laws of motion provided an insight into proper motion of the stars. Astronomy came to accept the existence of dark bodies that are not merely planetary in nature. The existence of huge amounts of gaseous nebular patches that belong to the stellar system with suns or stars not yet formed provides sufficient clues to the existence of other distant independent solar systems at a relative stage of development. Thus, the discipline of astronomy, the oldest of all natural sciences, started flourishing on stronger mathematical and theoretical foundations.

The theory of astronomy, later professional astronomy often considered to be synonymous with astrophysics, suggests that a sufficiently massive object like a star must inevitably collapse under its own gravitational self-attraction upon exhausting its nuclear fuel. It expels part of its material in a supernova explosion, but the core of these collapsed stars ends in a dense state of matter called a neutron star or they collapse infinitely toward a singularity, which causes the properties of the surrounding space to change due to the enormous strong gravitational fields. The gravitational field is so strong that it attracts everything into itself and swallows them and even light cannot escape from it, making it dark and invisible to the outside world. Such black voids in space are known as black holes (BHs). The phenomenon of a BH was first announced by John Michell in 1783 (Michell, 1784; Schaffer, 1979). A similar phenomenon was confirmed by Pierre-Simon Laplace in 1795 independently of Michell's work (Gillispie, 2000).

In 1915, Einstein published his famous and revolutionary general theory of relativity (Einstein, 1915, 1916). Space and time were no longer separate and independent entities. Instead, they were just different directions in a single object called space-time. This space-time is not flat, but is deformed and curved by the matter and energy in it. Thus, the idea of BH was rediscovered in 1916. Karl Schwarzschild (1873–1916) then solved Einstein's equations for the case of a BH (Schwarzschild, 1916), which he envisioned as a spherical volume of warped space surrounding a concentrated mass and completely invisible to the outside world. In the 1930s, Chandrasekhar worked out that a star 1.4 times the mass of the sun cannot support against its own gravity and collapses, which becomes a BH (Chandrasekhar, 1931a,b). A similar discovery was made by the Russian scientist Lev Davidovich Landau (Landau, 1932).

In 1939, Robert Oppenheimer (1904–1967) and Hartland Snyder (1913–1962) published the first detailed treatment of a gravitational collapse using Einstein's theory of gravitation (Oppenheimer and Snyder, 1939). This led to the idea that such an object might be formed by the collapse of a massive star that leaves behind a small and dense remnant core. If the core's mass is more than about three times the mass of the sun, the equations showed that the force of gravity overwhelms all other forces and produces a BH. The term BH was itself coined in 1968 by the Princeton physicist John Wheeler, who worked out further details of a BH's properties. The most common BHs are probably formed by the collapse of massive stars. Larger BHs are thought to be formed by the sudden collapse or gradual accumulation of a mass of millions or billions of stars. Many galaxies are believed to have such super massive BHs at their centers. There is a BH at the center of our Milky Way galaxy, known as Sagittarius A*, which has a mass 4 million times that of the solar mass. Most astronomers believe that the BHs in the centers of galaxies grew by swallowing stars and nebular gas and emitting light in the process. But there is not enough light coming out from the BHs in active galaxies to explain their growth. There are hints that much of the growth occurred behind a veil of dust (Davies, 1978; White and Diaz, 2004). Astrophysical discoveries such as quasi-stellar objects, pulsars, and the apparent gravitational radiation pulses being received from the galactic center indicate that unusual astrophysical objects containing intense gravitational fields may exist in the universe.

The form of motion of matter prevailing at first is heat. The heat, under certain conditions at high temperatures, can then be transformed into electricity or magnetism. It is postulated that the mechanical motion taking place in the sun arises solely from the conflict of heat with gravity (Engels, 1934). The theory of gravitation not only has a huge impact on many scientific inventions in the later centuries but also many other nature-inspired conceptual frameworks with roots in gravitational kinematics, a branch of physics in the recent years that models the motion of masses moving under the influence of gravity. On the basis of the laws of gravitation and natural physical forces, a number of computing algorithms have been proposed by researchers.

1.2.3 Transformation between Heat and Mechanical Energy

Christiaan Huygens (1629–1695) introduced an important quantity, which later became known as energy (Andriesse, 2005; Cercignani, 1998). When a point mass subjected to a force move along a path in the direction of the force, it increases a quantity called energy, which depends on the mass (m) and the velocity (v). This energy is known as kinetic energy, defined by $E_k = \frac{1}{2}mv^2$. Descartes introduced the concept of momentum and maintained that this quantity was conserved on grounds of metaphysical speculations. It was Huygens who defined momentum as the product of mass and square of the corresponding velocity. Leibnitz argued that there is no indication that momentum is conserved in nature. He used the term kinetic energy and also the term *potential matrix,* which includes for the first time potential energy* (Cercignani, 1998). Leibnitz considered the principle of equality between cause and effect, which was generalized systematically by Johannes Bernoulli (1667–1748) and Leonhard Euler (1707–1783). Daniel Bernoulli (1700–1782) was the first to reestablish the connection between Galileo and Huygens by avoiding the metaphysical consideration.

Benjamin Thompson (1753–1814) first observed the frictional heat generated while he was boring cannon at the arsenal in Bavaria. He also found that there was enormous amount of mechanical energy required for boring the cannons. He also observed that a huge amount of energy was lost through production of heat. Thompson concluded that there should be a direct relationship between the loss of mechanical energy and the heat produced. He proposed to assume that heat is a form of energy, which is transformed from mechanical energy (Thompson, 1798). Thus, the principle of conservation of energy was established.

During the same time, the concept and the term "work" emerged in connection with the development of applied mathematics used by Lazare Carnot (1753–1823) (Gillespie, 1971). Meanwhile, the industrial revolution occurred in England with the invention of the steam engine by James Watt (1736–1819) (Dickinson, 1939). This was another manifestation of conversion of heat energy into mechanical force. William Robert Grove† (1811–1896) published his paper on the correlation of physical forces in 1846 where he anticipated the general theory of the conservation of energy (Grove, 1874). He also proved that all so-called physical forces, mechanical forces, heat, light, electricity, magnetism, and indeed so-called chemical forces become transformed into one another under a definite condition without loss of force. Thus, Grove concluded along the line of Descartes that the quantity of motion present in the world is constant. Sadi Carnot‡ (1796–1832) argued that heat is the result of motion (see Carnot, 1824. He showed in his book published posthumously in 1878 that the motive power of heat is a universal function of temperature and proposed the Carnot process, which is the basic principle of thermodynamics. The basic point of the Carnot process is that at least two thermal sources are required at different temperatures to perform work and the maximum amount of work that can be performed is determined by the given temperatures. Carnot's work was also concerned with the fact that though energy can never be lost, it might be rendered unavailable. A debate emerged out at that time that the total heat must always be exactly proportional to the kinetic energy. Around the same time, Julius Robert Mayer (1814–1878) (Mayer, 1842) and James Prescott Joule (1818–1889) (Joule, 1850) made another milestone discovery in science by confirming the experimental proof of the transformation of heat into mechanical force and mechanical force into heat. It was not until the nineteenth century that the law of the conservation of energy was first recognized as a law of nature. Thus, the new discipline of thermodynamics was born.

The law of thermodynamics states that there are two kinds of energy in any physical system: mechanical and thermal. The sum may change because one performs work on the system or supplies heat to the system. The first law of thermodynamics states that the change in total energy equals the work performed plus the heat supplied to the system. This implies that there is a thermodynamic quantity "temperature," which alone decides the direction of heat transfer. The ideas of thermodynamics propagated very quickly among scientists of other disciplines. The ideas were extended to continuous media and the mechanics of continua developed earlier by Leonhard Euler and Augustin Louis Cauchy (1789–1857) (Truesdell, 1992). Hermann Ludwig Ferdinand von Helmholtz (1821–1894) showed that the conservation of energy, work in electrodynamics, and chemical thermodynamics are all based on the principle of mechanical thermodynamics (Helmholtz, 1871).

* Much later the term *potential energy* was introduced by physicist and engineer William Rankine (1820–1872) in the nineteenth century.
† Grove's book *The Correlation of Physical Forces* was published in 1846 based on the lectures he gave at the London Institute in January 1842.
‡ Sadi Carnot was a son of Lazare Carnot, who died of cholera at the age of 36.

By these interconnections and transitions of forces, it was possible to eliminate the existence of a number of physical forces that were commonly used those days. The theory was more famously put forward by Helmholtz on the conservation of force (*Über die Erhaltung der Kraft*) published the following year. The mass–energy equivalence theorem states that mass conservation is equivalent to total energy conservation, which is the first law of thermodynamics. The second law of thermodynamics is the equilibrium of entropy, which states that the entropy can never be decreased in a thermally isolated system (Waldram, 1987). Ludwig Boltzmann (1844–1906) introduced the statistical view of thermodynamics and the second law by defining entropy from a purely statistical standpoint. Boltzmann's entropy is a beautiful concept in physics, which is applicable to other disciplines and explanation of phenomena. In other words, entropy is a measure of disorder in the system (Waldram, 1987). The second law also states that there exists no process in which heat is transformed from a colder body to a warmer body without some other related change occurring at the same time.

The study of thermodynamics enriched the available tools, but another theory that of electromagnetism (EM) was still needed. Electric phenomena of attraction and repulsion that come out of friction are part of everyday experience, which were understood in terms of electric charges in the early seventeenth century. William Gilbert (1544–1603) attributed the electrification of a body to friction (Gilbert, 1600). Gilbert is credited for the term electricity used in modern times. The observations were organized in the form of a theory by Benjamin Franklin (1707–1790), which attributed "+" sign for a positive charge and "–" sign for a negative charge. Some materials are charged positively and some are negatively (Elliott, 1993; Inan and Inan, 1999). Charles-Augustin de Coulomb (1736–1806) performed experiments and demonstrated the law of electric force by employing his torsion balance in 1785, which was later known as Coulomb's law (Gillmor, 1971).

While the science of electrostatics had been progressing and fundamental laws were becoming clearer, Luigi Galvani (1737–1798) noticed that electrical charge produced by machines caused muscular contraction in a dissected frog (Marco, 1998). It was identified in further experiments that electric charge is transported by the metal from the nerve to the muscle causing the contraction. This simple fact inspired Alessandro Volta (1745–1827) to the invention of the electrical battery using chemicals in 1799, which was a milestone for this new field of electricity (Pancaldi, 2003). It was proved that electricity can be generated chemically. A relationship was thus established between physics and chemistry. This encouraged scientists to develop an atomistic view in chemistry. Heating effects caused by electrostatic discharges had been known for some time. There had been sporadic reports that iron objects become magnetized when struck by lightning. Hans Christial Oerstedt (1777–1851) was the first to observe that magnetic compass needle was influenced by currents in neighboring conductors in 1820 (Oerstedt, 1820). Around the same time, Andre Marie Ampere (1775–1836) showed that two parallel wires carrying current attracted each other if the current flowed in the same direction, while the wires repelled each other when the current flowed in opposite directions (Ampere, 1820). Jean-Baptiste Biot and Felix Savart in 1820 repeated Oersted's experiments and formulated a compact law of static magnetic fields generated by the current in a circuit, which later became known as the Biot–Savart law (Biot and Savart, 1820). The Biot–Savart law is the first quantitative analysis of this phenomenon and the most basic law of magnetostatics. In the following years, Ampere undertook the most thorough treatment of the subject, formulated this phenomenon in mathematical terms, and defined Ampere's law of magnetic force. He also postulated that magnetism itself was due to circulating current. Ampere's law founded the science of electric currents.

George Simon Ohm (1787–1854) investigated the conductivity of electricity and established a relationship between the potential difference, current flow, and resistance, published in 1827, which later became known as Ohm's law (Gupta, 1980). A definition of electric resistance of a wire was thus formalized. The relationship between electric currents and heat was investigated by James Prescott Joule (1818–1889) and formulated in its final form of theory in 1841 (Joule, 1841).

Many years after the establishment of Coulomb's law, the last seal to the experimental picture came from Michael Faraday (1791–1867) with a fundamental contribution to electromagnetic induction and diamagnetism (Faraday, 1855). Faraday understood electric and magnetic fields in terms of lines of force and concluded that an electric field is essentially a field of force. Faraday set up an experiment and deduced that a magnetic field could generate a steady current. These experiments enabled Faraday to formulate his famous law of electromagnetic induction widely known as Faraday's law. Faraday's law showed that steady current can be generated by a changing magnetic field, where mechanical energy is applied to change the magnetic field producing induced voltage. Thus, the world's first direct-current generator was invented. Wilhelm Weber (1804–1891) perfected the theory of electrodynamics such that all the electromagnetic phenomena known up to that point of time, that is, Coulomb's law, Ameper's law, and Faraday's law, could be explained (Hunt, 2003; Weber, 1848).

Faraday was not well-versed in the advanced mathematics of the time. It was James Clark Maxwell (1831–1879) who took the work of Faraday and others, and formulated them in a set of equations that came to be known as the complete classical theory of electromagnetism and considered as the axioms of electrodynamics (Maxwell, 1865). He described both electric and magnetic fields and the laws of their mutual changes in the mathematical form of partial differential equations. He showed the changes of differentials in several dimensions with respect to their neighbors. Electrodynamics thus proved to be local action theory. He predicted the existence of electromagnetic waves and inferred the electromagnetic nature of light (Maxwell, 1892). The existence of electromagnetic waves was experimentally confirmed 23 years later by Heinrich Rudolf Hertz (1857–1894) in 1887 (Hertz, 1888). Electrical engineering emerged as a discipline of study.

1.2.4 TRANSFORMATION BETWEEN MASS AND ENERGY

At the beginning of the twentieth century, Max Planck (1854–1947) introduced the quantum theory, which revolutionized human understanding of atomic and subatomic processes (Planck, 1900a,b). Later, Albert Einstein (1879–1955) extended the quantum theory to light. In the year 1905, known as *Annus Mirabilis*, Albert Einstein published four groundbreaking articles in the scientific journal *Annalen der Physik*. They pertain to the photoelectric effect that gave rise to quantum theory, Brownian motion, the special theory of relativity, and mass–energy equivalence ($E = mc^2$). The paper on electrodynamics of moving bodies resolved Maxwell's equations for electricity and magnetism with the laws of mechanics. It introduced major changes to mechanics close to the speed of light. This later became known as Einstein's special theory of relativity, which has further implications that include the time–space frame of a moving body appearing to slow down and contract. In the paper on mass–energy equivalence, Einstein proved $E = mc^2$ by applying the theory of special relativity. Einstein's 1905 work on relativity remained controversial for many years. The paper on the motion of small particles suspended in a stationary liquid explained the empirical evidence of atomic theory, known as Brownian motion, which has implications on statistical physics. The paper on the heuristic viewpoint of production and transformation of light, known as photoelectric effect, suggested that energy is exchanged only in discrete amounts or quanta, which has pivotal influence on the development of quantum theory. These four works contributed substantially to the foundation of modern physics and changed views on space, time, and matter.

It is now evident from the study of all sciences that energy is convertible. Motion had been linked to heat. Heat is linked to another form of motion—wave motion or light. Chemical change had been converted into current and back, current into magnetism and back, and heat into electricity and back, and electric current and chemical reaction into heat. In many cases, heat is the final form of energy.

There are three fundamental questions in physics. How did the universe begin? Does time have a beginning and an end? Does space have edges? Einstein's theory of relativity replied to these very old questions. According to Einstein's theory of relativity, space, time, and matter are related together. The universe is expanding from a big bang, BHs do distort space and time, and the dark energy could be pulling space apart, sending the galaxies forever beyond the edge of the visible universe (White and Diaz, 2004).

Most scientists believe that all the matter, energy, and space in the universe were once squeezed into an infinitesimally small volume. There was a cataclysmic explosion, which created the universe. Space, time, energy, and matter all came into being at an infinitely dense and infinitely hot gravitational singularity, and began expanding at once. The explosion became known as the big bang, which is considered to be the theory behind the birth of the universe. The big bang was neither big (the universe was incomparably smaller than the size of a proton), nor a bang (it was more of a snap or a sudden inflation). It is also believed that since the big bang, the universe is expanding. However, scientists also believe that this expansion will not continue forever. All matter will collapse into the biggest BH pulling everything within it, which is referred to as Big Crunch.

The big bang theory was finally proved by Stephen Hawking and Roger Penrose in 1970 (Hawking and Penrose, 1970). In other words, the big bang singularity refers to the birth of the universe and explains the subsequent evolution of the universe. The theory became very popular with the publication of the book *A Brief History of Time: From the Big Bang to Black Hole* by Stephen Hawking in 1987. The evolution of the universe resulted from the dialectics of the forces where the heavenly bodies in the universe such as moons, planets, suns, stars, galaxies, and nebular and stellar materials attract each other and this sets them in motion.

Galaxies are structures within the universe formed by the mutual gravitation of matter into bound systems of stars, stellar remnants, interstellar gas and dust, and dark matter (Hawking, 1998; Sandage, 1975). Galaxies range from dwarfs with just a few thousand stars to one hundred trillion stars. They are all orbiting around their galaxy's own center of mass. Our galaxy has hundreds of billions of stars, enough gas and dust to make billions of more stars, and at least 10 times as much dark matter as the sum of all the stars and gas together, which are all held together by gravity. The modern concept of the universe dates back to 1924, when Edwin Hubble showed that there are many other galaxies in the universe (Hubble, 1929). To date, approximately 170 billion galaxies are known in the observable universe. The space between galaxies is filled with interstellar gas and dust and dark matter. The majority of galaxies are gravitationally organized into associations known as galaxy groups, clusters, and super-clusters. Galaxies can be categorized according to their visual morphology into three groups: elliptical, spiral, and irregular (Sandage, 1975). Elliptical-shaped galaxies are early-type galaxies and exhibit a larger range of mass and less angular variation in brightness. A spiral galaxy is a certain kind of galaxy originally described by Hubble (1936). The characteristic beauty of this type of galaxy is its spiral structure. The spiral arms are sites of ongoing star formation and are brighter than their surroundings.

Albert Einstein determined that the laws of physics are the same for all nonaccelerating observers, and that the speed of light in a vacuum is independent of the motion of all observers. This theory is known as the special theory of relativity published in 1905 (Einstein, 1905). The special theory of relativity introduced a new framework and the concept of space and time in physics. Einstein worked for 10 years for the development of the theory of general relativity and presented his work to the Prussian Academy of Science in 1915 of what are now known as the Einstein field equations. These equations specify how the geometry of space and time is influenced by whatever matter and radiation are present, and form the core of Einstein's general theory of relativity. According to this theory, a massive object (e.g., an asteroid) causes curvature in the geometry of the space-time due to the gravitational field. This is felt as gravity (Einstein, 1915, 1916).

Einstein's general theory of relativity is the geometrical theory published during 1915–1916 (Einstein, 1915, 1916), which provides explanations to the distortion of space-time by mass, energy, and momentum. According to the general theory of relativity (also known as Einstein's equivalence principle), the curved geometry of space-time is seen as the geometrical distribution of gravity (Friedmann, 1922). The gravitational force can be observed by two reference frames: in a space free from gravitational fields and with uniform acceleration. These two frames are physically equivalent. In physics, gravitational waves are ripples in the curvature of space-time, which propagate as a wave outward from a moving object or system of objects transporting energy as gravitational radiation. Einstein (1915, 1918) first predicted the gravitational waves on the basis of his theory of general relativity (Einstein, 1916). Binary star[*] systems composed of white dwarfs[†], neutron stars[‡], or BHs[§] are sources of gravitational waves. Hulse and Taylor indirectly proved the existence of gravitational radiation in 1974. The measurement of the Hulse–Taylor binary system suggests that gravitational waves are more than mathematical anomalies. Stars orbiting each other are called a binary star system (also called pulsar such as the Hulse–Taylor binary star system). Hulse–Taylor also showed that the changing orbit of the binary star system matched with the loss of energy due to giving off gravitational radiation from the binary star system (Weisberg and Taylor, 2004). It is proved that gravitational radiation can occur only for an accelerating object. This means that gravitational radiation cannot occur for a static object or for a nonspinning object at constant velocity. Two general cases are considered for gravitational radiation: first, two or more stars orbiting each other and, second, an isolated nonaxis-symmetrical supernova[¶] expanding space. Since the strength of gravitational radiation depends on the mass of the star, the orbiting massive star and its companion in the binary star system will generate strong gravitational radiation. Hulse–Taylor detected pulsed radio emissions from a pulsar in a binary star system. The recent observation of gravitational waves was reported by NASA in 2016 (Abbott et al., 2016).

1.2.5 LIGHT AND OPTICS

The idea that light emitted from a source is reflected from an object and enters into the eye to create a sense of vision was given by Epicurus of Greece. Ptolemy in Alexandria measured the angles of incidence and of refraction of the light ray going from one transparent medium to another. He correctly deduced that the ray is bent toward the normal. Willebrord Snell (1580–1626) first established the so-called sine law that gives the index of refraction (Pledge, 1939). The index of refraction is a measure of the change in direction for light passing through a transparent medium. The laws of reflection and refraction were postulated by Pierre de Fermat (Pledge, 1939). It was stated that a ray of light traverses in a straight line taking the minimum time. While Isaac Newton supported the corpuscular theory, others supported wave theory of light in which light propagates through a universal medium called ether. The ether theory became void when James Maxwell proposed the electromagnetic theory of light. Maxwell's wave theory was based on a continuous medium. Max Planck demonstrated it is necessary to postulate that radiant-heat energy is emitted in quanta. In 1905, Einstein showed the wave–particle duality in the nature of light by the theory of the photoelectric effect (Einstein, 1905).

1.2.6 SOUND AND ACOUSTICS

Sounds are natural phenomena produced by various means. To perceive sounds created in the environment, the hearing organ in animal species has evolved, which is one of the most crucial means of survival for them. Speech, a form of sound, is one of the most distinctive features of humans.

---

[*] A binary star is a star system consisting of two stars orbiting around their common center of mass.

[†] A white dwarf is a stellar remnant composed mostly of electron-degenerate matter and very dense. A white dwarf's mass is comparable to that of the Sun with a volume comparable to that of the Earth.

[‡] A neutron star is a stellar remnant that results from the gravitational collapse of a massive star. Neutron stars are the densest and the smallest stars known to exist in the universe. They can have a radius of only about 12–13 km and a mass of about two times that of the Sun.

[§] A BH is a mathematically defined region of space-time exhibiting such a strong gravitational pull that no particle or electromagnetic radiation can escape from it.

[¶] A supernova is a stellar explosion that briefly outshines an entire galaxy. Supernovae can be triggered by a sudden re-ignition of nuclear fusion in a degenerate star or by the gravitational collapse of the core of a massive star. A supernova radiates as much energy as the Sun or any ordinary star is expected to emit over its entire lifespan. It then fades away from view within several weeks or months (Giacobbe, 2005).

Sound can propagate in air, liquids, and solids. The branch of physics that deals with the understanding of sound is called acoustics. Pythagoras (sixth century BC) showed that the pitch of a musical sound depends on the frequency of vibration of the sound-producing object. A fundamental feature of regular sounds, for example, speech, music, and so on, is their frequency (i.e., cycles per second) expressed in Hz (i.e., Hertz). Sound less than 20 Hz is called infrasonic, and sound greater than 20 kHz is called ultrasonic. Normal human ear can hear sound within the interval of 20 Hz to 20 kHz. Marcus Vitruvius Pollio (80–15 BC), a Roman architect, is believed to have understood the wave concept of sound. Little was reported in the middle ages except attempts for measurements of the velocity of sound. Isaac Newton suggested the theoretical measure of the velocity of sound and proposed that the velocity of sound in air should be the square root of the ratio of atmospheric pressure to density. Newton's formulation of sound propagation in air was based on constant temperature. Pierre-Simon Laplace gave a correction to Newton's formulation, assuming that compressions and rarefactions in sound propagation in air do not take place at constant temperature and proposed that the ratio of pressure to density should be multiplied by the specific heat. Jean Le Rond d'Alembert (1717–1783) proposed wave motion in 1747 (d'Alembert, 1747) on the basis of the mathematical expression developed by Leonhard Euler (1707–1783) (Euler, 1750) and later developed by Joseph-Louis Lagrange (1736–1813) (Lagrange, 1760). The development of the necessary mathematical tools for analyzing sound waves by Leonhard Euler, d'Alembert and Lagrange contributed to the understanding of the vibrations of major components of musical instruments such as strings, rods, membranes, plates, and organ pipes. Sound spreads across many facets of human life and society, for example, music, architecture, and industry. In music, the harmony of sounds is an important feature, which means playing multiple notes simultaneously or playing notes in a particular sequence that makes sounds pleasing to the ear. Harmony is not only a musical term; it also defines a relationship between musical tones and the perception of humans. Such relationships are expressible through elementary scientific investigations that lead to an optimal performance of the combination of the notes.

1.2.7 HYDROLOGY AND DYNAMICS

The concept of hydrologic cycle in antiquity, concerned with the waters of the earth, their circulation, distribution, and properties, was erroneous from a scientific point of view. The Greek philosopher Anaxagoras of Clazomenae (500–428 BC) possibly gave the first primitive concept of the hydrologic cycle (Koutsoyiannis and Angelakis, 2003). His concept was that the sun evaporates water from the sea into the atmosphere, which then falls as rain onto the earth. Rain water is deposited in underground reservoirs and feed the rivers to flow. An improved concept of the hydrologic cycle was given by Theophrastus (372–287 BC) (Koutsoyiannis and Angelakis, 2003). Further development was achieved by Roman architect and engineer Marcus Vitruvius (believed to have lived during the time of Jesus Christ) who conceived a theory which is accepted in the modern view (see Vitruvius, 1826). He suggested that ground water is derived from rain and snow through infiltration from the ground surface. The new concept of hydrologic cycle is based on the dynamical sequential system theory where the inputs are the rain and snowfall and ground water and the outputs are the evapotranspiration, infiltration, and run-off. The throughput is the water moving through the watersheds under the influence of gravitation toward the ocean. Applying system theoretic concepts, the hydrologic cycle can be better analyzed and its dynamic and optimal behavior can be better predicted. A number of approaches have been devised based on the principles of the hydrologic cycle such as the water drop algorithm (WDA) (Siddique and Adeli, 2014), river formation dynamics algorithm (RFDA), and the water cycle algorithm (WCA) (Siqueira et al., 2014).

The phenomena observed in particles, atoms, matter, solar systems, and galaxies seem to be similar and coherent—all in motion under the influence of gravitation and adhering to an order. The fundamental postulates of the dialectics of nature are based on the matter and its motion, on time and space, which seeks equilibrium of forces and energy to maintain stability. It appears that our lonely planet earth in the universe has been maintaining it ever since. These natural phenomena of gravitation, electrostatic, electromagnetic, thermodynamic, optical, and hydrodynamics can be seen as conditions of optimality in nature with its altruism. Researchers of different disciplines of science, engineering, and technology have been exploring these features to develop models, methods, algorithms, optimization, and computational procedures.

1.2.8 DEVELOPMENT IN CHEMISTRY

At the birth of the planet, the temperature was like that of the sun. Chemical compounds cannot form at such a high temperature. Everything was in a gaseous state. With progressive cooling, there occurred interplay of the physical forms of motion, which become transformed into one another. Chemical affinity started developing when the heat and motion reached a certain point during this process. The previously chemically indifferent elements became differentiated chemically one after another. They acquired chemical properties and formed compounds by combination with one another. These compounds changed continually with decreasing temperature, which affected differently not only each element but also each separate compound of the elements changing from the gaseous state into the liquid and then the solid state under the new conditions (Engels, 1934). Such chemical compounds are found everywhere, distributed in a random manner all over the planet earth.

A number of evidences proved that human civilizations started extracting chemical compounds by 1000 BC using different methods, for example, extracting metals from ores and making alloys like bronze, which eventually formed the basis of the various branches of chemistry. The history of chemistry dates back to the period of alchemy, which had been practiced for several millennia in various parts of the world. The alchemists considered four classical elements: fire, water, air, and earth. Johann Joachim Becher (1635–1682) published his book *Physica Subterranean* in 1667, which was the first mention of what became known as the phlogiston theory where he proposed three forms of earth by eliminating fire, water, and air from the classical element model and replacing them with three forms of earth: *terra lapidea*, *terra fluida*, and *terra pinguis* (Bowler, 2005). The element *terra pinguis* was oily, sulfurous, and had the properties that help combustion (Brock, 1993). Becher believed that *terra pinguis* was a key feature of combustion and was released when combustible substances were burned (Bowler, 2005). Later Georg Ernst Stahl proposed a variant of the theory and renamed Becher's *terra pinguis* to phlogiston in 1703. The theory probably had its greatest influence and dominated the chemistry of first half of the eighteenth century. Antoine Laurent Lavoisier (1743–1794) first disproved the phlogiston theory widely popular among scientists around that time and emancipated chemistry from alchemy (Bensaude-Vincent, 2003; see Lavoisier, 1787). The law of conservation of mass was also discovered by Lavoisier, which was of immense importance in the progress of alchemy. The law of conservation of mass states that the mass of a system must remain constant over time, that is, the mass can neither be created nor be destroyed. It may be changed from one form to another. John Dalton (1766–1844) first assumed the existence of elementary atoms (Dalton, 1808, 1810, 1827). He postulated that groups of atoms disassociate and then rejoin in new arrangements during chemical reactions. Atomic structure constrains different atoms to form groups in fixed ways as molecules. A molecule comprising several atoms is characterized by the atom type, bond, angle, and torsion, which is termed as molecular structure. With this information, a definition of chemical reaction can be provided.

In a chemical reaction, chemical bonds are broken by absorption of energy and new bonds are formed with the release of energy. A chemical reaction comprises different types of unimolecular and multimolecular elementary reactions, each of which releases or absorbs different level of energies. A reaction can be seen as a process of structural change in molecules. This definition includes all processes including the process of change of substance and change of forms. In that sense, freezing of water is also to be considered as a chemical reaction. On the basis of the exchange of energy, chemical reactions are classified into three types: exothermic (if, overall, energy is released), endothermic (if, overall, energy is absorbed), and athermic (if no energy is exchanged) (Waldram, 1987). In general, reactions require the introduction of energy in some form, for example, heat or light, from an external source. According to the theories of Max Planck and Albert Einstein, light is also energy comprising discrete particles called photons. It is now known that it is the energy relationship that accelerates or retards chemical reactions. The reason a chemical reaction occurs is attributed to entropy, a concept in thermodynamics, which is a measure of that energy and represents the measure of the disorder in the system. In a chemical reaction process, a population of reactants with a high energy level and unstable states (i.e., molecules of different structures) undergo a sequence of elementary chemical reactions, transform through different energy levels, and produce certain products (i.e., molecules of new structures) of low energy and stable states at the final stage. Above all of these, chemical reactions tend to achieve the equilibrium of energy, and an optimal or stable state. The process of the chemical reaction is seen as an optimization process.

1.2.9 DEVELOPMENT IN BIOLOGICAL SCIENCES

The most fundamental question and the least understood problem hitherto in science is the origin of life. Life introduces a degree of order, organization, and diversity on the earth which is not yet found elsewhere in the universe. There are many hypotheses on the origin of life. Many scientists and philosophers believed that the origin of life was a result of a supernatural event. Many believed that life arose spontaneously from nonliving matter in short periods of time.

In antiquity, Aristotle (384–322 BC) showed interest in zoology as fish was of extreme economic importance in Greece at the time. His view of classification was roughly near to that of the modern evolutionist. It ran from lower plants to higher plants through molluscs, arthropods, reptiles, birds, fishes, mammals, and men. His student Theophrastus (380–287 BC) did the same work in botany. The biological sciences emerged from traditions of medicine and natural history reaching back to Ayurveda, ancient Egyptian medicine, and the works of Aristotle and Galen (130–200 AD) in the ancient Greco-Roman world. In the middle ages, great work in biological science was done by the great Persian scholar Avicenna* (980–1037), which was later translated into Latin for Europeans (Pledge, 1939).

Biological science was revolutionized during the European Renaissance and by many discoveries in the early modern period. Cells are easiest to see in plants, which were discovered by Grew and Malpighi and thus cell theory started developing in the seventeenth century. Antonie Philips van Leeuwenhoek (1632–1723) became well known for his contribution to the development of the microscope (Pledge, 1939). He was able to see a thing much smaller than the cell. He was known for the discovery of the single-cell organism. Robert Brown's (1773–1858) contribution was the earliest description of the cell nucleus (Pledge, 1939).

The eighteenth century biologist was busy with collecting and sifting the vast material of botanical, zoological, anatomical, and physiological samples, and naming and classifying the dominating natural history. Scientists were busy with the essential investigation and comparison of various forms of life, geographical distribution, conditions of existence, and climate. In the years between 1750 and 1760, Carl Linnaeus (1707–1778) completed the vast classification tasks in botany and zoology (Linnaeus, 1758).

The efforts continued following improvements of the microscope in the early years of the nineteenth century. Matthias Jacob Schleiden (1804–1881) published his cell theory in 1838, applying it throughout the plant kingdom, which brought momentum in cell movement (Schleiden, 1838). Cells

---

* Ibn Sina was a Persian scholar and scientist known as Avicenna in the West. Among his famous works are the two books *The Book of Healing* and *The Canon of Medicine*, which became a standard medical text during the medieval age and as late as 1650. The book *Canon of Medicine* was reprinted in New York in 1973.

were seen as central to life. Theodor Schwann (1810–1882) announced the general theory of the cell as the basis of all life in 1839 (Schwann, 1839). Tissues (both plant and animal) are built up with various combinations of cells. The cell theory of Schleiden and Schwann had two important contributions to biological science: the first one was the microscopic creatures responsible for putrefaction, fermentation, and disease, and the second one was the cell division responsible for growth and reproduction. Jan Evangelista Purkyně (1787–1869) discovered the Purkinje cell in 1837 and coined the term protoplasm for the fluid substance in the cell (Pledge, 1939). Hugo von Mohl (1805–1872) distinguished the protoplasm from the sap of vegetable cells in 1846 and showed that the protoplasm is the source of those movements in cells (Pledge, 1939). This was the excitement of the time and attracted much attention. Thomas Huxley (1825–1895) referred to protoplasm as the physical basis of life, a view which considered that the properties of life resulted from the distribution of molecules within this substance (Harvey, 2004). Claude Barnard (1813–1878) insisted that all living creatures were bound by the same laws as inanimate matter. But he clearly gave the formulation that life is something more than the physical–chemical manifestations.

Jean-Baptiste Pierre Antoine de Monet, or Chevalier de Lamarck (1744–1829), also known as Lamarck, postulated that environmental influences such as climatic, nutritive, behavioral, and so on are transmitted to the next generation (Lamarck, 1809). Lamarck's contribution was the truly cohesive theory of evolution of inheritance of acquired characteristics. Gregor Johann Mendel (1822–1884) established many of the rules of heredity based on his pea plant experiments conducted between 1856 and 1863, which are now referred to as the laws of Mendelian inheritance (Mendel, 1866). The excellent scientific evidence behind this emergence of man was discerned by Charles Darwin (1809–1882) by publishing his groundbreaking work *The Origin of Species* in 1859 (Darwin, 1859). A modern rephrasing of his theory is that organisms slowly evolve toward greater complexity through adaptation and adaptation itself is a mechanism accomplished by natural selection. He established that all species of life have descended over time from common ancestors.

A genetic definition of life is said to be a system that is capable of evolution by natural selection, which means that a certain level of complexity cannot be achieved without adaptation via natural selection, which led to immense diversity of life and species on earth. Hence, it comes to a conclusion that man did not emerge on the bare planet but in an orderly and diverse biological world.

Much of the beauty and diversity of life on earth is due to sex. An asexual organism would need frequent mutations for adaptation and diversity to survive in the changing environment. The chance of an adaptation in an asexual organism then requires the mutations to wait for an accidental event, whereas sex solves this problem in an elegant way by reassorting the genetic material of the parents that produce completely new offspring with diverse combinations of genes. Some organisms developed more than two sexes. For example, paramecia have somewhere between 5 and 10 sexes and can combine their genetic material.

Energy must be supplied to living organisms. Organisms acquire this energy in two general ways. Some organisms (e.g., heterotrophs) acquire energy by controlled breakdown of organic molecules, that is, food, supplied by other organisms. Human and most animals are heterotrophs. Some organisms, known as autotrophs, acquire their energy from other sources, either from the energy of sunlight (in this case organisms are called photoautotrophs) or from controlled chemical reaction of inorganic materials (in this case organisms are called chemoautotrophs). Some organisms (e.g., photochemoautotrophs) acquire their energy from sunlight and chemical inorganic materials. These facts provide evidence for the direct relationship between energy and the very existence and survival of life. In our solar system, the sun is the largest of all providers of energy, which is not equally distributed over all the regions on the earth, causing different climatic and geographic conditions.

Georges-Louis Buffon (1707–1788) first noticed that the average patterns of distribution of the flora, fauna, and vegetation seemingly vary across geographic regions (Buffon, 1766). The present distribution patterns of plants and animals are the results of climatic and geographic conditions, the geological history of climates and landmasses, and the evolutionary history of the species. MacArthur and Wilson (1967) studied the geographical areas that are well suited as habitats for biological species. These geographical areas are said to have a high habitat suitability index (HSI). Features that correlate with the HSI include factors such as rainfall, diversity of vegetation, diversity of topographic features, land area, and temperature. The variables that characterize habitability are called suitability index variables (SIVs). SIVs can be considered the independent variables of the habitat, and the HSI the dependent variable (MacArthur and Wilson, 1967). MacArthur and Wilson (1967) tried to develop mathematical models of biogeography that describe the relationship between SIVs and the HSI. Biogeography studies the geographical distribution of biological species.

The twentieth century witnessed the remarkable achievements and discoveries of science such as the discovery of organic cells, the discovery of laws of conservation and transformation of energy, and Darwinian evolution that cover the major branches of natural science such as mathematics, mechanics, physics, chemistry, and biology. The major theories and hypothesis were established and new branches of natural science were brought into being. The philosophical significance of these discoveries was the exposure of the dialectical characteristics of nature and their development in a succinct form. The natural sciences had begun to flourish in the twentieth century in a remarkable way with applications in many dimensions for engineering, technology, industry, economy, society, and mankind overall. The activities of science have become multifarious and specialized as well as the scientific literature has become voluminous. Developing computing methods, algorithms, and ideas mimicking or imitating the behaviors or phenomena of nature for solving complex problems or achieving goal-oriented activities is termed as NIC (Shadbolt, 2004). Thus, nature-inspired concepts and ideas have great potential in developing methods, and algorithms to tackle engineering problems, especially those for which we lack adequate knowledge to design efficient solving methods.

1.3   TRADITIONAL APPROACHES TO SEARCH AND OPTIMIZATION

The aspiration of the preceding Section 1.2 was not only to look into the development of natural sciences over time but also to have some understanding of the laws of nature or, in other words, the philosophy of nature.* By nature, we mean all of the living and nonliving world, the planetary, galactic, stellar systems, and the heavenly bodies in the universe. One simple thing is clearly demonstrated in nature: be it physical, chemical, or biological, that nature maintains its equilibrium by any means known or unknown to us. A simplified explanation of the state of equilibrium possibly is the notion of optimum seeking in nature. There is optimum seeking in all spheres of life and nature (Adeli and Park, 1998; Arango et al., 2013; Chow, 2014). In all optimum seeking, there are goals or objectives to be achieved and constraints to be satisfied within which the optimum has to be found (Chen et al., 2014; Faturechi and Miller-Hooks, 2014; Jia et al., 2014). Eventually, the optimum seeking can be formulated as an optimization problem (Luo et al., 2013; Peng and Ouyang, 2014; Smith et al., 2014). That is, it is reduced to finding the best solution measured by a performance index often known as objective function in many areas of computing and engineering, which varies from problem to problem (Adeli, 1994; Adeli and Sarma, 2006; Aldwaik and Adeli, 2014; Gao and Zhang, 2013; Zhang and Wang, 2013). In general, a performance index can be described by

$$J(\theta) = \int_x Q(x,\theta) p(x) dx \qquad (1.1)$$

---

* In its German form, the term Naturphilosophie is chiefly identified with Friedrich Schelling and G. W. F. Hegel, both were early nineteenth century German philosophers.

where $Q(x,\theta)$ is the functional of the vector $\theta = (\theta_1, \theta_2, ..., \theta_N)$, which depends on the random sequence or process $x = (x_1, x_2, ..., x_N)$ with probability density function $p(x)$. The goal is to find the extremum of the functional $Q(x,\theta)$, that is, the minimum or the maximum depending on the problem. The expression in Equation 1.1 is generally known as the criterion of optimality. For the ease of application for certain problems, the criteria can also be defined based on the averaging of $Q(x,\theta)$ with respect to time depending on x. If x is a random sequence, that is, $x = \{x[n], n = 1, 2, ..., N\}$, then $J(\theta)$ is expressed as

$$J(\theta) = \lim_{N \to \infty} \frac{1}{N} \sum_{n=1}^{N} Q(x[n],\theta) \tag{1.2}$$

If x is a random process, that is, $x = \{x[t], 0 \le t < \infty\}$, then $J(\theta)$ is expressed as

$$J(\theta) = \lim_{T \to \infty} \frac{1}{T} \int_0^T Q(x[t],\theta)dt \tag{1.3}$$

The process or system for which the optimality is sought can be deterministic or stochastic in nature. For any system, be it deterministic or stochastic, the criterion of optimality (or the functional) $J(\theta)$ described by Equations 1.1 through 1.3 should be known *a priori* with sufficient information along with the constraints. If the functional $J(\theta)$ is differentiable, its extremum (i.e., maximum or minimum) can be obtained for the values of the parameter vector $\theta = (\theta_1, \theta_2, ..., \theta_N)$ when the partial derivatives $\partial J(\theta)/\partial \theta_v$, $v = 1, 2,..., N$ are simultaneously equal to zero; that is,

$$\nabla J(\theta) = \left(\frac{\partial J(\theta)}{\partial \theta_1}, \frac{\partial J(\theta)}{\partial \theta_2}, ..., \frac{\partial J(\theta)}{\partial \theta_N} \right) = 0 \tag{1.4}$$

The parameter vectors $\theta = (\theta_1, \theta_2, ..., \theta_N)$ which satisfy Equation 1.4 are called the stationary or singular vectors. The problem is that all stationary vectors are not optimal and they do not correspond to the desired extremum (i.e., solution) of the functional. Therefore, $\nabla J(\theta) = 0$ is only a necessary condition (Tsypkin, 1971). The sufficient conditions can be derived in the form of an inequality based on the determinant containing the partial derivatives of second order of the functional with respect to $\theta = (\theta_1, \theta_2, ..., \theta_N)$. However, it is not worth doing so, even in the case when the computation effort is not huge. If there is only one extremum, the stationary vector corresponding to the maximum or minimum can be found from the physical conditions of the problem. The conditions of optimality define only local extrema. Finding the global extremum becomes extremely difficult when the number of such extrema is large.

There are many methods reported in the literature for finding the unique optimal value of the vector θ^*. Gradient-based optimization techniques use derivative information in determining the search direction. A brief account of gradient calculations is provided in Appendix A. Among the gradient-based techniques, the steepest descent method (Knyazev and Lashuk, 2008) and Newton's method (Chong and Zak, 2008) are well-known methods. Conjugate gradient (Hestenes and Stiefel, 1952), Gauss–Newton (Hartley, 1961), quasi-Newton (Broyden, 1967), and Levenberg–Marquardt (More, 1977) are the well-known variants of these methods. There is no guarantee that a gradient-based descent algorithm will find the global optimum of a complex objective function within a finite time (Lin et al., 2012). The mathematical approaches, computational efforts, and difficulties thereof are discussed in the following sections, which will certainly lead to understanding the inevitability of NIC approaches to optimization problems.

1.3.1 LINE SEARCH

Very often, one-dimensional search methods apply a line search at every iteration in multidimensional optimization problems (Fletcher, 1987). If $f: R^n \rightarrow R$ is a function to be minimized, the iterative approach to minimize f can be defined as

$$x(k+1) = x(k) + \alpha_k d_k \qquad (1.5)$$

where $x(0)$ is a given initial point and $\alpha_k > 0$ is chosen to minimize $\phi_k(\alpha) = f(x(k) + \alpha d_k)$. The vector d_k is called the search direction. The first thing in line search method is to find a descent direction along which the objective function is to be reduced. It then computes a step size that determines how far it should move along that direction. The choice of α_k involves a one-dimensional minimization, which ensures under appropriate conditions defined below.

$$f[x(k+1)] < f[x(k)] \qquad (1.6)$$

The secant method (discussed in Section 1.3.5) can be used to find α_k. The descent direction d_k can be computed by various means such as gradient descent, Newton's method, and the quasi-Newton method. These methods are discussed later in this section. A good account of discussion on practical line search methods is provided in Fletcher (1987).

1.3.2 GOLDEN SECTION SEARCH

The Golden Section Search is a method of finding the minimum or maximum of a unimodal function within an interval by successively narrowing the range of values. The search algorithm is only applied when it is known that the minimum or maximum exists within that interval. A unimodal function contains only one minimum or maximum on the interval of $[a,b]$. Golden section search was developed by Kiefer (1953). Three points x_a, x_b, and x_c are chosen such that $x_a < x_b < x_c$, $f(x_b) > f(x_a)$, and $f(x_b) > f(x_c)$, where $f(x_a)$, $f(x_b)$, and $f(x_c)$ are the objective function values for x_a, x_b, and x_c, respectively, as shown in Figure 1.1. Therefore, the minimum or the maximum should lie between x_a and x_c. The fourth point x_d is chosen between the larger of the two intervals $[x_a,x_b]$ and $[x_b,x_c]$. It is assumed that $(x_c - x_b) > (x_b - x_a)$, that is, interval $[x_b,x_c]$ is larger. The new point x_d is chosen within the interval $[x_b,x_c]$. If $f(x_d) > f(x_b)$, then the three new points are $x_b < x_d < x_c$. If $f(x_d) < f(x_b)$, then the three new points are $x_d < x_b < x_c$. The process is repeated until the distance between the outer points is sufficiently small. The algorithm maintains the function values of three points

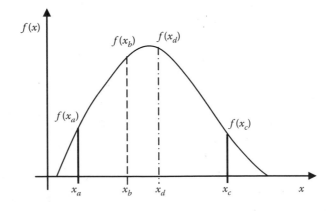

FIGURE 1.1   Cross section of the gutter.

forming a golden ratio (Bronshtein and Semendyyayev, 1998). The golden ratio is also known as the divine proportion, golden mean, or golden section. For a large number of function evaluations, the golden section search algorithm is the well-known Fibonacci search.

1.3.3 FIBONACCI SEARCH

Fibonacci search is a method for searching a sorted array by dividing the array into sub-arrays, which progressively narrows down possible locations with the aid of Fibonacci numbers. Fibonacci search was first introduced by Kiefer (1953) as a minimax search method for finding the maximum (or minimum) of a unimodal function. In binary search, the array is not split in the middle. Fibonacci search (Avriel and Wilde, 1966) is an elegant improvement to binary search where the search method splits the array corresponding to the Fibonacci number. A Fibonacci number is defined by

$$F_k = F_{k-1} + F_{k-2} \tag{1.7}$$

where $F_0 = 0$ and $F_1 = 1$ for $k \geq 2$. A sample of a Fibonacci sequence is shown in Figure 1.2.

If an array contains n elements where $n = F_k$, it will search at F_{k-1} first. If the sought value is greater than the found value, it will then search the sub-array from $F_{k-1} + 1$ to F_k. If the sought value is less than the found value, it will then search the sub-array from 1 to F_{k-1}. Interestingly, the length of the sub-arrays is also a Fibonacci number.

1.3.4 NEWTON'S METHOD

The problem is to find the minimum of the function f of a single real variable x. At each measurement of the point $x^{(k)}$, the values of $f(x^{(k)})$, $f'(x^{(k)})$, and $f''(x^{(k)})$ are available. A quadratic function can be fitted through $x^{(k)}$ that matches the first and second derivatives with that of the function f. The quadratic function has the form

$$q(x) = f(x^{(k)}) + f'(x^{(k)})(x - x^{(k)}) + \frac{1}{2}f''(x^{(k)})(x - x^{(k)})^2 \tag{1.8}$$

It is to be noted that $q(x) = f(x^{(k)})$, $q'(x) = f'(x^{(k)})$, and $q''(x) = f''(x^{(k)})$.

Newton's method for minimizing f uses second derivatives of f as follows:

$$x^{(k+1)} = x^{(k)} - \frac{f'(x^{(k)})}{f''(x^{(k)})} \tag{1.9}$$

Newton's method works well if $f''(x) > 0$ for all points (Chong and Zak, 2008).

1.3.5 SECANT METHOD

The secant method is a root finding algorithm. Succession of roots of secant lines is used to approximate the root of a function f. It is thought of as a finite difference approximation of Newton's method.

F_1	F_2	F_3	F_4	F_5	F_6	F_7	F_8
1	2	3	5	8	13	21	34

FIGURE 1.2 Sample of a Fibonacci sequence.

That is, Newton's method for minimizing f uses second derivatives of f as defined in Equation 1.9. If the second derivative is not available, then it can be approximated from the information of the first derivative, that is, $f''(x^{(k)})$ can be approximated by

$$\frac{f'(x^{(k)}) - f'(x^{(k-1)})}{x^{(k)} - x^{(k-1)}} \tag{1.10}$$

Using the approximation in Equation 1.10 of the second derivative, the following method is achieved:

$$x^{(k+1)} = x^{(k)} - \frac{x^{(k)} - x^{(k-1)}}{f'(x^{(k)}) - f'(x^{(k-1)})} f'(x^{(k)}) \tag{1.11}$$

The method described by Equation 1.11 is called the secant method. There is another equivalent form of the secant method described by

$$x^{(k+1)} = \frac{f'(x^{(k)})x^{(k-1)} - f'(x^{(k)})x^{(k)}}{f'(x^{(k)}) - f'(x^{(k-1)})} \tag{1.12}$$

The advantage of the secant method is that it requires only two initial points $x^{(-1)}$ and $x^{(0)}$. A good account of discussion on the secant method is provided in Press et al. (1992).

1.3.6   GRADIENT-BASED METHODS

Gradient-based optimization methods apply the derivative information of the objective function for determining the search direction. The preliminary concepts that sustain the descent algorithms are discussed for minimization problems in this section.

1.3.6.1   Descent Methods

To optimize a real-valued function, an objective function J is defined on an n-dimensional input space $x = [x_1, x_2, \ldots, x_n]^T$ in the descent method. Finding the minimum point $x = x^*$ that minimizes $J(x)$ is of primary concern. The next point $x(k + 1)$ of the iterative descent method is determined by a step down from the current point $x(k)$ in a direction vector d as follows:

$$x(k+1) = x(k) + \eta(k)d(k), \quad k = 1, 2, 3, \ldots \tag{1.13}$$

where η is a positive step size that regulates the direction of the next step and k represents the current iteration. $x(k)$ is expected to converge to local minimum x^* within the sequence of solution candidates $\{x(k)\}$. The computation of the k-th step requires determination of the direction d and the step size η. The next point $x(k + 1)$ must satisfy the following inequality:

$$J[x(k+1)] = J[x(k) + \eta d] < J[x(k)] \tag{1.14}$$

The optimum step size can be determined by

$$\eta^* = \arg \min_{\eta > 0} \phi(\eta) \tag{1.15}$$

where $\phi(\eta) = J[x(k) + \eta d]$. The search for η^* is accomplished by the line search mentioned earlier.

1.3.6.2   Gradient Methods

The straight downhill direction d can be determined based on the gradient of the objective function J. The gradient of a differentiable objective function $J: R^n \rightarrow R$ at x denoted as g (or $\nabla J(x)$) is given by a vector of the first derivatives of J as follows:

$$g(x) = \nabla J(x) \overset{\text{def}}{=} \left[\frac{\partial J(x)}{\partial x_1}, \frac{\partial J(x)}{\partial x_2}, \ldots, \frac{\partial J(x)}{\partial x_n} \right]^T \tag{1.16}$$

The direction d should satisfy the condition for feasible descent directions as follows:

$$\phi'(0) = \frac{dJ(x + \eta d)}{d\eta} \big|_{\eta=0} = g^T d = \left\| g^T \right\| \| d \| \cos(\xi(x)) < 0 \tag{1.17}$$

where $\xi(x)$ signifies the angle between g and d at point x.

A class of gradient-based descent methods has the following fundamental form, where the feasible descent directions can be determined by deflecting the gradients through multiplication by G:

$$\begin{aligned} x(k+1) &= x(k) - \eta Gg \\ x_{k+1} &= x_k - \eta Gg \end{aligned} \tag{1.18}$$

where η is some positive step size and G is some positive-definite matrix. If $d = -Gg$, the condition of descent direction in Equation 1.17 holds since $g^T g = -g^T Gg < 0$. There are many variants of gradient-based methods that have the similar form described in Equations 1.17 and 1.18 to bias the negative gradient direction $(-g)$.

1.3.6.3   Steepest Descent Method (or Gradient Descent)

Gradient descent is a first-order optimization algorithm. The basic idea of gradient descent algorithm is the use of steps proportional to the negative of the gradient (or of the approximate gradient) of the function at the current point that successively leads to the local minimum of the function defined on a multidimensional input space (Knyazev and Lashuk, 2008). If the positive of the gradient is used, it leads to the local maximum of the function. This simple procedure is known as gradient ascent or steepest descent (or the method of steepest descent) and is the most frequently used method due to its simplicity.

By setting $G = I$ with I as the identity matrix, Equation 1.18 becomes

$$\begin{aligned} x(k+1) &= x(k) - \eta g \\ x_{k+1} &= x_k - \eta g \end{aligned} \tag{1.19}$$

Equation 1.19 is the well-known gradient descent or the steepest descent method.

1.3.7   CLASSICAL NEWTON'S METHOD

The descent direction d can be determined using the second derivatives of the objective function J. If the starting point x_k is sufficiently close to a local minimum, the objective function J can be approximated by a quadratic function defined by

$$J(x) \approx J(x_k) + g^T(x - x_k) + \frac{1}{2}(x - x_k)^T H(x - x_k) \tag{1.20}$$

where H is the Hessian matrix consisting of the second derivatives of $J(x)$. The definition of the Hessian H of a function is provided in Appendix A.

The minimum point $x*$ can be obtained by differentiating the equation and setting it to zero as follows:

$$0 = g + H(x - x_k) \tag{1.21}$$

If the inverse of H exists, then there is a unique solution to Equation 1.21. The minimum point $x*$ is approximated by solving the equation as follows:

$$x^* = x_k - H^{-1}g \tag{1.22}$$

where $-H^{-1}g$ is called the Newton step and its direction is called the Newton direction. The gradient-based formulation in Equation 1.18 reduces to Newton's method when $G = -H^{-1}$ and $\eta = 1$. If H is a positive-definite matrix and $J(x)$ is quadratic, then the Newton's method gets a local minimum in a single step. If $J(x)$ is not quadratic, then the Newton's method has to be applied repeatedly.

1.3.8 MODIFIED NEWTON'S METHOD

If the current point x_k is far away from the local minimum point $x*$, the method may not provide a descent direction. Even if the Hessian H is positive definite, the quadratic approximation may not be satisfactory. This means that the direct Newton step with $\eta = 1$ in Equation 1.22 will take long to decrease $J(x)$. A simple modification is proposed by introducing an adaptive parameter as follows:

$$x_{k+1} = x_k - \eta H^{-1}g \tag{1.23}$$

where η is the adaptive parameter to minimize J and η is determined in such a way that it satisfies the condition $J(x_{k+1}) < J(x_k)$ in a heuristic manner. For example, an adaptive rule for η can be step-halving defined as follows:

$$\eta_{k+1} = \frac{1}{2}\eta_k \quad k = 0, 1, 2, 3, \ldots \tag{1.24}$$

where the starting value of η_0 can be set to 1.0 or smaller.

1.3.9 LEVENBERG–MARQUARDT MODIFICATION

If the Hessian matrix is not positive definite, the Newton direction may point to the local maximum. In such a case, the Hessian can be changed to positive definite by adding a positive-definite matrix P to the Hessian H. Equation 1.18 becomes

$$x_{k+1} = x_k - (H + P)^{-1}g \tag{1.25}$$

The approach was first used by Levenberg (1944) and Marquardt (1963) and later applied to Newton's method by Goldfeld et al. (1966). By applying $P = \lambda I$, Equation 1.25 becomes

$$x_{k+1} = x_k - (H + \lambda I)^{-1}g \tag{1.26}$$

where I is the identity matrix and λ is some nonnegative value. Depending on the value of λ, the method becomes Newton's method if $\lambda \to 0$ and the steepest descent method if $\lambda \to \infty$. There are a variety of Levenberg–Marquardt methods based on the selection of λ. Goldfeld et al. (1966)

computed eigenvalues of H and set the value of λ a little larger than the most negative eigenvalue. λ plays an important role, the same as an adjustable step length η in Equation 1.23. The step length η can also be introduced in Equation 1.26 as follows:

$$x_{k+1} = x_k - \eta(H + \lambda I)^{-1}g \tag{1.27}$$

1.3.10 Quasi-Newton Method

As can be seen in the previous sections, the Newton's method, modified Newton's method, and Levenberg–Marquardt method demand the computation of Jacobian or Hessian for finding the extrema (minima or maxima) of the function, which are not available or too expensive to compute at every iteration. Therefore, quasi-Newton methods require an alternative method for computing the Hessian to find local maxima or minima of functions.

Differentiating Equation 1.20 gives

$$H_k(x_{k+1} - x_k) = g_{k+1} - g_k \tag{1.28}$$

where k is the current iteration. This formulation in Equation 1.28 shows that the Hessian H can be interpreted as the rate of change of the gradients between g_{k+1} and g_k. Therefore, Hessian H can be approximated based on the information of $\{g_{k+1}, g_k\}$ and $\{x_{k+1}, x_k\}$.

This approach is known as quasi-Newton methods, also known as variable metric methods. Eventually, quasi-Newton methods construct an approximation of the Hessian matrix H or the inverse of the Hessian matrix H^{-1} in an iterative way. The approximations $\hat{H}$ will converge to H^{-1} near the solution point.

$$\hat{H}_k \approx \frac{(x_{k+1} - x_k)}{(g_{k+1} - g_k)} \tag{1.29}$$

$\hat{H}_k$ determines Δx_k and $\hat{H}_{k+1}$ is used to satisfy the condition of the quasi-Newton method described by

$$(x_{k+1} - x_k) = \hat{H}_{k+1}(g_{k+1} - g_k) \tag{1.30}$$

Equation 1.30 can be written in short form as

$$\Delta x_k = \hat{H}_{k+1}\Delta g_k \tag{1.31}$$

The initial $\hat{H}_0$ is often chosen as the identity matrix I. Two updating schemes Davidon–Fletcher–Powell (Chong and Zak, 2008) and Broyden–Fletcher–Goldfarb–Shanno (BFGS) (Chong and Zak, 2008) are widely used by researchers.

1.3.11 Conjugate Direction Methods

The conjugate direction methods have some advantages over the steepest descent and Newton's methods. They perform better than the method of steepest descent but not as well as Newton's method. Their features are stated as follows:

1. They can solve quadratics of n variables in n steps.
2. They do not need computation of the Hessian matrix.
3. They do not need inversion of a matrix or storage of an $n \times n$ matrix.

The quadratic function of n variables is defined by

$$f(x) = \frac{1}{2} x^T Q x - x^T b \qquad (1.32)$$

where $x \in R^n$ and Q is a symmetric positive-definite $n \times n$ matrix with $Q = Q^T > 0$.

The best search direction for the quadratic function in Equation 1.32 is the Q-conjugate direction. Two directions d_i and d_j in R^n are Q-conjugate if $d_i^T Q d_j = 0$ for all $i \neq j$. For a given starting point x_0, Q-conjugate directions d_0, d_1, d_2, ..., d_{n-1}, and $k \geq 0$, the conjugate direction algorithm can solve the equation $Qx = b$ in n steps as described by the following equations:

$$g_k = \nabla f(x_k) = Q x_k - b \qquad (1.33)$$

$$\alpha_k = -\frac{g_k^T d_k}{d_k^T Q d_k} \qquad (1.34)$$

$$x_{k+1} = x_k + \alpha_k d_k \qquad (1.35)$$

where α_k is the step size.

The basic conjugate direction algorithm converges to the unique x^* in n steps, that is, $x_n = x^*$.

1.3.12 Conjugate Gradient Methods

The conjugate gradient algorithm computes the conjugate directions at each stage of the algorithm. The direction is calculated as a linear combination of the previous direction and the current gradient such that all directions are mutually Q-conjugate. Therefore, it is called the conjugate gradient algorithm. It was mainly developed by Hestenes and Stiefel (1952). The conjugate gradient algorithm is used for the numerical solution of particular systems of linear equations with a symmetric and positive-definite matrix Q. The conjugate gradient algorithm is applicable to sparse systems that are too large to be handled by a direct implementation or other direct methods and also can be used to solve unconstrained optimization problems. The algorithm for the quadratic function of n variables given by $f(x) = (1/2)x^T Q x - x^T b$, $x \in R^n$, and $Q = Q^T > 0$ is implemented as an iterative procedure. The initial point x_0 is in the direction of steepest descent, that is,

$$d_0 = -g_0 \qquad (1.36)$$

The next point is computed as

$$x_1 = x_0 + \alpha_0 d_0 \qquad (1.37)$$

where

$$\alpha_0 = \arg \min_{\alpha \geq 0} f(x_0 + \alpha d_0) = -\frac{g_0^T d_0}{d_0^T Q d_0}.$$

This relationship leads to the generalization of α_k as follows:

$$\alpha_k = -\frac{g_k^T d_k}{d_k^T Q d_k} \qquad (1.38)$$

Equation 1.37 can be generalized for computation of x_{k+1} once α_k is known as follows:

$$x_{k+1} = x_k + \alpha_k d_k \tag{1.39}$$

The search is performed in the direction of d_1, which is Q-conjugate to d_0. d_1 is chosen as a linear combination of g_1 and d_0. Thus, it can be generalized as follows:

$$d_{k+1} = -g_{k+1} + \beta_k d_k, \quad k = 0, 1, 2,\dots \tag{1.40}$$

where β_k with $k = 0, 1, 2, \dots$ are the coefficients chosen in such a way that d_{k+1} is Q-conjugate to d_0, $d_1, \dots, d_k$. Computation of β_k can be performed using the expression as follows:

$$\beta_k = \frac{g_{k+1}^T Q d_k}{d_k^T Q d_k} \tag{1.41}$$

g_{k+1} is computed from the function $f(x_{k+1})$ as follows:

$$g_{k+1} = \nabla f(x_{k+1}) \tag{1.42}$$

The conjugate gradient algorithm comprises Equations 1.38 through 1.42. If $g_{k+1} = 0$, the algorithm stops. More details of the algorithm can be found in Chong and Zak (2008).

1.3.13 BFGS METHOD

BFGS method is a quasi-Newton method for solving unconstrained nonlinear optimization problems. The basic idea of BFGS is to replace the full Hessian matrix H with an appropriate matrix B in terms of an iterative updating formula with rank-one matrices (see Appendix A for rank-one matrix) as its increment. To minimize a function $f(x)$ with no constraint, the search direction d_k at each iteration is determined by

$$B_k d_k = -\nabla f(x_k) \tag{1.43}$$

where B_k is an approximation to the Hessian matrix at the k-th iteration.

Using the search direction d_k and an optimal step size β_k, the new trial solution is determined by

$$x_{k+1} = x_k + \beta_k d_k \tag{1.44}$$

Matrix B_k is updated using an estimate as follows:

$$B_{k+1} = B_k + \frac{v_k v_k^T}{v_k^T u_k} - \frac{(B_k u_k)(B_k u_k)^T}{u_k^T B_k u_k} \tag{1.45}$$

where u_k and v_k are two new variables defined by $u_k = x_{k+1} - x_k$ and $v_k = \nabla f(x_{k+1}) - \nabla f(x_k)$, respectively.

1.3.14 DETERMINISTIC VS STOCHASTIC ALGORITHMS

If the criterion of optimality $J(\theta)$ described by Equations 1.1 through 1.3 and their distributions are known, the approach for optimization is to be called ordinary. There exist many ordinary approaches and they are mainly analytic and algorithmic methods. These methods are suitable for simple problems of first and second order. Algorithmic methods are not very promising and hence approximations are to be used for higher-order problems. On the other hand, if the distribution is not known or sufficient *a priori* information is not available, then an adaptive approach is used for optimization. In an adaptive approach current information is actively used to compensate the insufficient *a priori* information. When a process is unknown (i.e., when not sure whether the process is deterministic or stochastic), also an adaptive approach is applicable (Siddique and Adeli, 2013). The adaptive approach is mainly an iterative method where a unique optimal vector is sought in a manner such that $\nabla J(\theta) = 0$. The algorithm is written as

$$\theta = \theta - \eta \nabla J(\theta) \tag{1.46}$$

The problem arises when the gradient of the functional in Equation 1.46 cannot be computed in an explicit form. There exist such situations when the functional $J(\theta)$ is discontinuous and nondifferentiable, or the dependence of the parameter vector cannot be expressed explicitly. In such situations, the algorithm in Equation 1.46 cannot be employed. Moreover, though all descent methods are deterministic, they require the initial points to be selected randomly by the users. If the initial points are to be chosen randomly, then the approach must be stochastic in nature. On the other hand, the objective function and the constraints in the real-world situations are often not analytically treatable or even not available in a closed form (Baeck, 1996). The only possible solution of the optimization problem under such conditions is possibly the search methods. Obviously, it suggests applying a stochastic method that is capable of searching a high-dimensional space.

Many methods have emerged for the solution of optimization problems of this kind, which can be divided into two categories based on the produced solutions (Weise et al., 2009), namely, deterministic algorithms and nondeterministic (stochastic) algorithms as shown in Figure 1.3. Deterministic algorithms in general follow more rigorous procedures repeating the same path every time and providing the same solution in different runs. Most conventional or classic algorithms are deterministic and based on mathematical programming. Many different mathematical programming methods have been developed in the past few decades. Examples of deterministic algorithms are linear programming (LP), convex programming, integer programming, quadratic programming, dynamic programming, nonlinear programming (NLP), and gradient-based (GB) and gradient-free (GF) methods. These methods usually provide accurate solutions for problems in continuous space. Most of these methods, however, need the gradient information of the objective function, constraints, and a suitable initial point.

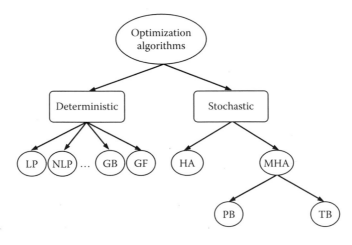

FIGURE 1.3 Classification of optimization algorithms.

On the other hand, nondeterministic or stochastic methods exhibit some randomness and produce different solutions in different runs. The advantage is that these methods explore several regions of the search space at the same time, and have the ability to escape from the local optima and reach a global optimum. Therefore, these methods are more capable of handling NP-hard problems (i.e., problems that have no known solutions in polynomial time) (Lin et al., 2012). A final conclusion can be drawn based on the transition from a derivative-based approach to derivative-free approach and the fact that even there are well-established analytical methods; however, the stochastic nature of all those approaches cannot be avoided. This assumption leads to the choice of derivative-free stochastic optimization methods. One school of stochastic search uses a single initial point and guides the search through the search space for the optimum, for example, simulated annealing (SA). The other school uses a population of initial points and guides the search according to relative efficiencies of the observed functional values, for example, genetic algorithm (GA). There are a variety of stochastic optimization algorithms, which are categorized into two types: heuristic algorithms (HAs) and meta-heuristic algorithms (MHAs), which are shown in Figure 1.3.

Heuristic means to find or discover by means of trial and error. Alan Turing was probably the first to use HAs during the Second World War and called his search methods heuristic search. Glover possibly revived the use of HAs in the 1970s (Glover, 1977). The general problem with HAs (e.g., scatter search) is that there is no guarantee that optimal solutions are reached though quality solutions are found in reasonable time. The second generation of the optimization methods are meta-heuristic that are proposed to solve more complex problems and very often provide better solutions than HAs. The 1980s and 1990s were the most exciting time for the proliferation of MHAs. Fred Glover was the first to use meta-heuristics in tabu search (TS) in the 1980s (Glover, 1986, 1989). The recent trends in MHAs are the stochastic algorithms with a certain trade-off of random and local search. Every meta-heuristic method consists of a group of search agents that explore the feasible region based on both randomization and some specified rules. The rules are usually inspired by natural phenomena or laws and the guidelines used are very simple. These methods rely extensively on repeated evaluations of the objective function and use heuristic guidelines for estimating the next search direction. MHAs are also well-known approximate algorithms and solve optimization problems with satisfying results. A general introduction to MHAs can be found in Manjarres et al. (2013). For a review on the field of meta-heuristics, the interested readers are referred to the book by Glover and Kochenberger (2003).

There are different classifications of MHAs reported in the literature (Fister et al., 2013; Manjarres et al., 2013). Two major types of meta-heuristics are common: one is population-based (PB) and the other is neighborhood-based or trajectory-based (TB), as shown in Figure 1.3. The neighborhood-based meta-heuristics such as SA (Glover, 1989) and TS (Kirkpatrick et al., 1983) only evaluate one potential solution at a time and the solution moves through a trajectory in the solution space. The steps or moves trace a trajectory in the search space, with nonzero probability that this trajectory can reach the global optimum. The PB meta-heuristics are very different from TB methods, where a set of potential solutions move toward goal simultaneously. For example, GA (Hejazi et al., 2013; Kociecki and Adeli, 2015) and particle swarm optimization (PSO) (Iacca et al., 2014; Shafahi and Bagherian, 2013) are PB algorithms and use a population of solutions. Some local search methods such as scatter search (Glover, 1977), TS (Glover, 1989), the random search method (Matyas, 1965), and the downhill simplex method (Nelder and Mead, 1965) are discussed briefly in the following sections.

1.3.15 LOCAL SEARCH METHODS

Local search is a meta-heuristic method for solving search and optimization problems. Local search algorithms move in the solution space by applying small changes to a candidate solution until it satisfies a criterion of optimality. The local search allows determining the extremum (minimum or maximum) of a function $f: R \rightarrow R$ over a closed interval, say, $[a,b]$. It is assumed that the objective function f is unimodal, meaning f has only one local minimum or maximum on the interval of $[a,b]$.

The local search methods have some common features such as local search, keep track of the single current state, move only to neighboring states, and use very little memory, and in this way, they very often are able to find reasonable solutions in large and infinite state spaces.

1.3.15.1   Scatter Search

Scatter search was introduced by Glover (1977) as a heuristic for integer programming based on the concept of surrogate constraints. The principle of the scatter search is that useful information about the global optima is stored in a diverse and elite set P, which is randomly generated purposely considering the characteristic features of the solution space. A reference set S_{ref} of candidate solutions from P is created such that $S_{ref} \subset P$. The reference set S_{ref} is partitioned into subsets S_1 and S_2 such that $S_{ref} = S_1 \cup S_2$ and $S_1 \cap S_2 = \varnothing$. New solutions are created by linear recombination of the solutions of the subsets S_1 and S_2 in an iterative way. The new solutions are refined using a heuristic and assessed as to whether or not they are selected for the next iteration, and thus the reference set S_{ref} is updated. The good solutions are defined by special criteria such as diversity that purposely go beyond the objective function value. Different improvements and designs of search algorithms originating from the basic scatter search algorithm can be found in Glover (1998) and Glover et al. (2000).

1.3.15.2   Tabu Search (TS)

TS is a meta-heuristic method originally proposed by Glover in the 1980s (Glover, 1989), which has found applications to various combinatorial problems and have appeared in the operations research literature. The basic principle of TS is to pursue local search whenever it encounters a local optimum by allowing nonimproving moves and preventing previously visited solutions by the use of memories, called tabu lists. TS algorithm starts with a feasible initial solution. It tries to improve the solution iteratively. It examines a set of candidate moves in the neighborhood of the solution. The neighborhood provides all feasible solutions obtainable from the current solution with a simple move. The neighboring solutions are evaluated and the current solution moves to its best neighboring solution. The best move is put on the tabu list so that it does not reverse the move for some time (i.e., for a number of iterations). A large-size tabu list explores the search space and prevents cycling. A small-size tabu list exploits the search space. The criterion for a tabu move is based on the fitness value if it is better than the best known solution. The move is remembered by a frequency list. The frequency list acts as a long-term memory and helps in detecting the most frequent moves. The algorithm stops on satisfaction of the criterion and the best solution is returned. Eventually, it turns out that TS is simply a combination of local search with short-term memories. Glover considered TS as a general strategy for guiding and controlling inner heuristics specifically tailored to problems. For more details on TS, see Gendreau and Potvin (2010), Glover (1994), and Glover and Laguna (1997).

1.3.15.3   Random Search (RS)

The RS technique was first proposed by Anderson (1953). RS methods are also known as direct-search or derivative-free methods. The advantage of RS methods is that they do not require any gradient information for the problem to be optimized. Therefore, it can also be applied on noncontinuous or nondifferentiable functions. Hooke and Jeeves (1961) suggested optimization via search at the early days of computational optimization. RS as a family of numerical optimization methods was proposed by Rastrigin (1963), which provided the basic mathematical analysis. Optimization by search was later used by Matyas (1965). It is an iterative procedure and progressively moves to better positions in the search space, which are sampled from the neighborhood of the current position.

1.3.15.4   Downhill Simplex (Nelder–Mead) Method

The downhill simplex method (also known as the Nelder–Mead method) is a commonly applied numerical method used to find the minimum or maximum of an objective function in a many-dimensional space, which was first proposed by Nelder and Mead (1965). The Nelder–Mead technique is a heuristic search method that can converge to nonstationary points and can be applied to nonlinear optimization problems where derivative information may not be known.

The strategy used in scatter search, TS, RS, and downhill simplex search methods are local search techniques and use a generate-and-test search, manipulating feasible solutions based on physical characteristics of the problem at hand. The selection of the initial values has a decisive effect on the final solution of these algorithms. In practice, knowing these initial values is nearly impossible. Also there is no known suitable heuristic approach to obtain these initial values, and they are selected randomly.

1.4 PARADIGM OF NIC

It is amply clear from the brief discussion on natural sciences in the earlier sections that the paradigm of NIC is fairly vast. Even though science and engineering have evolved over many hundred years with many clever tools and methods available for solutions, there is still a diverse range of problems to be solved, phenomena to be synthesized, and questions to be answered. Still there are many branches of science which have not been explored enough. In many situations there are alternative techniques to obtain the solution for a given problem, which may even provide superior solutions. However, natural computing approaches are not the only tools that provide solutions to these, nor are they always the most suitable and efficient approaches. Thus, it is important to carefully investigate the problem to be solved before choosing a specific approach for a solution, be it a natural computing tool or a traditional mathematical approach. In all cases, one or more of the natural computing approaches, briefly overviewed in the following section, can be used to solve the problem, synthesize the phenomenon, or answer the question. In general, natural computing tools or approaches should be considered when

- The problem is complex and nonlinear and involves a large number of variables or potential solutions or has multiple objectives.
- The problem to be solved cannot be suitably modeled using conventional approaches such as complex pattern recognition and classification tasks.
- Finding an optimal solution using traditional approaches is not possible, difficult to obtain, or cannot be guaranteed, but a quality measure exists that allows comparison of various solutions.
- The problem lends itself to a diversity of solutions or a diversity of solutions is desirable.
- The limits of current technology are reached or new computing approaches have to be sought.

NIC refers to a class of MHAs that imitates or is inspired by some natural phenomena explained by natural sciences, as discussed earlier. A common feature shared by all nature-inspired MHAs is that they combine rules and randomness to imitate some natural phenomenon. Many NIC paradigms have emerged in recent years. All these computing approaches (i.e., algorithms or methods) can be grouped into three broader classes: physics-based algorithms (PBAs), chemistry-based algorithms (CBAs), and biology-based algorithms (BBAs). Some researchers argue that society-, culture-, and civilization-based algorithms also belong to nature-inspired algorithms. These algorithms imitate phenomena observed in human behaviors that defined society, culture, and civilization. The society-, culture-, and civilization-based algorithms will be addressed under the broad category of BBAs. This broader class of search and optimization paradigms of NIC algorithms is shown in Figure 1.4. The discussion on NIC will be confined to only these three paradigms.

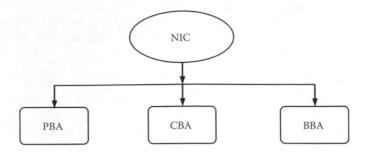

FIGURE 1.4 Broad classification of NIC.

1.5 PHYSICS-BASED ALGORITHMS

PBAs employ basic principles of physics, for example, Newton's laws of gravitation, laws of motion, Coulomb's force law of electrical charge, and so on, discussed in earlier sections of this chapter. They are all based on deterministic physical principles. These algorithms can be categorized broadly as follows:

1. Inspired by Newton's laws of motion, for example, colliding bodies optimization (CBO)
2. Inspired by Newton's gravitational force, for example, gravitational search algorithm (GSA), central force optimization (CFO), space gravitation optimization (SGO), and gravitational interactions optimization (GIO)
3. Inspired by celestial mechanics and astronomy, for example, big bang–big crunch (BB–BC) search, BH search (BHS), galaxy-based search algorithm (GbSA), artificial physics-based optimization (APO), and integrated radiation search (IRS)
4. Inspired by EM, for example, electromagnetism-like optimization (EMO), charged system search (CSS), and hysteretic optimization (HO)
5. Inspired by optics, for example, ray optimization (RO)
6. Inspired by acoustics, for example, harmony search algorithm (HSA)
7. Inspired by thermodynamics, for example, SA
8. Inspired by hydrology and hydrodynamics, for example, WDA, RFDA, and WCA

The earliest of all these algorithms was the SA algorithm proposed by Kirkpatrick et al. (1983) using the idea of the Metropolis and based on the principle of thermodynamics and applied it to optimization problems. The algorithm simulates the cooling process by gradually lowering the temperature of the system until it converges to a steady state. The idea to use SA to search for feasible solutions and converge to an optimal solution was very stimulating and led the researchers to explore other areas of physics.

An idea from the field of sound and acoustics led to the development of HSA inspired by a phenomenon commonly observed in music. The concept behind the HSA is to find a perfect state of harmony determined by esthetic estimation, which is like any esthetic quality widely popular in the music world (Geem, 2010; Geem et al., 2001). Reviews of HSAs, their variants, and applications are provided by Siddique and Adeli (2015a–c).

A number of recent MHAs are inspired by Coulomb and Gauss laws from physics and Newtonian laws from mechanics to guide the particles or masses to explore the search space and find the locations of the optimum. HO, EMO, APO, CFO, GSA, CSS, GIO, and CBO are based on Coulomb laws, Gauss laws, and Newton's law of gravity and laws of motion. Similarly, SGO, BB–BC, integrated radiation optimization (IRO), GbSA, spiral optimization (SpO), and BH introduced Einstein's general theory of relativity, space-time, and the spiral phenomenon. The difficulty in traditional HAs mentioned up to now is still persisting in these algorithms.

Zaránd et al. (2002) proposed a method of optimization inspired from these two processes of demagnetization. This is a process similar to SA where the material achieves a stable state by slowly decreasing the temperature, that is, finding the ground states of magnetic samples, which is to be compared to finding the optimal point in the search process. The new method is called HO. Zaránd et al. (2002) showed the similarity between magnetic hysteresis and the optimization process using the Ising model.

Based on the principles of EM, Birbil and Fang (2003) introduced the method of EMO. The EM-based algorithm imitates the attraction–repulsion mechanism of EM theory in order to solve unconstrained or bound-constrained global optimization problems. Therefore, it is called EMO algorithm. A solution in the EMO algorithm is seen as a charged particle in the search space and its charge relates to the objective function value.

Motivated by natural physical forces, William Spear and Diana Spear introduced APO framework (Spears and Spears, 2003). In APO, particles are seen as solutions sampled from a feasible region of the problem space. Particles move toward higher fitness regions and cluster at an optimal region over time. Heavier mass represents a higher fitness value and attracts other masses of lower fitness values. The individual with best fitness attracts all other individuals with lower fitness values. The individuals with lower fitness values repel each other. That means the individual with the best fitness has the biggest mass and moves with lower velocity than others. Thus, the attractive–repulsive rule can be treated as the search strategy in the optimization algorithm, which ultimately leads the population to search for the better fitness region of the problem. In the initial state, individuals are randomly generated within the feasible region. In APO, mass is defined as the fitness function for the optimization problem in question. A suitable definition of mass of the individuals is necessary.

CFO is a meta-heuristic optimization algorithm proposed and developed by Formato (2007a,b). CFO uses a population of probes that are distributed across a search space. The basic concept of the CFO is the search for the biggest mass that has the strongest force to attract all other masses distributed within a decision space toward it, which is to be considered as the global optimum of the problem at hand. CFO inherently conducts deterministic computation.

GSA proposed by Rashedi et al. (2009) is a PB search algorithm based on the law of gravity and mass interaction. The algorithm considers agents as objects consisting of different masses. All the agents move due to the gravitational attraction force acting between them and the progress of the algorithm directs the movements of all agents globally toward the agents with heavier masses (Rashedi et al. 2009).

CSS is inspired from electrostatics in physics, Coulomb's law, Gauss law, and Newtonian mechanics and was first proposed by Kaveh and Talatahari (2010). In CSS, a population of candidate solutions are generated within the parameter space of the optimization problem at hand. The candidate solutions within the search space are thought of as charged particles within the electric field. By applying electrostatics, Coulomb's law, Gauss law, and Newtonian mechanics, an optimal solution is to be found.

GIO is inspired from Newton's law. GIO has some similarities with GSA. GIO algorithm was first introduced by Flores et al. (2011) around the same time independently of GSA. The gravitational constant G in GSA decreases linearly with time, whereas GIO uses a hypothetical gravitational constant G as constant. GSA uses a set of best individuals to reduce computation time, while GIO allows all masses to interact with each other.

Through a process of continuous collision between bodies, some bodies will achieve better positions according to an optimality criterion defined by an objective function. The process continues for some time (i.e., for number of iterations) until the solutions meet the desired accuracy or condition. On the basis of this simple principle, Kaveh and Mahdavi (2014a,b) proposed CBO. In CBO, candidate solutions consist of problem variables and are considered as colliding bodies (CBs).

Hsiao et al. (2005) developed the SGO algorithm using the notion of space gravitational curvature based on the concept of Einstein's equivalence principle. SGO is a general purpose search technique for multidimensional optimization problems. In SGO, the search agents are asteroids moving around within the universe (i.e., search space). The agents travel through the search space by Newton's law of gravity, and the solution space is formulated as a curvature of the space-time according to the concept of Einstein's theory of general relativity. The search agents of SGO aggressively search undiscovered regions for heavier masses (i.e., a better solution) based on the geometrical variance around it. Technically speaking, SGO is an embryonic form of CFO (Kenyon, 1990; Yilmaz, 1965).

Based on the notion of the expansion phenomenon of big bang and the shrinking phenomenon of big crunch, an optimization algorithm has been introduced by Erol and Eksin (2006) called the BB–BC algorithm. The BB–BC algorithm has been improved by Kaveh and Talatahari (2009) over the years. There are two phases in the BB–BC algorithm, namely, the big bang phase and the big crunch phase. In the big bang phase, a population of masses is generated with respect to the center of mass. In Big Crunch phase, all masses collapse into one, that is, the center of mass. The center of mass has heavy gravitational force, which attracts all other masses. The center of mass is computed, which resembles a BH of the system. Thus, the big bang phase explores the solution space. The Big Crunch phase performs necessary exploitation as well as convergence. In the big bang phase, energy dissipation produces disorder and randomness, which is represented by the creation of a random population of individuals within the limitation of the search space, whereas in the Big Crunch phase, randomly distributed particles are brought into an order under gravitational attraction. Inspired by this theory, an optimization algorithm is constructed termed as the BB–BC method, which generates random points in the big bang phase and shrinks those points into a single representative point via a center of mass with minimal cost computation.

IRO algorithm was developed by Chuang and Jiang (2007) inspired by the gravitational radiation in the curvature of space-time. Based upon the theoretical analysis of gravitational radiation, there are two general cases: (i) two stars orbiting each other and (ii) an isolated nonaxis-symmetrical supernova expanding the space. Stars orbiting each other, also known as a binary star system or pulsar such as the Hulse–Taylor binary star system, generate strong gravitational radiation. The strength of gravitational radiation depends on the mass of the star. Since the orbiting star and its companion star are both massive, the gravitational radiation is also very strong.

Hosseini (2011) proposed the GbSA inspired by the spiral arm of spiral galaxies to search its surrounding. GbSA uses a spiral-like movement in each dimension of the search space with the help of chaotic steps and constant rotation around the initial solution. Gradually, the arm of the galaxy opens and covers the search space in order to find a better solution.

SpO is a multipoint search for continuous optimization problems. The SpO model is composed of plural logarithmic spiral models and their common center. The original work on the algorithm was first reported in Tamura and Yasuda (2011a,b).

Inspired by the phenomenon of the BH, Hatamlou (2013) proposed the new MHA called the BH algorithm, which starts with an initial population of candidate solutions for an optimization problem. In the BH algorithm, candidate solutions are considered as stars. The best candidate solution is selected to be the BH and the rest form the normal stars. After initialization, objective function values are calculated for all individuals in the population. At each iteration, the BH (i.e., the best solution) starts attracting other stars (i.e., candidate solutions) around it. If a star gets too close to the BH, it will be swallowed by the BH and is gone from the universe forever. In such a case, a new star (candidate solution) is randomly generated and placed in the search space and the algorithm starts a new search.

This basic idea of Snell's law is utilized in the RO algorithm developed by Kaveh and Khayatazad (2012, 2013). In the RO algorithm, a solution comprises a vector of variables, which is simulated by a ray of light passing through space consisting of media with different refractive indices.

A number of MHAs emerged based on the principles of hydro- and river dynamics and water cycles in the environment. On the basis of the idea of flow of water in rivers, which carries an amount of soil from one place to another, the WDA was introduced by Shah-Hosseini (2007, 2008). The novel algorithm, called "intelligent water drop (IWD)" imitates the dynamics of river systems and the behavior of water drops. The algorithm utilizes the features of river dynamics such as the variation of velocity, the change of soil in the river bed, the change of direction of the flow, and so on.

Water drops flow toward the sea over the land. As they flow, they change the landscape by eroding land at high altitudes and deposit carried sediments in flatter areas. By increasing or decreasing the altitude of nodes, gradients are modified, which in turn affects the flow of water drops. Eventually, decreasing gradients formed will depict paths from the points where it rains to the sea, and these paths can represent the solution of an optimization problem under consideration. Considering this natural phenomenon of a river forming by eroding the land and depositing sediments, Rabanal et al. (2007) proposed a meta-heuristic optimization algorithm called RFDA.

Eskandar et al. (2012) proposed a new MHA called the WCA for optimizing constrained functions and engineering problems. The WCA begins with an initial population of streams that give rise to rain or precipitation. The best individual (i.e., best stream) is chosen as a sea. A number of good streams (evaluated as per the cost function value) are chosen as a river and the rest of the streams flow to the rivers and sea directly. Depending on the magnitude of flow, each river takes in water from the streams. The amount of water in a stream joining rivers and/or the sea varies from other streams. All rivers flow to the sea, which is the ultimate destination and the optimal solution in terms of optimization.

Biswas et al. (2013) reported a survey on PBAs, which examines how natural phenomena can be used to solve a complex optimization problem with its excellent facts, functions, and phenomena. The survey is focused on inspirations that are originated from physics, their formulation of obtaining solutions, and their evolution with time.

1.6 CHEMISTRY-BASED ALGORITHMS

Compared to PBAs and BBAs, there are very few CBAs found in the literature. In the recent years, there have been some emerging chemistry-based optimization algorithms reported in the literature. These are broadly categorized into four classes of algorithms:

1. Artificial chemical process algorithm (ACPA)
2. Chemical reaction optimization (CRO)
3. Artificial chemical reaction optimization algorithm (ACROA)
4. Chemical reaction algorithm (CRA)
5. Gases Brownian motion optimization (GBMO)

ACPA was developed by Irizarry (2004, 2005a). In the ACPA, solutions are represented as vectors where the decision variables (or parameters) are encoded into a set of discrete variables called molecules. The ACPA is an iterative improvement methodology where a perturbation is applied to a randomly selected set called activation reactor (AR). If the trial state vector is the desired set, it is accepted as the best value; otherwise, an iterative procedure is applied to the selected set of molecules from AR. The ACPA has found applications in many domains (Irizarry, 2005b, 2006, 2011).

Lam and Li (2009) first reported the chemical-reaction-inspired MHA for optimization in a technical report in 2009. Shortly after that, they developed the meta-heuristic optimization algorithm based on the principles of chemical reactions and termed it CRO. Within a very short period of time, the CRO algorithm has found a wide range of applications in many domains (Lam and Li, 2010a,b; Lam et al., 2012).

ACROA was developed by Alatas (2011, 2012) based on the concept of chemical reactions that occur between reactants and their change to products by applying a set of operators such as synthesis, decomposition, single–double displacement, combustion, redox, and reversible reactions. In ACROA, molecules are encoded using an appropriate scheme for the optimization problems. The chemical system tends toward the highest entropy and lowest enthalpy. Enthalpy is used as a measure of performance or objective function for a minimization problem and entropy for a maximization problem. The observable properties of all participating reactants become stable as the chemical system reaches an equilibrium state when the process terminates. ACROA found applications in many domains (Alatas, 2012; Yang et al., 2011).

CRA was developed by Melin et al. (2013) based on the abstraction of chemical reactions where synthesis and decomposition reactions are applied as diversifying mechanisms and substitution and double-substitution reactions are applied as mechanisms of intensification. The CRA uses elements or compounds for parameter representation of the problem and can trigger single reaction or multiple reactions at a time, depending on the nature of the problem. The CRA has found applications in many domains (Melin et al., 2013).

GBMO algorithm was proposed by Abdechiri et al. (2013) based on the laws of Brownian motion and turbulent rotational motion of gas molecules. Each molecule has a position, mass, velocity, and radius of turbulence that represents parts of the solution. Molecules move in the solution space and toward an optimal solution. Mass and temperatures are updated. Higher temperature causes the molecules to move at higher velocity, that is, gases Brownian motion explores the solution space, while decreasing temperature causes the molecules to move at lower velocity, that is, gases turbulent rotational motion exploits the solution space. As the iterative process continues, velocity and position are updated, leading to a solution.

Xing and Gao (2014)* reported a survey on CBAs, which briefly describes the methods with useful references therein.

1.7 BIOLOGY-BASED ALGORITHMS

The paradigm of BBAs is huge and diverse in many aspects. It forms a broader class of search and optimization algorithms. These algorithms can be classified into three broad groups: evolutionary algorithms (EAs), bio-inspired algorithms (BIAs), and swarm intelligence-based algorithms (SIAs). The classification of these algorithms is shown in Figure 1.5.

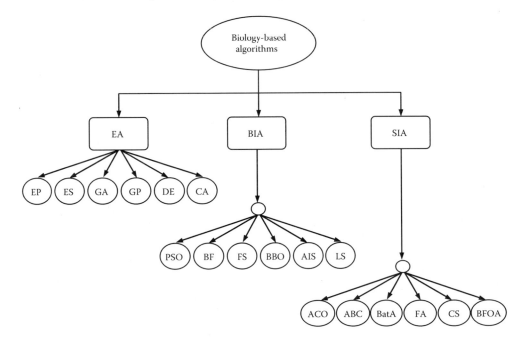

FIGURE 1.5 Classification of BBAs.

---

* The book of Xing and Gao (2014) published by Springer-Verlag is a reference book containing 134 clever algorithms classified into broad categories such as physics-, chemistry-, and biology-based algorithms. This is a very useful resource but details of algorithms and applications are not provided in the book.

EA is the emulation of the process of natural selection in a search procedure based on the seminal work on evolutionary theory by Charles Darwin (Darwin, 1859). The fundamental idea of the evolutionary process in nature as an optimization and search algorithm is based on Darwin's theory of evolution, which gained momentum in the late 1950s, nearly a century after publication of the book *Origin of Species*. Fraser (1957) was the first to conduct a simulation of genetic systems representing organisms by binary strings. Box (1957) proposed an evolutionary operation to optimizing industrial production. Friedberg (1958) proposed an approach to evolve computer programs. The fundamental works of Lowrence Fogel (Fogel, 1962) in evolutionary programming (EP), John Holland (Holland, 1962, 1975) in GA, and Ingo Rechenberg (Rechenberg, 1965, 1973) and Hans-Paul Schwefel (Schwefel, 1968) in evolution strategies (ES) had great influences on the development of EAs and computation as a general concept for problem solving and as a powerful tool for optimization. Since the developmental years of the 1960s, the field evolved into three main branches (De Jong, 2006): ES, EP, and GA. There were significant contributions to the field by many people. Among them are De Jong (1975), Goldberg (1989), and Fogel (1995). The 1990s have seen another set of developments in the EAs such as development of genetic programming (GP) by Koza (1992), development of cultural algorithms (CA) by Reynolds (1994, 1999), and development of DE by Storn and Price (1997).

EAs have now found widespread applications in almost all branches of science and engineering (Campomanes-Álvareza et al., 2013; Kim and Adeli, 2001; Kociecki and Adeli, 2013, 2014; Lin and Ku, 2014; Molina-García et al., 2014). Different variants of EAs such as EP, ES, GA, GP, DE, and CA are discussed in the book by Siddique and Adeli (2013).

BIAs are based on the notion of commonly observed phenomenon in some animal or insect societies and on the movement of organisms. Flocks of birds, herds of quadrupeds, and schools of fish are often shown as fascinating examples of self-organized coordination (Camazine et al., 2001). PSO proposed by Kennedy and Eberhart (2001) simulates social behavior of swarms such as birds flocking and fish schooling in nature. Particles make use of the best positions encountered and the best position of their neighbors to position themselves toward an optimal solution. The performance of each particle is measured according to a predefined fitness function. There are now as many as about 20 different variants of PSO (Bergh and Engelbrecht, 2006; Jiang et al., 2007; Tsai and Lin, 2011).

Bird flocking (BF) is seen as a feature of coherent maneuvering of a group of individuals in space. Natural flocks maintain two balanced behaviors: a desire to stay close to the flock and a desire to avoid collisions within the flock (Shaw, 1975). Joining a flock or staying with a flock seems to be the result of evolutionary pressure from several factors such as protecting from and defending against predators, improving the chances of survival of the (shared) gene pool from attacks from predators, profiting from a larger effective search for food, and advantages for social and mating activities (Shaw, 1962). Reynolds (1987) was the first to develop a model to mimic the flocking behavior of birds, which is described as a general class of polarized, noncolliding, aggregate motion of a group of individuals. Such flocking behaviors were simulated using three simple rules: namely, collision avoidance with flock mates, velocity matching with nearby flock mates, and flock centering to stay close to the flock (Momen et al., 2007; Turgut et al., 2008).

Fish schools (FS) show very interesting features in their behavior. About half the fish species are known to form FS at some stage in their lives. Fish can form loosely structured groups called shoals and highly organized structures called FS. FS are seen as self-organized systems consisting of individual autonomous agents (Shaw, 1962). FS also come in many different shapes, such as stationary swarms, predator avoiding vacuoles and flash expansions, hourglasses and vortices, highly aligned cruising parabolas, herds, and balls (Parrish et al., 2002), and can be of various sizes (Mackinson, 1999). Modeling the behavior of FS has been a subject of many researchers for a long time (Couzin et al., 2002; Niwa, 1996).

MacArthur and Wilson (1967) developed mathematical models of biogeography that describe how species migrate from one island to another, how new species arise, and how species become extinct. Since the 1960s, biogeography has become a major area of research that studies the geographical distribution of biological species. On the basis of the concept of biogeography, Simon (2008) derived a new family of algorithms for optimization called biogeography-based optimization (BBO).

The biological immune system defends the body from foreign pathogens. The innate immune system is an unchanging mechanism that detects and destroys certain invading organisms, while the adaptive immune system responds to previously unknown foreign cells and builds a response to them that can remain in the body over a long period of time. The remarkable features and information processing of immune systems have led to new computational techniques. On the basis of the principles of biological immune systems, models of artificial immune systems (AIS) were proposed. The first model of an AIS was formalized by Farmer et al. (1986) in the 1980s and stipulated the interaction between antibodies mathematically. The model was further interpreted by Bersini and Varela (1990) with numerous refinements, and the combination of these two approaches forms the cornerstone of AIS. The model was seen as having computationally useful properties, and provided a network-based approach distinct from both neural networks and GAs. Several models of the AIS have been proposed and applied for solution of real-world science and engineering problems (Hofmeyr and Forrest, 2000)

The beauty of the patterns observed in nature has attracted the attention of researchers for many years. In 1968, Lindenmayer (1968) introduced formalism for simulating the development of multicellular organisms, initially known as Lindenmayer systems and subsequently named L-systems, which attracted the immediate interest of theoretical computer scientists. The development of the mathematical theory of L-systems was followed by its applications to the modeling of plants. The central concept of L-systems is that of rewriting. In general, rewriting is a technique for defining complex objects by successively replacing parts of a simple initial object using a set of *rewriting rules* or *productions*. However, although a geometrical interpretation of strings was at the origin of L-systems, they were not applied to picture generation until 1984, when Aono and Kunii (1984) and Smith (1984) used them to create realistic-looking images of trees and plants.

There are many other bio-inspired search and optimization algorithms dispersedly reported in the literature, which have not attracted much attention in the research community such as atmosphere clouds model, dolphin echolocation, Japanese tree frogs calling, Egyptian vulture, flower pollination algorithm, great salmon run, invasive weed optimization, paddy field algorithm, roach infestation algorithm, shuffle frog-leaping algorithm, and many more. These algorithms are mostly unknown. All these algorithms will be explored and assimilated in Volume II.

SIAs are based on the idea of collective behaviors of insects living in colonies such as ants, bees, wasps, and termites, which have attracted the attention of researchers and naturalists for many years (Bonabeau et al., 1999). Close observation of an insect colony shows that the whole colony is very organized with every single insect having its own agenda. The seamless integration of all individual activities does not have any central control or any kind of supervision. Researchers are interested in this new way of achieving a form of collective intelligence called swarm intelligence. SIAs are advanced as a computational intelligence (CI) technique based around the study of collective behavior in decentralized and self-organized systems typically made up of a population of simple agents interacting locally with one another and with their environment. Although there is normally no centralized control structure dictating how individual agents should behave, local interactions between such agents often lead to the emergence of global behavior.

The inspiring source of ant colony optimization (ACO) is based on the foraging behavior of real ant colonies and was first formulated in 1990s by Dorigo et al. (1991, 1996). Ants are social insects that live in colonies and their behavior is governed by the goal of colony survival rather than the survival of individuals (Dorigo and Stützle, 2004; Dorigo et al., 2006). When searching for food, ants initially explore the surrounding area close to the nest in a random manner. While moving, ants leave a chemical pheromone trail on the ground. Ants can smell the pheromone (Jackson and Ratnieks, 2006). When choosing their way, they tend to choose, in probability, paths marked by strong pheromone concentrations. As soon as an ant finds a food source, it evaluates the quantity and the quality of the food and carries some of it back to the nest. During the return trip, the quantity of pheromone that an ant leaves on the ground may depend on the quantity and quality of the food. The pheromone trails will guide other ants to the food source. It has been shown that the indirect communication between the ants via pheromone trails enables them to find shortest paths between their nest and food sources. Ant colonies or societies in general can be compared to distributed systems, which present a highly structured social organization in spite of simple individuals (Blum, 2005). Researchers proposed simple stochastic models that adequately describe the dynamics of the ants' foraging behavior, and in particular, how ants can find shortest paths between food sources and their nest. ACO is a meta-heuristic optimization algorithm that can be used to find approximate solutions to difficult combinatorial optimization problems and has been applied successfully to an impressive number of optimization problems.

Honey bees live in colonies. They search for food sources and collect by foraging in promising flower patches. The foraging process begins by scout bees sent out to search for flower patches. Scout bees move randomly from one patch to another. When the scout bees return to the hive, those that found a patch that is rated above a certain quality threshold deposit their nectar or pollen and perform a waggle dance for colony communication. Waggle dance provides three pieces of information regarding a flower patch: the direction in which food source will be found, its distance from the hive, and its quality rating (Frisch, 1967). This information helps the colony to send its bees to flower patches precisely without using guides or maps. This simple mechanism of the honey bees inspired researchers to develop new search algorithms. The bee algorithm is a meta-heuristic optimization procedure that mimics bees' foraging behavior. The bee algorithm was first formulated by Nakrani and Tovey (2004). Later in 2005, a number of researchers independently developed the bees algorithm. The artificial bee colony (ABC) algorithm was proposed by Karaboga (2005), the virtual bee algorithm was proposed by Yang (2005), and the bee algorithm was proposed by Pham et al. (2005). The bee algorithm is becoming more popular among swarm intelligence researchers (Li et al., 2012). All of these algorithms will be denoted as ABC algorithms.

The bat algorithm (BatA) is based on the echolocation behavior of bats. The capability of micro-bats is fascinating as these micro-bats use a type of sonar, called, echolocation, to detect prey, avoid obstacles, and locate their roosting crevices in the dark. These bats emit a very loud sound pulse and listen for the echo that bounces back from the surrounding objects. Their pulses vary in properties and can be correlated with their hunting strategies, depending on the species. By comparing the outgoing pulse with the returning echoes, the brain and auditory system can produce detailed images of the bat's surroundings. This allows bats to detect, localize, and even classify their prey in complete darkness. If the echolocation characteristics of micro-bats are idealized in some way, various bat-inspired search algorithms can be developed. Yang (2010a, 2011) simulated echolocation behavior of bats and its associated parameters in a numerical optimization algorithm. The interesting simplification is that no ray tracing is used in estimating the time delay and three-dimensional topography.

Quite a number of cuckoo species engage in obligate brood parasitism by laying their eggs in the nests of the other host birds of different species. If the host bird discovers that the eggs are not its own, it either throws the eggs away or abandons its nest. Some cuckoo species specialize in mimicking the color and pattern of the eggs of the chosen host species and thus reduce the chances of their eggs being thrown out or abandoned. Yang and Deb (2010) developed a new meta-heuristic optimization algorithm, called cuckoo search (CS), which is based on the interesting breed behavior of certain cuckoo species.

The flashing of fireflies in the summer sky in the tropical regions has been attracting the attention of naturalists and researchers for many years. Most fireflies produce short and rhythmic flashes, which can be seen as a signaling system. The true function of such a signaling system is not completely known yet. Two fundamental functions of such flashes are to attract mating partners and to attract potential prey, or flashing may also serve as a warning mechanism. The rhythm, rate, and duration of flashing form part of the signaling system that brings two fireflies together. For example, females respond to a male's unique pattern of flashing. This unique feature of the fireflies can be formulated into new optimization algorithms. Yang (2009) proposed a new MHA, called the firefly algorithm (FA), based on some idealized rules such as fireflies attract one another by their flashing lights, attractiveness is proportional to their brightness, the less brighter firefly will move toward the brighter one, otherwise it will move randomly, and finally the brightness of a firefly is determined by the landscape of the objective function. For a maximization problem, a population of fireflies is generated and the brightness is simply proportional to the value of the objective function. The FA has found many applications in engineering and multiobjective optimization problems (Yang, 2010b).

Individual and groups of bacteria forage for nutrients, for example, chemotactic (foraging) behavior of *Escherichia coli* bacteria. The biology and physics of the underlying foraging behavior and strategy of the bacteria can be modeled as a distributed optimization process and the idea can be applied to solve real-world optimization problems. On the basis of this concept, researchers proposed an optimization technique known as the bacterial foraging optimization algorithm (BFOA). Passino (2002) first introduced the BFOA. The BFOA has been successfully applied to many real-world problems (Das et al., 2009; Liu and Passino, 2002).

There are many other swarm intelligence-based search and optimization algorithms dispersedly reported in the literature, which have not attracted much attention in the research community such as wolf search, cat swarm optimization, fish swarm optimization, eagle strategy, krill herd, monkey search, weightless swarm algorithms, and many more. These algorithms are mostly unknown. This chapter will make further investigation into these algorithms and make an attempt to assimilate those with some examples. These algorithms will be explored in Volume III.

1.8 CULTURE-, SOCIETY-, AND CIVILIZATION-BASED ALGORITHMS

There are a few more algorithms outside the paradigm of the PBAs, CBAs, and BBAs. These are also natural phenomena observed in human society. Culture is the sum total of the learned behavior of a population in the society. Culture generally is considered to be the tradition of a population that evolves from generation to generation. The cultural tradition can be seen as the memory of a population. Using this memory of tradition, Reynolds (1994) proposed the CA model where a belief space stores the generalizations of individuals' experiences. These experiences are updated at the macro-evolutionary level from time to time. The individual solutions are generated biased by existing beliefs. For a faster change in an individuals' performance, Reynolds and Chung (1997) used socio-behavioral dynamics whereby the individuals are described by a set of traits. These traits are passed down from a generation to another through operators called socially motivated operators. Sebag et al. (1997) developed a method based on civilized evolution, which learns from the past failures of evolution. Ursem (1999) introduced an idea based on the relationships between different nations and how countries are to interact in order to optimize a profit function.

A society can be seen as a collection of individuals in the parametric space. There are leaders in the society who are the best performing individuals that help others to improve through information exchange. Daneshyari and Yen (2004) proposed a novel optimization strategy based on a social algorithm and collective behaviors. The new algorithm incorporates the information of the individuals within the society introduced as their talent and the collective behavior of the society in the civilization called the liberty rate. Atashpaz-Gargari and Lucas (2007) proposed an algorithm that has been inspired by evolutionary development of human society. This algorithm has looked at the political process as a process of human sociopolitical evolution. They formulated a social-based algorithm by combining the evolutionary progress and the sociopolitical process, resulting in the imperialist competitive algorithm (ICA). People live in different types of communities, which leads to different styles of leadership development. This approach tries to capture several people involved in the characteristic of community development (Ramezani and Lotfi, 2013).

While a society is seen as a cluster of individuals in the parametric space, a civilization is seen as a set of all such societies based on the Euclidean closeness of the individuals. Individuals in a society interact with each other with the aim to improve themselves. The cooperative relationships among its societies define how a civilization will emerge and advance. Ray and Liew (2003) developed the concept of an optimization algorithm based on the intra- and intersociety interactions within a formal society and the civilization model to solve constrained optimization problems.

1.9 OVERVIEW

It is obvious from the above discussion that the field of NIC is huge. It is not possible to cover all of these topics in a single volume. Considering the constraints of time and limitation on page numbers, this volume focuses on the physics- and chemistry-based approaches and algorithms. Besides this, this volume comprises ten chapters: dialectics of nature, gravitational search algorithm, central force optimization, electromagnetism-like optimization, harmony search algorithm, water drop algorithm, spiral dynamics optimization, simulated annealing, chemical reaction optimization, and the final chapter covering miscellaneous algorithms.

Chapter 2: Gravitational Search Algorithms—The GSA is a newly developed stochastic search algorithm that is based on the law of gravity and mass interactions and was first introduced by Rashedi et al. in 2008. In the GSA, the search agents are the collection of masses which interact with each other based on Newtonian gravity and the laws of motion, completely different from other well-known PB optimization methods. In the GSA, agents are considered as objects and their performances are measured by their masses, and all these objects attract each other by the gravitational force, while this force causes a global movement of all objects toward the objects with heavier masses. In the GSA, each mass (agent) has four specifications: position, inertial mass, active gravitational mass, and passive gravitational mass. The position of the mass corresponds to a solution of the problem, and its gravitational and inertial masses are determined using a fitness function. Both the gravitational mass and the inertial mass, which control the velocity of an agent, are computed by fitness evolution of the problem. The positions of the agents in specified dimensions (solutions) are updated at each iteration, and the best fitness along with its corresponding agent is recorded. The termination condition of the algorithm is defined by a fixed number of iterations. In other words, each mass presents a solution, and the algorithm is navigated by properly adjusting the gravitational and inertia masses. By lapse of time, it is expected that masses be attracted by the heaviest mass. This mass will present an optimum solution in the search space.

This chapter investigates the physics of gravity, mass of moving objects, and acceleration of objects, and presents GSA, parameters of the GSA, and fitness evaluation. It also presents the different variants of GSA and a few applications to engineering problems.

Chapter 3: Central Force Optimization—CFO is a nature-inspired, gradient-like, gravity-based MHA for multidimensional search and optimization problems. CFO locates the global extrema of an objective function to be maximized. CFO searches by flying "probes" through the decision space at discrete time steps. Each probe's location is specified by its position vector computed from equations of motion under the influence of gravity. Each probe experiences an acceleration created by the gravitational pull of masses in the decision space. The acceleration causes the probe to move from an initial position to a new position. New probe distribution is computed by combining the acceleration and probe position. The value of the objective function is its fitness computed at each probe's location. The fitness is used in a user-defined function that becomes the CFO mass.

This chapter investigates the physics of the CFO and the CFO metaphor, presents the analysis on probe distribution and decision space, the CFO algorithm with a discussion on different parameters of CFO algorithms, and convergence analysis, variants and finally presents a few applications to engineering problems.

Chapter 4: Electromagnetism-Like Optimization—EMO is based on the principles of EM, a branch of physics. The EM-based optimization method imitates the attraction–repulsion mechanism of the EM theory in order to solve unconstrained or bound-constrained global optimization problems. A solution in the EMO algorithm is seen as a charged particle in the search space and its charge relates to the objective function value. The EMO algorithm is inherently stochastic, and as a consequence, EMO exhibits better diversity because all particles traverse throughout the decision space to an optimal point.

This chapter presents the principles of the EMO algorithm, different variants of EMO algorithms, and their applications to various domains.

Chapter 5: Harmony Search—HS is a music-inspired meta-heuristic search and optimization algorithm. Musical performances seek to find pleasing harmony (a perfect state) as determined by an esthetic standard, just as the optimization process seeks to find a global solution (a perfect state) as determined by an objective function. The pitch of each musical instrument determines the esthetic quality, just as the objective function value is determined by the set of values assigned to each decision variable. The concept behind the algorithm is to find a perfect state of harmony, which is like any esthetic quality. Harmony search can be formulated as finding a set of random notes, series of pitches, and adjustment of pitches. The HS meta-heuristic is an emerging optimization algorithm inspired by the underlying principles of music improvization. When musicians make up a harmony, they usually test various pitch combinations stored in their memories. The process of searching for optimal solutions to engineering problems is analogous to this efficient search for a perfect state of harmony. In music improvization, each player sounds any pitch within the possible range, together making one harmony vector. If all the pitches make a good harmony, that experience is stored in each player's memory, and the possibility to make a good harmony is increased the next time. Similarly, in engineering optimization, each decision variable initially chooses any value within the possible range, together making one solution vector. If all the values of decision variables make a good solution, that experience is stored in each variable's memory, and the possibility to make a good solution is also increased the next time.

This chapter investigates the music improvization and harmonic phenomena in music. It then investigates harmonic memory and harmony search and presents harmony search-based optimization methods, different variants and finally a few applications to engineering problems.

Chapter 6: Water Drop Algorithm—In nature, water has the general tendency of flowing toward the ocean. By virtue of this natural tendency, water drops flow into water bodies such as rivers, lakes, and seas. While flowing over the land, the water drops change the environment in many ways. The environment also has effects on the paths of the flowing water. In general, river flows from high terrain to lower terrain due to gravitational force and finally joins its destination like the sea or a lake. The gravitational force is straight toward the center of the earth. Therefore, the water of rivers would follow a straight path toward the destination, making the path shortest from the source to the destination. But due to natural obstacles and barriers, the river makes a number of bends and turns in its path. Even then, there is always a tendency of changing the real path to close to the ideal path. This continuous effort tries to make a better path over time. The velocity of water causes the river water to transfer an amount of soil from one place to another place. The velocity of water decreases as the river approaches the sea. This soil is usually transferred from the higher velocity parts of the river to the slower parts. The removed soils which are carried by the water drops are deposited in the beds of the river with slower flow. By taking these two properties into account, the WDA algorithm has been proposed by Hamed Shah Hosseini, which imitates the dynamics of river systems. The WDA algorithm changes the amount of soil on the paths that water drops traversed. This variation depends on the velocity and the soil carried by the water drop, and it can be increased or decreased to attract or obstruct other water drops.

This chapter investigates the natural water drops and principles of the intelligent WDA (IWDA). It presents the IWDA, the parameters of the IWDA, the convergence properties, variants and finally a few applications to engineering problems.

Chapter 7: Spiral Dynamics Optimization—Most recently, a new multipoint meta-heuristics search method has emerged for two-dimensional continuous optimization problems based on the analogy of spiral phenomena in nature, called two-dimensional SpO, first proposed by Tamura and Yasuda in 2010. Focused spiral phenomena are approximated to logarithmic spirals, which frequently appear in nature, such as whirling currents, low-pressure fronts, nautilus shells, and arms of spiral galaxies. A remarkable point about logarithmic spirals is that their discrete processes generating spirals can realize effective behavior in meta-heuristics. Two-dimensional SpO uses the feature of logarithmic spirals.

SpO is a multipoint search for continuous optimization problems. The SpO model is composed of plural logarithmic spiral models and their common center. The spiral model has two specific setting parameters: the convergence rate and the rotation rate whose values characterize its trajectory. The common center is defined as the best point in all search points. The search points moving toward the common center with logarithmic spiral trajectories can find better solutions and update the common center.

This chapter investigates the spiral phenomena in nature and presents logarithmic spiral models, two-dimensional spiral models, n-dimensional spiral models, and the stability analysis of spiral models. It also presents the SpO algorithm, its parameters, variants and finally a few applications to engineering problems.

Chapter 8: Simulated Annealing—SA is motivated by an analogy to annealing in solids. Metropolis developed a method for solving optimization problems that mimics the way thermodynamic systems go from one energy level to another. He thought of this after simulating a heat bath on certain chemicals. The method requires that a system of particles exhibit energy levels in a manner that maximizes the thermodynamic entropy at a given temperature value. Also, the average energy level must be proportional to the temperature, which is constant. This method is called SA. The structural properties of solids depend on the rate of cooling. If the liquid is cooled slowly enough, large crystals are formed. However, if the liquid is cooled quickly (quenched), the crystals contain imperfections. Metropolis's algorithm simulated the material as a system of particles. The algorithm simulates the cooling process by gradually lowering the temperature of the system until it converges to a steady, *frozen* state. In 1982, Kirkpatrick et al. used the idea of the Metropolis algorithm and applied it to optimization problems. The idea is to use SA to search for feasible solutions and converge to an optimal solution.

This chapter presents a detailed description of the thermodynamic process of annealing, the notion of optimization based on the annealing process, annealing and probability, choice of parameters, and the variants of SA-based algorithms and their parameters. This chapter also presents a few examples of benchmark optimization problems.

Chapter 9: Chemical Reactions Optimization—CRO is motivated by the analogy of chemical reactions that take place between chemical agents. A chemical reaction is a natural process of transforming unstable molecules to stable molecules through formation and destruction of chemical bonds. The process continues until it reaches the minimum free energy. This simple concept is utilized for optimization where molecules that take part in the reactions are representation of solutions of the problem at hand. The population of solutions undergoes a series of reactions and reaches a stable state, which is termed as the optimal solution.

This chapter presents a detailed description of the chemical reactions and the thermodynamic process of reactions' states, energy conservation, and the notion of optimization based on the chemical reaction process. The CRO algorithm is a recent addition to the meta-heuristic family; therefore, the number of variants is small. It discusses different choices of parameters, the variants of CRO-based algorithms, and their parameters. This chapter also presents examples of benchmark optimization problems and applications.

Chapter 10: Miscellaneous Algorithms—There are many other physics- and chemistry-based search and optimization algorithms dispersedly reported in the literature, which have not attracted much attention in the research community. These algorithms can be broadly categorized as motion-based algorithms, gravitational force-based algorithms, celestial mechanics and astronomy-based algorithms, EM-based algorithms, optics-based algorithms, thermodynamics-based algorithms, CBAs, and hydrology- and dynamics-based algorithms.

Among motion-based algorithms is CBO. Among gravitational force-based algorithms are SGO and GIO. BB–BC search, BHS, GbSA, APO, and IRS algorithms belong to celestial mechanics and astronomy-based algorithms. EM-based algorithms are CSS and HO algorithms. The optics-based algorithm is RO. Among CBAs are the ACPA, ACROA, CRA, and GBMO. RFDA and WCA are categorized as hydrology- and dynamics-based algorithms. These algorithms are mostly unknown. This chapter will make an attempt to assimilate those algorithms with some examples.

1.10 CONCLUSION

It is obvious from the above review that the field of NIC is large and expanding. Most of the algorithms reported in this volume have been implemented using MATLAB®. To keep the volume within an affordable size, it was not possible to provide MATLAB codes for all algorithms, but some links to valuable resources are provided, which will be very useful for the interested researchers. Moreover, there are huge resources available on the Internet for developing simulations and new applications using different programming languages.

A parallel development has been the emergence of the field of CI, mainly consisting of neural networks (Adeli and Karim, 1997; Adeli and Kim, 2001; Adeli and Park, 1995; Park and Adeli, 1997), and fuzzy logic (Siddique, 2014; Siddique and Adeli, 2013) in the past 20 years, starting with the seminal book of Adeli and Hung (1995), which demonstrated how a multiparadigm approach and integration of the three CI computing paradigms can lead to more effective solutions of complicated and intractable pattern recognition and learning problems. Memetic algorithms (Neri and Cotta, 2012) are also becoming increasingly popular (Boyd and Vandenberghe, 2004; Ong et al., 2010). It is observed that NIC and CI intersect. Some researchers have argued that swarm intelligence provides CI. The authors advocate and foresee more cross-fertilization of the two emerging fields. Evolving neural networks is an example of such cross-fertilization (Alexandridis, 2013; Cabessa and Siegelmann, 2014; Siddique and Adeli, 2013).

REFERENCES

Abbott, B. P., Abbott, R., Abbott, T. D., Abernathy, M. R., Acernese, F., Ackley, K., Adams, C., Adams, T., Addesso, P., Adhikari, R. X. et al. 2016. Observation of gravitational waves from a binary black hole merger, *Physics Review Letters*, 116(6), 061102.

Abdechiri, M., Meybodi, M. R., and Bahrami, H. 2013. Gases Brownian motion optimization: An algorithm for optimization (GBMO), *Applied Soft Computing*, 13, 2932–2946.

Adeli, H. 1994. *Advances in Design Optimization*, London, UK: Chapman and Hall.

Adeli, H. and Hung, S. L. 1995. *Machine Learning—Neural Networks, Genetic Algorithms, and Fuzzy Sets*, New York, NY: John Wiley & Sons.

Adeli, H. and Karim, A. 1997. Scheduling/cost optimization and neural dynamics model for construction, *Journal of Construction Management and Engineering, ASCE*, 123(4), 450–458.

Adeli, H. and Kim, H. 2001. Cost optimization of composite floors using the neural dynamics model, *Communications in Numerical Methods in Engineering*, 17, 771–787.

Adeli, H. and Park, H. S. 1995. A neural dynamics model for structural optimization—Theory, *Computers and Structures*, 57(3), 383–390.

Adeli, H. and Park, H. S. 1998. *Neurocomputing for Design Automation*, Boca Raton, FL: CRC Press.

Adeli, H. and Sarma, K. 2006. *Cost Optimization of Structures—Fuzzy Logic, Genetic Algorithms, and Parallel Computing*, West Sussex, UK: John Wiley & Sons.

Alatas, B. 2011. ACROA: Artificial chemical reaction optimization algorithm for global optimization, *Expert Systems with Applications*, 38, 13170–13180.

Alatas, B. 2012. A novel chemistry based meta-heuristic optimization method for mining of classification rules, *Expert Systems with Applications*, 39, 11080–11088.

Aldwaik, M. and Adeli, H. 2014. Advances in optimization of high-rise building structures, *Structural and Multidisciplinary Optimization*, 50(6), 899–919.

Alexandridis, A. 2013. Evolving RBF neural networks for adaptive soft-sensor design, *International Journal of Neural Systems*, 23(6), 1350029, 14 pp.

Ampere, A.-M. 1820. Memoir on the mutual action of two electric currents, *Annales de Chimie et Physique*, 15, 59.

Anderson, K. 1983. The mathematical techniques in Fermat's deduction of the law of refraction, *Historia Mathematica*, 10, 48–62.

Anderson, R. L. 1953. Recent advances in finding the best operating condition, *Journal of American Statistical Association*, 48, 147–169.

Andriesse, C. D. 2005. *Huygens: The Man Behind the Principle*, Cambridge, UK: Cambridge University Press.

Aono, M. and Kunii, T. L. 1984. Botanical tree image generation, *IEEE Computer Graphics and Applications*, 4(5), 10–34.

Arango, C., Cortés, P., Onieva, L., and Escudero, A. 2013. Simulation-optimisation models for the dynamic berth allocation problem, *Computer-Aided Civil and Infrastructure Engineering*, 28(10), 769–779.

Atashpaz-Gargari, E. and Lucas, C. 2007. Imperialist competitive algorithm: An algorithm for optimization inspired by imperialistic competition, *IEEE Congress on Evolutionary Computation (CEC 2007)*, September 25–28, Singapore, pp. 4661–4667.

Avriel, M. and Wilde, D. J. 1966. Optimality proof for the symmetric Fibonacci search technique, *Fibonacci Quarterly*, 4, 265–269.

Baeck, T. 1996. *Evolutionary Algorithms in Theory and Practice*, New York, NY: Oxford University Press.

Bensaude-Vincent, B. 2003. Languages in chemistry, in: M. J. Nye, ed., *The Cambridge History of Science, Vol. 5: The Modern Physical and Mathematical Sciences*, Cambridge, UK: Cambridge University Press, Chapter 9.

Bergh, F. V. D. and Engelbrecht, A. P. 2006. A study of particle swarm optimization particle trajectories, *Information Sciences*, 176, 937–971.

Bersini, H. and Varela, F. J. 1990. Hints for adaptive problem solving gleaned from immune networks, Parallel Problem Solving from Nature, First Workshop PPSW 1, October, Dortmund, Germany.

Biot, J.-B. and Savart, F. 1820. Note sur le magnétisme de la pile de Volta, *Annales de Chimie et de Physique*, 15, 222–223.

Birbil, I. and Fang, S. C. 2003. An electro-magnetism-like mechanism for global optimization, *Journal of Global Optimization*, 25, 263–82.

Biswas, A., Mishra, K. K., Tiwari, S., and Misra, A. K. 2013. Physics-inspired optimization algorithms: A survey, *Journal of Optimization*, 2013, 438152, 16 pp.

Blum, C. 2005. Ant colony optimisation: Introduction and recent trends, *Physics of Life Reviews*, 2, 353–373.

Bonabeau, E., Dorigo, M., and Theraulaz, G. 1999. *Swarm Intelligence: From Natural to Artificial Systems*. New York, NY: Oxford University Press.

Bounds, D. G. 1987. New optimization methods from physics and biology, *Nature*, 329, 215–219.

Bowler, P. J. 2005. *Making Modern Science: A Historical Survey*, Chicago, IL: University of Chicago Press.

Box, G. E. P. 1957. Evolutionary operation: A method for increasing industrial productivity, *Applied Statistics*, 6(2), 81–101.

Boyd, S. and Vandenberghe, L. 2004. *Convex Optimization*, Cambridge, UK: Cambridge University Press.

Boyer, C. B. 1991. *A History of Mathematics*, 2nd ed., Revised by U. C. Merzbach, New York, NY: John Wiley & Sons.

Brady, R. M. 1985. Optimization strategies gleaned from biological evolution, *Nature*, 317, 804–806.

Brasch, F. E. 1931. *Johann Kepler 1571–1630: A Tercentenary Commemoration of his Life and Works*, Baltimore, MD: Williams and Wilkins.

Brock, W. H. 1993. *The Norton History of Chemistry*, 1st American ed., New York, NY: W. W. Norton.

Bronshtein, I. N. and Semendyyayev, K. A. 1998. *Handbook of Mathematics*, Berlin, Heidelberg: Springer.

Broyden, C. G. 1967. Quasi-Newton methods and their applications to function minimisation, *Mathmatics of Computation*, 21, 368–381.

Buffon, G. L. L. 1766. *Histoire Naturelle, générale et particulière, avec la description du Cabinet du Roi*, Paris, France: Imprimerie Royale.

Cabessa, J. and Siegelmann, H. T. 2014. The super-turing computational power of evolving recurrent neural networks, *International Journal of Neural Systems*, 24(8), 1450029, 22 pp.

Camazine, S., Deneubourg, J.-L., Franks, N. R., Sneyd, J., Theraulaz, G., and Bonabeau, E. 2001. *Self-Organization in Biological Systems*, Princeton, NJ: Princeton University Press.

Campomanes-Álvareza, B. R., Cordón, O., and Damasa, S. 2013. Evolutionary multi-objective optimization for mesh simplification of 3D open models, *Integrated Computer-Aided Engineering*, 20(4), 375–390.

Carnot, S. 1824. *Reflections on the Motive Power of Fire and Other Papers on the Second Law of Thermodynamics*, R. H. Thurston (transl.), 1960, New York, NY: Dover.

Cercignani, C. 1998. *Ludwig Boltzmann—The Man Who Trusted Atom*, Oxford, UK: Oxford University Press.

Chandrasekhar, S. 1931a. The highly collapsed configurations of a stellar mass, *Monthly Notices of the Royal Astronomical Society*, 91, 456–466.

Chandrasekhar, S. 1931b. The maximum mass of ideal white dwarfs, *Astrophysical Journal*, 74, 81–82.

Chen, X., Zhang, L., He, X., Xiong, C., and Li, Z. 2014. Surrogate-based optimization of expensive-to-evaluate objective for optimal highway toll charging in a large-scale transportation network, *Computer-Aided Civil and Infrastructure Engineering*, 29(5), 359–381.

Child, J. M. 1920. *The Early Mathematical Manuscript of Leibnitz*, C. I. Gerhardt (ed. and transl.), Chicago, IL: Open Court.

Chong, E. K. P. and Zak, S. H. 2008. *An Introduction to Optimisation*, 3rd ed., Hoboken, NJ: John Wiley & Sons.

Chow, J. Y. J. 2014. Activity-based travel scenario analysis with routing problem reoptimization, *Computer-Aided Civil and Infrastructure Engineering*, 29(2), 91–106.

Chuang, C. and Jiang, J. 2007. Integrated radiation optimization: Inspired by the gravitational radiation in the curvature of space–time, *IEEE Congress on Evolutionary Computation (CEC)*, 25–28, 3157–3164.

Cohen, I. B. 1971. *Introduction to Newton's Principia*, Cambridge, UK: Cambridge University Press.

Couzin, I. D., Karause, J., James, R., Ruxton, G. D., Franks, N. R. 2002. Collective memory and spatial sorting in animal groups, *Journal of Theoretical Biology*, 218, 1–11.

d'Alembert, J. R. 1747. Recherches sur la courbe que forme une corde tenduë mise en vibration (Researches on the curve that a tense cord forms when set into vibration), *Histoire de l'académie royale des sciences et belles lettres de Berlin*, 3, 214–219.

Dalton, J. 1808. *A New System of Chemical Philosophy*, Vol. 1, Part 1, Manchester, UK: R. Bickerstaff.

Dalton, J. 1810. *A New System of Chemical Philosophy*, Vol. 1, Part 2, Manchester, UK: R. Bickerstaff.

Dalton, J. 1827. *A New System of Chemical Philosophy*, Vol. 2, Manchester, UK: R. Bickerstaff.

Daneshyari, M. and Yen, G. G. 2004. Talent based social algorithm for optimization, *Proceedings of the 2004 IEEE Congress on Evolutionary Computation*, June 19–23, Portland, OR, pp. 786–791.

Darwin, C. 1859. *The Origin of Species by Means of Natural Selection or the Preservation of Favoured Races in the Struggle for Life*, London, UK: John Murray.

Das, S., Dasgupta, S., Biswas, A., Abraham, A., and Konar, A. 2009. On the stability of the chemotactic dynamics in bacterial-foraging optimisation algorithm, *IEEE Transaction on Systems, Man and Cybernetics—Part A: Systems and Humans*, 39(3), 670–679.

Davies, P. C. W. 1978. Thermodynamics of black holes, *Reports on Progress in Physics*, 41(8), 1313–1355.

De Castro, L. N. 2007. Fundamentals of natural computing: An overview, *Physics of Life Reviews*, 4, 1–36.

De Jong, K. A. 1975. *Analysis of the Behaviour of a Class Genetic Adaptive Systems*, PhD thesis, Dept. of Computer and Communications sciences, University of Michigan, Ann Arbor, MI.

De Jong, K. A. 2006. *Evolutionary Computation: A Unified Approach*, Cambridge, MA: MIT Press.

Descartes, R. 1954. *The Geometry*, D. E. Smith and M. L. Latham (Paperback ed. and transl.), New York, NY: Dover.

Dickinson, H. W. 1939. *A Short History of the Steam Engine*, London, UK: Cambridge University Press.

Dorigo, M., Birattari, M., and Stutzle, T. 2006. Ant colony optimization, *IEEE Computational Intelligence Magazine*, 1(4), 28–39.

Dorigo, M., Maniezzo, V., and Colorni, A. 1991. *Positive Feedback as a Search Strategy*, Technical Report 91-016, Dipartimento di Elettronica, Politecnico di Milano, Milan.

Dorigo, M., Maniezzo, V., and Colorni, A. 1996. Ant system: Optimization by a colony of cooperating agents, *IEEE Transaction on System, Man and Cybernetics—Part B*, 26(1), 29–41.

Dorigo, M. and Stützle, T. 2004. *Ant Colony Optimization*, Cambridge, MA: MIT Press.

Drake, S. 1973. Mathematics and discovery in Galileo's physics, *Historia Mathematica*, 1, 129–150.

Einstein, A. 1905. Über einen die Erzeugung und Verwandlung des Lichtes betreffenden heuristischen Gesichtspunkt, *Annalen der Physik*, 17(6), 132–148.

Einstein, A. 1915. Die Feldgleichungun der gravitation, *Sitzungsberichte der Preussischen Akademie der Wissenschaften zu Berlin*, 844–847.

Einstein, A. 1916. The foundation of the general theory of relativity, *Annalen der Physik,* 354(7), 769–822. 10.1002/andp.19163540702

Einstein, A. 1918. Über Gravitationswellen, *Sitzungsberichte der Königlich Preussischen Akademie der Wissenschaften Berlin, Part 1*, 154–167.

Eisberg, J. 2003. Solar science and astrophysics, in: M. J. Nye, ed., *The Cambridge History of Science, Vol. 5: The Modern Physical and Mathematical Sciences*, Cambridge, UK: Cambridge University Press, Chapter 26.

Elliott, R. S. 1993. *Electromagnetics*, Piscatway, NJ: IEEE Press.

Engels, F. 1934. *Dialectics of Nature*, Moscow, Russia: Progress Publishers, 5th Print.

Erol, O. K. and Eksin, I. 2006. A new optimization method: Big bang–big crunch, *Advances in Engineering Software*, 37(2), 106–111.

Eskandar, H., Sadollah, A., Bahreininejad, A., and Hamdi, M. 2012. Water cycle algorithm—A novel meta-heuristic optimization method for solving constrained engineering optimization problems, *Computers & Structures*, 110–111, 151–166.

Euler, L. 1750. Sur la Vibration des Cordes (On the vibration of strings), *Memoires de l'academie des sciences de Berlin*, 4, 69–85.

Faraday, M. 1855. *Experimental Researches in Electricity*, Vol. 3, Art. 3249, London, UK: Bernard Quaritch.

Farmer, J. D., Packard, N., and Perelson, A. 1986. The immune system, adaptation and machine learning, *Physica D*, 2, 187–204.

Faturechi, R. and Miller-Hooks, E. 2014. A mathematical framework for quantifying and optimizing protective actions for civil infrastructure systems, *Computer-Aided Civil and Infrastructure Engineering*, 29(8), 572–589.

Fister, I. Jr., Yang, X.-S., Fister, I., Brest, J., and Fister, D. 2013. A brief review of nature-inspired algorithms for optimisation, *Elektrotehniski Vestnik*, 80(3), 1–7.

Fletcher, R. 1987. *Practical Methods of Optimisation*, 2nd ed., Chichester, UK: John Wiley & Sons.

Flores, J., Lopez, R., and Barrera, J. 2011. Gravitational interactions optimization, in: C.A. Coello-Coello, ed., *Learning and Intelligent Optimization*, Berlin, Germany: Springer, pp. 226–237.

Fogel, D. B. 1995. *Evolutionary Computation—Toward a New Philosophy of Machine Intelligence*, Piscataway, NJ: IEEE Press.

Fogel, L. J. 1962. Autonomous automata, *Industrial Research*, 4, 14–19.

Forbes, E. G. 1977. Descartes and the birth of analytical geometry, *Historia Mathematica*, 4, 141–151.

Formato, R. A. 2007a. Central force optimization: A new metaheuristic with applications in applied electromagnetics, *Progress in Electromagnetics Research, PIER*, 77(1), 425–491.

Formato, R. A. 2007b. Central force optimization: A new nature inspired computational framework for multidimensional search and optimization, in: N. Krasnogor, G. Nicosia, M. Pavone, and D. A. Pelta, eds., *NICSO, Studies in Computational Intelligence*, Vol. 129, Berlin, Germany: Springer-Verlag, 221–238.

Fraser, A. S. 1957. Simulation of genetic systems by automatic digital computers, I. introduction, *Australian Journal of Biological Sciences*, 10, 484–491.

Fraser, C. 1983. J. L. Langrange's early contributions to the principles and methods of mechanics, *Archive for History of Exact Sciences*, 28, 197–241.

Friedberg, R. M. 1958. A learning machine: Part I, *IBM Journal of Research and Development*, 2(1), 2–13.

Friedmann, A. 1922. Ueber die Kruemmung des Raumes, *Zeitschrift fuer Physik*, 10, 377–386.

Frisch, K. 1967. *The Dance Language and Orientation of Bees*. Cambridge, Mass.: The Belknap Press of Harvard University Press.

Galilei, G. 1960. *On Motion, On Mechanics*, Madison, WI: University of Wisconsin Press.

Gao, H. and Zhang, X. 2013. A Markov-based road maintenance optimization model considering user costs, *Computer-Aided Civil and Infrastructure Engineering*, 28(6), 451–464.

Geem, Z. W. 2010. *Recent Advances in Harmony Search Algorithm, Studies in Computational Intelligence*, Berlin, Germany: Springer-Verlag.

Geem, Z. W., Kim, J. H., and Loganathan, G. V. 2001. A new heuristic optimization algorithm: Harmony search, *Simulation*, 76(2), 60–68.

Gendreau, M. and Potvin, J.-Y. 2010. *Handbook of Metaheuristics*, New York, NY: Springer-Verlag.

Giacobbe, F. W. 2005. How a type II supernova explodes, *Electronic Journal of Theoretical Physics*, 2(6), 30–38.

Gilbert, W. 1600. *De Magnete*, P. Fleury Mottelay (transl.), *On the Lodestone and Magnetic Bodies*, 1893, New York, NY: John Wiley & Sons.

Gillespie, C. C. 1971. *Lazare Carnot Savant*, Princeton, NJ: Princeton University Press.

Gillispie, C. C. 2000. *Pierre-Simon Laplace, 1749–1827: A Life in Exact Science*, Princeton, NJ: Princeton University Press.

Gillmor, C. S. 1971. *Coulomb and the Evolution of Physics and Engineering in the Eighteenth Century France*, Princeton, NJ: Princeton University Press.

Glover, F. 1977. Heuristics for integer programming using surrogate constraints, *Decision Sciences*, 8(1), 156–166.

Glover, F. 1986. Future paths for integer programming and links to artificial intelligence, *Computing and Operational Research*, 13(5), 533–549.

Glover, F. 1989. Tabu search—Part I., *ORSA Journal on Computing*, 1(3), 190–206.

Glover, F. 1994. Tabu search for nonlinear and parametric optimization (with links to genetic algorithms), *Discrete Applied Mathematics*, 49, 231–255.

Glover, F. 1998. A template for scatter search and path relinking, in: J.-K. Hao, E. Lutton, E. Ronald, M. Schoenauer, and D. Snyers, eds., *Artificial Evolution, Lecture Notes in Computer Science*, Vol. 1363, Berlin, Germany: Springer, 3–51.

Glover, F. and Kochenberger, G. A. 2003. *Handbook of Metaheuristic*, New York, NY: Kluwer Academic Publishers.

Glover, F. and Laguna, M. 1997. *Tabu Search*, Boston, MA: Kluwer Academic Publishers.

Glover, F., Løkketangen, A., and Woodruff, D. L. 2000. Scatter search to generate diverse MIP solutions, in: M. Laguna and J. L. González-Velarde, eds., *Computing Tools for Modeling, Optimization and Simulation: Interfaces in Computer Science and Operations Research*, Boston, MA: Kluwer Academic Publishers, 299–317.

Goldberg, D. E. 1989. *Genetic Algorithms in Search, Optimization, and Machine Learning*, Boston, MA: Addison Wesley Publishing Company.

Goldfeld, S. M., Quandt, R. E., and Trotter, H. F. 1966. Maximisation by quadratic hill climbing, *Econometrica*, 34, 541–551.

Grattan-Guinness, I. 1971. *The Development of the Foundations of Mathematical Analysis from Euler to Riemann*, Cambridge, MA: MIT Press.

Grattan-Guinness, I. 1972. *Joseph Fourier 1768–1830*, Cambridge, MA: The MIT Press.

Grove, W. R. 1874. *The Correlation of Physical Forces*, 6th ed., London, UK: Longmans, Green.

Gupta, M. S. 1980. George Simon Ohm and ohm's law, *IEEE Transaction on Education*, 23(3), 156–162.

Hallyn, F. 1993. *The Poetic Structure of the World: Copernicus and Kepler*, D. M. Leslie (transl.), New York, NY: Zone Books.

Hartley, H. O. 1961. The modified gauss-newton method for the fitting of non-linear regression function by least squares, *Technometrics*, 3, 269–280.

Harvey, E. N. 2004. Some physical properties of protoplasm, *Journal of Applied Physics*, 9(2), 68.

Hatamlou, A. 2013. Black hole: A new heuristic optimization approach for data clustering. *Information Sciences*, 222, 175–184.

Hawking, S. 1998. *A Brief History of Time—From the Big Bang to Black Hole*, London, UK: Bantam Press.

Hawking, S. and Penrose, R. 1970. The singularities of gravitational collapse and cosmology, *Proceedings of the Royal Society A*, 314(1519), 529–548.

Hejazi, F., Toloue, I., Noorzaei, J., and Jaafar, M. S. 2013. Optimization of earthquake energy dissipation system by genetic algorithm, *Computer-Aided Civil and Infrastructure Engineering*, 28(10), 796–810.

Helmholtz, H. 1871. *Ueber die Wechselwirkung der Naturkraefte, Ein populaerwissenschaftlicher Vortrag in Koenigberg in Prussen, 2. Heft*, Braunschweig, Germany: Vieweg.

Hertz, H. R. 1888. Ueber die Ausbreitungsgeschwindigkeit der electrodynamischen Wirkungen, *Annalen der Physik*, 270(7), 551–569.

Hestenes, M. and Stiefel, E. 1952. Methods of conjugate gradients for solving linear systems, *Journal of Research of the National Bureau of Standards*, 49(6), 409–436.

Hobson, E. W. 1914. *John Napier and the Invention of Logarithms, 1614*, Cambridge, UK: The University Press.

Hofmeyr, S. A. and Forrest, S. 2000. Architecture for an artificial immune system, *Evolutionary Computing*, 8(4), 443–473.

Holland, J. 1962. Outline for a logical theory of adaptive systems, *Journal of ACM*, 3, 297–314.

Holland, J. 1975. *Adaptation in Natural and Artificial Systems*, Ann Arbor, MI: University of Michigan Press.

Hooke, R. and Jeeves, T. A. 1961. Direct search solutions of numerical and statistical problems, *Journal of ACM*, 8, 212.

Hosseini, H. S. 2011. Principal component analysis by galaxy-based search algorithm: A novel meta-heuristic for continuous optimisation, *International Journal of Computational Science and Engineering*, 6(1–2), 132–140.

Hsiao, Y. T., Chuang, C. L., Jiang, J. A. and Chien, C. C. 2005. A novel optimization algorithm: Space gravitational optimization, *Proceedings of 2005 IEEE International Conference on Systems, Man and Cybernetics*, October 10–12, 2005, Waikoloa, HI, USA, pp. 2323–2328.

Hubble, E. 1929. A relation between distance and radial velocity among extra-galactic nebulae, *Proceedings of the National Academy of Sciences of the United States of America (PNAS)*, 15(3), 168–173.

Hubble, E. P. 1936. *The Realm of the Nebulae*, New Haven, CT: Yale University Press.

Hunt, B. J. 2003. Electrical theory and practice in the nineteenth century, in: M. J. Nye, ed., *The Cambridge History of Science, Vol. 5: The Modern Physical and Mathematical Sciences*, Cambridge, UK: Cambridge University Press, Chapter 17.

Iacca, G., Caraffini, F., and Neri, F. 2014. Multi-strategy coevolving aging particle optimization, *International Journal of Neural Systems*, 24(1), 1450008, 19 pp.

Inan, U. S. and Inan, A. S. 1999. *Engineering Electromagnetics*, Menlo Park, CA: Addison-Wesley Longman Inc.

Irizarry, R. 2004. LARES: An artificial chemical process approach for optimization, *Evolutionary Computation*, 12(4), 435–459.

Irizarry, R. 2005a. A generalized framework for solving dynamic optimization problems using the artificial chemical process paradigm. Applications to particulate processes and discrete dynamic systems, *Chemical Engineering Science*, 60, 5663–5681.

Irizarry, R. 2005b. Fuzzy classification with an artificial chemical process, *Chemical Engineering Science*, 60, 399–412.

Irizarry R. 2006. Hybrid dynamic optimization using artificial chemical process: Extended LARES-PR, *Industrial & Engineering Chemistry Research*, 45, 8400–8412.

Irizarry, R. 2011. Global and dynamic optimization using the artificial chemical process paradigm and fast Monte Carlo methods for the solution of population balance models, in: I. Dritsas, ed., *Stochastic Optimization—Seeing the Optimal for the Uncertain*, Rijeka, Croatia: InTech, Chapter 16.

Jackson, D.E. and Ratnieks, F. L. W. 2006. Communication in ants, *Current Biology*, 16(15), 570–574.

Jia, L., Wang, Y., and Fan, L. 2014. Multiobjective bilevel optimization for production-distribution planning problems using hybrid genetic algorithm, *Integrated Computer-Aided Engineering*, 21(1), 77–90.

Jiang, M., Luo, Y. P., and Yang, S. Y. 2007. Stochastic convergence analysis and parameter selection of the standard particle swarm optimization algorithm, *Information Processing Letters*, 102, 8–16.

Joule, J. P. 1841. On the heat evolved by metallic conductors of electricity, *Philosophical Magazine*, 19, 260–265.

Joule, J. P. 1850. On the mechanical equivalent of heat, *Philosophical Transactions of the Royal Society of London*, 140, 61–82.

Jourdian, P. E. B. 1912. Note on Fourier's influence on the conceptions of mathematics, *International Congress of Mathematicians*, Cambridge, UK, Vol. 2, pp. 526–527.

Kant, I. 1781. *Universal Natural History and Theory of the Heavens*, translated by Stephen Palmquist in Kant's Critical Religion, 2000, Aldershot, UK: Ashgate.

Kari, L. and Rozenberg, G. 2008. Many facets of natural computing, *Communications of the ACM*, 51(10), 72–83.

Kaveh, A. and Khayatazad, M. 2012. A new meta-heuristic method: Ray Optimization, *Computers and Structures*, 112–113, 283–294.

Kaveh, A. and Khayatazad, M. 2013. Ray optimization for size and shape optimization of truss structures, *Computers and Structures*, 117, 82–94.

Kaveh, A. and Mahdavi, V. R. 2014a. Colliding bodies optimization: A novel meta-heuristic method, *Computers and Structures*, 139, 18–27.

Kaveh, A. and Mahdavi, V. R. 2014b. Colliding bodies optimization method for optimum discrete design of truss structures, *Computers and Structures*, 139, 43–53.

Kaveh, A. and Talatahari, S. 2009. Size optimization of space trusses using big bang–big crunch algorithm, *Computers & Structures*, 87(17–18), 1129–1140.

Kaveh, A. and Talatahari, S. 2010. A novel heuristic optimization method: Charged system search, *Acta Mechanica*, 213(3–4), 267–289.

Karaboga, D. 2005. *An Idea based on Honey Bee Swarm for Numerical Optimisation*, Technical Report TR06, Erciyes University, Turkey.

Kennedy, J. and Eberhart, R. 2001. *Swarm Intelligence*, San Francisco, CA: Morgan Kaufmann Publishers, Inc.

Kenyon, I. R. 1990. *General Relativity*, Oxford, UK: Oxford University Press.

Kiefer, J. 1953. Sequential minimax search for a maximum, *Proceedings of the American Mathematical Society*, 4(3), 502–506.

Kim, H. and Adeli, H. 2001. Discrete cost optimization of composite floors using a floating point genetic algorithm, *Engineering Optimization*, 33(4), 485–501.

Kirkpatrick, S., Gelatto, C. D., and Vecchi, M. P. 1983. Optimization by simulated annealing, *Science*, 220, 671–680.

Knyazev, A. V. and Lashuk, I. 2008. Steepest descent and conjugate gradient methods with variable preconditioning, *SIAM Journal on Matrix Analysis and Applications*, 29(4), 1267.

Kociecki, M. and Adeli, H. 2013. Two-phase genetic algorithm for size optimization of free-form steel space-frame roof structures, *Journal of Constructional Steel Research*, 90, 283–296.

Kociecki, M. and Adeli, H. 2014. Two-phase genetic algorithm for topology optimization of free-form steel space-frame roof structures with complex curvatures, *Engineering Applications of Artificial Intelligence*, 32, 218–227.

Kociecki, M. and Adeli, H. 2015. Shape optimization of free-form steel space-frame roof structures with complex geometries using evolutionary computing, *Engineering Applications of Artificial Intelligence*, 38, 168–182.

Koutsoyiannis, D. and Angelakis, A. N. 2003. Hydrologic and hydraulic sciences and technologies in ancient Greek times, in: B. A. Stewart and T. Howell, eds., *The Encyclopaedia of Water Science*, New York, NY: Markel Dekker Inc., 415–418.

Koza, J. R. 1992. *Genetic Programming: On the Programming of Computers by Means of Natural Selection*. Cambridge, MA: The MIT Press.

Lagrange, J. L. 1760. Nouvelles Recherches sur la Natuer et la Propagation du Son, *Miscellanea Taurenencia*, 2, 11–172.

Lagrange, J. L. 1901. *Lectures on Elementary Mathematics*, T. J. McCormack (transl.), Chicago, IL: Open Court.

Lam, A. Y. S. and Li, V. O. K. 2009. *Chemical-Reaction-Inspired Meta-Heuristic for Optimization*, Technical Report TR-2009-003.

Lam, A. Y. S. and Li, V. O. K. 2010a. Chemical reaction optimization for cognitive radio spectrum allocation, *Proceedings of the IEEE Global Communications Conference (GLOBECOM 2010)*, December 6–10, Miami, FL, pp. 1–5.

Lam, A. Y. S. and Li, V. O. K. 2010b. Chemical-reaction-inspired meta-heuristic for optimization, *IEEE Transaction on Evolutionary Computation*, 14(3), 381–399.

Lam, A. Y. S., Li, V. O. K., and Yu, J. J. Q. 2012. Real-coded chemical reaction optimization, *IEEE Transaction on Evolutionary Computation*, 16(3), 339–353.

Lamarck, J.-B. 1809. *Philosophie Zoologique (in French)*, New ed., Paris, France: Germer Baillière.

Landau, L. D. 1932. On the theory of stars, in: D. ter Haar, ed., *1965, Collected Papers of L. D. Landau*. New York, NY: Gordon and Breach, originally published in *Physikalische Zeitschrift der Sowjetunion*, 1, 285.

Laplace, P. S. 1966. *Mecanique Celeste*, N. Bowditch (ed. and transl.), 4 Vols, Reprint of the 1829–1839 ed., New York, NY: Chelsea.

Lavoisier A. L. 1787. Mémoire sur la necessite de reformer et de perfectionner la nomenclature de la Chimie, In: *Methode de Nomenclature Chimique*, De Morveau, Lavoisier, Betholet (sic), & De Fourcroy, Paris: Cuchet, pp. 1–25.

Le Bon, G. 1896. *The Crowd: A Study of the Popular Mind*, Translation published by Batoche Books in 2001, Kitchener, Ontario, Canada.

Leibnitz, G. W. 1975 [1686]. *Energy - historical development of the concept*. English Translation by R. B. Lindsay, Hutchinson & Ross, Dowden.

Levenberg, K. 1944. A method for the solution of certain problems in least squares, *Quarterly Applied Mathematics*, 2, 164–168.

Li, G., Niua, P., and Xiao, X. 2012. Development and investigation of efficient artificial bee colony algorithm for numerical function optimization, *Applied Soft Computing*, 12, 320–332.

Lin, D. Y. and Ku, Y. H. 2014. Using genetic algorithms to optimize stopping patterns for passenger rail transportation, *Computer-Aided Civil and Infrastructure Engineering*, 29(4), 264–278.

Lin, M.-H., Tsai, J.-F., and Yu, C.-S. 2012. A review of deterministic optimization methods in engineering and management, *Mathematical Problems in Engineering: Optimization Theory, Methods, and Applications in Engineering*, 2012, 756023, 15 pp.

Lindenmayer, A. 1968. Mathematical models for cellular interactions in development, Parts I and II, *Journal of Theoretical Biology*, 18, 280–315.

Linnaeus, C. 1758. *Systema Naturae*, 10th ed., Stockholm, Sweden: Laurentius Salvius.

Liu, Y. and Passino, K. M. 2002. Biomimicry of social foraging bacteria for distributed optimization: Models, principles, and emergent behaviors, *Journal of Optimization Theory and Application*, 115(3), 603–628.

Luo, D., Ibrahim, Z., Xu, B., and Ismail, Z. 2013. Optimization the geometries of biconical tapered fiber sensors for monitoring the early-age curing temperatures of concrete specimens, *Computer-Aided Civil and Infrastructure Engineering*, 28(7), 531–541.

MacArthur, R. and Wilson, E. 1967. *Theory of Biogeography*, Princeton, NJ: Princeton University Press.

Mackinson, S. 1999. Variation in structure and distribution of pre-spawning Pacific herring shoals in two regions of British Columbia, *Journal of Fish Biology*, 55, 972–989.

Manjarres, D., Landa-Torres, I., Gil-Lopez, S., Del Ser, J., Bilbao, M. N., Salcedo-Sanz, S., and Geem, Z. W. 2013. A survey on applications of the harmony search algorithm, *Engineering Applications of Artificial Intelligence*, 26(8), 1818–1831.

Marco, B. 1998. Medicine and science in the life of Luigi Galvani, *Brain Research Bulletin*, 46(5), 367–380.

Marquardt, D. W. 1963. An algorithm for least squares estimation of nonlinear parameters, *Journal of the Society of Industrial and Applied Mathematics*, 11, 431–441.

Matyas, J. 1965. Random optimization, *Automation and Remote Control*, 26, 244–251.

Maxwell, J. C. 1865. A dynamical theory of the electromagnetic field, *Philosophical Transaction of Royal Society, London*, 155, 450.

Maxwell, J. C. 1892. *A Treatise on Electricity and Magnetism*, Oxford, UK: Clarendon Press.

Mayer, J. R. 1842. Bemerkungen ueber die Kraefte der unbelebten Natur, *Liebig's Annalen der Chemie und Phramazie*, 42, 239.

Melin, P., Astudillo, L., Castillo, O., Valdez, F., and Valdez, F. 2013. Optimal design of type-2 and type-1 fuzzy tracking controllers for autonomous mobile robots under perturbed torques using a new chemical optimization paradigm, *Expert Systems with Applications*, 40, 3185–3195.

Mendel, J. G. 1866. Versuche über Pflanzenhybriden, in: *Verhandlungen des naturforschenden Vereines in Brünn*, Band IV für das Jahr, 1865 Abhandlungen, 3–47, English translation: Druery, C. T. and Bateson, W. 1901. Experiments in plant hybridization, *Journal of the Royal Horticultural Society*, 26, 1–32.

Michell, J. 1784. On the means of discovering the distance, magnitude, and of the fixed stars, *Philosophical Transactions of the Royal Society*, 74, 35–57.

Molina-García, M., Calle-Sánchez, J., González-Merino, C., Fernández-Durán, A., and Alonso, J. I. 2014. Design of in-building wireless networks deployments using evolutionary algorithms, *Integrated Computer-Aided Engineering*, 21(4), 367–385.

Momen, S., Amavasai, B. P., and Siddique, N. H. 2007. Mixed species flocking for heterogenous robotic swarms, in: *The International Conference on Computer as a Tool (EUROCON 2007)*, Piscataway, NJ: IEEE Press, pp. 2329–2336.

More, J. J. 1977. The Levenberg-Marquardt algorithm: Implementation and theory, in: G. A. Watson, ed., *Numerical Analysis, Lecture Notes in Mathematics*, Berlin, Germany: Springer-Verlag, 105–116.

Nakrani, S. and Tovey, C. 2004. On honey bees and dynamic server allocation in internet hosting centers, *Adaptive Behaviours*, 12, 223–240.

Napier, J. 1969. *A Description of the Admirable Table of Logarithms*, Amsterdam, Netherlands: Theatrum Orbis Terrarum; New York, NY: Da Capo Press.

Nelder, J. and Mead, R. 1965. The downhill simplex method, *Computer Journal*, 7, 308–313.

Neri, F. and Cotta, C. 2012. Memetic algorithms and memetic computing optimization: A literature review, *Swarm Evolutionary Computing*, 2, 1–14.

Newton, I. 1972. *Isaac Newton's Philosophiae Naturalis Mathematica*, 3rd ed., A. Koyre and I. B. Cohen, eds., with variant readings, 2 Vols., Cambridge, UK: Cambridge University Press.

Newton, I. 1999. *The Principia: Mathematical Principles of Natural Philosophy*, I. B. Cohen, J. Budenz, and A. M. Whitman (transl.), Berkeley, CA: University of California Press.

Niwa, H. S. 1996. Newtonian dynamical approach to fish schooling, *Journal of Theoretical Biology*, 181, 47–63.

Oerstedt, H. C. 1820. Experiments on the effect of current of electricity on the magnetic needle, English translation in *Annals of Philosophy*, 16, 273.

Ong, Y. S., Lim, M. H., and Chen, X. S. 2010. Research frontier: Memetic computation past, present and future, *IEEE Computational Intelligence Magazine*, 5(2), 24–36.

Oppenheimer, J. R. and Snyder, H. 1939. On continued gravitational contraction, *Physical Review*, 56(5), 455–459.

Packel, E. W. and Traub, J. F. 1987. Information-based complexity, *Nature*, 328, 29–33.

Pancaldi, G. 2003. *Volta: Science and Culture in the Age of Enlightenment*, Princeton, NJ: Princeton University Press.

Park, H.S. and Adeli, H. 1997. Distributed neural dynamics algorithms for optimization of large steel structures. *Journal of Structural Engineering ASCE*, 123, 880–888.

Parrish, J. K., Viscido, S. V., and Grunbaum, D. 2002. Self-organized fish schools: An examination of emergent properties, *The Biological Bulletin*, 202, 296–305.

Passino, K. M. 2002. Biomimicry of bacterial foraging for distributed optimization and control, *IEEE Control System Magazine*, 22(3), 52–67.

Peng, F. and Ouyang, Y. 2014. Optimal clustering of railroad track maintenance jobs, *Computer-Aided Civil and Infrastructure Engineering*, 29(4), 235–247.

Pham, D. T., Ghanbarzadeh, A., Koc, E., Otri, S., Rahim, S., and Zaidi, M. 2005. *The Bees Algorithm, Technical Note, Manufacturing Engineering Centre*, Cardiff, UK: Cardiff University.

Planck, M. 1900a. Ueber eine Verbesserung des Wienschen Spektralgleichung, *Verhandlungen der Deutschen Physikalische Gesellschaft*, 2, 202–204.

Planck M. 1900b. Zur Theorie des Gesetzen der Energieverteilung im Normalspektrum, *Verhandlungen der Deutschen Physikalische Gesellschaft*, 2, 237–245.

Pledge, H. T. 1939. *Science Since 1500: A Short History of Mathematics, Physics, Chemistry and Biology*, London, UK: Ten Shillings Net.

Press, W. H., Flannery, B. P., Teukolsky, S. A., and Vetterling, W. T. 1992. *Numerical Recipes in FORTRAN: The Art of Scientific Computing*, 2nd ed., Cambridge, UK: Cambridge University Press.

Prigogine, I. 1996. *The End of Certainty*, New York, NY: The Free Press.

Rabanal, P., Rodríguez, I., Rubio, F. 2007. Using river formation dynamics to design heuristic algorithms, in *Unconventional Computation, UC'07, LNCS 4618*, 13-17 August 2007, Berlin, Heidelberg: Springer, 163–177.

Ramezani, F. and Lotfi, S. 2013. Social-based algorithm (SBA), *Applied Soft Computing*, 13, 2837–2856.

Rashedi, E., Nezamabadi-pour, H., and Saryazdi, S. 2009. GSA: A gravitational search algorithm, *Information Sciences*, 179(13), 2232–2248.

Rastrigin, L. A. 1963. The convergence of the random search method in the extremal control of a many parameter system, *Automation and Remote Control*, 24(10), 1337–1342.

Ray, T. and Liew, K. M. 2003. Society and civilization: An optimization algorithm based on the simulation of social behaviour, *IEEE Transaction on Evolutionary Computing*, 7(4), 386–96.

Rechenberg, I. 1965. *Cybernetic Solution Path of an Experimental Problem*, Library Translation No. 1122, Farnborough, UK: Royal Aircraft Establishment.

Rechenberg, I. 1973. *Evolutionstrategie-Optimierung Technischer Systeme nach Prinzipien der Biologischen Information*, Freiburg, Germany: Fromman Verlag.

Reynolds, C. 1987. Flocks, herds, and schools: A distributed behavioural model, *Computer Graphics*, 21(4), 25–34.

Reynolds, R. G. 1994. Introduction to cultural algorithms, in: A. V. Sebald and L. J. Fogel, eds., *Proceedings of the Third Annual Conference on Evolutionary Programming*, Singapore: World Scientific Press, pp. 131–139.

Reynolds, R. G. 1999. *An Overview of Cultural Algorithms: Advances in Evolutionary Computation*, New York, NY: McGraw-Hill Press.

Reynolds, R. G. and Chung, C. J. 1997. A cultural algorithm framework to evolve multi-agent cooperation with evolutionary programming, in: *EP '97 Proceedings of the 6th International Conference on Evolutionary Programming VI*, London, UK: Springer-Verlag, 323–333.

Sandage, A. 1975. *Galaxies and Universe-Stars and Stellar Systems*, Chicago, IL: University of Chicago Press.

Schaffer, S. 1979. John Michell and black holes, *Journal for the History of Astronomy*, 10, 42–43.

Schleiden, M. J. 1838. Beiträge zur Phytogenesis. *Archiv für Anatomie, Physiologie und wissenschaftliche Medicin*, 137–176.

Schwann, T. 1839. *Mikroskopische Untersuchungen über die Uebereinstimmung in der Struktur und dem Wachsthum der Tiere und Pflanzen*, Berlin, Germany: Sander.

Schwarzschild, K. 1916. Über das Gravitationsfeld eines Massenpunktes nach der Einsteinschen Theorie, *Sitzungsberichte der Königlich-Preussischen Akademie der Wissenschaften*, Sitzung 3, Februar 1916, Deutsche Akademie der Wissenschaften zu Berlin, pp. 189–196.

Schwefel, H.-P. 1968. Projekt MHD-Strausstrhlrohr: Experimentelle Optimierung einer Zweiphasenduese, Teil I, Technischer Bericht 11.034/68, 35, AEG Forschungsinstitut, Berlin, Germany.

Sebag, M., Schoenauer, M., and Ravise, C. 1997. Toward civilized evolution: Developing inhibitions, in: T. Bäck, ed., *Proceedings of the Seventh International Conference on Genetic Algorithms*, San Francisco, CA: Morgan Kaufmann, pp. 291–298.

Shadbolt, N. 2004. Nature-inspired computing, *IEEE Intelligent Systems*, 19(1), 2–3.

Shafahi, Y. and Bagherian, M. 2013. A customized particle swarm method to solve highway alignment optimization problem, *Computer-Aided Civil and Infrastructure Engineering*, 28(1), 52–67.

Shah-Hosseini, H. 2007. Problem solving by intelligent water drops, *Proceedings of IEEE Congress on Evolutionary Computation (CEC 2007)*, September 25–28, Swissotel The Stamford, Singapore, pp. 3226–3231.

Shah-Hosseini, H. 2008. Intelligent water drops algorithm—A new optimisation method for solving the multiple knapsack problem, *International Journal of Intelligent Computing and Cybernetics*, 1(2), 193–212.

Shaw E. 1962. The schooling of fishes, *Scientific America*, 206, 128–138.

Shaw, E. 1975. Fish in schools, *Natural History*, 84(8), 40–46.

Siddique, N. 2014. *Intelligent Control: Hybrid Approach Using Fuzzy Logic, Neural Networks and Genetic Algorithms*, Heidelberg, New York, Dordrecht, London: Springer-Verlag.

Siddique, N. and Adeli, H. 2013. *Computational Intelligence: Synergies of Fuzzy Logic, Neural Networks and Evolutionary Computing*, Chichester, UK: John Wiley & Sons.

Siddique, N. and Adeli, H. 2014. Water drop algorithms, *International Journal on Artificial Intelligence Tools*, 23(6), 1430002 (22 pages).

Siddique, N. and Adeli, H. 2015a. Harmony search algorithm and its variants, *International Journal of Pattern Recognition and Artificial Intelligence*, 29(8), 1539001, 21 pp.

Siddique, N. and Adeli, H. 2015b. Hybrid harmony search algorithms, *International Journal on Artificial Intelligence Tools*, 24(6), 1530001, 16 pp.

Siddique, N. and Adeli, H. 2015c. Applications of harmony search algorithms in engineering, *International Journal on Artificial Intelligence Tools*, 24(6), 1530002, 15 pp.

Simon, D. 2008. Biogeography-based optimization, *IEEE Transaction on Evolutionary computation*, 12(6), 702–713.

Siqueira, H., Boccato, L., Attux, R., and Lyra, C. 2014. Unorganized machines for seasonal stream flow series forecasting, *International Journal of Neural Systems*, 24(3), 1430009, 16 pp.

Smith, A. R. 1984. Plants, fractals, and formal languages, *Proceedings of SIG-GRAPH '84 in Computer Graphics, ACM SIGGRAPH*, July 22–27, Minneapolis, MN, pp. 1–10.

Smith, R., Ferrebee, E., Ouyang, Y., and Roesler, J. 2014. Optimal staging area locations and material recycling strategies for sustainable highway reconstruction, *Computer-Aided Civil and Infrastructure Engineering*, 29(8), 559–571.

Spears, D. F. and Spears, W. M. 2003. Analysis of a phase transition in a physics-based multiagent system, *Lecture Notes in Computer Science*, 2699, 193–207.

Storn, R. and Price, K. 1997. Differential evolution—A simple and efficient heuristic for global optimisation over continuous space, *Journal of Global Optimisation*, 11(4), 431–459.

Tamura, K. and Yasuda, K. 2011a. Primary study of spiral dynamics inspired optimisation, *IEEJ Transactions on Electrical and Electronic Engineering*, 6(S1), 98–100.

Tamura, K. and Yasuda, K. 2011b. Spiral dynamics inspired optimisation, *Journal of Advanced Computational Intelligence and Intelligent Informatics*, 15(8), 1116–1122.

Thompson, B. 1798. An inquiry concerning the source of the heat which is excited by friction, *Philosophical Transactions of the Royal Society of London*, 88, 80–102.

Truesdell, C. A. 1992. Cauchy and the modern mechanics of continua, *Revue d'Histoire des Sciences*, 45(1), 5–24.

Tsai, H. and Lin, Y. 2011. Modification of the fish swarm algorithm with particle swarm optimization formulation and communication behavior, *Applied Soft Computing*, 11, 5367–5374.

Tsypkin, Y. Z. 1971. *Adaptation and Learning in Automatic Systems*, Z. J. Nikolic (transl.), New York, NY: Academic Press.

Turgut, A. E., Çelikkanat, H., Gökçe, F., and Sahin, E. 2008. Self-organized flocking in mobile robot swarms, *Swarm Intelligence*, 2, 97–120.

Ursem, R. K. 1999. Multinational evolutionary algorithms, *Proceedings of the Congress on Evolutionary Computation (CEC'99)*, 3, 1640–1645.

Vitruvius, M. 1826. *De Architectura*, J. Cwilt (transl.), London, UK: Priestley and Weale.

Waldram, J. R. 1987. *The Theory of Thermodynamics*. 1st ed. Cambridge University Press.

Weber, W. 1848. Elektrodynamische maassbestimmungen, *Annalen der Physik und Chemie*, 73(2), 193–240.

Weisberg, J. M. and Taylor, J. H. 2004. Relativistic binary pulsar B1913+16: Thirty years of observations and analysis, *Binary Radio Pulsars ASP Conference Series*, eds: F. A. Rasio and I. H. Stairs, Vol. 328, July 7–9, 2004, Tennessee, USA.

Weise, T., Zapf, M., Chiong, R., and Nebro, A. J. 2009. Why is optimization difficult? in: R. Chiong, ed., *Nature-Inspired Algorithms for Optimisation*, Berlin, Germany: Springer, 1–50.

White, N. E. and Diaz, A. V. 2004. Beyond Einstein: From the Big Bang to black holes, *Advances in Space Research*, 34, 651–658.

Xing, B. and Gao, W.-J. 2014. Emerging chemistry-based CI algorithms, in: *Innovative Computational Intelligence: A Rough Guide to 134 Clever Algorithms*, Intelligent Systems Reference Library 62, Switzerland: Springer International Publishing, Chapter 26.

Yang, X.-S. 2005. Engineering optimisation via nature-inspired virtual bee algorithms, IWINAC 2005, *Lecture Notes in Computer Science*, 3562, 317–323.

Yang, X.-S. 2009. Firefly algorithms for multimodal optimization, in: *Stochastic Algorithms: Foundations and Applications*, SAGA, *Lecture Notes in Computer Sciences*, 5792, 169–178.

Yang, X.-S. 2010a. A new metaheuristic bat-inspired algorithm, in: C. Cruz, J. Gonzalez, N. Krasnogor, and G. Terraza, eds., *Nature Inspired Cooperative Strategies for Optimization (NISCO 2010)*, *Studies in Computational Intelligence*, Berlin, Germany: Springer, 284, 65–74.

Yang, X.-S. 2010b. *Engineering Optimisation: An Introduction with Metaheuristic Applications*, Hoboken, NJ: John Wiley & Sons.

Yang, X.-S. 2011. Bat algorithm for multi-objective optimization, *International Journal of Bio-Inspired Computation*, 3(5), 267–274.

Yang, X.-S. and Deb, S. 2010. Engineering optimisation by cuckoo search, *International Journal of Mathematical Modelling and Numerical Optimisation*, 1(4), 330–343.

Yang, S.-D., Yi, Y.-L., and Shan, Z.-Y. 2011. G_{best}-guided artificial chemical reaction algorithm for global numerical optimization, *Procedia Engineering*, 24, 197–201.

Yilmaz, H. 1965. *Introduction to the Theory of Relativity and the Principles of Modern Physics*, New York, NY: Blaisdel Pub. Co.

Zaránd, G., Pázmándi, F., Pál, K. F., and Zimányi, G. T. 2002. Using hysteresis for optimization, *Physical Review Letters*, 89(15), 150201-1–150201-4.

Zhang, G. and Wang, Y. 2013. Optimizing coordinated ramp metering—A preemptive hierarchical control approach, *Computer-Aided Civil and Infrastructure Engineering*, 28(1), 22–37.

2 Gravitational Search Algorithm

2.1 INTRODUCTION

Scientists observed the natural motion of celestial bodies and the natural tendency of earthly bodies to move toward the center of Earth. These observations help scientists in understanding the principles of motion and conclude the cause to be universal gravitation. During the sixteenth and seventeenth centuries, several scientific contributions were made, which set the stage for Newton's gravitational theory. Gravitation plays the central role in the motion, shaping the structure and evolution of all celestial bodies, suns, stars, galaxies, and the entire universe. Galileo Galilei made major progress in understanding the properties of natural motion and measured the constant acceleration of bodies falling toward the Earth and also investigated the effect of gravity on projectiles (see Galilei, 1960). Isaac Newton formulated the law of universal gravitation as part of the general physical law derived from empirical observations. Newton later formulated his theories and published them in *Philosophiæ Naturalis Principia Mathematica* (or *Mathematical Principles of Natural Philosophy*) in 1687 (see Newton, 1999). Thus, he founded the modern quantitative science of gravitation and developed it as an important science in the seventeenth century. He assumed the presence of an attractive force between all bodies in the universe. This force does not need any bodily contact, and can act at a distance. Many of the physical and natural phenomena, for example, ocean tides, motion of the planets, and discovery of new planets (e.g., Neptune), have been explained by means of universal gravitation and classical mechanics. Newton explained that the circular motion of the Moon round the Earth is due to the gravitational force exerted by the Earth on the Moon. Acceleration due to gravity, denoted as *g*, is of importance to many disciplines of natural sciences. The absolute value of gravity provides a base for the standard of mass and establishes the derived standard of force.

Innumerable bodies are distributed all over in the universe. Forces are acting between bodies due to gravitation. These forces cause acceleration of the bodies that set them in motion. According to gravitational theory, bodies with bigger mass have higher attractive force and attract other bodies with smaller masses toward themselves and bodies change their motion and direction. As a result bodies change their position with certain velocity toward the body of bigger mass and often merge with the bigger mass. This natural phenomenon can be seen as an optimization process where the entire bodies move due to the gravitational attraction force acting between them and directs the movements of all bodies globally toward the body with heavier masses. At some stage of the process, the movements of the bodies come to an equilibrium state when no further position changes take place. The circular motion of a smaller mass around the bigger mass such as the motion of the Moon around the Earth and the motion of the planets around the Sun is the resultant equilibrium state. This equilibrium state can be seen as an optimal state.

Rashedi et al. (2009) used the notion of the universal law of gravitation for optimization problems and developed the gravitational search algorithm (GSA). GSA is a new stochastic search algorithm inspired by the law of gravity. In the GSA, a collection of objects* (or bodies) interact with each other based on the Newtonian gravity and the laws of motion, completely different from other well-known population-based optimization methods. Each object has a mass. The position of the mass of an object at specified dimensions represents a solution of the problem and the inertial mass of an object reflects its resistance to make its movement slow. The mass of an object is computed by using the mechanism of fitness evolution of the problem. The positions of the objects in specified dimensions are updated at every iteration, and the best fitness along with its corresponding object

---

* Objects or bodies will be used interchangeably. An object or body has a mass.

is stored. The algorithm terminates after a prespecified number of iterations after which the best fitness at the final iteration becomes the global fitness for a particular problem and the positions at the specified dimensions of the corresponding object becomes the global solution of that problem.

This chapter will present the physics of gravity, mass of moving objects, and acceleration of objects that encompass the GSA. The parameters of the GSA and fitness evaluation will also be discussed. It will also present the different variants of the GSA and finally a few applications to engineering problems.

2.2 PHYSICS OF GRAVITY

There are four fundamental forces in nature: gravity, electromagnetism, nuclear strong force, and nuclear weak force. Gravity is a natural phenomenon by which every physical mass is being attracted by others. It is commonly recognized as the force that causes a physical body to fall from a height and gives weight to a mass. According to the Newton's law of universal gravitation, the gravitational force acts along the line joining the centers of the two objects that is proportional to the product of their masses and inversely proportional to the square of the distance between the two objects. Newton's law of gravity can be expressed mathematically as a relation described by

$$F = G \frac{m_1 m_2}{r^2}$$ (2.1)

where F is the magnitude of the gravitation force, the constant of proportionality G is known as the universal gravitation constant, m_1 and m_2 are the two masses of objects, and r is the distance between the two masses as shown in Figure 2.1. G is termed a universal constant because it is same at all places and all times, and thus universally characterizes the intrinsic strength of the gravitational force.

Gravity is the weakest of the four fundamental forces, and appears to have unlimited range (unlike the strong or weak force). The gravitational force is approximately 10^{-38} times the strength of the strong force (i.e., gravity is 38 orders of magnitude weaker), 10^{-36} times the strength of the electromagnetic force, and 10^{-29} times the strength of the weak force. Gravity is responsible for causing the Earth and the other planets to orbit the Sun; and for causing the Moon to orbit the Earth. Gravity is the only force acting on all objects with mass; has an infinite range; is always attractive and never repulsive; and cannot be absorbed, transformed, or shielded against. This means gravity acts between objects without any intermediary and without any delay (Holliday et al., 1993; Schutz, 2003). According to Equation 2.1, there is an attracting gravitational force among all bodies of the universe and the gravitational force is greater for bigger and closer bodies. Increasing the distance between two bodies will decrease the gravitational force between them. This is also illustrated in Figure 2.2. The total gravitational force F on a body of mass m produced by objects with masses m_i, $i = 1, 2, \ldots, n$, is given by

$$F = Gm \sum_{i=1}^{n} \frac{m_i r_i}{r_i^3}$$ (2.2)

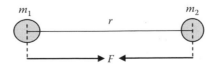

FIGURE 2.1   Gravitational force between two masses.

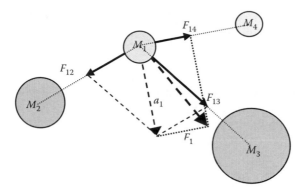

FIGURE 2.2 Acceleration vector and force vector.

where r_i is the special vector from the mass m to the masses m_i. The force on m due to masses m_i becomes the sum of the forces due to each mass separately. The individual forces are added together vectorially to yield the direction and magnitude of the force. The symbol Σ represents summation. The factor r_i/r_i^3 is used to yield the direction, which is numerically equivalent to division by r_i^2. Equation 2.2 is Newton's gravitational law in its vector form.

2.2.1 Acceleration of Objects

According to Newton's second law of motion, gravitational attraction force F acting on an object of mass m causes acceleration, denoted as a, which only depends on the force and its mass. Therefore, acceleration is defined by Newton's law as

$$a = \frac{F}{m} \tag{2.3}$$

The acceleration acting on objects changes the position and velocity of the objects. In general, objects attract each other toward themselves. An example of forces acting in different directions with accelerations is shown in Figure 2.2. The forces acting on M_1 from the masses M_2, M_3, and M_4 are denoted as F_{12}, F_{13}, and F_{14}, respectively. The overall resultant force F_1 is acting on M_1, which causes the mass M_1 to move with an acceleration of a_1 toward the mass M_3. Here, F_1 represents the direction and magnitude of the force.

Thus, the larger mass has a stronger attraction force, attracting the smaller masses toward it. Smaller masses merge with the larger mass and the larger mass becomes even larger.

2.2.2 Mass of Moving Objects

There are three kinds of masses considered in theoretical physics (Kenyon, 1990; Yilmaz, 1965) that a body can have:

1. *Active gravitational mass denoted as M_{ai}*: Active gravitational mass M_{ai} is a measure of the strength of the gravitational field due to a particular body. Gravitational field of a body with a small active gravitational mass is weaker than that of a body with larger active gravitational mass.
2. *Passive gravitational mass denoted as M_{pi}*: Passive gravitational mass M_{pi} is a measure of the strength of a body's interaction with the gravitational field. Within the same gravitational field, a body with a smaller passive gravitational mass experiences a smaller force than a body with a larger passive gravitational mass.

3. *Inertial mass denoted as M_{ii}:* Inertial mass M_{ii} is a measure of an object's resistance to changing its state of motion when a force is applied. Therefore, an object with a large inertial mass will move slowly, while an object with a small inertial mass will move rapidly.

At the turn of the twentieth century, the Hungarian physicist Lorant Eötvös found that different materials accelerate in the Earth's field at an identical rate. His experiments show the equality of gravitational and inertial mass for different substances (Eötvös, 1890). Recent experiments also observed the equality of accelerations. Newtonian theory of gravitation also states that the gravitational force is proportional to an object's mass. In order to simplify the computation of masses, it is assumed that the active gravitational mass and the passive gravitational mass are equal to the inertial mass. That is

$$M_{ai} = M_{pi} = M_{ii} = M_i, \quad i = 1, 2, \ldots, N \tag{2.4}$$

In the GSA, a system of masses is considered. Each object in the GSA has four features by which it is specified (Rashedi et al., 2009): position of the mass in the *d*-th dimension, inertial mass, active gravitational mass, and passive gravitational mass. It is, like a small artificial world of masses, obeying the Newtonian laws of gravitation and motion. More precisely, masses obey the following two laws:

1. *Law of gravity:* Each object attracts every other object and the gravitational force between two objects is directly proportional to the product of their masses and inversely proportional to the distance (R) between them as defined in Equation 2.1.
2. *Law of motion:* The current velocity of any mass is equal to the sum of the fraction of its previous velocity and the variation in the velocity. Variation in the velocity or acceleration of any mass is equal to the force acted on the system divided by the mass of inertia defined by Equation 2.3.

2.3 GRAVITATIONAL SEARCH ALGORITHM

The GSA is based on the metaphor of gravitational interaction between masses. Inspired by the theory of Newton's laws of motion and the laws of mass interactions, Rashedi et al. (2009, 2010) proposed the GSA for optimization problems. A hypothetical universe with a finite number of objects is considered in the GSA where every object in the universe attracts every other object with a gravitational force. Gravity minimizes the total gravitational potential energy within this system of objects assumed in the hypothetical universe. The performances of objects are measured by their masses. All these objects attract each other by the gravitational force, while this force causes a global movement of all objects toward the objects with heavier masses (i.e., objects with higher performances). The position of the object corresponds to a solution of the problem, and its gravitational and inertial masses are determined using a fitness function. The fitness function is problem specific, which measures the performance of the object that belongs to the system to be optimized. Both the gravitational mass and the inertial mass, which control the velocity of an object, are computed by fitness evolution of the problem. Objects with heavier masses represent good solutions and lighter ones represent worse solutions. Heavier masses move more slowly than lighter ones. The positions of the objects in specified dimensions are updated at each iteration and the best fitness along with its corresponding object is stored. The termination condition of the algorithm is defined by a fixed number of iterations. In other words, each object at a position presents a solution, and the algorithm is driven by properly adjusting the gravitational and inertial masses. Over time, it is expected that objects be attracted by the heaviest mass (object with the highest fitness function value). This object with its current position presents an optimal solution in the search space.

In the GSA, the masses that exist in the hypothetical universe are of theoretical importance. Virtually, they do not have any real mass. In order to compute the mass, an elegant fitness mechanism is devised in the GSA so that they can be ranked only with respect to their fitness. Therefore, the mass assigned to the objects must be a function of their fitness value. As referred to in Rashedi et al. (2009), the mass is calculated through comparison of the fitness assigned to each object defined by Equations 2.5 through 2.7. In the GSA, the inertial mass is calculated according to the fitness function value of the object. It is easy to understand that an object with a heavier mass (i.e., better performance) has a strong gravitational field and moves slowly, because it has larger inertial mass. According to the fitness value, the inertial mass of the i-th object is defined as

$$m_i(t) = \frac{\text{fit}_i(t) - \text{worst}(t)}{\text{best}(t) - \text{worst}(t)} \tag{2.5}$$

where $\text{fit}_i(t)$ is the fitness function value of the i-th object, and $\text{best}(t)$ and $\text{worst}(t)$ represent the strongest and weakest object, respectively, within the population with respect to their fitness function value. For a minimization problem, the $\text{best}(t)$ and the $\text{worst}(t)$ are defined as

$$\begin{cases} \text{best}(t) = \min_{j \in 1,\dots,N} \text{fit}_j(t) \\ \text{worst}(t) = \max_{j \subset 1,\dots,N} \text{fit}_j(t) \end{cases} \tag{2.6}$$

This means that objects with higher fitness values have higher attractions and move more slowly. The mass of the objects can be distributed over a wide range, which affects their mutual gravitational interaction. It has been observed that assigning the mass after normalizing them provides a better simulation of the force acting between the objects. The normalized mass M_i of the objects are expressed as

$$M_i(t) = \frac{m_i(t)}{\sum_{j=1}^{N} m_j(t)} \quad \text{with} \quad 0 \le M_i(t) < 1 \tag{2.7}$$

All masses attract each other by a gravitational force, and this force causes a global motion of all masses toward the heavier masses. The heavier masses have higher fitness values and they offer better optimal solution to the problem and they move more slowly than lighter ones representing poorer solutions. In the GSA, a mass has four particular features: position, inertial mass, active gravitational mass, and passive gravitational mass (Rashedi et al., 2009, 2010). The position of the mass corresponds to a solution of the problem. The gravitational and inertial masses are determined by evaluating a fitness function. In other words, each mass contributes to a solution and the algorithm steers to solutions through appropriately adjusting the gravitational and inertial masses (Ceylan et al., 2010; Rashedi et al., 2009, 2010).

Let us consider a system with N objects in the free space. Each object has a mass and all the masses are attracted toward the heavier ones and the collection of masses ultimately converges to the optimal mass. From Newton's theory of gravity, the gravitational force acting between two masses is defined as F in Equation 2.1 (Rashedi et al., 2009). To develop an efficient GSA, the first thing is to distribute the objects randomly over the entire search space and observe their gravitational interaction. At a specific time t, masses are calculated according to the fitness function.

According to Newton's gravitation theory, a gravitational force from mass j that acts on mass i is specified as follows:

$$F_{ij}^d = G(t)\frac{M_i(t)\times M_j(t)}{R_{ij}(t)}\left(x_j^d(t)-x_i^d(t)\right) \tag{2.8}$$

where M_i is the normalized mass of the object i, M_j is the normalized mass of the object j, $G(t)$ is the gravitational constant at time t, and $R_{ij}(t)$ is the distance between object i and object j. Rashedi et al. (2009) used $R_{ij}(t)$ instead of $R_{ij}^2(t)$. It was found in experimental studies for a number of benchmark optimization problems that R provided better results than R^2. Initially, the objects are distributed all over the search space and the distance between objects is nonzero. As the objects move toward the heavier ones under the gravitational force F_{ij}^d, the distance between objects become smaller and the distance $R_{ij}(t)$ tends to zero at an optimal mass. To avoid division by zero caused by the term $R_{ij}(t)$ in the denominator, a small constant term ε is added to $R_{ij}(t)$ in Equation 2.8. Thus, the gravitational force becomes

$$F_{ij}^d = G(t)\frac{M_i(t)\times M_j(t)}{R_{ij}(t)+\varepsilon}\left(x_j^d(t)-x_i^d(t)\right) \tag{2.9}$$

Most of the researchers used a constant small value for ε (Chatterjee et al., 2012; David et al., 2013; Duman et al., 2012a,b; Li et al., 2012b; Rashedi et al., 2009). Precup et al. (2012) suggested a decreasing value over time for ε with an initial starting value of ε_0.

$$\varepsilon(t) = \varepsilon_0 - \varepsilon_0\frac{(t-0.15t_{\max})}{0.85t_{\max}} \tag{2.10}$$

where t is the current iteration time and $t_{\max}$ is the maximum iteration time.

The position of the i-th object is given by $\mathbf{X}_i = \left(x_i^1,\ldots,x_i^d,\ldots,x_i^n\right)$ for $i = 1, 2, \ldots, N$, where x_i^d and x_j^d are the positions of the i-th and j-th objects at the d-th dimension, respectively. $R_{ij}(t)$ is the distance between object i and object j. Therefore, the distance $R_{ij}(t)$ between two objects is defined as the Euclidian distance expressed by

$$R_{ij}(t) = \left\|\mathbf{X}_j(t)-\mathbf{X}_i(t)\right\| \tag{2.11}$$

where $\mathbf{X}_i(t)$ and $\mathbf{X}_j(t)$ are the position vectors of the objects, and $\left\{x_i^d,x_j^d\right\}$ are the d-th elements in $\mathbf{X}_i(t)$ and $\mathbf{X}_j(t)$, respectively. It has been found that use of R_{ij} instead of R_{ij}^2 gives better convergence. There are also criticisms about the use of the term R_{ij} as it contradicts the basic principle of the Newtonian law of gravity (Gauci et al., 2012). To perform a good trade-off between exploration and exploitation, one strategy would be to reduce the number of objects and increase the number of generations (Pal et al., 2013).

The gravitational constant $G(t)$ in the real world is a constant with respect to time. However, in order to enhance the convergence and exploration of the approach, the gravitational constant $G(t)$ is a decreasing constant with the age of the universe (Mansouri et al., 1999). Rashedi et al. (2009) defined $G(t)$ for GSA as follows:

$$G(t) = G(t_0)\cdot\left(\frac{t_0}{t}\right)^{\beta} \tag{2.12}$$

where $G(t_0) = G_0$ is the gravitational constant at the first cosmic quantum-interval of time t_0, t is the time of the current iteration, and $\beta < 1$. The parameter β is used to control the value of $G(t)$. A lower value of β yields a higher value of $G(t)$. Since the value of β is less than 1, the initial value of $G(t)$ is very high. This helps in exploring a larger search space and guarantees a faster convergence (Pal et al., 2013). To extend the search space, some researchers such as Precup et al. (2012) proposed linear decrease for the gravitational constant $G(t)$ defined by

$$G(t) - G_0 \cdot \left(\frac{1 - \gamma \cdot t}{t_{\max}} \right) \tag{2.13}$$

where $\gamma > 0$ is an arbitrary parameter, which controls the convergence and search accuracy, and $t_{\max}$ is the maximum iteration time. In the binary version of the GSA, $G(t)$ is defined as a linearly decreasing function of time (Han et al., 2013, 2014) as

$$G(t) = G_0 \cdot \left(1 - \frac{t}{t_{\max}} \right) \tag{2.14}$$

Some applications may demand more aggressive or exponential decrease of the gravitational constant. Therefore, some researchers introduced an exponential decrease of $G(t)$ (Chatterjee et al., 2012; David et al., 2013; Duman et al., 2012a,b; Li et al., 2012b; Precup et al., 2012), defined by

$$G(t) = G_0 \cdot \exp\left(-\alpha \frac{t}{t_{\max}} \right) \tag{2.15}$$

where $\alpha > 0$ is an arbitrary parameter. It is important to explore the entire search space for modified or newly optima after a change in the environment. At the last stage of the search, when exploitation is preferred over exploration, convergence of $G(t)$ tends to G_0 asymptotically. The parameter α is used to control the convergence and search accuracy of the GSA. Chatterjee et al. (2012) reported $G_0 = 100$ and $\alpha = 20$, yielding good results in their experiments.

To give a stochastic characteristic to the GSA, the total force acting on the i-th object in dimension d is a weighted sum of the forces exerted from other objects (Chaterjee et al., 2012; David et al., 2013; Li et al., 2012b; Rashedi et al., 2009):

$$F_i^d(t) = \sum_{j=1, j \neq i}^{N} \lambda_j F_{ij}^d(t) \tag{2.16}$$

where λ_j is a random number within the interval of [0,1].

To strike a good compromise between exploration and exploitation, Rashedi et al. (2009) has proposed to reduce the number of attractive masses in Equation 2.16. This way, only a set of heavier masses exert force on the others. Exploration must be preferred at the beginning to avoid convergence toward a local optimum. Exploitation is introduced after a certain number of iterations and gradually dominating over exploration. In the GSA, therefore, only the $K_{\text{best}}(t)$ of the moving objects will attract the others. To implement this scheme, only K_{best} objects are allowed to attract other objects (Duman et al., 2012a,b; Pal et al., 2013).

$$K_{\text{best}} = N + \left\lceil \frac{1 - N}{t_{\max} - 1} \right\rceil (t - 1) \tag{2.17}$$

where $\lceil \cdot \rceil$ denotes the ceiling of the argument, t is the current iteration number, $t_{\max}$ is the maximum number of iterations, and N is the number of objects. The number of objects N has a direct impact on the computation of total force F_i^d and K_{best}. The value of N mainly depends on the problem and should be found empirically.

$K_{\mathrm{best}}(t)$ is the set of the first K objects that have the best fitness value (i.e., biggest gravitational mass), K is a function of time initialized with a value that decreases with time, and $K_{\mathrm{best}}(t)$ is chosen as a decreasing function of time. In the original GSA, the first iteration K equals the number of objects N and this value decreases to 1 linearly over time. At a final epoch, there will be just one mass applying force on the other masses. On the basis of this scheme, the total force exerted on a single object is given by the modified version of Equation 2.16 as follows:

$$F_i^d(t) = \sum_{j \in K_{\mathrm{best}}, j \neq i}^{N} \lambda_j F_{ij}^d(t) \tag{2.18}$$

The masses exist in the hypothetical universe. In order to compute the mass, an elegant fitness mechanism is devised in the GSA so that they can be ranked only with respect to their fitness. Therefore, the mass assigned to the objects must be a function of their fitness value. As referred to in Rashedi et al. (2009), the mass is calculated through a comparison of the fitness assigned to each object defined by Equations 2.3 through 2.7.

The gravitational attraction force acting on each object creates acceleration a, which is defined in Equation 2.2. This acceleration causes changes in the position and velocity of the objects. Consequently, these objects tend to move closer to each other eventually. In the GSA, the parameters are defined in such a way that the object with the best fitness value among the population attracts the whole collection of objects toward itself. In that course, some of the moving objects come closer to the optima and become the best object (evaluated as best mass). On the basis of the law of motion, the acceleration of the i-th object at the d-th dimension is calculated by

$$a_i^d(t) = \frac{F_i^d(t)}{M_i(t)} \tag{2.19}$$

where $M_i(t)$ is the mass of an object evaluated as per the fitness function defined in Equation 2.7. The searching strategy would be to compute the change of velocity v_i^d and the position x_i^d of the object. The position x_i^d is changed due to the velocity change of each object. For updating the position in the proposed optimization algorithm, it should calculate the velocity of each object. The velocity update is calculated according to the classical mechanics as follows:

$$v_i^d(t+1) = \rho_i \cdot v_i^d(t) + a_i^d(t) \times T \tag{2.20}$$

where ρ_i is a random number within the interval of [0,1] to give a random characteristic to the search algorithm.

The new position of the object is calculated by

$$x_i^d(t+1) = x_i^d(t) + v_i^d(t+1) \times T \tag{2.21}$$

In these relations, the time duration of the motion (T) is considered 1 such that acceleration in Equation 2.20 is converted into velocity and velocity in Equation 2.21 is converted into position for addition operation. In the rest of this chapter, time duration (T) will not be used just for avoiding repetition of the same information. To prevent the objects going out of the search space, the positions are bounded to the limits of each variable. The GSA can now be described by the following steps:

Step 1: Create N objects and make randomized initialization of n-dimensional positions
Set iteration $t = 1$
Step 2: Calculate fitness $\text{fit}_i(t)$ for $i = 1, 2, \ldots, N$
Calculate best fitness $\text{best}(t) = \min_{j \in 1, \ldots, N} \text{fit}_j(t)$
Calculate worst fitness $\text{worst}(t) = \max_{j \in 1, \ldots, N} \text{fit}_j(t)$
Calculate the mass using Equation 2.5
Update normalized mass using Equation 2.7
Step 3: Calculate gravitational constant using any of the methods from Equations 2.12 through 2.15
Step 4: Calculate distance between objects using Equation 2.11
Calculate the total force on the i-th object using Equation 2.18
Step 5: Calculate accelerations using Equation 2.19
Step 6: Update velocity using Equation 2.20
Update positions using Equation 2.21
Step 7: If (termination condition not satisfied), Goto Step 2
Step 8: Return solution

The algorithm of GSA is presented by the flow chart in Figure 2.3.

The basic advantage of the GSA is that it is a memoryless algorithm, that is, the best position of each object does not require storing. In the GSA, the gravitational force guides the masses and leads the masses to be absorbed by each other depending on the strength of the force. The disadvantage of the GSA is that if there is an occurrence of premature convergence, there is no recovery for the algorithm. In other words, the algorithm has no ability to explore and becomes inactive once it has converged. Therefore, a disruption mechanism has to be incorporated into the algorithm (Sarafrazi et al., 2011). There are some other distinct features of the GSA noted by Rashedi et al. (2009, 2010):

1. Gravitational force is an information transferring tool as each object in the GSA can observe the performance of the others.
2. A heavy mass has a larger effective attraction radius and attracts other masses with a higher intensity of attraction.
3. In the GSA, the objects tend to move toward the best objects with greater gravitational mass defined in terms of higher performance.
4. An object with heavy inertial mass provides a slower motion in the search space meaning exploring the search space more locally. Some researchers consider it as an adaptive learning rate. Conversely, a bigger gravitational mass causes a higher attraction of objects, which permits a faster convergence.
5. A large absolute value of the velocity means, in general, that the current position of the mass is not proper and large movement will be required to reach the optimum position.
6. A small absolute value of the velocity indicates that the current position of the mass is close to the optimum position and a small movement or small distance to be travelled is necessary to reach the optimum position. As it reaches the optimum position, the velocity becomes close to zero.
7. The gravitational constant decreases with time as it adjusts the accuracy of the GSA.

In the above discussion, it is assumed that the gravitational and inertial masses are the same. However, for some applications different values for each can also be used.

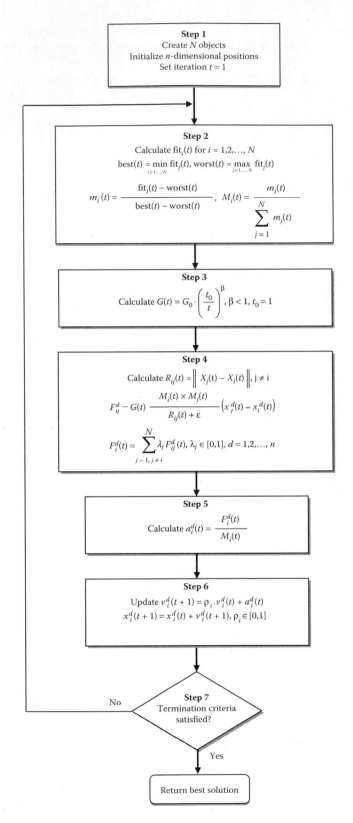

FIGURE 2.3   Flow chart of GSA.

2.4 PARAMETERS OF GSA

The general issue in the GSA is how exactly the parameter values can be estimated. There are, in general, nine parameters in the GSA that are required to be initialized, controlled, or tuned by the user to have a desired optimal performance of the algorithm. These parameters are N, K, K_{best}, G, ε, α, β, γ, and t_{max}. Some parameters appear to be much more important than others such as N, K, K_{best}, G, and ε (Table 2.1).

There is no systematic investigation carried out on the approximate value for the number of objects N required for an application. It is mainly problem dependent though but it has a direct impact on the computation of total force F_i^d and K_{best}. Most of the researchers set the value of N mainly from their experience or find it out empirically. In the GSA, K_{best} and G are two parameters that balance the exploration and exploitation during the search. To avoid trapping in a local optimum, any meta-heuristic algorithm should explore the search space at an early stage of iterations. In the GSA, this is accomplished by assigning high values to parameters $G_{min} \leq G \leq G_{max}$ and $K_{min} \leq K_{best} \leq K_{max}$ at the beginning, that is, the value of $K_{initial}$ and G_0 (or $G_{initial}$) must be high (Dowlatshahi et al., 2014). It is difficult to find a suitable value of G_0 for various test functions. Nobahari et al. (2011) suggested $G_0 = \beta \max_{d \in \{1,..,n\}} \left| x_U^d - x_L^d \right|$, where β is a parameter and $\left\{ x_U^d, x_L^d \right\}$ are, respectively, the upper and lower boundary of x_i^d for $i = 1, 2, \ldots, N$. The high value for parameter K allows an object to move in the solution space based on the position of other objects and consequently increases the exploration ability of the algorithm.

Also, a high value for parameter G increases the mobility of objects in the solution space and hence the exploration capability of the algorithm is increased. With high values for $K_{initial}$ and G_0, the good regions of the solution space are probably recognized in the early iterations of the algorithm. Over the iterations, the exploration of the GSA must fade out and its exploitation must fade in. Exploitation of the search space is accomplished by reducing the values of parameters K and G over time. The low value for parameter K causes objects to move in the solution space based on the position of few objects, increasing the exploitation ability of the algorithm consequently. The parameter $\beta < 1$ is used to control the value of $G(t)$ (as per Equation 2.12). $\alpha > 0$ is an arbitrary parameter, which controls the convergence and search accuracy of the GSA. By choosing suitable values for G_0 and α (as per Equation 2.15), the exploration can be guaranteed, which also ensures slow movement of heavier objects. $\gamma > 0$ is an arbitrary parameter to control the value of $G(t)$ (Equation 2.13), which eventually controls the convergence and search accuracy.

The gravitational constant $G(t)$ in the real world is a constant with respect to time. However, in order to enhance the convergence and exploration of the approach, the gravitational constant $G(t)$ is decreasing with the age of the universe as suggested by a number of researchers (Chatterjee et al., 2012; David et al., 2013; Duman et al., 2012a,b; Li et al., 2012b; Mansouri et al., 1999; Pal et al., 2013; Precup et al., 2012; Rashedi et al., 2009) in Equations 2.12 through 2.14. Also, the low value for parameter G decreases the mobility of each object in the solution space and hence the exploitation of the algorithm is increased. By choosing suitable values for parameters G_0 and α (as per

TABLE 2.1
Parameters of GSA

Parameters	Description
N	Total number of objects
K, K_{best}	K objects, best K objects
G	Gravitational constant
ε	A small constant term ε is added to $R_{ij}(t)$ to avoid division by zero
α, β, γ	Parameters to control the value of $G(t)$
t_{max}	Maximum number of iterations

Equation 2.12), the exploration can be guaranteed. Slow movement of heavier objects and the reduction of the participating objects can guarantee the exploitation ability of the GSA (Rashedi et al., 2009). Therefore, the good regions of the solution space are exploited in the final stage of iterations of the algorithm. The parameter ε is a small constant used to prevent division by zero while computing the force in Equation 2.9. It is chosen mostly arbitrarily (Chatterjee et al., 2012; David et al., 2013; Duman et al., 2012a,b; Li et al., 2012b; Rashedi et al., 2009). Some researchers also suggested a decreasing value for $\varepsilon(t)$ over time (Precup et al., 2012). t_{max} is the maximum number of iterations, which affects $G(t)$ (Equations 2.13 through 2.15), and is generally found empirically.

2.5 FITNESS FUNCTION

Each object in the GSA is specified by two parameters: position of the object in the d-th dimension and mass (Rashedi et al., 2009). The position of an object at specified dimensions represents a solution of the problem and the inertial mass of an object reflects its resistance to make its movement slow. Both the gravitational mass and the inertial mass, which control the velocity of an object in specified dimensions, are computed by fitness evolution of the problem. The positions of the objects in specified dimensions (solutions) are updated at every iteration and the best fitness along with its corresponding objects is recorded.

A fitness function is a criterion of optimality, also called the objective function, which is used to evaluate as a single figure of merit to see how close a given setting of parameters is to achieving the set of aims or the desired performance. In other words, this is a performance measure for individual objects, which is assigned as a fitness value $fit_i(t)$ to each individual at time t. Depending on the fitness value, it is decided whether the individual will be selected or not. This eventually defines the basis for improvement over iterations. The only thing that the fitness function must do is to rank the individuals in some way by producing the fitness value. The fitness function $fit_i(t)$ must be defined by the user, which is very problem dependent. $fit_i(t)$, $i = 1, 2, \ldots, N$, is the fitness of the solution defined within the scope of the problem, and best(t) and worst(t) represent the strongest and weakest mass within the population with respect to their fitness defined by best$(t) = \min_{j\in 1,\ldots,N} fit_j(t)$ and worst$(t) = \max_{j\in 1,\ldots,N} fit_j(t)$, respectively.

2.5.1 FITNESS SCALING

Proportionate selection of K_{best} individuals can cause premature convergence in earlier stages of the process due to the presence of an individual with a high fitness value that eventually dominates a population. Selection pressure decreases in the later generations as most of the individuals achieve the same fitness value. Scaling or readjustment of fitness values is sometimes useful to sustain a steady exploration or selective pressure on the population (Siddique and Adeli, 2013). The use of fitness scaling is common among researchers. The calculated raw fitness $fit_i(t)$ is transformed into a scaled fitness of $\hat{fit}_i(t)$, which is used for selection. There have been a variety of methods of scaling proposed by researchers (Cordon et al., 2001). In linear scaling, the scaled fitness $\hat{fit}_i(t)$ is defined as

$$\hat{fit}_i(t) = c_0 fit_i(t) + c_1 \tag{2.22}$$

where c_0 and c_1 are parameters, which can be static or dynamic to be adjusted based on the raw fitness distribution over the current population. In sigma truncation scaling, the scaled fitness is defined as

$$\hat{fit}_i(t) = \frac{fit_i(t) - \overline{fit}(t) - \sigma}{\sigma} \tag{2.23}$$

where $\overline{\text{fit}}(t)$ is the average fitness at time (iteration) t and σ is the standard deviation of the fitness values in the population.

2.6 VARIANTS OF GSA

The GSA has been an effective optimization algorithm compared with other meta-heuristic algorithms. However, there are two major issues in terms of the search performance:

1. Premature convergence at local optima occurs in the standard GSA due to rapid reduction of diversity.
2. Rapid convergence occurs at the beginning of the search process while very slow convergence near the optimum of the local search space, which induces more ineffective iterations to get an accurate estimation of the local optima.

A good meta-heuristic algorithm is one which can strike a balance between exploration and exploitation. Researchers have made much effort to improve the GSA to overcome the two major problems mentioned above. A number of variants have been reported in the literature recently since the publication of the GSA in 2009.

2.6.1 BINARY GSA

As discussed in Section 2.3, a large absolute value of the velocity indicates that a large movement is required to reach the optimum position. This means that a high probability value is required for the change in position of the mass to happen. On the other hand, a low probability value of the velocity will ensure that a mass close to the optimum position will move less frequently. In other words, a mass in a good position close to the optimum should have a velocity close to zero. On the basis of this idea, Rashedi et al. (2010) proposed the binary GSA (BGSA) by introducing a probability function for the absolute value of velocity such that the probability of changing the position x_i^d is low for small values of $\left|v_i^d\right|$ and the probability of changing the position x_i^d is high for large values of $\left|v_i^d\right|$. The probability function $P\left(v_i^d\right)$ is defined as

$$P\left(v_i^d(t)\right) = \left|\tanh\left(v_i^d(t)\right)\right| \tag{2.24}$$

$P\left(v_i^d\right)$ is bounded within the interval of [0,1] by using the absolute value of the hyperbolic function $\tanh(\cdot)$, which satisfies all the requirements mentioned earlier. The hyperbolic function $\tanh(\cdot)$ is defined as

$$\tanh(v) = \frac{e^v - e^{-v}}{e^v + e^{-v}} \tag{2.25}$$

The shape of the probability function $P\left(v_i^d\right)$ can also be controlled by introducing a shape parameter a to the hyperbolic function defined as

$$\tanh(v) = \frac{e^{av} - e^{-av}}{e^{av} + e^{-av}} \tag{2.26}$$

The parameter a controls the slope of the curve. Figure 2.4 illustrates the distribution of the probability function $P\left(v_i^d\right)$ with different values of slope parameter a. The parameter a will provide

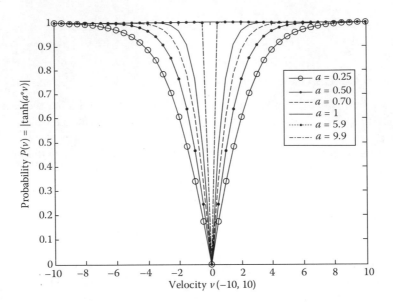

FIGURE 2.4 Probability function $P(v_i^d)$.

an extra flexibility to the probability distribution of the BGSA. Certainly, the choice of the shape function will again depend on the application itself. It can be seen from Figure 2.4 that the shape of the function is wider for smaller values of a. The shape of the function gets narrower for higher values of a and it is narrowest at $a = 5.9$. Interestingly, it does not get narrower any further for $a > 5.9$. Rashedi et al. (2010) investigated their BGSA for $a = 1$.

According to this simple rule of probability, the position x_i^d is updated as follows:

$$x_i^d(t+1) = \begin{cases} \overline{x_i^d(t)} & \text{if } \left(\text{rand} < P\left[v_i^d(t+1)\right]\right) \\ x_i^d(t) & \text{else} \end{cases} \tag{2.27}$$

where $\overline{x_i^d(t)}$ is the complement of $x_i^d(t)$. To achieve good convergence, Rashedi et al. (2010) limit the velocity $\left|v_i^d\right| < v_{\max}$. The main difference between the standard GSA and the BGSA is that in the BGSA the update of position switches between values of 0 and 1 with a given probability and all other steps are the same as the standard GSA. The updating of force, acceleration, and velocity are considered continuous similar to the standard GSA. In the BGSA, parameters such as G_0 and K_{best} are required to be tuned to achieve the best results.

Rashedi et al. (2010) applied the BGSA to a range of unimodal, multimodal, and discrete benchmark functions and reported improved results. They compared the performance of the BGSA with that of genetic algorithms and PSO. Their performances are almost the same and show a good convergence rate. Yuan et al. (2014) applied the BGSA to a unit commitment problem (UCP) with the decision variable representing on/off status of power-generating units. The simulation results for solving several UCP instances with the number of units in the range of 10–100 shows the competitive performance of the BGSA.

There are a few variants of the BGSA proposed by researchers to fit their own applications. Ibrahim et al. (2012) and Han et al. (2013) proposed the quantum-inspired BGSA (QBGSA) for solving the optimal power quality monitor (PQM) placement problem in power systems, which is discussed in later in this section.

2.6.2 Chaotic GSA

Though the GSA has been successful to a wide range of application domains, there are some obvious drawbacks of the algorithm in terms of performance issues. In general, a random number is used in updating the position of objects in the standard GSA, which is mostly responsible for premature convergence and getting stuck in a local optimum. Han and Chang (2012a) introduced chaotic dynamic operator to overcome these drawbacks of the standard GSA. In the modified GSA, a d-dimensional random vector $c_1^d \in [0,1]$ is produced. The chaotic sequence is then generated with iteration using the logistic function defined as

$$c_i^d = 4c_{i-1}^d \left(1 - c_{i-1}^d\right), \quad i = 2,3,\ldots,N \tag{2.28}$$

where c_i^d is now the chaotic vector and N is the number of objects. A logistic map is used to generate chaotic variables. These chaotic vectors are then added to the search space.

The velocity update rule is modified using chaotic vector c_i^d as follows:

$$v_i^d(t+1)' = \left[\text{rand}_i \times v_i^d(t) + \xi(c_i^d - 0.5) \right] + a_i^d(t), \quad i = 1,2,3,\ldots,N \tag{2.29}$$

where rand_i is a uniform random variable within the interval of [0,1] and ξ is a control factor for adjusting the scope of the chaos.

The positions are updated using $v_i^d(t+1)'$ according to

$$x_i^d(t+1)' = x_i^d(t) + v_i^d(t+1)' \tag{2.30}$$

The main steps of the chaotic GSA are as follows:

Step 1: Initialize positions $x_i^d(t)$, $i = 1, 2, \ldots, N, d = 1, 2, \ldots, n$
Step 2: Compute fitness
Step 3: Update $G(t)$, best(t), worst(t), $M_i(t)$
Step 4: Compute total force $F_i^d(t)$ and acceleration $a_i^d(t)$
Step 5: Generate chaotic vectors c_i^d using Equation 2.28
      Update velocity $v_i^d(t)$ using Equation 2.29
      Update position $x_i^d(t)$ using Equation 2.30
Step 6: If (termination condition not satisfied), Goto Step 2
Step 7: Return solution

The effectiveness and performance of the chaotic GSA were verified on digital filter modeling for reducing channel noise. The performances were compared with those of the standard GSA and PSO.

Chaos-based approaches have been successful in many meta-heuristic algorithms as well as in the GSA. Chaos has been used in the GSA in two ways: in the first approach, chaotic sequences are used to replace random numbers in parameters for the GSA; in the second approach, a chaotic search is employed as a local search technique to improve solution quality by exploiting the search space. Gao et al. (2014) proposed a chaotic GSA by combining the GSA with logistic map function to generate chaotic sequences to substitute random numbers and chaotic search to act as a local search approach (discussed later in this section). In the chaotic GSA, Gao et al. (2014) used the chaotic sequence c_j^d defined by logistic map function (shown in Equation 2.28) and replaced the random numbers with chaotic sequence c_i^d in the computation of the total force in Equation 2.18 and chaotic sequence c_i^d in computation of velocity in Equation 2.20 as follows:

$$F_i^d(t) = \sum_{j \in K_{\text{best}}, j \neq i}^{N} c_j^d F_{ij}^d(t) \tag{2.31}$$

$$v_i^d(t+1) = c_i^d \cdot v_i^d(t) + a_i^d(t) \tag{2.32}$$

They also generated the initial positions of objects using a chaotic sequence. All the steps in the algorithm of chaotic GSA remain the same as those of the standard GSA except the computation of force and velocity. To verify the effectiveness and performance, the chaotic GSA approach was applied to a set of benchmark functions of numerical optimization problems.

2.6.3 Piece-Wise Linear Chaotic Map and Sequential Quadratic Programming with GSA

Due to fast reduction of diversity and slow convergence when the GSA approaches close to the optimum of the local search space, the GSA encounters premature convergence. To counteract the premature convergence of the GSA, Han et al. (2014) proposed incorporation of a piece-wise linear (PWL) chaotic map into the GSA for increasing the population diversity and then integration of sequential quadratic programming (SQP) for accelerating local exploitation for the global best solution found by the combined chaotic PWL-GSA. While the PWL chaotic map helps in increasing the diversity of the search space (Baranovsky and Daems, 1995), SQP speeds up local search and improves the precision of the GSA. A simple PWL chaotic map is defined by

$$M(x,p) = \begin{cases} x/p & x \in [0,p] \\ (x-p)/\left(\frac{1}{2}-p\right) & x \in \left[0,\frac{1}{2}\right] \\ M[(1-x),p] & x \in \left[\frac{1}{2},1\right] \end{cases} \tag{2.33}$$

where $p \in \left[0,\frac{1}{2}\right]$ is a control parameter. The piece-wise chaotic map $M(x,p)$ has some distinct properties: (i) the system behaves chaotic within the interval of [0,1] satisfying the ergodicity and determination of certainty in the defined interval; (ii) the system is subject to uniform invariant distribution; and (iii) self-correlation function of the output trajectory is a δ-like function. These properties of $M(x,p)$ are exploited to enhance the diversity of the search space in this proposed modified GSA. Considering the fact that chaotic search is efficient within a small range of radius (Mendel et al., 2011), the position x_i^d is computed as follows:

$$x_i^d(t) = \text{best}(t)\left[1 + \omega \cdot M(-1,1)\right] \tag{2.34}$$

where $M(-1,1)$ is defined in terms of $M(0,1)$ as $M(-1,1) = 2M(0,1) - 1$, ω is a scaling parameter that can be chosen arbitrarily, $M(-1,1)$ is the chaotic map within the range $[-1,1]$, $x_i^d(t)$ denotes the position of the i-th object in the d-th direction at time t, and best(t) is the best fitness of all objects. The new position $x_i^d(t)$ can go out of bound and it needs to be brought back within the problem space of $[x_{\min}, x_{\max}]$ using the following equation:

$$x_i^d(t) = \max\left\{\min\left\{x_i^d(t), x_{\max}\right\}, x_{\min}\right\} \tag{2.35}$$

The position modification in Equations 2.34 and 2.35 is performed on $x_i^d(t)$ only if the Euclidean distance $R\left(x_i^d(t), \text{best}(t)\right) \geq 1$, that is, $x_i^d(t)$ is far from the best solution, and if $R\left(x_i^d(t), \text{best}(t)\right) < 1$, that is, $x_i^d(t)$ is close to the best solution. Before the next iteration, the global best solution found so far is used to begin a local search using the SQP method (Han et al., 2014) to exploit around the best solution. SQP is a local search method used for constrained nonlinear optimization problems. It provides fast convergence speed and high precision. If the local search SQP produces a better solution (denoted as $x_{\text{SQP}}(t)$), then best(t) is replaced with $x_{\text{SQP}}(t)$; otherwise, best(t) is retained. Han et al. (2014) applied the PWL-GSA-SQP approach to a feature subset selection (FSS) problem. Since many FSS problems are inherently set in binary space, a BGSA is finally employed in this hybrid approach of the GSA. Thus, the proposed new method is proved to be very successful for solving the FSS problem. The performance of the method has been verified on a set of benchmark problems.

2.6.4 GSA WITH CHAOTIC LOCAL SEARCH

During the exploration phase of the standard GSA, the algorithm experiences rapid reduction of diversity, leading to premature convergence. During the exploitation phase, the convergence becomes slow near the optimum of the local search space, leading to ineffective iterations in searching for the local optima. To improve the diversity and to make the local search more effective, a chaotic system is incorporated into the standard GSA as a kind of chaotic local search procedure that helps in improving global performance.

Chaos is typically a mathematical property of a dynamical system, which exhibits dynamic, unstable, pseudo-random, ergodic, and nonperiodic behavior. The behavior is sensitive to the initial value and can be controlled using a set of parameters (Elaydi, 1999). There are many chaotic maps used in chaotic search such as the logistic map, Chebyshev map, and iterative chaotic map with infinite collapse (ICMIC). A number of chaotic maps are presented in Appendix C. Logistic map, a discrete chaotic system with any dimensionality, can exhibit strange attractors. He et al. (2000, 2009) and Li et al. (2012b, 2013) reported that the ICMIC map with asymmetrical region $[-1,0) \cup (0,1]$ shows an advantage in chaotic search. The ICMIC map is defined by the following equation (Ott, 2002):

$$cx^{(k+1)} = \sin\left(\frac{a}{cx^{(k)}}\right) \quad \text{for} \quad a > 0, cx^{(k)} \in [-1,0) \cup (0,1] \tag{2.36}$$

where k denotes the iteration number at local search. Randomness and ergodicity of chaos variables make chaos optimization possible to achieve the optimum quickly. The local search is carried out for the current best solution X_g found by the GSA. The range around X_g could be the most promising area to exploit by using chaotic local search. The basic procedure of chaotic search is to generate chaos variables using a chaotic map such as the one defined in Equation 2.36, utilize chaos variables for determining variable intervals of the problem, search for an optimal solution, and then reduce the search radius $\delta(k)$ for an accelerated convergence defined as

$$\delta(k+1) = w * \delta(k) \tag{2.37}$$

where $w < 1$ is to be chosen arbitrarily. Li et al. (2012b) found that $w = 0.98$ gives the best result in an application for parameter optimization of a chaotic system. Chaotic local search is applied around the current best solution X_g. The chaotic local search procedure is described as follows:

Step 1: Set $k = 0$,
   Initialize randomly n chaotic variables $cx_i^{(k)} \in [-1,0) \cup (0,1]$, $i = 1, 2, \ldots, n$
   Set search radius $\delta_i(k)$
   Set optimization variables $x_{i\max} = X_g + \delta_i/2$, $x_{i\min} = X_g - \delta_i/2$, $X_{\text{best}} = X_g$, and $F_{\text{best}} = J$

Step 2: Determine chaotic variables $cx_i^{(k+1)}$ using Equation 2.36

Step 3: Calculate optimization variables

$$x_i^{(k+1)} = \frac{x_{i\max} + x_{i\min}}{2} + \frac{x_{i\max} - x_{i\min}}{2} cx_i^{(k+1)}$$

Step 4: Calculate new solutions $X_{k+1} = \left[x_1^{(k+1)}, \dots, x_i^{(k+1)}, \dots, x_n^{(k+1)} \right]$ by evaluation of objective function J_{k+1}

Step 5: Set $X_{best} = X_{k+1}$, if $J_{k+1} < F_{best}$, then set $F_{best} = J_{k+1}$

Step 6: Reduce search radius by using Equation 2.37

Step 7: If (max iteration/termination condition not reached), Goto Step 2

Step 8: Return solution

The main steps of the GSA with chaotic local search are as follows:

Step 1: Initialize positions $x_i^d(t)$, $i = 1, 2, \dots, N$, $d = 1, 2, \dots, n$

Step 2: Compute fitness

Step 3: Update $G(t)$, best(t), worst(t), $M_i(t)$

Step 4: Compute total force $F_i^d(t)$ and acceleration $a_i^d(t)$

Step 5: Update velocity $v_i^d(t)$ and position $x_i^d(t)$

Step 6: If (max iteration not reached), Goto Step 2

Step 7: Current best solution X_g

Step 8: Perform chaotic local search on X_g for a fixed number of iterations

Step 9: Return solution

Li et al. (2012b) applied a coarse GSA with a fine chaotic local search based on the current best solution found by the GSA, while chaotic local search seeks an optimal solution further. Their proposed approach shows more effectiveness and efficiency than the genetic algorithm, PSO, and the standard GSA.

The GSA with chaotic local search has been applied by a number of researchers over the past few years. Ju and Hong (2013) proposed a chaotic GSA by incorporating a chaotic sequence for improving local search. The chaotic sequence is generated using logistic map functions such as Equation 2.28. Chaotic local search is performed around the current best solution to improve the global convergence and to avoid the local optimum. The algorithm was applied to improve the performance of electricity load forecasting using a support vector regression model. Li et al. (2013) proposed a novel GSA to search for better parameters of the coarse antecedent membership function around the coarse results, in which chaotic search is embedded in the iteration of the standard GSA to search and replace the current best solution of the GSA. In an approach of using chaos in the GSA, Gao et al. (2014) employed chaotic search as a local search technique to improve solution quality by exploiting the search space. The chaotic local search is applied to the current global best solution X_g found by the GSA. The search neighborhood of X_g is constructed in a hypercube of $[X_g - r, X_g + r]^n$, where r is the radius of the neighborhood. To verify the effectiveness and performance of the GSA with chaotic local search, the algorithm was applied to a set of benchmark functions representative of numerical optimization problems.

2.6.5 DISCRETE GSA

The GSA was originally designed for solving continuous optimization problems. It is now well established that the efficiency of the standard GSA depends on how the algorithm exploits the local solution space. In multidimensional continuous space, local search is performed by moving the i-th object toward all members of the K_{best} objects in all dimensions in the search space. In discrete space, moves of the i-th object toward all members of the K_{best} objects are defined by the neighborhood space and represented by an undirected graph $G = (V, E)$ associated with the solution space of the problem. The nodes (or vertices) in V correspond to candidate solutions and the edges correspond to the moves in the neighborhood structure. A neighbor denoted by $N(x)$ for a solution $x \in V$ is generated by applying a small move m to the solution x using the move operator $\oplus$. The connectivity within the neighborhood space is important. For any two arbitrary solutions x_i and x_{i+1} in graph G, there should be a definite path of solutions from x_i to x_{i+1}. The concept of path relinking plays a role to generate and explore the trajectory in the neighborhood space connecting a starting solution x_i to a target solution x_n. This path relinking was originally proposed by Glover (1996) within the framework of scatter search. Using this path relinking strategy, Dowlatshahi et al. (2014) proposed a discrete GSA. Instead of using the classic way of moving from the current position to the target position, they introduced movement operators. A template of the basic path relinking strategy is shown for a movement length of l as follows:

While dist$(x_i, x_n) \neq 0$ Do
{
For $i = 1$ to l
    $x_{i+1} = x_i \oplus m$
}

where dist(.) is distance function.

Two movement operators are introduced: independent movement operator (IMO) and dependent movement operator (DMO). Dowlatshahi et al. (2014) applied the discrete GSA to a traveling salesman problem (TSP). They reported that the discrete GSA provided satisfactory results compared to 54 different algorithms and it ranked ninth in terms of performance.

2.6.6 MASS-DISPERSED GSA

The practical problems with some population-based meta-heuristic algorithms are that they tend to perform best when the optimum is located at or near the center of the search space (Davarynejad et al., 2012). In general, such meta-heuristic algorithms show various types of search bias. Understanding such biases is crucial to designing and implementing meta-heuristic algorithms. In order to address such biasness, Davarynejad et al. (2014) introduced two metrics: measuring center-seeking bias (CSB) and initialization region bias (IRB). An approach known as center offset was proposed by Monson and Seppi (2005), which changes the search space of the original problem by reducing it on one side and expanding it on the other. The optimal solution then moves away from the center of the search space.

In the case of the GSA, this kind of biasness is caused by the increase in the number of objects, which, in turn, smooth out the differences between the masses of the objects while computing the mass by using Equation 2.7. This is caused by the increase of the denominator term in the equation. In absolute terms, the exerting forces (attracting or repulsing) become almost equal for all objects to make any significant change in their position as a result of the applied gravitational force. The population of objects then can be seen as one object with a uniform mass distribution. Under the Newtonian gravitational force, the objects tend to move closer to the center of the population, resulting in an increase in the density of the population. As a result, they move more quickly toward the center of the search space (Davarynejad et al., 2012). This may explain the center-seeking behavior of the standard GSA.

Davarynejad et al. (2014) proposed a mass-dispersed GSA (mdGSA) that can recover from the center biasness. In this approach, based on their fitness, the objects are assigned a mass within an upper and lower bound, for example, within a range of $[M_U, M_L]$. This helps in dispersing the mass M_i well distributed over the whole search space. The calculation of the mass M_i in Equation 2.7 is now modified according to Davarynejad et al. (2014) as follows:

$$M_i = M_L + (M_U - M_L)\frac{\text{fit}_i(t) - \text{worst}(t)}{\text{best}(t) - \text{worst}(t)} \tag{2.38}$$

With the new definition of M_i, the algorithm recovers from the center biasness. The only change to the standard GSA shown in Figure 2.3 is the calculation of M_i in Step 2 and the rest of the algorithm remains the same. The main steps of the mdGSA are as follows:

Step 1: Initialize positions $x_i^d(t)$, $i = 1, 2, \dots, N$, $d = 1, 2, \dots, n$
Step 2: Compute fitness
Step 3: Update $G(t)$, best(t), worst(t)
Step 4: Compute $M_i(t)$ using Equation 2.38
Step 5: Compute total force $F_i^d(t)$ and acceleration $a_i^d(t)$
Step 6: Update velocity $v_i^d(t)$ and position $x_i^d(t)$
Step 7: If (termination condition not satisfied), Goto Step 2
Step 8: Return solution

Davarynejad et al. (2014) applied the mdGSA to 14 unimodal and multimodal benchmark functions and a gene regulatory network model identification problem and verified the performance of the mdGSA. The mdGSA showed improved performance in terms of accuracy and robustness.

2.6.7 OPPOSITION-BASED GSA

The notion of opposition is familiar to us all, for example, opposition party in politics, antiparticle in physics, and antithesis in dialectical philosophy. The concept of opposition-based learning was first introduced by Tizhoosh (2005). In the event of no *a priori* information about the solution, a random initialization is used in meta-heuristic algorithms. If the distribution of the initial guess is far away from the optimal solution, then the opposite should be close to the optimal solution. The notion of opposition-based learning is the simultaneous consideration of an estimate and its corresponding opposite estimate (i.e., guess and opposite guess) in order to achieve a better approximation for the current candidate solution. This idea has been applied to algorithms to accelerate learning (Shokri et al., 2006; Tizhoosh, 2006). Rahnamayan et al. (2007a,b) showed how an opposition-based initial population can accelerate EAs better than a random initial population. Rahnamayan et al. (2008) also investigated the effectiveness of opposition numbers over random numbers in soft computing techniques. The opposition number for a real $x \in [a,b]$ is defined by

$$\breve{x} = (a + b) - x \tag{2.39}$$

where a and b are the lower and upper bounds of the real value x, respectively. Similarly, the above definition can be extended to higher dimensions. Let $X = (x_1, x_2, \dots, x_n)$ be a point in n-dimensional space and $(x_1, x_2, \dots, x_n) \in R$, $x_i \in [a_i, b_i]$ for $\forall i \in \{1, 2, \dots, n\}$. The opposite of X is defined as

$$\breve{X} = (a_i + b_i) - x_i \quad \text{for} \quad \forall i \in \{1, 2, \dots, n\} \tag{2.40}$$

A recent survey on opposition-based learning can be found in Xu et al. (2014). Opposite numbers are employed in the initialization of the population of objects' positions in the opposition-based GSA. In opposition-based optimization, if the fitness of $\breve{X}$, i.e. $f(\breve{X})$, is better than the fitness of X, i.e. $f(X)$, then X is replaced with $\breve{X}$, otherwise X is retained. Shaw et al. (2012) first proposed the use of opposition-based numbers in the GSA and utilized opposition-based learning for population initialization and also for generation jumping, which also accelerates the performance and improves the convergence rate of the GSA. The standard GSA can now be modified to an opposition-based GSA by converting the positions into corresponding opposite numbers using Equation 2.40. The main steps of opposition-based GSA are as follows:

Step 1: Initialize positions $x_i^d(t)$, $i = 1, 2, \ldots, N$, $d = 1, 2, \ldots, n$
   Generate opposite population $\breve{x}_i^d(t)$ from initial population $x_i^d(t)$
Step 2: Compute fitness
Step 3: Update $G(t)$, best(t), worst(t), $M_i(t)$
Step 4: Compute total force $F_i^d(t)$ and acceleration $a_i^d(t)$
Step 5: Update velocity $v_i^d(t)$
Step 6: Update positions $x_i^d(t)$
Step 7: Perform opposition-based generation jumping
   Choose generation jumping rate J_r
   If $(rand < J_r)$
   $$\breve{x}_i^d(t) = \left[x_i^{d-\min}(t) + x_i^{d-\max}(t) \right] - x_i^d(t)$$

   Select N fittest individuals from $\left\{ x_i^d, \breve{x}_i^d \right\}$
Step 8: If (termination condition not satisfied), Goto Step 2
Step 9: Return solution

Shaw et al. (2012) verified the effectiveness and performance of an opposition-based GSA on 7 unimodal and 16 multimodal benchmark functions. They also applied the opposition-based GSA to different test systems of a combined economic and emission dispatch (EED) problem. Shaw et al. (2014) applied an opposition-based GSA to an optimal reactive power dispatch (RPD) problem of a power generation system and reported that the opposition-based GSA yields optimal settings of the control variables and also shows robustness and superiority of the proposed approach. Niknam et al. (2013) proposed a new approach by introducing a self-adaptive learning mechanism to the opposition-based GSA. The self-adaptive mechanism is eventually a mutation strategy where two mutation strategies are used adaptively to increase diversity in the solution space. All objects in the population have the same chance of taking part in the mutation controlled by a mutation probability. The proposed approach was verified on the multiobjective RPD and voltage control problem in a power generation system. Coelho et al. (2014) introduced quasi-opposition-based learning (Rahnamayan et al., 2007a,b) to the GSA. A quasi-opposition-based GSA utilizes opposite numbers to describe the mass. The opposite of a random number is likely to be closer to the mass than the original number. The effectiveness of quasi-opposition-based GSA was verified onto six-hump-camel-back function, Griewank function, Rosenbrock function, and pole shape of the magnetizer design problem. In all cases, the quasi-opposition-based GSA achieved an optimal solution.

2.6.8 GSA with Wavelet Mutation

In order to explore the solution space more efficiently for obtaining a better solution, Chatterjee et al. (2012) proposed a new variant of the GSA utilizing the wavelet theory for a finer mutation of the solutions. The mutation process postulates that the mutating search in hyperspace is limited. It is proposed that every particle will undergo mutation based on the mutation probability $p_m \in [0,1]$. If a generated random number is greater than or equal to p_m, the mutation will be performed on a solution; otherwise, it will keep the solution without mutation. The d-th element of a randomly selected object i will undergo mutation at time t after the following rule:

$$\bar{x}_i^d(t) = \begin{cases} x_i^d(t) + \sigma \cdot \left(x_i^{d-\max} - x_i^d(t) \right) & \text{if } \sigma > 0 \\ x_i^d(t) + \sigma \cdot \left(x_i^d(t) - x_i^{d-\min} \right) & \text{if } \sigma \leq 0 \end{cases} \quad (2.41)$$

where $\bar{x}_i^d(t)$ is the mutated new object at time t, and $x_i^{d-\min}$ and $x_i^{d-\max}$ represent the minimum and maximum value of the d-th element of the i-th object, respectively. σ is the wavelet function defined by $\sigma = \psi_{(a,0)}(\phi) = (1/\sqrt{a})\psi(\phi/a)$, where $\psi(\phi)$ is the Morlet wavelet, a is the dilation parameter, and $\psi_{(a,0)}(\phi)$ is the amplitude. For an increasing value of a, the amplitude of the wavelet will be scaled down defining the precision of the search space. Equation 2.41 represents the mutation strategy in the GSA. The new GSA with wavelet mutation is called GSAWM. The main steps of the GSAWM are as follows:

Step 1: Initialize population of positions $x_i^d(t)$, $i = 1, 2, \ldots, N$, $d = 1, 2, \ldots, n$
Step 2: Compute fitness
Step 3: Update $G(t)$, best(t), worst(t), $M_i(t)$
Step 4: Compute total force $F_i^d(t)$ and acceleration $a_i^d(t)$
Step 5: Update velocity $v_i^d(t)$ and position $x_i^d(t)$
Step 6: Perform wavelet mutation using Equation 2.41 with probability p_m
Step 7: If (termination condition not reached), Goto Step 2
Step 8: Return solution

Chatterjee et al. (2012) verified the effectiveness and performance of the GSAWM on an economic load dispatch (ELD) problem. The GSAWM provides a better-quality near-optimal solution with faster convergence. Saha et al. (2015) investigated the accuracy, speed of convergence, and stability of the GSAWM by applying the algorithm to design an eighth-order low-pass infinite impulse response (IIR) filter and compared the performance with that of a real-coded GA, PSO, and standard GSA.

2.6.9 QUANTUM-INSPIRED GSA

The quantum-inspired computing method is a numerical computation technique first introduced by Moore and Nayaranan (1995). In the standard GSA, only after calculating the velocity, it can obtain the position of each object according to the classical mechanics as described in Equations 2.19 and 2.20. According to Heisenberg principle of uncertainty (Schurmann and Hoffmann, 2009), it is impossible to measure the velocity and position of a particle simultaneously with an acceptable degree of accuracy and certainty. Soleimanpour-Moghadam et al. (2014), therefore, proposed a quantum-inspired GSA (QGSA) where no velocity and position term will be used for updating Equations 2.19 and 2.20, since the quantum state function can only provide a probability density function for the position y of the object. The position of an object is measured accurately according to the formulation below.

$$|y| = \frac{h}{2\sqrt{-2mE}} \ln\left(\frac{1}{\mathrm{rand}}\right) \tag{2.42}$$

$$\begin{cases} y = x - c = \pm \dfrac{h}{2\sqrt{-2mE}} \ln\left(\dfrac{1}{\mathrm{rand}}\right) \\ x = c \pm \dfrac{h}{2\sqrt{-2mE}} \ln\left(\dfrac{1}{\mathrm{rand}}\right) \end{cases} \tag{2.43}$$

where h is the Planck's constant, m is the mass of the object, E represents the energy of the object, and rand $\in [0,1]$ is a random value uniformly distributed over [0,1]. The term $h/(2\sqrt{-2mE})$ in Equation 2.43 is replaced with the term $\xi \cdot |c - x|$, where ξ is a constant, $|c - x|$ is used as a scaling factor for the variation of the new object around a solution from the K_{best} members, and c is the position center of an object selected at random from K_{best} set at each iteration. Equation 2.43 now can be written as

$$x = c \pm \xi \cdot |c - x| \cdot \ln\left(\frac{1}{\mathrm{rand}}\right) \tag{2.44}$$

Soleimanpour-Moghadam et al. (2014) provided a very good account of description and detailed derivation of the formulation, which is suggested for an interested reader. Equation 2.44 can be used for the computation of position of an object in an n-dimensional search space defined by

$$x_i^d(t+1) = \begin{cases} c_i^d + \xi \cdot |c_i^d - x_i^d(t)| \cdot \ln\left(\dfrac{1}{\text{rand}}\right) & \text{if } \rho \geq 0.5 \\[2ex] c_i^d - \xi \cdot |c_i^d - x_i^d(t)| \cdot \ln\left(\dfrac{1}{\text{rand}}\right) & \text{otherwise} \end{cases} \quad (2.45)$$

where $i = 1, 2, \ldots, N$, $d = 1, 2, \ldots, n$, and ξ is a parameter that determines the balance between exploration and exploitation, and $\rho \in [0,1]$ is a random number. Thus, the QGSA remains the same, except that the position update uses Equation 2.45. The main steps of the QGSA are described as follows:

Step 1: Initialize randomly n-dimensional positions x_i^d of N objects
Step 2: Evaluate fitness
Step 3: Update K_{best}, best(t), worst(t), $M(t)$
Step 4: Choose position center c by a probabilistic procedure
       Update position $x_i^d(t)$ using Equation 2.45
Step 5: If (termination condition not satisfied), Goto Step 2
Step 6: Return solution

The proposed QGSA was applied to 25 standard benchmark functions by Soleimanpour-Moghadam et al. (2014) and the effectiveness of the algorithm was confirmed by providing detailed experimental results and comparison with other meta-heuristic algorithms from the literature.

2.6.10 QUANTUM-INSPIRED BGSA

In order to avoid premature convergence and improve the efficiency of the BGSA (Han and Kim, 2002; Vlachogiannis and Lee, 2008), Ibrahim et al. (2012) proposed a QBGSA. The QBGSA integrates the concept and principles of quantum computing into the BGSA. The positions of decision variables (x_{ij}) in the QBGSA are represented by a string of Q-bits or Q-bit individuals. In quantum computing, a Q-bit individual is updated by a quantum gate (Q-gate). A Q-gate is a unitary operator U defined in the form of a rotation gate (Hey, 1999). The rotation gate is defined by

$$U(\Delta\theta) = \begin{bmatrix} \cos(\Delta\theta) & -\sin(\Delta\theta) \\ \sin(\Delta\theta) & \cos(\Delta\theta) \end{bmatrix} \quad (2.46)$$

where $\Delta\theta$ is the rotation angle.

Rotation angle ($\Delta\theta$) is used in the QBGSA proposed by Ibrahim et al. (2012). Since the concept of acceleration update is modified to obtain a rotational angle, the gravitational mass is replaced to the magnitude of the rotation angle. The rotation angle ($\Delta\theta$) is calculated according to

$$\Delta\theta_{ij}(t) = \sum_{k \in K_{\text{best}}, k \neq i} \left[\theta \times \gamma_i^k \times \left(x_{kj}(t) - x_{ij}(t) \right) \right] \quad (2.47)$$

where θ is the magnitude of the rotation angle within the interval of $[\theta_{\max}, \theta_{\min}]$ and γ_i^k is defined by

$$\gamma_i^k = \begin{cases} \lambda_i^k + 1 & \text{if } f(X_j) = f(X_{\text{best}}) \\ \lambda_i^k & \text{otherwise} \end{cases} \quad (2.48)$$

λ_i^k is defined by

$$\lambda_i^k = \begin{cases} 1 & \text{if } f(M_j) > f(M_i) \text{ and } R_{ij} \leq \tau \\ 0 & \text{otherwise} \end{cases} \qquad (2.49)$$

where τ is the maximum of the different numbers of bits between two objects obtained from the percentage of total bits.

The Q-bits are updated based on the rotation angle according to

$$\begin{bmatrix} \alpha_{ij}(t+1) \\ \beta_{ij}(t+1) \end{bmatrix} = U\left[\Delta\theta_{ij}(t)\right] \begin{bmatrix} \alpha_{ij}(t) \\ \beta_{ij}(t) \end{bmatrix} \qquad (2.50)$$

The position vector of the object is updated based on the probability of $|\beta|^2$ according to

$$x_{ij} = \begin{cases} 1 & \text{if rand} < \left|\beta_{ij}(t+1)\right|^2 \\ 0 & \text{otherwise} \end{cases} \qquad (2.51)$$

where rand is a random number uniformly distributed over [0,1].

Ibrahim et al. (2012) verified the performance of the QBGSA for solving the optimal PQM placement problem in power systems. The performance of the QBGSA was compared with the BGSA, quantum-inspired binary PSO (QBPSO) (Jeong et al., 2010), and binary PSO (BPSO). The experimental results show that the QBGSA is more effective and precise than the BGSA, QBPSO, and BPSO.

Han et al. (2013) also proposed a new QBGSA, where the positions of decision variables (x_{ij}) represented by a string of Q-bits (also called Q-bit individuals). The Q-bit (i.e., the state of each decision variable) is equal to 0 or 1 determined by the probability of two corresponding complex numbers $|\alpha|^2$ or $|\beta|^2$ defined by $|\psi\rangle = \alpha|0\rangle + \beta|1\rangle$ such that the probability of state can be normalized to unity as $|\alpha|^2 + |\beta|^2 = 1$. Q-bit individuals are updated using the rotation gate defined in Equation 2.46. The object's new position is determined by the rotation angle $(\Delta\theta)$ defined in Equation 2.47. The random variables for the calculation of force in Equation 2.18 and velocity in Equation 2.20 are removed to avoid random exploration. The small positive coefficient ε is avoided in acceleration calculation. Han et al. (2013) applied the QBGSA to feature selection in a classification problem along with the K-nearest neighbor (K-NN) algorithm (discussed later in this section), which shows improved accuracy with an appropriate feature subset for binary problems.

2.6.11 PIECE-WISE FUNCTION-BASED GSA

In a standard GSA, gravitational constant $G(t)$ plays an important role in determining the performance of the GSA by having an impact on the force. In general, an exponential form of equation, for example, Equation 2.15, is used for calculation of $G(t)$. Moreover, exploration and exploitation of the search space is also dictated by the value of $G(t)$. Li et al. (2014) proposed a piece-wise function-based GSA (PFGSA), where a PWL function is used to compute the gravitational constant by replacing the traditional exponential form of the equation. The piece-wise function provides a more rational gravitational constant and has more control over the convergence of the algorithm. The proposed PFGSA is applied to parameter identification of an automatic voltage regulator system, which demonstrated competitive performance compared to the standard GSA. The PWL function for $G(t)$ is defined as

$$G(t) = \begin{cases} \dfrac{G_1 - G_0}{t_1} t + G_0 & 0 \le t < t_1 \\[2ex] \dfrac{G_1 - G_2}{t_1 - t_2} t + \dfrac{t_1 G_2 - t_2 G_1}{t_1 - t_2} & t_1 < t \le t_2 \\[2ex] \dfrac{G_2 - G_3}{t_2 - N} t + \dfrac{t_2 G_3 - N G_2}{t_2 - N} & t_2 < t \le N \end{cases} \qquad (2.52)$$

The main steps of the PFGSA are as follows:

Step 1: Initialize positions $x_i^d(t)$, $i = 1, 2, \ldots, N$, $d = 1, 2, \ldots, n$
Step 2: Compute fitness
Step 3: Update $G(t)$ using PWL function
Step 4: Update K_{best}, best(t), worst(t), $M(t)$
Step 5: Compute total force $F_i^d(t)$ and acceleration $a_i^d(t)$
Step 6: Update velocity $v_i^d(t)$ and position $x_i^d(t)$
Step 7: If (termination condition not satisfied), Goto Step 2
Step 8: Return solution

The parameter selection of the piece-wise function in Equation 2.52 is critical to performance. Li et al. (2014) suggested $G_0 \in [20, 100]$, $G_1 = G_0/10$, $G_2 - G_1/100$, $G_3 \in [1e-4, 1e-10]$, $t_1 = N/4$, and $t_2 = N/25$, which yielded acceptable performance. Li et al. (2014) applied the PFGSA to parameter identification of an automatic voltage regulator used for maintaining the voltage of a synchronous generator. Eight key parameters were chosen for identification and the performances were compared with those of the GSA and PSO.

2.6.12 Adaptive GSA

The variants of GSA reported in the literature are mainly based on updates of the position or velocity vector of the GSA. It is found that the GSA is sensitive to a number of parameters such as gravitational constant $G(t)$ and ε. Precup et al. (2012) proposed adjustment of the two parameters $G(t)$ and ε during execution of the original GSA (Rashedi et al., 2009) and called it an adaptive GSA. The adaptive GSA procedure is divided into five stages: engagement, exploration, exploitation, elaboration, and evaluation.

In the engagement stage, N objects are generated randomly, the position of which is given by $X_i = \left(x_i^1, \ldots, x_i^d, \ldots, x_i^n \right)$, $i = 1, 2, \ldots, N$. Each object's fitness is calculated at this stage.

In the exploration stage, the search space is extended by deploying a linearly decreasing gravitational constant $G(t)$ defined by Equation 2.13. The exploration stage is carried out for 15% of the total iterations. The position and velocities are updated at this stage.

In the exploitation stage, a more aggressive decrease of the gravitational constant is deployed using an exponential decrease of $G(t)$ defined by Equation 2.15. ε is decreased from a starting value of ε_0 defined by Equation 2.10, which helps reducing distances between objects. It is suggested that the exploitation stage should be carried out for 45% of the total iterations.

In the elaboration stage, refining of the solutions is performed by replacing worst solutions with best solutions. The decreasing gravitational constant $G(t)$ is fixed to G_0 and $\varepsilon(t)$ continues with a decrease in value till t_{max}. It is suggested that the elaboration stage is carried out for the remaining 40% of the total iterations. In the evaluation stage, optimized parameters obtained from the adaptive GSA are applied to a real-world system.

The effectiveness and performance of the adaptive GSA were verified on the optimal tuning of Takagi–Sugeno proportional-integral (PI) fuzzy controllers (Precup et al., 2011a,b, 2012). Experimental results confirmed the efficiency of the proposed method.

Ibrahim et al. (2014) also proposed an adaptive QBGSA by incorporating an artificial immune system into the QBGSA discussed earlier in this section. The effectiveness and performance of an adaptive QBGSA were verified on optimal PQM placement in power systems, which demonstrated competitive performance compared to the QBGSA and the GSA.

2.6.13 MUTATION-BASED GSA

In general, premature convergence in the GSA still remains as an outstanding problem to be addressed. Nobahari et al. (2011) found for the case of complex multimodal problems that the close moving objects have an impact on each other and consequently the direction of their movement does change. To solve this problem, a mutation operator has been introduced to the original GSA to escape the local optima. Therefore, to simulate the irregularities that exist in the movement of the real objects, they proposed two new mutation operators: sign and reordering mutation.

2.6.13.1 Sign Mutation

The sign mutation was first proposed by Ho et al. (2005) in PSO where the sign of velocity $v_i^d(t)$ is changed. The idea is borrowed by Nobahari et al. (2011) to update the position of objects by changing the sign of the velocity vector temporally as follows:

$$v'^d_i(t+1) = s_i^d v_i^d(t+1), \quad i = 1,\ldots,N \quad d = 1,\ldots,n \tag{2.53}$$

where $v'^d_i(t+1)$ is the mutated velocity, and s_i^d is the sign mutation operator, which changes the direction of the velocity vector with a predefined probability p_s, defined by

$$s_i^d = \begin{cases} -1 & \text{rand} < p_s \\ 1 & \text{otherwise} \end{cases} \tag{2.54}$$

where rand is a uniform random number generated in the interval [0,1].

There is a distinct difference between the two mutations. The sign mutation in Ho et al. (2005) changes the direction of the velocity, whereas the sign mutation in Nobahari et al. (2011) mutates the position of the object by temporally changing the sign of $v'^d_i(t+1)$ to update the position, and then the nonmutated velocity $v_i^d(t+1)$ is used to calculate $v_i^d(t+2)$ in the later iteration. Figure 2.5 shows different possible mutations of a velocity vector in two-dimensional (2-D) space.

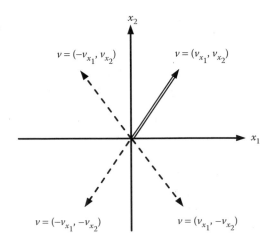

FIGURE 2.5 Sign mutation.

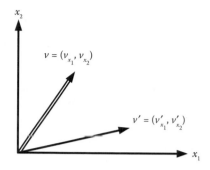

FIGURE 2.6   Reordering mutation.

2.6.13.2   Reordering Mutation

In this mutation, some objects are chosen randomly to be mutated according to reordering mutation probability p_r. Figure 2.6 shows the reordering mutation of a velocity vector in 2-D space. Then, elements of the velocity vector are rearranged randomly. The reordering mutation is performed on the velocity vector as follows:

$$v''^d_i(t+1) = r^d_i v'^d_i(t+1), \quad i = 1,...,N, \quad d = 1,...,n \tag{2.55}$$

where r^d_i is the reordering mutation operator, which changes the value of the velocity vector with a predefined probability p_r. r^d_i can be additive or multiplicative to velocity vector $v'^d_i(t+1)$ for performing the mutation.

   Sign and reordering mutation can also be applied together. Figure 2.7 shows the sign and reordering mutation of a velocity vector applied together. As can be seen, when the sign and the reordering mutations are applied together, the exploration space is increased with respect to the case when only a single operator is applied.

   Once the mutated velocity $v''^d_i(t+1)$ is obtained, the position is updated using the updated velocity vector $v''^d_i(t+1)$. A uniform mutation on the position vector is also applied to position $x'^d_i(t+1)$. The main steps of the mutation-based GSA are as follows:

Step 1: Initialize positions $x^d_i(t)$, $i = 1, 2, ... , N, d = 1, 2, ... , n$
Step 2: Compute fitness
Step 3: Update $G(t)$, K_{best}, best(t), worst(t), $M(t)$
Step 4: Compute total force $F^d_i(t)$ and acceleration $a^d_i(t)$
Step 5: Update velocity $v^d_i(t+1)$
Step 6: Compute $v'^d_i(t+1)$ by performing sign mutation
        Compute $v''^d_i(t+1)$ by performing reordering mutation

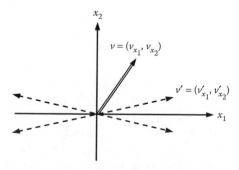

FIGURE 2.7   Combination of sign and reordering mutations.

Step 7: Update position $x'^d_i(t+1)$ using $v'^d_i(t+1)$
       Update $x^d_i(t+1)$ by performing uniform mutation
Step 8: If (termination condition not satisfied), Goto Step 2
Step 9: Return solution

Nobahari et al. (2011) applied a mutation-based GSA for a multiobjective optimization problem (MOOP) using nondominated sorting concept. The proposed GSA approach is applied to nine benchmark functions and the performances are compared with NSGA-II (Deb et al., 2002).

2.6.14   DISRUPTION-BASED GSA

When a swarm of gravitationally bound objects with a total mass of m approaches very close to a massive object M, the swarm of objects is torn apart due to gravitational disruption. Thus, the disruption is a sudden inward fall of gravitationally bound objects under the influence of gravitational force. The same thing can happen to a solid body when it approaches a massive object (Harwit, 1998). In the GSA, the mass of the best solution M becomes massive, which turns M into the star of the system. Thus, the best solution M can potentially disrupt and scatter other solutions in the GSA. Sarafrazi et al. (2011) introduced a new operator borrowed from astrophysics to the GSA, called disruption. To prevent such a divergence in solutions and increasing complexity of the algorithm, a mechanism is devised by controlling the ratio of $R_{i,j}$ (distance between the masses i and j) to $R_{i,\text{best}}$ (distance between the mass i and the best mass or solution), defined by

$$\frac{R_{i,j}}{R_{i,\text{best}}} < C \tag{2.56}$$

where C is a threshold variable. For nonconverging masses, C should be large enough to ensure exploration of the search space. For converging masses, C should be small, enabling masses to get closer and perform exploitation of the search space. If the ratio of the distances in Equation 2.56 is smaller than the threshold C, then the solution i is disrupted. The idea is that if two solutions are close to each other, they are not useful. It would be useful to move one of them for further exploration. On the basis of this simple notion, the following steps are incorporated into the search procedure.

1. The threshold condition in Equation 2.56 is checked for all solutions except for the best solution.
2. The solution that satisfies the condition is disrupted, that is, the position of the solution is changed, according to

$$x'^d_i = x^d_i \cdot D \tag{2.57}$$

The factor D is defined as follows:

$$D = \begin{cases} R_{i,j} \cdot U(-0.5,0.5) & \text{if } R_{i,\text{best}} \geq 1 \\ 1 + \rho \cdot U(-0.5,0.5) & \text{otherwise} \end{cases} \tag{2.58}$$

where $U(-0.5,0.5)$ is a random number uniformly distributed over $[-0.5,0.5]$ and ρ is a small number. Thus, when $R_{i,\text{best}} \geq 1$, the solution is far away from the best solution, the solutions are not converging, and the algorithm needs to explore the search space thoroughly. When $R_{i,\text{best}} < 1$, the solution is close to the best solution, solutions are converging, and the algorithm needs to exploit the search space around the best solution. The value of D is set to a region with radius $\rho/2$ to provide exploitation around the old solution x^d_i.

3. The old solutions x^d_i are replaced by new solutions x'^d_i.

The disruption operator improved the search ability of the GSA for exploration at the initial stage and over time the operator switches to exploitation as the solution moves closer to the best solution in the search space. The main steps of a disruption-based GSA are as follows:

Step 1: Initialize positions $x_i^d(t)$, $i = 1, 2, \ldots, N, d = 1, 2, \ldots, n$
Step 2: Compute fitness
Step 3: Update $G(t)$, K_{best}, best(t), worst(t), $M(t)$
Step 4: Compute total force $F_i^d(t)$ and acceleration $a_i^d(t)$
Step 5: Update velocity $v_i^d(t)$
Step 6: Update position $x_i^d(t)$
Step 7: If disruption condition satisfied, apply disruption operator using Equations 2.57 and 2.58
Step 8: If (termination condition not satisfied), Goto Step 2
Step 9: Return solution

The proposed disruption-based GSA has been verified on 23 nonlinear benchmark functions and compared with the standard GSA, GA, and PSO. The obtained results confirm the high performance of the disruption-based GSA method in solving various nonlinear functions.

2.6.15 RANDOM LOCAL EXTREMA-BASED GSA

Chakraborti et al. (2014) proposed three modifications to the GSA for applying it to 2-D solution space. The first variation is the 2-D GSA where the position of an object is represented by an $n \times n$-dimensional matrix. The second variation introduces a novel random local extrema-based GSA (RLE-GSA), which employs a stochastic local neighborhood-based search instead of a global search generally performed in the basic GSA. The third variation is an automated selection of solution vectors incorporated in the 2-D RLE-GSA, called the modified RLE-GSA (MRLE-GSA), for improving performance.

In the 2-D GSA, the position variable for the i-th object $x_i = \left\{ x_i^{kl}(t) \right\}$, $i = 1, 2, \ldots, N, k = 1, 2, \ldots, n$, $l = 1, 2, \ldots, n$ is an $n \times n$ matrix. The modified force, acceleration, velocity, and updated position calculations are based on the 2-D position variables. The mass update equations remain the same as in the standard GSA.

In a 2-D RLE-GSA, instead of using the global best and global worst values in the mass update equation, the local best and local worst values are used, that is, these are the best and worst fitness values in the neighborhood of that object. The method proposes incorporation of a random element in the determination of neighborhood to boost the rate of convergence. A normalized Euclidean distance of the i-th object is calculated between n-dimensional position vector $x_i = \left\{ x_i^{kl}(t) \right\}$ and any other object defined as

$$d_{ij} = \frac{|x_i - x_j|}{\max_j(|x_i - x_j|)} \text{ with } i, j = 1, 2, \ldots, N \quad \text{and} \quad i \neq j \tag{2.59}$$

The corresponding distance vector for the i-th object is created as $D_i = \{d_{ij}\}, j = 1, 2, \ldots, N$. The vector D_i is rearranged in ascending order and denoted as D_i^{asc}. At iteration t, the objects in the neighborhood of object i having distances corresponding to the first $q_i(t)$ number of elements of D_i^{asc}, where $q_i(t)$ is defined as

$$q_i(t) = \text{round}\left(\left[\kappa \times \frac{t}{t_{max}} \right] \times N \times \text{rand}(0,1) \right) \tag{2.60}$$

where κ is a scaling constant. If $q_i(t) > N$, then $q_i(t) = N$. Once the neighborhood of the i-th object is found, the best and worst fitness values are calculated over the neighborhood. The neighborhood of each object grows in size with the iterations and finally approaches the traditional extrema case, where the entire search space becomes the neighborhood of each object.

In an MRLE-GSA, an automated selection of projection vectors $S_i(t) = \left\{ s_i^l(t) \right\}$, $i = 1, 2, \dots, N$, $l = 1, 2, \dots, n$, is incorporated in the 2-D RLE-GSA where each element $s_i^l(t) \in S_i(t)$ is initialized with a random number uniformly distributed over [0,1]. $s_i^l(t)$ undergoes modification in its velocity and position in subsequent iterations. If the value of the position of $s_i^l(t)$ is greater than 0.5, then the l-th column of the position matrix $x_i = \left\{ x_i^{kl}(t) \right\}$ is selected. The process is repeated for each l-th dimension of the vector $S_i(t)$ in each iteration. In the MRLE-GSA, both the candidate solution matrix x_i and selection vector $S_i(t)$ are evolved, that is, force, distance, velocity, and position are computed for both $x_i^{kl}(t)$ and $s_i^l(t)$ at each iteration. Finally, the vector of solution is selected according to selection vector $S_i(t)$ at the end of the MRLE-GSA.

The effectiveness and performance of the proposed 2-D-GSA, RLE-GSA, and MRLE-GSA are verified on 2-D image processing. Experimental results based on benchmark face databases, namely, Yale A and ORL database, show that the proposed methods outperform the standard GSA and several other existing schemes. The RLE-GSA requires several trial runs, whereas the MRLE-GSA requires a single run.

2.6.16 MODIFIED GSA

In the standard GSA, objects move to new positions based on the updated values of velocities determined by the direction and distance. By changing the values of velocities, the objects are likely to reach the global optima. At the start of the search process, the objects are thought to be far away from the global optima and hence large values for the velocities are required to explore the search space. As the objects move closer to the global optima, velocities are gradually reduced to exploit the search space; that is, small values are required for small movements. Khajehzadeh et al. (2012) proposed a modified GSA by introducing a controlled trajectory for the velocity updates that limits the velocity within an interval of $-v_{\max} \leq v_i^d \leq v_{\max}$, where $|v_{\max}|$ is the designated maximum velocity allowed. It is also found that if the value of $|v_{\max}|$ is too large, then the object may move erratically. If the value of $|v_{\max}|$ is too small, then the movement will be too slow to reach an optimal solution. To strike a balance between exploration and exploitation of the search space, Khajehzadeh et al. (2012) proposed a time-varying velocity profile defined by

$$v_{\max} = \left[1 - \left(\frac{t}{t_{\max}} \right)^h \right] \cdot v_{\max 0} \tag{2.61}$$

where h is a positive parameter to be chosen arbitrarily by trial and error, $t_{\max}$ is the maximum number of iterations, and $v_{\max 0}$ is the initial velocity to be considered as a fraction of the search space defined by

$$v_{\max 0} = \alpha (x_{\max} - x_{\min}) \tag{2.62}$$

where $0 < \alpha < 1$ to be chosen arbitrarily. By limiting the velocity $v_i^d(t)$ within the allowable range, the algorithm controls the unnecessary oscillations in the trajectory of objects in the search space. The main steps of the modified GSA are as follows:

Step 1: Initialize positions $x_i^d(t)$, $i = 1, 2, \dots, N$, $d = 1, 2, \dots, n$
Step 2: Compute fitness

Step 3: Update $G(t)$, K_{best}, best(t), worst(t), $M(t)$
Step 4: Compute total force $F_i^d(t)$ and acceleration $a_i^d(t)$
Step 5: Update velocity $v_i^d(t)$
Step 6: Determine max velocity v_{max} using Equation 2.61
      If $v_i^d(t) > v_{max}$
      Replace $v_i^d(t)$ with representative limits
Step 7: Update position $x_i^d(t)$
Step 8: If (termination condition not satisfied), Goto Step 2
Step 9: Return solution

The modified GSA approach is seen as an adaptive maximum velocity constraint to the standard GSA. Khajehzadeh et al. (2012) verified the performance of the proposed GSA on a slope stability analysis problem and reported excellent performance in terms of accuracy, convergence rate, stability, and robustness.

2.7 HYBRID GSA

Some variants of GSA have been discussed in Section 2.6. These modified GSA methods are developed for enhancing solution quality, exploration and exploitation of search space, and accelerating convergence. Modifications to the standard GSA can be categorized as (1) Parameters such as $G(t)$ and $\varepsilon(t)$ are dynamically adjusted for exploration and exploitation of the search space; (2) New operations such as mutation are introduced from other algorithms; (3) Different techniques such as chaotic, quantum and wavelet etc. are used to update velocity and positions.

There is still scope to improve exploration or global search ability, exploit local search, increase convergence speed, improve solution quality, and minimize computational cost. In general, different meta-heuristic algorithms have their own advantages as well as disadvantages. The common disadvantages of meta-heuristic algorithms are the premature convergence at local minima, slow convergence demanding more computational cost, low quality solution meaning poor exploitation or local search ability, and lack of diverse operations. In order to further improve the algorithm, the standard GSA algorithm has been hybridized with other meta-heuristic algorithms such as PSO, EA, K-NN, K-harmonic means (K-HM), and fuzzy logic. The hybridization of GSA with other algorithms can be implemented in two ways: (i) incorporation of some techniques of other meta-heuristic algorithms into GSA procedure and (ii) incorporation of some GSA techniques into other meta-heuristic algorithms. There are some good hybrid variants reported in the literature in recent years which are discussed in this section.

2.7.1 HYBRID PSO-GSA

A good heuristic algorithm tries to strike a balance between exploration and exploitation. In the standard GSA, the movement of each object is determined based on the total force obtained by all other objects acting on it. The GSA is memoryless and only the current velocity and position of the object plays a role in the updating procedure. The velocity update depends on the acceleration and the position update depends on the velocity terms as per Equations 2.20 and 2.21 mentioned in Section 2.3. Some researchers attempted to improve the standard GSA by introducing the idea of memory and social information, a concept borrowed from PSO. PSO is a population-based optimization algorithm developed by Kennedy and Eberhart (1995, 2001) based on the social behavior of swarms such as flocking of birds and schooling of fish seen in nature. Each particle, referred to as an individual in the swarm, representing a candidate solution to the optimization problem, flies through the multidimensional search space, adjusting its position in the search space according to its own experience and that of neighboring particles. Particles make use of the best positions encountered and the best position of its neighbors to position themselves toward an optimal solution. The performance of each particle is measured according to a predefined fitness function related to the problem at hand. Applications of PSO include function approximation, clustering, optimization of mechanical structures, and solving systems of equations. There are now as many as 20 different variants of PSO.

PSO uses a kind of memory to represent the best previous position of an individual. This is done by introducing a velocity update mechanism in PSO. Li and Zhou (2011) proposed an improved GSA (IGSA) by introducing a novel moving strategy in the search space, obeying the law of gravity, and utilizing memory and social information from PSO. The velocity-updating rule of the IGSA is thus defined by

$$v_i^d(t+1) = \rho_{i1} \cdot v_i^d(t) + a_i^d(t) + c_1 \cdot \rho_{i2} \cdot \left(x_{p\text{best}}^d(t) - x_i^d(t)\right) + c_2 \cdot \rho_{i3} \cdot \left(x_{g\text{best}}^d(t) - x_i^d(t)\right) \qquad (2.63)$$

where ρ_{i1}, ρ_{i2}, and ρ_{i3} are random variables within the range of [0,1], $c_1, c_2 \in [0,1]$ are learning rates, $x_{p\text{best}}^d(t)$ is the best previous position that the i-th individual has ever encountered until time t, and $x_{g\text{best}}^d(t)$ is the previous best global position. Once the velocity is obtained, the position is updated according to Equation 2.21. By adjusting the value of the learning rate of c_1 and c_2, the balance of the effectiveness of "law of gravity" and the effectiveness of "memory and social information" are maintained. If c_1 and c_2 are set to zero, the IGSA becomes the standard GSA. If c_1 and c_2 are set to 1, the IGSA becomes PSO. The main steps in the IGSA are the same as those of the standard GSA except that the velocity update is performed using Equation 2.63.

Li and Zhou (2011) applied the IGSA to the nonlinear model parameter identification problem of hydraulic turbine governing systems (HTGSs) and reported the improvement of the proposed algorithm. They showed that the IGSA performs the best in experiments with high accuracy and stability, which means the improvement made to the GSA is effective compared to the standard GSA and PSO.

Mirjalili et al. (2012) combined the ability of social thinking in PSO with the local search capability of the GSA and proposed a hybridized PSO-GSA where the velocity update method is defined as follows:

$$v_i^d(t+1) = w * v_i^d(t) + c_1 \cdot \rho_1 \cdot a_i^d(t) + c_2 \cdot \rho_2 \cdot \left[x_{p\text{best}}^d(t) - x_i^d(t)\right] \qquad (2.64)$$

where w is a weighting function, c_1 and c_2 are coefficients, $\rho_1, \rho_2 \in [0,1]$ are random numbers, and $x_{p\text{best}}^d(t)$ is the best solution up to time t. The proposed hybrid PSO-GSA was applied to train a feedforward neural network (NN) and it was able to avoid local minima and speed up convergence.

Mallick et al. (2013) proposed a hybrid PSO-GSA by combining the ability of social thinking of PSO with the local search capability of the GSA. The velocity update has been modified as follows:

$$v_i^d(t+1) = w(t+1) * v_i^d(t) + c_1 \cdot \rho_1 \cdot a_i^d(t) + c_2 \cdot \rho_2 \cdot \left[x_{p\text{best}}^d(t) - x_i^d(t)\right] \qquad (2.65)$$

where $w(t+1)$, c_1, and c_2 are the weighting factors, $\rho_1, \rho_2 \in [0,1]$ are random numbers, and $x_{p\text{best}}^d(t)$ is the best solution up to time t. The velocity update rule is the same as the rule in Mirjalili et al. (2012) except that $w(t+1)$ varies with time.

Mirjalili et al. (2012) and Mallick et al. (2013) used a mechanism of combining PSO and GSA for calculation of velocity updates. The approach of Tsai et al. (2013) is different where they first calculated the velocity using the formulation of PSO and GSA separately and then combined the velocities of PSO and GSA. They call it gravitational particle swarm. The velocity updates are formulated as follows:

$$v_i^d(t+1)_{\text{PSO}} = w(t) \cdot v_i^d(t) + a_i^d(t) + c_1 \cdot \rho_2 \cdot \left(x_{p\text{best}}^d(t) - x_i^d(t) \right) + c_2 \cdot \rho_2 \cdot \left(x_{g\text{best}}^d(t) - x_i^d(t) \right) \quad (2.66)$$

$$v_i^d(t+1)_{\text{GSA}} = \rho_i \cdot v_i^d(t) + a_i^d(t) \quad (2.67)$$

$$v_i^d(t+1)_{\text{GPS}} = c_3 \cdot \rho_{i3} v_i^d(t+1)_{\text{PSO}} + c_4 (1 - \rho_{i3}) \cdot v_i^d(t+1)_{\text{GSA}} \quad (2.68)$$

where c_3 and c_4 are weighting factors and $\rho_{i3} \in [0,1]$ is a random number with uniform distribution.

In the standard GSA, communications between objects are not considered in the calculation of velocity updates. Chen et al. (2014) proposed an IGSA by introducing communication and memory characteristics of PSO. The new velocity is computed as follows:

$$v_i^d(t+1) = \rho_1 \cdot v_i^d(t) + a_i^d(t) + c_1 \cdot \rho_2 \cdot \left(x_{p\text{best}}^d(t) - x_i^d(t) \right) + c_2 \cdot \rho_3 \cdot \left(x_{g\text{best}}^d(t) - x_i^d(t) \right) \quad (2.69)$$

where $\rho_1, \rho_2, \rho_3 \in [0,1]$ are random numbers, $c_1, c_2 \in [0,2]$ are learning rates, $x_{p\text{best}}^d(t)$ is the best position that the i-th object has ever encountered until time t, and $x_{g\text{best}}^d(t)$ is the previous global best position. The position is updated according to Equation 2.21. The IGSA proposed by Chen et al. (2014) is the same as the IGSA proposed by Li and Zhou (2011) except that Chen et al. used a wider range of learning rates. Chen et al. has applied the IGSA to a water turbine regulation system (WTRS) for identification of the system parameters, which shows promising results with higher accuracy and stability than the standard GSA and PSO.

Jiang et al. (2014) proposed a hybrid PSO and GSA (HPSO-GSA). The HPSO-GSA embodies interesting features from both PSO and the GSA, incorporating the social essence of PSO with the motion mechanism of the GSA. The proposed HPSO-GSA adopts a technique to simultaneously update particle positions with PSO velocity and GSA acceleration.

$$v_i^d(t+1)_{\text{PSO}} = w(t) \cdot v_i^d(t) + c_1 \cdot \rho_1 \cdot \left(x_{p\text{best}}^d(t) - x_i^d(t) \right) + c_2 \cdot \rho_2 \cdot \left(x_{g\text{best}}^d(t) - x_i^d(t) \right) \quad (2.70)$$

$$v_i^d(t+1)_{\text{GSA}} = \rho_i \cdot v_i^d(t)_{\text{GSA}} + a_i^d(t)_{\text{GSA}} \quad (2.71)$$

$$v_i^d(t+1)_{\text{HPSO-GSA}} = c_3 \cdot \rho_3 \cdot v_i^d(t+1)_{\text{PSO}} + c_4 \cdot (1 - \rho_3) \cdot v_i^d(t+1)_{\text{GSA}} \quad (2.72)$$

The two terms of the velocity-updating formulation in the HPSO-GSA includes the cooperative contributions of PSO velocity and GSA acceleration; c_3 and c_4 are two acceleration coefficients which adjust the degree of influence of PSO velocity and GSA acceleration on the HPSO-GSA. If c_3 or c_4 is set to zero, the HPSO-GSA becomes independent of PSO or the GSA. If c_3 and c_4 are set to 1, the shared impact of PSO and the GSA on the HPSO-GSA depends stochastically on the value of ρ_3. Finally, the position of the HPSO-GSA is updated based on the HPSO-GSA velocity at every iteration by

$$x_i^d(t+1) = x_i^d(t) + v_i^d(t+1)_{\text{HPSO-GSA}} \quad (2.73)$$

The HPSO-GSA is expected to obtain an efficient balance between exploration and exploitation. From results of canonical benchmark test functions, the HPSO-GSA shows significant improvement over PSO and the GSA in terms of performance. It appears that the velocity computation in the IGSA involves a good deal of effort compared to memoryless characteristics of the standard GSA, which may restrict application of the IGSA to a wide range of problem domains.

Chatterjee and Mahanti (2010) made a comparative study on PSO and the GSA, which shows that there are features in both the methods that can be combined for an improved algorithm. Mirjalili et al. (2012) combined the PSO with the local search capability of the GSA and proposed a hybridized PSO-GSA where the velocity update method is modified using Equation 2.64 presented earlier in this section. The GSA becomes slow due to the presence of heavier masses in the population toward the end of the run, requiring more time to reach the optimal solution. In order to overcome this tardiness, Mallick et al. (2013) proposed a coevolutionary heterogeneous hybrid PSO-GSA by combining the PSO with the GSA and modified the velocity update procedure as defined in Equation 2.65. The hybrid PSO-GSA has the following features, which make it efficient over the standard GSA.

The quality of solutions (fitness) is considered in the PSO-GSA in the updating procedure.

The agents near good solutions are attracted by other agents, which are exploring different parts of the search space.

When all agents are near a good solution, they move very slowly and $g$best helps them in exploiting the global best.

$g$best is the best solution found so far, which serves as a memory for the PSO-GSA.

Each agent can compare itself with $g$best and move toward it.

The global searching and local searching are balanced by adjusting the coefficients c_1 and c_2.

2.7.2 HYBRID GSA-EA

Khatibinia and Khosravi (2011) proposed a hybrid of IGSA and orthogonal crossover (OC), a crossover operator widely used in EAs, to improve the local search by preserving the previous best solution. The proposed hybrid approach is called IGSA-OC, which improves the global exploration ability of the IGSA method and increases its convergence speed.

Research works have shown that OC outperforms existing crossover operators used in EAs. An orthogonal array is integrated into the classical crossover operator so that two parents can be used to generate a set of sampling points (i.e., children) based on the orthogonal array. A detailed study on the construction of an orthogonal array is reported in Fang and Wang (1994). A new version of OC based on combination of the quantization technique and the OC operator, called QOC, is proposed by Leung and Wang (2001). The search space defined by any two parents $P_1 = \left\{ p_1^1, p_2^1, \ldots, p_d^1 \right\}$ and $P_2 = \left\{ p_1^2, p_2^2, \ldots, p_d^2 \right\}$ has the range of $\left[\min\left(p_i^1, p_i^2 \right), \max\left(p_i^1, p_i^2 \right) \right]$ for the i-th variable. In QOC, the search range $\left[\min\left(p_i^1, p_i^2 \right), \max\left(p_i^1, p_i^2 \right) \right]$ is quantized into Q levels such as $\left\{ l_i^1, l_i^2, \ldots, l_i^Q \right\}$ for variable x_i using the following scheme proposed by Leung and Wang (2001):

$$ l_i^j = \min\left(p_i^1, p_i^2 \right) + \frac{j-1}{Q-1} \left(\max\left(p_i^1, p_i^2 \right) - \min\left(p_i^1, p_i^2 \right) \right), \quad j = 1, 2, \ldots, Q \tag{2.74} $$

An orthogonal array for K factors with Q levels and M combination is often denoted by $L_M \left(Q^K \right)$. After quantization of the search space, Q^d points are produced. Since d is much greater than K, the orthogonal array $L_M \left(Q^K \right)$ cannot be applied. To apply QOC, Leung and Wang (2001) proposed to divide $\left\{ x_i^1, x_i^2, \ldots, x_i^d \right\}$ into K subvectors as

$$ \begin{cases} \bar{H}_1 = \left\{ x_i^1, x_i^2, \ldots, x_i^{n_1} \right\} \\ \bar{H}_2 = \left\{ x_i^{n_1+1}, x_i^{n_1+2}, \ldots, x_i^{n_2} \right\} \\ \vdots \\ \bar{H}_K = \left\{ x_i^{n_{K-1}+1}, x_i^{n_{K-1}+2}, \ldots, x_i^d \right\} \end{cases} \tag{2.75} $$

where $1 < n_1 < n_2 < \ldots < n_{K-1} < d$. Here, $\bar{H}_i$ is treated as a factor such that $L_M(Q^K)$ orthogonal arrays can be applied to construct M solutions. In order to eliminate the pitfalls of the IGSA and explore promising search space, a QOC operator is utilized in the hybrid IGSA-OC. The QOC operator is used as local search and helps in improving the best previous position among all agents (i.e., gbest) in each iteration of the IGSA. The IGSA-QOC procedure consists of six steps as follows:

Step 1: Three numbers $r_1, r_2, r_3 \in \{1, 2, \ldots, N\}$ are selected randomly
Step 2: Compute mutant vector $U = pbest_{r1} + \text{rand}(pbest_{r2} - pbest_{r3})$
Step 3: Generate M trial solutions by combining gbest and U using QOC based on $L_M(Q^K)$
Step 4: Evaluate the fitness function values $\bar{f}$ of the M trial solutions
Step 5: Select one U_g with the smallest fitness function
Step 6: Set gbest $= U_g$ if $f(U_g) < f(\text{gbest})$

The procedure can increase the global search capability of the IGSA with the help of QOC to search the region defined by these two vectors U and gbest. To verify the robustness and efficiency of the proposed IGSA-OC, it has been applied to four well-known benchmark functions from the literature and its performances are compared with those of the standard GSA and other modified GSA methods. IGSA-OC is then applied to find the optimal shape of concrete gravity dams. The numerical results demonstrate that the proposed IGSA-OC significantly outperforms the standard GSA, IGSA, and PSO.

2.7.3 Hybrid Fuzzy GSA

The multi-objective GSA (MO-GSA) provides a set of Pareto optimal solutions. A major difficulty in MO-GSA is to select a competitive global optimal solution from a set of optimal solutions for each agent in the population. For practical applications, it is required to select one solution which should be the best compromise from the Pareto front that satisfies the different conflicting goals. It is difficult for a decision maker to select the best possible solution. Therefore, a fuzzy logic-based mechanism is used for selecting the best Pareto optimal solution. Ghasemi et al. (2013) applied a fuzzy-based mechanism into the MO-GSA to figure out this kind of selection problem. The experience and the intuitive knowledge of the decision maker are used to define a linear membership function for each of the objectives of the optimization problem. To have the best compromise of solutions in the interest of the decision maker, a fitness-sharing scheme is employed. The proposed method is called a fuzzy GSA (FGSA). To verify the performance of the FGSA, it has been applied to damping control of electromechanical modes of oscillations for a wide range of operating conditions. Experimental results show that the FGSA is able to provide better and faster solutions.

Kumar et al. (2013) analyzed the quantitative and qualitative relationship between the parameters, especially the gravitational constant, and the performance of the GSA. The gravitational constant $G(t)$ has momentous influence in the computation of force and consequently on the acceleration; therefore, a selection mechanism for gravitational constant $G(t)$ can strike a balance between global exploration and local exploitation as well as reduce the total number of iterations. Kumar et al. (2013) proposed an adaptive GSA where the gravitational constant is being dynamically adjusted using a Mamdani-type fuzzy controller. Normalized fitness and the gravitational constant are the inputs and the change of gravitational constant is the output of the fuzzy controller. Three membership functions are used for each input and output variables. The main steps of the fuzzy adaptive GSA are as follows:

Step 1: Initialize positions $x_i^d(t)$, $i = 1, 2, \ldots, N$, $d = 1, 2, \ldots, n$
Step 2: Compute fitness for each object
Step 3: Compute $G(t)$ using fuzzy controller

Step 4: Update best(t), worst(t), $M(t)$
Step 5: Compute total force $F_i^d(t)$ with updated $G(t)$ and acceleration $a_i^d(t)$
Step 6: Update velocity $v_i^d(t)$ and position $x_i^d(t)$
Step 7: If (termination condition not satisfied), Goto Step 2
Step 8: Return solution

The effectiveness of the fuzzy adaptive GSA has been verified on an optimal bidding problem and the performance is compared with the standard GSA, PSO, GA, and golden section search.

2.7.4 HYBRID QBGSA-K-NN

Han et al. (2013) proposed a hybrid of the QBGSA (discussed earlier in this section) with KNN method with leave-one-out cross-validation to improve classification accuracy with an appropriate feature subset in binary problems. The K-NN (Fix and Hodges, 1989) method is a nonparametric method, which requires only one parameter K (the number of nearest neighbors) to be determined. The classification procedure is based on the minimum Euclidean distance. If an object is close to K-NN, the object then belongs to the K-th object category. The advantage of K-NN is that it is easy to implement. The main steps of the QBGSA-K-NN algorithm are as follows:

Step 1: Initialize Q-bit individual and position population
Step 2: Select feature subset using QBGSA
Step 3: Train 1-NN classifier with selected feature subset using training set
Step 4: Evaluate fitness
Step 5: Update Q-bit individual of the i-th object
        Modify position of the i-th object
        Update the i-th object
Step 6: If (termination condition not satisfied), Goto Step 2
Step 7: Return solution

In order to improve classification accuracy by appropriate feature selection in binary problems, the hybrid QBGSA-K-NN method was applied to several machine learning benchmark examples from the University of California, Irwine (UCI) data sets over 20 runs. The experimental results show that the proposed method is able to select the discriminating input features correctly and achieve high classification accuracy, which is comparable to or better than that of well-known similar classifier systems such as BPSO-K-NN and GA-K-NN. As the fitness evaluation in the QBGSA is proportional to the force unlike in BPSO or the GA, it provides better values of solutions. This showed that the QBGSA was able to optimize the feature subset and escape the local optimum.

2.7.5 HYBRID K-HARMONIC MEANS AND GSA

K-means clustering is sensitive to initialization and its winner-take-all strategy, which makes the association between data points and their nearest cluster centers very strong, the result of which prevents cluster centers from moving out from a local density of data and makes the algorithm converge to the local optimum. K-HM clustering algorithm gives higher weights to data points that are not close to any cluster centers. The K-HM strategy thus distributes the association between data points and cluster centers. The K-HM algorithm was first proposed by Zhang et al. (2000) and modified by Hammerly and Elkan (2002).

Yin et al. (2011) integrated an IGSA with the K-HM clustering algorithm to help K-HM escape from local minima. In this approach, a solution vector comprises $d \times k$ elements in the vector, where d is the dimensionality of the data to be clustered and k is the number of clusters. The fitness function is the objective function of the K-HM algorithm.

The IGSA is a simple modification to the GSA when objects go out of bound. The IGSA brings the objects back within the solution space by assigning boundary values. The following strategy is applied to objects out of bound:

If $[(x_i^d(t) > x_U^d) \text{ or } (x_i^d(t) < x_L^d)]$

$\quad v_i'^d(t) = -\text{rand} * v_i^d(t)$

$\quad x_i^d(t) = x_i^d(t) + v_i'^d(t)$

The steps of the IGSA are the same as those of the GSA except the modification to position recalculation. The main steps of the IGSA-K-HM algorithm are

Step 1: Initialize positions $x_i^d(t)$, $i = 1, 2, \ldots, N$, $d = 1, 2, \ldots, n$
Step 2: Compute fitness
Step 3: Apply IGSA
Step 4: Assign x_i as cluster center of K-HM
Step 5: Recalculate cluster center using K-HM
Step 6: If (termination condition not satisfied), Goto Step 2
Step 7: Return solution

The proposed approach was tested on seven data sets, namely, ArtSet1, ArtSet2, Wine, Glass, Iris, Breast cancer Wisconsin, and Contraceptive Method Choice, which cover examples of data. Standard parameter setting is used for the GSA. The K-HM algorithm requires fewer function evaluations and tends to converge faster than the GSA. In all tests, the results of the proposed approach demonstrated superior performance in terms of efficiency and effectiveness compared to that of K-HM and PSO-K-HM.

2.8 APPLICATION TO ENGINEERING PROBLEMS

The GSA is a recent meta-heuristic search and optimization algorithm, which has been applied successfully to many domains of computing, engineering, manufacturing, and management science. Since its inception, the research communities from different fields have been showing interest in the GSA and applying it to their own fields for optimization purposes. The GSA has found a wide range of applications and the domains of application are comparatively broad, which are classified under different headings discussed in the following.

2.8.1 BENCHMARK FUNCTION OPTIMIZATION

Rashedi et al. (2009) applied the GSA to 23 benchmark functions for minimization tasks to demonstrate the usefulness, power, and performance of the GSA. Their experimental investigations provided superior results compared to other meta-heuristic algorithms. Soleimanpour-Moghadam et al. (2014) applied QGSA to 25 standard benchmark functions. The effectiveness of the algorithm was confirmed by providing a detailed description of the experimental results and by comparison with other available results from the literature. Coelho et al. (2014) verified the effectiveness of the quasi-opposition-based GSA on six-hump-camel-back function, Griewank function, and Rosenbrock function. These functions are discussed in later chapters. In all cases, the quasi-opposition-based GSA achieved an optimal solution with 15 masses and within 50 iterations. Shaw et al. (2012) applied an opposition-based GSA to a comprehensive set of 23 complex benchmark test functions including unimodal and multimodal high-dimensional functions, which confirms the potential and effectiveness of the proposed GSA. Yazdani et al. (2014) also applied GSA to multimodal optimization problems.

2.8.2 Combinatorial Optimization Problems

A combinatorial optimization problem is the problem of finding the best possible grouping, ordering, or assignment of a finite set of objects satisfying certain conditions or constraints. Combinatorial optimization problems are very common in many areas of computer science and other disciplines such as applied mathematics, artificial intelligence, operations research, bioinformatics, electronic commerce, engineering, and industry where an exhaustive search is not feasible. Very often the set of feasible solutions is discrete or can be reduced to be discrete. In all cases, the goal is to find the best possible solution. There are combinatorial optimization problems which are NP-hard and there are no algorithms that can solve them in polynomial time. Such combinatorial optimization problems are very common in engineering and industry, which need novel solutions using new meta-heuristic algorithms. In the following sections, application of the GSA and its variants to some engineering- and industry-type problems is discussed.

2.8.3 Economic Load Dispatch (ELD) Problem

ELD is a method of determining the most efficient, low-cost, and reliable operation of a power system by dispatching available electricity generation resources to supply load on the system most economically. The problem of ELD is multimodal, nondifferentiable, and highly nonlinear. The objective function of the ELD problem can be stated as follows:

$$F_T = \text{Min} \sum_{i=1}^{N_g} C_i(P_i) \tag{2.76}$$

where F_T is the minimized total electricity generation cost, C_i is the cost function of the i-th generator, and N_g is the number of power-generating units each loaded to P_i in MW. $C_i(P_i)$ in terms of cost coefficients without and with valve-point loading as follows:

$$C_i(P_i) = \begin{cases} \alpha_i + \beta_i P_i + \gamma_i P_i^2 & \text{without valve-point loading} \\ \alpha_i + \beta_i P_i + \gamma_i P_i^2 + \left| \delta_i \cdot \sin\left(e_i\left(P_i^{\min} - P_i\right)\right) \right| & \text{with valve-point loading} \end{cases} \tag{2.77}$$

where α_i, β_i, and γ_i are the cost coefficients of the i-th power-generating unit with power output P_i in MW. δ_i and e_i are valve-point effects taken into account when used. A constrained optimization problem must be feasible. Such methods usually employ a concept of penalty function where it penalizes the unfeasible solutions. The objective is to find an optimal combination of power generation systems that minimizes the total cost of generation while satisfying different equality and inequality constraints. The equality constraint is defined by

$$\sum_{i=1}^{N_g} P_i = P_D + P_L \tag{2.78}$$

where P_D is the total load demand and P_L is the transmission loss. The power output P_i of each generating unit should be within the minimum and maximum limits. The inequality constraint is defined by

$$P_i^{\min} \le P_i \le P_i^{\max}, \quad i = 1, 2, \ldots, N_g \tag{2.79}$$

where P_i^{min} and P_i^{max} are the minimum and maximum power output of the i-th generating unit, respectively. A detailed mathematical model and description of the ELD problem are provided in Chatterjee et al. (2012). A number of meta-heuristic algorithms have been applied to solve the ELD problem by many researchers. Chatterjee et al. (2012) proposed a GSAWM (discussed in Section 2.6.8) and applied it to an ELD problem. The objective function for the ELD problem, described by Equations 2.76 through 2.79, is defined incorporating the penalization factors that require to be minimized.

$$F_{ELD} = \sum_{i=1}^{N_g} C_i(P_i) + 100 \cdot P_L + 1000 \cdot abs \left(\sum_{i=1}^{N_g} P_i - P_D - P_L \right) \tag{2.80}$$

The penalization factors 100 and 1000 are used for loss and power mismatch, respectively. The objective function is taken as a minimization problem. A detailed sensitivity analysis of the weighing/penalization factors can be found in Shaw et al. (2011). Swain et al. (2012) investigated the effectiveness and performance of the standard GSA by applying it to ELD systems with 3 and 13 thermal generating units, taking into account valve-point effects, and showed the effectiveness of the GSA by performing two test cases. In both cases, competitive solutions were achieved compared to evolutionary programming. In another study, Chatterjee et al. (2012) applied the GSAWM approach to the ELD problem and carried out a very comprehensive study onto four different ELD systems with 3, 6, 13, and 140 generating units incorporating valve-point effects. In all test cases, the GSAWM succeeds in finding a near-optimal solution. The minimum costs achieved were 8233 for the ELD system with 3 generating units, 15,441 with 6 generating units, 17,817 with 13 generating units, and 1,573,200 with 140 generating units (for Case 1). In all cases, the GSAWM converged to a better-quality near-optimal solution within 100 iterations.

2.8.4 ECONOMIC AND EMISSION DISPATCH (EED) PROBLEM

The consumption of electrical energy is increasing everyday demanding cost-effective production of energy. The total production cost of electricity can be reduced by the scheduling generators, which produce energy to meet the load demand and to operate at minimum cost. The thermal power plants using fossil fuels produce economic energy but release toxic gases such as carbon dioxide (CO_2), sulfur dioxide (SO_2), nitrogen oxide (NO_x), and some other particles, which cause huge pollution to the environment. The objective of EED of power generation is to schedule the committed generating unit outputs to meet the load demand at minimum operating cost while satisfying all the environmental constraints of minimum emission levels. The EED problem thus becomes one of the most important optimization problems in power system operation and forms the basis of a benchmark problem for optimization algorithms. Thus, the EED poses a bi-objective optimization problem formulated as Min[F,E]. F is the total fuel cost of generators $i = 1, 2, \ldots,$ N_g, defined as

$$Min\, F = \left[\sum_{i=1}^{N_g} F_i(P_{G_i}) \right] \tag{2.81}$$

where $F_i(P_G) = a_i + b_i P_{G_i} + c_i P_{G_i}^2 + \left| d_i \sin \left\{ e_i \left(P_{G_i}^{min} - P_{G_i} \right) \right\} \right|$, $i = 1, 2, \ldots, N_g$; P_{G_i} is the real power output; N_g is the number of generators; a_i, b_i, and c_i are the cost coefficients; and d_i and e_i are valve-point effects taken into account when used. E is the total emission dispatch from the generators expressed as the sum of all types of emission considered, such as NO_x, SO_2, CO_2, particles, and thermal emissions:

$$\text{Min} E = \left[\sum_{i=1}^{n_g} E_i(P_{G_i}) \right] \tag{2.82}$$

where, $E_i(P_{G_i}) = \alpha_i + \beta_i P_{G_i} + \gamma_i P_{G_i}^2 + \eta_i \exp(\delta_i P_{G_i})$, $i = 1, 2, \ldots, N_g$; α_i, β_i, and γ_i are emission coefficients; and δ_i and η_i are valve-point effects taken into account when used. Rather than handling the EED problem with two objectives, many researchers converted the bi-objective problem into a single objective optimization function using a price penalty factor (Bharathi et al., 2007; Sayah et al., 2014), defined by

$$\text{Min} C_{\text{EED}} = [h \cdot F + (1 - h) \cdot P_f \cdot E] \tag{2.83}$$

where C_{EED} is the total operating cost, $h \in [0,1]$ is the weighting factor, and P_f is the price penalty factor defined as $P_f = (F(P_{G_i}^{\text{max}}))/(E(P_{G_i}^{\text{max}}))$, which blends the emission cost with the normal fuel costs. Kumarappan et al. (2002) suggested a four-step procedure to find out the value of P_f. The objective is now to minimize the total operating cost C_{EED} of the EED problem. Further details of problem formulation of EED can be found in Wood and Wollenberg (1994). Although the GSA is memoryless, it works efficiently like algorithms with memory, and it can also be considered as an adaptive learning algorithm (Rashedi et al., 2010). On account of all these advantages, the GSA has been applied to the EED problem by many researchers (Güvenç et al., 2012; Jiang et al., 2014; Mondal et al., 2013; Shaw et al., 2012).

To demonstrate the effectiveness of the GSA, Güvenç et al. (2012) applied the algorithm to four different test cases for the EED problem with quadratic cost and emission function. The test systems consist of 6, 10, 11, and 40 generating units. They used standard GSA parameter settings, for example, $G_0 = 100$, $\alpha = 20$ (as per Equation 2.14), and the total number of iterations was set to 1000. For the test system with 6 generating units and the total power demand of 1000 MW, the GSA achieved a very competitive solution with a minimum cost of 51255.78 and emission of 827.13 kg/h. For the test system with 11 generating units, a total power demand of 2500 MW, and without loss coefficients, the GSA achieved a very competitive solution with a minimum cost of 12422.66 and emission of 2002.94 kg/h. For the test system with 10 generating units with nonsmooth cost function, a total power demand of 2000 MW, and with loss coefficients, the GSA achieved a very competitive solution with a minimum cost of 1.1349×10^5 and emission of 4111.4 lb. For a test system with 40 generating units with nonsmooth cost function and a total power demand of 10,500 MW, the GSA achieved a very competitive solution with a minimum cost of 1.2578×10^5 and emission of 12.1093×10^5 ton. In all test cases, the GSA provided better solutions than other stochastic algorithms.

Shaw et al. (2012) first proposed the use of an opposition-based GSA to solve a EED problem. To assess the proposed opposition-based GSA, four test case studies of the EED are carried out utilizing the objective function defined as

$$F_{\text{EED}} = \sum_{i=1}^{N_g} C_{\text{EED}(i)} + 100 \cdot P_L + 1000 \cdot \text{abs} \left(\sum_{i=1}^{N_g} P_i - P_D - P_L \right) \tag{2.84}$$

where P_D is the total load demand, P_L is the transmission loss, and P_i is the power output of each generating unit. In order to measure the effectiveness of the proposed algorithm, it has been implemented on four different test systems. The Test System 1 consists of an IEEE 30-bus system with six generators without P_L. The fuel cost and emission are individually optimized. The Test System 2 consists of an IEEE 30-bus system with six generators with P_L. The fuel cost and emission are individually optimized. The Test System 3 consists of a Taiwan power system of 40 generating units with valve-point loading effects. The fuel cost and emission are individually optimized. The Test System 4 consists of a Korean power system of 140 generating units with valve-point loading effects, prohibited operating zones, and ramp rate, and without network loss. The results obtained have confirmed the potential and effectiveness of an opposition-based GSA for the application to an EED problem. Both the near-optimal solution and convergence speed of the opposition-based GSA are promising.

The EED problem seeks a balance between minimum fuel cost and minimum emission, which is actually an MOOP. Mondal et al. (2013) applied the GSA to a bi-objective EED problem. An IEEE 30-bus system having six conventional thermal generators has been considered as a test system. Two extra wind power sources have been placed on those two load buses which are selected based on their voltage stability limits. After placing the wind power sources, those buses have been converted to a generator bus. The minimum fuel cost, minimum emission cost, and the best compromising solution have been calculated applying the GSA. The objective function defined in Equation 2.83 and the P_f value determined by Kumarappan et al. (2002)'s method are used. The initial value of h in Equation 2.83 is 0, and then increases in steps of 0.05 up to 1. Equation 2.83 is minimized for different weights of each objective subject to constraints. Out of many sets of Pareto optimal solutions, the best compromise solution is found out using a fuzzy logic-based selection technique. It is found that the GSA has the ability to converge to quality solutions and it has better convergence characteristics, which help in achieving a competitive solution.

Jiang et al. (2014) proposed HPSO-GSA to solve EED problems with five various features such as with/without losses, with/without valve-point effects, with/without prohibited operating zones, with/without multiple fuels, and ramp rate limits. The test systems consist of 6 generating units with smooth cost function, 6 generating units with transmission losses, ramp rate limits, and prohibited operating zones, 10 generating units with transmission losses, 10 generating units with multiple fuels, and 40 generating units with nonsmooth cost function. The experiments carried out on the different power systems with various constraints provide competitive results and confirmed the potential and effectiveness of the HPSO-GSA approach compared to PSO, GSA, and other algorithms published in the recent state-of-the art works in the literature for the solution of the EED problems.

2.8.5 Optimal Power Flow (OPF) Problem

OPF has become the heart of modern economic power systems and markets. The problem is inherently complex in terms of economy, reliable power supply, and computational effort. Efficient market equilibrium requires multipart nonlinear pricing to be economical. As the power flow is an alternating current, it induces additional nonlinearities. The problem poses nonconvexities, which include both binary variables and continuous functions making the optimization problem computationally difficult to solve (Carpentier, 1979). Moreover, the power system must be able to withstand the loss of any generator or transmission unit, and the system operator must make binary decisions to start up and shut down generation and transmission units in response to system events. There are mainly two objectives that the electrical power system must achieve besides the consideration of the operational constraints: minimization of total generating cost and minimization of active transmission losses. The OPF problem is a large-scale, highly nonlinear control optimization problem, which seeks the most favorable settings of a given power system that minimizes the total fuel cost, active power loss, and bus voltage deviation, and enhances voltage stability while at the same time satisfying a number of equality and inequality constraints.

The first objective is the minimization of the total fuel cost F, which is defined in Equation 2.81. The second objective is the minimization of active power loss P_{APL}, defined by

$$Min\,P_{APL} = g_k \sum_{k=1}^{N_{TL}} \left(V_i^2 + V_j^2 - 2V_iV_j \cos(\theta_{ij}) \right) \qquad (2.85)$$

where N_{TL} is the number of transmission lines, g_k is the conductance of branch k connected between buses i and j, θ_{ij} is the voltage angle difference between buses i and j, and V_i and V_j are the voltages of the buses i and j, respectively. To improve the voltage profile, it is important to minimize the bus voltage deviation, which is considered as the third objective defined by

$$Min\,V_{BVD} = \sum_{i=1}^{N_B} |V_i - V_{ref}| \qquad (2.86)$$

where V_{BVD} is the bus voltage deviation, N_B is the number of buses, and V_{ref} is the reference voltage of the buses. For a power system it is important to enhance the voltage stability, which is also considered as an objective. The voltage stability of the total system is described by the L-index of the global power system. Mathematically, it is defined as

$$L = \max(L_j) \qquad (2.87)$$

The four objective functions (Equations 2.81 and 2.85 through 2.87) should satisfy a set of equality and inequality constraints. Further details can be found in Bhowmik and Chakraborty (2014).

Conventional optimization techniques have the disadvantages due to inherent nonlinear characteristics of OPF. A number of population-based optimization methods have been applied to the OPF problem such as PSO (Abido, 2002), DE (Abou El Ela et al., 2010), BBO (Bhattacharya and Chattopadhyay, 2011), GA (Deveraj and Yegnanarayana, 2005), improved GA (Lai et al., 1997), and SA (Roa-Sepulveda and Pavez-Lazo, 2003).

Duman et al. (2012a) applied the GSA to a standard IEEE 30-bus test system with six generators. The proposed GSA approach has been applied to solve the OPF problem for different cases with various objective functions such as quadratic cost function, voltage profile improvement, voltage stability enhancement, voltage stability enhancement during contingency, piece-wise quadratic fuel cost functions, and quadratic cost curve with valve-point loading. The gravitational constant $G(t)$ is set according to Equation 2.14 with $G_0 = 100$, $\alpha = 10$, and $t_{max} = 200$ for all cases. In each case, 50 test runs are conducted. In all cases, the GSA outperformed BBO, DE, or PSO. To evaluate the performance and effectiveness of the GSA on larger power systems, it is applied to an IEEE 57-bus test system with 80 transmission lines and seven generators to determine the optimal settings of control variables of the OPF problem with quadratic cost function. After 50 test runs, the GSA provided the best fuel cost comparable to the best results reported in the literature.

2.8.6 Reactive Power Dispatch (RPD) Problem

The OPF problem discussed in the earlier section can be divided into two subproblems: optimal RPD and optimal real power dispatch (Shi et al., 2012). The electric power loads in a power-generating system vary from hour to hour. The change in load causes variation in the reactive power requirement, which again depends on the voltage, so that the variation of load causes the variation of voltage. Therefore, the important operating task is to maintain the voltage level within the allowable range for high-quality consumer service and minimize real power transmission losses.

The RPD problem is a nonlinear optimization problem with a number of equality and inequality constraints. The objective of RPD is to minimize the objective functions, which are real power losses, voltage profile improvement, and voltage stability enhancement, while satisfying a number of constraints such as load flow, generator bus voltages, load bus voltages, switchable reactive power compensations, reactive power generation, transformer tap setting, and transmission line flow. In general, the objective functions are the minimization of real power loss and maximization of the static voltage stability margin subject to satisfying a set of equality and inequality constraints.

Active power losses: Minimization of the active power losses in the transmission lines can be formulated as follows:

$$\text{Min} F_{\text{loss}} = \sum_{k=1}^{\text{NTL}} g_k \left[V_i^2 + V_j^2 - 2V_i V_j \cos(\delta_i - \delta_j) \right] \qquad (2.88)$$

where NTL is the number of transmission lines, g_k is the conductance of the i-th line, V_i and V_j are the voltage magnitude of the i-th and j-th buses, and δ_i and δ_j are the voltage phase angles of the i-th and j-th buses, respectively.

Voltage deviation: Minimization of the sum of voltage deviations at load buses can be formulated as follows:

$$\text{Min } F_{\text{deviation}} = \sum_{j=1}^{N_{\text{Load}}} \left| V_j - V_j^{\text{ref}} \right| \qquad (2.89)$$

where N_{Load} is the number of load buses, and V_j and V_j^{ref} are the voltage magnitude and reference voltage of the j-th bus, respectively.

Voltage stability index: A fast indicator of voltage stability is the L-index presented by Kessel and Glavitsch (1986). L-index is chosen as an objective function described by

$$L_j = \left| 1 - \sum_{i=1}^{N_g} F_{ji} \frac{V_i}{V_j} \right|, \quad j = N_g + 1, \ldots, n \qquad (2.90)$$

where N_g is the number of generation units. The matrix F_{ji} is computed using the two submatrices $[Y_{LL}]$ and $[Y_{LG}]$ defined as $[F_{ji}] = -[Y_{LL}]^{-1}[Y_{LG}]$. L-index gives the proximity of the system to voltage collapse. The condition for stable voltage is $0 \le L_j \le 1$. A global indicator L describes the stability of the whole system defined by

$$L = \max(L_j), \quad j \in \alpha_L \qquad (2.91)$$

A value of L lower than a threshold ensures a stable system; therefore, minimization of L is considered as an objective function. Further description on the RPD problem can be found in Duman et al. (2012a,b), Niknam et al. (2013), and Shaw et al. (2014). The RPD problem has some constraints, which are discussed in Section 6.7.7 of Chapter 6.

A number of traditional mathematical algorithms have been applied to the RPD problem. Due to its nonlinearity and multimodality, these algorithms have not been quite successful. Recently, researchers have been applying meta-heuristic algorithms. Abou El Ela et al. (2011) and Varadarajan and Swarup (2008) applied a differential evolution (DE) algorithm, Bhattacharya and Chattopadhyay (2011) used a BBO approach, Lee and Park (1995) used a modified simple genetic algorithm, Lai and Ma (1997) applied evolutionary programming, Liang et al. (2007) used a cooperative coevolutionary DE algorithm, Mahadevan and Kannan (2010) used PSO, Subbaraj and Rajnarayanan (2009) used a self-adaptive real-coded genetic algorithm, Wu et al. (1998) applied an adaptive genetic algorithm (AGA), and Zhao et al. (2005) used PSO based on multiagent systems (MAPSO) to solve the RPD problem. Generally, these techniques suffer from algorithmic complexity and insecure convergence as well as being sensitive to the initial search point.

Duman et al. (2012b) applied the GSA to the RPD problem. The test system consists of 19 control variables of a standard IEEE 30-bus system with six generators, four transformers with an off-nominal tap ratio, and selected buses as shunt VAR compensation buses. The proposed GSA approach has been applied to solve the RPD problem for different cases with various objective functions. $G(t)$ is set according to Equation 2.14 with $G_0 = 100$, $\alpha = 10$, and $t_{max} = 200$ for the case of minimization of power loss and $t_{max} = 500$ for the cases of improvement of the voltage profile and enhancement of voltage stability. In each case, 30 test runs are conducted. In all cases, the GSA outperformed BBO, DE, or PSO. In order to evaluate the performance and effectiveness of the GSA on larger power system, it is applied to an IEEE 57-bus test system with 80 transmission lines including 7 generators, 15 transformer taps, and 3 reactive power sources (i.e., 25 dimensions) and to an IEEE 118-bus test system with 186 transmission lines including 54 generators, 9 transformer taps, and 14 reactive power sources (i.e., 77 dimensions). The results obtained from the GSA approach provide a better-quality solution compared to the best results reported in the literature.

Niknam et al. (2013) applied an opposition-based self-adaptive modified GSA to optimal RPD and voltage control in power system operation. The RPD problem is formulated as a mixed-integer nonlinear optimization problem. In order to demonstrate the efficiency of the proposed GSA, four case studies are investigated. In the first three cases, active power loss, voltage deviation, and voltage stability index are optimized individually. In the fourth case, multiobjective optimization has been carried out using two and three objectives. The proposed GSA approach has been verified on the IEEE 30-bus test system. The test system consists of six generators, four transformer taps, two capacitor banks, and two tuning parameters, that is, the GSA has a 14-dimensional space to search. An opposition-based population initialization (discussed earlier) and self-adaptive parameter tuning GSA (also discussed earlier) were applied. In Case 1, minimization of voltage deviation is carried out. It clearly demonstrates the applicability to a real system. It converges to a lower value compared to EAs, PSO, and the standard GSA within 25 iterations. In Case 2, minimization of active power loss is carried out by optimizing compensator capacitors, transformer taps, and the voltage of generator buses. Again, the proposed GSA approach demonstrates superior performance and converges to a lower value compared to the EA, PSO, and standard GSA. In Case 3, minimization of the voltage stability index is carried out. The obtained results show superior performance over the EGA, PSO, and standard GSA. In Case 4, multiobjective optimization is carried out and nondominated Pareto optimal solutions are obtained for proposed objective functions.

Shaw et al. (2014) used opposition-based population initialization and generation jumping in the GSA to further improve the performance of the GSA and applied the approach to an optimal RPD problem. In order to verify the proposed GSA and to solve the RPD problem, three test systems, namely, IEEE 30-, 57-, and 118-bus standard power systems, were considered. The three test systems were used for different cases with various objective functions. The GSA parameter $G(t)$ is set according to Equation 2.14 with $G_0 = 100$ and $\alpha = 10$. The Test System 1 consists of 19 control variables of a standard IEEE 30-bus system with six generators, four transformers with an off-nominal tap ratio, and nine shunt VAR compensation devices. Three cases of the RPD problem are verified on Test System 1, namely, the minimization of power loss, minimization of voltage deviation, and minimization of the voltage stability index as objective functions. The results obtained by all these test cases were compared to those reported in the literature, such as the standard GSA, BBO, DE, self-adaptive GA, and PSO, and found to be very competitive. The convergence profile was promising for the proposed GSA-based approach.

To verify the effectiveness of the proposed GSA, the approach was tested on two larger power systems. The Test System 2 consists of 25 control variables of a standard IEEE 57-bus power system with 7 generators, 15 transformer taps, and 3 reactive power sources. The Test System 3 consists of 77 control variables of a standard IEEE 118-bus power system with 54 generators, 9 transformer taps, and 14 reactive power sources. The three cases of the RPD problem, the minimization of power loss, minimization of voltage deviation, and minimization of the voltage stability index as objective functions, are tested on Systems 2 and 3. In all cases, the proposed GSA outperformed BBO, DE, GA, or PSO. The results obtained from the proposed GSA approach also provide a better-quality solution compared to the best results reported in the literature.

2.8.7 Energy Management System (EMS)

An EMS is an integrated computer-based system, which is used by operators of electric utility grids to monitor, control, and optimize the performance of the generation and/or transmission system. An EMS also used for optimal use of distributed energy sources. The EMS is also used for the auto-mated control and monitoring of electromechanical facilities in a building, which yield significant energy consumption such as heating, ventilation, and lighting installations. The scope of use of the EMS may span from a single building to a group of buildings such as university campuses, office buildings, retail stores networks, or factories. Failure of load feeding is likely to occur in the power distribution system if the total demand is higher than the maximum capacity of the power genera-tion sources. Both performance optimization and scheduling of the distributed generation (DG) are relevant, implementing an EMS within microgrid (MG). Furthermore, optimization methods need to be applied to achieve maximum efficiency, improve economic dispatch, and attain the best performance. In the recent years, researchers have been applying population-based algorithms such as the GA (Chen et al., 2011), and PSO (Hassan and Abido, 2011) and also simulated annealing (Zhuang and Galiana, 1990).

Marzband et al. (2014) applied a GSA-based optimization method to a real-time EMS problem in an MG including different types of DG units with particular attention to the technical constraints. The system under investigation consists of a stand-alone wind turbine (WT), photovoltaic (PV), microturbine (MT), and energy storage (ES) system. For the optimization problem, the voltage level in all of the points of the MG is considered the same. The power loss and reactive power flow are neglected. The objective function for the optimization of the EMS is then to minimize the total general cost defined as

$$\min \sum_{t=1}^{m} \left(C_t^{rg} + C_t^{ng} + C_t^{ES-} - C_t^{l} - C_t^{ES+} + \Omega_t \right) \times \Delta T \qquad (2.92)$$

where m is the number of time periods in the scheduling time horizon T; C_t^{rg} and C_t^{ng} are the cost of renewable and nonrenewable energy production units at time t, respectively; C_t^{ES-} and C_t^{ES+} are the cost of energy produced by ES units during the charging and discharging operation mode at time t, respectively; C_t^{l} is the cost of energy consumed by responsive load demand; and Ω_t is the penalty cost resulting from undelivered power at time t. The further definition of the different production costs and constraints can be found in Marzband et al. (2014). The application includes the imple-mentation of some variation in the load consumption model considering accessibility to the ES and demand response (DR). The GSA method considers a space of dimension $nT \times D \times N$, where nT rep-resents the number of periods, D is the dimension of independent variables, and N is the number of masses. The position of the masses are determined by the vector $XM_j = \left[P_j^{WT}, P_j^{PV}, P_j^{MT}, E_j^{ES}, E_j^{DR} \right]$. These variables $P_j^{WT}, P_j^{PV}, P_j^{MT}, E_j^{ES}, E_j^{DR}$ are vectors of power of WT, PV, MT, ES, and DR, respec-tively. The variables are influenced by the forces between the masses. The parameters P_j^{ES+} and P_j^{ES-} can be derived when E_j^{ES} is known.

The optimization procedure using the GSA is validated experimentally, which shows improved performance in the isolated MG, in comparison with the conventional EMS. Moreover, this approach, which is feasible from a computational viewpoint, has many advantages as peak consumption reduction and electricity generation cost minimization among others. The GSA also achieves a good compromise between computation time and the precision of the solution. It is found that the GSA provides a better-quality solution (i.e., minimum generation cost) with a faster convergence rate within 100 iterations than PSO. The convergence characteristic is more obvious when applied to a large-scale system such as multiple MGs with a lot of variables.

2.8.8 CLUSTERING PROBLEM

Clustering is a search process of discovering hidden patterns or relationships between data objects in large data sets. The process involves partitioning data sets into homogeneous subgroups, generally termed as clusters, subject to satisfying mainly two objectives: data items within one cluster should be similar to each other and those within different clusters should be dissimilar based on a similarity measure. It is generally acknowledged that clustering is an ill-posed problem when prior information about the underlying data distributions is not well defined. Clustering relies on the use of certain criteria that attempt to capture those aspects that humans perceive as the properties of a good clustering solution. The clustering algorithms mainly aim to minimize within-cluster variation (that is intracluster distance) and maximize the between-cluster variation (that is intercluster distance). Let the set of data points be $X = \{x_1, x_2, \ldots, x_n\}$, where $x_j = (x_{j1}, x_{j2}, \ldots, x_{jd}) \in R^d$ and each x_{ji} represents a feature. A clustering algorithm tries to find K partitions with clusters $C = \{C_1, C_2, \ldots, C_K\}$ of the data set X such that

$$C_i \neq \varnothing, \quad i = 1,2,\ldots,K \tag{2.93}$$

$$C_i \cap C_j = \varnothing, \quad i,j = 1,2,\ldots,K \quad \text{and} \quad i \neq j \tag{2.94}$$

$$\bigcup_{i=1}^{K} C_i = X \tag{2.95}$$

In general, data points belong to a cluster based on a similarity measure. The most common similarity measure is the Euclidean distance between the data point x_j and cluster center m_i of cluster C_i defined as

$$d_{ij} = \sqrt{\sum_{k=1}^{d} (x_{jk} - m_{ik})^2} \tag{2.96}$$

A candidate solution $P_j = \{m_1, m_2, \ldots, m_k\}, j = 1, 2,\ldots, S$, is represented by a one-dimensional array encoding centroids of the desired clusters comprising $d \times k$ elements in the array, where d is the dimensionality of the data set or number of features, k is the number of clusters, S is the size of the population or the number of candidate solutions, and $m_i = \{m_{i1}, m_{i2}, \ldots, m_{id}\}, i = 1, 2, \ldots, K$, is the i-th cluster center of the j-th candidate solution.

There are many popular clustering methods reported in the literature to date such as K-means clustering (MacQueen, 1967), fuzzy C-means clustering (Bezdek, 1981), mountain clustering (Yager and Filev, 1994), subtractive clustering (Chiu, 1994), and K-HM clustering (Zhang et al., 2000). While a number of clustering and classification algorithms have been developed in recent years, the search for effective algorithms still continues. Finding a high-performance search method for mining of huge data known as *big data* is of great current interest. A requisite for mining of big data is an effective clustering algorithm. There are many applications of clustering such as data mining, pattern recognition, machine learning, information retrieval, and bioinformatics.

The performance of K-means and fuzzy C-means algorithms depends on the initial selection of the cluster centers. The computation in the mountain method grows exponentially with the dimension of the problem as it requires the evaluation of mountain function over all grid points. Similarly, the computation in subtractive clustering is proportional to data points since each data point is considered as the initial cluster center. Thereby, it is advisable either to employ some front-end methods for finding good initial cluster centers or run the algorithm several times with a different set of initial center values each time.

In K-HM clustering, data objects are partitioned into K clusters and the objective function is to minimize the harmonic mean average from all points in the data set to cluster centers. Therefore, K-HM is not sensitive to the initial selection of cluster centers. But the K-HM algorithm easily runs into local optima. Therefore, researchers exploited the features of the GSA in finding the optimal cluster centers for these clustering algorithms. Yin et al. (2011) integrated an IGSA into the K-HM clustering algorithm to help K-HM escape from local minima. In this approach, a solution vector comprises $d \times k$ elements in the vector, where d is the dimensionality of the data to be clustered and k is the number of clusters. The fitness function is the objective function of the K-HM algorithm. The proposed approach was tested on seven data sets, namely, ArtSet1, ArtSet2, Wine, Glass, Iris, Breast cancer Wisconsin, and Contraceptive Method Choice, which cover examples of data. Standard parameter setting is used for the GSA. The K-HM algorithm requires fewer function evaluations and tends to converge faster than the GSA. In all tests, the results of the proposed approach demonstrated superior performance in terms of efficiency and effectiveness compared to K-HM and PSO-K-HM.

Hatamlou et al. (2012) applied a hybrid GSA-K-means approach to solve the optimal clustering problem. The K-means algorithm is used in generating the initial population, that is, cluster centers, and then the GSA is employed as an improvement method to find the optimal solution. The quality of the resulting clusters and the convergence speed of the GSA have been enhanced by incorporating the K-means algorithm. Five real data sets, namely, Iris, Wine, Glass, Breast cancer Wisconsin, and Contraceptive Method Choice with different number of clusters, data objects, and features, were used to validate the proposed clustering approach. The GSA-K-means provided the highest quality solution compared to the standard GSA, PSO, GA, SA, ACO, HBMO, and K-means algorithm.

Clustering algorithms require the number of clusters to be specified, but in most of the cases, it is unknown or difficult to estimate. Kumar et al. (2014) proposed an automatic clustering using GSA (ACGSA) where it finds the number of clusters in a data set and optimal partitioning. It uses three metrics, namely, the number of clusters, intercluster, and intracluster distances. It uses threshold value setting for each cluster center. A weighted Euclidean distance measure has been preferred to the simple Euclidean distance that helps in assigning data points to appropriate clusters, which is defined by

$$d_{ij}^w = \sqrt{\sum_{k=1}^{d} w_i^2 (x_{jk} - m_{ik})^2} \qquad (2.97)$$

where w_i is the threshold assigned to the cluster center m_i. In this approach, a solution vector comprises real numbers of $k_{max} + d \times k_{max}$ elements in the vector. The first k_{max} elements are positive real numbers within the range of [0,1] representing the weights (i.e., threshold values) and the second k_{max} elements are the cluster centers having d dimensions of the data to be clustered. If the threshold is greater than a specified value, then the center m_{ij} is selected for partitioning; otherwise it is inactive. Most of the fitness evaluation for clustering is based on intercluster and intracluster distance matrices. The fitness used by Kumar et al. (2014) is defined by

$$F = \text{trace}\left(S_W^{-1}S_B\right) \cdot \frac{1}{(k-1)} \tag{2.98}$$

where S_W is the intercluster variation, S_B is the intracluster variation, and $1/(k-1)$ is a penalty function. Five real data sets, namely, Iris, Wine, Glass, Breast cancer, and Vowel with different number of clusters, data objects, and features, were used to validate the proposed clustering approach, and the performances were compared to those of three well-known algorithms such as ACDE (automatic clustering using DE), GCUK (genetic clustering with unknown number of clusters), and DCPSO (dynamic clustering using PSO). In all cases, the proposed approach was able to determine the correct number of clusters, the clusters were clearly separated from each other, and clusters were more compact.

2.8.9 CLASSIFICATION PROBLEM

Classification is the grouping of objects into categories. Mathematically, it is a mapping from input features into a set of labels or classes. There are many popular techniques for the classifier system such as binary classifiers (Garcia-Predajas and Ortiz-Boyer, 2011), decision tree classifiers (Kurzynski, 1983), NN classifiers (Ozyildirim and Avci, 2013), Bayesian classifiers (Hernández-González et al., 2013), support vector machines (Suykens and Vandewalle, 1999), and instance-based classifiers (Lil et al., 2000). Some researchers also applied a meta-heuristic approach such as PSO (De Falco et al., 2007) and ABC (Karaboga and Ozturk, 2011) to classification problems. Bahrololoum et al. (2012) applied a GSA-based approach to an instance-based classifier system. Three fitness functions were used to measure the performance of the GSA-based classifier system applied onto D samples of data represented by an ordered pair consisting of vectors I_j.

$$\min(\text{fit}_1) = \frac{100}{D_T} \sum_{j=1}^{D} m(I_j) \tag{2.99}$$

$$\min(\text{fit}_2) = \frac{1}{D_T} \sum_{j=1}^{D} d(I_j, P_j) \tag{2.100}$$

$$\min(\text{fit}_3) = \frac{1}{2}\left(\frac{\text{fit}_1}{100} + \text{fit}_2\right) \tag{2.101}$$

where $m(I_j)$ is the misclassification, P_j is the prototype to which the class belongs, and $d(\cdot)$ is the Euclidean distance between two I_j. $D_T \subset D$ is the training sample used for training the classifier system. The term $m(I_j)$ provides a value equal to 1 for a misclassification; otherwise, it is equal to 0. The value 100 is used for percentile such that the fitness function value ranges within the interval of [0,100]. The proposed GSA-based approach was tested on 12 data sets, and 20 objects were generated randomly. The GSA parameter $G(t)$ is set according to

$$G(t) = G_0\left(1 - \frac{t}{t_{\max}}\right) \tag{2.102}$$

where $G_0 = 1$ and the number of iterations $t_{max} = 50$. The results were compared to other meta-heuristic algorithms such as PSO and ABC algorithms. The experimental results confirm the effectiveness and performance of the GSA on classification problems.

Chakraborti et al. (2014) applied a local extrema-based GSA, which employs a stochastic local neighborhood-based search instead of global search, to a face recognition problem. They carried out several experiments using benchmark databases, which demonstrated comparable recognition accuracy and outperform many other algorithms.

2.8.10 Feature Subset Selection (FSS)

Pattern recognition, data mining, or knowledge discovery problems require FSS to represent the patterns to be classified. The performance of classifier systems and the computational cost of classification are sensitive to the features set. The FSS refers to the task of identifying and selecting a useful subset of features to be used to represent a pattern from a larger set of features, which is very often mutually redundant and contains possible irrelevant features with different associated measurement costs and/or risks. Such large data sets increase the size of the search space as well as increase the computational cost. The problem is then how to select the minimum subset of features which can preserve the meaning of the original pattern or knowledge without affecting the information represented by the entire set of features. If there are N features in a data set, then FSS can be seen as a search process over a search space of 2^N possible subsets of features. Thus, FSS leads to a search or optimization problem, where each point in the search space specifies a subset of features. Evaluation of all possible feature subsets will be very exhaustive and will lead to a large amount of computational effort. The practical and feasible approach would be to eliminate the redundant, uninformative, and noisy features from the features set, which will reduce the dimension of the data set and the search space for the classification or learning algorithm.

There are two categories of FSS algorithms: filter approach and wrapper approach. In the filter approach, the feature selection is done independently of the learning algorithm. In the wrapper approach, the feature selection is done dependent on the learning algorithm. The filter approach is generally computationally more efficient than the wrapper approach but its major drawback is that an optimal FSS may not be independent of the learning algorithm (Dash and Liu, 1997). On the other hand, the wrapper approach involves the computational overhead of evaluating candidate feature subsets by executing a selected learning algorithm on the data set. The wrapper method selects the subset of features using learning algorithms which provides more promising results than the filter method in terms of classification accuracy (Kohavi and John, 1997).

When the problem of FSS is considered as a search or optimization problem in a search space of 2^N, there is a number of stochastic and evolutionary search algorithms available. Many search techniques have been proposed to solve the FSS problem when there is no *a priori* knowledge about the nature of the task, carrying out an intelligent search in the space of possible solutions. Raymer et al. (2000) suggested using genetic algorithms (GAs) to tackle the problem. Wang et al. (2007) proposed BPSO for feature selection. Zhang and Sun (2002) applied tabu search to this problem. Han et al. (2013, 2014) first proposed to apply the GSA-based approach to the FSS problem. Since the FSS problem poses itself in binary space, Han et al. (2013, 2014) introduced a binary modified GSA referring to the original BGSA of Rashedi et al. (2010).

Binary encoding is generally used in FSS. The length of the encoding represents the number of features. The position of an object is represented by a binary string as shown in Figure 2.8 where 1 represents a selected feature and 0 represents a nonselected feature.

Once a set of features is selected, the classification accuracy is verified on the classifier system or a learning system. K-NN or NN-based classifiers are generally popular methods. The common

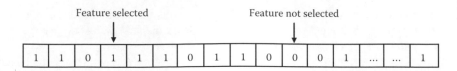

FIGURE 2.8   Encoding of FSS problem.

fitness function considers higher classification accuracy and minimum number of features. Han et al. (2013, 2014) used a fitness function for FSS defined as follows:

$$\text{fit}_i = w_1 \cdot ac_i + w_2 \cdot \left[1 - \frac{\sum_{j=1}^{p} f_j}{p} \right] \tag{2.103}$$

where $f_j \in \{0,1\}$ is the binary value of the feature mask, w_1 is the weight for the classification accuracy, w_2 is the weight for the number of selected features, p is the set of features, and ac_i is the accuracy of the classification at the i-th iteration, which is defined as

$$ac_i = \frac{\text{corr}}{\text{corr} + \text{incorr}} \times 100\% \tag{2.104}$$

where corr and incorr represent the number of correctly classified samples and the number of incorrectly classified samples, respectively.

Han et al. (2013) applied the QBGSA to an FSS problem and used the K-NN method as the classifier system with the aim of improving the classification accuracy in FSS. The method was verified on several machine learning benchmark examples from the University of California Machine Learning Repository. The following parameter settings are used: initially 20 objects are generated, linearly decreasing $G(t)$ is used with G_0 set to 1, the maximum number of iterations is set to 100, and K_{best} is monotonically decreased from 100% to 2.5%. The parameters $\{w_1, w_2\}$ are set to 0.8 and 0.2, respectively. The experimental results are compared with the binary PSO-K-NN (Chung et al., 2011) and GA-K-NN (Li et al., 2001) methods. The results show that the proposed method is able to select the discriminating input features correctly and achieve high classification accuracy, which is comparable to that of well-known similar classifier systems.

Han et al. (2014) applied the modified GSA to an FSS problem using the wrapper methods to evaluate the goodness of the classification. Firstly, a PWL chaotic map is combined with the GSA (called PWL-GSA). Then, SQP is applied to accelerate the local search on the global best solution found by the PWL-GSA. Thus, the exploitation of the search space was enhanced. This integration forms a new method, which enables exploring for a global optimum with more local optima while speeding up the convergence. The method was verified on several machine learning benchmark examples with the same parameter setting as before. Four learning algorithms from different families are selected, such as the ID3 classification tree algorithm, Naïve–Bayes algorithm, K-NN classifier and support vector machine, and they are used as the classifier system with the aim of improving the classification accuracy in FSS. The experimental results are compared with the BPSO (Chung et al., 2011) and GA (Li et al., 2001) methods, both wrapped by K-NN. The results showed higher accuracy compared to that of BPSO and the GA. It also proved that the number of evaluations required by the PWL-GSA method is less than that by the standard GSA. Overall, the results show that the proposed method has higher performance in exploring the search space as evident from the high classification accuracy which is comparable to other well-known meta-heuristic classifier systems.

2.8.11 PARAMETER IDENTIFICATION

A dynamical system is described by a mathematical function with a set of parameters that can represent the dynamic behavior of the system. System identification is a general mathematical procedure to build a dynamical model from measured input–output data. That means the system identification needs to deal with analysis, determination of order, and determination and estimation of parameters of the dynamic system. In other words, system identification is to find the mapping of the functions $\Psi: X \rightarrow Y$ for $\forall x \in X$ and $\forall y \in Y$. Thus, the system identification problem is then transformed into constructing a suitable model of the system from input–output data which when subjected to the same input $u(k)$ produces an output $\hat{y}(k)$ such that $e(k) = \| y(k) - \hat{y}(k) \| < \varepsilon$ for some desired $\varepsilon > 0$, and $\|\cdot\|$ is a suitably defined norm, where $e(k) = y(k) - \hat{y}(k)$ and the norm $\| y(k) - \hat{y}(k) \|$ is the absolute error or squared error.

Parameter identification is to find the best possible values for the set of parameters by minimizing an objective function. The objective function is defined in terms of the parameter set, for example as a function of errors between the system's actual output $y(k)$ and model's evaluated output $\hat{y}(k)$. If the error $e(k) > \varepsilon$, then the parameters are re-estimated and fed back to the model. The process continues until $e(k) < \varepsilon$ is reached. The problem of parameter identification simply leads to a problem of optimization of the parameter set with an objective function J to be minimized (Ljung, 1999). The objective function J can be defined as

$$\min_{\min} J = \frac{1}{M} \sum_{k=1}^{M} \| y(k) - \hat{y}(k) \|^2 < \varepsilon \qquad (2.105)$$

Such an optimization-based general system identification process is illustrated in Figure 2.9. The model in Figure 2.9 is described in terms of a set of parameters, which are estimated by the optimizer. The estimated model output $\hat{y}(k)$ is compared with system output $y(k)$ and the error is fed back to the optimizer. There are many optimization algorithms available that can be applied as an optimizer in this case. For example, the GA and PSO have been used for power system parameter identification problems (Quispe and Graciela, 2008; Carlos and Schirru, 2008).

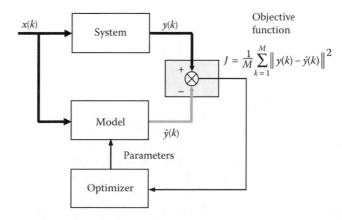

FIGURE 2.9 Parameter identification process.

Li and Zhou (2011) applied the GSA-based optimization algorithm to the solution of parameter identification of HTGS. The HTGS model is described by the following differential equations:

$$
\begin{cases}
\dfrac{dx}{dt} = e_g x + \dfrac{1}{T_a} m_t - \dfrac{1}{T_a} m_{g0} \\[2ex]
\dfrac{dy}{dt} = \dfrac{k_P}{T_y} x + \dfrac{1}{T_y} x_I + \dfrac{1}{T_y} x_D - \dfrac{1}{T_y} y + \dfrac{k_P}{T_y} c \\[2ex]
\dfrac{dx_I}{dt} = -k_I x + k_I b_p y + k_I c \\[2ex]
\dfrac{dx_D}{dt} = \dfrac{k_D e_g}{T_D T_a} x - \dfrac{1}{T_D} x_D - \dfrac{k_D}{T_D T_a} m_t + \dfrac{k_D}{T_D T_a} m_{g0} + \dfrac{k_D}{T_D} c \\[2ex]
\dfrac{dq}{dt} = -\dfrac{1}{T_w} h \\[2ex]
u = x_P + x_I + x_D, \; u_{\min} \le u \le u_{\max}, \; y_{\min} \le y \le y_{\max}
\end{cases}
\tag{2.106}
$$

where $[x,y,x_I,x_D,q]$ is the vector of state variables and $[m_t,h,c,m_{g0}]$ are input variables with $m_t = f_1(y,x,h)$, $q = f_2(y,x,h)$, $h = f_3(y,x,q)$. The model parameter vector is $\theta = [k_P,k_I,k_D,T_y,T_w,T_a,e_g]$, which is to be identified through an optimization process using the IGSA. A detailed model description of the HTGS is provided in Li and Zhou (2011). The fitness function is defined as

$$
F(\theta) = \sum_{k=1}^{N} \sum_{j=1}^{n} w_j [z_j(k) - \hat{z}_j(k)]^2
\tag{2.107}
$$

where $z = [u,y,m_t,x]$ is the system output, $\hat{z} = [\hat{u},\hat{y},\hat{m}_t,\hat{x}]$ is the estimated model output, $w = [w_1,\dots,w_4]$ is the weight vector, N is the number of samples, and n is the number of parameters or the dimension of the system, that is, $n = 4$ for the case of the HTGS.

The GSA and the IGSA, discussed in the previous section, have been applied to identification of parameters $\theta = [k_P,k_I,k_D,T_y,T_w,T_a,e_g]$ of the HTGS. The following parameter settings of the GSA and the IGSA are used: 30 objects are randomly generated, the gravitational constant $G(t)$ defined in Equation 2.15 is used, with $G_0 = 30$, $\alpha = 10$, and $c_1 = c_2 = 0.5$ for the IGSA. The maximum number of iterations used is 500. The results show that the GSA achieved better identification accuracy than PSO and the GA. It also shows that the IGSA achieves the best performance in terms of computational cost, convergence, identification accuracy, and stability compared to the GSA, PSO, and GA.

Chen et al. (2014) also applied the IGSA-based optimization algorithm to the parameter identification problem of WTRS where they introduced three improvements to the IGSA: velocity impacted by the best object, off-boundary agents brought back with a novel strategy, and the chaotic mutation operator incorporated into the search space. A parametric model of a WTRS is developed using $[x,y,x_I,x_D,h]$ as the vector of state variables and $[c,m_g]$ as the input variables. The model is similar to the model in Equation 2.106. The model parameter vector is $\theta = [k_P,k_I,k_D,T_y,T_w,T_a,e_g]$, which is to be identified using the GSA. The fitness function used for the WTRS is the same as Equation 2.107, where $z = [x,y,m_t]$ is the system output, $\hat{z} = [\hat{x},\hat{y},\hat{m}_t]$ is the estimated model output, $w = [w_1,w_2,w_3]$ is the weight vector, N is the number of samples, and n is the number of parameters or the dimension of the system, that is, $n = 3$ for the case of the WTRS. The following parameter settings of the GSA and the IGSA are used: 80 objects are randomly generated, the gravitational constant $G(t)$ defined in Equation 2.15 is used with $G_0 = 30$, $\alpha = 10$, and $c_1 = c_2 = 0.5$ for the IGSA. The maximum number of iterations used is 100. Thirty trial runs were carried out for this case. The results show that the IGSA achieved better identification accuracy than PSO and the standard GSA. The IGSA also shows faster convergence than the other two methods.

Parameter identification of a chaotic system is a multidimensional optimization problem. Different meta-heuristic algorithms have been applied to the identification of chaotic systems. Li et al. (2012b) applied the GSA and the chaotic GSA to identify the parameters of a Lorenz system described by

$$\begin{cases} \dot{x} = \sigma(y - x) \\ \dot{y} = \rho x - xz - y \\ \dot{z} = xy - \beta z \end{cases} \tag{2.108}$$

where $4 < \sigma < 14$, $24 < \rho < 90$, and $1.5 < \beta < 4.5$ are the parameters of the Lorenz system, which defines the behavior of the chaotic system. The task of identification of the chaotic parameters $[\sigma,\rho,\beta]$ of the Lorenz system will be to optimize the objective function defined in Equation 2.105. The position vector is then described as

$$x_i = [\sigma,\rho,\beta] \tag{2.109}$$

The parameter settings for the GSA and the chaotic GSA used by Li et al. (2012b) are as follows: 20 objects, the gravitational constant $G(t)$ defined in Equation 2.15 is used with $G_0 = 20$, $\alpha = 8$, and max iterations 100 for the GSA; and 20 objects, the gravitational constant $G_0 = 20$, $\alpha = 8$, and max iterations 50 for the chaotic GSA. The experimentation was carried out on three groups of initial settings of $[\sigma,\rho,\beta]$. Thirty trial runs were carried out for each case. The results show that the GSA and the chaotic GSA achieved better identification accuracy than PSO and the GA. The chaotic GSA also shows faster convergence.

There are a lot of inherent nonlinearities and uncertainties in a real-world system, which cannot be incorporated into a model described by a set of differential equations. Many researchers, therefore, attempt to model such real-world systems using a Takagi–Sugeno-type fuzzy, also called T-S fuzzy, modeling approach (Siddique and Adeli, 2013; Takagi and Sugeno, 1985).

Li et al. (2013) applied the chaotic GSA to a T-S fuzzy model of the HTGS for the mathematical model of the HTGS described by Equation 2.105 to incorporate unknown nonlinearities. Assuming N_I inputs and N_R rules, there are $N_I \times N_R$ variables in the T-S fuzzy HTGS system. Parameterized Gaussian membership functions with c_k^j as the center and σ_k^j as the width, $j = 1$, ... , N_I, $k = 1$, ... , N_R are used to represent the input membership functions and the rule base of the antecedent part of the model. The parameters of the consequent part are estimated using the least squares method, whereas the parameters of the antecedent part are estimated using an optimization method such as the chaotic GSA method. To apply the GSA, the position of the i-th agent is represented by

$$x_i = \left[c_1^1,...,c_1^{N_I},\sigma_1^1,...,\sigma_1^{N_I},...,c_{N_R}^1,...,c_{N_R}^{N_I},...,\sigma_{N_R}^1,...,\sigma_{N_R}^{N_I} \right] \tag{2.110}$$

The objective function used for the three-rule T-S fuzzy model is the mean squared error (MSE) defined in Equation 2.105. The parameter settings for the chaotic GSA used by Li et al. (2013) are as follows: 30 objects, the gravitational constant $G(t)$ defined in Equation 2.15 is used with $G_0 = 20$, $\alpha = 8$, and max iterations 200 for the chaotic GSA. The experimentation was carried out on three groups of the frequency disturbance process. The results show that the T-S fuzzy model optimized by the chaotic GSA can predict the system output accurately. Hyperplane clustering is a useful approach widely used in fuzzy space partitioning and structure identification of the T-S fuzzy model. Li et al. (2012a) also used GSA-based hyperplane clustering for T-S fuzzy model identification. Experimental results show that the accuracy of model identification is slightly better than that of the fuzzy c-means regression model.

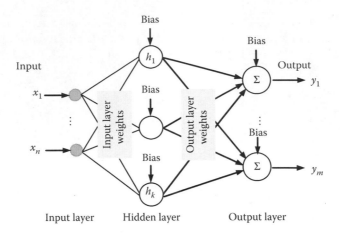

FIGURE 2.10 Feedforward NN architecture.

2.8.12 TRAINING NNS

In general, a feedforward NN consists of one input layer, one or more hidden layers with nonlinear activation function, and one output layer neuron with linear activation function. Biases can be set to nonzero or zero. A generic architecture of a feedforward NN is shown in Figure 2.10.

The problem is to find the weights of the connectivity of the network for the given architecture to produce the correct output for the function for each corresponding input. Thus, for the given architecture, the NN for a given function has to optimize a set of weights as shown in Figure 2.10. The weights training of NN can be carried out by minimizing the MSE of the network error function defined by

$$\text{MSE}(w,b) = \min_{w,b}\left\{ \frac{1}{m} \sum_{i=1}^{m} (y_i - \hat{y}_i)^2 \right\} \qquad (2.111)$$

where w and b are the weights and biases of the NN. The minimum of $MSE(w,b)$ leads to optimum weights and biases of the NN where y_i is the target output and $\hat{y}_i$ is the actual output of the network. Most of NN applications use the standard or a variant of the backpropagation (BP) algorithm for training NNs. The BP method requires computation of the gradient of error function with respect to weights and biases, which again needs the error function to be differentiable. Very often, the error function is not differentiable or is very costly. When global minima are hidden among the local minima, the BP algorithm can get trapped in local minima without reaching the global optimum and end up bouncing between local minima without much overall improvement, which leads to very slow training (Siddique and Adeli, 2013; Siddique and Tokhi, 2001). As a result, BP cannot handle discontinuous optimality criteria or discontinuous neuron (or node) transfer functions. BP's speed and robustness are sensitive to parameters such as learning rate, momentum and acceleration constant, and the best parameters to use seem to vary from problem to problem.

Therefore, researchers have been applying various heuristic optimization algorithms to the training of NNs such as PSO (Settles et al., 2003), the harmony search algorithm (Kattan and Abdullah, 2011), central force optimization (Green et al., 2012), and spiral dynamics optimization (Khan and Sahai, 2013). Mirjalili et al. (2012) proposed a hybrid method combining PSO and the GSA, called the hybrid PSO-GSA, to overcome the slow searching speed of the GSA during the last iterations. The GSA and the hybrid PSO-GSA (described in Section 2.6) approach have been applied to finding optimal combination of weights and biases that will provide the minimum error for the NN defined in Equation 2.111 for a generic NN architecture with n inputs, k hidden layer neurons, and m outputs as shown in Figure 2.10. Mirjalili et al. (2012) used learning error E_k as the fitness function for their application defined by

$$E_k = \sum_{i=1}^{m} (y_i - \hat{y}_i)^2 \tag{2.112}$$

The fitness function is calculated by averaging E_k over q training samples. The objective function for the GSA and the hybrid PSO-GSA used is the typical MSE defined as

$$F = \frac{1}{q} \sum_{k=1}^{q} E_k \tag{2.113}$$

The GSA and the hybrid PSO-GSA were applied to train the NN with predefined architecture representing three benchmark problems: three-bit parity, function approximation, and Iris classification. The standard parameter settings were used for the GSA, for example, $G_0 = 1$, $\alpha = 20$ (as per Equation 2.14), and a population size of 200. For the PSO-GSA, c_1 and c_2 were set to 1 and a population size of 200 was used.

For a three-bit parity problem, a 3-h-1 NN is used where h is the number of neurons in the hidden layer ranging from 5 to 30. For all training samples over 30 independent runs were made. For approximation of the function $y = \sin(2x)e^{-x}$, a 1-h-1 NN is used where h ranges from 3 to 7. A total of 150 training data samples were used for over 30 independent runs to estimate the performance. The PSO-GSA has the best performance for the 1-7-1 NN and the best convergence rate for all 1-h-1 NNs. For the Iris classification problem, a 4-h-3 NN is used where h ranges from 3 to 15. A total of 150 training data samples containing three classes with four features were used for over 30 independent runs to estimate the performance. The PSO-GSA has the best performance and the best convergence rate for all 4-h-3 NNs. The experimental results are promising compared to those of PSO, GSA, and PSO-GSA. For all benchmark problems, the PSO-GSA shows better performance in terms of convergence rate and avoiding local minima.

2.8.13 Traveling Salesman Problem (TSP)

The TSP was mathematically formulated in the 1800s by the Irish mathematician W. R. Hamilton and by the British mathematician Thomas Kirkman. The general form of the TSP appears to have been first studied by mathematicians during the 1930s in Vienna and at Harvard, notably by Karl Menger, who defines the problem. Hassler Whitney at Princeton University introduced the name TSP (Schrijver, 2005). In the 1950s and 1960s, the TSP became very popular in scientific communities in Europe and the United States. TSP is modeled as an undirected weighted graph $G = (V,E)$, where the set of vertices V denotes the n cities and the edge set E denotes the edges between cities, such that cities are the graph's vertices, paths are the graph's edges, and a path's distance is the edges' length. In TSP, a map (i.e., a graph) of n cities is given to the salesman who is required to visit every city only once, starting from an initial city and returning to the starting city once the tour through all cities is completed. The goal of TSP is to find a tour with the minimum total length among all such possible tours for the given graph. A solution to the TSP represented by the graph $G = (V,E)$ is an ordered set of n distinct cities. For such a TSP with n cities, there are $(n-1)!/2$ feasible solutions in which the global optimum(s) is sought. In fact, many real-world problems can be formulated as instances of the TSP, which can lead to solutions to many other combinatorial optimization problems such as scheduling, routing, placement of goods in the warehouse, placement of machines in workshops, and printed circuit design (Dariane and Sarani, 2013). Therefore, it is probably the most widely and extensively studied problem in the field of combinatorial optimization.

A TSP is called a symmetric TSP when the distance from city i to city j (denoted as C_{ij}) is the same as the distance from city j to city i (denoted as C_{ji}), that is, $C_{ij} = C_{ji}$. The symmetric TSP is formulated as follows:

$$\min \sum_{i=1}^{n} \sum_{j=1}^{n} C_{ij} x_{ij} \qquad (2.114)$$

where $x_{ij} \in \{0,1\}$, x_{ij} is 1 if the salesman travels from city i to city j and x_{ij} is 0 otherwise, and $\sum_{j=1}^{n} x_{ij} = 1$, $i, j = 1, 2, \dots, n$, $\sum_{i \in P} \sum_{j \in P, i \neq j} x_{ij} \leq |P| - 1$, $\forall P \subset \{1,2,\dots,n\}$, $2 \leq |P| \leq n - 2$. If the cities can be represented as points in the plane such that the distance between two cities i and j is equal to the Euclidean distance between two points, then the corresponding TSP is called Euclidean TSP.

Dowlatshahi et al. (2014) applied a discrete GSA (discussed in an earlier section) to a set of 54 Euclidean benchmark instances of TSP with sizes ranging from 51 to 2392 nodes. The TSP is encoded as a permutation problem and the small move operator is used based on the swap operator. In order to give a simple representation of the solution and easy manipulation of operators, standard array representation is used for the candidate tour. The array representation reduced the time and space complexity of the operations to be performed on the representation. A greedy randomized algorithm is used to generate the initial population where each member of the population is a tour. The discrete GSA parameters are as follows: the number of objects is set to 10, K is set to 5, the value of G_0 is set to 0.5, and the end value of $G(t)$ is set to 0.1. The stopping criterion is satisfied after 200 runs. The results were satisfactory and in the majority of the instances, the results were equal to the best known solution. Balachandar and Kannan (2007) also applied the randomized gravitational emulation search algorithm to the symmetric travelling salesman problem.

2.8.14 FILTER DESIGN AND COMMUNICATION SYSTEMS

A digital filter transforms input signals into desired signals by keeping the frequency contents of the desired band and eliminating others. Two important types of linear filters are finite impulse response (FIR) filters and infinite impulse response (IIR) filters (Mitra, 2002). Nonlinear filters are another type of digital filters used in many systems with nonlinear behavior. The output of an FIR filter depends on the inputs, whereas the output of an IIR filter depends on the inputs as well as the outputs. IIR filters have the advantage that they require fewer coefficients to compute. The transfer function $H(z)$ of an IIR filter is described by

$$H(z) = \frac{\sum_{i=0}^{m} \hat{a}_i z^{-i}}{1 + \sum_{i=1}^{n} \hat{b}_i z^{-i}} \qquad (2.115)$$

where m and n are the order of the filter, and $\hat{a}_i$ and $\hat{b}_i$ are the filter coefficients to be estimated using a suitable algorithm. A nonlinear filter is defined by the ratio of two polynomials expressed as

$$\hat{y}(k) = \frac{\hat{a}\left[x(k-1),\dots,x(k-n),\hat{y}(k-1),\dots,\hat{y}(k-m)\right]}{\hat{b}\left[x(k-1),\dots,x(k-n),\hat{y}(k-1),\dots,\hat{y}(k-m)\right]} \qquad (2.116)$$

where $x(k-1) \dots x(k-n)$ are the inputs to the filter, $\hat{y}(k)$ is the estimated output of the filter, m and n are the order of the filter, and $\hat{a}_i$ and $\hat{b}_i$ are the filter coefficients to be estimated using a suitable algorithm. Rashedi et al. (2011) are possibly the first to have applied the GSA to modeling digital filters where they estimated the filter parameters $\hat{a}_i$ and $\hat{b}_i$ using the GSA-based optimization procedure. The objective function is defined as

$$J = \frac{1}{L} \sum_{k=1}^{L} \left[\hat{y}(k) - y(k) \right]^2 \tag{2.117}$$

where $y(k)$ is the actual output of the filter. Rashedi et al. estimated the parameters of a second-order IIR filter and a third-order nonlinear filter. In all cases, a randomly generated initial population of 50 objects and 200 data samples were used. The GSA achieved a global solution using the gravitational constant function (Equation 2.14) with $G_0 = 100$ and $\alpha = 10$ within 1000 iterations. The performance was compared with those of the GA and PSO for all cases. Considering the promising results in those experiments, it can be concluded that the GSA is a new tool for filter modeling and design. Han and Chang (2012b) employed a secure communication scheme that consists of encoding, GSA-based filtering, chaotic receiver, and decoding. A useful signal is encoded as a parametric continuous-time carrier signal in the unified chaotic system. The chaotic states recovered in the receiver are corrupted with noise, which requires a GSA-based filtering to be used to estimate the states. The estimated states are used in the decoding scheme to achieve a useful message. To verify the effectiveness of the GSA-based filter modeling, different sets of initial populations with the presence of noise are given and tested with a chaotic signal of length 1024. In experiments, the number of objects is set to 100, the gravitational constant is set to 200, ε is set to 3, and the number of iterations is set to 500. The performance of the GSA-based filter design is compared with PSO- and GA-based filters, which demonstrates superior performance. Channel noise is very often a problem, which degrades the performance of chaos-based secure communication systems. Most chaotic systems fail to compensate such channel noise. Han and Chang (2012a) proposed a chaotic secure communication scheme based on the modified GSA (MGSA), which minimizes the risk of premature convergence of the GSA and employed the MGSA-based nonlinear rational filter modeling, where the MGSA is used to achieve the global optimal solution. To verify the effectiveness of the MGSA-based filter, different sets of initial populations are given and tested with a chaotic signal of length 1024. In experiments, the gravitational constant is set to 200, ε is set to 3, and the number of iterations is set to 300. The results are compared with similar runs using GSA- and PSO-based filters. Saha et al. (2015) applied the GSA with wavelet mutation (GSAWM), discussed in Section 2.6, to the design of an eighth-order IIR filter. An extensive simulation study of low-pass, high-pass, band-pass, and band-stop IIR filters demonstrated the potential of the GSAWM in achieving better cut-off frequency sharpness, smaller pass band and stop band ripples, smaller transition width, and higher stop band attenuation with assured stability. Doraghinejad et al. (2014) applied the IGSA to wireless mesh networks for a channel assignment problem.

2.8.15 Unit Commitment Problem (UCP)

The UCP in power systems is a scheduling procedure for combining power-generating units in a cost-effective way, satisfying the generator and transmission constraints and meeting the forecasted load and reserve requirements. Generally, the UCP is completed over a time horizon, for example, over 1 day to 1 week. It determines which generators will be operating during which hours. This commitment schedule takes into account the intertemporal parameters of each generator (minimum run time, minimum down time, notification time, etc.) but does not specify production levels. Production levels are determined over a very short period, for example, 5 minutes before delivery. The determination of these levels is known as economic dispatch and it is the least-cost usage of the committed assets during a single period to meet the demand.

The objective is to minimize the total system cost F of generating power from N units over a time horizon T. The total cost from each generator at a given period of time is the fuel cost FC_i plus any start-up costs S_i that may be incurred during the period (Happ et al., 1971).

$$F = \sum_{t=1}^{T} \sum_{i=1}^{N} \left[FC_i(P_i(t)) + S_i(h_i(t), u_i(t)) \right] \qquad (2.118)$$

The fuel costs FC_i are dependent on the level of power generation $P_i(t)$. The start-up costs S_i are dependent on the state of the unit h_i, which indicates the number of hours the unit has been on (positive) or off (negative). The discrete decision variable u_i denotes if power generation of the unit at time t is up (denoted as 1) or down (denoted as -1) from that of the unit at time $t + 1$. The start-up costs S_i can be split into start-up and shut-down costs. The objective function in Equation 2.118 can be rewritten as

$$F = FC_T + SU_T + SD_T + \sum_{j=1}^{NC} PF_j \qquad (2.119)$$

where $PF_j = \mu_j \cdot |V_j|$, V_j is the amount of violation of constraint j, PF_j is the penalty associated with violated constraint j, μ_j is the penalty factor associated with constraint j, and NC is the number of violated operating constraints. FC_T is the total fuel cost, SU_T is the unit start-up cost, and SD_T is the unit shut-down cost. The total fuel cost FC_T is computed as the sum of the hourly fuel costs by economically dispatching the load demand to the operating units for every hour t of the scheduling period. If the power balance equation is not satisfied for some time t, a penalty term PF_j multiplied by the factor μ_j proportional to power balance violation is added to the fitness function. The factor μ_j is chosen sufficiently large to discourage selection of solutions with violated constraints. The time-dependent start-up and shut-down costs can be computed from the schedule to be optimized. Depending on the nature of the power system under study, there are a number of equality and inequality constraints in the UCP (Happ et al., 1971).

Binary numbers are to be used in the UCP to indicate the units' status (i.e., committed or not committed). Roy (2013) applied the GSA to the UCP using binary numbers to deal with units' status presented in binary form. Roy (2013) used a large number of solutions generated according to their initial status, that is, flat start. Depending upon the agent size, several numbers of feasible solutions are selected, which is represented by the agent matrix.

$$X = \begin{bmatrix} x_1 \\ x_2 \\ \vdots \\ x_N \end{bmatrix} = \begin{bmatrix} x_1^1 & x_1^2 & \cdots & x_1^n \\ x_2^1 & x_2^2 & \cdots & x_2^n \\ \vdots & \vdots & \ddots & \vdots \\ x_N^1 & x_N^2 & \cdots & x_N^n \end{bmatrix} \qquad (2.120)$$

where x_i is the position of the i-th agent represented by x_i^j, $i = 1, \ldots, N, j = 1, \ldots, n$, n is the dimension of the agent, and N is the number of agents. The fitness value of each agent is calculated using the objective function in Equation 2.119, and acceleration, velocity, and position are updated. The units' status is updated using a sigmoid function $f(x_i^j) = 1/(1 + \exp(x_i^j))$, which provides 1 or 0. Roy (2013) used a population size of 50, a maximum number of iterations 1000, $G_0 = 0.165$, and $\alpha = 20$ as parameter values for the GSA, which was applied to small-scale and large-scale UCPs. In each test case, the GSA appears to be a robust and reliable optimization algorithm and showed competitive performance over many other heuristic algorithms. Yuan et al. (2014) and Ji et al. (2014) also applied the BGSA to the UCP, which also showed competitive results in all experiments.

2.8.16 MULTIOBJECTIVE OPTIMIZATION PROBLEM (MOOP)

Problems requiring simultaneous optimization of more than one objective functions are known as MOOPs. They can be defined as the problems consisting of multiple objectives, which are to be minimized or maximized while maintaining some constraints. Formally, it can be defined as

$$\text{Minimize/maximize } f(x) \tag{2.121}$$

$$\text{Subject to } g_j(x) \geq 0, \quad j = 1, 2, 3, \ldots, J \tag{2.122}$$

$$h_k(x) = 0, \quad k = 1, 2, 3, \ldots, K \tag{2.123}$$

where $f(x) = \{f_1(x), f_2(x), \ldots, f_n(x)\}$ is a vector of objective functions, $x = \{x_1, x_2, \ldots, x_p\}$ is a vector of decision variables, n is the number of objectives, and p is the number of decision variables. Here, the problem optimizes n objectives and satisfies J inequality and K equality constraints. This type of problem has no unique perfect solution. In traditional multiobjective optimization, it is very common to simply aggregate all the objectives together to form a single (scalar) fitness function. However, the obtained solution using a single scalar is sensitive to the weight vector used in the scaling process. This requires knowledge about the underlying problem which is not known *a priori* in most of the cases. Moreover, the objectives can interact or conflict with each other. Therefore, trade-offs are sought when dealing with such MOOPs rather than a single solution. Most MOOPs do not provide a single solution, rather they offer a set of solutions. Such solutions are the "trade-offs" or good compromises among the objectives. In order to generate these trade-off solutions, an old notion of optimality called the Pareto-optimum set (Ben-Tal, 1980) is normally adopted.

Hassanzadeh and Rouhani (2010) first proposed the multiobjective GSA (MOGSA). They used an external archive to store the nondominated solutions. Their approach is the same as in simple multiobjective PSO (SMOPSO) of Cagnina et al. (2005). In the MOGSA, only the stored agents apply gravitational forces to the moving ones and after each movement, the well-known uniform mutation is applied to the new positions. The main disadvantage of the MOGSA approach by Hassanzadeh and Rouhani (2010) is that no equation was presented which relates the mass value to the distance value. In the MOGSA, there is no chance of establishing a relationship between the mass of each object and its multiple objectives.

Hadi et al. (2012) proposed the nondominated sorting GSA (NSGSA) utilizing the nondominated sorting approach. In Hadi et al. (2012), the nondominated sorting concept is used to divide the agents to several layers within the performance space. The NSGSA utilizes a limited-length external archive to store the last found nondominated solutions. A number of researchers also applied the multiobjective GSA to a variety of applications (Ganesan et al., 2013; Ghasemi et al., 2013; Mondal et al., 2013; Tian et al., 2014).

2.8.17 FUZZY CONTROLLER DESIGN

There are many other applications of the GSA reported in the literature such as control design. The performance of any fuzzy controller depends on the rule base and the shape of the membership functions. The rule base and membership functions are constructed using expert knowledge and then optimized or tuned. A detailed description of different approaches and methods on the design, optimization, and tuning of the fuzzy controller using EAs can be found in Siddique and Adeli (2013) and Siddique (2014). One way to improve the performance of fuzzy controllers is to optimize or tune the parameters of membership functions and scaling factors using meta-heuristic algorithms. Roy and Sharma (2013) employed the GSA to fine-tune the membership function parameters and scaling factors of an adaptive fuzzy controller. David et al. (2013) applied the GSA to optimize the parameters of the Takagi–Sugeno-type PI fuzzy controller. Integral squared error between the esti-mated output and reference output was used as the objective function to be minimized.

There are also many other applications of GSA reported in the literature such as vertex covering problem (Balachandar and Kannan, 2009), set covering problem (Balachandar and Kannan, 2010), forecasting (Behrang et al., 2011; Zhang et al., 2013), and image enhancement (Zhao, 2011).

2.9 CONCLUSION

The GSA algorithm is an effective and efficient search and optimization algorithm based on gravi-tational mechanics. Even though in its infancy, and not nearly as highly evolved as many other algorithms since its inception in 2009, the GSA nevertheless appears to hold considerable promise. Because it is deterministic, the GSA is fast and reproducible, characteristics which lend themselves well to implementations that react to how well the algorithm is performing, which are already evident from the applications in various domains. The GSA has already attracted significant atten-tion from the research community, which will help GSA's further development and hopefully will involve both empirical algorithmic improvements and theoretical refinements. This is again evident from the different variants that have already come into being in the literature in the recent past. Some researchers have made some valuable source code in MATLAB available at http://www. cs.bgsu.edu/greenr/cfo/, which will help new researchers in applying the GSA algorithm to new applications. Despite all of these successes so far, there are many open research questions.

Most optimization algorithms come in different variants. The GSA has so far a very good num-ber of variants available in the literature. The GSA still needs further research in other variants like multiobjective formulations. Any application of the GSA should be explored and compared with other population-based meta-heuristic methods.

Another important issue is the convergence of the GSA. The GSA is very likely to stick in local minima. There have been many variants published so far to overcome this situation. This is the general belief, but there has been little theoretical analysis done so far. In addition, premature avoidance, convergence analysis, estimation of convergence rate, searching behaviors explanation, accelerating convergence, and parameter selection are important issues to be addressed.

REFERENCES

Abido, M. A. 2002. Optimal power flow using particle swarm optimization, *Electrical Power Energy Systems*, 24, 563–571.

Abou El Ela, A. A., Abido, M. A., and Spea, S. R. 2010. Optimal power flow using differential evolution algo-rithm, *Electrical Power Systems Research*, 80, 878–885.

Abou El Ela, A. A., Abido, M. A., and Spea, S. R. 2011. Differential evolution algorithm for optimal reactive power dispatch, *Electrical Power Systems Research*, 81, 458–464.

Bahrololoum, A., Nezamabadi-pour, H., Bahrololoum, H., and Saeed, M. 2012. A prototype classifier based on gravitational search algorithm, *Applied Soft Computing*, 12(2), 819–825.

Balachandar, S. R. and Kannan, K. 2007. Randomized gravitational emulation search algorithm for symmet-ric travelling salesman problem, *Applied Mathematics and Computation*, 192, 413–421.

Balachandar, S. R. and Kannan, K. 2009. A meta-heuristic algorithm for vertex covering problem based on gravity, *International Journal of Computational and Mathematical Sciences*, 3(7), 332–336.

Balachandar, S. R. and Kannan, K. 2010. A meta-heuristic algorithm for set covering problem based on grav-ity, *International Journal of Computational and Mathematical Sciences*, 4(5), 223–228.

Baranovsky, A. and Daems, D. 1995. Design of one-dimensional chaotic maps with prescribed statistical properties, *International Journal of Bifurcation and Chaos*, 5, 1585–1598.

Behrang, M. A., Assareh, E., Ghalambaz, M., Assari, M. R., and Noghrehabadi, A. R. 2011. Forecasting future oil demand in Iran using GSA (gravitational search algorithm), *Energy*, 36(9), 5649–5654.

Ben-Tal, A. 1980. Characterization of Pareto and lexicographic optimal solutions, in: G. Fandel and T. Gal, eds., *Multiple Criteria Decision Making: Theory and Application, Lecture Notes in Economics and Mathematical Systems*, Vol. 17, Berlin, Germany: Springer, pp. 1–11.

Bezdek, J. C. 1981. *Pattern Recognition with Fuzzy Objective Function Algorithms*, New York, NY: Plenum Press.

Bharathi, R., Kumar, M. J., Sunitha, D., and Premalatha, S. 2007. Optimisation of combined economic and emission dispatch problem—A comparative study, *Proceedings of the Power Engineering Conference*, December 3–6, Singapore, pp. 134–139.

Bhattacharya, A. and Chattopadhyay, P. K. 2011. Application of biogeography-based optimisation to solve different optimal power flow problems, *IET Generation Transmission & Distribution*, 5(1), 70–80.

Bhowmik, A. R. and Chakraborty, A. K. 2014. Solution of optimal power flow using non-dominated sorting multi objective gravitational search algorithm, *Electrical Power & Energy Systems*, 62, 323–334.

Cagnina, L., Esquivel, S., and Coello, C. A. 2005. A particle swarm optimizer for multi-objective optimization, *Journal of Computer Science and Technology*, 5(4), 204–210.

Carlos, C. M. J. A. and Schirru, R. 2008. Identification of nuclear power plant transients using the particle swarm optimisation algorithm, *Annals of Nuclear Energy*, 35(4), 576–582.

Carpentier, J. 1979. Optimal power flows, *International Journal of Electrical Power & Energy Systems*, 1(1), 3–15.

Ceylan, O., Ozdemir, A., and Dag, H. 2010. Gravitational search algorithm for post-outage bus voltage magnitude calculations, International Universities Power Engineering Conference, August 31–September 3, Wales, UK.

Chakraborti, T., Sharma, K. D., and Chatterjee, A. 2014. A novel local extrema based gravitational search algorithm and its application in face recognition using one training image per class, *Engineering Applications of Artificial Intelligence*, 34, 13–22.

Chatterjee, A., Ghoshal, S. P., and Mukherjee, V. 2012. A maiden application of gravitational search algorithm with wavelet mutation for the solution of economic load dispatch Problems, *International Journal of Bio-Inspired Computation*, 4(1), 33–46.

Chatterjee, A., Mahanti, G. K., and Pathak, N. 2010. Comparative performance of gravitational search algorithm and modified particle swarm optimisation algorithm for synthesis of thinned scanned concentric ring array antenna, *Progress in Electromagnetics Research B*, 25, 331–348.

Chen, C., Duan, S., Cai, T., Liu, B., and Hu, G. 2011. Smart energy management system for optimal microgrid economic operation, *IET Renewable Power Generation*, 5(3), 258–267.

Chen, Z., Yuan, X., Tian, H., and Ji, B. 2014. Improved gravitational search algorithm for parameter identification of water turbine regulation system, *Energy Conversion and Management*, 78, 306–315.

Chiu, S. L. 1994. Fuzzy model identification based on cluster estimation, *Journal of Intelligent and Fuzzy Systems*, 2(3), 267–278.

Chung, L. Y., Yang, C. H., and Li, J. C. 2011. Chaotic maps based on binary particle swarm optimisation for feature selection, *Applied Soft Computing*, 11, 239–248.

Coelho, L. S., Mariani, V. C., Tutkun, N., and Alotto, P. 2014. Magnetizer design based on a quasi-oppositional gravitational search algorithm, *IEEE Transaction on Magnetics*, 50(2), 7017404.

Cordon, O., Herrera, F., Hoffmann, F., and Magdalena, L. 2001. *Genetic Fuzzy Systems: Evolutionary Tuning and Learning of Fuzzy Knowledge Bases*, Singapore: World Scientific.

Dariane, A. B. and Sarani, S. 2013. Application of intelligent water drop algorithm in reservoir operation, *Water Resource Management*, 27, 4827–4843.

Dash, M. and Liu, H. 1997. Feature selection for classification, *Intelligent Data Analysis*, 1(3), 131–156.

Davarynejad, M., Berg, J., and Rezaei, J. 2014. Evaluating center-seeking and initialization bias: The case of particle swarm and gravitational search algorithms, *Information Sciences*, 278, 802–821.

Davarynejad, M., Forghany, Z., and Berg, J. 2012. Mass-dispersed gravitational search algorithm for gene regulatory network model parameter identification, *Simulated Evolution and Learning*, 7673, 62–72.

David, R.-C., Precup, R.-E., Petriu, E. M., Radac, M.-B., and Preitl, S. 2013. Gravitational search algorithm-based design of fuzzy control systems with a reduced parametric sensitivity, *Information Sciences*, 247, 154–173.

De Falco, I., Cioppa, A. D., and Tarantino, E. 2007. Facing classification problems with particle swarm optimisation, *Applied Soft Computing*, 7(3), 652–658.

Deb, K., Pratap, A., Agarwal, S., and Meyarivan, T. 2002. A fast and elitist multi-objective genetic algorithm: NSGA-II, *IEEE Transactions on Evolutionary Computation*, 6(2), 182–197.

Deveraj, D. and Yegnanarayana, B. 2005. Genetic algorithm based optimal power flow for security enhancement, *IEE Proceedings: Generation, Transmission and Distribution*, 152(6), 899–905.

Doraghinejad, M., Nezamabadi-pour, H., and Mahani, A. 2014. Channel assignment in multi-radio wireless mesh networks using an improved gravitational search algorithm, *Journal of Network and Computer Applications*, 38, 163–171.

Dowlatshahi, M. B., Nezamabadi-pour, H., and Mashinchi, M. 2014. A discrete gravitational search algorithm for solving combinatorial optimization problems, *Information Sciences*, 258, 94–107.

Duman, S., Güvenç, U., Sönmez, Y., and Yörükeren, N. 2012a. Optimal power flow using gravitational search algorithm, *Energy Conversion and Management*, 59, 86–95.

Duman, S., Sonmez, Y., Guvenc, U., and Yorukeren, N. 2012b. Optimal reactive power dispatch using a gravitational search algorithm, *IET Generation, Transmission & Distribution*, 6(6), 563–576.

Elaydi, S. N. 1999. *Discrete Chaos*, Boca Raton, FL: Chapman & Hall/CRC Press.

Eötvös, R. V. 1890. Über die Anziehung der Erde auf verschiedene Substanzen, *Mathematische und Naturwissenschaftliche Berichte aus Ungarn*, 8, 65 (in German).

Fang, K. T. and Wang, Y. 1994. *Number-Theoretic Methods in Statistics*, New York, NY: Chapman and Hall.

Fix, E. and Hodges, J. 1989. Discriminatory analysis, nonparametric discrimination: Consistency properties, *International Statistical Review*, 57, 238–247.

Galilei, G. 1960. *On Motion, On Mechanics*, Madison, WI: University of Wisconsin Press.

Ganesan, T., Elamvazuthi, I., Shaari, K. Z. K., and Vasant, P. 2013. Swarm intelligence and gravitational search algorithm for multi-objective optimization of synthesis gas production, *Applied Energy*, 103, 368–374.

Gao, S., Vairappan, C., Wang, Y., Cao, Q., and Tang, Z. 2014. Gravitational search algorithm combined with chaos for unconstrained numerical optimization, *Applied Mathematics and Computation*, 231, 48–62.

Garcia-Predajas, N. and Ortiz-Boyer, D. 2011. An empirical study of binary classifier fusion methods for multi-class classification, *Information Fusion*, 12(2), 111–130.

Gauci, M., Dodd, T. J. and Groß, R. 2012. Why 'GSA: A gravitational search algorithm' is not genuinely based on the law of gravity, *Natural Computing*, 11(4), 719–720.

Ghasemi, A., Shayeghi, H., and Alkhatib, H. 2013. Robust design of multi-machine power system stabilizers using fuzzy gravitational search algorithm, *International Journal of Electrical Power*, 51, 190–200.

Glover, F. 1996. Tabu search and adaptive memory programming—Advances, applications and challenges, in: R. Barr, R. Helgason, and J. Kennington, eds., *Interfaces in Computer Science and Operations Research*, Norwell, MA: Kluwer Academic Publishers, 1–71.

Green, R. C., Wang, L., and Alam, M. 2012. Training neural networks using central force optimization and particle swarm optimization: Insights and comparisons, *Expert Systems with Applications*, 39, 555–563.

Güvenç, U., Sönmez, Y., Duman, S., and Yörükeren, N. 2012. Combined economic and emission dispatch solution using gravitational search algorithm, *Scientia Iranica*, 19(6), 1754–1762.

Hadi, N., Mahdi, N., and Patrick, S. 2012. A multi objective gravitational search algorithm based on non-dominated sorting, *International Journal of Swarm Intelligence Research*, 3(3), 32–49.

Hammerly, G. and Elkan, C. 2002. Alternatives to the k-means algorithm that find better clustering, *Proceedings of the 11th International Conference on Information and Knowledge Management*, November 4–9, McLean, VA, pp. 600–607.

Han, X. and Chang, X. 2012a. A chaotic digital secure communication based on a modified gravitational search algorithm filter, *Information Sciences*, 208(15), 14–27.

Han, X.-H. and Chang, X.-M. 2012b. Chaotic secure communication based on a gravitational search algorithm filter, *Engineering Applications of Artificial Intelligence*, 25(4), 766–774.

Han, X.-H., Chang, X.-M., Quan, L., Xiong, X.-Y., Li, J.-X., Zhang, Z.-X., and Liu, Y. 2014. Feature subset selection by gravitational search algorithm optimization, *Information Sciences*, 281, 128–146.

Han, K. H. and Kim, J. H. 2002. Quantum-inspired evolutionary algorithm for a class of combinatorial optimization, *IEEE Transaction on Evolutionary Computing*, 6, 580–593.

Han, X.-H., Quan, L., Xiong, X.-Y., and Wu, B. 2013. Facing the classification of binary problems with a hybrid system based on quantum-inspired binary gravitational search algorithm and K-NN method, *Engineering Applications of Artificial Intelligence*, 26(10), 2424–2430.

Happ, H. H., Johnson, R. C., and Wright, W. J. 1971. Large scale hydrothermal unit commitment method and results, *IEEE Transaction on Power Apparatus Systems*, PAS-90(3), 1373–1384.

Harwit, M. 1998. *The Astrophysical Concepts*, 3rd ed., New York, NY: Springer.

Hassan, M. A. and Abido, M. A. 2011. Optimal design of microgrids in autonomous and grid connected modes using particle swarm optimization, *IEEE Transaction on Power Electronics*, 26(3), 755–769.

Hassanzadeh, H. R. and Rouhani, M. 2010. A multi-objective gravitational search algorithm, *International Conference on Computational Intelligence, Communication Systems and Networks*, July 28–30, Liverpool, UK, pp. 117–122.

Hatamlou, A., Abdullah, S., and Nezamabadi-pour, H. 2012. A combined approach for clustering based on K-means and gravitational search algorithms, *Swarm and Evolutionary Computation*, 6, 47–52.

He, D., He, C., Jiang, L. G., Zhu, H., Hu, G. 2000. A chaotic map with infinite collapses, *Proceedings of 2000 IEEE TENCON*, September 24–27, Kuala Lumpur, Malaysia, pp. 95–99.

He, Y., Zhou, J., Xiang, X., Chen, H., and Qin, H. 2009. Comparison of different chaotic maps in particle swarm optimization algorithm for long-term cascaded hydro-electric system scheduling, *Chaos, Solutions & Fractals*, 42(5), 3169–3176.

Hernández-González, J., Inza, I., and Lozano, J. A. 2013. Learning Bayesian network classifiers from label proportions, *Pattern Recognition*, 46(12), 3425–3440.

Hey, T. 1999. Quantum computing: An introduction, *Journal of Computing and Control Engineering*, 10(3), 105–112.

Ho, S. L., Shiyou, Y., Guangzheng, N., Lo, E. W. C., and Wong, H. C. 2005. A particle swarm optimization-based method for multi-objective design optimizations, *IEEE Transactions on Magnetics*, 41(5), 1756–1759.

Holliday, D., Resnick, R., and Walker, J. 1993. *Fundamentals of Physics*, New York, NY: John Wiley & Sons.

Ibrahim, A. A., Mohamed, A., and Shareef, H. 2012. A novel quantum-inspired binary gravitational search algorithm in obtaining optimal power quality monitor placement, *Journal of Applied Sciences*, 12(9), 822–830.

Ibrahim, A. A., Mohamed, A., and Shareef, H. 2014. Optimal power quality monitor placement in power systems using an adaptive quantum-inspired binary gravitational search algorithm, *International Journal of Electrical Power & Energy Systems*, 57, 403–413.

Jeong, Y. W., Park, J. B., Jang, S. H., and Lee, K. Y. 2010. A new quantum-inspired binary PSO: Application to unit commitment problems for power systems, *IEEE Transaction on Power System*, 25, 1486–1495.

Ji, B., Yuan, X., Chen, Z., and Tian, H. 2014. Improved gravitational search algorithm for unit commitment considering uncertainty of wind power, *Energy*, 67, 52–62.

Jiang, S., Ji, Z., and Shen, Y. 2014. A novel hybrid particle swarm optimization and gravitational search algorithm for solving economic emission load dispatch problems with various practical constraints, *International Journal of Electrical Power & Energy Systems*, 55, 628–644.

Ju, F.-Y. and Hong, W.-C. 2013. Application of seasonal SVR with chaotic gravitational search algorithm in electricity forecasting, *Applied Mathematical Modelling*, 37, 9643–9651.

Karaboga, D. and Ozturk, C. 2011. A novel clustering approach: Artificial bee colony (ABC) algorithm, *Applied Soft Computing*, 11(1), 652–657.

Kattan, A. and Abdullah, R. 2011. Training of feed-forward neural networks for pattern-classification applications using music inspired algorithm, *International Journal of Computer Science and Information Security*, 9, 44–57.

Kennedy, J. and Eberhart, R. 1995. Particle swarm optimization, *Proceedings of the IEEE International Conference on Neural Networks*, November 27–December 1, Perth, Australia, pp. 1942–1948.

Kennedy, J. and Eberhart, R. 2001. *Swarm Intelligence*, San Francisco, CA: Morgan Kaufmann Publishers, Inc.

Kenyon, I. R. 1990. *General Relativity*, Oxford, UK: Oxford University Press.

Kessel, P. and Glavitsch, H. 1986. Estimating the voltage stability of a power system, *IEEE Transactions on Power Delivery*, 1(3), 346–354.

Khajehzadeh, M., Taha, M. R., El-Shafie, A., and Eslami, M. 2012. A modified gravitational search algorithm for slope stability analysis, *Engineering Applications of Artificial Intelligence*, 25, 1589–1597.

Khan, K. and Sahai, A. 2013. Spiral dynamics optimization-based algorithm for human health improvement, *GESJ: Computer Science and Telecommunications*, 37(1), 31–38.

Khatibinia, M. and Khosravi, S. 2011. A hybrid approach based on an improved gravitational search algorithm and orthogonal crossover for optimal shape design of concrete gravity dams, *Applied Soft Computing*, 16, 223–233.

Kohavi, R. and John, G. H. 1997. Wrappers for feature selection, *Artificial Intelligence*, 97, 273–324.

Kumar, V., Chhabra, J. K., and Kumar, D. 2014. Automatic cluster evolution using gravitational search algorithm and its application on image segmentation, *Engineering Applications of Artificial Intelligence*, 29, 93–103.

Kumar, J. V., Kumar, D. M. V., and Edukondalu, K. 2013. Strategic bidding using fuzzy adaptive gravitational search algorithm in a pool based electricity market, *Applied Soft Computing*, 13(5), 2445–2455.

Kumarappan, N., Mohan, M. R., and Murugappan, S. 2002. ANN approach applied to combined economic and emission dispatch for large scale system, *Proceedings of the 2002 International Joint Conference on Neural Networks (IJCNN)*, May 12–17, Honolulu, HI, Vol. 2(1), pp. 323–327.

Kurzynski, M. W. 1983. The optimal strategy of a tree classifier, *Pattern Recognition*, 16, 81–87.

Lai, L. L. and Ma, J. T. 1997. Application of evolutionary programming to reactive power planning approach, *IEEE Transaction on Power Systems*, 12(1), 198–206.

Lai, L. L., Ma, J. T., Yokoyama, R., and Zhao, M. 1997. Improved genetic algorithms for optimal power flow under normal and contingent operation states, *International Journal of Electrical Power Energy Systems*, 19(5), 287–292.

Lee, K. Y. and Park, Y. M. 1995. Optimization method for reactive power planning by using a modified simple genetic algorithm, *IEEE Transaction on Power Systems*, 10(4), 1843–1850.

Leung, Y. W. and Wang, Y. 2001. An orthogonal genetic algorithm with quantization for global numerical optimization, *IEEE Transaction on Evolutionary Computation*, 5, 41–53.

Li, L., Darden, T. A., Weingberg, C. R., Levine, A. J., and Pedersen, L. G. 2001. Gene assessment and sample classification for gene expression data using a genetic algorithm-K-nearest neighbor method, *Combinatorial Chemistry & High Throughput Screening*, 4, 727–739.

Li, C. S., Li, H. S., and Kou, P. 2014. Piece-wise function based gravitational search algorithm and its application on parameter identification of AVR system, *Neurocomputing*, 124, 139–148.

Li, C. and Zhou, J. 2011. Parameters identification of hydraulic turbine governing system using improved gravitational search algorithm, *Energy Conversion and Management*, 52(1), 374–381.

Li, C., Zhou, J., Fu, B., Kou, P., and Xiao, J. 2012a. T-S fuzzy model identification with a gravitational search-based hyperplane clustering algorithm, *IEEE Transaction on Fuzzy Systems*, 20(2), 305–317.

Li, C., Zhou, J., Xiao, J., and Xiao, H. 2012b. Parameters identification of chaotic system by chaotic gravitational search algorithm, *Chaos, Solutions & Fractals*, 45(4), 539–547.

Li, C., Zhou, J., Xiao, J., and Xiao, H. 2013. Hydraulic turbine governing system identification using T-S fuzzy model optimized by chaotic gravitational search algorithm, *Engineering Applications of Artificial Intelligence*, 26(9), 2073–2082.

Liang, C. H., Chung, C. Y., Wong, K. P., and Duan, X. Z. 2007. Parallel optimal reactive power flow based on cooperative co-evolutionary differential evolution and power system decomposition, *IEEE Transaction on Power Systems*, 22(1), 249–257.

Lil, J., Dong, G., and Ramamohanarao, K. 2000. Instance-based classification by emerging patterns, principles of data mining and knowledge discovery, *Lecture Notes in Computer Science*, 1910, 191–200.

Ljung, L. 1999. *System Identification—Theory for the User*, 2nd ed., Upper Saddle River, NJ: PTR Prentice Hall.

MacQueen, J. 1967. Some methods for classification and analysis of multivariate observations, in: L. LeCam and J. Neyman, eds., *Proceedings of the Fifth Berkeley Symposium on Mathematics Statistics and Probability*, Berkeley, CA: University of California Press, Vol. 1, pp. 281-297.

Mahadevan, K. and Kannan, P. S. 2010. Comprehensive learning particle swarm optimization for reactive power dispatch, *Applied Soft Computing*, 10(2), 641–652.

Mallick, S., Ghoshal, S. P., Acharjee, P., and Thakur, S. S. 2013. Optimal static state estimation using improved particle swarm optimization and gravitational search algorithm, *Electrical Power and Energy Systems*, 52, 254–265.

Mansouri, R., Nasseri, F., and Khorrami, M. 1999. Effective time variation of G in a model universe with variable space dimension, *Physics Letters*, 259, 194–200.

Marzband, M., Ghadimi, M., Sumper, A., and Domínguez-García, J. L. 2014. Experimental validation of a real-time energy management system using multi-period gravitational search algorithm for microgrids in islanded mode, *Applied Energy*, 128, 164–174.

Mendel, E., Krohling, R. A., and Campos, M. 2011. Swarm algorithms with chaotic jumps applied to noisy optimization problems, *Information Science*, 181, 4494–4514.

Mirjalili, S., Hashim, S. Z. M., and Sardroudi, H. M. 2012. Training feed forward neural networks using hybrid particle swarm optimisation and gravitational search algorithm, *Applied Mathematics and Computation*, 218, 11125–11137.

Mitra, S. K. 2002. *Digital Signal Processing—A Computer-Based Approach*, Boston, MA: McGraw-Hill.

Mondal, S., Bhattacharya, A., and Dey, S. H. 2013. Multi-objective economic emission load dispatch solution using gravitational search algorithm and considering wind power penetration, *International Journal of Electrical Power & Energy Systems*, 44(1), 282–292.

Monson, C. K. and Seppi, K. D. 2005. Exposing origin-seeking bias in PSO, Proceedings of the 2005 Conference on Genetic and Evolutionary Computation, June 25–29, Washington, DC, pp. 241–248.

Moore, M. and Nayaranan, A. 1995. Quantum-Inspired Computing, Technical Report, Department of Computer Science, University of Exeter, UK.

Newton, I. 1999. *The Principia: Mathematical Principles of Natural Philosophy*, I. B. Cohen, J. Budenz, and A. M. Whitman (trans.), Berkeley, CA: University of California Press.

Niknam, T., Narimani, M. R., Azizipanah-Abarghooee, R., and Bahmani-Firouzi, B. 2013. Multiobjective optimal reactive power dispatch and voltage control: A new opposition-based self-adaptive modified gravitational search algorithm, *IEEE Systems Journal*, 7(4), 742–753.

Nobahari, H., Nikusokhan, M., and Siarry, P. 2011. Non-dominated sorting gravitational search algorithm, ICSI 2011: International Conference on Swarm Intelligence, June 14–15, Cergy, France, paper id-1.

Ott, E. 2002. *Chaos in Dynamical Systems*, 2nd ed., Cambridge, UK: *Cambridge University Press*.

Ozyildirim, B. M. and Avci, M. 2013. Generalized classifier neural network, *Neural Networks*, 39, 18–26.

Pal, K., Saha, C., Das, S., and Coello Coello, C. A. 2013. Dynamic constrained optimization with offspring repair based gravitational search algorithm, 2013 IEEE Congress on Evolutionary Computation, June 20–23, Cancún, México, pp. 2414–2421.

Precup, R.-E., David, R.-C., Petriu, E. M., Preitl, S., and Paul, A. S. 2011a. Gravitational search algorithm-based tuning of fuzzy control systems with a reduced parametric sensitivity, in: A. Gaspar-Cunha, R. Takahashi, G. Schaefer, and L. Costa, eds., *Advances in Intelligent and Soft Computing*, Berlin, Germany: Springer-Verlag, Vol. 96, pp. 141–150.

Precup, R.-E., David, R.-C., Petriu, E. M., Preitl, S., and Radac, M.-B. 2011b. Gravitational search algorithms in fuzzy control systems tuning, Proc. 18th World Congress of Int. Federation of Automatic Control (IFAC'11), August 28–September 2, Milano, Italy, pp. 13624–13629.

Precup, R.-E., David, R.-C., Petriu, E. M., Preitl, S., and Radac, M.-B. 2012. Novel adaptive gravitational search algorithm for fuzzy controlled servo systems, *IEEE Transaction on Industrial Informatics*, 8(4), 791–800.

Quispe, P. J. and Graciela, C. D. 2008. Parameters identification of excitation system models using genetic algorithm, *IET Generation, Transmission & Distribution*, 2(3), 456–467.

Rahnamayan, S., Tizhoosh, H. R., and Salama, M. M. A. 2007a. Quasi-oppositional differential evolution, Proceedings of the IEEE Congress on Evolutionary Computation, September 25–28, Singapore, pp. 2229–2236.

Rahnamayan, S., Tizhoosh, H. R., and Salama, M. M. A. 2007b. A novel population initialisation method for accelerating evolutionary algorithms, *Computers and Mathematics with Applications*, 53(10), 1605–1614.

Rahnamayan, S., Tizhoosh, H. R., and Salama, M. M. A. 2008. Opposition versus randomness in soft computing techniques, *Applied Soft Computing*, 8, 906–918.

Rashedi, E., Nezamabadi-pour, H., and Saryazdi, S. 2009. GSA: A gravitational search algorithm, *Information Sciences*, 179(13), 2232–2248.

Rashedi, E., Nezamabadi-pour, H., and Saryazdi, S. 2010. BGSA: Binary gravitational search algorithm, *Natural Computing*, 9, 727–745.

Rashedi, E., Nezamabadi-pour, H., and Saryazdi, S. 2011. Filter modelling using gravitational search algorithm, *Engineering Applications of Artificial Intelligence*, 24, 117–122.

Raymer, M. L., Punch, W. F., Goodman, E. D., Kuhn, L. A., and Jain, A. K. 2000. Dimensionality reduction using genetic algorithms, *IEEE Transaction on Evolutionary Computation*, 4, 164–171.

Roa-Sepulveda, C. A. and Pavez-Lazo, B. J. 2003. A solution to the optimal power flow using simulated annealing, *Electrical Power Energy Systems*, 25, 47–57.

Roy, P. K. 2013. Solution of unit commitment problem using gravitational search algorithm, *International Journal of Electrical Power & Energy Systems*, 53, 85–94.

Roy, A. and Sharma, K. D. 2013. Gravitational search algorithm and Lyapunov theory based stable adaptive fuzzy logic controller, *Procedia Technology*, 10, 581–586.

Saha, S. K., Kar, R., Mandal, D., and Ghoshal, S. P. 2015. Optimal IIR filter design using gravitational search algorithm with wavelet mutation, *Journal of King Saud University—Computer and Information Sciences*, 27, 25–39.

Sarafrazi, S., Nezamabadi-pour, H., and Saryazdi, S. 2011. Disruption: A new operator in gravitational search algorithm, *Scientia Iranica, Transactions D: Computer Science & Engineering and Electrical Engineering*, 18(3), 539–548.

Sayah, S., Hamouda, A., and Bekrar, A. 2014. Efficient hybrid optimization approach for emission constrained economic dispatch with non-smooth cost curves, *Electric Power Energy Systems*, 56, 127–139.

Schrijver, A. 2005. On the history of combinatorial optimization (till 1960), in: K. Aardal, G. L. Nemhauser, and R. Weismantel, eds., *Handbooks in Operations Research and Management Science: Discrete Optimization*, Amsterdam, Netherlands: Elsevier.

Schurmann, T. and Hoffmann, I. 2009. A closer look at the uncertainty relation of position and momentum, *Foundations of Physics*, 39, 958.

Schutz, B. 2003. *Gravity from the Ground Up*, Cambridge, UK: Cambridge University Press.

Settles, M., Rodebaugh, B., and Soule, T. 2003. Comparison of genetic algorithm and particle swarm optimizer when evolving a recurrent neural network, in: E. Cantú-Paz, J. A. Foster, K. Deb, D. Lawrence, R. Roy, U.-M. O'Reilly, H.-G. Beyer, R. Standish, G. Kendall, S. Wilson, et al., eds., *Genetic and Evolutionary Computation, GECCO'2003*, Vol. 2723, Berlin, German: Springer, pp. 148–149.

Shaw, B., Mukherjee, V., and Ghoshal, S. P. 2011. Seeker optimisation algorithm: Application to the solution of economic load dispatch problems, *IET Generation, Transmission & Distribution*, 5(1), 81–91.

Shaw, B., Mukherjee, V., and Ghoshal, S. P. 2012. A novel opposition-based gravitational search algorithm for combined economic and emission dispatch problems of power systems, *Electrical Power and Energy Systems*, 35(1), 21–33.

Shaw, B., Mukherjee, V., and Ghoshal, S. P. 2014. Solution of reactive power dispatch of power systems by an opposition-based gravitational search algorithm, *International Journal of Electrical Power & Energy Systems*, 55, 29–40.

Shi, L., Wang, C., Yao, L., Ni, Y., and Bazargan, M. 2012. Optimal power flow solution incorporating wind power, *IEEE Systems Journal*, 6(2), 233–241.

Shokri, M., Tizhoosh, H.R., and Kamel, M. 2006. Opposition-based Q(λ) algorithm, Proceedings of IEEE International Joint Conference on Neural Networks part of IEEE World Congress on Computational Intelligence, July 16–21, Vancouver, Canada, pp. 646–653.

Siddique, N. and Adeli, H. 2013. *Computational Intelligence: Synergies of Fuzzy Logic, Neural Networks and Evolutionary Computing*, Chichester, UK: John Wiley & Sons.

Siddique, N. H. and Tokhi, M. O. 2001. Training neural networks: Backpropagation vs genetic algorithms, Proceedings of the IEEE International Joint Conference on Neural Networks (IJCNN-2001), July 15–19, Washington, DC, pp. 2673–2678.

Soleimanpour-Moghadam, M., Nezamabadi-pour, H., and Farsangi, M. M. 2014. A quantum inspired gravitational search algorithm for numerical function optimization, *Information Sciences*, 267, 83–100.

Subbaraj, P. and Rajnarayanan, P. N. 2009. Optimal reactive power dispatch using self-adaptive real coded genetic algorithm, *Electrical Power Systems Research*, 79(2), 374–381.

Suykens, J. A. K. and Vandewalle, J. 1999. Least squares support vector machine classifiers, *Neural Processing Letters*, 9(3), 293–300.

Swain, R. K., Sahu, N. C., and Hota, P. K. 2012. Gravitational search algorithm for optimal economic dispatch, *Procedia Technology*, 6, 411–419.

Takagi, T. and Sugeno, M. 1985. Fuzzy identification of systems and its applications to modelling and control, *IEEE Transaction on Systems, Man and Cybernetics*, 15(1), 116–132.

Tian, H., Yuan, X., Ji, B., and Chen, Z. 2014. Multi-objective optimization of short-term hydrothermal scheduling using non-dominated sorting gravitational search algorithm with chaotic mutation, *Energy Conversion and Management*, 81, 504–519.

Tizhoosh, H. R. 2005. Opposition-based learning: A new scheme for machine intelligence, Proceedings of International Conference on Computational Intelligence for Modelling Control and Automation, CIMCA 2005, November 28–30, Vienna, Austria, Vol. I, pp. 695–701.

Tizhoosh, H. R. 2006. Opposition-based reinforcement learning, *Journal of Advanced Computational Intelligence and Intelligent Informatics*, 10(5), 578–585.

Tsai, H. C., Tyan, Y. Y., Wu, Y. W., and Lin, Y. H. 2013. Gravitational particle swarm, *Applied Mathematics and Computation*, 219(17), 9106–9117.

Varadarajan, M. and Swarup, K. S. 2008. Differential evolution approach for optimal reactive power dispatch, *Applied Soft Computing*, 8(4), 1549–1561.

Vlachogiannis, J. G. and Lee, K. Y. 2008. Quantum-inspired evolutionary algorithm for real and reactive power dispatch, *IEEE Transaction on Power System*, 23, 1627–1636.

Wang, X., Yang, J., Teng, X., Xia, W., and Jensen, R. 2007. Feature selection based on rough sets and particle swarm optimization, *Pattern Recognition Letters*, 28, 459–471.

Wood, A. J. and Wollenberg, B. F. 1994. *Power Generation, Operation and Control*, New York: John Wiley.

Wu, Q. H., Cao, Y. J., and Wen, J. Y. 1998. Optimal reactive power dispatch using an adaptive genetic algorithm, *International Journal of Electrical Power Energy Systems*, 20(8), 563–569.

Xu, Q., Wang, L., Wang, N., Hei, X., and Zhao, L. 2014. A review of opposition-based learning from 2005 to 2012, *Engineering Application of Artificial Intelligence*, 29, 1–12.

Yager, R. R. and Filev, D. P. 1994. Approximate clustering via the mountain method, *IEEE Transaction on Systems, Man and Cybernetics*, 24, 1279–1284.

Yazdani, S., Nezamabadi-pour, H., and Kamyab, S. 2014. A gravitational search algorithm for multimodal optimization, *Swarm and Evolutionary Computation*, 14, 1–14.

Yilmaz, H. 1965. *Introduction to the Theory of Relativity and the Principles of Modern Physics*, New York, NY: Blaisdel Publications Co.

Yin, M., Hu, Y., Yang, F., Li, X., and Gu, W. 2011. A novel hybrid K-harmonic means and gravitational search algorithm approach for clustering, *Expert Systems with Applications*, 38, 9319–9324.

Yuan, X., Ji, B., Zhang, S., Tian, H., and Hou, Y. 2014. A new approach for unit commitment problem via binary gravitational search algorithm, *Applied Soft Computing*, 22, 249–260.

Zhao, W. 2011. Adaptive image enhancement based on gravitational search algorithm, *Procedia Engineering*, 15, 3288–3292.

Zhao, B., Guo, C. X., and Cao, Y. J. 2005. A multi-agent-based particle swarm optimization approach for optimal reactive power dispatch, *IEEE Transaction on Power Systems*, 20(2), 1070–1078.

Zhang, B., Hsu, M., and Dayal, U. 2000. K-harmonic means, International Workshop on Temporal, Spatial, and Spatio-temporal Data Mining, TSDM2000, September 2, Lyon, France.

Zhang, W., Niu, P., Li, G., and Li, P. 2013. Forecasting of turbine heat rate with online least squares support vector machine based on gravitational search algorithm, *Knowledge-Based Systems*, 39, 34–44.

Zhang, H. and Sun, G. 2002. Feature selection using tabu search method, *Pattern Recognition*, 35, 701–711.

Zhuang, F. and Galiana, F. D. 1990. Unit commitment by simulated annealing, *IEEE Transaction on Power Systems*, 5(1), 311–318.

3 Central Force Optimization

3.1 INTRODUCTION

According to classical mechanics, a central force acting on an object is defined by the force that points radially, the magnitude of which depends only on the distance of the object from the origin and is directed along the line joining them (Whittaker, 1944). The central force problem is an important problem in classical physics. Many naturally occurring forces are central (Goldstein, 1980) such as gravitational force described by Newton's law of gravity and electromagnetic force described by Coulomb's law. The concept of CFO is based on gravitational force where masses move under the influence of gravity. In CFO, a set of "probes" fly through a decision space (DS) according to two simple equations derived from the gravitational metaphor. Probes are analogous to agents or particles used in many other meta-heuristic algorithms. CFO is inherently deterministic, unlike other widely used meta-heuristics.

For a problem at hand, it is assumed that there is no *a priori* information about the maxima. The objective function is defined on a DS of unknown topology that is searched by the algorithm. CFO searches by flying probes through the DS at discrete time steps (iterations). Probes are agents similar to ants used in ACO and particles used in PSO (Formato, 2007a,b). Each probe's location is specified by its position vector computed from equations of motion that are analogous to their real-world counterparts for material objects moving through physical space under the influence of gravity without energy dissipation. CFO comprises two simple "equations of motion" drawn from its metaphor of gravitational kinematics. Gravity is deterministic, and so is CFO because it holds Newton's mathematically precise laws of gravity and motion. Each probe experiences an acceleration created by the gravitational pull of masses in the DS. The acceleration causes the probe to move from an initial position to the next position according to the trajectory equation. By combining the acceleration and the probe position, a new probe distribution is computed. In CFO, the initial probe distribution is based on a deterministic rule and has a significant effect on the algorithm's convergence. The value of the objective function to be maximized, called fitness, is computed step by step at each probe's location, and then input to a user-defined function that becomes the CFO's "mass." Mass in CFO space is analogous to real mass in the Universe moving under Newton's law of gravity.

This chapter will present the CFO metaphor, analysis of probe distribution and DS, the CFO algorithm with discussion on different parameters of CFO algorithms, convergence analysis, variants of CFO, and finally a few applications to engineering problems.

3.2 CENTRAL FORCE OPTIMIZATION METAPHOR

CFO is a relatively new meta-heuristic optimization algorithm proposed by Formato (2009b, 2010a–e). CFO uses a population of probes. The probes are distributed over the entire search space. The main concept behind the CFO is the search for the biggest mass that has the strongest force to attract all other masses toward it within a DS and converge toward the optimal probe that achieves the highest mass measured in terms of fitness, which is considered to be the global optimum of the problem at hand.

In contrast to other population-based algorithms where the initial population is generated randomly such as genetic algorithms (GAs) (Adeli and Hung, 1995), CFO uses probes as its basic population. The movement of the probes is based on the theory of gravitational kinematics that describes the force between two objects as defined by the following equation where force F is proportional to two masses m_1 and m_2 and inversely proportional to the distance r between them, and g is the gravitational constant (Marion, 2013):

$$F = g \frac{m_1 m_2}{r^2} \tag{3.1}$$

The force F acts along the line connecting the centers of gravity of the two masses. According to Newton's law of motion, $F = m_1 a$, where a is the acceleration and m_1 is the mass. The acceleration $\vec{a}$ of mass m_1 toward the mass m_2 is given by the following equation where $\hat{e}$ denotes the unit vector acting along the line joining the centers of gravity of the masses m_1 and m_2:

$$\vec{a} = -g \frac{m_2 \hat{e}}{r^2} \tag{3.2}$$

CFO is based on three basic kinematic equations in terms of the force F between masses, acceleration a, and change of position of the mass. The new position can be calculated from the old position and the distance traveled by the mass with an initial velocity V_0 and acceleration $\vec{a}$ over time Δt. Thus, the new position after Δt is calculated using the following equation:

$$\vec{X}(t + \Delta t) = \vec{X}_0 + \vec{V}_0 \Delta t + \frac{1}{2} \vec{a} \Delta t^2 \tag{3.3}$$

where $\vec{X}(t + \Delta t)$ is the new position at time $t + \Delta t$, $\vec{X}_0$ is the position at time t, and $\vec{V}_0$ is the velocity at time t. A position vector $\vec{X}$ in a 3-D space described by the Cartesian coordinate system is defined by $\vec{X} = x\hat{e}_i + y\hat{e}_j + z\hat{e}_k$, where $\hat{e}_i$, $\hat{e}_j$, and $\hat{e}_k$ are the unit vectors along the x, y, and z axes, respectively.

CFO is basically a maximization algorithm applied to an optimization problem where the criterion of optimality is defined by an objective function. The objective function is again problem dependent. The search procedure is implemented by flying a limited number of probes through the DS. Probes in CFO are equivalent to chromosomes in the GA (Siddique, 2014), that is, each probe position represents a solution to the problem at hand. Each probe p is a feasible solution to the problem in an N_d-dimensional space (i.e., with N_d coordinates). The vector $\vec{X}_j^p = \sum_{k=1}^{N_d} x_{k,j}^p \hat{e}_k$ represents its position vector at time step (or iteration) j, where $x_{k,j}^p$ is the k-th coordinate (decision variable) of probe p's coordinates and $\hat{e}_k$ is the unit vector along the x_k-axis. A fitness value, the mass M in CFO, is calculated by evaluating the objective function of the optimization problem and assigned to each probe. Smaller probes are attracted by bigger probes within the DS like a larger mass attracts a smaller mass in the universe. The attraction force F defined by Equation 3.1 causes the probes to fly with an acceleration a defined by Equation 3.2 through the space over time. As a consequence, the probe position vectors are updated by applying the equation of motion, and all probes tend to settle around the larger probes. In order to represent a solution to a problem, CFO defines each probe as having a position vector X, an acceleration vector a, and a fitness value M. The position vector is a representation of the probes' current coordinates with regard to each dimension of the search space.

With these theoretical underpinnings, the representation of an optimization algorithm can now be pertained to Equations 3.1 through 3.3 and reformulated based on Newton's universal laws of gravitation. Let us consider two probes $k, p \in N_p$ with position vectors $X_{j-1}^k, X_{j-1}^p \in X^{N_d}$ at time step $j - 1$. An attraction force that will act on the probes creating an acceleration a_{j-1}^p is described by (Formato, 2007a)

$$a_{j-1}^p = g \sum_{\substack{k=1 \\ k \neq p}}^{N_p} U(M_{j-1}^k - M_{j-1}^p) \cdot (M_{j-1}^k - M_{j-1}^p)^\alpha \cdot \frac{(X_{j-1}^k - X_{j-1}^p)}{\left| X_{j-1}^k - X_{j-1}^p \right|^\beta} \tag{3.4}$$

where $j = 1,\ldots, N_t$, N_p and N_t are the total number of probes and the total number of time steps, respectively. Here, $\alpha > 0$ and $\beta > 0$ are constant parameters of the CFO model to be chosen. In the physical space, α and β are 1 and 3, respectively. In the CFO space, the user can choose $\alpha > 0$ and $\beta > 0$ depending on the problem and based on experience with the problem at hand. That is, α and β are free parameters, and a user is free to assign a completely different variation of the gravitational acceleration defined by Equation 3.4 in terms of mass and distance. In other words, the gravity in CFO is different from real gravity. Simulation results reveal that the convergence of the CFO algorithm is sensitive to the choice of values of α and β. The term $U(\cdot)$ in Equation 3.4 is the unit step function defined as follows:

$$U(M_{j-1}^k - M_{j-1}^p) = \begin{cases} 1 & \text{if } (M_{j-1}^k - M_{j-1}^p) \geq 0 \\ 0 & \text{otherwise} \end{cases} \tag{3.5}$$

The position vector of probe k at step j is defined by

$$X_j^k = \sum_{l=1}^{N_d} x_{l,j}^k \hat{e}_l \tag{3.6}$$

where $x_{l,j}^k$ are probe k's coordinates, $\hat{e}_l$ is the unit vector along the x_l-axis, and N_d is the number of dimensions (or axes) of the DS.

There are a total of N_p probes within the N_d-dimensional DS, which are flying through the space as a function of time along the trajectories of acceleration and the position vector. At each time step, probes move to new positions, thus creating a new probe distribution. Because the CFO algorithm is based on a metaphor gravitational space, a physically realizable mass does not exist in this space. The fitness function value of each probe is called the mass in CFO. The mass is calculated from an appropriate fitness function defined for the optimization problem in question. The fitness at the location of the k-th probe at time step $j - 1$ is defined by

$$M_{j-1}^k = f(x_{1,j-1}^k, x_{2,j-1}^k, \ldots, x_{N_d,j-1}^k) \tag{3.7}$$

where M_{j-1}^k, $k = 1, \ldots, p - 1, p, p + 1, \ldots, N_p$ are the fitness function values or the masses in CFO space; $f(\cdot)$ is the fitness function; and $x_{l,j-1}{}^k$ are the decision variables of the optimization problem. The probes that are close to each other in the DS are likely to have similar fitness values, which will lead to an excessive gravitational force on the subject probe. In practice, the difference between fitness values is used as the mass, for example, $(M_{j-1}^k - M_{j-1}^p)$. The advantage of the mass calculation based on fitness values is that it avoids excessive gravitational force when a probe is close to another probe. According to gravitational theory, real mass must be positive. But there is no such mass in CFO space. Due to the difference used, the term $(M_{j-1}^k - M_{j-1}^p)$ can be negative or positive depending on which fitness is greater. Therefore, the unit step function $U(M_{j-1}^k - M_{j-1}^p)$ is included to avoid the possibility of negative mass in CFO. It forces CFO to create only positive masses that are consequently attractive in nature. If negative masses were allowed, the corresponding accelerations would be repulsive instead of attractive. The effect of a repulsive gravitational force is to fly probes away from large masses (i.e., fitness values) instead of causing attraction toward them. Thus, the mass in CFO is defined as the difference of fitness raised to the power α multiplied by the unit step function $U(\cdot)$:

$$U(M_{j-1}^k - M_{j-1}^p)(M_{j-1}^k - M_{j-1}^p)^\alpha \tag{3.8}$$

The term $U(M_{j-1}^k - M_{j-1}^p)(M_{j-1}^k - M_{j-1}^p)^\alpha$ in the numerator in Equation 3.4 is the fitness function that corresponds to the mass. As explained earlier, α is a free parameter and it is the designers' choice to select any value for the parameter $\alpha > 0$. It is usually set to 1 (Formato, 2007a).

The distance between two masses M_{j-1}^k and M_{j-1}^p is given by the position distance between the two probes k and p at time step $j - 1$ defined by the following relation:

$$\left| X_{j-1}^k - X_{j-1}^p \right| = \sqrt{\sum_{l=1}^{N_d} (X_{j-1}^{k,l} - X_{j-1}^{p,l})^2} \tag{3.9}$$

The term $\left| X_{j-1}^k - X_{j-1}^p \right|^\beta$ in the denominator in Equation 3.4 is used to represent the distance where β is a free parameter in CFO and it is the designers' choice to select any value for β.

The update of the position vector for probe p at time step j is calculated by adding the distance to the previous position, that is,

$$X_j^p = X_{j-1}^p + S_j^p \tag{3.10}$$

The distance S_j^p traveled by the probe p due to initial velocity V_0^p and acceleration a_{j-1}^p from time step $j - 1$ to time step j is given by

$$S_j^p = V_0^p \Delta t + \frac{1}{2} a_{j-1}^p \Delta t^2 \tag{3.11}$$

The position of probe p at time step j is given by

$$X_j^p = X_{j-1}^p + V_0^p \Delta t + \frac{1}{2} a_{j-1}^p \Delta t^2, \quad j \geq 1 \tag{3.12}$$

In Equations 3.3 and 3.12, the initial velocity V_0^p and the time increment Δt have been used primarily as a formalism that preserves the analogy to gravitational kinematics. The initial velocity V_0^p is considered 0, and the time step increment Δt is actually the time difference between two time steps, that is, $j - (j - 1) = 1$; hence, Δt is unity here. Thus, Equation 3.12 becomes

$$X_j^p = X_{j-1}^p + \frac{1}{2} a_{j-1}^p, \quad j \geq 1 \tag{3.13}$$

The new positions of probes are calculated using Equation 3.13. The factor 1/2 in Equation 3.13 comes from Equation 3.12 after setting the values of initial velocity $V_0^p = 0$ and the time difference $\Delta t = 1$. There is no specific significance of the factor. Therefore, some researchers dropped the factor 1/2 in Equation 3.13 (Chao et al., 2014). The possible problem with the calculation of new position vectors using Equation 3.13 is that some new positions may fall outside of the defined DS and CFP may search regions beyond the DS, which may significantly degrade the performance of CFO and waste valuable computation time. Formato (2007a) suggested a simple deterministic repositioning scheme for avoiding unallowable search space and repairing unfeasible solutions as follows:

$$X_{j,i}^p = x_i^{\min} + F_{\text{rep}} \cdot (X_{j,i}^p - x_i^{\min}) \quad \text{if } X_{j,i}^p < x_i^{\min} \tag{3.14}$$

$$X_{j,i}^p = x_i^{\max} - F_{\text{rep}} \cdot (x_i^{\max} - X_{j,i}^p) \quad \text{if } X_{j,i}^p > x_i^{\max} \qquad (3.15)$$

where $x_i^{\min}$ and $x_i^{\max}$ are the lower and upper bounds of the i-th decision variables, respectively, and F_{rep} is an arbitrary repositioning factor specified by the user within the range of $0 \le F_{\text{rep}} \le 1$. The unfeasible probe retrieval scheme can be implemented using the following pseudocode.

For $i = 1$ *to* N_d

    If $X_{j,i}^p < x_i^{\min}$ Then

$$X_{j,i}^p = x_i^{\min} + F_{\text{rep}} \cdot \left(X_{j,i}^p - x_i^{\min} \right)$$

    If $X_{j,i}^p > x_i^{\max}$ Then

$$X_{j,i}^p = x_i^{\max} - F_{\text{rep}} \cdot \left(x_i^{\max} - X_{j,i}^p \right)$$

end

Another relocation mechanism would be to reposition the probes randomly. While CFO has been shown to be promising in terms of solution quality and functional evaluations, the computational time required to solve optimization problems is often very high when compared with other algorithms. Studies have shown that this increased computational time is due to the computations used to update the acceleration of each probe (Green et al., 2012a).

Figure 3.1 shows a sample 3-D DS with three probes k, p, and q; their position; distances between them; and their masses. N_p, α, β, G, and F_{rep} are the parameters of the CFO. While CFO has been shown to be promising in terms of solution quality and functional evaluations, the computational time required to solve optimization problems is often very high when compared with other algorithms. Studies have shown that this increased computational time is due to the computations used to update the acceleration of each probe (Green et al., 2012a,c). Another issue with the CFO is that the global optimum depends on the initial probe distribution within the search space. There are a

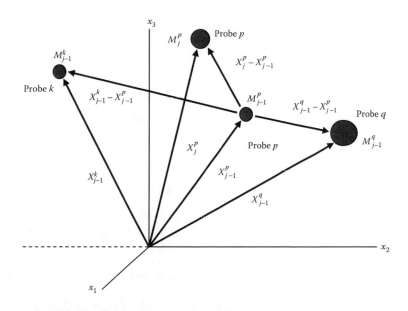

FIGURE 3.1 DS of CFO in 3-D with three probes k, p, and q.

number of approaches to a pertinent search space topology and probe distribution, discussed in a later section, which may help in improving the solution quality.

3.3 CFO ALGORITHM

The CFO algorithm is a deterministic and meta-heuristic algorithm for solving multidimensional optimization problem. Let us consider an N_d-dimensional search space defined by $\Omega = \left\{ X \mid x_d^{\min} < x_d < x_d^{\max}, d = 1, 2, \ldots, N_d \right\}$, $1 \le d \le N_d$, where x_d is the decision variable, d is the coordinate number, N_d is the space's dimensionality (number of coordinates), and $x_d^{\min}$ and $x_d^{\max}$ are the lower and upper bounds of the dimension d, respectively. The problem is to determine the locations of the global maximum of the objective function $f(x_1, x_2, \ldots, x_N)$ defined within the search space Ω. The term fitness refers to the value of $f(x)$ at point $x \in \Omega$. It is to be noted that most optimization problems are formulated as minimization problems, but that CFO usually performs maximization, which, of course, is the same as minimizing $-f(x)$. There is no *a priori* information about the maximum of the objective function available, that is, the topology or landscape of the objective function $f(x)$ is unknown, which may be continuous or discontinuous, unimodal or highly multimodal and possibly subject to a set of constraints among the decision variables.

The CFO algorithm comprises two simple equations, (3.4) for the probe's acceleration and (3.13) for its position vector in the search space. These two equations are explained in detail in Section 3.2. It is important to note that this terminology is chosen solely to reflect the CFO's gravitational metaphor. While G and Δt have direct analogs in the equations of motion for real masses moving under the influence of gravity, there is no analogy in nature for the CFO exponents α and β. They are free parameters included to provide some added flexibility for implementation. They are perfectly acceptable in metaphorical CFO space, where the user is free to change how gravity varies with mass or distance or both in order to achieve a more effective algorithm.

As explained in Section 3.2, the mass in CFO space is a user-defined objective function's value, not necessarily the fitness value itself. For example, the mass in CFO is defined as $U(M_{j=1}^k - M_{j=1}^p) \cdot (M_{j=1}^k - M_{j=1}^p)^\alpha$, that is, the difference in fitness values raised to the α power multiplied by the unit step function $U(\cdot)$. The user is free to define some other function as the mass in CFO if the alternative definition provides better results. The unit step function $U(\cdot)$ solves the negative mass problem by creating a positive-definite mass. $U(\cdot)$ also is the basis for defining a new type of hyperspace directional derivative. In nature, the force of gravity is a conservative vector force field resulting in action at a distance, and accordingly its basic equations are cast in terms of vectors. But real gravity can also be calculated as the gradient of a scalar gravitational potential function, just as the electric field can be derived from an electric potential. The notion of a gradient applies in CFO space as well; but to some degree, its utility depends on how mass is defined.

With the above introduction to the principles of the CFO, implementation of the CFO algorithm can be stated by the following steps:

Step 1: Initialize CFO parameters

Step 2: Generate initial probes and assign initial accelerations

Step 3: Compute initial probe positions, evaluate fitness function values, and compute masses

Step 4: Compute each probe's new position using Equation 3.13 based on previously computed accelerations using Equation 3.4

Step 4.1: Verify that each probe is located inside the search space, making corrections as required using Equations 3.14 and 3.15

Step 5: Update mass at each new probe position
Step 6: Update acceleration for the next time step using new positions
Step 7: If (termination condition not met), Goto step 3
Step 8: Return solution

The initial acceleration is set to zero. A nonzero value flies probes in the directions of the accelerations at the initial step, which may or may not improve performance. Because CFO may fly probes from the DS domain into regions of unfeasible solutions, these probes should be returned to the DS. There are many possible retrieval schemes discussed in Section 3.2. The repositioning factor, F_{rep}, which is used to repair probes that fall outside the search space, is an important parameter of the CFO algorithm. The flowchart of the CFO algorithm is presented in Figure 3.2.

The pseudocode of CFO is given below, which may be helpful in developing simulation in different programming environments.

Step 1: Initialization of population
  a. Create arrays of position vectors $X(p,i,j)$ and acceleration vectors $a(p,i,j)$, where p, $1 \le p \le N_p$, represents the probe number, i, $1 \le i \le N_d$, is the coordinate number, and j, $0 \le j \le N_t$, is the time step
  b. Define fitness $M(p,j) = f[X(p,i,j)]$
  c. Create array of last saved N_{saved}, best fitness $M_{best}(q)$, $1 \le q \le N_{saved}$
Step 2: Initialization of algorithm
  a. At time step $j = 0$:
    i. Uniform probes on each coordinate axis
      For $i = 1$ to N_d, $n = 1$ to $\dfrac{N_p}{N_d}$
      {

$$p = n + \frac{(i-1)N_p}{N_d}$$

$$X(p,i,0) = x_i^{min} + \frac{(n-1)(x_i^{max} - x_i^{min})}{\dfrac{N_p}{N_d} - 1}$$

      }
    ii. Probes slightly off DS diagonal
      For $p = 1$ to N_p, $i = 1$ to N_d
      {

$$X(p,i,0) = x_i^{min} + \frac{(n-1)(x_i^{max} - x_i^{min})[N_d(p-1)+i-1]}{N_p N_d - 1}$$

      }
    iii. Uniform 2-D grid

$$\Delta x_1 = \frac{\left(x_1^{max} - x_1^{min}\right)}{\dfrac{N_p}{N_d} - 1}, \quad \Delta x_2 = \frac{\left(x_2^{max} - x_2^{min}\right)}{\dfrac{N_p}{N_d} - 1}$$

      For $k = 1$ to $\dfrac{N_p}{N_d}$, $m = 1$ to $\dfrac{N_p}{N_d}$
      {

$$p = \frac{N_p}{N_d}(k-1) + m$$

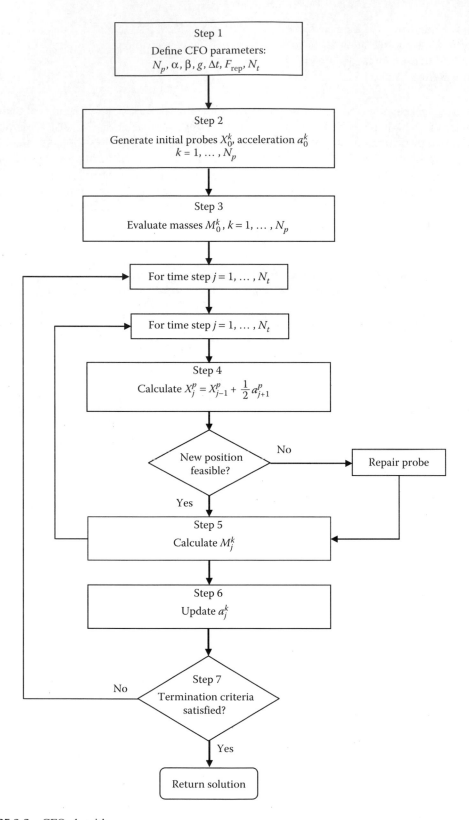

FIGURE 3.2 CFO algorithm.

$$X(p,1,0)=x_1^{\min}+(k-1)\Delta x_1$$
$$X(p,2,0)=x_2^{\min}+(m-1)\Delta x_2$$
}

    iv. Other user-specific probe distribution

  b. Initial acceleration

$$a(p,i,0),\quad 1\le p\le N_p,\quad 1\le i\le N_d$$

  c. Initial fitness

$$M(p,0)=f[X(p,i,0)],\quad 1\le p\le N_p,\quad 1\le i\le N_d$$

$$\text{Best fitness}=\text{Max}(M(p,0),1\le p\le N_p)$$

  d. Save fitness values and tolerance
  e. Initialize reposition factor and increment

$$F_{\text{rep}}=F_{\text{rep}}^{\text{init}},\quad \Delta F_{\text{rep}}$$

Step 3: Time loop $(0\le j\le N)$
  a. Saved fitness array index: $s=j\ \text{MOD}\ N_{\text{saved}}$; If $s=0$, Then
     New probe positions
     For $p=1$ to N_p, $i=1$ to N_d

    i. $X(p,i,j)=X(p,i,j-1)+1/2A(p,i,j-1)$

    ii. Retrieve unfeasible probes as required for repositioning factor $0\le F_{\text{rep}}\le 1$
      If $X(p,i,j)<x_i^{\min}$, Then $X(p,i,j)=x_i^{\min}+F_{\text{rep}}\cdot(X(p,i,j)-x_i^{\min})$
      If $X(p,i,j)>x_i^{\max}$, Then $X(p,i,j)=x_i^{\max}-F_{\text{rep}}\cdot(x_i^{\max}-X(p,i,j))$
  b. Update fitness matrix
     For $p=1$ to N_p do
     {
       $M(p,j)=f[X(p,i,j)]$
       If $M(p,j)\ge \text{Best\_fitness}$, Then
         {
           $\text{Best\_fitness}=M(p,j)$
           $M_{\text{best}}(s)=M(p,j)$
         }
     }
  c. Update reposition factor

     If $\left|M_{\text{best}}(N_{\text{saved}})-(1/N_{\text{sat}})\sum_{q=s}^{N_{\text{saved}}}M_{\text{best}}(q)\right|\le Tol_{\text{fit}}$, $s=N_{\text{saved}}-N_{\text{sat}}+1$, Then
     {
       $F_{\text{rep}}=F_{\text{rep}}+\Delta F_{\text{rep}}$
     }
     If $F_{\text{rep}}\ge 1$, Then $F_{\text{rep}}=F_{\text{rep}}^{\text{init}}$
  d. Update accelerations
     For $p=1$ to N_p, $i=1$ to N_d
     {

$$a(p,i,j)=g\sum_{k=1,k\ne p}^{N_p}U(M(k,j)-M(p,j))\cdot(M(k,j)-M(p,j))^{\alpha}\cdot\frac{X(k,i,j)-X(p,i,j)}{\left|X(k,i,j)-X(k,i,j)\right|^{\beta}}$$

}
Where

$$\left| X_j^k - X_j^p \right| = \sqrt{\sum_{m=1}^{N_d} (X(k,m,j) - X(p,m,j))^2}$$

    e. Increment $j = j + 1$

Repeat from Step 3(a) until $j = N_t$ or other stopping criterion has been met.

3.4 PARAMETERS OF THE CFO ALGORITHM

The main issue in CFO is the selection or estimation of the parameter values. There are, in general, seven parameters in CFO that are required to be initialized, controlled, or tuned by the user in order to achieve an optimal performance for the algorithm. They are the total number of probes N_p, repositioning factor F_{rep}, gravitational constant G, parameters α and β required for computation of acceleration, time interval Δt, and the maximum number of iterations N_t. Some of these appear to be more important than others. The parameters are given in Table 3.1.

Formato (2009a–c) empirically found that CFO's performance mostly depends on the number of probes N_p and its initial distribution. If the initial probes inadvertently sample only points at which the fitness is zero, then CFO has no information to begin its search; it will simply lead to nonzero initial acceleration as the solution. This issue has been discussed further in Section 3.5. There is no physical correspondence to the exponents α and β, though they are considered as parameters. Parameters α and β are included to provide added flexibility for the implementation of the CFO algorithm. In physical space, α and β would take any value. In general, α and β are found empirically. In most of the CFO applications reported in the literature, α and β are set to 2 across a wide range of test functions and applications. The parameter values for α and β seem to provide good results for many test functions and reveal that the CFO's convergence is sensitive to the values of α and β.

The repositioning factor $0 \le F_{rep} \le 1$ is another important parameter of CFO with significant influence on the convergence of the algorithm. Formato (2009a–c) used a simple and deterministic approach to set values of F_{rep} from a starting value and increment it by an arbitrary value ΔF_{rep}, to a final value of 1. At each step, the current and previous four fitness values are stored in a five-element array. It is seen that F_{rep} starts at a value of 0.5 and is incremented by 0.005 whenever the absolute value of the difference between the fifth array element and elements 3, 4, and 5 is less than 0.0005.

TABLE 3.1
Parameters of the CFO Algorithm

Parameters	Description
N_p	Total number of probes
F_{rep}	Repositioning factor
G	Gravitational constant
α, β	Arbitrary parameters required to compute acceleration
Δt	Time interval
N_t	Maximum number of iterations

If incrementing F_{rep} in this way results in $F_{rep} \geq 1$, then F_{rep} is reset to the starting value. The starting value of F_{rep}, increment ΔF_{rep}, and the fitness tolerance are determined empirically. Certain values of F_{rep} can also help avoid local trapping of the algorithm. Local trapping of the CFO algorithm can be measured by the normalized average distance, denoted as D_{avg}. Oscillation in the D_{avg} curve is an indication of local trapping and a convergence difficulty (further discussed in Section 3.8). Changing F_{rep} at each step by some small increment appears to mitigate local trapping effectively in most cases.

The parameter G is the gravitational constant. Very often G is chosen the same as α or β. The parameter Δt is the time interval between steps during which the acceleration is constant. This is chosen to reflect the CFO's gravitational metaphor, and Δt is usually set to 1. It is seen that G and Δt have direct analogs in the equations of motion for real masses moving under gravity. Constant values for G and Δt combine multiplicatively in a single coefficient (Formato, 2009a–c), but still there is the possibility of varying these two parameters individually. The maximum number of iterations N_t has been chosen between 250 and 300 for most of the applications reported in the literature. Another possible parameter is the initial probe acceleration a_0 which is usually set to zero. Some researchers consider it as a parameter and tune it for improving performance (Formato, 2009a–c).

3.5   DECISION SPACE AND PROBE DISTRIBUTION

The efficiency of a meta-heuristic algorithm is mainly evident by the effectiveness of exploitation of the search space. Otherwise, the algorithm gets stuck at local optima. In CFO, probes fly through the DS as a function of time, and at each step, a new probe distribution is computed, which are distributed all over the search space. The probes gradually move toward the probe with the highest fitness. It is apparently clear that the global optimum in CFO depends on the probe distribution within the DS. The CFO algorithm is sensitive to the initial probe distribution. The initial probe distribution is defined by two variables: the total number of probes N_p used for the CFO algorithm and where the probes are placed inside the search space Ω. Different suggestions have been made on how to distribute the probes within the search space such as uniform distribution across the axis of each dimension, uniform distribution on the diagonals of the problem space, or random distribution across the search space. Researchers have shown that the CFO algorithm is sensitive to distinctive topological distributions and these topologies can be mapped to certain mathematical functions (Formato, 2009c). Researchers have used topologies to improve the local search and avoid the trapping in local maxima. Toscano-Pulido et al. (2011) first studied the effect of neighborhood topologies in the behavior of the PSO algorithm. They investigated distinct neighborhood topologies such as ring, fully connected, mesh, star, toroidal, and static tree. Green et al. (2012a) show that CFO and PSO share some features and some neighborhood topologies used in the PSO algorithm provide clues to the application of such neighborhood topologies in the CFO algorithm. Green et al. (2012a) note that the neighborhood topologies used by Toscano-Pulido et al. (2011) for PSO can also be applied to CFO to ensure a good distribution of search points within the DS. The different topologies that can be applied are standard or fully connected, linear, mesh, ring, star, static tree, and toroidal. Figure 3.3 shows the graphical illustration of these topologies. Each of these topologies is described in brief in the sequel.

3.5.1   STANDARD OR FULLY CONNECTED

In this topology, the probes are fully connected to each other. Standard topology is the classic implementation where every probe influences every other probe in the CFO algorithm. A standard topology is shown in Figure 3.3a. There is no special consideration given to the probe with the best fitness. By the definition of mass, the influence on a probe by another probe is scaled by its fitness

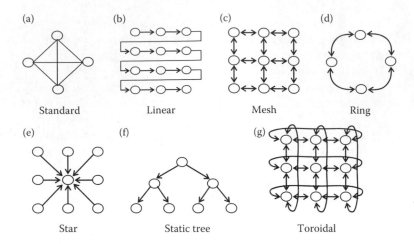

FIGURE 3.3   Various neighborhood topologies used in CFO.

value. This topology is also known as the gbest version in PSO, that is, every particle is attracted by the best solution found by any member of the population.

3.5.2   LINEAR

In linear topology, all probes are ordered in a linear fashion as shown in Figure 3.3b. When calculating the acceleration, the influence on the current probe only includes the previous probe. For example, the arrow in Figure 3.3b indicates the influence of Probe 1 on Probe 2.

3.5.3   MESH

The mesh topology arranges all probes in an $M \times N$ grid, with each probe connected to four neighboring probes located directly to north, south, east, and west of the probe, as shown in Figure 3.3c. Probes located on the border of the mesh are connected with its three adjacent probes, and probes located in the corners are connected with its two adjacent probes. There is a disadvantage of using the mesh as there are overlapping neighbors in each probe causing redundancy in the search effort.

3.5.4   RING

In this topology, all probes are connected to and influenced by those probes that are immediately before and after them. Each probe is affected by the best performance of its k immediate neighbors in the topological population. The last probe in the population is also connected to the first probe in the population as shown in Figure 3.3d. For example, when $N_p = 4$, Probe 1 is connected to and influenced by Probes 2 and 4, Probe 2 is connected to and influenced by Probes 1 and 3, Probe 3 is connected and influenced by Probes 2 and 4, and Probe 4 is connected to and influenced by Probes 3 and 1. This close connection reacts when one probe has an increase in fitness. This reaction decreases proportionally with respect to distance. Thus, each probe is affected by the best fitness of its immediate neighbors in the topology. This topology is also known as the lbest version in PSO.

3.5.5   STAR

In star topology, one central probe influences all others and it is influenced by all others. In other words, each probe is influenced only by the probe with the best overall fitness. The probe with the best fitness has no influence on itself and changes position only when another probe achieves a superior fitness value. The star topology is shown in Figure 3.3e. The central probe is selected randomly. At each step, all probes direct their move toward the central probe. The central probe compares the performance of every individual probe in the population and adjusts its own trajectory toward the best probe.

3.5.6 STATIC TREE

This topology is also known as hierarchical topology. In this topology, all probes are arranged into a hierarchical tree structure with a central probe acting as the root probe. The root probe is connected to one or more probes that are one level lower in the hierarchy. This pattern continues until the entire population of probes is structured to form a tree. In the current implementation, the structure of the tree remains static and each probe is influenced only by its children probes. All probes are also influenced by the probe with the overall best fitness value. The static tree topology is shown in Figure 3.3f.

3.5.7 TOROIDAL

The structure of the toroidal topology is similar to that of the mesh topology except that each probe is connected to four adjacent neighboring probes. The toroidal topology is shown in Figure 3.3g. The toroidal topology connects every corner probe with its symmetrical neighbor. The same occurs with the toroidal boundaries.

The topologies shown in Figure 3.3 and discussed may also be classified into two groups: lbest and gbest (Kennedy and Mendes, 2002). In the lbest-type topology, each individual is influenced by a smaller portion of the population. The linear, mesh, ring, and toroidal topologies are considered to fall into the lbest topology. In the gbest-type topology, every individual in a population is influenced by the probe with the best fitness in the population. The standard, static tree, and star topologies are considered to fall into the gbest topology. Kennedy and Mendes (2002) suggested that populations that use the gbest strategy tend to converge faster to optima than those that use the lbest strategy, but they are more prone to converge to local optima. The probes in the static tree could be randomly or dynamically restructured (Janson and Middendorf, 2005). The static tree and star topologies appear to be members of the lbest classification; the implementations use the probe with the best fitness value to influence all other probes in the population. A typical modification that would allow the star topology to be placed in the lbest category would be choosing the probe that influences other probes randomly (Toscano-Pulido et al., 2011).

Green et al. (2011) investigated the performance of the CFO algorithm using different topologies on four benchmark functions, namely Sphere, Rastrigin, Griewank, and Rosenberg functions. They tested each of these functions using each topology discussed above with the parameter-free (PF)-CFO algorithm. On repeated testing, they tried to find out the implications and impacts of topologies and the performance of CFO in terms of computation time, functional evaluations, probe corrections, solution quality, convergence characteristics, and parameter selection that will help in designing an efficient CFO algorithm with a predefined topology. In all cases of the seven topologies discussed above, there was significant reduction in computation time, which is about 80%–90% lesser than that of the standard topology. It is due to the reduced complexity encountered during updating the acceleration values for each topology. In terms of function evaluation, the mesh, toroidal, and static tree topologies perform the best. The linear and star topologies show good performance except with the Griewank function, on which they require excessive evaluations. The ring topology shows poor performance except with the Rosenbrock function, on which it outperforms all other topologies. Probe correction is an important issue when applying the CFO algorithm, which is required when the probe goes outside the given bound. The experiments show that all the topologies reduce the number of probe corrections for all cases of the functions.

For example, probe corrections were reduced roughly by 30% for the Rosenberg function and by 96% for the Rastrigin function. In every case, each probe is connected to a smaller number of probes than in the standard topology, which results in slower acceleration and prevents probes going outside the search space. In terms of solution quality, each combination of topology and function achieves good results comparable to those of the standard topology. In terms of convergence, all topologies except the ring topology outperform the standard topology, though the linear and star topologies converge more slowly. The results suggest that mesh, toroidal, and static tree topologies not only improve performance in terms of computation, evaluation, and probe corrections but also converge faster in fewer time steps. In terms of parameter selection, namely the number of probes N_p and gravitational constant G, the experiments show varying values for N_p and G for use with each topology and function.

Every CFO algorithm needs an initial probe distribution that plays an important role in finding the quality of the optimal solution. Formato presented a number of examples of the initial probe distribution for 2-D and 3-D DS with varying values of a parameter γ, $0 \leq \gamma \leq 1$ in Formato (2010b–d). The parameter γ, further discussed in Equation 3.21 in Section 3.6.5, determines where the diagonal the probe lines intersect. These results are very promising and useful for defining the DS landscape for an efficient search.

3.6 VARIANTS OF CFO

There have been many implementations and successful applications of CFO since the first publication of the algorithm by Formato in 2007 (Formato, 2007a,b). Despite many successes, the CFO algorithm suffers from some well-known problems like premature convergence at local optima. The quality and accuracy of the optimal solution mainly depend on how the algorithm explores and exploits the search space. Several different formulations for calculation of the mass, acceleration, and position have been proposed to improve the quality of solution and convergence speed. A number of modifications of parameters have been proposed to improve the exploration and exploitation of the search space. Some of the variants of the CFO algorithm are discussed in this section.

3.6.1 SIMPLE CFO

In simple CFO (SCFO), objects are attracted by gravitation force based on the defined mass (Rashedi et al., 2009; Xie et al., 2011). The objects are probes, and their performances are measured by the fitness function. In other words, position of each object represents a solution, which moves toward an object with optimal mass (i.e., an optimal solution) by adjusting its position according to the position update rules described by Equation 3.13. The SCFO algorithm considers three basic procedures:

1. Initialization of the probe distribution
2. Calculation of acceleration
3. Consideration of motion

A population of probes in N_d-dimensional space is initialized. An initial probe distribution is formed by deploying N_p/N_d objects uniformly on each probe line determined by distribution factor γ (Formato, 2010b), where N_p denotes the number of probes. The initial acceleration is set to zero. Under a predefined distribution, SCFO searches the optimum by a deterministic method. The acceleration is calculated using Equation 3.4 described in Section 3.2. The compound acceleration of one object from components in each direction is calculated according to the metaphorical principle of Newton's law of gravity. Mass is a user-defined function from the fitness function to be maximized. To avoid missing landscapes of problems, the simplest definition of mass in Xie et al. (2011) is used. In the gravitational field, the mass of object p is defined by Equation 3.8, where the unit step function $U(\cdot)$ used in the definition of the CFO's mass is described by Equation 3.5.

By $M_{j-1}^p = f\left(x_1^{pj-1}, x_2^{pj-1}, \ldots, x_{N_d}^{pj-1}\right)$ is denoted the fitness function for object p at $j - 1$, $j \in \{1, \ldots, N_t\}$, N_t being the total number of iterations. For other mass definitions with different performances, the reader is referred to Xie et al. (2011).

The final consideration is the motion. According to the acceleration calculated previously, the positions and velocities of probes are updated based on Newton's laws of motion (Green et al., 2012a; Rashedi et al., 2009; Xie et al., 2011). If acceleration a_{j-1}^p is applied to a probe p, it will move from position X_{j-1}^p to position X_j^p according to the motion equation described in Equation 3.13. The mass is based on the relative fitness value of the probes in SCFO. If the fitness is larger than the mass of the probe, it then only attracts the probe. This causes the SCFO algorithm to converge quickly to local optima as the information on the overall fitness is lost. A mechanism is to be found that considers a larger computation incorporating initial distribution to reach global optima.

3.6.2 Extended CFO

To enhance the global search ability of CFO and speed up its convergence, Ding et al. (2012) proposed an extended CFO (ECFO) algorithm by defining a new mass function which is updated in an adaptive fashion based on historical information (using values from previous iterations) and added to the acceleration term.

The optimal contraction theorem indicates that no optimizers can be optimal for arbitrary problems (Chen et al., 2009) unless there is a balance between exploitation and exploration in CFO for general problems. In CFO, the mass is defined as positive by using the unit step function. In ECFO, a new landscape of mass is introduced by defining a new unit step function based on an adaptive mean threshold. The total relative masses will adjust to different probe distributions adaptively. The adaptive mean threshold is defined as follows:

$$M_{j-1}^{\mathrm{amt}} = \frac{1}{N_p - 1} \sum_{\substack{k=1 \\ k \neq p}}^{N_p} \left(M_{j=1}^k - M_{j=1}^p\right) \tag{3.16}$$

The unit step function $U(.)$ is then defined based on the value of the adaptive mean threshold:

$$U(z) = \begin{cases} 1 & \text{if } z \geq -M_{j-1}^{\mathrm{amt}} \\ 0 & \text{else} \end{cases} \tag{3.17}$$

This definition of the unit step function in Equation 3.17 expands the gravitational range of both larger and smaller probes and helps in exploiting the search space globally, and, in addition, the initial distribution of each process is harnessed to its full potential.

A second modification included in ECFO is the addition of a weighted historical experience of acceleration term to the calculation of position in Equation 3.13. Although Newton's motion law says that the velocity term is necessary, it is avoided for simplicity in most of the CFO implementations (Formato, 2010a,b; Green et al., 2012a). The historical velocity information, that is, last initial velocity, is the same as the inertia term in PSO. The larger inertia weight changes the dynamic searching process intrinsically to achieve better global exploration (Bergh and Engelbrecht, 2006; Jiang et al., 2007). Therefore, a weighted historical experience term $a_{j-2}^p \Delta t$ is added to the original CFO, and the position information is calculated according to the following equation:

$$X_j^p = X_{j-1}^p + a_{j-2}^p \Delta t + \frac{1}{2} a_{j-1}^p \Delta t^2 \tag{3.18}$$

By this approach, the exploitation of overall fitness information is expanded and useless searching attempts are avoided at a large scale. Meanwhile, the cost of finding the optimal initial distribution is reduced accordingly. A mathematically elegant and rigorous convergence proof of ECFO is provided by Ding et al. (2012).

3.6.3   Pseudo-Random CFO

The CFO algorithm is inherently deterministic where it searches the DS by using a population of probes governed by the laws of motion. It is evident from the empirical investigations of PSO and ACO that these algorithms fail completely if the randomness is removed from their implementation. Therefore, in order to improve the basic algorithm and introduce some kind of stochasticity into CFO, the basic CFO algorithm has been modified by Formato (2010e). Randomness is introduced into the algorithm indirectly without affecting the deterministic nature of the algorithm. This new version of CFO with adjustment to include near-stochastic characteristics is called pseudo-random CFO (PR-CFO) (Formato, 2013).

PR-CFO extends the basic CFO algorithm through the addition of variations of random behavior, yet it is completely calculable and allows the algorithm to remain deterministic. The pseudo-randomness is introduced to the classic CFO algorithm through injecting randomness in three ways:

- Initial probe distribution
- Repositioning factor
- Dynamic DS bounds

A variable initial probe distribution is an effective way to inject randomness into the CFO algorithm and to provide a better sampling of the decision space Ω than a static distribution. In the original CFO algorithm, Formato (2007a) suggested a simple mechanism of deterministic repositioning and repairing unfeasible solutions using Equations 3.14 and 3.15, where the repositioning factor F_{rep} is an empirical parameter with values in the range of $0 \leq F_{rep} \leq 1$ and specified by the user. In PR-CFO, a variable F_{rep} is used defined by $\Delta F_{rep} \leq F_{rep} \leq 1$, where ΔF_{rep} is the step increment. The variable F_{rep} has the effect of a pseudo-random distribution of probes throughout the decision space Ω. In order to achieve a better convergence speed, the decision space is gradually reduced around the location of the best probe by using F_{rep}. This eventually redistributes the probes within a smaller decision space Ω. The main steps of the PR-CFO algorithm are given as follows:

Step 1: Initialize position and acceleration vectors;
Step 2: Evaluate fitness
Step 3: Update probe positions
     Retrieve probes that moved out of the search space
     Calculate fitness values
     Compute new accelerations
     Increment F_{rep} by ΔF_{rep}
     If $F_{rep} > 1$, Then
         $F_{rep} \leftarrow F_{rep}^{min}$
Step 4: Reduce the search space
Step 5: If (Termination condition not met), Goto Step 3
Step 6: Return solution

3.6.4   PARAMETER-FREE CFO

The most troublesome part in implementing meta-heuristic algorithms is the selection of values for a number of parameters considering the following unsettling facts:

1. There is no methodology for choosing good values.
2. Parameter values are very often problem specific.
3. Solutions are often sensitive to small changes.
4. Exact setting of the same parameters never yields the same results due to the inherently stochastic nature of the algorithms.

Therefore, the simple strategy would be to eliminate or reduce the set of parameters, which demands substantial design time. Parameter-free CFO (PF-CFO) is a further modification of the standard CFO to reduce a number of parameters that must be tweaked in order to generate sufficiently good results and was first proposed by Formato (2010d). In PF-CFO, the attraction force of two probes $k, p \in N_p$ with position vectors $X_{j-1}^k, X_{j-1}^p \in X^{N_d}$ at time step $j-1$ will cause an acceleration a_{j-1}^p on the probe p. The acceleration a_{j-1}^p is described in a parameter-free form as follows:

$$a_{j-1}^p = \sum_{\substack{k=1 \\ k \neq p}}^{N_p} U\left(M_{j-1}^k - M_{j-1}^p\right) \cdot \left(M_{j-1}^k - M_{j-1}^p\right)^{\alpha=1} \cdot \frac{\left(X_{j-1}^k - X_{j-1}^p\right)}{\left|X_{j-1}^k - X_{j-1}^p\right|^{\beta=1}} \tag{3.19}$$

The three parameters, the gravitational constant G and the constant arbitrary parameters α and β, are eliminated by setting fixed values. This simplified PF-CFO equation is obtained by setting fixed values for the basic parameters $G = 1$, $\alpha = 1$, and $\beta = 1$. The unit step function $U(\cdot)$ remains the same as for the standard CFO. It can be seen by comparing the definition of acceleration in Equation 3.19 with that of Equation 3.4 that these parameters are set to unity. Further discussion on these parameters can be found in Formato (2007a, 2010b). The second equation of motion in PF-CFO is the position vector $X_j^p \in X^{N_d}$ which is now defined as follows:

$$X_j^p = X_{j-1}^p + a_{j-1}^p, \quad j \geq 1 \tag{3.20}$$

It is important to note that the new position of the probe is calculated using Equation 3.20, where the factor 1/2 in Equation 3.13 is also dropped as there is no real significance of this factor in implementation of the CFO algorithm (Chao et al., 2014). Thus, the PF-CFO can now be described by the two simplified equations of motions (3.19) and (3.20). The internal parameters such as the number of probes N_p, repositioning factor F_{rep}, ΔF_{rep}, and the maximum number of iterations N_t remain the same. To verify the performance of the PF-CFO, only a CFO run is required and no statistical description is needed as CFO is inherently deterministic. Formato (2010d) applied the PF-CFO algorithm to 23 unimodal and multimodal benchmark functions to demonstrate its effectiveness. In the first set, there were seven high-dimensional unimodal functions. PF-CFO produced the best fitness on all seven functions. In the second set, there were six high-dimensional multimodal functions with many local maxima. PF-CFO provided the best fitness on three of them. In the last set of functions, there were 10 multimodal functions with few local maxima. PF-CFO returned the best fitness on four of them, equal fitness on four of them, and slightly lower fitness on four of them. Formato (2010d) has verified the performance of PF-CFO on a number of benchmark problems. Once again, the CFO's performance was competitive against the standard CFO, PSO, or GA.

Green et al. (2011) also applied the PF-CFO algorithm using different topologies to four benchmark functions such as Sphere, Rastrigin, Griewank, and Rosenberg functions in order to demonstrate the impact of topology as well as to determine the optimal parameters. The parameters are chosen by the PF-CFO algorithm for each function and topology combination. It can be concluded that the mesh, toroidal, and static tree topologies tend to perform the best in terms of functional evaluation. The linear and star topologies show good performance demanding excessive functional evaluations for the Griewank function. The ring topology shows poor performance in terms of functional evaluations except for the Rosenbrock function, on which it outperforms all other topologies. In all cases except for the Rastrigin function, the number of probes N_p used in the standard topology is less than or equal to the number of probes used for other topologies. The Rastrigin function is particularly interesting as the function is nonlinear and multimodal, and contains multiple local optima requiring a higher number of probes. This investigation also confirms the fact that lbest topologies tend to work better than gbest topologies on multimodal functions (Janson and Middendorf, 2005).

3.6.5 Improved CFO

Formato (2010b) studied the performance of the standard CFO algorithm by applying it to 23 benchmark functions. From these empirical investigations, he found out that the performance of the CFO algorithm depends on the initial probe distribution. In other words, the initial probe distribution is a measure of the well-sampled DS topology that enables the CFO algorithm to explore the search space efficiently. The DS reconfiguration or adaptation helps the CFO algorithm in exploiting the search space and improving convergence speed. Formato (2010c) proposed an improved CFO (ICFO) by introducing variable initial probe distribution and DS adaptation. The initial probe distribution is defined by two variables: (i) the total number of probes N_p to be considered within the DS and (ii) the dimension of the DS N_d. A well-distributed initial probe distribution can be formed by an orthogonal array of N_p/N_d probes per dimension deployed uniformly on each probe line determined by distribution factor γ described in Equation 3.21. Formato (2010c) presented an example of initial probe distribution in a 2-D DS as shown in Figure 3.4. The probe lines are parallel to x_1 and x_2 axes (shown in dotted lines) and intersect at a point within the decision space on the principal diagonal. By $\hat{e}_1$ and $\hat{e}_2$, one denotes the unit vectors along the axes x_1 and x_2, respectively. The number of probes per axis can vary. Only six probes per axis are shown on each probe line in Figure 3.4. The position vector X is defined as follows:

$$X = X_{\min} + \gamma\left(X_{\max} - X_{\min}\right)$$

(3.21)

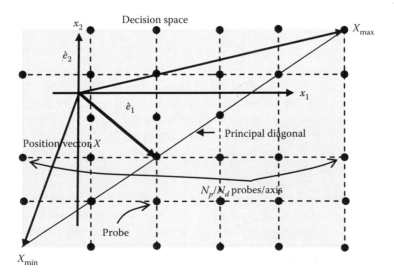

FIGURE 3.4 Probe distribution on a 2-D DS.

where $X_{min} = \sum_{i=1}^{N_d=2} x_i^{min} \hat{e}_i$ and $X_{max} = \sum_{i=1}^{N_d=2} x_i^{max} \hat{e}_i$ are the positions of the diagonal's endpoint. Here, $N_d = 2$ for 2-D DS. The DS is defined by $x_i^{min} \le x_i \le x_i^{max}$, $1 \le i \le N_d$, where x_i is the decision variable. The parameter $\gamma \in [0,1]$ determines where the probe lines are to be intersected along the diagonal. Using this simple notion, the initial probe distribution can be generalized for N_d-dimensional decision space Ω, and N_d number of probe lines can be drawn in the decision space Ω parallel to N_d coordinates.

Formato presented a number of examples of the initial probe distribution for 2-D and 3-D DS with varying values of parameter γ in Formato (2010b,c). These results are very promising and useful for defining the DS landscape for an efficient search strategy.

An efficient local search mechanism would better exploit the DS locally, avoid unnecessary iterations, and improve the overall convergence speed. Adaptive reconfiguration of the DS would be a suitable mechanism that will reduce the size of the decision space Ω around the location of the probe with the best fitness. The DS boundary is reduced by one-half the distance from the best probe's position to the boundary of each coordinate, defined as follows:

$$x'^{min}_i = x_i^{min} + \frac{X_{best} \cdot \hat{e}_i - x_i^{min}}{2} \tag{3.22}$$

$$x'^{max}_i = x_i^{max} + \frac{x_i^{max} - X_{best} \cdot \hat{e}_i}{2} \tag{3.23}$$

where x_i is the decision variable and x'_i is the new boundary of the DS.

Figure 3.5 illustrates how the DS reduces over every n time steps adaptively based on the distance between the best probe's position X_{best} and each boundary. It is to be noted that X_{best} may also change in each iteration. Formato (2010b) found out that changing the DS boundary by $n = 10$ steps provides a good result. It is also suggestive that the other interval is arbitrary as well.

The performance of the ICFO algorithm has been assessed against recognized antenna benchmark problems. Results also are presented for a standard 23-function suite of analytic benchmarks. The ICFO implementation exhibits excellent performance (Formato, 2010c).

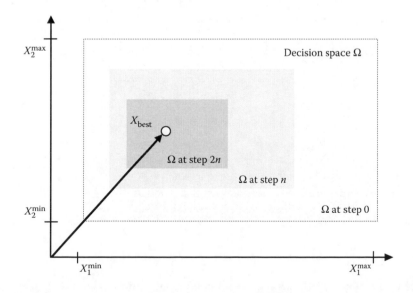

FIGURE 3.5   Adaptive DS.

3.6.6 CFO with Acceleration Clipping

There are a number of run parameters in CFO such as acceleration, which are required to set to obtain the global optimal solution as well as to prevent probes being trapped into the local optimum. The run parameters very often become troublesome as they need to be adjusted and tuned for the application at hand. Improper values for the run parameters may cause the probes' motion to go out of control. Qubati and Dib (2010) introduced acceleration clipping to standard CFO to enhance CFO's global search ability while maintaining its simplicity. Acceleration clipping is also a modification to release the CFO from its dependency on the run parameters. Very often probes fly out of the DS due to their motion, which requires a repositioning procedure to be carried out to bring them back to the DS. When the length of the acceleration vector A_j^p is greater than the length of the diagonal of the search space multiplied by a predefined factor A_{max}, the acceleration vector A_j^p is clipped by this same factor. After clamping, the probes keep the original direction of the acceleration. The clipping procedure is given by

$$\text{If } A_j^p > A_{max} \times \text{DSDL, Then } A_j^p = A_{max} \times A_j^p \qquad (3.24)$$

where $A_{max} \in [0.001, 0.5]$ is a constant and the default value is 0.01. In standard CFO, $A_{max} = 1$. DSDL is the length of the diagonal of the DS defined by

$$\text{DSDL} = \sqrt{\sum_{i=1}^{N_d} \left(X_i^{max} - X_i^{min} \right)^2} \qquad (3.25)$$

It is seen that A_{max} is an important parameter that refines the probes' motion and decreases the number of probes flying outside the DS. Therefore, it reduces the computation for the probes' repositioning scheme and enhances the global search ability. A comprehensive comparison between CFO and CFO with acceleration clipping (CFO-AC) is provided by Qubati (2009). The effectiveness and performance CFO-AC were verified on many mathematical benchmark functions and on linear and circular array antenna design problems. In every case, CFO-AC outperformed standard CFO.

There are possibly many other variants of CFO, which may have been reported in the literature but did not attract much attention from the scientific community and remained unknown. Some of them are really simple and small modifications to standard CFO that were required to fit specific problems. Qubati et al. (2010) introduced a new selection scheme to standard CFO to enhance CFO's global search ability while maintaining its simplicity. The selection mechanism is described by the following condition:

$$\text{If } M_j^p < M_{j-1}^p, \text{ then } X_j^p = \text{rand}\left(X_j^p + X_{j-1}^p \right) \qquad (3.26)$$

where rand() is a uniform random number distributed over the interval of [0,1]. The selection scheme helps in mitigating a local optimum problem when CFO is trapped into the local optimum. Unfortunately, the scheme does not guarantee that the probe will not fly outside the DS. The proposed scheme has been applied to design linear array and circular array antenna, which produced competitive results.

3.6.7 Binary CFO

CFO has been very successful for a number of continuous optimization problems demonstrating competitive and often superior performance compared to other meta-heuristic algorithms in terms of functional evaluations and solution quality. But it has not been applied to discrete optimization problems. Green et al. (2012b) proposed a binary version of the CFO algorithm called binary CFO (BCFO) by accommodating binary values into CFO and maintaining the deterministic features and natural basis inherent in the CFO. In order to create BCFO, the PR-CFO algorithm (discussed earlier) has been modified to accommodate binary representation. This involves four steps: the modification of acceleration initialization, calculation of the initial probe distribution, update of position vectors, and update of acceleration vectors. The four steps are discussed in the following.

Step 1: Acceleration initialization—In order to ensure determinism and repeatability, the acceleration vector A_i^p is initialized with values in the range of [0,1] using a low-discrepancy sequence. The acceleration vector for the probe p in the i-th dimension is initialized according to the following equation:

$$A_i^p = \mathrm{LDS}_i^p \tag{3.27}$$

where LDS_i^p is the low-discrepancy sequence. The Halton sequence (Halton, 1964) is a low-discrepancy sequence, which was applied by Green et al. (2012b).

Step 2: Calculation of initial probe distribution—Once the acceleration vector is initialized, the position vector X_i^p can be initialized according to a given probe distribution. Green et al. (2012b) proposed an initial probe distribution based on a combination of the previously initialized acceleration vector A_i^p and the Van der Corput sequence (Corput, 1935). The position vector X_i^p for probe p in the i-th dimension is determined according to the following equation:

$$X_i^p = \begin{cases} 0 & \text{if } \mathrm{sigmoid}\left(A_i^p\right) \le k \\ 1 & \text{otherwise} \end{cases} \tag{3.28}$$

where k is the i-th value in the Van der Corput sequence using the prime number P for each probe p in dimension i.

Step 3: Update of position vectors—The position vector X_i^p is updated at each iteration j to reflect the changes of the current acceleration. This is done again using the Van der Corput sequence, where k is assigned the $(N_p \times j + i)$th value of the Van der Corput sequence using the prime number P for each probe p in dimension i.

Step 4: Update of acceleration vectors—In BCFO, the acceleration is bounded within the range of [0,1]. To ensure acceleration in the range of [0,1], the value of A_i^p is scaled according to

$$A_i^p = \frac{A_i^p - A_i^{\min}}{A_i^{\max} - A_i^{\min}} \tag{3.29}$$

where $A_i^{\max}$ and $A_i^{\min}$ are the maximum and minimum acceleration in a given dimension i, respectively.

BCFO has been applied to eight test functions of varying difficulty. The results of BCFO are then compared with BPSO. Results demonstrate that the BCFO algorithm is capable of delivering results that are as good as or better than those of the BPSO algorithm in terms of fitness value.

3.6.8 Multistart CFO

Since CFO is inherently deterministic, it cannot overcome the local minimum if it gets trapped into it. Some kind of diversification is needed for CFO to overcome the local minimum and explore the search region. Liu and Tian (2015) proposed a multistart strategy for the CFO algorithm to overcome the problem of premature convergence. The CFO algorithm with a multistart strategy is called the multistart CFO or modified CFO (MCFO). The MCFO has two stages. In the first stage, the solutions are produced by two different initialization methods. In the second stage, the initial solutions are improved by CFO. In other words, the proposed MCFO is obtained by simply applying the CFO algorithm twice, and it returns the best solution over all starts. A trial-and-error-based initialization method is widely used in CFO, which incurs a higher computational cost. Therefore, two different initialization methods are used in MCFO. In the first approach, uniform distribution of probes on each coordinate axis is used. In the second approach, all probes are uniformly spaced on the diagonals of a specific problem. The algorithm of the MCFO is simple and can be described as follows:

Step 1: $f^* \leftarrow$ initialize random value $+\infty$
Step 2: $X \leftarrow$ initialize population
Step 3: $f \leftarrow$ Perform CFO on X
      If $(f < f^*)$
         $f = f^*$
Step 4: If (termination condition not met), Goto step 2
Step 5: Return f^*

In a new start of MCFO, an initial population is generated using one of the two initialization methods. In a new start, the best value obtained so far is stored. In the second run, the population is generated using the other initialization method. Once the two initializing methods are used, the MCFO is terminated. The performance and effectiveness of the MCFO approach are verified on a set of 23 benchmark functions and compared with other well-known evolutionary algorithms such as CMA-ES (Hansen and Ostermeier, 2001), CLPSO (Liang et al., 2006), and CoDE (Wang et al., 2011).

3.7 HYBRID CFO

In general, different meta-heuristic algorithms have their own advantages and disadvantages. The most common disadvantage of all meta-heuristic algorithms is the premature convergence at local minima. Among other detrimental features are the slow convergence demanding more computational cost, low-quality solution meaning poor exploitation or local search ability, lack of powerful operations, for example, crossover or mutation operations. In order to minimize the impact of detrimental features and improve exploration or global search ability, exploit local search, increase convergence speed, improve solution quality, and minimize computational cost, the standard CFO algorithm has been hybridized with other meta-heuristic algorithms. Some of these hybrid CFO algorithms are discussed in the following.

3.7.1 Hybrid CFO-Nelder–Mead (CFO-NM)

CFO is a global meta-heuristic algorithm, which is efficient for exploring the whole search space and localizing the best area of a search space for the global optimum. Once a promising area containing the global optimum is found, an efficient local search method would be appropriate to apply for exploiting the region to obtain the optimum solution. This will ensure a better solution in terms of quality and computation time. Nelder–Mead (NM) simplex search is a local search method developed by Nelder and Mead (1965). The NM method belongs to the general class of direct search methods. The NM method has four possible steps during an iteration: reflection, contraction in one dimension, contraction around a low vertex, and expansion of the DS, which are specified by the parameters reflection (ρ), expansion (χ), contraction (δ), and shrinkage (σ). The standard NM parameter values are $\rho = 1$, $\chi = 2$, $\delta = 0.5$, and $\sigma = 0.5$. A detailed description of the algorithm can be found in the original work of Nelder and Mead (1965).

Mahmoud (2011) proposed a hybrid method combining CFO with the NM algorithm for improving the local search. This hybrid method is called CFO NM, which performs exploration with CFO and exploitation with NM. Once a global near-optimum solution is found by CFO, the NM method is applied for fine-tuning the solution by local search. After every k-th step, the DS size is adaptively reduced around the probe's location with the best fitness X_{best}. The DS boundary coordinates are reduced by half coordinate-by-coordinate as defined by Equations 3.22 and 3.23. The adaptive DS update procedure is shown in Figure 3.5. The main steps of the CFO-NM algorithm are as follows:

Step 1: Set CFO parameters $N_p, \alpha, \beta, g, \Delta t, F_{\text{rep}}, N_t$
Step 2: Initialize probe position and acceleration vectors
Step 3: Evaluate fitness M_j^p
Step 4: Select the best probe fitness
Step 5: Update probe positions and acceleration
      Retrieve errant probes
      Update fitness
      If ($M_j^p >$ best fitness)
         best fitness = M_j^p
Step 6: If (Probe <= Np), Goto step 5
Step 7: Shrink DS around the best probe using Equations (3.22)–(3.23)
Step 8: If (Time step<=Nt), Goto step 4
Step 9: Perform NM algorithm
      Start with probe position $M_{j\max}^p$
Step 10: Return solution, i.e. probe position of global best fitness

The effectiveness of the CFO-NM algorithm was verified by applying it to a set of 13 benchmark functions first and then designing rectangular microstrip patch antennas (Mahmoud, 2011). The benchmark functions are classified into three groups: five unimodal functions, five multimodal functions, and three low-dimension multimodal functions. The CFO-NM outperformed the GA, PSO, group search optimizer (GSO), and CFO on three unimodal functions, on four multimodal functions, and on all three low-dimensional multimodal functions. Mahmoud (2011) and Montaser et al. (2012) applied the CFO-NM algorithm on antenna design applications, which is discussed later in Section 3.8 in detail.

3.7.2 HYBRID CFO AND INTELLIGENT STATE SPACE PRUNING

A probabilistic methodology called Intelligent State Space Pruning (ISSP) (Green et al., 2013) has recently been reported. The ISSP has two advantages: firstly, it reduces the computational resources necessary for convergence and, secondly, it improves the process of state space decomposition for nonsequential Monte Carlo simulation (MCS). The algorithm consists of three steps:

Step 1: Prune original state space using population-based meta-heuristics.
Step 2: Run MCS using the pruned state space until it converges.
Step 3: Perform calculation to introduce pruned states again during evaluation.

The ISSP algorithm improves the MCS process in two ways. Firstly, intelligence present in ISSP guides the algorithm to sample states as determined by the fitness function. Second, when ISSP samples a state, it is stored after classification. ISSP has been shown to perform differently when implemented using different population-based meta-heuristic algorithms.

Green et al. (2012b) proposed a hybrid CFO-ISSP algorithm that integrates the binary version of the CFO algorithm with the ISSP algorithm and applied the hybrid approach to a composite power system reliability problem.

3.7.3 Multistart or Modified CFO (MCFO)

The convergence speed of CFO depends on the distance between the probes. As the distance between probes decreases, convergence speed decreases. To improve the convergence speed toward the end of the search process, Chen et al. (2016) proposed a MCFO algorithm by applying a position update mechanism from the PSO algorithm and a mutation operation of the GA. The mutation operation of the GA adds diversity to solutions. The MCFO is actually a hybrid of CFO, PSO, and GA. CFO is a memoryless method, and the position update is based on the current position. MCFO applies PSO's memory mechanism for the position updating rule as follows:

$$\tilde{X}_j^p = X_j^p + c_1 r_{j1}^p \left[p\text{best}_j - X_j^p \right] + c_2 r_{j2}^p \left[g\text{best}_j - X_j^p \right] \tag{3.30}$$

where c_1 and c_2 are positive constants, r_{j1} and r_{j2} are random numbers, $p\text{best}_j$ is the best position of the jth probe, and $g\text{best}_j$ is the global best probe.

A mutation operation is applied on the new position $\tilde{X}_j^p$ obtained by applying PSO's position updating rule. In this operation, a small value m_j^p is added to $\tilde{X}_j^p$ where m_j^p is defined by

$$m_j^p = \begin{cases} \xi e^{-cj} & \tau \le c_3 \\ 0 & \tau > c_3 \end{cases}, \tag{3.31}$$

where $c > 0$ and c_3 are constant values, τ is a threshold value, and $\xi \in [-0.5, 0.5]$ is a coefficient.

The updated position $\hat{X}_j^p$ using mutation operation is obtained as follows:

$$\hat{X}_j^p = \begin{cases} \tilde{X}_j^p + m_j^p & \text{if } \left[f(\hat{X}_j^p) > f(\tilde{X}_j^p) \right] \\ \tilde{X}_j^p & \text{else} \end{cases} \tag{3.32}$$

The main steps of the MCFO algorithm are as follows:

Step 1: Define CFO parameters
Step 2: Generate initial probes and acceleration
Step 3: Compute fitness
Step 4: Calculate M_j^k
Step 5: Calculate a_j^k
Step 6: Update position $X_j^p = X_{j-1}^p + 1/2\, a_{j-1}^p$
Step 7: Update position using PSO operator using Equation 3.30
Step 8: Update position using GA's mutation operator using Equation 3.32
Step 9: If (termination condition not satisfied), Goto step 3
Step 10: Return solution

The performance of MCFO was verified on a 3-D unmanned aerial vehicle (UAV) path-planning problem. The 3-D UAV is formulated as a path-optimization problem. The objective is to minimize the multiobjective cost function that includes minimizing the path length, minimizing the fuel cost, and ensuring the smoothness of the path of the UAV.

3.7.4 HYBRID CFO AND HILL-CLIMBING

The CFO algorithm is a global meta-heuristic method, and it is good at localizing promising areas of the search space. To improve the solution quality, Abdel-Rahman et al. (2012) proposed an efficient global hybrid optimization method by combining the CFO as a global search algorithm with an efficient local search hill-climbing (HC) method. HC is an iterative local search algorithm. It starts with a given initial solution and attempts to improve the solution by incrementally changing one element of the solution. If the change results in an improved solution evaluated by an objective function, the new solution is accepted. The process continues until no further improvements are found. The advantage of the HC is that it requires a limited amount of memory, that is, only the current state. The disadvantage of HC is that it does not guarantee the best possible solution (Russell and Norvig, 1995). The proposed CFO-HC approach was applied to design a bandpass filter with microstrip resonator-loaded capacitors. The CFO-HC algorithm is used to optimize the dimension of the microstrip and values of the capacitance.

3.8 APPLICATIONS TO ENGINEERING PROBLEMS

Due to CFO's deterministic nature, easy implementation, and effectiveness, it has become quite popular in optimization techniques. Furthermore, only one run is needed to assess the performance of CFO and even no statistics are required (Formato, 2009c). The CFO algorithm has been applied to a variety of application domains and has been a great success in estimating parameter sets and optimizing the performance of various complex systems. Initially, the CFO algorithm has been applied to various well-known benchmark functions to verify its performance as an optimization method. Then, CFO has been successfully applied in many real-world and engineering applications, such as the Fano load equalizer and linear array synthesis problems (Formato, 2007a), antenna design (Qubati and Dib, 2010), leakage detection for water networks (Haghighi and Ramos, 2012), neural network training (Green et al., 2012a), and nonlinear circuits (Roa et al., 2012). This section briefly reports on different applications of CFO.

3.8.1 ELECTRONIC CIRCUIT DESIGN

Linear circuits with simple and ideal models work fine for basic digital circuits. As the number of components in a nonlinear circuit increases, it becomes difficult to solve the equations analytically. Therefore, numerical approaches become popular among researchers. Some researchers are interested to apply heuristic algorithms such as CFO. Formato (2007a) first applied the CFO algorithm to a standard electronic network problem such as the Fano load equalizer. The Fano load comprises an inductor L_f in series with a parallel combination of capacitor C_f and resistor R_f. The equalizer comprises parallel capacitors C_1 and C_2 connected in series with inductor L_1. The power is delivered to the load from a generator with internal impedance R_g. A generic Fano load equalizer circuit is shown in Figure 3.6.

The components of the equalizer are the capacitors C_1 and C_2 and inductor L_1. The problem is to match the source power to the load. The objective of the optimization problem is to determine the component values of the equalizer to optimally match the Fano load. Two different CFO optimizers were applied to this problem: 2-D algorithm and 3-D algorithm. The 2-D algorithm optimizes capacitor C_2 and inductor L_1 with fixed C_1. In this case, $N_d = 2$, and $C_2 \in [0.1,10]$ and $L_1 \in [0.1,10]$ are optimized for a fixed value of $C_1 = 0.386$. With an initial uniform probe distribution of 25 probes (i.e., $N_p = 25$), CFO achieved a maximum fitness of 0.853 within 50 time steps (i.e., $N_t = 50$), giving the optimal values for $C_2 = 0.961$ Farad and $L_1 = 3.041$ Henry. The 3-D algorithm optimizes three equalizer components: the capacitors C_1 and C_2 and inductor L_1. In this case, $N_d = 3$, and C_1, C_2,

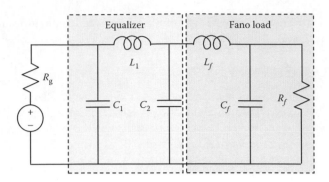

FIGURE 3.6 Fano load and equalizer circuit.

and L_1 are optimized. With an initial uniform probe distribution of 210 probes (i.e., $N_p = 210$), CFO achieved a maximum fitness of 0.852 within 40 time steps (i.e., $N_t = 40$), giving the optimal values for $C_1 = 0.460$ Farad, $C_2 = 1.006$ Farad, and $L_1 = 2.988$ Henry. In both cases, CFO achieved competitive results compared to other heuristic methods such as the GA-simplex method, real frequency technique (RTE), and recursive stochastic optimization (RSE).

Roa et al. (2012) applied the CFO algorithm to design a simple electronic circuit consisting of one nonlinear element such as a diode. A diode generally shows exponential behavior in an electronic circuit. In this case, the CFO algorithm was unable to find an adequate solution of the basic circuit with two unknowns due to the exponential behavior. Roa et al. (2012) also applied the CFO algorithm to the Buck converter. It was found that after including a mapping of the search domain, and varying the reposition factor and the gravitational constant, the results improved. It was able to find the exact solution of the basic circuit with two unknowns but could not provide a good solution for higher-order systems with three or more unknowns.

3.8.2 ANTENNA DESIGN

Due to wide use of radio frequency identification (RFID) technology in industry, design of antennas for low, high, and ultra-high frequency and microwave is becoming important. There are various types of antennas such as linear array, bow-tie, microstrip, and dipole antennas. Formato (2007a) was the first to apply the CFO algorithm to the synthesis of a 32-element linear array antenna design to compare the performance of CFO to other algorithms. The linear array comprises $2N_d$ elements equally spaced by a half wavelength ($\lambda/2$) and positioned symmetrically about the origin along the x-axis. The array factor is simplified to

$$F(\varphi, x'_i) = 2 \sum_{i=1}^{N_d} \cos\left(\pi x'_i \cos\varphi\right) \tag{3.33}$$

where $x'_i = x_i/(\lambda/2)$ is the normalized x_i, $i = 1, 2, \ldots, N_d$ and x_i are the element coordinates uniformly spaced. The array factor $F(\varphi, x'_i)$ has a maximum value of $2N_d$. The CFO algorithm has to determine the array element coordinates x_i where $x_i^{min} \leq x_i \leq x_i^{max}$, $i = 1, 2, \ldots, N_d$, $N_d = 16$ so as to maximize a user-defined fitness function $g(\varphi, x'_i)$ that best suits an antenna designers' choice to use as a measure of the array's merit. This is a constrained optimization problem as x_i must meet the requirement that no array elements occupy the same position, that is, $x_i \neq x_j$ for $i \neq j$ and $i, j = 1, 2, \ldots, N_d$. The CFO algorithm optimized the 32-element array using 1° pattern resolution in a DS defined by $0.1 \leq x'_i \leq 32.5$, $i = 1, 2, \ldots, 16$ with parameters $N_p = 48$ (3 probes per dimension), $N_t = 7$, $G = 2$, $\alpha = 2$, $\beta = 2$, and initial acceleration $a_0 = 0$.

Soon after the success of linear array antenna design using the CFO algorithm, CFO was then subsequently applied to many other antenna design problems reported by many researchers (Asi and Dib, 2010; Formato, 2010c, 2011; Mahmoud, 2011; Mohammad and Dib, 2009; Montaser et al., 2012, 2013; Qubati, 2009; Qubati and Dib, 2010; Qubati et al., 2010).

Qubati (2009) did a detailed study into the CFO algorithm and its application to antenna design. Qubati et al. (2010) also applied the CFO algorithm to the same problem with a different fitness function that yields a better result than the result in Formato (2007a). A random initial probe distribution was used in this application. The CFO algorithm optimized the 32-element array using $1°$ pattern resolution with parameters $N_p = 48$, $N_t = 500$, $G = 2$, $\alpha = 0.3$, and $\beta = 1$ with negative fitness to be maximized.

Qubati and Dib (2010) applied the MCFO algorithm to design a microstrip patch antenna. Microstrip patch antennas are widely used in wireless and mobile communication systems because of their low profile, light weight, and ease of fabrication. The objective is to determine the geometric parameters of the antenna to achieve the best design that satisfies certain performance criteria. They demonstrated further examples of the effectiveness of CFO by introducing a new scheme called acceleration clipping to enhance CFO's global search ability. The ICFO algorithm is applied to two different wideband microstrip patch antennas: microstrip line-fed E-shaped and coaxial line-fed double E-shaped patch antennas. The algorithm is applied for one trial with acceleration $a_{max} = 0.1$ using 20 probes. The other parameters are $F_{rep} = 0.5$, $N_t = 500$, $G = 2$, $\alpha = 0.3$, and $\beta = 1.5$. The E-shaped patch antenna achieved a frequency range of 2.4 GHz $\leq f_r \leq$ 2.484 GHz, while the double E-shaped patch antenna achieved the frequency range of 1.7 GHz $\leq f_r \leq$ 2.5 GHz.

Qubati et al. (2010) applied the CFO algorithm to design optimization of a 32-element linear array and a 10-element nonuniform circular array antenna. In the first case, CFO optimized the 32-element linear array using $1°$ pattern resolution and random initial probe distribution according to

$$X_1^p(i) = \text{RU}(i) + 0.5 r_1 \left(-1^{U(r_2 - 0.5)} \right) \tag{3.34}$$

where RU is the uniformly spaced reference array vector, RU $= (\lambda/2)[0.5, 1.5, 2.5, \ldots, 15.5]$, and $r_1 \in [0,1]$ and $r_2 \in [0,1]$ are uniform random numbers. An initial random probe distribution of 48 probes (i.e., $N_p = 48$) and parameter values $G = 2$, $\alpha = 0.3$, and $\beta = 1$ were used. A random repositioning scheme for errant probe retrieval was used as follows:

If $X_j^p(i) < \text{RU}(i) - 0.5(\lambda/2)$, Then

$\quad X_j^p(i) = \text{RU}(i) - \text{rand}(\lambda/2)$

If $X_j^p(i) > \text{RU}(i) + 0.5(\lambda/2)$, Then

$\quad X_j^p(i) = \text{RU}(i) + \text{rand}(\lambda/2)$

where rand($\cdot$) is a uniform random number between [0,1] and ($\lambda/2$) is a placing factor. CFO achieved a substantially better result within 500 time steps (i.e., $N_t = 500$) compared to other methods.

In the second design case, CFO optimized a 10-element nonuniform circular array antenna. The experiment provides further examples of its effectiveness and performance with a new selection scheme, defined by Equation 3.26, and random initial probe distribution with zero initial acceleration, $N_p = 20$, $G = 2$, $\alpha = 2$, and $\beta = 2$, $N_d = 20$. Errant probes were retrieved using a reposition factor of $F_{rep} = 0.9$. CFO achieved the null objective of around -65 dB with 7500 time steps (i.e., $N_t = 7500$). The best fitness increased monotonically through about time step 6350. CFO implementation onto a circular array antenna shows quite a significant result because CFO returns a design with uniform amplitude excitation.

Bluetooth technology (2.4–2.484 GHz) has been widely used in portable devices, and Ultra-Wideband (UWB) (3.1–10.6 GHz) has been widely used in various radars and communication systems as well as indoor and handheld devices. To design lightweight consumer products, it is necessary to integrate UWB with Bluetooth wireless technology. The simple solution would be to have a single antenna to work in both UWB and Bluetooth. The problem is that some existing narrow-band communication systems, such as wireless local area network (WLAN) (5.15–5.825 GHz), interfere with UWB systems. To minimize potential interference, researchers try to design antennas with band-notched characteristic (Kim and Kwon, 2004). Montaser et al. (2013) applied the CFO algorithm to optimize an E-shaped patch antenna for Bluetooth and UWB applications with WLAN band-notched characteristics. When the optimized antenna is embedded in a laptop computer, it slightly affects the antenna return loss without disturbing the resonant frequency and the impedance bandwidth. But when embedded in the screen of the laptop computer, it affects the antenna radiation patterns. The CFO parameters are not provided in Montaser et al. (2013). Finally, the CFO-optimized antenna showed satisfactory characteristics with different angles between the base unit and the screen of the laptop.

Pantoja, Bretones, and Martin developed a suite of antenna benchmark problems known as the PBM suite (Pantoja et al., 2007). The PBM suite comprises five antenna problems designed to test the effectiveness of evolutionary algorithms. Of the five problems, the problem number one is by far the most difficult unimodal problem with a single global maximum. It is lumpy with strong local maxima. Formato (2010c) applied the ICFO algorithm to verify its performance on PBM benchmark problem one. The objective is to maximize a center-fed dipole's directivity $D(L,\theta)$ as a function of its total length L and the polar angle θ. That means to determine the best fitness of the objective function $D(\cdot)$ over the DS domain $0.5\lambda \leq L \leq 3\lambda$ and $0 \leq \theta \leq (\pi/2)$. The ICFO produced a maximum directivity of $D(L,\theta)_{max} = 3.2584$ at $(L,\theta) = (2.5815\lambda, 0.60697)$. It appears to have determined the global maximum with a value slightly higher than the value computed by the numerical electromagnetic code. A standard CFO was also applied to the antenna problem. Four probes with symmetrical initial probe distribution achieved a fitness value of 3.02691 at $(L,\theta) = (2.23044\lambda, 1.07992)$ within 100 time steps. These results clearly show the effectiveness of CFO for PBM problems.

Asi and Dib (2010) applied the CFO algorithm to the optimal design of multilayer microwave absorbers in a specific frequency range. Multilayer microwave absorbers are important elements of many items of civil and military electronic equipment that can be used for minimizing electromagnetic reflection from metal plates such as aircrafts, ships, tanks, and many other electronic appliances. Optimal characteristics can be obtained by varying different parameters of the absorbers. These parameters are the number of layers, thickness of layers, dielectric constant, permeability, frequency, angle of incidence, and wave polarization (Macedo et al., 2005). The challenge in designing an absorber is the minimization of the reflection coefficient of an incident wave on a multilayer structure for a range of frequencies and incidence angles. CFO has been applied to absorber design since it has been successful in many antenna optimization problems. CFO is run for 20 independent trials. The number of iterations is set to 1000. The values of the CFO parameters depend on the initial probe distribution. For random initial probe distribution, the values used are as follows: $N_t = 1000$, $A_{max} = 0.1$, $N_p = 20$, $G = 1.7$, $\alpha = 0.6$, $\beta = 0.9$, and $F_{rep} = 0.9$, which achieved a maximum reflection coefficient of −25.698 dB. For uniform initial probe distribution, the values used are as follows: $N_t = 1000$, $N_p = 20$, $G = 1.2$, $\alpha = 0.5$, $\beta = 0.8$, and $F_{rep} = 0.5$, which achieved a maximum reflection coefficient of −22.773 dB.

The notion of increasing antenna bandwidth by adding impedance loading has been around for long time. Formato (2011) applied the parameter-free CFO algorithm to optimize a loaded monopole antenna for increasing its bandwidth. This application of CFO demonstrates its usefulness in broadband antenna design.

Mahmoud (2011) used the hybrid CFO-NM algorithm to optimize rectangular microstrip patch antennas. Montaser et al. (2012) also applied the hybrid CFO-NM algorithm to optimize a triple-band dual bow-tie slot antenna for RFID applications. The antenna showed accepted bandwidths. Variable Z_0 is a new concept in antenna design and optimization problems. Dib et al. (2014) used the CFO algorithm to optimize an UWB meander monopole antenna. An improved performance is observed for variable Z_0 design using CFO.

3.8.3   BENCHMARK FUNCTION OPTIMIZATION

In addition to testing the CFO algorithm in typical engineering applications, Formato (2007a, 2008, 2009a–c) also tested it against a variety of standard benchmark functions whose extrema are known so that its effectiveness can be evaluated and compared precisely.

1. Schwefel problem
2. Rastrigin's function
3. Ackley's problem
4. Griewank function
5. Step function
6. Sphere function
7. Rosenbrock's function
8. Colville function
9. Six-hump camel-back function
10. Branin function
11. Shekel's inverted foxholes function
12. Keane's bump function
13. Goldstein–Price function

The Schwefel problem is defined as

$$f(\overline{x}) = \sum_{i=1}^{30} x_i \sin\left(\sqrt{|x_i|}\right), \quad -500 \le x_i \le 500 \tag{3.35}$$

The Schwefel problem is highly multimodal, and its maximum is 12,569.5 at $x_i = 420.9687$ with $i = 1,2, \dots, 30$. Formato (2007a, 2009a–c) applied the CFO algorithm to the Schwefel problem with an initial acceleration of 0, 8 probes uniformly distributed along 30 coordinate axes, that is, $N_d = 30$ and $N_p = 240$. The fitness increase was extremely rapid, reaching 12,569.09 in 8 steps.

Rastrigin's function is defined as

$$f(\overline{x}) = \sum_{i=1}^{30} (x_i - x_0)^2 - 10\cos(2\pi(x_i - x_0) + 10), \quad -5.12 \le x_i \le 5.12 \tag{3.36}$$

where $x_0 = 1.123$ and $i = 1,2, \dots, 30$. The function has its maximum at $x_i = 1.123$. The Rastrigin function is usually tested for the maximum at the origin. But the maximum can be offset to a point on the diagonal in order to avoid biasing CFO. Formato (2007a, 2009a–c) applied the CFO algorithm to Rastrigin's problem with an initial distribution of 6 uniformly spaced probes along 30 coordinate axes, that is, $N_d = 30$ and $N_p = 180$. Formato (2009a–c) found that fitness saturated at -0.0577637 at Step 43. The fitness increase was extremely rapid reaching 12,569.09 in 8 steps, and the CFO algorithm converged remarkably quickly.

Ackley's problem is defined as

$$f(\overline{x}) = -20 - e + 20\exp\left(-0.2\sqrt{\frac{1}{N_d}\sum_{i=1}^{N_d}(x_i - x_0)^2}\right) + \exp\left(\frac{1}{N_d}\sum_{i=1}^{N_d}\cos(2\pi(x_i - x_0))\right) \tag{3.37}$$

where $-32.768 \leq x_i \leq 32.768$, $i = 1,2, \ldots, 30$, and $x_0 = 4.321$. Its global maximum is zero at $x_i = 4.321$. The maximum is offset to avoid biasing the CFO with an on-axis probe distribution. Formato (2007a, 2009a–c) applied the CFO algorithm to Ackley's problem with 780 probes ($N_p = 780$), which returned coordinates between 4.27333 and 4.38386 for the global maximum with the best fitness value of -0.2067925 at Step 220. The CFO algorithm achieved an absolute accuracy of 0.2067925 with 39,780 iterations.

The Griewank function is a 30-D function defined as

$$f(x) = \frac{1}{400}\sum_{i=1}^{30}(x_i - x_0)^2 + \prod_{i=1}^{30}\cos\left(\frac{(x_i - x_0)}{\sqrt{i}}\right) - 1, \quad -600 \leq x_i \leq 600 \tag{3.38}$$

where $x_0 = 75.123$. The Griewank function is continuous and very highly multimodal. Formato (2007a) applied the CFO algorithm to the Griewank function with 780 probes ($N_p = 780$) and parameter values of $G = 2$, $\alpha = 2$, $\beta = 2$, and $a_0 = 0$, which returned coordinates between 74.9 and 75.2653 for the global maximum with the best fitness value of -0.0459 at iteration 6.

The Step function is defined as

$$f(x) = \sum_{i=1}^{30}\left(\lfloor x_i - x_0 + 0.5 \rfloor\right)^2, \quad -100 \leq x_i \leq 100 \tag{3.39}$$

where $x_0 = 75.123$. The Step function value varies over a wide range from a maximum of 0 at 75.123 to a minimum below $-60,000$. Formato (2007a) applied the CFO algorithm to the Step function with 600 probes ($N_p = 600$) and parameter values of $G = 2$, $\alpha = 2$, $\beta = 2$, and $a_0 = 0$, which returned coordinates between 73.6842 and 75 for the global maximum with the best fitness value of -1 at iteration 4.

Sphere function is defined as

$$f(x) = \sum_{i=1}^{30}(x_i - x_0)^2, \quad -100 \leq x_i \leq 100 \tag{3.40}$$

where $x_0 = 75.123$. Formato (2007a) applied the CFO algorithm to the Sphere function with 15,000 probes ($N_p = 15,000$) and parameter values of $G = 2$, $\alpha = 2$, $\beta = 2$, and $a_0 = 0$, which returned coordinates between 75.0854 and 74.9168 for the global maximum with the best fitness value of -0.0836 at time step 2. CFO required 30,000 iterations with 15,000 probes to reach the solution, which demands a substantial amount of run time.

Rosenbrock's function is defined as

$$f(x) = \sum_{i=1}^{29}[100((x_{i+1} - x_0) - (x_i - x_0)^2)^2 + ((x_i - x_0) - 1)^2] \tag{3.41}$$

where $x_0 = 25.123$. The function has a maximum value of 0 at $x_i = 26.123$, $i = 1,2, \ldots, 30$. Formato (2007a) applied the CFO algorithm to Rosenbrock's function with 60 probes ($N_p = 60$) and parameter values of $G = 2$, $\alpha = 2$, $\beta = 2$, and $a_0 = 0$, which returned coordinates between 26.0078 and 26.1289 for the global maximum with the best fitness value of -3.8 at iteration 250.

The Colville function is a 4-D function defined as

$$
\begin{cases}
f(x_1, x_2, x_3, x_4) = -100 \cdot (x'_2 - x'^2_1)^2 - (1 - x'_1)^2 \\
f(x_1, x_2, x_3, x_4) = -90 \cdot (x'_4 - x'^2_3)^2 - (1 - x'_3)^2 \\
f(x_1, x_2, x_3, x_4) = -10.1 \cdot ((x'_2 - 1)^2 - (x'_4 - 1)^2) \\
f(x_1, x_2, x_3, x_4) = -19.8 \cdot (x'_2 - 1) \cdot (x'_4 - 1)
\end{cases}
\tag{3.42}
$$

where $-10 \le x_i \le 10$, $x'_i = x_i - 7.123$, $i = 1, \ldots, 4$. The function has a maximum value of 0 at $x_i = 8.123$. Formato (2007a) applied the CFO algorithm to the Colville function with 56 probes ($N_p = 56$) and parameter values of $G = 2$, $\alpha = 2$, $\beta = 2$, and $a_0 = 0$, which returned coordinates between 7.74637 and 7.83799 for the global maximum with the best fitness value of -19.387 at iteration 15.

The Six-hump camel-back function is a 2-D function defined as

$$
f(x) = \quad -4(x_1 - 1)^2 + 2.1(x_1 - 1)^4 - \frac{1}{3}(x_1 - 1)^6
$$
$$
-(x_1 - 1)(x_2 - 1) + 4(x_2 - 1)^2 - 4(x_2 - 1)^4
\tag{3.43}
$$

where $-5 \le x_1, x_2 \le 5$. The maximum is 1.0316285 at (1.08983,0.2874) and (0.91017,1.7126). Formato (2007a) applied the CFO algorithm to the Six-hump camel-back function with 220 probes ($N_p = 220$) and parameter values of $G = 2$, $\alpha = 2$, $\beta = 2$, and $a_0 = 0$, which returned coordinates between 1.11294 and 0.28745 for the global maximum with the best fitness value of 1.02956 at time step 5. The camel-back function has two global maxima. Simply changing the number of the initial probes toggles the CFO between these maxima.

The Branin function is a 2-D function defined as

$$
f(x) = \left(x_2 - \frac{5.1}{4\pi^2} x_1^2 + \frac{5}{\pi} x_1 \right)^2 - 10 \left(1 - \frac{1}{8\pi} \right) \cos x_1 - 10
\tag{3.44}
$$

where $-5 \le x_1, x_2 \le 15$. The function has a maximum value of -0.398 at $(-3.142, 12.275)$, $(3.142, 2.275)$, and $(9.425, 2.425)$. Formato (2007a) applied the CFO algorithm to the Branin function with 400 probes ($N_p = 400$) and parameter values of $G = 2$, $\alpha = 2$, $\beta = 2$, and $a_0 = 0$, which returned coordinates between 3.142 and 2.275 for the global maximum with the best fitness value of -0.398689 at iteration 18. The Branin function has global maxima at three points. The CFO toggles between two of them depending on the number of initial probes.

Shekel's inverted foxholes function is 2-D function defined as

$$
f(x) = \left[\frac{1}{500} + \sum_{j=1}^{25} \frac{1}{j + \sum_{i=1}^{2} (x_i - a_{ij})^6} \right]^{-1}
\tag{3.45}
$$

where $-65.536 \le x_i \le 65.536$ and

$$
a_{ij} = \begin{bmatrix}
-32 & -16 & 0 & 16 & 32 & -32 & \cdots & 0 & 16 & 32 \\
-32 & -32 & -32 & -32 & -32 & -16 & \cdots & 32 & 32 & 32
\end{bmatrix}.
$$

The function has a maximum values of ≈ -1 at $(-32, -32)$. Formato (2007a) applied the CFO algorithm to Shekel's inverted foxholes function with 240 probes ($N_p = 240$) and parameter values of $G = 2$, $\alpha = 2$, $\beta = 2$, and $a_0 = 0$, which returned coordinates between -31.9419 and -32.768 for the global maximum with the best fitness value of -1.2023 at iteration 2.

Keane's bump function is a constrained objective function defined as

$$f(x_1, x_2) = \begin{cases} 0 & \text{for } x_1 + x_2 \geq 15 \quad \text{or} \quad x_1 x_2 \leq 0.75 \\ \dfrac{\cos^4(x_1) + \cos^4(x_2) - 2\cos^2(x_1)\cos^2(x_2)}{\sqrt{x_1^2 + 2x_2^2}} & \text{otherwise} \end{cases} \tag{3.46}$$

where $-5 \leq x_1, x_2 \leq 5$. The precise locations and values of the maxima are unknown. Formato (2007a) applied the CFO algorithm to Keane's bump function with 196 probes ($N_p = 196$) and parameter values of $G = 2$, $\alpha = 2$, $\beta = 2$, and $a_0 = 0$, which returned coordinates between 1.60267 and 0.46804 for the global maximum with the best fitness value of 0.364915 at iteration 20.

The Goldstein–Price function is a 2-D benchmark function defined as

$$\begin{aligned} f(x_1, x_2) = &-\left[1 + (x_1 + x_2 + 1)^2 \cdot \left(19 - 14x_1 + 3x_1^2 - 14x_2 + 6x_1 x_2 + 3x_2^2 \right) \right] \\ &\times \left[30 + (2x_1 - 3x_2)^2 \cdot \left(18 - 32x_1 + 12x_1^2 + 48x_2 - 36x_1 x_2 + 27x_2^2 \right) \right] \end{aligned} \tag{3.47}$$

where $-100 \leq x_1, x_2 \leq 100$. The global maximum of the function is -3 at $(0, -1)$. This function has far fewer local maxima. Formato (2009a–c) applied the CFO algorithm to the Goldstein–Price function with a uniform grid of 100 probes, which returned coordinates as $(-0.000719142, -1.00412)$ with a maximum fitness of 3.0068739 at iteration 1500.

3.8.4 TRAINING NEURAL NETWORK

The CFO algorithm has also been successfully applied to a variety of other problems. Green et al. (2012a) were the first to apply the CFO algorithm to train a basic neural network that represents a logical exclusive-OR (XOR) function. The XOR problem has historically been considered a good test of a network model and learning algorithm. The XOR problem has been chosen as one of the benchmark problems for testing training algorithms. There are many reasons for this choice. Firstly, the XOR problem is one of the simplest problems, which is not linearly separable and complex enough for backpropagation learning algorithm (Siddique and Adeli, 2013) to be trapped in local minima without reaching the global optimum. Secondly, there is a significant number of research works reported on analytical work on XOR, which claim that the XOR problem exhibits local minima, a view that is widely accepted in the neural network literature (Dayhoff, 1990). The network consists of one input layer with two neurons, a single hidden layer containing three neurons with a sigmoidal activation function, and one output layer neuron with a linear activation function. All biases are set to zero. The architecture of an XOR neural network is shown in Figure 3.7.

The problem is to find the weights of the connectivity of the network for the given architecture to produce the correct output for the XOR function for each corresponding input. Thus, for the given architecture, the neural network for the XOR function has to optimize nine weights, as shown in Figure 3.7. The weights training of the artificial neural network (ANN) can be carried out by minimizing (optimizing) the MSE of the network error function defined by

$$\text{MSE}(w) = \min_{w} \left\{ \frac{1}{N} \sum_{i=1}^{N} (y_i - \hat{y}_i)^2 \right\} \tag{3.48}$$

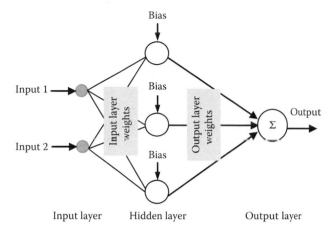

FIGURE 3.7 XOR neural network.

The minimum of MSE(w) leads to optimum behavior of the ANN where y_i is the target output and $\hat{y}_i$ is the actual output of the network. An ANN can be trained using the backpropagation learning algorithm with the possible risk of being trapped in local minima without reaching the global optimum (Siddique and Adeli, 2013). Green et al. (2012a) then extended the approach to train two different neural network architectures for classification of an Iris data set using three and five neurons in the hidden layer. The Iris data set consists of 150 samples. The objective function for the CFO algorithm used is the typical MSE. The stopping criterion in CFO is defined as

$$\text{HFS}_j = \left(\frac{1}{25} \sum_{k=j-25}^{j} F_k \right) - F_j \tag{3.49}$$

where j is the current time step and F_j is the best fitness at time step j. If the value of HFS$_j$ after time step 25 becomes less than a tolerance value, the algorithm is stopped. A more typical stopping criterion for the training neural network would be MSE ≤ 0.2. PF-CFO was applied to train the neural networks where initial velocities and accelerations were set to zero. Uniform-on-diagonal and uniform-on-axis methods were used for the initial probe distribution. PF-CFO resulted in a fitness value of zero for XOR networks. CFO is found to be sensitive to the initial probe distribution for Iris data classification, and the uniform-on-axis initial probe distribution approach shows good results. Chao et al. (2014) applied a distributed multiobjective CFO (DM-CFO) algorithm to optimize individual network components of a neural network ensemble and showed that the CFO is capable of achieving better solutions in terms of convergence speed and local minima.

3.8.5 WATER PIPE NETWORKS

Leak detection, reduction, and calibration for friction factors in water supply systems and distribution network systems are an important engineering issue since leaks and ruptures in such network systems cause major physical damage, inadequate operation, and high operating pressure level, and ultimately involves a huge economic cost. Efficient detection of leaks and locations is thus required in order to effectively control water losses and to quickly repair the system. A hypothetical water pipe network was originally introduced by Pudar and Liggett (1992). The benchmark network system consists of 11 pipes and 7 nodes with a reservoir at Node 1 which feeds the water into the

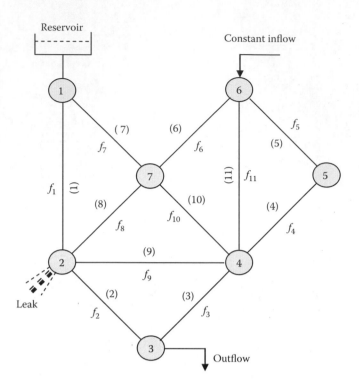

FIGURE 3.8   A generic pipe network used as benchmark problem.

network system at constant inflow under gravitation. A constant inflow also supplied water into the network at Node 6. All pipes have uniform diameter, length, and wave speed. A generic pipe network is shown in Figure 3.8 with the network's layout, node numbers {1,2,...,7}, pipes named by numbers {(1),(2), ... , (11)}, and their respective friction factors {$f_1, f_2, ... , f_{11}$}, which has been used as the benchmark problem for many optimization algorithms (Liggett and Chen, 1994; Pudar and Liggett, 1992). A small leak is located at Node 2. Initially, there is a steady outflow at Node 3. Inverse transient analysis (ITA) is a powerful tool and has been used for leak detection, location identification, and calibration of friction in pressurized pipe networks by many researchers (Liggett and Chen, 1994; Soares et al., 2011). The method is based on the minimization of MSEs between measured and calculated state variables (pressures or flow rates). Based on this concept, the water pipe network is then modeled as a function of unknown variables. The leakage detection can now be formulated as a minimization problem. The indirect approach to solve the ITA problem of parameter optimization can be set as the minimization of weighted MSEs between observed and computed pressure at a number of measurement sites (i.e., piezometric heads) in the pipe network defined by

$$\min_{Z} F = \sum_{t=1}^{m_H} \left[\sum_{j-1}^{n_H} \left\{ \frac{(H_{t,j}^d - H_{t,j}^o)^2}{\left(\sum_{i=1}^{n_H} H_{t,i}^o / n_H \right)^2} \right\} \right] \tag{3.50}$$

where F is the objective function to be minimized, z is the set of decision variables; m_H is the number of observed hydraulic transient events t; $H_{t,j}^d$ is the desired (or computed) pressure of the piezometric heads for transient event t at site j, and $H_{t,j}^o$ is the observed pressure of the piezometric heads (collected data from the sites); $(\sum_{i=1}^{n_H} H_{t,i}^o / n_H)^2$ is the weight term used for the function; and n_H

is the number of pressure observation sites in the system. Haghighi and Ramos (2012) used a simplified version of the objective function by dropping the weight term in Equation 3.50 and defined the objective function as

$$\min_{Z} F = \sum_{t=1}^{m_H} \sum_{j-1}^{n_H} \left(H_{t,j}^d - H_{t,j}^o \right)^2 \tag{3.51}$$

Applying a nonlinear optimization method, the objective function F is minimized, resulting in optimum leak areas and pipe friction factors in the network under investigation. Theoretically, F is close to zero. Researchers applied various optimization techniques. For example, Soares et al. (2011) and Vítkovsky et al. (2000) introduced the GA as an optimization technique to the ITA problem. Haghighi and Ramos (2012) used the CFO algorithm for the problem of the ITA-based approach to the minimization problem of the pipe network system. Nodes 4 and 7 are chosen as the measurement sites where pressures are sampled. All nodes except the reservoir node and friction factors are considered unknown, resulting in 17 unknown variables to be optimized. CFO was applied to the pipe network problem with 36 probes ($N_p = 36$) and parameter values of $G = 2$, $\alpha = 2$, $\beta = 2$, and $F_{rep} = 0.26$, which returned the best fitness value of 0.031 m² at iteration 75. CFO is found computationally more efficient than the GA.

3.8.6 Multiobjective CFO Algorithm

In order to address the problem of overfitting in the training procedure of an ensemble neural network (ENN), Chao et al. (2014) applied the DM-CFO algorithm to train individual network components of the ENN. For a given training set $\{x_k, y_k\}$, $k = 1, 2, \ldots, M$, the output of the ENN is the combination of n component networks' $f_i(x)$ defined as

$$f_{enn}(x_k) = w_i^{gen} \sum_{i=1}^{n} f_i(x_k) \tag{3.52}$$

where w_i^{gen} is the combining weight for the i-th component network of the ENN. The objective of the DM-CFO is to minimize the MSE of the empirical training defined by

$$MSE = \frac{1}{M} \sum_{k=1}^{M} (f_{enn}(x_k) - y_k)^2 \tag{3.53}$$

This objective function eventually leads to overfitting. To enhance the generalization ability, Chao et al. (2014) proposed the DM-CFO algorithm. In DM-CFO, three objective functions are defined. The performance objective is defined by

$$\psi_1 = \sum_{k=1}^{M} \left(f_i(x_k) - y_k \right)^2 \tag{3.54}$$

The correlation objective is to measure the variability in ENN component networks and defined by

$$\psi_2 = -\sum_{k=1}^{M} \left(f_i(x_k) - f_{enn}(x_k) \right)^2 \tag{3.55}$$

And, the regularization objective is to punish large networks and is defined by

$$\psi_3 = \mathbf{w}_i^T \mathbf{w}_i \tag{3.56}$$

where $\mathbf{w}_i$ is the weight vector of the i-th network. Rather than handling the ENN problem with three objectives, the three objectives can be combined into a single objective optimization function using a combining weight, reward, and punishment factor (Chen and Yao, 2009). The cost function e_i for the network i is defined by

$$e_i = w_i^{\text{gen}} \psi_1 - \lambda w_i^{\text{gen}} \psi_2 + \alpha_i \psi_3 \tag{3.57}$$

The DM-CFO algorithm is used to train the component networks with 18 hidden nodes. CFO used 240 probes ($N_p = 240$), which returned the best fitness. Extensive experiments were carried out on eight different benchmark data sets to compare the performance of DM-CFO. DM-CFO is found computationally more efficient than PSO, BP, and standard CFO. DM-CFO showed that the CFO is capable of achieving better solutions in terms of convergence speed and local minima.

3.9 CONVERGENCE OF CFO

An appropriate measure of efficiency of the meta-heuristic algorithm demands convergence analysis. Like any other meta-heuristic algorithm, there is no doubt that CFO can be stuck in local optima or end up with premature solutions. In spite of random movements, CFO exploits deterministic strategies to escape from local solutions. Convergence is thus an important issue for any heuristic algorithm. How well CFO performs is still an open question and needs more developments and discussions in future. CFO's convergence is measured against the normalized average distance between the probe with the best fitness and all other probes at each time step. The normalized average distance, denoted as D_{avg}, is defined as

$$D_{\text{avg}} = \frac{1}{L(N_p - 1)} \sum_{p=1}^{N_p} \sqrt{\sum_{i=1}^{N_d} \left(x_i^{p,j} - x_i^{p*,j}\right)^2} \tag{3.58}$$

where $p*$ is the number of the probe with the best fitness and L is the length of DS Ω's principal diagonal. Here, L is defined as

$$L = \sqrt{\sum_{i=1}^{N_p} \left(x_i^{\text{max}} - x_i^{\text{min}}\right)^2} \tag{3.59}$$

The D_{avg} curve should be monotonically decreasing throughout the epochs indicating good convergence. Oscillation in CFO's D_{avg} curve is a clear and reliable indication of local trapping and would indicate local maxima (Formato, 2011). Changing F_{rep} at each step by a small increment appears to effectively mitigate the local trapping in most of the cases and bring a momentum to the probe stuck in local minima. Therefore, F_{rep} can have significant influence on the convergence of CFO. CFO has been introduced much the same way that ACO was in 1996. CFO is still not based on a deep theoretical foundation. Therefore, a rigorous convergence analysis is not yet available. Some mathematical and theoretical analyses and proof of convergence of the CFO algorithm have been provided in Ding et al. (2011, 2012) and Xie et al. (2011).

3.10   CONCLUSION

The CFO algorithm is an effective and efficient deterministic search and optimization algorithm based on gravitational mechanics. Even though it is still in its infancy, and not nearly as highly evolved as many other algorithms since its inception in 2007, CFO nevertheless appears to hold considerable promise. Because it is deterministic, CFO is fast, characteristics which lend themselves well to implementations that react to how well the algorithm is performing. The CFO algorithm has already attracted significant interest from the research community, which will help CFO's further development and hopefully will involve both empirical algorithmic improvements and theoretical refinements. Some researchers have made some valuable source code in Matlab available on the Internet, which will help new researchers apply the CFO algorithm to new applications. Despite all of these successes so far, there are still many open research questions.

CFO is computationally intensive and demands large computational resources, especially computation time, which is the major obstacle to new applications of CFO. Further research is needed for developing new methods for reducing this complexity while continuing to harvest the benefits of CFO. Most optimization algorithms come in different variants. CFO has so far a good number of variants available in the literature. CFO still needs further research in hybridizing with local search methods to further improvements in exploitation of the search space. Any application of CFO should be explored and compared with other population-based meta-heuristic methods.

CFO is a novel deterministic physics-inspired heuristic optimization algorithm, which has found a number of successful applications. But there is still some outstanding issue of convergence in it. Since it is a deterministic search and optimization algorithm, such a deterministic mechanical process converges rapidly to the optima. This is the general belief, but there has been little theoretical analysis done so far. In addition, premature avoidance, convergence analysis, searching behaviors explanation, accelerating convergence, and parameter selection are important issues to be addressed.

REFERENCES

Abdel-Rahman, A. B., Montaser, A. M., and Elmikati, H. A. 2012. Design a novel bandpass filter with microstrip resonator loaded capacitors using CFO-HC algorithm, *2012 Middle East Conference on Antennas and Propagation (MECAP)*, December 29–31, 2012, Cairo, Egypt, pp. 1–6.

Adeli, H. and Hung, S. L. 1995. *Machine Learning—Neural Networks, Genetic Algorithms, and Fuzzy Sets*, John Wiley and Sons, New York.

Asi, M. J. and Dib, N. I. 2010. Design of multilayer microwave broadband absorbers using central force optimisation, *Progress in Electromagnetics Research B*, 26, 101–113.

Bergh, F. V. D. and Engelbrecht, A. P. 2006. A study of particle swarm optimization particle trajectories, *Information Sciences*, 176, 937–971.

Chao, M., Xin, S. Z., and Min, L. S. 2014. Neural network ensembles based copula methods and distributed multi-objective central force optimisation algorithm, *Engineering Applications of Artificial Intelligence*, 32, 203–212.

Chen, H. and Yao, X. 2009. Regularised negative correlation learning for neural network ensembles, *IEEE Transaction on Neural Networks*, 20(12), 1962–1979.

Chen, J., Xin, B., Peng, Z., Dou, L., and Zhang, J. 2009. Optimal contraction theorem for exploration–exploitation trade-off in search and optimization, *IEEE Transaction on System, Man and Cybernetics – Part A: Systems and Humans*, 39(3), 680–691.

Chen, Y., Yu, J., Mei, Y., Wang, Y., and Su, X. 2016. Modified central force optimisation (MCFO) algorithm for 3D UAV path planning, *Neurocomputing*, 171, 878–888.

Corput, J. V. D. 1935. Verteilungsfunktionen I, *Akademische Wetenschaft*, 38, 813–821.

Dayhoff, J. E. 1990. The exclusive-OR: A classic problem, in: *Neural Network Architectures: An Introduction*, New York: Van Nostrand Reinhold, pp. 76–79.

Dib, N., Sharaqa, A., and Formato, R. A. 2014. Variable Z_0 applied to optimal design of multi-stub matching network and meander monopole, *International Journal of Microwave and Wireless Technologies,* 6(5), 505–514.

Ding, D., Luo, X., Chen, J., Wang, X., Du, P., and Guo, Y. 2011. A convergence proof and parameter analysis of central force optimization algorithm, *Journal of Convergence Information Technology*, 6(10), 16–23.

Ding, D., Qi, D., Luo, X., Chen, J., Wang, X., and Du, P. 2012. Convergence analysis and performance of an extended central force optimisation algorithm, *Applied Mathematics and Computation*, 219(4), 2246–2259.

Formato, R. A. (2007a). Central force optimization: A new metaheuristic with applications in applied electromagnetics, *Progress in Electromagnetics Research, PIER*, 77(1), 425–491.

Formato, R. A. (2007b). Central force optimization: A new nature inspired computational framework for multidimensional search and optimization, in: N. Krasnogor, G. Nicosia, M. Pavone, D. A. Pelta, eds., *NICSO, Studies in Computational Intelligence*, Vol. 129, Berlin: Springer-Verlag, pp. 221–238.

Formato, R. A. 2008. Central force optimization: A new nature inspired computational framework for multidimensional search and optimization, in: N. Krasnogor, G. Nicosia, M. Pavone, D. A. Pelta, eds., *NICSO*, Berlin, Heidelberg: Springer-Verlag, pp. 221–238.

Formato, R. A. (2009a). Central force optimization: A new deterministic gradient-like optimization metaheuristic, *Journal of the Operations Research Society of India*, 46(1), 25–51.

Formato, R. A. (2009b). Are near earth objects the key to optimization theory? Computing Research Repository, No. arXiv:0912.1394, December 2009.

Formato, R. A. (2009c). Central force optimisation: A new gradient-like meta-heuristic for multidimensional search and optimisation, *International Journal of Bio-Inspired Computation*, 1(4), 217–238.

Formato, R. A. (2010a). Central force optimization applied to the PBM suite of antenna benchmarks, Computing Research Repository, Vol. abs/1003.0221, [Online]. Available: http://arxiv.org/abs/1003.0221.

Formato, R. A. (2010b). Comparative results: group search optimizer and central force optimization, Computing Research Repository, Vol. abs/1002.2798, [Online]. Available: http://arxiv.org/abs/1002.2798.

Formato, R. A. (2010c). Improved CFO algorithm for antenna optimization, *Progress in Electromagnetics Research. PIER B*, 19, 405–425.

Formato, R. A. (2010d). Parameter-free deterministic global search with central force optimization, Computing Research Repository, Vol. abs/1003.1039, [Online]. Available: http://www.springerlink.com/content/d45u6135702015wq/.

Formato, R. A. (2010e). On the utility of directional information for repositioning errant probes in central force optimization, Computing Research Repository, Vol. abs/1005.5490, 2010. [Online]. Available: http://arxiv.org/abs/1005.5490.

Formato, R. A. 2011. Central force optimization with variable initial probes and adaptive decision space, *Applied Mathematics and Computation*, 217(21), 8866–8872.

Formato, R. A. 2013. Pseudorandomness in central force optimization, *British Journal of Mathematics and Computing*, 3(3), 241–264.

Goldstein, H. 1980. The two-body central force problem, in: *Classical Mechanics*, 2nd ed. Reading, MA: Addison-Wesley, pp. 70–127.

Green, R. C., Wang, L., and Alam, M. (2012a). Training neural networks using central force optimization and particle swarm optimization: Insights and comparisons, *Expert Systems with Applications*, 39, 555–563.

Green, R. C., Wang, L., and Alam, M. (2012b). Intelligent state space pruning with local search for power system reliability evaluation, *2012 3rd IEEE PES International Conference and Exhibition on Innovative Smart Grid Technologies (ISGT Europe)*, October 14–17, 2012, pp. 1–8.

Green, R. C., Wang, L., Alam, M., and Formato, R. 2011. Central force optimization on a GPU: A case study in high performance metaheuristics using multiple topologies, *Proceedings of the IEEE Congress on Evolutionary Computation*, June 2011, New Orleans, Los Angeles, pp. 550–557.

Green, R. C., Wang, L., Alam, M., and Formato, R. A. (2012c). Central force optimization on a GPU: A case study in high performance metaheuristics, *Journal of Supercomputing*, 62, 378–398.

Green, R. C., Wang, L., Alam, M., and Singh, C. 2013. Intelligent state space pruning for Monte Carlo simulation with applications in composite power system reliability, *Engineering Applications of Artificial Intelligence*, 26(7), 1707–1724.

Haghighi, A. and Ramos, H. M. 2012. Detection of leakage freshwater and friction factor calibration in drinking networks using central force optimization, *Water Resources Management*, 26(8), 2347–2363.

Halton, J. H. 1964. Algorithm 247: Radical-inverse quasi-random point sequence, *Communications of the ACM*, 7, 701–702.

Hansen, N. and Ostermeier, A. 2001. Completely derandomized self adaptation in evolution strategies, *Evolutionary Computing*, 9(2), 159–195.

Janson, S. and Middendorf, M. 2005. A hierarchical particle swarm optimizer and its adaptive variant, *IEEE Transactions on Systems, Man, and Cybernetics – Part B: Cybernetics*, 35(6), 1272–1282.

Jiang, M., Luo, Y. P., and Yang, S. Y. 2007. Stochastic convergence analysis and parameter selection of the standard particle swarm optimization algorithm, *Information Processing Letters*, 102, 8–16.

Kennedy, J. and Mendes, R. 2002. Population structure and particle swarm performance, IEEE Congress on Evolutionary Computation, Vol. 2, Honolulu, Hawaii, pp. 1671–1676.

Kim, Y. and Kwon, D.-H. 2004. CPW-fed planar ultra wideband antenna having a frequency band notch function, *Electronics Letters*, 40(7), 403–405.

Liang, J. J., Qin, A. K., Suganthan, P. N., and Baskar, S. 2006. Comprehensive learning particle swarm optimizer for global optimization of multimodal functions, *IEEE Transactions on Evolutionary Computation*, 10(3), 281–295.

Liggett, J. A. and Chen, L. C. 1994 Inverse transient analysis in pipe networks, *Journal of Hydraulic Engineering*, 120(8), 934–955.

Liu, Y. and Tian, P. 2015. A multi-start central force optimization for global optimization, *Applied Soft Computing*, 27, 92–98.

Macedo, J., de Sousa, M., and Dmitriev, V. 2005. Optimization of wide band multilayer microwave absorbers for any angle of incidence and arbitrary polarization, *2005 SBMO/IEEE MTT-S International Conference on Microwave and Optoelectronics*, July 25–25, 2005, 558–561.

Mahmoud, K. R. 2011. Central force optimization: Nelder–Mead hybrid algorithm for rectangular microstrip antenna design, *Electromagnetics*, 31(8), 578–592.

Marion, J. B. 2013. *Classical Dynamics of Particles and Systems*, Int. Edt., New York, London: Academic Press Inc.

Mohammad, G. and Dib, N. 2009. Synthesis of antenna arrays using central force optimization, *The Mosharaka International Conference on Communications, Propagation and Electronics*, March 6–8, 2008, Amman, Jordan.

Montaser, A. M., Mahmoud, K. R., and Elmikati, H. A. 2012. Tri-band slotted bow-tie antenna design for RFID reader using hybrid CFO-NM algorithm, *29th National Radio Science Conference (NRSC 2012)*, April 10–12, 2012, Cairo University, Cairo, Egypt, pp. 119–126.

Montaser, A. M., Mahmoud, K. R., Abdel-Rahman, A. B., and Elmikati, H. A. 2013. Design bluetooth and notched-UWB e-shape antenna using optimization techniques, *Progress in Electromagnetics Research B*, 47, 279–295.

Nelder, J. A. and Mead, R. 1965. A simplex method for function minimisation, *Journal of Computing*, 7, 308–313.

Pantoja, M. F., Bretones, A. R., and Martin, R. G. 2007. Benchmark antenna problems for evolutionary optimisation algorithms, *IEEE Transactions on Antennas and Propagation*, 55(4), 1111–1121.

Pudar, R. S. and Liggett, J. A. 1992. Leaks in pipe networks, *Journal of Hydraulics Engineering*, 118(7), 1031–1046.

Qubati, G. 2009. Central force optimization method and its application to the design of antennas, Master's thesis, Jordan University of Science and Technology.

Qubati, G., Formato, R., and Dib, N. I. 2010. Antenna benchmark performance and array synthesis using central force optimisation, *Microwaves, Antennas & Propagation, IET*, 4(5), 583–592.

Qubati, G. M. and Dib, N. I. 2010. Microstip patch antenna optimization using modified central force optimization, *Progress in Electromagnetics Research B*, 21, 281–298.

Rashedi, E., Nezamabadi-pour, H., and Saryazdi, S. 2009. GSA: A gravitational search algorithm, *Information Sciences*, 179(13), 2232–2248.

Roa, O., Amaya, I., Ramirez, F., and Correa, R. 2012. Solution of nonlinear circuits with the central force optimization algorithm, *Proceedings of the IEEE 4th Colombian Workshop on Circuits and Systems*, November 2012, Barranquilla, Colombia, pp. 1–6.

Russell, S. and Norvig, P. 1995. *Artificial Intelligence: A Modern Approach*, Englewood Cliffs, NJ: Prentice-Hall.

Siddique, N. 2014. *Intelligent Control: Hybrid Approach Using Fuzzy Logic, Neural Networks and Genetic Algorithms*, Heidelberg, New York, London: Springer-Verlag.

Siddique, N. and Adeli, H. 2013. *Computational Intelligence: Synergies of Fuzzy Logic, Neural Networks and Evolutionary Computing*, Chichester: John Wiley & Sons.

Soares, A. K., Covas, D. I. C., and Reis, L. F. R. 2011. Leak detection by inverse transient analysis in an experimental PVC pipe system, *Journal of Hydraulics*, 13(2), 153–166.

Toscano-Pulido, G., Reyes-Medina, A. J., and Ramirez-Torres, J. G. 2011. A statistical study of the effects of neighbourhood topologies in particle swarm optimisation, in: K. Madani, A. Correria, A. Rosa, J. Filipe, eds. *Computational Intelligence, Studies in Computational Intelligence*, Berlin, Heidelberg: Springer Verlag, Vol. 343, pp. 179–192.

Vıtkovský, J. P., Simpson, A. R., and Lambert, M. F. 2000. Leak detection and calibration issues using transient and genetic algorithms, *Journal of Water Resource Planning and Management*, 126(4), 262–265.

Wang, Y., Cai, Z., and Zhang, Q. 2011. Differential evolution with composite trial vector generation strategies and control parameters, *IEEE Transactions on Evolutionary Computation*, 15(1), 55–66.

Whittaker, E. T. 1944. Central forces in general: Hamilton's theorem, §47, in: *A Treatise on the Analytical Dynamics of Particles and Rigid Bodies: With an Introduction to the Problem of Three Bodies.* New York: Dover, pp. 77–80.

Xie, L., Zeng, J., and Formato, R. A. 2011. Convergence analysis and performance of the extended artificial physics optimization algorithm, *Applied Mathematics and Computation*, 218, 4000–4011.

4 Electromagnetism-Like Optimization

4.1 INTRODUCTION

Electromagnetism (EM) is the study of the electromagnetic force that causes a type of physical interaction between electrically charged particles. According to electromagnetic theory, each particle has a charge in an electromagnetic field and there is an electromagnetic force acting between two particles, which follow Coulomb's law. The theory was first published by French physicist Charles Augustin de Coulomb in 1785 (Coulomb, 1785a,b). This was an essential invention to the development of the theory of EM. He determined that the magnitude of the electric force between two point charges is directly proportional to the product of the charges and inversely proportional to the square of the distance between them. Mathematically, Coulomb's law can be expressed in scalar form as follows:

$$|F| = k_e \frac{|q_i \cdot q_j|}{r^2} \tag{4.1}$$

where k_e is Coulomb's constant, q_i and q_j are the signed magnitudes of the charges, and r is the distance between the charges. The force acts along the straight line joining q_i and q_j, as shown in Figure 4.1. If the two charges have the same sign, the electrostatic force between them is repulsive. If they have different signs, the force between them is attractive.

The vector form of Equation 4.1 can be expressed as

$$F_i = k_e \frac{q_i \cdot q_j}{|r_{ij}|^2} \hat{r}_{ij} \tag{4.2}$$

where the vector $r_{ij} = r_i - r_j$ is the distance between the particles and $\hat{r}_{ij} = r_{ij} / |r_{ij}|$ is the unit vector acting from q_j to q_i. The vector form of the equation calculates the force F_i applied on q_i by q_j. In other words, the electrostatic force is proportional to the product of the charges with the same distance dependence. The charge determines the magnitude of attraction or repulsion. The particle with higher charge will attract the other, whereas the particle with lower charge will repulse the former. The total force acting on a particle exerted by other particles defines the direction of movement in order to reach equilibrium in the space from where the particle will not move any further. The equilibrium point is thus considered as an optimal point.

In any search or optimization algorithm, points move toward the optimum region and move away from a nonoptimum or steeper region. This concept is analogous to attraction and repulsion forces in EM theory. In a similar disposition, sample points can be thought of as charged particles distributed over a search space where the charge itself represents the fitness value defined by an objective function and the point represents the particle.* In other words, a particle represents a point in n-dimensional space. The charge also determines the magnitude of attraction or repulsion. Points (or particles) with higher charges have higher attraction forces that act on other points and determine the direction of movements. In the search process, the points gradually move toward the

---

* Particle and point will be used interchangeably throughout this chapter. A particle in EM represents a solution in an n-dimensional space.

FIGURE 4.1   Force in Coulomb's law.

optimum region of the search space with increasing fitness values and attract the other points. The point in the optimum region with the highest fitness value yields the solution. This notion is utilized in EMO.

This chapter presents the principles of the EMO algorithm, variants of EMO algorithms, and their applications to various domains.

4.2   EMO ALGORITHM

Based on the principles of EM, Birbil and Fang (2003) first introduced the method of EMO. The EMO method imitates the attraction–repulsion mechanism of the EM theory (Cowan, 1968) in order to solve unconstrained or bound-constrained global optimization problems. Therefore, it is called the EMO algorithm. A solution in the EMO algorithm is seen as a charged particle in the search space, and its charge relates to the objective function value. The better the objective function value, the higher is the magnitude of the attraction or repulsion force. EMO is inherently stochastic; as a consequence, EMO exhibits better diversity (ability to fully explore the decision space) because all particles traverse throughout the decision space to an optimal point.

A random population of particles x_i^k, $i = 1, \dots, N$ in n-dimensional space (i.e., $k = 1, \dots, n$) is generated within the search space defined by the problem at hand. The objective function values for each particle $f(x_i)$ for all $k = 1, \dots, n$ are computed. A suitable objective function $f(\cdot)$ is defined for the problem at hand, and it is very much problem dependent. For an initial population $S = x_1, x_2, \dots, x_N$, the best particle is computed by

$$x_{\text{best}} = \arg \min_{x_i \in S}\{f(x_i)\} \tag{4.3}$$

A charge-like value q_i is defined and assigned to each particle x_i. The charge q_i of x_i depends on the fitness value $f(x_i)$ and $f(x_{\text{best}})$. The charge determines the strength of attraction or repulsion to other particles. The charge q_i for the particle x_i is computed as follows:

$$q_i = \exp\left[-n\frac{f(x_i) - f(x_{\text{best}})}{\displaystyle\sum_{j=1}^{N}\left[f(x_j) - f(x_{\text{best}})\right]}\right], \quad \forall i = 1, \dots, N \tag{4.4}$$

where n is the maximum number of dimensions.

Unlike electrical charges, no signs are attached to the charge of an individual particle in Equation 4.4. It is assumed that q_i is always positive. Particles with higher values of q_i attract others. The force $F_{i,j}$ between two particles x_i and x_j is computed according to Equation 4.2:

$$F_{i,j} = \begin{cases} (x_j - x_i)\dfrac{q_i q_j}{\left\|x_j - x_i\right\|^2} & \text{if } f(x_i) > f(x_j) \\[4mm] (x_i - x_j)\dfrac{q_i q_j}{\left\|x_j - x_i\right\|^2} & \text{if } f(x_i) \le f(x_j) \end{cases}, \tag{4.5}$$

where $\|x_j - x_i\|$ is the Euclidean distance between x_j and x_i. The term $f(x_i) > f(x_j)$ represents attraction, and the term $f(x_i) \leq f(x_j)$ represents repulsion.

The total force F_i corresponding to particle x_i is computed as

$$F_i = \sum_{j=1, j \neq i}^{N} F_{i,j} \tag{4.6}$$

The direction of a particular force between two particles will be determined after comparing their objective function values. Figure 4.2 shows the direction of the total force acting on a candidate solution.

Attraction or repulsion between two particles x_j and x_i depends on the objective function values $f(x_j)$ and $f(x_i)$. The computation of the force F_i, $\forall i, j = 1$ to N (attraction or repulsion) is performed as follows:

$$F_i = \begin{cases} F_i + (x_j - x_i)\dfrac{q_i q_j}{\left\| x_j - x_i \right\|^2} \text{(attraction)} & \text{if } [f(x_j) < f(x_i)] \\[4mm] F_i - (x_j - x_i)\dfrac{q_i q_j}{\left\| x_j - x_i \right\|^2} \text{(repulsion)} & \text{else} \end{cases} \tag{4.7}$$

Due to the total force F_i acting on the particle x_i, it will move along the direction of the force F_i by a random step. The movement of the particle x_i is computed according to

$$x_i = x_i + \lambda \frac{F_i}{\left\| F_i \right\|}(\text{rng}), \quad i \neq \text{best} \tag{4.8}$$

where $\|F_i\|$ is the normalized force vector and $\lambda \in [0,1]$ is a random number uniformly distributed over $[0,1]$. rng denotes a vector that represents allowable movement within the lower bound L^k or upper bound U^k of the corresponding dimension (for $k = 1, \ldots, n$):

$$\text{rng} = \begin{cases} U^k - x_i^k & \text{if } F_i^k > 0 \\ x_i^k - L^k & \text{else} \end{cases}, \quad k = 1, \ldots, n \tag{4.9}$$

It is to be noted that the best particle x_{best} (based on fitness value) does not move and it is directly carried over to the next iteration. All other particles' movements are adjusted within the lower and upper bounds according to

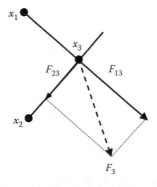

FIGURE 4.2   Direction of total force.

$$x_i^k = \begin{cases} x_i^k + \lambda F_i \left(U^k - x_i^k \right) & \text{if } F_i > 0 \\ x_i^k + \lambda F_i \left(x_i^k - L^k \right) & \text{else} \end{cases} , \quad i \neq \text{best}, \quad k = 1, \dots, n \qquad (4.10)$$

A local search is performed for each x_i. A maximum of m particles are generated in each direction in the δ neighborhood of x_i. It continues the local search until it finds a better point than x_i within the maximum trial runs. A simple local search procedure can be described as follows:

Step 1: Compute $\Delta = \delta \left[\max_k (U^k - L^k) \right], \forall k = 1, \dots, n$

Step 2: $x_i' = x_i, \forall i = 1$ to N

$$x_i' = \begin{cases} x_i' + \lambda_2 \times \Delta & \text{if } (\lambda_1 > 0.5) \\ x_i' - \lambda_2 \times \Delta & \text{else} \end{cases}$$

If $f(x_i') < f(x_i)$
$x_i = x_i'$

Step 3: If (max trial runs not reached), Goto Step 2

Step 4: Return $x_{\text{best}} = \arg \min\{f(x_i), \forall i\}$

where $\delta \in [0,1]$ is a local search parameter, and λ_1 and λ_2 are random numbers uniformly distributed over $[0,1]$.

A number of local search methods are widely in use such as great deluge (GD), SA, and TS. The local search procedure can also be omitted or applied to all particles or only to selected particles. Applying local search to all points will be very exhaustive. Therefore, the local search procedure can be applied only to the current best point to reduce computation time. The main steps of a standard EMO algorithm are as follows:

Step 1: Initialize a random population x_i, $i = 1, \dots, N$ in n-dimensional space
Step 2: Evaluate $f(x_i)$ and determine x_{best}
Step 3: Calculate charge q_i and total force F_i
Step 4: Calculate movement
Step 5: Perform local search
Step 6: If (termination condition not satisfied), Goto Step 2
Step 7: Return solution

The termination condition can be the maximum number of iterations or any suitable condition. Some researchers use $(f(x_{\text{best}}) - f(x_{\text{gbest}}))/|x_{\text{gbest}}| \leq 10^{-4}$ as the termination condition (Birbil and Fang, 2003; Huyer and Neumaier, 1999), where $f(x_{\text{gbest}})$ is the fitness of the known global best solution or optimum. The only important parameter of the EMO algorithm is the population size. The population size should be fairly large to be representative of the whole search space. Too large a population will be computationally exhaustive. Birbil and Fang (2003) tested the EMO algorithm without the local search procedure, with the local search procedure, and with the local search procedure applied only to the current best points. The performance of the EMO algorithm was verified on general test functions.

4.3 VARIANTS OF EMO

In order to improve the performance of the EMO algorithm and to apply it to different optimization problems, different modifications and hybridizations with other meta-heuristic algorithms have been proposed to the original EMO algorithm by researchers. A number of variants of the EMO algorithm have been reported in the literature since its inception. Variants of EMO can be classified into two broad groups: variants based on parameters and variants based on hybridization with other methods. Some of the well-known variants are briefly discussed in the following sections.

4.3.1 EMO Variants Based on Parameters

A number of modified variants of EMO have been reported in the literature. They are firstly based on the modified calculation of charge and force for enhancing convergence speed. Secondly, they are based on initialization of single or multiple populations for enhancing solution quality.

4.3.1.1 Revised EMO

Though the original EMO algorithm proved to be promising, it has the disadvantage of premature convergence and can get stuck in the local optimum. In order to improve premature convergence, Birbil et al. (2004) proposed a revised version of the EMO where a modification to the force computation is introduced. In the revised EMO, a perturbed point x_p is selected, which is the farthest point from x_{best}. The perturbed point x_p is determined by

$$x_p = \arg\max \{\| x_{\text{best}} - x_i \|\}, \quad i = 1, \dots, N \tag{4.11}$$

Computation of the total force remains the same except that the force acting on x_p is computed using the modified force definition given by

$$F_{p,j} = \begin{cases} (x_j - x_p) \dfrac{\lambda q_p q_j}{\| x_j - x_p \|^2} & \text{if } f(x_j) > f(x_p) \\[3mm] (x_p - x_j) \dfrac{\lambda q_p q_j}{\| x_j - x_p \|^2} & \text{if } f(x_p) \le f(x_j) \end{cases} \tag{4.12}$$

The parameter $\lambda \in [0,1]$ is a random value uniformly distributed over $[0,1]$. The direction of the component forces acting on x_p is perturbed. If the parameter $\lambda < v \in (0,1)$, the direction of the component force is also reversed. Since the local search procedure in standard EMO is considered not very significant, the local search procedure is removed from the revised EMO algorithm. The main steps of the revised EMO are as follows:

Step 1: Initialize random population x_i, $i = 1, \dots, N$ in n-dimensional space
Step 2: Evaluate $f(x_i)$ and determine x_{best} and perturbed point x_p using Equation 4.11
Step 3: Calculate charge q_i and total force F_i using Equation 4.12
Step 4: Calculate movement using Equation 4.8
Step 5: If (termination condition not satisfied), Goto Step 2
Step 6: Return solution

Birbil et al. (2004) provided detailed analytical proofs of the convergence, which show that the revised EMO exhibits global convergence with higher probability.

Guan et al. (2011) proposed a revised EMO algorithm for flow path design of an automated guided vehicle (AGV) system where a variable neighborhood search strategy is employed as the local search technique. Variable neighborhood search (Mladenovic and Hansen, 1997) explores distant neighborhoods of the current incumbent solution and moves from there to a new solution if and only if an improvement is made. This simple procedure is applied repeatedly until the local optimum is reached. In the revised EMO, individuals are encoded discretely and the Hamming distance is used to reduce computation of variable neighborhood search. In order to reduce computational cost, only the best individual is considered in the local search procedure. In the movement procedure, each element of the individual is moved according to the ordering probability strategy. Guan et al. (2012) applied a version of revised EMO algorithm to the layout design problem of a reconfigurable manufacturing system. The revised EMO has been verified on several cases of path flow design and the layout design problem of an AGV system.

4.3.1.2 Discrete EMO (DEMO)

Originally the EMO algorithm was developed for searching a continuous space and restricted to real numbers. As a result it cannot be applied directly to problem domains featuring discrete or qualitative values between variables such as scheduling, knapsack, and permutation flow shop problems. Adaptations were required for applying the algorithm to the discrete domain. Those adaptations are mostly made by applying a random-key representation to limit the required modifications of the original algorithm. Javadian et al. (2008) proposed the discrete binary version of EMO for solving a TSP followed by another discrete binary version of EMO for solving combinatorial optimization problems reported in Javadian et al. (2009). In order to apply EMO to discrete domains, two distinct features have to be incorporated into EMO:

1. Each point should comprise binary variables. For example, $x_i = [x_i^1, x_i^2, \ldots, x_i^n]$ is the i-th point (solution) in n-dimensional space where x_i^k can have only discrete binary values, that is, $x_i^k \in \{0,1\}$ for $k = 1, \ldots, n$.
2. The total force vector $F_i = [F_i^1, F_i^2, \ldots, F_i^n]$ is the total force vector acting on the point i. The largest positive component $F_i^{\max} \in F_i$ is selected. This selected component of $F_i^{\max} \in F_i$ has a corresponding component in x_i, denoted as $x_i^{F\max}$. This corresponding component $x_i^{F\max} \in \{0,1\}$ is switched between the binary values 0 and 1.

In order to apply the EMO algorithm to a discrete domain such as discrete permutation flow shop scheduling, Liu and Gao (2010) redefined the calculation of the charge of particles, forces acting on them, and their movement in discrete space. The new approach is called DEMO. According to DEMO, the charge of a particle is calculated as

$$q_i = r_c \cdot \exp\left[1 - \frac{f(x_i)}{f(x_{p\text{best}})}\right] \tag{4.13}$$

where r_c is a predefined constant, $f(x_i)$ is the objective function value of particle x_i, and $f(x_{p\text{best}})$ is the objective function value of the best personal memory $x_{p\text{best}}$. The best particle is recorded as personal memory x_i^p. The charge of the personal memory is calculated as

$$q_i^p = r_c \cdot \exp\left[1 - \frac{f(x_i^p)}{f(x_{p\text{best}})}\right] \tag{4.14}$$

where $f\left(x_i^p\right)$ is the objective function value of the personal memory x_i^p.

The force acting on particle i by particle j is calculated according to

$$F_1 = q_i q_j \frac{H_1 - 1}{H_1} \tag{4.15}$$

where H_1 represents the Hamming distance between x_i and x_j. It is the distance between two binary strings x_i and x_j. It is the number of positions at which the corresponding bits between x_i and x_j differ.

Similarly, the force acting on personal memory x_i^p by particle x_i is calculated as follows:

$$F_2 = q_i q_i^p \frac{H_2 - 1}{H_2},\qquad(4.16)$$

where H_2 represents the Hamming distance between x_i and x_i^p.

The proposed DEMO algorithm can be described as follows:

Step 1: Initialize particles and personal memory
Step 2: Calculate distance H_1 and calculate force F_1 using Equation 4.15 in discrete space
    Perform Swap() operation
    Calculate distance H_2 and calculate force F_2 using Equation 4.16 in discrete space
    Perform Insertion() operation
Step 3: Repeat Step 2 N times
Step 4: If rand() $< p_{mu}$
    Perform swap operation

$$\begin{cases} \text{Randon-swap()} & \text{if } [\text{rand}() > 0.5] \\ \text{Reverse-swap()} & \text{else} \end{cases}$$

Step 5: Repeat Steps 2–4 for N times
Step 6: Evaluate objective function for each solution
Step 7: Perform local search on the best personal memory
Step 8: If (termination condition not met), Goto Step 2
Step 9: Return solution

Swap() and Insertion() are two operators that perform the movement of one permutation (i.e., solution) toward another by reducing the Hamming distance (H_1 and H_2) between them. The Swap() operation swaps the positions of two elements ξ_k in x_i. The Insertion() operation directly inserts the differing element ξ_k at position k from permutation x_j into permutation x_i. Random-swap() operation randomly swaps two positions of a permutation representation of x_i. Reverse-swap() operation reverses the sequence of all the positions between two randomly generated positions in solution x_i. The rand() is a function that generates a random value between [0,1], and p_{mu} is a predefined value. The variant neighborhood search (VNS) is used as the local search on the best personal memory at every iteration.

In order to demonstrate the efficiency of the algorithm, DEMO has been applied to the TSP and the single machine scheduling problem. The experimental results show that DEMO is capable of solving such well-known benchmark problems more efficiently than the standard EMO algorithm. The proposed DEMO algorithm was developed for the distributed permutation flow scheduling problem, and its effectiveness and performance were verified on a large instance set of benchmark suite. The experimental results show that the proposed DEMO algorithm outperforms the continuous EMO algorithm (Yuan et al., 2006).

Chao and Liao (2012) proposed a DEMO for minimizing the single machine total weighted tardiness (TWT) problem with sequence-dependent setup times. The proposed DEMO algorithm employs an attraction–repulsion mechanism involving crossover and mutation operators to avoid the calculation of real-valued numbers. The general algorithm consists of the following four steps: initialization of the population, calculation of charges of particles (i.e., permutation representation of solutions), movement of permutations according to the attraction–repulsion force, and local search using reference local search. A permutation of n jobs of the form $x_i = \{\xi_1, \xi_2, \dots, \xi_k, \dots, \xi_n\}$, $\forall i = 1, \dots, N$ of a population is considered where ξ_k is the k-th job and N is the size of the population. A destruction and construction procedure is applied to generate permutations. In the destruction phase, random d jobs are removed from a permutation. In the construction phase, the d jobs are reinserted into permutation. The proposed DEMO algorithm has been tested on a set of benchmark problem instances from the literature. The performance has been compared to the best-performing algorithms from the recent literature and found very competitive with the best-performing meta-heuristics such as discrete DE.

Bonyadi and Li (2012) proposed a DEMO to solve the multidimensional knapsack problem (MKP) where GA operators are utilized to work in discrete spaces. The proposed DEMO works in the same way as the EMO except that the operators work in discrete space. The vector calculations in the EMO are replaced by specific types of GA operators to determine the effects that particles have on one another. A new operator based on the principles of quantum mechanics is introduced. The operator is called the annihilation/creation (A/A') operator, which helps the algorithm in sampling the search space and also improves the performance of the method further. The general algorithm consists of four steps: calculation of the charge of particles, calculation of force, movement of particles applying the A/A' operator on particles, and local search. All tests are done on standard problems of the MKP, and the results are compared with several stochastic population-based optimization methods. Experiments showed that the proposed algorithm was found not only comparable but even better for the standard MKP.

4.3.1.3 Opposition-Based EMO

In the event of no *a priori* information about the solution, a random initialization is used in meta-heuristic algorithms. Very often the initial guess is far away from the optimal solution, which incurs huge computational costs for the algorithm. The notion of opposition-based learning is the simultaneous consideration of an estimate and its corresponding opposite estimate (i.e., guess and opposite guess) in order to achieve a better approximation for the current candidate solution and reduce computational costs. The concept of opposition-based learning introduced by Tizhoosh (2005) has been applied to meta-heuristic algorithms to accelerate learning (Shokri et al., 2006; Tizhoosh, 2006). Rahnamayan et al. (2007a–c) showed that EAs with an opposition-based population are faster than EAs with a random population. Rahnamayan et al. (2008) also investigated the effectiveness of opposition numbers over random numbers in soft computing techniques.

Cuevas et al. (2012a) proposed an opposition-based (OB) EMO (OBEMO) applying opposition-based numbers to optimization algorithms for better exploration of the search space and enhancing the convergence speed. The notions of opposition numbers and opposition-based learning have been introduced in Section 2.6.7 of Chapter 2.

The main steps of the OBEMO algorithm are as follows:

Step 1: Initialize a random population x_i, $i = 1, \dots, N$ in n-dimensional space
Step 2: Apply OB strategy
Step 3: Evaluate $f(x_i)$ and determine x_{best}
Step 4: Calculate charge q_i and total force F_i
Step 5: Calculate movement
Step 6: Perform local search
Step 7: Apply OB strategy
Step 8: If (termination condition not satisfied), Goto Step 3
Step 9: Return solution

The performance of OBEMO has been verified on a comprehensive set of benchmark functions (e.g., 14 different global optimization test problems).

4.3.1.4   Improved EMO

Computation of the total force in standard EMO is complicated, the precision is not satisfactory, and the running cost is very high and increases with the population size. The other disadvantage is when summing up the force components, which may cancel each other out and result in small force. Zhang et al. (2013a,b) proposed an improved EMO algorithm by introducing simplification to force vector calculation and incorporating movement probability to movement operation. Firstly, when calculating the total force of a particle, only a part of the particles are used for summing up. The number of particles is set to three, and they are selected randomly. Secondly, the distance quadratic term in the total force definition in Equation 4.5 is removed as the term does not contribute toward the total force significantly. Thus, the total force computation becomes

$$F_i = \sum_{j=R_1,R_2,R_3} \begin{cases} (x_j - x_i)q_iq_j & \text{if } [f(x_i) > f(x_j)] \\ (x_i - x_j)q_iq_j & \text{else} \end{cases} \tag{4.17}$$

where R_1, R_2, and R_3 denote three randomly selected particles.

The movement of particle defined in Equation 4.8 is modified by introducing a new parameter called move probability denoted as p_{move}. Thus, the movement of particle is defined by

$$x_i = \begin{cases} x_i + \lambda F_i & \text{if } (\text{rand} < p_{\text{move}}) \\ x_i & \text{else} \end{cases} \tag{4.18}$$

where rand and λ are random numbers distributed over [0,1].

The main steps of the improved EMO algorithm remain the same as those of the standard EMO. Only the calculation of the total force and movement of particle are replaced by Equations 4.17 and 4.18, respectively. The main steps are as follows:

Step 1: Initialize random population x_i, $i = 1, \ldots, N$ in n-dimensional space
Step 2: Evaluate $f(x_i)$ and determine x_{best}
Step 3: Calculate charge q_i and total force F_i using Equation 4.17
Step 4: Calculate movement using Equation 4.18
Step 5: Perform local search
Step 6: If (termination condition not satisfied), Goto Step 2
Step 7: Return solution

The effectiveness of the improved EMO has been verified on 13 classical functions, 3 engineering design problems, and 22 benchmark functions taken from CEC'06.

4.3.1.5   Multipopulation EMO

In order to handle dynamic optimization problems where the problem parameters change during the search or optimization process, Turky and Abdullah (2014) introduced multipopulation EMO. The population is divided into several subpopulations, which are then allocated for different regions of the search space. Each subpopulation is used for exploration (i.e., diversification) or exploitation (i.e., intensification) of the region by a separate EMO algorithm, which runs for a number of times until a change is seen when the subpopulations are merged together and partitioned again for exploration and exploitation. If a subpopulation is allocated for exploitation, local search will accept only an improved solution. If a subpopulation is allocated for exploration, local search will accept any solution regardless of quality. The mechanism of multipopulation EMO is better illustrated by the flow diagram in Figure 4.3.

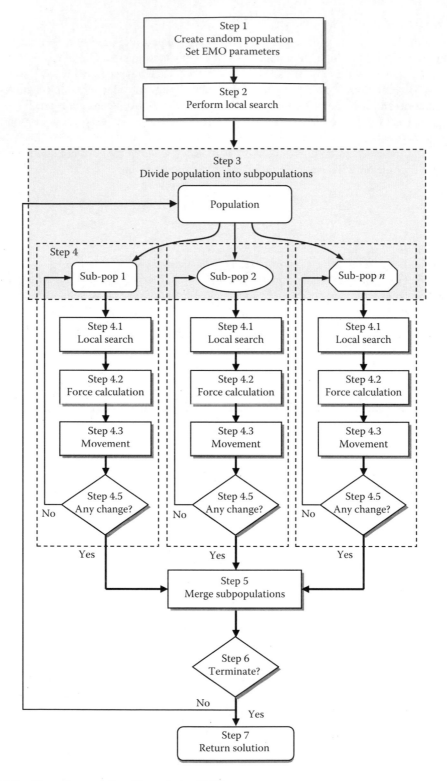

FIGURE 4.3   Flow diagram of multipopulation EMO.

To handle the changing solution landscape in dynamic problems, a population diversity mechanism is to be in place. Two of such population diversity mechanisms are as follows: immigration mechanism and memory-based mechanism. Turky and Abdullah (2014) proposed three diversity mechanisms for EMO: random immigration mechanism, memory-based mechanism, and memory-based immigration mechanism. In the random immigration mechanism, a small subset of reasonable size (e.g., *population_size**0.2) of solutions are randomly generated at each iteration and replace the worse solutions in the population. In the memory-based mechanism, an explicit EM memory of reasonable size (e.g., *population_size**0.1) is used where a subset of the best solutions are kept and reinserted in the EM memory replacing worse solutions once the changes are detected. In the memory-based immigration mechanism, immigration and memory-based mechanisms are combined. At each iteration, a set of solutions (e.g., *population_size**0.1) are selected from EM memory, which undergoes mutation with a mutation probability (e.g., $p_m = 0.01$). The mutated solutions then replace worse solutions in the population.

The population-based EMO has been tested on the well-known moving peak benchmark problem.

4.3.1.6 Memory-Based EMO

The redundancy allocation problem (RAP) is a combinatorial optimization problem where it aims to find the optimal number of proper components in the system. Teimouri et al. (2016) proposed a memory-based EMO (MBEMO) for the RAP. The main idea of using a memory matrix is to separate positive variations from the negative variations, use the information for local search, and improve the quality of solutions. In MBEMO, a particle is represented by an $s \times z$ matrix, where s is the number of subsystems and z is the maximum number of component types that can be used in subsystems for the RAP. The memory matrix is updated by the best particle found so far by the local search method. The main steps of MBEMO are the same as those of the standard EMO except for the memory-based local search. The simulation tests show that the memory matrix in local search has improved efficiency and robustness of EMO.

4.3.2 Hybrid EMO with Other Meta-Heuristics

In EMO, exploration of the search space is performed by EMO itself and exploitation is performed by the local search method. In general, different meta-heuristic algorithms and local search techniques have their own advantages and disadvantages. The strategy is to combine EMO with other meta-heuristic and local search techniques to overcome the individual disadvantages. Birbil and Fang (2003) showed experimentally that EMO was able to approximate the optimum solution. But the accuracy of average function values was not good enough. In pursuit of improving the accuracy, they applied local search to all points to examine thoroughly the attractive part of the feasible region. They found that the performance of EMO improved at the cost of the number of added evaluations required by the local search procedure. Further they applied the local search only to the current best point to reduce the number of evaluations. This idea eventually strikes a balance between the average number of evaluations and the accuracy of solutions, which gave a clue to hybridization of EMO and other local search methods. Moreover, EMO does not have any powerful operations such as crossover and mutation to enhance population diversity, which is another reason for slow convergence and low quality of the solution. In order to improve exploration or global search ability, exploit local search, increase convergence speed, improve solution quality, and minimize computational cost, the standard EMO algorithm has been hybridized with other meta-heuristic algorithms. There have been a number of hybrid variants of EMO with other heuristic and local search algorithms reported in the literature in the recent years, which are discussed in this section.

4.3.2.1 Hybrid Modified EMO and Scatter Search

Scatter search was introduced by Glover (1977) as a heuristic for integer programming and widely accepted as a local search method. The concept of the scatter search is to store useful information about the global optima in a set P, which is a randomly generated set of solutions such that recombining samples from the set P can exploit this information. A reference set S_{ref} of candidate solutions is created from P such that $S_{ref} \subset P$. The reference set S_{ref} comprises subsets S_1 and S_2 such that $S_{ref} = S_1 \cup S_2$ and $S_1 \cap S_2 = \emptyset$. New solutions are created by linear recombination of the subsets S_1 and S_2 in an iterative way, and the solutions are refined using heuristic criteria and assessed as to whether or not they are selected for the next iteration followed by updating of the reference set S_{ref}. Further detailed description on scatter search can be found in Glover et al. (2000).

In order to apply EMO to resource-constrained project scheduling (RCPS), Debels et al. (2006) proposed hybridization of EMO and scatter search (EMO-SS) with modification to EMO where point charges are not computed independently but also based on the point on which where the forces act. In other words, a predetermined number of particles act on any given particle. Thus, the computation of charge is simplified. The charge q_{ij} depends on the relative difference between fitness values between particle i and j. We compute q_{ij} as

$$q_{ij} = \frac{f(x_i) - f(x_j)}{f(x_{worst}) - f(x_{best})} \tag{4.19}$$

where x_{worst} and x_{best} are the worst and best solutions of the reference set of the scatter search procedure, respectively. Since $q_{ij} \in [-1,1]$, better solutions have a higher value on q_{ij}; that is, q_{ij} is positive and solution j attracts i when $f(x_i) > f(x_j)$. The opposite happens when q_{ij} is negative and solution j repulses i when $f(x_i) < f(x_j)$. No force acts when $f(x_i) = f(x_j)$. The force F_{ij} that acts on i from j is defined as

$$F_{ij} = (x_j - x_i) \cdot q_{ij} \tag{4.20}$$

The perturbation of point x_i is performed by moving the particle x_i toward the direction of force F_{ij}, and x_i is then updated according to

$$x_i = x_i + F_{ij} \tag{4.21}$$

The main steps of the EMO-SS algorithm are as follows:

Step 1: Initialize a random population x_i, $i = 1, \ldots, N$ in n-dimensional space
Step 2: Evaluate $f(x_i), f(x_j)$ and determine x_{best} and x_{worst}
Step 3: Calculate charge q_{ij} using Equation 4.19 and force F_{ij} using Equation 4.20
Step 4: Calculate movement using Equation 4.21
Step 5: Perform scatter search to generate a new reference set
Step 6: If (termination condition not satisfied), Goto Step 2
Step 7: Return solution

Experimental results of the hybrid approach to the RCPS problem show that the procedure outperforms other state-of-the-art heuristics in the literature and it is competitive with other procedures.

4.3.2.2 Hybrid EMO and Restarted Arnoldi Algorithm

Calculation of eigenvalues and eigenvectors of a large nonsymmetric matrix A is very challenging. Among different methods, the Arnoldi method (Arnoldi, 1951) is a widely used method where the idea is to make the matrix similar to the Hessenberg form (Press et al. 1992). The method starts with an initial vector v_1. After m-steps, it produces an $m \times m$ Hessenberg matrix H_m and an orthogonal $n \times m$ matrix $V_m = [v_1, v_2, \ldots, v_m]$ such that $V_m^H A V_m \approx H_m$ for $m \leq n$ so that the eigenvalues of H_m are an approximation of the eigenvalues of nonsymmetric matrix A. In most of the restarting methods for the Arnoldi algorithm, the best initial eigenvector $v_1 \in V_m$ is selected for the next iteration to increase the accuracy.

The difficulty with the restarted Arnoldi method is that the algorithm starts diverging before getting a good approximation of the eigenvalues and eigenvectors. Even after many iterations, the solution does not improve to the required level of accuracy as it depends on the initial vector. Taheri et al. (2007) proposed the hybrid approach combining EMO and the restarted Arnoldi (EMO-RA) algorithm to improve the initial vector. The main steps of the EMO-RA are as follows:

Step 1: Initialize x_0, $x_{\text{best}} = x_0$
Step 2: Perform the restarted Arnoldi method
Step 3: Perform local search
Step 4: Calculate charge q_i and total force F_i
Step 5: Calculate movement
Step 6: Determine $x_{\text{best}} = \arg \min\{f(x_i), i = 1, \dots, N\}$ where $f(x_i) = \|Ax_i - \lambda_i x_i\|$
Step 7: If (termination condition not satisfied), Goto Step 2
Step 8: Return solution

The local search is a neighborhood search around the solution obtained by the restarted Arnoldi method where m_1 points are generated using random direction v. The termination condition could be $\|Ax_{\text{best}} - \lambda x_{\text{best}}\| \leq \varepsilon$ or the maximum number of iterations. The EMO-RA approach is an interesting method for matrix computation, especially eigenvalues and eigenvectors, which has huge implication in mathematics and engineering. The approach has been applied to a number of test cases from quantum chemistry.

4.3.2.3 Hybrid EMO and Iterated Swap Procedure

Yurtkuran and Emel (2010) proposed a hybrid of EMO and the iterated swap procedure (EMO-ISP) approach for solving the capacitated vehicle routing (CVR) problem. The iterated swap procedure (ISP) (Ho and Ji, 2003) is used as a local search technique. ISP is a simple procedure described by the following steps:

Step 1: Select two components randomly from the vector of selected particle
Step 2: Exchange the values in the selected components to form a child
Step 3: Swap neighbors of the exchanged components to form four children
Step 4: Evaluate children
Step 5: If the best child is better than parent, replace parent with the best child
Step 6: Return solution

A Roulette Wheel mechanism is used to select the particle that will undergo ISP. Yurtkuran and Emel used EMO of Debels et al. (2006) where the charge q_{ij} and the force F_{ij} are calculated using Equations 4.19 and 4.20, respectively. In the EMO-ISP approach, the perturbation of x_i is calculated using the modified equation given by

$$x_i = x_i + \frac{F_{ij}}{t} \tag{4.22}$$

where the increment is a decreasing value with the iteration number t.

After movement operation, particles can go out of bound and they need to be brought inside the search space. To control the search space within the limits of n-dimensional space, the upper bound U_i and the lower bound L_i are checked at every iteration as follows:

$$x_i^k = \begin{cases} U_i^k & \text{if}\left(x_i^k > U_i^k\right) \\ L_i^k & \text{if}\left(x_i^k < L_i^k\right) \end{cases}, \quad \forall k = 1, \dots, n \qquad (4.23)$$

The main steps of the EMO-ISP algorithm are as follows:

Step 1: Initialize a random population x_i, $i = 1, \dots, N$ in n-dimensional space
Step 2: Evaluate $f(x_i)$, $f(x_j)$ and determine x_{best} and x_{worst}
Step 3: Perform local search using ISP
    Select x_i using the Roulette Wheel mechanism
    Apply ISP on x_i
Step 4: Calculate charge q_{ij} using Equation 4.19 and force F_{ij} using Equation 4.20
Step 5: Calculate movement using Equation 4.22
Step 6: Check boundary

$$x_i^k = \begin{cases} U_{i.}^k & \text{if}\left(x_i^k > U_i^k\right) \\ L_i^k & \text{if}\left(x_i^k < L_i^k\right) \end{cases}, \quad \forall k = 1, \dots, n$$

Step 7: If (termination condition not satisfied), Goto Step 2
Step 8: Return solution

EMO-ISP has been applied to the CVR problem to verify the effectiveness and performance. The CVR problem involves finding the best sequence that satisfies certain constraints. A real-coded population is used to represent solutions of sequences in most of the popular methods like the GA. To find a set of routes for CVR, real codes are required for decoding. The random-key procedure (RKP) (Bean, 1994) is used for solving sequencing problems. In order to apply EMO to CVR with discrete variables, an extra procedure for objective value calculation is required, which involves three phases such as decoding, routing, and coding. RKP is applied in the decoding phase to uncover the corresponding routes.

4.3.2.4 Hybrid EMO and SA

The EMO algorithm is good at global search but it is not efficient at local search. To improve solution quality, a local search technique is employed as an inherent part of EMO. SA is an efficient and the oldest local search technique among the meta-heuristic algorithms with an explicit searching strategy to avoid local optima. In SA, a perturbation generates a new solution in the neighborhood of the current solution. The new solution is accepted or rejected depending on a transition probability. The mechanism of the SA algorithm is presented in detail in Chapter 8. Another issue with EMO is that the EMO's performance very much depends on the initial solution. Tavakkoli-Moghaddam et al. (2009) proposed to hybridize EMO with a local search-based meta-heuristic. SA is more or less independent of the initial solution. Therefore, the hybrid EMO-SA brings some advantages by combining SA for obtaining the initial solution and EMO for global search. In EMO-SA, a modified movement operation is employed, which is defined using normalized total force $\bar{F}_i = \dfrac{F_i}{\|F_i\|}$ as follows:

$$x_i = \begin{cases} x_i + \beta \times \bar{F}_i(1 - x_i) & \text{if } (\bar{F}_i > 0) \\ x_i + \beta \times \bar{F}_i(x_i) & \text{else} \end{cases} \qquad (4.24)$$

where $\beta \in [0,1]$ is a random number.

EMO-SA proposed by Tavakkoli-Moghaddam et al. (2009) comprises two phases: initialization phase and improvement phase. In the initialization phase, SA starts from an initial solution and produces an improved solution. In the improvement phase, EMO is applied for global search to improved solutions produced by SA.

The main steps of the EMO-SA algorithm are as follows:

Initial phase:
    Step 1: Generate an initial solution (using suitable heuristic or analyzing the problem)
    Step 2: Apply SA to produce an improved initial solution x_i for EMO
Improvement phase:
    Step 3: Evaluate $f(x_i)$ and determine x_{best}
    Step 3: Calculate charge q_i and total force F_i
    Step 4: Calculate movement using Equation 4.24
    Step 6: If (termination condition not satisfied), Goto Step 3
    Step 7: Return solution

Tavakkoli-Moghaddam et al.'s version of EMO-SA has been applied to job shop scheduling where operation-based representation is used as the encoding scheme and TWT is used as the objective function to be minimized.

Naderi et al. (2010) applied the hybrid of the EMO-SA algorithm to a flow shop scheduling problem using permutation coding of jobs, with jobs sequenced using random keys. SA is employed as a local search technique that uses two operators, for example, swap and relocation, to improve the solution. In the EMO, movement is calculated using Equation 4.24. The hybrid EMO-SA has been applied to the scheduling flow shop problem with objective function of minimizing the makespan and the TWT. The effectiveness and performance of the proposed EMO-SA approach have been verified on some randomly constructed instances. Jamili et al. (2011) also proposed a similar EMO-SA approach for the periodic job shop scheduling problem (JSSP) using random keys as the encoding scheme. SA is used to provide the initial solution to the EMO algorithm. The movement operation is performed using Equation 4.24. The EMO-SA algorithm comprises two phases. In the first phase, SA obtains a good initial solution. In the second phase, $N - 1$ solutions are randomly generated and the EMO combined with the SA algorithm is applied to the population followed by a feasibility check for valid solutions. The performance and effectiveness of the proposed hybrid EMO-SA algorithm have been verified on randomly generated instances of the periodic JSSP with an objective function of minimizing the TWT.

4.3.2.5 Hybrid EMO and Solis–Wets Search

EMO algorithm is inherently a global search method capable of exploring the search space, which has been demonstrated in many applications. Further improvement of the performance of EMO eventually rests upon clever use of local search methods. A point $x \in S$ in the search space of S is to be found, which minimizes $f(x)$ or at least yields an acceptable approximation of the infimum of $f(x)$ on S where $f(x)$ is a function defined by $f: R^n \to R$ and $S \subset R^n$. Solis and Wets (1981) proposed a local search technique known as the Solis–Wets search. The conceptual algorithm is described by the following steps:

    Step 1: Find an initial point $x^0 \in S$ and $k = 0$
    Step 2: Generate $\xi^k \in R^n$ randomly from distribution μ_k
    Step 3: Set $x^{k+1} = D(x^k, \xi^k)$, $D(x, \xi) \in S \times R^n$ is a map that satisfies $f[D(x, \xi)] \le f(x)$ and $f[D(x, \xi)] \le f(\xi)$
       if $\xi \in S$
    Step 4: Choose distribution μ_{k+1}, set $k = k + 1$, and Goto Step 1.

Alikhani et al. (2009) proposed a novel hybrid approach for EMO by employing the Solis–Wets local search method discussed above. The main steps of the EMO-Solis–Wets algorithm are as follows:

Step 1: Initialize random population x_i, $i = 1, \ldots, N$ in n-dimensional space
Step 2: Perform the Solis–Wets local search
Step 3: Evaluate $f(x_i)$, determine x_{best}, and perturbed point x_p using Equation 4.11
Step 4: Calculate charge q_i and total force F_i using Equation 4.12
Step 5: Calculate movement using Equation 4.8
Step 6: If (termination condition not satisfied), Goto Step 2
Step 7: Return solution

To demonstrate the effectiveness of the hybrid EMO-Solis–Wets approach, a number of experiments are carried out on a set of well-known benchmark problems. The experimental results confirmed improved performance compared to EMO and the revised EMO algorithm.

4.3.2.6 Hybrid EMO and Great Deluge (GD)

Most authors consider hybridization of the EMO algorithm with another meta-heuristic in order to benefit from the advantages of the individual approaches. The GD algorithm is a local search meta-heuristic algorithm introduced by Dueck (1990, 1993) used for optimization problems. The basic principle of GD is simple. The name GD algorithm comes from the analogy of the mythical story of the great flood where a man climbs up a hill in the event of the GD. The water level keeps rising with the continuous falling of rain. The man makes a move in any direction such that he does not get his feet wet in water. The man keeps going on his way up as the water level rises until he reaches the top of the hill. The GD algorithm starts with an initial guess of solution x. A new solution is generated in the neighborhood of x, which is perturbed once again. The quality of solution E is estimated using a function such as $EQx(\cdot)$. The basic principle of the GD algorithm can be described by the following steps:

Step 1: Guess an initial solution x
       Set rate of rain $r_r > 0$ and water level $w_l > 0$
Step 2: Generate a new solution x' in any direction from x
       $x' = N(x)$
Step 3: Perturb x
Step 4: Estimate quality of solution
       $E = EQx(x')$
Step 5: If $(E > w_l)$
       $x = x'$
       Update water level $w_l = w_l + r_r$
Step 6: If (termination condition not met), Goto Step 2
Step 7: Return solution x

The operator $EQx(\cdot)$ is a measure for the quality of the solution. The simplest form of $EQx(\cdot)$ would be the fitness function for the problem at hand.

There is another known implementation of the GD algorithm based on the basic principle. The algorithm starts with an initial guess of the optimum solution x. A numerical value termed badness $\beta(x)$ is calculated for the solution. It is seen that $\beta(x)$ measures the undesirability of the initial approximation and it is compared with a tolerance value Δ. The tolerance $\Delta = \psi(\Delta)$ is defined as a decaying function $\psi(\cdot)$ that decreases the value of tolerance with rising water levels. The GD method approximates a new solution x' from the neighborhood $N(x)$. If $\beta(x') > \Delta$, then $x = x'$ is set. If $\beta(x') < \Delta$, then a different x' is selected from the neighborhood $N(x)$. The process is repeated until all the solutions in the neighborhood of $N(x)$ are beyond tolerance Δ and x is returned as the best approximate solution. This version of the GD method can be described by the following steps:

Step 1: Guess an initial solution x
Step 2: Create a new solution $x' = N(x)$
Step 3: Estimate badness $\beta(x')$
Step 4: Define tolerance $\Delta = \psi(\Delta)$
Step 5: If $\beta(x') > \Delta$
    $x = x'$
Step 6: If (termination condition not met), Goto Step 2
Step 7: Return solution

Abdullah and Burke (2006) developed a force decay rate β for the DG algorithm. The force decay rate β is defined by

$$\beta = \frac{f(x) - EQx}{t_{max}} \qquad (4.25)$$

where EQx is the estimated quality of solution that a user wants to achieve. We define EQx as $EQx = f(x) - F$ where F is the total force. This is the total force computed by the EMO algorithm. The initial water level w_l is set to $f(x)$. The force decay GD algorithm is described by the following steps:

Step 1: Set an initial solution x
Step 2: Calculate fitness $f(x)$
    $x_{best} = x$
Step 3: Estimated quality of solution EQx
    $EQx = f(x) - F$
Step 4: Set initial water level
    $w_l = f(x)$
Step 5: Set force decay rate β using Equation 4.25

$$\beta = \frac{f(x) - EQx}{t_{max}}$$

Step 6: Generate a new solution from neighborhood
    $x' = N(x)$
Step 7: Calculate fitness $f(x')$
Step 8: If $[f(x') < f(x_{best})]$
        $x = x'$
        $x_{best} = x'$
      Else if $[f(x') \leq w_l]$
        $x = x'$
      $w_l = w_l - \beta$
Step 9: If (t_{max} not reached), Goto Step 6
Step 10: Return solution

Abdullah et al. (2009) proposed a hybridization of the EMO algorithm with the force decay rate GD described above. The main steps of the EMO-GD algorithm are the same as those of the standard EMO where the local search method is replaced with GD or the force decay rate GD method as follows:

Step 1: Initialize a random population x_i, $i = 1, \ldots, N$ in n-dimensional space
Step 2: Evaluate $f(x_i)$ and determine x_{best}
Step 3: Calculate charge q_i and total force F_i
Step 4: Calculate movement
Step 5: Perform force decay rate GD search (or basic GD search)
Step 6: If (termination condition not satisfied), Goto Step 2
Step 7: Return solution

The hybrid EMO-GD approach has been verified on a university timetabling problem (TP). Combination of the population-based approach with GD search has provided very good results for a variety of scheduling problems. A penalty value representing the degree to which various constraints are satisfied is measured that reflects the quality of the solution. Turabieh et al. (2009) applied hybrid EMO-GD algorithm with the force decay rate to a university TP using established data sets and compared it with state-of-the-art techniques from the literature.

4.3.2.7 Hybrid EMO and GA

The local search procedure of EMO is stochastic, which incurs high computational cost. In order to improve the computation of EMO, the random neighborhood local search is replaced by the reproduction operation, that is, competitive selection, crossover, and mutation of the GA, or GA operators are applied along with local search. The GA is a population-based optimization algorithm based on the principle of Darwinian evolution. Details of the GA and its operators can be found in Siddique and Adeli (2013). Chang et al. (2009) proposed a hybrid framework integrating modified EMO with the GA for faster convergence of the search procedure. In this hybrid EMO-GA implementation, the algorithm starts with a local search method used by original EMO; GA applies binary tournament selection and uniform crossover operator, and the final steps are performed by EMO, that is, charge, force, and movement calculations described by Equations 4.19 through 4.21 are carried out. The hybrid EMO-GA starts with a selected set of solutions which undergo GA operation or EMO operation based on the average fitness value. The main steps of EMO-GA are as follows:

Step 1: Initialize a random population x_i, $i = 1, \ldots, N$ in n-dimensional space
Step 2: Perform local search
Step 3: Evaluate fitness $f(x_i)$, calculate average fitness f_{avg}, and determine x_{best}
Step 4: If $[f(x_i) < f_{\text{avg}}(x_i)$ and $x_i \neq x_{\text{best}}]$
    Select x_j using binary tournament selection
    Apply uniform crossover between x_j and x_i
Step 5: If $[f(x_i) > f_{\text{avg}}]$
    Calculate charge q_{ij} using Equation 4.19 and force F_{ij} using Equation 4.20
    Calculate movement using Equation 4.21
    Check boundary of solutions using Equation 4.23

$$x_i^k = \begin{cases} U_i^k & \text{if } \left(x_i^k > U_i^k\right) \\ L_i^k & \text{if } \left(x_i^k < L_i^k\right) \end{cases}, \quad \forall k = 1,\ldots,n$$

Step 6: If (termination condition not satisfied), Goto Step 2
Step 7: Return solution and transform to appropriate representation

The hybrid EMO-GA is verified on single machine earliness/tardiness problem where random-key method is used to represent solutions for scheduling problems.

Lee and Chang (2010) proposed hybridization of the improved EMO algorithm with the GA (IEMO-GA) for optimization of a proportional-integral-differential (PID) controller. In the IEMO-GA, the random neighborhood local search is replaced with the GA. After determination of the best particle, the competitive selection procedure is applied to generate new solutions for the next generation of the population. From this competitive selection procedure, 50% of the front solutions are selected. The GA performs crossover and mutation operation on the population to generate a new population. Of the GA generated solutions, 50% are selected. Finally, 50% of the GA-generated solutions and 50% of the front solutions are combined to form a new population. EMO operations are applied to the new population. The main steps of the IEMO-GA are as follows:

Step 1: Initialize a random population x_i, $i = 1, \ldots, N$ in n-dimensional space
Step 2: Evaluate fitness $f(x_i)$, do ranking, and determine x_{best}
Step 3: Select 50% solutions using competitive selection based on x_{best}
Step 4: Applying GA on selected solutions by employing crossover and mutation
Step 5: Create new population by selecting 50% of the GA and 50% of the competitive selection
Step 6: Calculate fitness $f(x_i)$, do ranking, and determine x_{best}
Step 7: Calculate charge q_i and total force F_i
Step 8: Calculate movement
Step 9: If (termination condition not satisfied), Goto Step 2
Step 10: Return solution

The performance of the IEMO-GA has been tested on fractional-order PID controller design. The advantage of the IEMO-GA is that it does not need any gradient information and it is capable of multiple searches to achieve global optimization with less computational complexity. Lee et al. (2010) applied the IEMO-GA to designing a recurrent fuzzy neural network system for control purposes. In this implementation, the random neighborhood local search is replaced by competitive selection and the GA.

4.3.2.8 Species-Based Improved EMO

Lee et al. (2011) proposed species-based improved EMO (SIEMO) by combining EMO and the gradient-descent technique. SIEMO has faster convergence and lower computational complexity. SIEMO consists of the following four phases: initialization, evaluation, species, and IEMO operation. When using random initialization, the population may be crowded in a region. Therefore, the initial population is generated using a uniform distribution method. The population is distributed evenly over the high-dimensional solution space. The uniform method has a lower probability of producing outliers, which affect the results significantly. The population is divided into two nonredundant subpopulations by applying similarity measurement. In the evaluation phase, evaluation of particles is carried out using a fitness function defined for the problem at hand. Particles with the same fitness values and locations are removed, and particles with better fitness values among two subsequent generations are retained. In the species phase, multiple species in a population are identified followed by identification of the best particle in each species. The dominant particle in each species is called the species seed, and it is the fittest individual in the same species. The particles that fall within a certain distance r_s from the species seed are classified as the same species. If r_s is small, there are many isolated species in each generation. If there are an inadequate number of particles in each species, species do not evolve. If r_s is large, there may not be enough isolated species in each generation and the species technique has no use. The IEMO operation phase is carried out in the following three steps: calculation of the total force for each species, calculation of movement for each species, and performance of local search for the best particle by gradient descent.

The main steps of the SIEMO algorithm are as follows:

Step 1: Set parameters,
   Initialize uniform population x_i, $i = 1, \ldots , N$
Step 2: Evaluate $f(x_i)$
   Rank particles based on fitness
   Determine the best particle x_{best}
Step 3: Determine species
Step 4: Calculate total force F_i for each subspecies
   Calculate movement for each subspecies
   Perform local search for the best particle by gradient descent
Step 5: If (all species not done), Goto Step 3
Step 6: If (termination condition not satisfied), Goto Step 2
Step 7: Return solution

The proposed hybrid algorithm has been applied to train an interval-valued neural fuzzy system. Simulation results show that SIEMO has faster convergence.

4.3.2.9 Hybrid EMO and Davidon–Fletcher–Powell Search

To obtain an approximate solution close to the true optimal solution, EMO requires a large number of iterations. In an attempt to achieve a near-optimal solution within a reasonable number of iterations, Yin et al. (2011) proposed to combine EMO with the Davidon–Fletcher–Powell (DFP) (Chong and Zak, 2008) method to boost the convergence speed. The DFP method is a quasi-Newton method, inherently a generalization of the secant method, for application to multidimensional problems. A brief description of the quasi-Newton method and its variants is provided in Chapter 1. The mechanism of the DFP method is simple. The search direction d_k of an initial solution x_k at iteration k is computed from the gradient of $\nabla f(x_k)$ and Hessian matrix H_k according to

$$d_k = -H_k \nabla f(x_k) \tag{4.26}$$

where H_k is a positive-definite symmetric matrix. Further detailed description of the gradient of function and the Hessian matrix is provided in Appendix A. H_k is initialized with a unit matrix at the beginning and calculated iteratively as follows:

$$H_{k+1} = H_k + \frac{p_k (p_k)^T}{(p_k)^T q_k} - \frac{H_k q_k (q_k)^T H_k}{(q_k)^T H_k q_k} \tag{4.27}$$

where $p_k = x_{k+1} - x_k$ and $q_k = \nabla f(x_{k+1}) - \nabla f(x_k)$. Once the search direction d_k is obtained, the next solution x_{k+1} is calculated according to

$$x_{k+1} = x_k + \lambda d_k \tag{4.28}$$

where $\lambda \in [0,1]$ is a random search step size used in the modified DFP (MDFP) method proposed by Yin et al. (2011). The DFP search continues until an error $e = \|\nabla f(x_k)\|$ is less than a small termination value $\varepsilon > 0$. In order to apply the MDFP method in an appropriate manner, it is to be ensured that the initial value close to the optimum solution is provided to DFP by EMO, the search space for DFP is restricted to small area, and finally the search direction is known per iteration. The main steps of the hybrid EMO-DFP algorithm are as follows:

Step 1: Initialize a random population x_i, $i = 1, \ldots , N$ in n-dimensional space
Step 2: Evaluate $f(x_i)$ and determine x_{best}

Step 3: Calculate charge q_i and total force F_i
Step 4: Calculate movement
Step 5: If (termination condition not satisfied), Goto Step 2
Step 6: Perform MDFP local search using Equations 4.26 through 4.28
Step 7: Return solution

The above algorithm can be run in two ways: firstly, running the DFP method as a local search for a small number of iterations, and secondly, running EMO without local search to obtain a good approximate solution and then running DFP to converge to the optimal solution as shown in the algorithm. The EMO-DFP has been applied for solving inverse kinematics for the PUMA robot, which confirms optimal solutions. It also reveals that MDFP is computationally efficient and numerically stable.

4.3.2.10 Hybrid EMO and PSO

A number of researchers have attempted to enhance the EMO algorithm by replacing random neighborhood local search methods with improved search methods. These improved search methods do not incorporate cognitive or social information. Lee and Lee (2012) proposed a hybrid approach by combining EMO with PSO, incorporating the idea of cognitive and social information used in PSO. PSO is a population-based optimization algorithm developed by Kennedy and Eberhart (1995, 2001) based on the social behavior of swarms. The hybrid approach is called EMO-PSO, and it replaces the local search method of EMO to help in improving the search efficiency. Each particle in PSO represents a candidate solution of the optimization problem. Particles fly through the multi-dimensional search space adjusting their velocity and position in the search space according to the best positions encountered and the best position of its neighbors toward an optimum solution. Each particle is attracted by the global best position x^{gbest} and its own best position x_i^{pbest}. It is seen that x^{gbest} is the best position for all particles defined as

$$x^{gbest} = \min\{f(x_i)\}, \quad i = 1, \ldots, n \qquad (4.29)$$

and $x_i^{pbest}(t)$ is the personal best position for particle i up to current iteration t. It can be determined using simple codes as follows:

$$\text{if } f[x_i(t)] > f\left[x_i^{pbest}(t)\right]$$

$$x_i^{pbest}(t) = x_i(t)$$

The velocity of the particles drives the optimization process of the PSO algorithm. The velocity update is computed using x^{gbest} and $x_i^{pbest}(t)$ according to

$$v_i(t+1) = v_i(t) + c_1 r_1 \left[x^{gbest} - x_i(t)\right] + c_2 r_2 \left[x_i^{pbest}(t) - x_i(t)\right] \qquad (4.30)$$

where c_1, $c_2 \in [0,1]$ are learning rates representing cognitive and social information of swarms, and r_1, $r_2 \in [0,1]$ are random values. The standard PSO uses both global best x^{gbest} and personal best x_i^{pbest}. A simplified version that could speed up the convergence of the PSO is to drop the personal best $x_i^{pbest}(t)$ term in Equation 4.30. A brief discussion on different velocity update mechanisms is presented in Chapter 2. Once the velocity is obtained, the position is updated according to

$$x_i(t+1) = x_i(t) + v_i(t+1) \times \Delta t \qquad (4.31)$$

where Δt is the time taken to reach the new position and used to convert velocity into position. Usually, Δt is set to unity, that is, $\Delta t = 1$. The main steps of the PSO algorithm are as follows:

Step 1: Initialize position x_i and velocity v_i for particles $i = 1, \ldots, n$

Step 2: Evaluate fitness $f(x_i)$ and find the global best x^{gbest}

Step 3: For all $i = 1, \ldots, n$

    Calculate velocity $v_i(t + 1)$ using Equation 4.30

    Update position $x_i(t + 1)$ using Equation 4.31

    Evaluate fitness $f(x_i[t + 1])$ at new positions

    Update the personal best $x_i^{\text{pbest}}(t + 1)$

Step 4: Update the global best x^{gbest}

Step 5: If (termination condition not met), Goto Step 3

Step 6: Return solution

The standard EMO algorithm uses the local search procedure, which very often demands huge computation time. The PSO algorithm is simple, which can reduce some computational time. After evaluation by EMO, each particle updates its velocity and position in PSO and provides improved solutions. The main steps of the EMO-PSO are as follows:

Step 1: Initialize a random population x_i, $i = 1, \ldots, N$ in n-dimensional space

Step 2: Evaluate $f(x_i)$ and determine x_{best}

Step 3: Calculate charge q_i and total force F_i

Step 4: Calculate movement

Step 5: Perform PSO local search

Step 6: If (termination condition not satisfied), Goto Step 2

Step 7: Return solution

Lee and Lee (2012) verified the performance of EMO-PSO on a recurrent fuzzy neural system with a finite-impulse-response filter where EMO-PSO is used as a learning algorithm. Simulation results show that the proposed EMO-PSO algorithm is effective and shows faster convergence.

4.3.2.11 Hybrid EMO and TS

To enhance the performance of any search technique for an optimization problem, it is crucially important to strike a balance between exploration and exploitation. A good combination of a global search and local search methods is essential for such a strategy. Sels and Vanhoucke (2014) proposed a hybrid approach combining EMO and TS. TS is a meta-heuristic method originally proposed by Glover in the 1980s (Glover, 1989; Glover and Laguna, 1997). TS is simply a combination of local search with short-term memories and generally applied as a local search technique to exploit the local search space around the neighborhood of the solution provided by EMO using memory structures and by restricting the visited solutions during the search. A brief description of TS is provided in Chapter 1. The TS stops after a specified number of iterations and returns the best solution found back to the EMO algorithm. The hybrid of the EMO and TS method balances the trade-off between exploration and exploitation. The main steps of the EMO-TS algorithm are as follows:

Step 1: Initialize a random population x_i, $i = 1, \ldots, N$, for all $k = 1, \ldots, n$

Step 2: Evaluate $f(x_i)$ and determine x_{best}

Step 3: Calculate charge q_i and total force F_i, and update position x_i

Step 4: Calculate movement

Step 5: Perform TS

Step 6: If (termination condition not satisfied), Goto Step 2

Step 7: Return solution

The EMO-TS approach was applied to JSSP on a single machine where random-key representation is used for job sequences. Compared to other meta-heuristic algorithms such as the GA, TS, and dual-population GA, the hybrid EMO-TS approach showed improved performance.

4.3.2.12   Hybrid EMO and DE

Though successful on a wide variety of applications, the EMO algorithm suffers from slow convergence because of its inability to exploit the local search space due to the local search method used. A powerful explorative algorithm may improve this weakness. DE is a population-based direct search algorithm and a fast and simple implementation of the EA developed by Storn and Price (1997). DE involves a few parameters (Storn, 1999) and has better explorative power (Zaharie, 2002). DE uses differences of two selected individuals as the source of generating a new population. It performs mutation using the distribution information of the population, that is, standard deviation, and then applies a crossover operator. A detailed description of DE and its operators can be found in Siddique and Adeli (2013). The basic algorithm of DE is also provided in Chapter 5. Muhsen et al. (2015) proposed a hybrid approach by combining EMO and DE with adaptive mutation (DEAM). DEAM comprises four basic steps: initialization, mutation, crossover, and selection, like any standard DE. Muhsen et al. (2015) proposed two types of mutation: classical DE-type mutation and EMO-type mutation. In the classical DE-type mutation, three solution vectors (i.e., position of particles) x_j, x_k, and x_l are randomly selected from the population and the mutated vector $\hat{x}_i$ is computed as follows:

$$\hat{x}_i(t) = x_j(t) + \lambda \left[x_k(t) - x_l(t) \right] \tag{4.32}$$

where $\lambda \in [0.5,1]$ is a scaling parameter, $i, j, k, l = 1, \ldots, N$, and N is the population size.

In the EMO-type mutation, three vectors x_i, x_j, and x_k are randomly selected from the population, charges between the particles, and total force exerted on them are calculated and finally the total force is used to perform mutation operation on the vector. The charges between $\{x_i$ and $x_j\}$ and $\{x_i$ and $x_k\}$ are calculated using the worst fitness $f[x_w(t)]$ and the best fitness $f[x_b(t)]$ by

$$q_{ij}(t) = \frac{f[x_i(t)] - f[x_j(t)]}{f[x_w(t)] - f[x_b(t)]} \tag{4.33}$$

$$q_{ik}(t) = \frac{f[x_i(t)] - f[x_k(t)]}{f[x_w(t)] - f[x_b(t)]} \tag{4.34}$$

The charge calculation in Equations 4.33 and 4.34 uses the same formulation as in Equation 4.19. Forces exerted on x_i, x_j, and x_k are computed as follows:

$$F_{ij}(t) = [x_j(t) - x_i(t)] \times q_{ij}(t) \tag{4.35}$$

$$F_{ik}(t) = [x_k(t) - x_i(t)] \times q_{ik}(t) \tag{4.36}$$

Force calculation in Equations 4.35 and 4.36 uses the same formulation as in Equation 4.20. The total force exerted on x_i, x_j, and x_k is given by

$$F_i(t) = F_{ij}(t) + F_{ik}(t) \tag{4.37}$$

EMO-type mutation is carried out using the total force as follows:

$$\hat{x}_i(t) = x_i(t) + F_i(t) \tag{4.38}$$

The EMO-type mutation operation in Equation 4.38 has the same form of the movement operation in Equation 4.21. This EMO-type mutation operation eventually represents a proper hybridization of the EMO and DE algorithms. The type of mutation to be carried out is decided based on the standard deviation of the population as follows:

$$\text{Mutation type} = \begin{cases} \text{EMO type} & \text{if } \left(\|\sigma(t)\| < \varepsilon \|\sigma_0\| \right) \\ \text{DE type} & \text{else} \end{cases} \tag{4.39}$$

where $\sigma(t)$ is the standard deviation of the population at iteration (or generation) t, σ_0 is the standard deviation of the initial population, and $\varepsilon \in [0,1]$ is the parameter for switching between EMO-type and DE-type mutations.

Crossover operation is performed in a similar way to that of classical DE using the mutated vector $\hat{x}_i(t)$ and a target vector $x_i(t)$, which is randomly picked from the population. The crossover operation is described by

$$x'_{i,\xi}(t) = \begin{cases} \hat{x}_{i,\xi}(t) & \text{if } (r \le p_c) \\ x_{i,\xi}(t) & \text{else} \end{cases} \text{ for } \xi = 1, \ldots, n \tag{4.40}$$

where $r \in [0,1]$ is a random number and $p_c \in [0.5,1]$ is the crossover rate. A boundary check is useful to bring the vectors within the search space back if they go out of the search space after crossover operation.

The selection operation is performed after it has carried out crossover on all N vectors in the population. It selects vectors having better fitness values that will go to next generation. It is described as follows:

$$x_i(t+1) = \begin{cases} x'_i(t) & \text{if } \left[f(x'_i(t)) < f(x_i(t)) \right] \\ x_i(t) & \text{else} \end{cases} \tag{4.41}$$

The main steps of the EMO-DE algorithm are as follows:

Step 1: Initialize population x_i, $i = 1, \ldots, N$ in n-dimensional space
Step 2: Compute $\|\sigma_0\|$, fitness $f(x_i)$, $f(x_b)$, and $f(x_w)$
Step 3: Compute $\|\sigma(t)\|$
Step 4: Decide on mutation type
    If $(\|\sigma(t)\| < \varepsilon \|\sigma_0\|)$
        Perform EMO-type mutation using Equations 4.33 through 4.38
    Else
        Perform DE-type mutation using Equation 4.32
Step 5: Perform crossover using Equation 4.40
    Do boundary check
Step 6: Perform selection using Equation 4.41
Step 7: If (max iteration not reached), Goto Step 3
Step 8: Return solution

The validity of the proposed EMO-DE has been verified on experimental data and optimization of model parameters of a photovoltaic module model.

4.3.2.13   Opposite Sign Test-Based EMO (EMO-OST)

Wang et al. (2015) proposed an improved EMO for the optimum feature selection problem to achieve high classification accuracy where the local search method is replaced with the opposite sign test (OST) method. The OST method involves testing the current state of a feature (i.e., 1 denotes that a feature is selected, and 0 denotes that a feature is not selected) in the feature selection vector and changes it to the opposite state (i.e., from 1 to 0 or 0 to 1). The OST method can be described by the following codes:

For all $k = 1, \ldots, n$

Set $x_i' = x_i$

$$x_{ik}' = \begin{cases} 0 & \text{if } (x_{ik} = 1) \\ 1 & \text{if } (x_{ik} = 0) \end{cases}$$

Calculate fitness $f(x_i)$ and $f(x_i')$

If $[f(x_i) < f(x_i')]$

   $x_{ik} = x_{ik}'$

Implementing the OST as a local search into EMO, the main steps of the EMO-OST algorithm become the following:

Step 1: Initialize a random population x_i, $i = 1, \ldots, N$, for all $k = 1, \ldots, n$

Step 2: Perform OST search

Step 3: Calculate charge q_i and total force F_i

Step 4: Calculate movement

Step 5: If (termination condition not satisfied), Goto Step 2

Step 6: Return solution

The proposed EMO-OST algorithm has been verified on prediction systems using 54 public data sets of diabetes mellitus.

Generally, local search and perturbed points are two widely accepted methods for improving solution quality and escaping local optima. Lin et al. (2012) provided a survey and comparative study of some widely used EMO algorithms reported in the literature.

4.4   APPLICATIONS TO ENGINEERING PROBLEMS

The EMO algorithm has found many applications in different domains within a very short period of time since it was introduced by Birbil and Fang (2003). They applied the EMO algorithm to a range of unimodal, easy, and difficult global optimization functions and verified the performance of the EMO algorithm on 15 test functions. The revised EMO algorithm was applied to a similar set of global optimization problems by Birbil et al. (2004). The EMO algorithm converges rapidly to the optimum when the number of function evaluations is included into performance measure. Brief discussions on selected applications are presented in the following sections.

4.4.1   Constrained Optimization Problem

Constrained global optimization problems are found in many application domains in science and engineering. Such problems can be mathematically formulated as follows: given a real objective function $f(\cdot)$ defined on a feasible set $\Omega \subset R^n$, find a point $x^* \in \Omega$ such that $f(x^*) = \min\{f(x)\}$, $\forall x \in \Omega$ for constraints $g_i(x) \leq 0$, $i = 1, 2, \ldots, m$ and for bounded x within lower and upper limits. In constrained optimization problems, optimal solutions lie on the boundary of the feasible region. Therefore, some researchers convert the constrained problems into unconstrained problems with penalty function. The advantage of this approach is that the existing implementation of unconstrained optimization can easily be applied to those problems by defining a suitable penalty function and modifying the objective function. A number of researchers applied the EMO algorithm to constrained optimization problems (Ali and Golalikhani, 2010; Han and Han, 2010; Rocha and Fernandes, 2008).

4.4.2 TRAVELING SALESMAN PROBLEM

For a TSP with n cities, there are $(n-1)!/2$ feasible solutions in which the global optimum is sought. In general, the TSP is an NP-hard problem that attracted researchers from the combinatorial optimization domain to apply algorithms to the TSP as test problems. Detailed mathematical treatment of the TSP is given in Chapter 2. In fact, many real-world problems can be formulated as instance of the TSP, which can lead to solutions to many other combinatorial optimization problems such as scheduling, routing, rostering, and timetabling. The EMO algorithm has been applied to the TSP in a number of research works (Bonyadi et al., 2008; Javadian et al., 2008; Wu and Chiang, 2005; Wu et al., 2006), which demonstrate competitive results and the effectiveness of the approach.

4.4.3 TIMETABLING PROBLEM

The TP includes assigning a set of events to a set of given rooms each with particular features and timeslots on a certain period of time. The rostering problem (RP) is an optimization problem of finding an optimal way to assign staffs with various skills to a set of shifts over a predefined period subject to hard and soft constraints. All valid solutions must follow hard constraints, whereas soft constraints generally reflect the preferences of staffs and organizational requirements. Mathematical descriptions of the TP and the RP are presented in Chapter 5. Abdullah et al. (2009) and Turabieh et al. (2009) applied the hybrid EMO-GD algorithm to TP, which demonstrates the effectiveness of EMO. Also, the EMO algorithm has been applied to a nurse RP by Maenhout and Vanhoucke (2007), which shows competitive performance over other meta-heuristic methods reported in the literature.

4.4.4 JOB SHOP SCHEDULING PROBLEM

In the JSSP, n jobs have to be processed on m machines satisfying a number of constraints. The constraints are namely splitting of a job is not allowed, interruption of an operation is not permitted, each machine can perform one operation at a time, and each operation is performed once on a unique machine. JSSP is a typical NP-hard problem. Detailed mathematical description of job shop scheduling is presented in Chapter 5 and also in Chapter 8. The EMO algorithm has been applied to JSSP by a number of researchers (Chao and Liao, 2012; Roshanaei et al., 2009). Roshanaei et al. (2009) use the EMO algorithm with random-key representation to solve the JSSP with sequence-dependent setup times in order to minimize the makespan. Computational results show the effectiveness and competitiveness of the EMO algorithm over other meta-heuristic algorithms. Davoudpour and Molana (2008) used the EMO algorithm with discrete variables to flow shop scheduling with deteriorating jobs. The distributed permutation flow shop scheduling problem is a generalization of the permutation flow shop scheduling problem, which is considered an NP-hard problem. The i-th permutation representation at time t of the distributed permutation flow scheduling problem is described by

$$x_i^t = \{\xi_1, \xi_2, \dots, \xi_k, \dots, \xi_n, \xi_{n+1}, \dots, \xi_{n+M-1}\}, \quad \forall i = 1, \dots, N \qquad (4.42)$$

where ξ_k represents the k-th job, $1 \leq k \leq n + M - 1$, $\xi_k \in Z$, Z is the set of all integer numbers, and ξ_{n+1} represents the separation point after which jobs belong to other factories, that is, when $k > n$.

Mirabi et al. (2008) reported a hybrid EMO approach with SA for flow shop scheduling with sequence-dependent setup times with the objective of minimizing the makespan. Chang et al. (2009) proposed a hybrid EMO algorithm to solve the single machine earliness/tardiness problem. Naderi et al. (2009) present an EMO algorithm for the flexible flow shop scheduling problem with sequence-dependent setup times and transportation times with the objective of minimizing the TWT. Naderi et al. (2010) applied a similar approach to the flow shop problem with stage-skipping in order to minimize the makespan and the TWT. In order to improve performance, Liu and Gao (2010) proposed a DEMO algorithm for the distributed permutation flow shop scheduling problem by introducing a redefinition of distance and movement and of charge and force calculation. Jamili et al. (2011) applied a hybrid EMO and SA (EMO-SA) algorithm to periodic JSSP and verified the performance on randomly constructed instances. Khalili and Tavakkoli-Moghaddam (2012) applied multiobjective EMO to a bi-objective flow shop scheduling problem for minimizing the makespan and TWT where all jobs may not be processed by all machines. There are many other scheduling problems reported in the literature where the EMO algorithm has been applied for solutions demonstrating promising results (Chang et al., 2009; Chen et al., 2007; Debels et al., 2006; Debels and Vanhoucke, 2004; Gilak and Rashidi, 2009; Jolai et al., 2012; Maenhout and Vanhoucke 2007; Naderi et al., 2010).

4.4.5 Knapsack Problem (KP)

The KP is to select a subset of items with each item yielding a profit and resource requirement such that all the selected items fit into the knapsack of certain capacity and maximize the sum of profits. The multidimensional knapsack problem (MKP) is a well-known discrete programming problem, which is considered as an NP-hard combinatorial optimization problem. MKP has been chosen as the benchmark problem for many optimization algorithms. The MKP is defined as follows: there are n objects; each of them has a price p_j and m knapsacks, and each of which has a capacity of b_i. Each object j occupies w_{ij} unit space of each knapsack m_j. The goal is to pick a subset of the objects to fill the knapsack such that the sum of the prices of selected objects is maximized. A more detailed description on MKP is provided in Chapter 6. Chang et al. (2010) and Chou et al. (2010) applied quantum-inspired EMO to a 0/1 KP problem. Bonyadi and Li (2012) applied the DEMO algorithm to solving the MKP and a set of test problems of MKP.

4.4.6 Set Covering Problem

The set covering problem (SCP) is to identify the smallest subset of S comprising n sets, that is, $S = \{s_1, s_2, \ldots, s_n\}$ whose union equals the universal set U consisting of elements $\{1, 2, \ldots, m\}$. The SCP is a classical problem in combinatoric optimization, and it is known to be an NP-hard problem. Naji-Azimi et al. (2010) applied EMO to SCP and tested the performance on 80 instances from the literature. For a fixed set of parameters, the EMO algorithm found the best-known solutions and improves the current best solutions for 12 instances.

4.4.7 Feature Subset Selection

Eliminating the redundant, uninformative, and noisy features reduces the dimension of the data set and reduces the search space to be used in many applications such as classification and learning algorithms. FSS is seen as a search problem. A detailed mathematical description on the FSS problem is presented in Chapter 2. Su and Lin (2011) applied the EMO algorithm combined with 1-nearest-neighbor (1NN) for feature selection and classification. To verify the effectiveness of the proposed approach, numerical experiments were carried out on several data sets with diverse sizes, features, separability, and classes. Experimental results show that the proposed method outperforms other well-known algorithms in terms of balanced classification accuracy and efficiency of feature selection.

4.4.8 INVERSE KINEMATICS PROBLEM IN ROBOTICS

The solution to the arm equation of a serial-chain manipulator is a complicated task. The problem is known as inverse kinematics where it has to find the values of the joint positions of the robot arm for a given position and orientation of the end effector. The solution to inverse kinematics has a huge impact on industrial application, for example, robot trajectory planning, motion control, and workspace analysis. Traditional mathematical approaches to inverse kinematics demand enormous computation when higher precision is sought. Yin et al. (2011) proposed a novel hybrid algorithm by combining EMO and MDFP called EMO-MDFP, and applied it to an inverse kinematics problem. Upon obtaining the approximate solution provided by the EMO algorithm, the MDFP algorithm is applied to improve the solution to the desired precision. Unlike the traditional algorithms, the MDFP algorithm randomly chooses the search step size between 0 and 1, which helps in reducing the computational complexity to a significant amount. The EMO-MDFP is a powerful and easy algorithm for solving the inverse kinematics problem. The EMO-MDFP approach was verified on 10 general test functions and the PUMA 560 robot, which demonstrated that the new near-real-time hybrid method can produce the best performance.

4.4.9 VEHICLE ROUTING PROBLEM

The vehicle routing problem (VRP) has been extensively analyzed within the past few decades since it plays a central role in optimization of distribution networks. The VRP is a set of tours for a subset of vehicles such that all customers are served within a time window. The VRP has been an important problem in the field of distribution and logistics. Further detailed mathematical description of the VRP is presented in Chapter 6. Wu et al. (2007) applied revised EMO to the VRP. Yurtkuran and Emel (2010) applied a hybrid approach by combining EMO with the iterated swap procedure utilizing the random-key procedure to a capacitated VRP. The efficiency of the proposed algorithm has been verified on 14 classical benchmark instances.

4.4.10 MAXIMUM BETWEENNESS PROBLEM

The betweenness problem is a well-known combinatorial optimization problem. Let a set S of n objects be defined by $S = \{x_1, x_2, \ldots, x_n\}$. A set C of triples defined by $C = \{x_i, x_j, x_k\} \in S \times S \times S$ is to be determined from the set S such that triples from C satisfy the betweenness constraints, that is, the element x_j is between the elements x_i and x_k. The betweenness problem is defined as a problem of determination of the total ordering of the elements from S. The maximum betweenness problem (MBP) is defined as finding the total ordering that maximizes the number of satisfied constraints. The MBP and variants of betweenness problem belong to a class of discrete optimization problems, which have many applications in various fields such as physical mapping problems in molecular biology and the process of gene mapping in bioinformatics. Filipovic et al. (2013) applied the EMO algorithm to the MBP by employing special representation of the individuals, which enables the EMO operators to explore the search space and reach high-quality solutions. Filipovic et al. also employed an effective 1-swap-based local search procedure improved by the specific caching technique on each point obtained by EMO. The algorithm is tested on real and artificial instances from the literature. The results show that the proposed EMO approach achieves all previously known optimal solutions.

4.4.11 REDUNDANCY ALLOCATION PROBLEM (RAP)

In industry, design of any hardware system involves discrete choices among available many components based on cost, reliability, performance, weight, etc. In general, the objective is the minimization of design cost for a certain reliability requirement. Finding the optimal combination of components and/or design configuration becomes very challenging. Furthermore, when there are many functionally similar components available, it becomes difficult to identify the optimal solution. This also leads to redundancy of components in any combination or design configuration. Rather than reducing redundancy, it can be used as a means of improving reliability. The problem of finding the optimal combination of components based on cost, reliability, performance, and weight is known as the RAP. The RAP is a combinatorial optimization problem where the reliability objective is achieved by discrete choices made from available hardware components. For a series–parallel system with s subsystems in series, each having n_i, $i = 1, 2, 3, \ldots, s$ components in parallel, the goal is to design a system such that the system reliability within the system-level constraints on cost and weight is maximized. The RAP can be formulated as a maximization problem of system reliability given restrictions on system cost and weight. The RAP can be mathematically defined as

$$\max R = \prod_{i=1}^{s} R\left(x_i \middle| k_i\right) \tag{4.43}$$

subject to

$$\begin{cases} \sum_{i=1}^{s} C_i(x_i) \leq C \\ \sum_{i=1}^{s} W_i(x_{ij}) \leq W \end{cases} \tag{4.44}$$

where x_{ij} is an integer for $i = 1, 2, \ldots, s$ and $j = 1, 2, \ldots, T_i$, C is the limit of total cost, and W is the limit of the weight.

It is assumed that the states of a component and its subsystem are one of the True or False states. Failed components do not lead to system damage or failure and are not repaired either. The supply of components is unlimited. The component reliabilities are assumed to be known and deterministic. The state of components is assumed to be independent of each other. Active redundancy strategies are considered. Teimouri et al. (2016) applied EMO to the RAP. In order to suit it for the RAP, the proposed algorithm employs a memory matrix in local search to save the features of good solutions and feed it back to the algorithm. The inclusion of the memory matrix to the local search procedure also increases efficiency.

4.4.12 Uncapacitated Multiple Allocation p-hub Median Problem

Generally in networks, the hubs are fully interconnected and the nonhub nodes route all of their traffic via one or more hubs. Hub location problems arise when given a set of nodes with pairwise traffic demands, p of them have to be chosen as hub locations and all the traffic has to be routed through these hubs at a minimal cost. Hubs are uncapacitated when there are no capacity limitations on the hubs or on the flow between them. Multiple allocations allow the nonhub nodes to be allocated to several hubs such that the overall cost of satisfying the flow demand is minimized. The setup costs are not considered in such a case. It is assumed that there are known nonnegative flows associated with each origin–destination pair of nodes. The transportation cost between a pair of nodes depends on the distance between the nodes in the network, the amount of flow to be moved across these distances, and the type of link between the nodes whether the flow is translated between hubs, collected from, or distributed to a nonhub node. The total flow cost is computed as the sum of transportation costs between all pairs of nodes. This problem is known as the uncapacitated multiple allocation p-hub median problem (UMApHMP), proposed by O'Kelly (1987). The UMApHMP is known to be an NP-hard problem. Kratica (2013) applied a hybrid approach combining EMO with the local search technique for solving the UMApHMP. The local search procedure interchanges one of the hubs with one of the nonhub nodes to improve the solution. The approach was tested on large-scale instances of nodes, and expected performance was confirmed.

4.4.13 Resource-Constrained Project Scheduling (RCPS) Problem

A RCPS problem considers resources of limited availability and tasks of known durations and resource requests linked by precedence relations. The RCPS problem has been a challenging problem due to its strong NP-hard nature. The objective is to find a schedule of minimal duration by assigning a start time to each activity such that the precedence relations and the resource availabilities are satisfied. These features of RCPS restrict the effectiveness of exact optimization to relatively small instances. RCPS is seen as a generalization of the JSSP. A generic JSSP is discussed earlier in Section 4.4.4. A set of tasks $T = \{\tau_0, \tau_1, \tau_2, \ldots, \tau_n, \tau_{n+1}\}$ are to be scheduled on a set of finite resources R with each $r \in R$ with its capacity limit. By convention, t_0 represents the start of schedule and t_{n+1} represents the end of schedule. The start and end tasks τ_0 and τ_{n+1} are dummy, have zero duration, and do not require any resource. The precedence relation among the tasks is that one task cannot be started until all of its predecessors have been finished. Each task puts some demand on resources described by a capacity function $C: R \rightarrow N$, a duration function $D: T \rightarrow N$, an assignment start time $S: T \rightarrow N$, a deadline δ, and a utilization function $U: T \times R \rightarrow N$. A partial ordering P on the tasks T is also given, specifying that some task must precede others. The goal is to minimize the makespan without violating the precedence constraints or overutilizing the resources. If the task $\tau_1 \in T$ precedes $\tau_2 \in T$ in the partial ordering P, then the precedence constraints are defined as

$$S(\tau_1) + D(\tau_1) \leq S(\tau_2) \tag{4.45}$$

For any time τ and running time $running(\tau) = \{\tau | S(\tau) \leq \tau \leq S(\tau) + D(\tau)\}$, the resource constraints are defined as

$$\sum_{t \in running(\tau)} U(\tau, r) \leq C(r), \quad \forall \tau, \forall r \in R \tag{4.46}$$

For all tasks, $\tau \in T$, $S(\tau) \geq 0$, and $D(\tau)$ must satisfy the deadline condition as follows:

$$S(\tau) + D(\tau) < \delta \tag{4.47}$$

Slowinski (1981) described the multiobjective RCPS framework and provided a list of objectives. The primary objectives include makespan, tardiness, resource investment, and robustness. The RCPS problem has been widely expanded over the past several decades. A solution for the RCPS problem is feasible if and only if the precedence and resource constraints are satisfied. In general, a solution of RCPS is represented as a schedule or a list of start times that implies a corresponding finishing time. To apply heuristic algorithms like EMO to the RCPS problem, a schedule representation scheme, a schedule generation scheme, and an evaluation procedure are required so that EMO operators can be applied to the representation and solutions can be evaluated. Debels et al. (2006) and Debels and Vanhoucke (2004) used the random-key scheme for solution representation. In random-key representation, a solution corresponds to a point in $(n + 1)$-dimensional Euclidean space such that the i-th vector element functions as a priority value for the i-th task. The advantage of random-key representation is that EMO operators or mathematical operations can easily be performed. Debels et al. (2006) applied the hybrid EMO-scatter search method to the RCPS problem. The computational results show that the proposed approach outperforms other heuristic algorithms reported in the literature.

4.4.14 Multiobjective Optimization Problem

EMO has been applied to many single-objective optimization problems with great success. Real-world problems involve simultaneous optimization of multiple objectives. MOOPs are defined as the problems consisting of multiple objectives, which are to be minimized or maximized while maintaining some constraints. Most MOOPs do not provide a single solution; rather, they offer a set of solutions. Therefore, trade-offs are sought when dealing with such MOOPs. It is also a common approach to aggregate all objectives together to form a single fitness function, but the solution using a single scalar is sensitive to the weight vector used in the scaling process and the individual objectives may conflict with each other. EMO has been applied to MOOPs by a number of researchers (Jiekang et al., 2014; Khalili and Tavakkoli-Moghaddam, 2012; Tsou and Kao, 2006, 2008).

Tsou and Kao (2006) first proposed a framework of multiobjective EMO for generating non-dominated Pareto front solutions for optimization problems. The framework was validated on three well-known benchmark functions. Tsou and Kao (2008) applied multiobjective EMO to an inventory control problem with two objectives, namely minimization of cost and storage without surrogate measure and prior preference information. The simple outranking method is used to prioritize nondominated solutions. Khalili and Tavakkoli-Moghaddam (2012) applied multiobjective EMO to a flow shop scheduling problem. Random-key representation is used for solutions. The two objectives used are the makespan and total tardiness. To measure the goodness of a solution, three approaches, namely the nondominated sorting method, strength Pareto method, and the weighted sum of them are used. The simulation results were compared with multiobjective SA. Jiekang et al. (2014) applied multiobjective EMO to short-term scheduling for dynamic generation flow of cascaded hydroelectric plants, which is known as the complex and nonlinear optimization problem. The multiple objectives considered for dynamic generation flow are maximization of energy production over a period, maximization of storage water, minimization of the volume of water used for energy production, minimization of water consumption, and minimization of spillage flow subject to satisfaction of a number of constraints. EMO was applied as the global search technique and data envelopment analysis (DEA) as the local search method. Simulation results confirmed the efficiency of the hybrid EMO-DEA approach. Xiao et al. (2016) extended EMO to multiobjective EMO for applying to the RCPS problem (discussed above) with the aim of optimizing two objectives, namely makespan and total tardiness. A solution in RCPS is a schedule, on which multiobjective EMO operations cannot be applied. Standard random-key (SRK) representation is used for the schedule so that EMO operations can be applied. A schedule generation scheme is employed for converting the schedule into an SRK.

4.4.15 Other Applications

EMO algorithm has been applied to many other applications in several domains such as fuzzy systems (Lee et al., 2009; Wu et al., 2005), matrix optimization (Taheri et al., 2007), neural networks (Wang et al., 2008; Wu et al., 2010), neural fuzzy controller design and optimization (Lee and Chang, 2008; Lee et al., 2007, 2009, 2011), antenna array design application (Jhang and Lee, 2009; Lee and Jhang, 2008), motion stabilization in robotics application (Santos et al., 2009), PID controller design (Lee and Chang, 2010), wireless communication networks optimization (Tsai et al., 2010), flow path designing (Guan et al., 2011), circle detection (Cuevas et al., 2012b) design of the angular position sensor (Zhang et al., 2013a,b), exploration of isospectral cantilever beams (Dutta et al., 2013), milling process optimization (Wu et al., 2013), intelligent forecasting system (Wu et al., 2014), multilevel thresholding in image segmentation (Oliva et al., 2014), tool path planning (Kuo et al., 2015), prediction of diabetes mellitus (Wang et al., 2015), parameter extraction of the photovoltaic module model (Muhsen et al., 2015), and wireless sensor networks (Özdag and Karcı, 2016).

4.5   CONCLUSIONS

Though a very recent addition to the meta-heuristic family, EMO has received considerable attention, and it has been successfully applied to various real-world problems. EMO nevertheless appears to hold considerable promise and has already attracted significant interest from the research community which is evident from a good number of improved, modified and hybrid versions of EMO and applications reported in the literature. This will help EMO's further development and hopefully will involve both empirical algorithmic improvements and theoretical refinements.

A recent trend has been the hybridization of EMO with other algorithms, in particular nature-inspired approaches such as GAs, SA, TS, and scatter search. There has been little theoretical analysis done on different issues of convergence analysis. In addition, premature avoidance, estimation of the convergence rate, searching behaviors explanation, accelerating convergence, and parameter selection are important issues to be addressed.

REFERENCES

Abdullah, S. and Burke, E. K. 2006. A multi-start large neighbourhood search approach with local search methods for examination timetabling, *The International Conference on Automated Planning and Scheduling (ICAPS 2006)*, June 6–10, 2006, Cumbria, UK, pp. 334–337.

Abdullah, S., Turabieh, H., and McCollum, B. 2009. A hybridization of electromagnetic-like mechanism and great deluge for examination timetabling problems, *Lecture Notes in Computer Science*, 5818, 60–72.

Ali, M. M. and Golalikhani, M. 2010. An electromagnetism-like method for nonlinearly constrained global optimization, *Computers and Mathematics with Applications*, 60, 2279–2285.

Alikhani, M. G., Javadian, N., and Tavakkoli-Moghaddam, R. 2009. A novel hybrid approach combining electromagnetism-like method with Solis and Wets local search for continuous optimization problems, *Journal of Global Optimization*, 44(2), 227–234.

Arnoldi, W. E. 1951. The principle of minimised iterations in the solution of the matrix eigenvalue problem, *Quarterly Applied Mathematics*, 9, 17–29.

Bean, J. C. 1994. Genetic algorithms and random keys for sequencing and optimisation, *ORSA Journal of Computing*, 6(2), 154–160.

Birbil, S. I. and Fang, S. C. 2003. An electromagnetism-like mechanism for global optimization, *Journal of Global Optimization*, 25(3), 263–282.

Birbil, S. I., Fang, G., and Sheu, R.-L. 2004. On the convergence of a population-based global optimization algorithm, *Journal of Global Optimization*, 30, 301–318.

Bonyadi, M. R., Azghadi, M. R., and Shah-Hossaini, H. 2008. Population-based optimisation algorithms for solving travelling salesman problem, in: F. Greco, ed., *Travelling Salesman Problem, Chapter 1*, Vienna, Austria: In-Tech, pp. 1–34.

Bonyadi, M. R. and Li, X. 2012. A new discrete electromagnetism-based meta-heuristic for solving the multidimensional knapsack problem using genetic operators, *Operational Research*, 12(2), 229–252.

Chang, C.-C., Chen, C.-Y., Fan, C.-W., Chao, H.-C., and Chou, Y.-H. 2010. Quantum-inspired electromagnetism-like mechanism for solving 0/1 knapsack problem, *Second IEEE International Conference on Information Technology Convergence and Services (ITCS)*, August 11–13, 2010, Cebu, Philippines, pp. 1–6.

Chang, P.-C., Chen, S.-H., and Fan, C.-Y. 2009. A hybrid electromagnetism-like algorithm for single machine scheduling problem, *Expert Systems with Applications*, 36(3), 1259–1267.

Chao, C.-W. and Liao, C.-J. 2012. A discrete electromagnetism-like mechanism for single machine total weighted tardiness problem with sequence-dependent setup times, *Applied Soft Computing*, 12, 3079–3087.

Chen, S.-H., Chang, P.-C., Chan, C.-L., and Mani, V. 2007. A hybrid electromagnetism-like algorithm for the single machine scheduling problem, *Lecture notes in Computer Sciences*, 4682, 543–552.

Chong, E. K. P. and Zak, S. H. 2008. *An Introduction to Optimisation*, 3rd edn. New York: John Wiley & Sons.

Chou, Y.-H., Chang, C.-C., Chiu, C.-H., Lin, F.-J., Yang, Y.-J., and Peng, Z.-V. 2010. Classical and quantum-inspired electromagnetism-like mechanism for solving 0/1 knapsack problems, *IEEE International Conference on Systems, Man and Cybernetics*, October 10–13, 2010, Istanbul, Turkey, pp. 3211–3218.

Coulomb, C. A. (ed.). 1785a. Premier mémoire sur l'électricité et le magnétisme, in: *Histoire de l'Académie Royale des Sciences*, Pairs: Imprimerie Royale, pp. 569–577.

Coulomb, C. A. (ed.). 1785b. Second mémoire sur l'électricité et le magnétisme, in: *Histoire de l'Académie Royale des Sciences*, Pairs: Imprimerie Royale, pp. 578–611.

Cowan, E. W. 1968. *Basic Electromagnetism*, New York: Academic Press.

Cuevas, E., Oliva, D., Zaldivar, D., Perez-Cisneros, M., and Pajares, G. 2012a. Opposition-based electromagnetism-like for global optimisation, *International Journal of Innovative Computing, Information and Control*, 12, 8181–8198.

Cuevas, E., Oliva, D., Zaldivar, D., Perez-Cisneros, M., and Sossa, H. 2012b. Circle detection using electromagnetism optimization, *Information Sciences*, 182(1), 40–55.

Davoudpour, H. and Molana, M. 2008. Solving flow shop sequencing problem for deteriorating jobs by using electro magnetic algorithm, *Journal of Applied Science*, 8, 4121–4128.

Debels, D. De Reyck, B., Leus, R., and Vanhoucke, M. 2006. A hybrid scatter search/electromagnetism metaheuristic for project scheduling, *European Journal of Operational Research*, 169, 638–653.

Debels, D. and Vanhoucke, M. 2004. The electromagnetism meta-heuristic applied to the resource-constrained project scheduling problem, *Lecture Notes in Computer Science*, 3871, 259–270.

Dueck, G. 1990. New optimization heuristics: The great deluge algorithm and the record-to-record travel, Technical Report, IBM Germany, Heidelberg Scientific Center, 1990.

Dueck, G. 1993. New optimization heuristics: The great deluge algorithm and the record-to-record travel, *Journal of Computational Physics*, 104(1), 86–92.

Dutta, R., Ganguli, R., and Mani, V. 2013. Exploring isospectral cantilever beams using electromagnetism inspired optimization technique, *Swarm and Evolutionary Computation*, 9, 37–46.

Filipovic, V., Kartelj, A., and Matíc, D. 2013. An electromagnetism metaheuristic for solving the maximum betweenness problem, *Applied Soft Computing*, 13, 1303–1313.

Gilak, E. and Rashidi, H. 2009. A new hybrid electromechanism algorithm for job shop scheduling, *Third UKSim European Symposium on Computer Modelling and Simulation*, David Al-Dabass, Sokratis K. Katsikas, Ioannis Koukos, Richard N. Zobel: EMS 2009, November 25–27, 2009, Athens, Greece, pp. 327–332.

Glover, F. 1977. Heuristics for integer programming using surrogate constraints, *Decision Sciences*, 8(1), 156–166.

Glover, F. 1989. Tabu search – Part I, *ORSA Journal on Computing*, 1(3), 190–206.

Glover, F. and Laguna, M. 1997. *Tabu Search*, Boston: Kluwer Academic Publishers.

Glover, F., Løkketangen, A., and Woodruff, D. L. 2000. Scatter search to generate diverse MIP solutions, in: M. Laguna and J. L. González-Velarde, eds., *OR Computing Tools for Modeling, Optimization and Simulation: Interfaces in Computer Science and Operations Research*, Kluwer Academic, Publishers, 2000, pp. 299–317.

Guan, X., Dai, X., and Li, J. 2011. Revised electromagnetism-like mechanism for flow path design of unidirectional AGV systems, *International Journal of Production Research*, 49(2), 401–429.

Guan, X., Dai, X., Qiu, B., and Li, J. 2012. A revised electromagnetism-like mechanism for layout design of reconfigurable manufacturing system, *Computers and Industrial Engineering*, 63, 98–108.

Han, L. and Han, Z. 2010. Electromagnetism-like method for constrained optimisation problems, *2010 International Conference on Measuring Technology and Mechatronics Automation*, IEEE Computer Society, pp. 87–90.

Ho, W. and Ji, P. 2003. Component scheduling for chip shooter machines: A hybrid genetic algorithm approach, *Computers and Operations Research*, 30, 2175–2189.

Huyer, W. and Neumaier, A. 1999. Global optimisation by multilevel coordinate search, *Journal of Global Optimization*, 14, 331–355.

Jamili, A., Shafia, M. A., and Tavakkoli-Moghaddam, R. 2011. A hybridization of simulated annealing and electromagnetism-like mechanism for a periodic job shop scheduling problem, *Expert Systems with Applications*, 38(5), 5895–5901.

Javadian, N., Alikhani, M., and Tavakkoli-Moghaddam, R. 2008. A discrete binary version of the electromagnetism-like heuristic for solving travelling salesman problem, *Lecture Notes in Computer Science*, 5227, 123–130.

Javadian, N., Golalikhani, M., and Tavakkoli-Moghaddam, R. 2009. Solving a single machine scheduling problem by a discrete version of electromagnetism-like methods, *Journal of Circuits, Systems, and Computers*, 18(8), 1597–1608.

Jhang, J.-Y. and Lee, K.-C. 2009. Array pattern optimization using electromagnetism-like algorithm, *International Journal of Electronics and Communications*, 63, 491–496.

Jiekang, W., Zhuangzhi, G., and Fan, W. 2014. Short-term multi-objective optimisation scheduling for cascaded hydroelectric plants with dynamic generation flow limit based on EMA and DEA, *Electrical Power and Energy Systems*, 57, 189–197.

Jolai, F., Tavakkoli-Moghaddam, R., Golmohammadi, A., and Javadi, B. 2012. An electromagnetism-like algorithm for cell formation and layout problem, *Expert Systems with Applications*, 39, 2172–2182.

Kennedy, J. and Eberhart, R. 1995. Particle swarm optimization, *Proceedings of the IEEE International Conference on Neural Networks*, November 27–December 1, 1995, Perth, WA, pp. 1942–1948.

Kennedy, J. and Eberhart, R. 2001. *Swarm Intelligence*, San Francisco, CA: Morgan Kaufmann Publishers, Inc.

Khalili, M. and Tavakkoli-Moghaddam, R. 2012. A multi-objective electromagnetism algorithm for a bi-objective flow shop scheduling problem, *Journal of Manufacturing Systems*, 31, 232–239.

Kratica, J. 2013. An electromagnetism-like metaheuristic for the uncapacitated multiple allocation p-hub median problem, *Computers & Industrial Engineering*, 66, 1015–1024.

Kuo, C.-L., Chu, C.-H., Li, Y., Li, X., and Gao, L. 2015. Electromagnetism-like algorithms for optimized tool path planning in 5-axis flank machining, *Computers & Industrial Engineering*, 84(C), 70–78.

Lee, C. H. and Chang, F. K. 2008. Recurrent fuzzy neural controller design for nonlinear systems using electromagnetism-like algorithm, *Far East Journal of Experimental and Theoretical Artificial Intelligence*, 1(1), 5–22.

Lee, C. H. and Chang, F. K. 2010. Fractional-order PID controller optimization via improved electromagnetism-like algorithm, *Expert Systems with Applications*, 37(12), 8871–8878.

Lee, C. H., Chang, F. K., and Chen, C. W. 2007. A modified electromagnetism-like algorithm for training neural fuzzy systems in control applications, *CACS International Automatic Control Conference*, November 2007, Taichung, Taiwan.

Lee, C.-H., Chang, F.-K., and Lee, Y.-C. 2010. An improved electromagnetism-like algorithm for recurrent neural fuzzy controller design, *International Journal of Fuzzy Systems*, 12(4), 280–290.

Lee, C. H., Kuo, C., Chang, H., Chien, J., and Chang, F. 2009. A hybrid algorithm of electromagnetism-like and genetic for recurrent neural fuzzy controller design, *Proceedings of the International Multi Conference of Engineers and Computer Scientists*, Hong Kong, March 18–20, 2009, pp. 224–229.

Lee, C. H. and Lee, Y. C. 2012. Nonlinear systems design by a novel fuzzy neural system via hybridization of electromagnetism-like mechanism and particle swarm optimisation algorithms, *Information Sciences*, 186(1), 2012, 59–72.

Lee, C.-H., Li, C.-T., and Chang, F. Y. 2011. A species based improved electromagnetism like mechanism algorithm for TSK-type interval-valued neural fuzzy system optimization, *Fuzzy Sets and Systems*, 171, 22–43.

Lee, K. C. and Jhang, J. Y. 2008. Application of electromagnetism-like algorithm to phase-only syntheses of antenna arrays, *Progress in Electromagnetics Research*, 83, 279–291.

Lin, J.-L., Wu, C.-H., and Chung, H.-Y. 2012. Performance comparison of electromagnetism-like algorithm for global optimisation, *Applied Mathematics*, 3, 1265–1275.

Liu, H. and Gao, L. 2010. A discrete electromagnetism-like mechanism algorithm for solving distributed permutation flow scheduling problem, *Proceedings of the 2010 International Conference on Manufacturing Automation*, ICMA '10, December 13–15, 2010, Hong Kong, pp. 156–163.

Maenhout, B. and Vanhoucke, M. 2007. An electromagnetic metaheuristic for the nurse scheduling problem, *Journal of Heuristics*, 13, 359–385.

Mirabi, M., Fatemi Ghomi, S., Jolai, F., and Zandieh, M. 2008. Hybrid electromagnetism-like algorithm for the flow scheduling with sequence dependent setup times, *Journal of Applied Sciences*, 8, 3621–3629.

Mladenovic, N. and Hansen, P. 1997. Variable neighborhood search, *Computers and Operations Research*, 24(11), 1097–1100.

Muhsen, D. H., Ghazali, A. B., Khatib, T., and Abed, I. A. 2015. Extraction of photovoltaic module model's parameters using an improved hybrid differential evolution/electromagnetism-like algorithm, *Solar Energy*, 119, 286–297.

Naderi, B., Tavakkoli-Moghaddam, R., and Khalili, M. 2010. Electromagnetism-like mechanism and simulated annealing algorithms for flow shop scheduling problems minimizing the total weighted tardiness and make-span, *Knowledge-Based Systems*, 23(2), 77–85.

Naderi, B., Zandieh, M., and Shirazi, M. 2009. Modeling and scheduling a case of flexible flow: Total weighted tardiness minimization, *Computers & Industrial Engineering*, 57, 1258–1267.

Naji-Azimi, Z., Toth, P., and Galli, L. 2010. An electromagnetism metaheuristic for the unicost set covering problem, *European Journal of Operational Research*, 205, 290–300.

O'Kelly, M. 1987. A quadratic integer program for the location of interacting hub facilities, *European Journal of Operational Research*, 32, 393–404.

Oliva, D., Cuevas, E., Pajares, G., Zaldivar, D., and Osuna, V. 2014. A multilevel thresholding algorithm using electromagnetism optimization, *Neurocomputing*, 139, 357–381.

Özdag, R. and Karcı, A. 2016. Probabilistic dynamic distribution of wireless sensor networks with improved distribution method based on electromagnetism-like algorithm, *Measurement*, 79, 66–76.

Press, W. H., Flannery, B. P., Teukolsky, S. A., and Vetterling, W. T. 1992. Reduction of a general matrix to Hessenberg form, Section 11.5, in: William H. Pres, ed., *Numerical Recipes in FORTRAN: The Art of Scientific Computing*, 2nd ed., Cambridge, UK: Cambridge University Press, pp. 476–480.

Rahnamayan, S., Tizhoosh, H. R., and Salama, M. M. A. 2007a. Quasi-oppositional differential evolution, *Proceedings of the IEEE Congress on Evolutionary Computation*, September 2007, pp. 2229–2236.

Rahnamayan, S., Tizhoosh, H. R., and Salama, M. M. A. 2007b. A novel population initialisation method for accelerating evolutionary algorithms, *Computers and Mathematics with Applications*, 53(10), 1605–1614.

Rahnamayan, S., Tizhoosh, H. R., and Salama, M. M. A. 2007c. Oppositional-based differential evolution, *IEEE Transaction on Evolutionary Computation*, 12(1), 64–79.

Rahnamayan, S., Tizhoosh, H. R., and Salama, M. M. A. 2008. Opposition versus randomness in soft computing techniques, *Applied Soft Computing*, 8, 906–918.

Rocha, A. M. A. C. and Fernandes, E. M. G. P. 2008. Feasibility and dominance rules in the electromagnetism-like algorithm for constrained global optimization, *Lecture Notes in Computer Science*, 5071, 768–783.

Roshanaei, V., Balagh, A., Esfahani, M., and Vahdani, B. 2009. A mixed integer linear programming model along with an electromagnetism-like algorithm for scheduling job shop production system with sequence-dependent set-up times, *International Journal of Advanced Manufacturing Technology*, 47, 783–793.

Santos, C., Oliveira, M., Matos, V., Maria, A., Rocha, A. C., and Costa, L. A. 2009. Combining central pattern generators with the electromagnetism-like algorithm for head motion stabilization during quadruped robot locomotion, in: *Proceedings of the 2nd International Workshop on Evolutionary and Reinforcement Learning for Autonomous Robot Systems*, Missouri, USA, October 15, 2009, pp. 1–9.

Sels, V. and Vanhoucke, M. 2014. A hybrid electromagnetism-like mechanism/tabu search procedure for the single machine scheduling problem with a maximum lateness objective, *Computers and Industrial Engineering*, 67, 44–55.

Shokri, M., Tizhoosh, H. R., and Kamel, M. 2006. Opposition-based Q(λ) algorithm, *Proceedings of IEEE International Joint Conference on Neural Networks part of IEEE World Congress on Computational Intelligence*, Vancouver, Canada, pp. 646–653.

Siddique, N. and Adeli, H. 2013. *Computational Intelligence: Synergies of Fuzzy Logic, Neural Networks and Evolutionary Computing*, Chichester, UK: John Wiley and Sons.

Slowinski, R. 1981. Multi-objective network scheduling with efficient use of renewable and non-renewable resources, *European Journal of Operation Research*, 7(3), 265–273.

Solis, F. J. and Wets, R. J. B. 1981. Minimisation by random search techniques, *Mathematics of Operations Research*, 6(1), 19–30.

Storn, R. 1999. System design by constraint adaptation and differential evolution, *IEEE Transaction on Evolutionary Computation*, 3(1), 22–34.

Storn, R. and Price, K. V. 1997. Differential evolution – A simple and efficient heuristic for global optimization over continuous spaces, *Journal of Global Optimization*, 11(4), 341–359.

Su, C.-T. and Lin, H.-C. 2011. Applying electromagnetism-like mechanism for feature selection, *Information Sciences*, 181(5), 972–986.

Taheri, S. H., Ghazvini, H., Saberi-Nadjafi, J., and Biazar, J. 2007. A hybrid of the restarted Arnoldi and electromagnetism meta-heuristic methods for calculating eigenvalues and eigenvectors of a non-symmetric matrix, *Applied Mathematics and Computation*, 191, 79–88.

Tavakkoli-Moghaddam, R., Khalili, M., and Nasiri, M. 2009. A hybridization of simulated annealing and electromagnetic-like mechanism for job shop problems with machine availability and sequence-dependent setup times to minimize total weighted tardiness, *Soft Computing*, 13, 995–1006.

Teimouri, M., Zaretalab, A., Niaki, S. T. A., and Sharifi, M. 2016. An efficient memory-based electromagnetism-like mechanism for the redundancy allocation problem, *Applied Soft Computing*, 38, 423–436.

Tizhoosh, H. R. 2005. Opposition-based learning: A new scheme for machine intelligence, *Proceedings of International Conference on Computational Intelligence for Modelling Control and Automation, CIMCA'2005*, Vienna, Austria, Vol. I, pp. 695–701.

Tizhoosh, H. R. 2006. Opposition-based reinforcement learning, *Journal of Advanced Computational Intelligence and Intelligent Informatics*, 10(5), 578–585.

Tsai, C., Hung, H., and Lee, S. 2010. Electromagnetism-like method based blind multi-user detection for MC-CDMA interference suppression over multi-path fading channel, *2010 International Symposium on Computer, Communication, Control and Automation (3CA '10)*, May 5–7, 2010, Tainan, China, pp. 470–475.

Tsou, C.-S. and Kao, C.-H. 2006. An electromagnetism-like meta-heuristic for multi-objective optimization, *Proceedings of the IEEE Congress on Evolutionary Computation*, IEEE, July 16–21, 2006, Vancouver, BC, Canada, pp. 1172–1178.

Tsou, C.-S. and Kao, C.-H. 2008. Multi-objective inventory control using electromagnetism-like meta-heuristic, *International Journal of Production Research*, 46(14), 3859–3874.

Turabieh, H., Abdullah, S., and McCollum, B. (eds.). 2009. Electromagnetism-like mechanism with force decay rate great deluge for the course timetabling problem, in: *Rough Sets and Knowledge Technology*, Berlin, Germany: Springer, pp. 497–504.

Turky, A. M. and Abdullah, S. 2014. A multi-population electromagnetic algorithm for dynamic optimisation problems, *Applied Soft Computing*, 22, 474–482.

Wang, K.-J., Adrian, A. M., Chen, K.-H., and Wang, K.-M. 2015. An improved electromagnetism-like mechanism algorithm and its application to the prediction of diabetes mellitus, *Journal of Biomedical Informatics*, 54, 220–229.

Wang, X., Gao, L., and Zhang, C. 2008. Electromagnetism-like mechanism based algorithm for neural network training, in: *Advanced Intelligent Computing Theories and Applications with Aspects of Artificial Intelligence*, Berlin, Germany: Springer, pp. 40–45.

Wu, P. and Chiang, H. 2005. The Application of electromagnetism-like mechanism for solving the traveling salesman problems, *Proceeding of the 2005 Chinese Institute of Industrial Engineers Annual Meeting*, December 2005, Taichung, Taiwan.

Wu, P., Hung, Y.-Y., and Lin, Z.-P. 2014. Intelligent forecasting system based on integration of electromagnetism-like mechanism and fuzzy neural network, *Expert Systems with Applications*, 41, 2660–2677.

Wu, P., Yang, K., and Fang, H. 2006. A revised EM-like algorithm + KOPT method for solving the travelling salesman problem, *1st International Conference on Innovative Computing, Information and Control 2006 (ICICIC '06)*, August 2006, IEEE, Beijing, China. Vol. 1, pp. 546–549.

Wu, P., Yang, K.-J., and Huang, B.-Y. 2007. A revised EM-like mechanism for solving the vehicle routing problem, *2nd International Conference on Innovative Computing, Information and Control 2007 (ICICIC '07)*, pp. 1–4.

Wu, P., Yang, K., and Hung, Y. 2005. The study of electromagnetism-like mechanism based fuzzy neural network for learning fuzzy if-then rules, in: *Knowledge-Based Intelligent Information and Engineering Systems*, Berlin, Germany: Springer, pp. 907–907.

Wu, Q., Gao, L., Li, X., Zhang, C., and Rong, Y. 2013. Applying an electromagnetism-like mechanism algorithm on parameter optimisation of a multi-pass milling process, *International Journal of Production Research*, 51, 1777–1788.

Wu, Q., Zhang, C., Gao, L., and Li, X. 2010. Training neural networks by electromagnetism-like mechanism algorithm for tourism arrivals forecasting, *Proceedings of the IEEE 5th International Conference on Bio-Inspired Computing: Theories and Applications (BIC-TA '10)*, Liverpool, UK, September 8–10, 2010, pp. 679–688.

Xiao, J., Wu, Z., Hong, X.-X., Tang, J.-C., and Tang, Y. 2016. Integration of electromagnetism with multi-objective evolutionary algorithms for RCPSP, *European Journal of Operational Research*, 251, 22–35.

Yin, F., Wang, Y.-N., and Wei, S.-N. 2011. Inverse kinematic solution for robot manipulator based on electro-magnetism-like and modified DFP algorithms, *Acta Automatica Sinica*, 37(1), 74–82.

Yuan, K., Henequin, S., Wang, X. J., and Gao, L. 2006. A new heuristic-EM algorithm for permutation flow-shop scheduling, *Proceedings of the 12th IFAC Symposium on Information Control Problems in Manufacturing (INCOM06)*, May 2006, Saint-Etienne, France, pp. 17–19.

Yurtkuran, A. and Emel, E. 2010. A new hybrid electromagnetism-like algorithm for capacitated vehicle routing problems, *Expert Systems with Applications*, 37(4), 3427–3433.

Zaharie, D. 2002. Critical values for the control parameters of differential evolution algorithms, in: R. Matousek, P. Osmera, eds., *Proceedings of Mendel 2002, 8th International Conference on Soft Computing*, Brno, Czech Republic, June 2002, pp. 62–67.

Zhang, C., Li, X., Gao, L., and Wu, Q. 2013a. An improved electromagnetism-like mechanism algorithm for constrained optimisation, *Expert Systems with Applications*, 40, 5621–5634.

Zhang, Z., Ni, F., Dong, Y., Jin, M., and Liu, H. 2013b. A novel absolute angular position sensor based on electromagnetism, *Sensors and Actuators A: Physical*, 194, 196–203.

5 Harmony Search

5.1 INTRODUCTION

The term harmony is derived from the Greek word harmonía, meaning joint, agreement, or concord. Harmony is based on repeating patterns in the sound waves, and therefore firing in the neurons in our brains. In music, harmony means the effect of simultaneous pitches (tones, notes) or chords. The study of harmony involves chords, their construction, progressions, and the principles of connection that govern them. Harmony is often said to refer to the vertical aspect of music, as distinguished from melodic line, or the horizontal aspect. Some other researchers also consider harmony as the base of consonance, a concept whose definition has changed various times during the history of Western music. In a psychological approach, consonance is a continuous variable that can vary across a wide range. A chord may sound consonant for various reasons. It is important to remember that much of the progress made in music has been in dissonance, and how to use it in a way that is still a pleasing sound to us despite its neurological disadvantage.

Harmony is also seen in the nature in various forms, especially in waveforms of sound. For example, mosquitoes can impress potential mates by harmonizing the high-pitched whine of their tiny wings. Now, scientists have discovered how this musical matchmaking helps the insects to pick their perfect partner (Borrell, 2009). In practice, this broad definition of harmony can also include some instances of notes sounded one after the other. If the consecutively sounded notes call to mind the notes of a familiar chord (a group of notes sounded together), the ear creates its own simultaneity in the same way that the eye perceives movement in a motion picture. In such cases, the ear perceives the harmony that would result if the notes had sounded together. In a narrower sense, harmony refers to the extensively developed system of chords and the rules that allow or forbid relations between chords that characterize Western music.

We see harmony in nature and in the environment around us to which we are used to, and we react in response to any disharmony seen in any natural or artificial system. All we look for is a harmony in all natural and artificial systems comprising different components or individuals acting and interacting together in a group or in a society. Thus, harmony means that each individual element in a group or in a system does their best to contribute to the harmony. Therefore, harmony is seen as an optimal performance or behavior of individuals or elements working together in a group or in a system. Therefore, harmony seeking can be seen as a process of search and optimization.

Based on the notion of harmonic phenomena in musical performance, a new meta-heuristic algorithm was derived called harmony search (HS). HS is a population-based search and optimization algorithm, which was developed by Z.W. Geem in 2001 (Geem, 2010a,b; Geem et al., 2001). In computer science and operations research, HS is a phenomenon-mimicking algorithm inspired by the improvisation process of musicians. The concept behind the algorithm is to find a perfect state of harmony of different elements or variables working together within a system where the state of harmony is determined by an aesthetic estimation or aesthetic quality widely used in Western music. In musical performances, a pleasing harmony (a perfect state) is always sought. Likewise in optimization, a perfect state of combination of different variables is sought, which leads to a feasible solution of a problem. The feasibility or quality of the solution is determined by an objective function. The pitch of each musical instrument determines the aesthetic quality, just as the objective function value is determined by the set of values assigned to each decision variable. The HS metaheuristic is an emerging optimization algorithm inspired by the principles of music improvisation.

In music improvisation, each musician plays sounds of any pitch within the possible range, together making one harmony vector. If all the pitches make a good harmony, they save the experience and build up their memory of good harmonies. The process continues by testing various pitch combinations stored in their memories, and the possibility of improving harmony is increased in the following trials. The process of searching for optimal solutions for any problems is analogous to this efficient search technique where a perfect state of harmony of variables is sought (Lee and Geem, 2005). In engineering optimization, each decision variable initially chooses any value within the possible range, together making one solution vector. If the combination of all the values of decision variables makes a good solution, the values of the decision variables pertaining to that solution are stored in memory. By this process, the possibility of finding a good solution is also increased in the next iteration.

The HS has several advantageous features compared to other meta-heuristic algorithms. The HSA demands less mathematical requirements and iteratively generates new solutions (Coelho and Bernert, 2009; Mahdavi et al., 2007; Saka, 2007). Programming HSA is simple compared to other meta-heuristic algorithms (Ayvaz, 2009a,b). Furthermore, instead of a gradient search, the HSA uses a stochastic random search that is based on the harmony memory (HM), HM consideration rate, and the pitch adjustment so that derivative information is unnecessary. It uses a stochastic mechanism that reduces the number of iterations (Geem, 2008). In the HSA, each musician plays a note, and all musicians try to find a pleasing harmony collectively. The features of HS as a meta-heuristic algorithm can be summarized as follows:

- HS does not require differential gradient information, and therefore, it can be applied to discontinuous functions as well as continuous functions.
- HS can handle discrete variables as well as continuous variables.
- HS can avoid the drawback of building block theory used in other EAs.
- HS has a novel stochastic mechanism applied to discrete variables, which uses musicians' experiences as a searching direction.

This chapter will investigate the harmonic phenomena and improvisations in music that are being applied by musicians. This chapter will also investigate the memory mechanism the musicians apply for harmony and its application to HS. This chapter will finally present HSA and different variants of HSA as well as applications of the HSA to engineering problems.

5.2   HARMONY IN MUSIC

The aim of any musical performance is to discover some kind of pleasing harmony, which is considered as a perfect state of combination of different musical instruments determined by an aesthetic quality. It is the same process that an optimization looks for a perfect combination of values of different decision variables determined by an optimality criterion leading to a global optimal solution. The estimation of aesthetic quality of a musical performance is very subjective and leads to the difficulty of being exactly defined by a set of mathematical equations or any kind of quantitative description. Practically, the aesthetic quality of a musical instrument is essentially determined by its pitch (or frequency), timbre (or sound quality), and amplitude (or loudness). Timbre is largely determined by the harmonic content that is in turn determined by the waveforms or modulations of the sound signal. However, the harmonics that it can generate will largely depend on the pitch or frequency range of the particular instrument (Lee and Geem, 2005).

There have been efforts put forth by many theorists for estimation of musical aesthetic quality. The Greek philosopher and mathematician Pythagoras (582–497 BC) is credited for being the first to relate musical harmony to mathematics by working out the frequency ratios (or string length ratio with equal tension). He noted that pleasing intervals correspond to simple frequency ratios and formulated the mathematical relationship between different notes and showed which notes sound pleasant together (Lee and Geem, 2005; Yang, 2009). The octave was found at 1:2 ratio. Galileo (1564–1642) echoed this idea when he said that these proportionate ratios do not keep the eardrum in perpetual torment (Ball, 2008). The French composer Leonin (1135–1201) is the first known composer of polyphonic "organum" involving a simple doubling of the chant at an interval of a fifth or fourth above or below. The French composer Jean-Philippe Rameau (1683–1764) developed the classical harmony theories in his book *Treatise on Harmony*.

Harmony occurs in music when two pitches vibrate at frequencies in small integer ratios. Different notes have different frequencies. For example, the note A above middle C (or standard concert A4) has a fundamental frequency (f_0) of 440 Hz. The speed of sound (v) in dry air is about $331 + 0.6T$ m/s, where T is the temperature. At room temperature, say $T = 20°C$, the A4 note has a wavelength $l = v/f_0 \gg 0.7795$ m. Adjusting the pitch means trying to change the frequency. In music theory, pitch P_n in musical instrument digital interface (MIDI) is often represented as a numerical scale (a linear pitch space) using the following formula (Yang, 2009):

$$P_n = 69 + 12\log_2\left(\frac{f}{f_0}\right) \tag{5.1}$$

or

$$f = f_0 \times 2^{(P_n - 69)/12} \tag{5.2}$$

Equations 5.1 and 5.2 mean that the A4 notes have a pitch number of 69. On this scale, octaves correspond to size 12 (denominator in Equation 5.2), while semitone corresponds to size 1, which leads to the fact that the ratio of frequencies of two notes that are an octave apart is 2:1 (i.e., the ratio of f_0 to f). Thus, the frequency of a note is doubled (halved) when it is raised (lowered) an octave. For example, note A2 has a frequency of 110 Hz, A3 has a frequency of 220 Hz, A4 has a frequency of 440 Hz, and A5 has a frequency of 880 Hz.

The measurement of harmony where different pitches occur simultaneously, like any aesthetic quality, is subjective to some extent. However, it is possible to use some standard estimation for stylistic regularities identifiable in music (Krumhansl, 1990). Parncutt (1989) gave an excellent explanation to the theory of harmony in Western music by taking a psychoacoustical approach and defined harmony in terms of quantitative relationships between sound (acoustics) and sensation (psychology). The pitch properties of tones, chords, and chord sequences are calculated on the assumption that the human auditory system is familiar with the pitch pattern of the audible harmonics of complex tones. The frequency ratio, pioneered by the ancient Greek mathematician Pythagoras, is also a good way for such estimation. For example, the octaves with a ratio of 1:2 sounds pleasant when played together, so are the notes with a ratio of 2:3. However, it is unlikely for playing any random notes to produce a pleasant harmony.

5.3   MUSICAL IMPROVISATION

It is obvious that improvisation in music requires a greater creative effort on the part of the performer than does composition of music. Indeed, improvisation is measured by the degree to which the performer is creatively involved during the performance (Nettl, 1974). Traditionally, improvisation also combines performance with communication of emotions and instrumental technique as well as spontaneous response to other comusicians. This is also known as musical extemporization in musical studies. The notion of improvisation mainly involves chord changing in classical music and indeed in some genres of Western music. One definition is a performance given extempore without planning or preparation. This means to play or sing (music) extemporaneously, especially by devising variations on a melody or creating new melodies in accordance with a set progression of chords. Encyclopedia of Britannica defines it as the extemporaneous composition or free performance of a musical passage. Thus, musical improvisation is seen as a technique of creating a melodious piece of music.

Musical improvisation can be seen as a search strategy for finding a pleasing or optimal harmony where individual instruments or music players try out various combinations of chords or pitches of their instrument to go with other instruments within a musical context (Al-Betar et al., 2012). This musical improvisation is made up by adjusting the pitches of musical instruments from repeated practice trials and evaluated by their own musical aesthetic standard. In general, there are three options open to a musician to improvise a pitch from his musical instrument:

1. Improvising a pitch from his memory
2. Improvising an adjacent pitch of one from his memory
3. Improvising a pitch from a possible pitch range

Let us consider a Jazz bandstand consisting of five musicians playing guitar, trumpet, drum, saxophone, and double bass. The five musicians playing the instruments with the following sets of pitches from their choices randomly stored in their memory:

Guitarist = {Do, Mi, Sol},
Saxophonist = {Fa, Do, La},
Trumpeter = {La, Si, Sol},
Drummer = {Re, Sol, Si},
Double bassist = {Re, Sol, Mi}

Suppose in a practice session the Guitarist randomly improvises {Do} from his memory, Saxophonist improvises {La} from his memory, Trumpeter improvises {Sol} from his memory, Drummer adjusts {Re} from his memory to {Fa}, and Double bassist improvises {Si} from the available range {Do, Re, Mi, Fa, Sol, Si}. All these pitches together create a new harmony of (Do, La, Sol, Fa, Si), which is estimated by an audio-aesthetic standard to be, for example, sounding pleasing to the ear. The new harmony substitutes the worst harmony stored in the HM if it is better. This process is repeated until a pleasing harmony is considered to have been reached. The process of musical improvisation is shown in Figure 5.1.

The process of musical improvisation described above can be applied to an optimization process in a very similar way. Let us consider a function $f(x)$ with a set of decision variables $\{x_1, x_2,..., x_5\}$ for an optimization problem. The variables $\{x_1, x_2, ... , x_5\}$ are assigned with initial values. An optimal solution can be reached by repeatedly modifying the variables subject to evaluation of an objective function. In the process of optimization, the value of each decision variable is determined according to one of the following options: (i) assigning a value stored in memory, (ii) modifying an existing value in memory, or (iii) assigning a value from its feasible range. These three options are formalized into three operators: memory consideration, pitch adjustment, and random consideration.

In terms of optimization, let us consider the function $f(x) = f(x_1, x_2, x_3, x_4, x_5)$. Each of the five decision variables has stored experience values in the HM as follows:

x_1: {101,302,504}
x_2: {200,300,690}
x_3: {100,90,600}
x_4: {100,200,300}
x_5: {0,250,300}

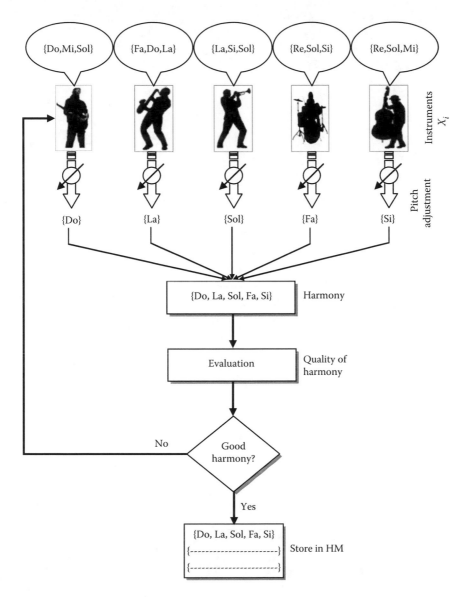

FIGURE 5.1   Process of musical improvisation.

Suppose in an iteration x_1 is assigned 101 from its memory, x_2 is assigned 690 from its memory, x_3 is adjusted from the value 90 stored in its memory to 120, x_4 is assigned 200 from its memory, and x_5 is assigned 320 from its feasible range $x_5 \in [0,600]$. The new constructed solution $f(x^*) = \{101,690,120,200,320\}$ is evaluated by an objective function suitable for the function $f(x)$. If the new solution $f(x^*)$ is better than the worst solution stored in the HM, it is stored in HM, replacing the worst solution. The process of improvisation is performed repeatedly until an optimal solution is reached. Figure 5.2 shows the optimization process inspired from the musical improvisation process where a_i, b_i, and c_i, $i = 1, 2, 3$, represent the values of the decision variables.

When musicians make up a harmony, they usually test various pitch combinations stored in their memories. In an engineering optimization problem, decision variables are assigned values from the full range of values and sometimes modified, together making one solution. If the set of values of decision variables makes a good solution based on the evaluation of objective function, it is stored in each variable's memory. The process of searching for optimal solutions to engineering problems

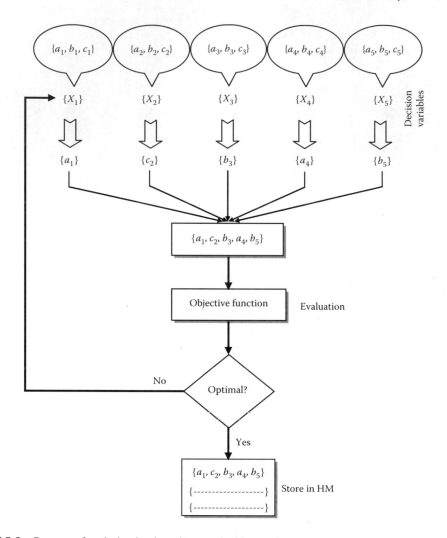

FIGURE 5.2 Process of optimization based on musical improvisation.

is analogous to this efficient search for a harmony (Xu et al., 2010). Table 5.1 shows equivalence between optimization and HS in musical terms.

5.4   HARMONY MEMORY

In the improvisation process discussed in Section 5.3, a musician has three possible choices:

1. Play a piece of music (a series of pitches in harmony) exactly from his memory
2. Play something similar to a known piece (thus adjusting the pitch slightly)
3. Compose new or random notes

These three options have been formalized into a quantitative optimization process by Geem et al. (2001) using three corresponding mechanisms:

1. Choosing a value from HM, which is defined as memory consideration
2. Select a nearby value of a chosen value from HM, which is defined as pitch adjusting
3. Choosing completely a random value from within the range of the pitch, which is defined as randomization

TABLE 5.1

HSA Terminologies

In Music		In Optimization
Instrument	⟺	Decision variable
Pitch	⟺	Value
Pitch range	⟺	Range of values
Improvisation	⟺	Generation/construction
Harmony	⟺	Solution vector
Aesthetic quality	⟺	Objective function
Trials	⟺	Iterations
Pleasing harmony	⟺	Optimal solution

HM is a matrix that stores in its rows a selection of the current best harmonies (or solutions), which is the principal HS data structure. The HM matrix contains as many randomly generated solution vectors as the harmony memory size (HMS), and the solution vector consists of N decision variables. The HM matrix has a dimension of HMS $\times N$. The HM matrix is constructed from the randomly generated decision variables x_i^j using the following equation:

$$x_i^j = L_i + r \cdot (U_i - L_i) \tag{5.3}$$

where $r \in [0,1]$ is a random number, $i = 1, \ldots, N$ and $j = 1, \ldots$, HMS.

$$HM = \begin{bmatrix} x_1^1 & x_2^1 & \cdots & x_{N-1}^1 & x_N^1 \\ x_1^2 & x_2^2 & \cdots & x_{N-1}^2 & x_N^2 \\ \vdots & \vdots & \ddots & \vdots & \vdots \\ x_1^{HMS-1} & x_2^{HMS-1} & \cdots & x_{N-1}^{HMS-1} & x_N^{HMS-1} \\ x_1^{HMS} & x_2^{HMS} & \cdots & x_{N-1}^{HMS} & x_N^{HMS} \end{bmatrix} \tag{5.4}$$

The process of producing a new harmony is called improvisation. The new harmony vector $x' = [x'_1, x'_2, \ldots, x'_N]$ is improvised from the HM matrix HM by applying the three options mentioned above. These options are implemented using explicitly two parameters for the HSA:

1. Harmony memory consideration rate (HMCR)
2. Pitch adjustment rate (PAR)

The usage of HM will ensure that the best harmonies will be carried over to the new HM. In order to use the HM effectively, it is typically assigned as a parameter called the HMCR, usually HMCR $\in [0,1]$. The decision variable x'_i is selected from the vectors of variables $x_i^j, j = 1, 2, \ldots$, HMS existing in the HM with a probability of the HMCR, that is

$$x'_i = \begin{cases} x_i \in \left\{ x_i^1, x_i^2, \ldots, x_i^{HMS} \right\} & \text{with probability HMCR} \\ x_i \in X_i & \text{with probability } (1 - HMCR) \end{cases} \tag{5.5}$$

Here, X_i is the decision space of all decision variables. $(1 - HMCR)$ is the probability of randomly selecting a fresh value from possible range of X_i. If the HMCR is too low, only a few best harmonies are selected and the HS will progress too slowly. If the HMCR is extremely high (near 1), almost all the harmonies are used in the HM, then other harmonies are not explored well, leading to potentially poor solutions (Das et al., 2011; Geem et al., 2001; Lee and Geem, 2005; Yang, 2009). Therefore, the HMCR should typically be chosen within the range of HMCR $\in [0.7, 0.95]$ (Ricart et al., 2011; Yang, 2009).

Every component obtained from the HM is examined whether it should be pitch-adjusted or not. This operation uses the parameter PAR. The PAR is the pitch adjustment determined by an arbitrary pitch bandwidth (*bw*). Though in music, pitch adjustment means to change the frequencies, it corresponds to generate a slightly different solution in the HSA (Geem et al., 2001). Theoretically, the pitch can be adjusted linearly or nonlinearly, but in practice, linear adjustment is used, which is defined as

$$x_i' = \begin{cases} x_i + \text{rand}(-1,1) * bw & \text{with probability PAR} \\ x_i & \text{with probability } (1 - \text{PAR}) \end{cases} \qquad (5.6)$$

where x_i is the existing pitch or solution from the HM, and x_i' is the new pitch after the pitch-adjusting action. This essentially produces a new solution around the existing solution by varying the pitch slightly by a small uniformly distributed random value rand(−1,1) (Das et al., 2011; Geem et al., 2001; Lee and Geem, 2005; Yang, 2009). The parameter PAR is used to control the degree of the adjustment. A low value of the PAR with a narrow *bw* can slow down the convergence of HS because of the limitation in the exploration of only a small subspace of the whole search space. On the other hand, a very high value of the PAR with a wide *bw* may cause the solution to scatter around some potential optima as in a random search. Thus, most applications use the PAR within the range of [0.1, 0.5] (Ricart et al., 2011; Yang, 2009).

The third component of the improvisation process is randomization, which generates a completely random new pitch within the range of the pitch and also increases the diversity of the solutions. Although adjusting pitch has a similar role, it is limited to certain local pitch adjustments and thus corresponds to a local search. The use of randomization can drive the system further to explore various diverse solutions so as to find the global optimality. Geem (2012) did extensive experiments with the initial HM and found out that generating more initial harmonies in the HM did not affect the solution quality, but limiting the number of identical harmonies in the HM enhances the solution quality.

5.5 HARMONY SEARCH ALGORITHM

By implementing the three components *memory consideration*, *pitch adjustment*, and *randomness* discussed in the earlier section, Geem et al. (2001) formulated the new meta-heuristic algorithm in 2001, called HSA. In the HSA, an initial population of harmonies are randomly generated and stored in the HM. A new candidate harmony is improvised in each iteration by using memory consideration, pitch adjustment, and random generation. This newly generated harmony is compared with the worst harmony at the current iteration. The worst harmony vector is replaced by the new candidate harmony vector if the generated harmony is better than the worst harmony and the HM is updated. This process is repeated until the predetermined number of improvisations (NI) is reached. The three rules in the HSA are effectively directed using the parameters HMCR, PAR, and *bw*. The HS meta-heuristic optimization algorithm can be described using the five main steps as follows (Das et al., 2011; Geem et al., 2001, 2002; Lee and Geem, 2005):

1. Initializing the optimization problem and algorithm parameters
2. Initializing HM
3. Improvising harmony from the HM
4. Updating HM
5. Stopping condition

5.5.1   Initializing the Optimization Problem and Algorithm Parameters

In this step, the optimization problem is formulated and parameters of HS are initialized. In general, the optimization problem can be specified as follows:

$$\begin{cases} \text{Minimize} & f(x) \\ \text{subject to} & g_i(x) \geq 0, i = 1,2,\ldots,M \\ & h_j(x) = 0, j = 1,2,\ldots,P \\ & x_k^L \leq x_k \leq x_k^U, k = 1,2,\ldots,N \end{cases} \tag{5.7}$$

where $f(x)$ is the objective function, M is the number of inequality constraints, P is the number of equality constraints, x_k is the set of decision variables, and N is the number of decision variables. The lower and upper bounds for each decision variable are x_k^L and x_k^U, respectively. The HSA parameters that are required to solve the optimization problem defined by Equation 5.7 are also specified in this step. The control parameters are as follows: the HMS, the number of solution vectors in the HM, HMCR, PAR, NI, and the stopping criterion. The HM is an array where all the solution vectors (sets of decision variables) are stored. This is discussed in greater detail in Section 5.4. HMCR, PAR, and NI are parameters that are used to improve the solution vector. These parameters are further explained later.

5.5.2   Initializing HM

In this step, the IIM matrix, shown in Equation 5.8, is initialized with randomly generated solution vectors. The individual components of the HM matrix are initialized using Equation 5.3:

$$\text{HM} = \begin{bmatrix} x^1 \\ x^2 \\ \vdots \\ x^{\text{HMS}} \end{bmatrix} = \begin{bmatrix} x_1^1 & x_2^1 & \cdots & x_{N-1}^1 & x_N^1 \\ x_1^2 & x_2^2 & \cdots & x_{N-1}^2 & x_N^2 \\ \vdots & \vdots & \ddots & \vdots & \vdots \\ x_1^{\text{HMS}-1} & x_2^{\text{HMS}-1} & \cdots & x_{N-1}^{\text{HMS}-1} & x_N^{\text{HMS}-1} \\ x_1^{\text{HMS}} & x_2^{\text{HMS}} & \cdots & x_{N-1}^{\text{HMS}} & x_N^{\text{HMS}} \end{bmatrix} \tag{5.8}$$

The $f(x^i)$ values are calculated and sorted by the values of the objective function. The solutions that are not feasible, that is, those that do not satisfy the constraints in Equation 5.7, can have a chance to be included in the HM matrix to enforce the search toward the feasible solution space. Some researchers use a static function for calculating the penalty cost for an unfeasible solution. The total cost for each solution vector is evaluated using the following fitness function:

$$\text{fit}(x) = f(x) + \sum_{i=1}^{M} \alpha_i \times \min[0, g_i(x)]^2 + \sum_{j=1}^{P} \beta_j \times \min[0, h_j(x)]^2 \tag{5.9}$$

where $\text{fit}(x)$ denotes the fitness value of x, $g_i(x)$ and $h_j(x)$ are the penalty functions, and α_i and β_j are the penalty coefficients. There is no specific rule for determining the values of the penalty coefficients. The penalty coefficients α_i and β_j are considered parameters and remain problem dependent.

5.5.3   Improvising Harmony from HM

Generating a new harmony is termed as improvisation (Lee and Geem, 2004). In this step, a new harmony vector $x' = [x'_1, x'_2, \ldots, x'_N]$ is improvised (or constructed) using the vectors stored in the HM using three operators: (i) HM consideration, (ii) random consideration, and (iii) pitch adjustment.

5.5.3.1 Harmony Memory Consideration

The decision variable x'_1 is randomly chosen from the historical values $\{x_1^1, x_1^2, \ldots, x_1^{HMS}\}$ stored in the HM. Values of the other decision variables $\{x'_2, x'_3, \ldots, x'_N\}$ can be chosen in the same way with a probability of HMCR $\in$ [0,1]. The HMCR is the probability of choosing one value from historical values stored in the HM. Recently, Al-Betar et al. (2013b) and Awadallah et al. (2014) reported some other types of memory consideration rates. These memory consideration rates seem to be very promising.

Random memory consideration selection: This is the original selection method introduced by Geem et al. (2001). A vector $x = [x_1, x_2, \ldots, x_N]$ is randomly selected from the HM. In this selection method, each vector in the HM has the same chance (or probability) of being selected.

Global best memory consideration selection: This selection method is derived from PSO introduced by Kennedy and Eberhart (1995). At each iteration, the best vector x^{best} in the HM has the highest probability (i.e., close to 1) and the rest has the probability 0.

Proportional memory consideration selection: This selection method comes from the GA introduced by Holland (1975). At each iteration, the fittest vectors (individuals in the HM) have the higher probability of being selected. The probability of the individual vector is based on the relative fitness defined as

$$p(x^i) = \frac{\text{fit}(x^i)}{\sum_{i=1}^{HMS} \text{fit}(x^i)} \tag{5.10}$$

where fit(x^i) is the fitness value of x^i and $p(x^i)$ is the probability of an individual being selected, which is directly proportional to the relative fitness value of that individual.

This may cause an individual to dominate the production of new harmony vectors, thereby limiting diversity in the new HM. In the *roulette wheel* method, the relative fitness values are calculated (or fitness values are normalized by dividing each fitness value by the maximum fitness value or the sum of fitness values). The probability distribution can then be thought of as a roulette wheel, where each slice has a width corresponding to the selection probability of an individual. Selection can then be visualized as the spinning of the wheel. A detailed description of proportional selection can be found in Siddique and Adeli (2013).

Tournament memory consideration selection: This is tournament selection method introduced by Goldberg (1989) used in the GA. In tournament selection, k number of individuals are randomly chosen from the HMS individuals. The k individuals then take part in a tournament. The performance of the k individuals is carried out. The individual with the best fitness is selected from the tournament participants. The advantage of tournament selection is that the worse individuals of the HM will not be selected for the next iteration, and the best individual will not dominate in the improvisation process. As a result, the tournament selection has a low selective pressure that ensures an optimum solution. It is very important to remember that the size of k individuals is directly related to selective pressure. It causes a very high selective pressure for $k = $ HMS, and it results in a low selective pressure for $k = 1$. Further detailed description of tournament selection can be found in Siddique and Adeli (2013).

Linear ranking memory consideration selection: This selection method was introduced by Baker (1985). Rank-based selection uses the rank ordering of the fitness values to determine the probability of selection and not the fitness values itself. This means that the selection probability is independent of the actual fitness. Ranking, therefore, has the advantage that a highly fit individual will not dominate in the selection process as a function of the magnitude of its fitness. At each iteration, each vector in the HM is ranked, where the best vector is denoted as x^1 and the worst vector is denoted as x^{HMS} in the HM. The selection probability is computed using

$$p(x^i) = \frac{1}{\text{HMS}}\left(\min + \frac{(\max - \min)[\text{rank}(x^i) - 1]}{\text{HMS} - 1}\right) \tag{5.11}$$

where $1 \leq \max \leq 2$ and $\max + \min = 2$.

Ensemble consideration: Geem (2006a,b) added a new operation called *ensemble consideration* to the original HSA structure. The new operation considers the relationship among decision variables, and the value of each decision variable can be determined from the strong relationship with other variables.

5.5.3.2 Random Consideration

The decision variables that have not been assigned values according to memory consideration are assigned values randomly with a probability of (1-HMCR). The random consideration helps in choosing one feasible value not limited to those stored in the HM. The new value is chosen using the HMCR parameter according to the following rule:

$$x_i' \leftarrow \begin{cases} x_i' \in \{x_i^1, x_i^2, \ldots, x_i^{\text{HMS}}\} & \text{if probability HMCR} \\ x_i' \in X_i & \text{if probability } (1 - \text{HMCR}) \end{cases} \tag{5.12}$$

where X_i is the set of the possible range of values for each decision variable. For an HMCR value of 0.90, only 10% of the entire feasible range will undergo random selection with a probability of (1-HMCR). An HMCR value of 1.0 is not recommended because it does not allow improving the solution by values not stored in the HM (Lee and Geem, 2005).

5.5.3.3 Pitch Adjustment

The pitch-adjusting process is performed only after a value is chosen from the HM. Every decision variable x_i' of the new harmony vector $x' = \{x_1', x_2', \ldots, x_N'\}$ is examined to determine whether it should be pitch-adjusted or not. Pitch adjustment is controlled by the parameter called the pitch adjustment rate, PAR $\in [0,1]$. The PAR sets the rate of adjustment for the pitch chosen from the HM according to the following rule:

$$\text{Pitch adjustment for } x_i' \leftarrow \begin{cases} \text{Yes} & \text{if probability PAR} \\ \text{No} & \text{if probability } (1 - \text{PAR}) \end{cases} \tag{5.13}$$

A value for PAR = 0.3 indicates that the HSA modifies the neighboring values of the decision variables with a probability of (PAR*HMCR), while the other values of the decision variables remain unchanged. If the pitch adjustment decision for x_i' is Yes and x_i' is assumed to be $x_i(k)$, that is, the k-th element in X_i, the pitch-adjusted value of x_i' is defined as

$$x_i' = x_i' + \text{rand}(-1,1) * bw \tag{5.14}$$

where bw is an arbitrary distance bandwidth for the continuous decision variable and rand(−1,1) is a uniform random distribution. If the pitch adjustment decision for x'_i is No, it means that the value of $(1 - \text{PAR})$ sets the rate for doing no change. The improvisation procedure, which is the main mechanism of the HSA that drives the search toward an optimal solution, can now be summarized as

$$x_i' \leftarrow \begin{cases} x_i' \in \{x_i^1, x_i^2, \ldots, x_i^{\text{HMS}}\} & \text{if probability HMCR} * (1 - \text{PAR}) \\ x_i' = x_i' \pm \text{rand}() * bw & \text{if probability HMCR} * \text{PAR} \\ x_i' \in \{v_{i,1}, v_{i,2}, \ldots, v_{i,Ki}\} & \text{if probability } (1 - \text{HMCR}) \end{cases} \tag{5.15}$$

where x_i' is assigned a value $v_{i,k}$, that is, the k-th element in X_i. A stochastic derivation of the value $v_{i,k}$ of the variable x_i can be found in Al-Betar et al. (2012) and Geem (2008).

5.5.4 UPDATING HM

At this step, the HM is updated. If the new harmony vector is better than the worst harmony vector in the HM in terms of the objective function value, the new harmony vector is included in the HM and the existing worst harmony vector is excluded from the HM. This is actually a selection process of the algorithm where the objective function value is evaluated to determine if the new variation should be included in the HM. The HM is then sorted by the objective function value.

5.5.5 STOPPING CONDITION

At this step, the HSA is repeated until the termination criterion is satisfied. In general, the termination condition is the maximum number of improvisations or iterations. Achieving the desired objective function value could also be considered as the termination condition. The computations are terminated when either of the two conditions is met.

The basic HSA can be summarized in the following pseudocodes in five steps:

Step 1: Initialize HSA parameters HMS, HMCR, PAR, and NI
Step 2: Initialize HM and find x^{worst}
Step 3: Improvise a new harmony
     If $(r \leq \text{HMCR})$
       Select $x_i' \in \left\{ x_i^1, x_i^2, \ldots, x_i^{\text{HMS}} \right\}$, $\forall i = 1, \ldots, N$ {HM consideration}
       If $(r \leq \text{PAR})$
         $x_i' = x_i' + \text{rand}(-1,1) * bw$ {pitch adjustment}
       Else $x_i' = L_i + \text{rand}(0,1) \times (U_i - L_i)$ {random generation from the range}
Step 4: Update the HM
     If $[f(x_i') < f(x^{\text{worst}})]$
       HM $\leftarrow$ HM $\cup \{x_i'\}$ and HM $\leftarrow$ HM$\backslash\{x^{\text{worst}}\}$
     {i.e., include x_i' to HM and exclude x^{worst} from HM}
Step 5: If (termination condition not satisfied), Goto Step 3
Step 6: Return solution

where $r \in [0,1]$ is a random number uniformly distributed over the interval [0,1].

Figure 5.3 represents the flow chart of the optimization procedure of the HSA consisting of five steps.

5.6 CHARACTERISTIC FEATURES OF PARAMETERS IN THE HSA

The HSA has a number of parameters that are required to be adjusted or modified in order to control the optimization process, improve the diversification of the search and convergence speed of the algorithm, manipulate local search, and ensure the quality of the optimal solution. HSA parameters that can be varied are as follows:

- HMS
- HMCR
- PAR
- Distance bandwidth (bw)
- Termination condition or number of iterations (NI)

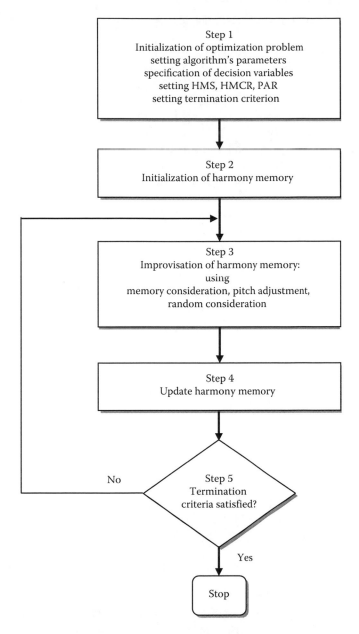

FIGURE 5.3 Flow chart of HSA.

The HMS is important that defines the spread of the search space. HMS represents the number of solution vectors in the HM. The HMS is an analogous construct similar to population size of GAs. It generally varies from 1 to 100, but it is typically set to 30 (Kim et al., 2001; Geem and Choi, 2007; Geem et al., 2002, 2005a,b; Lee and Geem, 2004; Panchal, 2009). Panchal (2009) investigated different values of HMS between 1 and 20 in his experiments in therapeutic medical physics and found that convergence gets worse with increase in the HMS. With an increased HMS, the chance of a better solution decreases, and also there is a demand for more time as a new vector is constructed by random selection from the HM. Fesanghary et al. (2008) also made a systematic study with the frequently used HMS values ranging between 1 and 10 reported in the literature (Geem, 2006b; Geem et al., 2001, 2002, 2005a; Kim et al., 2001; Mahdavi et al., 2007). It appears from this investigation that, for most moderate-sized problems, a typical value for the HMS is in the range of 4–10.

The HMCR is the rate of choosing a vector from the HM and is a parameter that determines whether the value for the current decision variable in the new vector should come from the HM or is to be generated randomly. The parameter HMCR also controls the balance between exploration and exploitation. This is a mechanism for preventing the HSA from getting trapped in local minima (or maxima). It generally varies from 0.7 to 0.99, but the typical value for HMCR is 0.95. Panchal (2009) investigated different values for HMCR within the range of [0.3,0.95] in his experiments in therapeutic medical physics and found that convergence improves with increasing values of the HMCR for optimization problems with increased iterations.

PAR is another parameter that determines the rate at which the decision variables should be modified (adjusted). The typical value for the PAR is set to 0.9 (Ricart et al., 2011; Yang, 2009). Panchal (2009) investigated different values for the PAR within the range of [0.3,0.9] in his experiments and found that solution quality improves with increasing values of the PAR, but the average time for iterations increases. Increasing the PAR continuously in the HSA might be a questionable strategy for some reasons. Since the PAR controls the probability of either selecting a pitch from the HM randomly or further adjusting a selected pitch, it is believed that a successful search should be proceeded with progressively in the beginning and then should gradually settle down. Thus, the PAR should be decreased with time to prevent overshooting and oscillation.

Distance bandwidth (bw) is the amount of maximum change in pitch adjustment. This can be $(0.01 \times$ allowed range) to $(0.001 \times$ allowed range). If the memory consideration rule is to be used, the PAR determines whether further adjustment is required according to the bw parameter and can be visualized as local search. Here, bw is to be considered as the step size of the PAR parameter; bw has a considerable influence on the precision of solutions and should be problem dependent. Therefore, it seems reasonable that decreasing bw with iteration number could fine-tune the final solutions.

Termination condition or number of iterations (NI): NI is the maximum number of improvisations or iterations. In other words, it is used as the termination condition.

It is possible to vary the parameter values as the search progresses, which gives an effect similar to SA. Parameter-setting-free researches have also been performed. In those researches, algorithm users do not need a tedious parameter-setting process. Lee and Geem (2005) have recommended the HSA parameter value range. The HMCR takes a range between 0.7 and 0.95, PAR values between 0.2 and 0.5, and HMS takes a range between 10 and 50 to produce good performance of the HSA.

5.7 VARIANTS OF THE HSA

Since the first publication of the HSA in 2001, it has attracted huge attention from the meta-heuristic research community as well as from other fields, which contributed to the improvement of the algorithm and its application to different domains. There have been many variants of the HSA reported in the literature since its inception. These variants can be categorized considering two aspects of the improvements: performance improvements by parameter setting and performance improvements by hybridization of the HSA with other meta-heuristic algorithms. The discussion in this section is divided into two parts: variants based on parameters of HSA and variants based on hybridization with other meta-heuristic algorithms.

5.7.1 HSA VARIANTS BASED ON PARAMETERS

The main parameters of the HSA are those that constitute the algorithm, such as HMS, HMCR, PAR, and bw, and that control its performance and convergence. The initialization of the HM also plays a crucial role in the algorithm. The operations involved in HM consideration and pitch adjustment are also vital for the performance. The proper selection of the HSA parameter values is considered as one of the challenging tasks in HSA design. It has been difficult as there is no general rule for selecting these parameter values. The parameter values are also problem dependent. Therefore, an empirical trial is the only guide to this difficult task, which has led to research into new variants of the HSA. There have been numerous variants reported in the literature. The important variants are discussed in this section. The following are the few variants among them:

1. Binary HS (Geem, 2005)
2. Improved HS (IHS) (Mahdavi et al., 2007)
3. Global best HS (GHS) (Omran and Mahdavi, 2008)
4. Adaptive HS (Wang and Huang, 2010)
5. Self-adaptive GHS (SGHS) (Pan et al., 2010)
6. Discrete HS (Geem, 2008; Wang et al., 2010a,b)
7. Chaotic HS (CHS) (Alatas, 2010)
8. Gaussian HS (Duan and Li, 2014)
9. Innovative global HS (IGHS) (Askarzadeh and Rezazadeh, 2012)
10. Dynamic/parameter adaptive HS (Hasancebi et al., 2009; Saka and Hasancebi, 2009)
11. Explorative HS (EHS) (Das et al., 2011; Mukhopadhyay et al., 2008)
12. Quantum-inspired HS (Layeb, 2013)
13. Opposition-based HS (Banerjee et al., 2014)
14. Cellular HS (cHS) (Al-Betar et al., 2013a,b)
15. Design-driven HS (DDHS) (Murren and Khandelwal, 2014)
16. Island-based HS (Al-Betar et al., 2015)
17. HM initialization (Degertekin, 2008)
18. Grouping HS (Linda-Torres et al., 2012)
19. Multiple PAR HS (Geem et al., 2005a)
20. Geometric selective HS (Castelli et al., 2014)
21. Adaptive binary HS (Wang et al., 2013)

5.7.1.1 Binary Harmony Search (HS)

The standard HSA has been applied to the optimization problem with a continuous decision variable problem with significant success. Application of HS to the optimization problem with a binary decision variable was not reported until 2005. Geem (2005) first adopted the standard HS with binary coding, called BHS. In this proposed BHS, the decision variables x_{ij}, $i = 1, 2, \ldots,$ HMS, $j = 1, 2, \ldots, N$ are coded into binary values. The HM then comprises binary decision variables $x_{ij} \in \{0,1\}$ as defined below:

$$HM = \begin{bmatrix} x_{1,1} & x_{1,2} & \cdots & x_{1,N} \\ x_{2,1} & x_{2,2} & \cdots & x_{2,N} \\ \vdots & \vdots & \ddots & \vdots \\ x_{HMS,1} & x_{HMS,2} & \cdots & x_{HMS,N} \end{bmatrix} \tag{5.16}$$

The binary decision variable $x_{ij} \in \{0,1\}$ can only be on or off. Since the decision variables are encoded into binary form, there is no need for the pitch adjustment operator, that is, pitch adjustment operator is discarded from BHS. The main steps of BHS are as follows:

Step 1: Initialize parameters HMS, HMCR, and NI
Step 2: Initialize HM and find x^{worst}

Step 3: Improvise new harmony
    If ($r \leq$ HMCR)
        Apply HM consideration
        Apply random generation from the range
Step 4: Update the HM
    If $[f(x') < f(x^{worst})]$
        Replace x^{worst} with x' in HM
Step 5: If (termination condition not satisfied), Goto Step 3
Step 6: Return solution

Geem (2005) applied BHS to a water pump switching system consisting of m water pumps in n stations, where the sum of energy functions E_{ij} is to be minimized for distribution of water from the source to places of demand. In this particular case, the total electrical energy E_{ij} is the energy required for a water pump $j, j = 1, 2, \ldots, m$ based in station $i, i = 1, 2, \ldots, n$. The binary decision variable $x_{ij} \in \{0,1\}$ decides on/off scheduling of the pump at each station. Each row of the HM represents m pumps of n pumping stations, that is, $n \times m$ pumps, which are to be switched on or off, that is, if $x_{ij} = 1$, the pump is on and if $x_{ij} = 0$, the pump is off. Within 3500 improvisations, the HSA found the global minimum of energy. Later on, Geem and Williams (2007, 2008) utilized BHS to tackle an ecologic optimization problem and achieved better results than those using the SA algorithm. Greblicki and Kotowski (2009) analyzed the properties of BHS on one-dimensional binary knapsack problems, and the experimental results indicate that the performance of BHS was not ideal.

5.7.1.2 Improved HS (IHS)

It is found from the empirical studies that the performance of the HSA depends on the choice of the values for the PAR and bw, which are mainly fixed during the run of the algorithm. The PAR and bw are two important parameters in the standard HSA that help in fine-tuning solution vectors and adjusting the convergence rate. But the standard HSA applies fixed values for the PAR and bw, which incurs prolonged iterations to reach an optimal solution. Small PAR values with large bw values can cause poor performance and considerable increase in iterations. To improve the performance of the HSA and eliminate the drawbacks, Mahdavi et al. (2007) proposed an improved HSA (IHSA) by introducing dynamic adaptation for PAR and bw values during improvisation. The value of PAR is linearly increased from a minimum value to a maximum value according to

$$\text{PAR}(t) = \text{PAR}_{min} + \frac{(\text{PAR}_{max} - \text{PAR}_{min})}{\text{NI}} \cdot t \tag{5.17}$$

where PAR(t) is the pitch-adjusting rate at current iteration t, PAR_{min} is the minimum adjusting rate, and PAR_{max} is the minimum adjusting rate. NI is the maximum number of iterations (or improvisations). The value of bw is exponentially decreased from a maximum value to a minimum value according to the following expression:

$$bw(t) = bw_{max} \exp\left(\frac{\ln(bw_{min}/bw_{max})}{\text{NI}} \cdot t \right) \tag{5.18}$$

where bw_{max} and bw_{min} are the maximum and minimum bandwidths, respectively.

The main steps of the IHSA are the same as those of the standard HSA except the dynamic computation of PAR(t) and $bw(t)$.

Step 1: Initialize HSA parameters
Step 2: Initialize HM

Step 3: Improvise a new harmony
    If ($r \leq$ HMCR)
        Select $x_i' \in \left\{ x_i^1, x_i^2, \ldots, x_i^{\text{HMS}} \right\}$
    If ($r \leq$ PAR(t))
        $x_i' = x_i' + \text{rand}(-1,1) * bw(t)$
        $x_i' = L_i + \text{rand}(0,1) \times (U_i - L_i)$ {random generation from the range}
Step 4: Update the HM
Step 5: If (termination condition not satisfied), Goto Step 3
Step 6: Return solution

A large value of *bw* in early generations enforces diversity, and a small value of *bw* in final generations helps in increasing the fine-tuning of the solutions. In the final generations, a large value for the PAR with a small value for *bw* usually provides improved solutions. The IHSA has been applied to a number of test problems, which demonstrated good performance in terms of fitness and quality of solutions compared to other meta-heuristic algorithms. A major shortcoming of the IHSA is the need to specify the values of $bw_{\max}$ and $bw_{\min}$, which are very problem dependent and difficult to guess. Of the two parameters, *bw* is more difficult to tune as it may take any value between $[0,\infty]$.

Coelho and Bernert (2009) proposed another version of the IHSA where only the parameter PAR is dynamically updated. The concept is borrowed from dispersed PSO introduced by Cai et al. (2008). The fitness function value is incorporated into the PAR(t) updating mechanism defined by the following equation:

$$\text{PAR}(t) = \text{PAR}_{\min} + (\text{PAR}_{\max} - \text{PAR}_{\min}) \cdot grade \qquad (5.19)$$

where PAR(t) is the pitch-adjusting rate of the current iteration *t*, and $\text{PAR}_{\min}$ and $\text{PAR}_{\max}$ are the minimum adjusting rate and the minimum adjusting rate, respectively. The factor *grade* is updated according to the following expression:

$$grade = \frac{(F_{\max}(t) - \bar{F})}{(F_{\max}(t) - F_{\min}(t))} \qquad (5.20)$$

where $F_{\max}(t)$ and $F_{\min}(t)$ are the maximum and minimum fitness values at iteration *t*, respectively, and $\bar{F}$ is the mean of fitness values of the HM. The proposed approach has been tested on a discrete-time chaotic system, and the effectiveness was confirmed against the standard HSA and the global best HSA.

Jaberipour and Khorram (2010) proposed two improvements for the IHSA by introducing new definitions of *bw*. The first definition of *bw* is as follows:

$$bw(i) = \frac{df(x)/dx_i}{x_i^u - x_i^l} \qquad (5.21)$$

Calculation of the derivative term $df(x)/dx_i$ is complicated and sometimes very expensive. To simplify and reduce the cost of computation, the derivative term $df(x)/dx_i$ is approximated to

$$\frac{df(x)}{dx_i} \approx \frac{f(x + \varepsilon \cdot e_i) - f(x - \varepsilon \cdot e_i)}{2\varepsilon}$$

by applying the finite difference approach where e_i is the *i*-th unit vector and $\varepsilon = 10^{-8}$. With this approximation, *bw* becomes

$$bw(i) = \frac{f(x + 2\varepsilon \cdot e_i) - f(x - 2\varepsilon \cdot e_i)}{4\varepsilon(x_i^u - x_i^l)} \qquad (5.22)$$

The pitch adjustment is performed based on *bw* calculated in Equation 5.22 according to

$$x_i' = \begin{cases} x_i + \text{rand}(-0.5, 1.5) \cdot \dfrac{f(x + 2\varepsilon \cdot e_i) - f(x - 2\varepsilon \cdot e_i)}{4\varepsilon(x_i^u - x_i^l)} & \text{with probability HMCR} \times \text{PAR} \\ x_i & \text{with probability HMCR} \times (1 - \text{PAR}) \end{cases} \tag{5.23}$$

The second definition of *bw* is based on the Yang bandwidth (Yang, 2008) as follows:

$$bw(i) = \begin{cases} \dfrac{x_u - x_l}{1000} & \text{with rand}(0,1) < \text{PAR} - A \\ \dfrac{f(x + 2\varepsilon \cdot e_i) - f(x - 2\varepsilon \cdot e_i)}{4\varepsilon(x_i^u - x_i^l)} & \text{with rand}(0,1) \geq \text{PAR} - A \end{cases} \tag{5.24}$$

where $A \in [0, \text{PAR}]$.

The pitch adjustment is performed based on *bw* calculated in Equation 5.24 according to

$$x_i' = \begin{cases} x_i + \text{rand}(-0.5, 1.5) \cdot \dfrac{f(x + 2\varepsilon \cdot e_i) - f(x - 2\varepsilon \cdot e_i)}{4\varepsilon(x_i^u - x_i^l)} & \text{with probability HMCR} \times \text{PAR} \times A \\ x_i + \text{rand}(-0.5, 1.5) \cdot \dfrac{x_u - x_l}{1000} & \text{with probability HMCR} \times (\text{PAR} - A) \\ x_i & \text{with probability HMCR} \times (1 - \text{PAR}) \end{cases} \tag{5.25}$$

The efficiency of the two proposed methods has been validated by applying it to 13 well-known benchmark optimization problems.

In order to explore and exploit the search space, Degertekin (2012) proposed an efficient HSA by introducing dynamic update rule for the PAR and *bw*. The PAR is increased linearly for exploration of the search space in the initial stage and linearly decreased for exploitation in the final stage. Two strategies are used for the PAR, defined as

$$\text{PAR}(t) = \begin{cases} \text{PAR}_{\min} + \dfrac{(\text{PAR}_{\max} - \text{PAR}_{\min})}{NI} \times t & \text{for linearly increasing} \\ \text{PAR}_{\max} - \dfrac{(\text{PAR}_{\max} - \text{PAR}_{\min})}{NI} \times t & \text{for linearly decreasing} \end{cases} \tag{5.26}$$

where *t* is the current iteration.

A large value for *bw* is used in the initial stage of the search to increase diversity, and a small value is used for local search in the final stage. Therefore, the *bw* update rule is expressed by the same mechanism defined in Equation 5.18. The proposed approach has been verified on a weight minimization problem of two planar and two spatial bar truss structures, which demonstrated improved convergence and robust behavior compared to the standard HSA.

In the previous IHSA implementations, NI is fixed and NI is to be found empirically. Contreras et al. (2014) proposed a new variant to make the algorithm independent of prior knowledge required for maximum iterations. The proposed approach is based on variable parameters; that is, the parameters HMCR and PAR should be changed through time, starting at a minimum value and ramping up to a maximum value in a given number of iterations. After reaching the maximum value, it is kept constant until the end of the run. Once the maximum tolerable iterations are reached, defined as a saturation parameter, the parameter bw is increased. The proposed approach has been applied to a weight minimization problem of two planar and two spatial bar truss structures for a saturation parameter value of 100, which demonstrated improved convergence and robust behavior compared to the standard HSA. There are other improved HSAs reported in the literature by Ashrafi and Dariane (2013) and Yousefi et al. (2013).

5.7.1.3 Global Best HS (GHS)

Inspired by the concept of the global best position in PSO, Omran and Mahdavi (2008) proposed a new variant of HS, called the GHS, which modifies the pitch adjustment step of the HSA such that the new harmony can mimic the best harmony x^{best} in the HM. GHS incorporates a social dimension to the HSA by completely removing the parameter bw from the HSA. It, thus, eliminates the limitation of determining the lower and upper bounds of the bandwidth bw used in IHS and further includes a randomly selected decision variable from the best harmony vector in the HM. Step 3 of the standard HSA is modified with the pseudocode shown below:

> If rand(0,1) ≤ HMCR, then
> $\quad x'_i = x_i^j$ where $j \in \{1, \ldots, HMS\}$
> $\quad$ If rand(0,1) ≤ PAR(t), then
> $\quad\quad x'_i = x_k^{best}$ where *best* is the index of best harmony in HM, $k = 1, \ldots, n$
> $x'_i = L_i + \text{rand}(0,1) \times (U_i - L_i)$ {random generation from the range}

where x_k^{best}, $k = 1, \ldots, n$ are the k-best harmonies in HM and U_i and L_i are the upper and lower bounds of x_i, respectively.

The main steps of the GHS algorithm are as follows:

> Step 1: Initialize HSA parameters
> Step 2: Initialize HM
> Step 3: Improvise a new harmony
> $\quad\quad$ If (rand(0,1) ≤ HMCR)
> $\quad\quad\quad x'_i = x_i^j$ for $j \in \{1, \ldots, HMS\}$ and $i = 1, \ldots, N$
> $\quad\quad\quad$ If (rand(0,1) ≤ PAR(t))
> $\quad\quad\quad\quad x'_i = x_k^{best}$ where *best* is the index of best harmony in HM, $k = 1, \ldots, n$
> $\quad\quad x'_i = L_i + \text{rand}(0,1) \times (U_i - L_i)$ {random generation from the range}
> Step 4: Update the HM
> Step 5: If (termination condition not satisfied), Goto Step 3
> Step 6: Return solution

The performance of GHS has been verified on 10 benchmark functions where it shows the best results when applied to high-dimensional problems.

In order to improve the convergence and accuracy of the GHS algorithm, Cobos et al. (2011) proposed a modification to the GHS algorithm using the concepts of a learnable evolution model (LEM). The LEM can locate promising areas where the global optimum is to be found and work with discrete and continuous variables (Michalski, 2000). The LEM is used to determine which individuals in a population or a set of individuals from a previous population are better than others. Thus, the LEM helps in generating new populations. The approach is called GHS+LEM. A set of conjunctive rules is used, which delineate the regions about which there is a greater chance of finding a better value for each x_i. Rules are selected using the parameter RCR (rule consideration rate). The procedure of the improvisation step of the GHS+LEM algorithm is presented below:

If rand(0,1) < HMCR, then
{
$x_i' = x_k^j, j \in \{1, \dots, \text{HMS}\}$ and for all $i = 1, \dots, N$
If rand(0,1) ≤ PAR(t), then
    $x_i' = x_k^{\text{best}}$ where *best* is the index of the best harmony in HM and $k = 1, \dots, n$
}
Else
{
If rand(0,1) ≤ RCR, then
    $x_i' = (L_i^{\text{best}}, U_i^{\text{best}})$ where *best* is the best set of rules
Else
    $x'_i = L_i + \text{rand}(0,1) \times (U_i - L_i)$ {random generation from the range}
}

The performance of the algorithm is verified on 15 optimization functions commonly used by the optimization community. The results show that the GHS+LEM algorithm improves the accuracy of the solutions obtained in relation to the other options, producing better results in most situations, but more specifically in problems with high dimensionality, where it offers a faster convergence with fewer iterations.

El-Abd (2013) proposed an improved GHS with a view to investigate the search space by going through the stages of exploration with large steps at the beginning and exploitation with smaller steps toward the end of the search. In the pitch selection step, a randomly selected pitch x_i^r is adjusted according to a Gaussian distribution defined as

$$x_i' = x_i' + \text{Gauss}(0,1) \cdot bw \tag{5.27}$$

Gauss(0,1) is a Gaussian random number with zero mean and standard deviation 1. In the pitch adjustment step, the current pitch of the best harmony is selected and adjusted at the same time using uniform distribution:

$$x_i' = x_k^{\text{best}} + \text{rand}(-1,1) \cdot bw \tag{5.28}$$

where $x_k^{\text{best}}, k = 1, 2, \dots, n$ are the k-best harmonies from the HM and rand(−1,1) is a random number uniformly distributed over the range of [−1,1]. Thus, the continuous perturbation even around the best harmony will enable the improved GHS to escape the local minima. Gaussian distribution will provide a wide area of search around the pitch. The proposed improved GHS algorithm has been applied to 14 benchmark functions and compared against 8 previously modified HSAs, which show superior performance.

Xiang et al. (2014) proposed an improved GHS search algorithm improved GHS to further enhance the convergence of GHS. In the improved GHS algorithm, the HM is initialized using opposition-based learning for improving the solution quality of the initial HM; a new improvisation scheme based on DE for enhancing the local search ability and a modified random consideration based on the ABC algorithm for reducing the randomness of the GHS algorithm, as well as two perturbation schemes for avoiding premature convergence, are integrated. In addition, two parameters of improved GHS, the HM consideration rate and the pitch-adjusting rate, are dynamically updated based on a composite function composed of a linear time-varying function, a periodic function, and a sign function in view of approximate periodicity of evolution in nature. Experimental results from

28 benchmark functions indicate that improved GHS is far better than the basic HSA and GHS. Though improved GHS shows improved performance over the HSA and GHS, the computational cost in this approach is huge.

5.7.1.4 Adaptive HS

As discussed with regard to the improved HSA earlier in this section, the PAR and bw have significant influence on the quality of the optimal solution. Therefore, by modifying the pitch adjustment step of the HS, better harmony is achievable. Wang and Huang (2010) proposed a new variation of the HSA called adaptive HS where they applied dynamic selection of the parameters bw and PAR to the HSA. In adaptive HS, the parameter PAR is decreased linearly from 1.0 to 0.0 over the iterations:

$$PAR(t) = 1 - \frac{1}{NI} \times t \qquad (5.29)$$

The parameter bw is replaced with a new idea that is based on using the maximal and minimal values in the HM. The pitch adjustment can be done through using the following equations:

$$\begin{cases} x_i' = x_i + (\max(HM_i) - x_i) \cdot \text{rand}[0,1) \\ x_i' = x_i + (x_i - \min(HM_i)) \cdot \text{rand}[0,1) \end{cases} \qquad (5.30)$$

where $\max(HM_i)$ and $\min(HM_i)$ denote the maximal and minimal values in of the i-th variable in the HM, respectively; x_i is the selected pitch from the HM; and $\text{rand}[0,1)$ is a random number uniformly distributed over the range of [0,1] without 1. Spontaneously, $\max(HM_i)$ and $\min(HM_i)$ would help the HSA in approaching the optimum gradually. Thus, the adaptive mechanism does finer adjustment to harmony progressively over the interactions. Dynamic PAR(t) uses Equation 5.29, and pitch adjustment uses Equation 5.30. The main steps of the adaptive HSA are as follows:

Step 1: Initialize HSA parameters
Step 2: Initialize HM
Step 3: Improvise a new harmony
    If $(r \le HMCR)$
      Select $x_i' \in \left\{ x_i^1, x_i^2, \ldots, x_i^{HMS} \right\}$
    If $(r \le PAR(t))$
      $\begin{cases} x_i' = x_i + (\max(HM_i) - x_i) \cdot \text{rand}[0,1) \\ x_i' = x_i + (x_i - \min(HM_i)) \cdot \text{rand}[0,1) \end{cases}$
   $x_i' = L_i + \text{rand}(0,1) \times (U_i - L_i)$ {random generation from the range}
Step 4: Update the HM
Step 5: If (termination condition not satisfied), Goto Step 3
Step 6: Return solution

The adaptive HS has been applied to well-known benchmark problems. The experimental results reveal the superiority of the proposed method over the standard HSA and other recently developed variants.

Degertekin (2012) also proposed an extension of adaptive HSA where pitch adjustment is carried out according to the following formulation:

$$x_i' = \begin{cases} x_i + (\max(HM_i) - x_i) \cdot \text{rand}[0,1) & \text{if } \text{rand}(0,1) \le 0.5 \\ x_i + (x_i - \min(HM_i)) \cdot \text{rand}[0,1) & \text{if } \text{rand}(0,1) > 0.5 \end{cases} \qquad (5.31)$$

The proposed approach has been applied to a weight minimization problem of two planar and two spatial bar truss structures, which demonstrated improved convergence and robust behavior compared to the standard HSA.

Kattan and Abdullah (2013) proposed a new adaptive HSA where the parameters PAR and *bw* vary dynamically rather than monotonically using an auto-tuning technique. The new quality measure computes the current best-to-worst (BtW) ratio from the fitness function values of the HM. A high value of BtW indicates that the quality of the current solution is as good as that of the best solution. Based on the value of BtW, the PAR value is adjusted dynamically. That is, for a decreasing BtW, the PAR value should be increasing to reflect local exploitation ability, and for an increasing BtW, the PAR value should be decreasing to diversify the search area. The strategy is to increase the BtW as the search progresses owing to higher quality of the solution. The BtW value is defined as

$$\text{BtW} = \frac{f(\overline{x}_{\text{best}})}{f(\overline{x}_{\text{worst}})} \tag{5.32}$$

where $f(\overline{x}_{\text{best}})$ and $f(\overline{x}_{\text{worst}})$ are the mean of the best and worst fitness function values, respectively. The PAR value is not monotonic and also not limited by the value of NI. The dynamic determination of the PAR is based on the strategy described by Figure 5.4.

Therefore, the PAR value is determined as a function of BtW defined as

$$\text{PAR} = slope \cdot \text{BtW} + \text{PAR}_{\text{max}} \tag{5.33}$$

where *slope* is the slope of the line between PAR_{max} and PAR_{min} expressed by

$$slope = \frac{\text{PAR}_{\text{max}} - \text{PAR}_{\text{min}}}{0-1} = \text{PAR}_{\text{min}} - \text{PAR}_{\text{max}} \tag{5.34}$$

PAR_{max} is set to 1.0, and PAR_{min} is set to a small value greater than 0, as shown in Figure 5.4. Thus, the variation in the PAR is not monotonic, and change in the PAR depends on the quality of the current solutions in the HM. The change in the PAR is reflected by the *bw* parameter, which represents the actual adjustment of each dimension variable. The value of *bw* is computed independently of each dimension variable and is determined dynamically based on the standard deviation of the dimension variable in the HM. An active bandwidth, denoted as bw^a, is defined considering each harmony vector component separately by

$$bw_i^a = C \cdot \sigma(x_i) \tag{5.35}$$

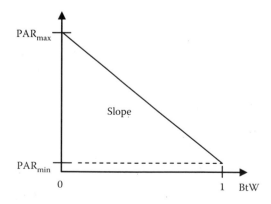

FIGURE 5.4 Determination of PAR.

where $\sigma(\cdot)$ is the standard deviation for each x_i in the HM. The value of C is decided by the following rule:

$$C = \begin{cases} 2 & \text{if } A_r \geq A_t \\ 2 \cdot \dfrac{A_r}{A_t} & \text{if } A_t > A_r > 1 \\ 0.1 & \text{if } A_r \leq 1 \end{cases} \qquad (5.36)$$

A new parameter A_r is introduced, which is the percentage of improvisation acceptance rate. A_t denotes the threshold value for A_r, which is found to yield a good result at $A_t = 20\%$. The dynamic bandwidth, denoted as bw^d, is then calculated as

$$bw_i^d = \text{rand}(-bw_i^a, bw_i^a) \qquad (5.37)$$

A value for $C > 1.0$ will extend the value of bw^d to avoid eliminating the benefits of exploration of the search space outside the current boundary, and a value for $C = 2.0$ will render the performance of the proposed algorithm compared to other algorithms. A minimum value for $C = 0.1$ can be used for low acceptance rates.

The proposed dynamic adaptive HSA has been applied to nine well-known benchmark functions and compared against five other recent methods with adaptive features. It demonstrated superior performance in all cases.

5.7.1.5 Self-Adaptive GHS

Pan et al. (2010) proposed the self-adaptive mechanism to GHS by employing a new improvisation scheme and adaptive parameter tuning. Their approach has some modifications to the GHS algorithm: firstly, self-adaptation of the parameters HMCR and PAR, and secondly, modification of pitch adjustment. In the parameter adaptation step, the HMCR and the PAR are dynamically adapted over time according to a normal distribution within a suitable range defined by the recorded historic values, for example, the range of HMCR $\in [0.9, 1.0]$ and PAR $\in [0.0, 1.0]$, with mean $\overline{\text{HMCR}}$ and $\overline{\text{PAR}}$ and standard deviation $\sigma_{\text{HMCR}} = 0.01$ and $\sigma_{\text{PAR}} = 0.05$, respectively. Initially, $\overline{\text{HMCR}}$ is set to 0.98, and $\overline{\text{PAR}}$ is set to 0.90. After a specified number of iterations, for example, 100 iterations, the mean and standard deviations are recalculated, and the procedure is repeated. As a result, appropriate values for the HMCR and the PAR are learnt. To balance the exploration and exploitation, bw is decreased dynamically with increasing iteration from bw_{max} to bw_{min} using the following formula:

$$bw(t) = \begin{cases} bw_{\text{max}} - \dfrac{bw_{\text{max}} - bw_{\text{min}}}{NI} \cdot 2t & \text{if } t < NI/2 \\ bw_{\text{min}} & \text{if } t \geq NI/2 \end{cases} \qquad (5.38)$$

In the improvisation step, the current pitch is adjusted using uniform distribution given by

$$x_i' = x_i^j \pm \text{rand}(-1,1) \times bw(t) \qquad (5.39)$$

where $x_i^j, j \in \{1, 2, \ldots, N\}$ are the k-best harmonies from the HM and rand$(-1,1)$ is the random number uniformly distributed in the range of $[-1,1]$.

The SGHS algorithm does not require a precise setting of the parameters in accordance with the problem's characteristics and complexity as the parameters will self-adapt themselves by learning over the generations. The main steps of the SGHS algorithm are as follows:

Step 1: Initialize HSA parameters
Step 2: Initialize HM
Step 3: Improvise a new harmony
        Calculate $\text{HMCR} = f\{\overline{\text{HMCR}}, \sigma_{\text{HMCR}}\}$ and $\text{PAR} = \{\overline{\text{PAR}}, \sigma_{\text{PAR}}\}$
        If $(\text{rand}(0,1) < \text{HMCR})$
            $x_i' = x_i^j \pm \text{rand}(-1,1) \times bw(t)$
        If $(\text{rand}(0,1) < \text{PAR})$
            $x_i' = x_k^{\text{best}}$ where *best* is the index of the best harmony in HM, $k = 1, \ldots, n$
        $x_i' = L_i + \text{rand}(0,1) \times (U_i - L_i)$ {random generation from the range}
Step 4: Update the HM
Step 5: Recalculate $\overline{\text{HMCR}}$ and $\overline{\text{PAR}}$ every 100 iterations
Step 6: If (termination condition not satisfied), Goto Step 3
Step 7: Return solution

Extensive computational simulations and comparisons are carried out by employing a set of 16 benchmark functions from the literature. The computational results show that the proposed SGHS algorithm is more effective in finding better solutions than the standard HS variants.

Poursalehi et al. (2013a,b) proposed an SGHS algorithm for reactor core fuel management optimization. Their approach has some modifications on GHS algorithms such as self-adaptation of the HMCR and the PAR that dynamically change over time and normal distribution within the range of $\text{HMCR} \in [0.7, 1.0]$ and $\text{PAR} \in [0.0, 0.5]$, with mean $\text{HMCR} = 0.85$ and $\text{PAR} = 0.25$ and standard deviation $\sigma_{\text{HMCR}} = 0.01$ and $\sigma_{\text{PAR}} = 0.05$, respectively. After a specified number of generations, for example, 100 in this case, the mean and standard deviations are recalculated and the procedure is repeated. To balance the exploration and exploitation, bw is decreased dynamically over generations from $bw_{\max}$ to $bw_{\min}$ using the following formula:

$$bw(t) = bw_{\max} - \frac{bw_{\max} - bw_{\min}}{\text{NI}} \cdot t \tag{5.40}$$

The proposed SGHS has been applied to a reactor core fuel management optimization problem. In every experiment, the search approached a semi-optimal solution in a reasonable computational time.

5.7.1.6   Discrete HS

There are many real-world situations where design variables contain discrete values by nature. As a result of this, a derivative-based optimization technique cannot be applied to these types of problems. In order to consider these realistic design situations, Geem (2008) proposed a variant of the HSA for discrete design variables.

The value of design variable x_i, $i = 1, 2, \ldots, N$ can be randomly selected from the set of all candidate discrete values $\{x_i(1), x_i(2), \ldots, x_i(K_i)\}$ with a probability of p_{random}. It can be selected from the set of good values $\{x_i^1, x_i^2, \ldots, x_i^{\text{HMS}}\}$ stored in the HM with a probability of p_{HM}. The value can be adjusted by moving to neighboring values (a value up or down from the current value) denoted as $N(x_i(k \pm m))$ once $x_i(k)$ is selected from the HM with a probability of p_{PA} for pitch adjustment. Based on the HM at a certain iteration, Geem (2008) introduced the partial stochastic derivative for the discrete variables, defined as

$$\frac{\partial f}{\partial x_i} = \frac{1}{K_i} \cdot p_{\text{random}} + \frac{N(x_i(k))}{\text{HMS}} \cdot p_{\text{HM}} + \frac{N(x_i(k \mp m))}{\text{HMS}} \cdot p_{\text{PA}} \tag{5.41}$$

The stochastic derivative for discrete variables provides information on the probability for a certain value $x_i(k)$ to be selected. The first term on the right-hand side yields the probability for random selection, the second term the probability of memory consideration, and the third term the probability of pitch adjustment. The performance of the new stochastic derivative has been verified on a benchmark function and on a two-loop water distribution network (WDN) optimization problem. Detailed analysis has been carried out to demonstrate the validity of the proposed discrete HS.

When the problem under consideration consists of integer decision variables, a discrete optimization algorithm is preferred to handle the problem effectively. Askarzadeh (2013a,b) proposed a discrete variant of the HSA, which can effectively solve the optimization problem. Initially, N_h number of feasible solutions is generated for the HM. The new harmony is then produced by the following pseudocode:

If $(r_1 > \text{HMCR})$
$\quad x_{\text{new}}(k) = x' \in \{X\}, \forall k = 1 \text{ to } N_h$
$x_{\text{new}}(k) = hm' \in \{\text{HM}\}$
If $(r_2 < \text{PAR})$
$\quad x_{\text{new}}(k) = x_{\text{new}}(k) + r_w$

where $x_{\text{new}}(k)$ is the improvised harmony, $x' \in X$ is randomly selected from the feasible subset of integer numbers $\{X\}$, $hm' \in \text{HM}$ is a value randomly selected from the HM, and r_1 and r_2 are uniformly distributed random numbers within the interval of [0,1]. The parameter r_w is defined as follows:

$$r_w = \begin{cases} 1 & r_3 < 0.5 \\ -1 & \text{otherwise} \end{cases} \tag{5.42}$$

Here, r_3 is a uniformly distributed random number within the interval of [0,1]. The proposed discrete HS has been applied to size optimization of a wind–photovoltaic hybrid energy system. The obtained results show that the proposed approach converges to an optimal solution after 150 iterations. The discrete HS approach seems promising for integer optimization problems.

5.7.1.7 Chaotic HS (CHS)

Chaos is typically a mathematical property of a dynamical system, which exhibits dynamic, unstable, pseudo-random, ergodic, and nonperiodic behavior. Simulating complex phenomena, sampling, numerical analysis, decision-making, and heuristic optimization algorithms need random sequences with good uniformity (Schuster, 1988). Recently, chaotic sequences have been adopted instead of random sequences, and very interesting and somewhat good results have been achieved in many heuristic algorithms and applications. The behavior of a chaotic system is sensitive to the initial value and can be controlled using a set of parameters (Elaydi, 1999). There are many chaotic maps used in chaotic search such as the logistic map (May, 1976), Tent map (Peitgen et al., 1992), sinusoidal iterator (May, 1976), Gauss map (Peitgen et al., 1992), circle map (Zheng, 1994), sinus map (Alatas, 2010), and Henon map (Hénon, 1976). A detailed mathematical description of a number of chaotic maps is presented in Appendix C. Chaotic maps can also be used for parameter adaptation in order to improve the convergence characteristics and to prevent the HS from getting stuck on local solutions. Alatas (2010) proposed CHS using different types of chaotic maps. Chaotic maps have been iteratively used to generate chaotic sequences of numbers that are then mapped to parameters such as the HM, HMCR, PAR, and bw in the HSA. Alatas (2010) investigated the CHS algorithm by applying seven chaotic maps (logistic, Tent, sinusoidal, Gauss, circle, sinus, and

Henon) for generating the initial HM, modifying the PAR and *bw*. The pseudocode for generating the initial HM can be expressed as follows:

```
j = 0;
while (j < HMS)
  {
  cm(0,0) = random_initialisation;
  i = 0;
  while (i < N)
    {
    cm(j,i) = generate(chaotic_map);
    x_{j,i} = x_i^{min} + cm(j,i) · (x_i^{max} − x_i^{min})
    i = i + 1;
    }
  j = j + 1;
  }
```

where $cm(j,i)$ represents the chaotic map, x_i^{max} and x_i^{min} are the max and the min value of x_i, respectively.

The parameter PAR is modified using the selected chaotic map defined as

$$\text{PAR}(t+1) = f(\text{PAR}(t)) \quad \text{where } 0 < \text{PAR}(t) < 1 \tag{5.43}$$

Here, $f(\cdot)$ is the selected chaotic map mentioned above beginning with a value within the range of constraints. The *bw* value is modified using the selected chaotic map defined as

$$bw(t+1) = f(bw(t)) \quad \text{where } 0 < bw(t) < 1 \tag{5.44}$$

The steps of the CHS are the same as the standard HSA except that the initial HM is generated using the chaotic map, PAR value is modified using the chaotic map, and *bw* value is modified using the chaotic map. Alatas (2010) investigated the CHS algorithm for seven different combinations of the HM, PAR, and *bw*, all generated and modified using the seven chaotic maps. The CHS algorithms have been applied to well-known multimodal benchmark problems such as the Griewangk and Rastrigin functions, showing better performance and convergence than other HSAs. Pan et al. (2011a,b) developed a chaotic local search scheme and embedded it in the chaotic HSA to enhance the local search ability of HS. The proposed CHS has been applied to a flow shop scheduling problem with limited buffers. Computational simulations show better results compared to some improved HSAs. A hybrid parallel chaos algorithm with the HSA is also reported by Yuan et al. (2014).

5.7.1.8 Gaussian HS

The foremost features of an ideal optimization algorithm are faster convergence to the global optimum and accuracy of the solution. A number of variants of HS reported in the literature demonstrated acceptable convergence and accuracy of the solution. The parameter *bw* plays a crucial role in the HSA. A large *bw* may result in premature convergence in local minima, whereas a small *bw* causes excessive local search resulting in slow convergence. Duan and Li (2014) introduced a Gaussian factor into the improvisation process of HS and called the procedure Gaussian HS. A parameter α is defined to facilitate decreasing *bw* with iterations.

$$\alpha = (1-\lambda)\frac{t_{\text{current}}}{t_{\text{max}}} \tag{5.45}$$

where $t_{current}$ and t_{max} are the current iteration and max number of iterations, respectively. The parameter λ decides the speed of decrease of bw. The Gaussian factor is introduced by the expression defined below:

$$\xi = g \cdot \alpha = g \cdot (1 - \lambda)\frac{t_{current}}{t_{max}} \tag{5.46}$$

Here, g is a Gaussian distribution function with variance σ. In Gaussian HS, the pitch adjustment is performed according to the following definition:

$$x_i' = \begin{cases} x_i \pm \xi \cdot bw & rand < PAR \\ x_i & rand < (1 - PAR) \end{cases} \tag{5.47}$$

The combination of the parameters g and α could improve the convergence and solution quality. The main steps of the Gaussian HS are as follows:

Step 1: Initialize HSA parameters
Step 2: Initialize HM
Step 3: Improvise a new harmony
      If ($r \leq$ HMCR) where $r \in [0,1]$ is random number
        Select $x_i' \in \left\{ x_i^1, x_i^2, \ldots, x_i^{HMS} \right\}$
      If ($r <$ PAR)
        $x_i' = x_i \pm \xi \cdot bw$
      If ($r <$ 1 − PAR)
        $x_i' = x_i$
        $x_i' = L_i + rand(0,1) \times (U_i - L_i)$ {random generation from the range}
Step 4: Update the HM
Step 5: If (termination condition not satisfied), Goto Step 3
Step 6: Return solution

The Gaussian HS has been applied to solving Loney's solenoid problem. Comparative results show that the Gaussian HS outperforms HS and PSO in both convergence speed and efficiency. Li and Duan (2014) also investigated the feasibility and effectiveness of Gaussian HS. In this approach, the pitch adjustment is modified with the following formulation:

If ($r <$ PAR)
$$x_i' = \begin{cases} x_i + bw \times (r_2 - 0.5) \times \beta \times \alpha & \text{for } r_1 > p \\ x_i + bw \times (r_2 - 0.5) \times 2 \times \alpha & \text{for } r_1 \leq p \end{cases}$$
If ($r <$ 1 − PAR)
$$x_i' = x_i + bw \times (r_2 - 0.5) \times 2 \times \alpha$$

where β is the random number with Gaussian distribution, $\{r_1, r_2\}$ are two random numbers with uniform distribution over $[0,1]$, p is a parameter that determines the balance between Gaussian and uniform distribution, and α is the search step defined by

$$\alpha = 0.5 - 0.25\frac{t_{current}}{t_{max}} \tag{5.48}$$

The efficiency of the proposed approach was verified on Itti's visual attention model. The simulation results again demonstrated the expected performance.

5.7.1.9 Innovative GHS

The problem in HS is that the best harmony is not the global best, which leads the algorithm to premature convergence. Askarzadeh and Rezazadeh (2012) proposed an innovative GHS that uses a predefined number of elite harmonies with the best fitness values using the roulette wheel selection mechanism. A probabilistic approach is used to select the interesting elite harmony for the improvisation process. The selection probability of the k-th elite harmony is given by the following formula:

$$p_k = \frac{1/f_k}{\sum\limits_{k=1}^{E} 1/f_k} \tag{5.49}$$

where f_k is the fitness of the k-th elite harmony and E is the number of elite harmonies. MSE is defined as the fitness value of f_k. Thus, the probability of generating a harmony with higher quality increases as the new harmony is improvised using the best harmonies. The improvisation of a new harmony is carried out using the following algorithm:

If $r_1 \geq$ HMCR
$\quad x_i' = L_i + r_2 \times (U_i - L_i)$ for $i = 1$ to N
Else
$\quad$ Select elite harmonies
$\quad$ Calculate selection probability
$\quad$ Select the interesting elite harmony e_i with probability p_k
$\quad x_i' = e_i$
If $r_3 <$ PAR(t)
$\quad x_i' = x_i' + (r_4 - r_5) \times bw(t) \times |U_i - L_i|$

where r_1, r_2, r_3, r_4, and r_5 are random numbers distributed over the interval [0,1]. Linear increasing PAR(t) and exponentially decreasing $bw(t)$ are used in innovative GHS. The main steps of the innovative GHS are as follows:

Step 1: Initialize HSA parameters
Step 2: Initialize HM
Step 3: Improvise a new harmony
$\quad\quad$ If $r_1 \geq$ HMCR
$\quad\quad\quad x_i' = L_i + r_2 \times (U_i - L_i)$ for $i = 1$ to N
$\quad\quad$ Else
$\quad\quad\quad$ Select elite harmonies
$\quad\quad\quad$ Calculate selection probability
$\quad\quad\quad$ Select the interesting elite harmony e_i with probability p_k
$\quad\quad\quad x_i' = e_i$
$\quad\quad$ If $r_3 <$ PAR(t)
$\quad\quad\quad x_i' = x_i' + (r_4 - r_5) \times bw(t) \times |U_i - L_i|$
Step 4: Update the HM
Step 5: If (termination condition not satisfied), Goto Step 3
Step 6: Return solution

The innovative GHS algorithm has been applied to parameter identification problems, and its performance is compared against that of the HSA, PSO algorithm, bee swarm optimization algorithm, and seeker optimization algorithm. Simulation results demonstrate better and robust results than the other studied algorithms.

5.7.1.10 Dynamic/Parameter Adaptive HS

Initially, HMCR and PAR values are set to HMCR$^{(0)}$ and PAR$^{(0)}$, respectively, in the initialization step of the HM. Some researchers have proposed adaptive parameter theories that enable HS to automatically acquire the best parameter values at each iteration (Geem, 2009a). By dynamically changing both the HMCR and the PAR during the improvisation process, the performance of HS can be improved. Suggesting this adaptive strategy, Hasancebi et al. (2009) and Saka and Hasancebi (2009) proposed a new adaptation mechanism for the HMCR and the PAR during the improvisation process of HS. The dynamic calculation of the parameters is defined by

$$HMCR_k = \frac{1}{\left(1 + ((1 - \overline{HMCR})/\overline{HMCR}) \cdot e^{-\gamma \cdot N(0,1)}\right)} \tag{5.50}$$

$$PAR_k = \frac{1}{\left(1 + ((1 - \overline{PAR})/\overline{PAR}) \cdot e^{-\gamma \cdot N(0,1)}\right)} \tag{5.51}$$

where $HMCR_k$ and PAR_k are updated parameters for the new harmony vector. Here, $N(0,1)$ is a normally distributed random number, γ is the learning rate of adapted parameters and should be chosen within the interval of $\gamma \in [0.25, 0.50]$. $\overline{HMCR}$ and $\overline{PAR}$ are the mean of all the corresponding values of all the solution vectors within the HM matrix defined by

$$\overline{HMCR} = \frac{1}{HMS} \left(\sum_{i=1}^{HMS} HMCR_i \right) \tag{5.52}$$

$$\overline{PAR} = \frac{1}{HMS} \left(\sum_{i=1}^{HMS} PAR_i \right) \tag{5.53}$$

In fact, the new set of values of $HMCR_k$ and PAR_k is obtained by probabilistic selection from average values observed within the current HM matrix. The effectiveness of the adaptive parameter HS has been verified on large-scale steel frameworks design for minimum weight. The simulation results have been compared with standard algorithms as well as with other meta-heuristic algorithms, demonstrating improved performance.

Kumar et al. (2014) proposed the linear and exponential parameter adaptation mechanism for the HMCR and the PAR. The linear form of the parameter adaptation for the HMCR and the PAR is defined as

$$HMCR(k) = HMCR_{min} + \frac{HMCR_{max} - HMCR_{min}}{NI} \cdot t \tag{5.54}$$

$$PAR(k) = PAR_{min} + \frac{PAR_{max} - PAR_{min}}{NI} \cdot (NI - t) \tag{5.55}$$

The exponential form of parameter adaptation for the HMCR and the PAR is defined as

$$HMCR(k) = HMCR_{min} \cdot \exp\left(-\frac{\ln(HMCR_{min}/HMCR_{max})}{NI} \cdot t \right) \tag{5.56}$$

$$PAR(k) = PAR_{min} \cdot exp\left(-\frac{ln(PAR_{min}/PAR_{max})}{NI} \cdot t \right) \tag{5.57}$$

where t is the current iteration and NI is the maximum iteration.

The parameter adaptive HS has been verified on seven unimodal and eight multimodal benchmark functions and finally on a data-clustering application.

5.7.1.11 Explorative HS (EHS)

To avoid any premature convergence and to ensure exploration of the search space, the variation operators must adjust the population variance from generation to generation. When the population variance is decreased, the variation operators must increase the balance between exploration and exploitation. An algorithm is said to have good exploratory power when the population variance is increasing over generations. Let us consider an initial population of scalar variables $x = \{x_1, x_2, \ldots, x_m\}$ with $x_i \in R$ and HMS = m. The variance of the population x is given by

$$Var(x) = \frac{1}{m} \sum_{k=1}^{m} (x_k - \bar{x})^2 = \overline{x^2} - \bar{x}^2 \tag{5.58}$$

where $\bar{x}$ is the population mean and $\overline{x^2}$ is the quadratic population mean. When the elements in the population are perturbed with random values, then the variance of the population will also be random. The explorative power of the population can be measured by $E(Var(x))$. Das et al. (2011) and Panigrahi et al. (2010) mathematically analyzed the evolution of population variance for the HSA and proposed a small but effective amendment to the HSA, increasing its explorative power. Das et al. (2011) analytically showed that the expected variance of the t-th population without selection operation becomes

$$E\left[Var(x_t)\right] = \left[\frac{(m-1)}{m} \cdot HMCR \cdot \left(1 + \frac{1}{3} \cdot k^2 \cdot PAR\right)^t \right] \cdot Var(x_0) \tag{5.59}$$

where x_0 is the population at iteration t. If the parameters HMCR, PAR, and k in Equation 5.59 are chosen in such a way that the term within the brackets becomes greater than 1, then the expected variance $E[Var(x_t)]$ becomes exponential, causing exponential growth of population variance. Das et al. (2011) suggested that bw will be the standard deviation of the current population when the HMCR is close to 1. When the HMCR is very high (i.e., very close to 1) and the distance bandwidth bw is chosen to be proportional to the standard deviation $\sigma(x)$ of the current population, then it can be written as

$$bw \propto \sigma(x) = \sqrt{Var(x)} \tag{5.60}$$

In the case of Equation 5.60, the expected population variance without selection can grow exponentially over iterations. Equation 5.60 can be written with a proportionality constant c:

$$bw = c \cdot \sqrt{Var(x)} \tag{5.61}$$

The proportionality constant c now provides a mechanism of controlling the exponential growth of the population variance. By choosing a suitable value of c for a given set of values of the HMCR and the PAR, the population variance of the HSA and its explorative power can be controlled over iterations without selection. The new scheme of tuning bw (by making it proportional to the current population variance) provides HS with high explorative power and yields very good results on a wide variety of objective functions. The trade-off between exploration and exploitation can also be controlled by proper selection of the proportionality constant c. After estimating bw, the main steps of the EHS are as follows:

> Step 1: Initialize HSA parameters
> Step 2: Initialize HM
> Step 3: Improvise a new harmony
>> If ($r \leq$ HMCR)
>>> $x_i' \in \left\{ x_i^1, x_i^2, \ldots, x_i^{HMS} \right\}, \forall i = 1, \ldots, N$ {HM consideration}
>>> If ($r \leq$ PAR)
>>>> $x_i' = x_i' + r * bw$ {pitch adjustment using bw given by (5.61)}
>>> $x_i' = L_i + \text{rand}(0,1) \times (U_i - L_i)$ {random generation from the range}
> Step 4: Update the HM
> Step 5: If (termination condition not satisfied), Goto Step 3
> Step 6: Return solution

The EHS has been applied to 15 unconstrained and 5 constrained benchmark functions. The EHS algorithm has always outperformed the other HS variants over all of the tested problems in terms of solution quality, speed of convergence, and frequency of hitting the optimum.

5.7.1.12 Quantum-Inspired HS

The quantum-inspired HS is the integration of quantum representation scheme in the basic HSA that allows applying some kind of quantum-inspired operators like measurement and interference. The origin of quantum computing goes back to the 1940s when Richard Feynman observed that some quantum mechanical effects cannot be efficiently simulated on a computer (Feynman, 1942). A quantum representation deemed necessary. The smallest unit of information storage is the binary bit with two possible values {0,1}. In quantum computing, a bit (or qubit) will be in the superposition of those two values. The state of a qubit can be represented by following bracket notation:

$$|\psi\rangle = a|0\rangle + b|1\rangle \tag{5.62}$$

where $|\psi\rangle$ denotes more than a vector ψ in some vector space. Here, $|0\rangle$ and $|1\rangle$ represent the classical bit values 0 and 1, respectively. It is seen that a and b are complex numbers that specify the probability amplitudes of the corresponding states such that

$$|a|^2 + |b|^2 = 1 \tag{5.63}$$

Equation 5.63 further clarifies that the qubit's state "0" has a probability of $|a|^2$ and the state "1" has a probability of $|b|^2$. The salient feature is that a system with n-qubits can represent $2n$ states and quantum computers can perform computations on all these values at the same time. This suggests exponential speedup of quantum operations. Researchers have been trying to implement some of the features of quantum computing onto the classical algorithms. The quantum computing facilitates parallelism that obviously reduces the algorithmic complexity where combinatorial optimization problems require parallel processing to explore large search spaces. For example, quantum computing features have been incorporated into the EA, which allow representing the individual, the evaluation function, and potential solutions for a given problem. The quantum EA has been successfully applied to many problems (Draa et al., 2010; Han and Kim, 2004; Layeb and Saidouni, 2008).

Interested readers are directed to Jaeger (2006) and Yanofsky and Mannucci (2008) for in-depth theoretical insights and a description on quantum computing.

To apply quantum techniques to computing problems of binary nature, a mapping of solutions onto quantum representation is to be established so that the quantum operators can be applied to solutions for manipulation. Therefore, each binary representation of solutions is mapped to a quantum register of length N as shown in Figure 5.5. Each column represents a qubit corresponding to binary digit 0 or 1. The upper and lower bounds of a value $x_i \in [x_{iL}, x_{iU}]$ are -1 and 1, respectively, which can also mean that -1 and 1 represent, respectively, the minimum and maximum values of the qubit components a and b. In other words, a binary value is computed according to its probabilities $|a|^2$ and $|b|^2$ for each qubit.

Quantum measurement: The binary values are computed according to its probabilities $|a|^2$ and $|b|^2$ for each qubit. The quantum measurement is an operation that transforms the quantum vector (shown in Figure 5.5) into a binary vector. The operation is performed according to the following rule:

$$r = \text{rand}(0,1)$$

$$x_i = \begin{cases} 1 & \text{if } |b_i|^2 > r \\ 0 & \text{otherwise} \end{cases}$$

Here, rand(0,1) is a random number uniformly distributed over [0,1]. Applying the rule, the quantum vector in Figure 5.5 will look like Figure 5.6. The binary vector is then translated into the solution vector of the problem.

Another interesting feature of the measure operation is that two successive measurement operations do not give the same solution, which increases the diversification capability of the quantum approach.

Quantum interference: A good heuristic algorithm strikes a balance between exploration (global search) and exploitation (local search). The interference operation plays the role of a local search method and intensifies the search around the best solution. The operation also amplifies the amplitude of the best harmony and decreases the amplitudes of the worst ones. The operation is performed using a unit transformation which attains a rotation angle $\delta\theta$ as shown in Figure 5.7. The value of rotation angle $\delta\theta$ is a function of the amplitudes of a_i and b_i. The value of $\delta\theta$ is an indicator of convergence, where a big value of $\delta\theta$ can lead to premature convergence or divergence, and a small value of $\delta\theta$ can slow down the convergence speed. In order to avoid premature convergence, a well-estimated value of $\delta\theta$ is essential. The value of $\delta\theta$ is determined experimentally, and its direction is determined as a function of the values of a_i and b_i. The corresponding binary value depends on the sign of $\delta\theta$ as shown in Figure 5.7.

Quantum mutation: The notion of mutation operation is adopted from evolutionary computation, which introduces diversity to the population and allows exploration of new solutions in the

$$\begin{pmatrix} a_1 & a_2 & a_3 & \cdots & a_n \\ b_1 & b_2 & b_3 & \cdots & b_n \end{pmatrix}$$

FIGURE 5.5   Quantum representation of a binary vector.

$$\begin{pmatrix} a_1 & a_2 & a_3 & \cdots & a_n \\ b_1 & b_2 & b_3 & \cdots & b_n \end{pmatrix} \xrightarrow{\text{measurement}} \{0,1,0,\dots,1\} \Rightarrow \text{Solution}$$

FIGURE 5.6   Transformation of qubits into binary vector.

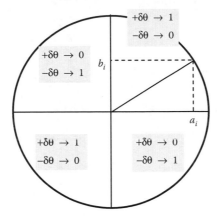

FIGURE 5.7 Interference operation.

neighborhood. A variety of mutation operators can be applied on to quantum-based representation of binary vectors shown in Figure 5.5. Among them, interqubit and intraqubit mutation operations are simple and easy to implement.

Interqubit mutation: It performs permutation between two qubits within a vector by selecting a register in the quantum matrix randomly. The pairs of qubits are chosen randomly according to a predefined probability and swapped with another pair of qubits, as shown in Figure 5.8.

Intraqubit mutation: It performs permutation between amplitudes of a qubit, that is, a_i and b_i, by selecting a qubit randomly according to a predefined probability, as shown in Figure 5.9.

The problem with the interqubit and intraqubit mutation operations is that they are not useful when a decision variable goes out of the bound of the search space after pitch adjustment. A new mutation operation is introduced, which depends on the bandwidth (bw_{min}, bw_{max}), iteration, and the maximum number of improvisations (or iterations) NI. The mutation operator is defined as follows:

$$x_i'(t) = \begin{cases} x_i \cdot \cos(\alpha \times t) & \text{if } x_i > 1 \\ \sin(\alpha \times t) \times \sqrt{(1 - x_i^2)} & \text{if } x_i < -1 \end{cases} \tag{5.64}$$

where t is the current iteration and α is the coefficient defined as

$$\left(\begin{array}{c|c|c|c|c} a_1 & a_2 & a_3 & \cdots & a_n \\ b_1 & b_2 & b_3 & \cdots & b_n \end{array} \right)$$

Swap

FIGURE 5.8 Interqubit mutation.

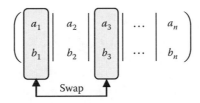

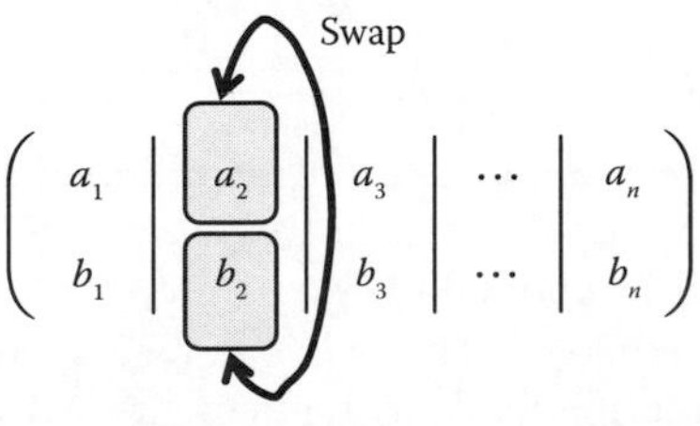

Swap

FIGURE 5.9 Intraqubit mutation.

$$
\left[
\begin{pmatrix}
a_1^1 & a_2^1 & a_3^1 & \cdots & a_n^1 \\
b_1^1 & b_2^1 & b_3^1 & \cdots & b_n^1
\end{pmatrix} \\
\begin{pmatrix}
a_1^2 & a_2^2 & a_3^2 & \cdots & a_n^2 \\
b_1^2 & b_2^2 & b_3^2 & \cdots & b_n^2
\end{pmatrix} \\
\vdots \\
\begin{pmatrix}
a_1^{\text{HMS}} & a_2^{\text{HMS}} & a_3^{\text{HMS}} & \cdots & a_n^{\text{HMS}} \\
b_1^{\text{HMS}} & b_2^{\text{HMS}} & b_3^{\text{HMS}} & \cdots & b_n^{\text{HMS}}
\end{pmatrix}
\end{array}
\right]
$$

FIGURE 5.10   Quantum representation of HM.

$$
\alpha = \frac{\log(bw_{\min}/bw_{\max})}{(0.7 \times \text{NI})} \tag{5.65}
$$

The quantum representation of binary vector shown in Figure 5.5 can be used to form harmony vectors for the HM. An HM of size HMS is shown in Figure 5.10. The quantum measurement, interference, or mutation operations can be easily performed on the HM matrix in Figure 5.10. The quantum vectors in the HM can be seen as a probabilistic representation of problem solutions, which is not only a powerful way of representing the solution space but also a way of reducing the size of the HM (Layeb, 2013).

Layeb (2013) proposed a quantum-inspired HSA using the quantum operations like measurement, quantum interference, and quantum mutation. With the theoretical underpinning discussed above, the main steps of the quantum-inspired HS can be described as follows:

Step 1: Quantum representation of HM
Step 2: Improvisation procedure
          Apply memory consideration
             Apply pitch adjustment
             Apply quantum mutation operation
          Apply random selection
          Apply quantum measure operation
          Compute fitness
Step 3: Update HM
          Apply quantum interference
Step 4: If (termination condition not satisfied), Goto Step 2
Step 5: Return solution

Layeb (2013) carried out an extensive experimentation on the proposed quantum-inspired HSA by applying it to widely used benchmark problems such as 0–1 and multidimensional knapsack problems. A brief introduction to the knapsack problem is provided in Chapter 6. The experimental studies prove the feasibility and effectiveness of the proposed quantum-inspired HSA.

5.7.1.13   Opposition-Based HS

Meta-heuristic algorithms start the search process with an initial solution or a population of solutions. Such algorithms apply some heuristic rules to improve the solutions toward optimal solution(s) and terminate when some predefined optimality criteria are satisfied. In most of the cases, *a priori* information about the initial solution(s) is not available for the meta-heuristic algorithm to start with. In the absence of *a priori* information, only random guesses are used. The computation time, quality of solutions, convergence, and overall performance of the meta-heuristic algorithms mainly depend on the distance of the initial solutions from the optimal solution. This is the intrinsic feature of any meta-heuristic algorithm. The only way to improve the performance of such meta-heuristic algorithms that involve such guesses is to have a better guess about the initial solution. Tizhoosh (2005a) introduced the notion of opposition-based guess, which helps in improving the initial guess in meta-heuristic algorithms and learning. The idea is to improve the guess by checking the opposite guess (Tizhoosh, 2005a,b, 2006). According to probability theory, 50% of the time a guess is further away from the solution than its opposite guess (Tizhoosh, 2005a; Upadhyay et al., 2014). By applying a guess and opposite guess, a better initial solution can be found. The notion of opposition-based learning is the simultaneous consideration of an estimate and its corresponding opposite estimate (i.e., guess and opposite guess) in order to achieve a better approximation for the initial candidate solution. More details of opposition-based learning can be found in Tizhoosh (2005a,b, 2006). This idea has been applied to other learning algorithms to accelerate learning (Ergezer and Simon, 2014; Rahnamayan et al., 2008; Shokri et al., 2006; Tizhoosh, 2006; Ventresca and Tizhoosh, 2006). The opposition-based number and opposite number in n-dimensional space are discussed in Chapter 2.

For a point $X = (x_1, x_2, \ldots, x_n)$ in n-dimensional space with $(x_1, x_2, \ldots, x_n) \in R$, $x_i \in [a_i,b_i]$, $\forall i \in \{1,2, \ldots, n\}$, a and b being the lower and upper bounds, respectively, of the real value x_i, the opposite of X is defined as

$$\breve{X} = (a_i + b_i) - x_i \quad \text{for} \quad \forall i \in \{1,2,\ldots,n\} \tag{5.66}$$

In the opposition-based HSA, opposite numbers defined by Equation 5.66 are employed in the initialization of the HM. The standard HSA is then modified to the opposition-based HSA by converting the harmonies into corresponding opposite harmony vectors. Banerjee et al. (2014) proposed opposition-based HS to optimization of engineering problems. In opposition-based HS, both X and $\breve{X}$ are evaluated. The main steps the opposition-based HS are as follows:

Step 1: Initialize HSA parameters
Step 2: Initialize HM
            Generate opposition HM from HM
Step 3: Improvise a new harmony
            If ($r \le$ HMCR)
                $x_i' \in \left\{ x_i^1, x_i^2, \ldots, x_i^{\text{HMS}} \right\}$, $\forall i = 1, \ldots, N$ {HM consideration}
                If ($r \le$ PAR)
                    $x_i' = x_i' + r \times bw$ {pitch adjustment}
                $x_i' = L_i + \text{rand}(0,1) \times (U_i - L_i)$ {random generation from the range}
Step 4: Update the HM
            If $f(\breve{X}) \le f(X)$, replace X with $\breve{X}$
Step 5: Form new HM by selecting the fittest vectors from HM and opposition HM
Step 6: If (termination condition not satisfied), Goto Step 3
Step 7: Return solution

The effectiveness of opposition-based HS has been verified on a suite of 16 benchmark function optimization problems and on a reactive power system model. Upadhyay et al. (2014) applied the opposition-based HSA to optimize nine benchmark IIR filter design coefficients. The simulation results were compared with GA, PSO, and DE.

5.7.1.14 Cellular HS

A challenging problem in population-based meta-heuristic algorithms is to maintain a high level of diversity and counteract the genetic drift over the generations. There have been many techniques for maintaining population diversity, for example, mutation. Cellular automata, a concept introduced by Neumann (1966), have been embedded into the process of the EA in order to provide a decentralized method (Alba and Dorronsoro, 2005). The idea is to partition the population into multiple sets of small populations with common features such that the population structure can be preserved. The cellular structures have been applied to other structured EAs, which have proved to be useful in preserving diversity and shown improved performance (Alba and Dorronsoro, 2008; Shi et al., 2011).

Al-Betar et al. (2013a) introduced the concept of cellular automata into the HSA and called the new approach cellular harmony search (cHS). In cHS, the HM is arranged as a two-dimensional toroidal grid, where each cell in the grid represents an individual of the HM. The structure is arranged in such a way that each cell has exactly the same number of neighbors. The cellular structure of the HM is shown in Figure 5.11.

A neighborhood matrix NM of dimension HMS × HMS is generated with binary values. Binary values are assigned to each element $a_{i,j}$ of NM based on neighborhood parameters. The NM matrix is expressed as

$$NM = \begin{bmatrix} a_{1,1} & a_{1,2} & \cdots & a_{1,HMS} \\ a_{2,1} & a_{2,2} & \cdots & a_{2,HMS} \\ \vdots & \vdots & \ddots & \vdots \\ a_{HMS,1} & a_{HMS,2} & \cdots & a_{HMS,HMS} \end{bmatrix} \tag{5.67}$$

where $a_{i,j}$ is defined as

$$a_{i,j} = \begin{cases} 1 & \text{if } x^j \in N(x^i) \\ 0 & \text{otherwise} \end{cases} \quad \forall i,j \in \{1,\ldots,HMS\} \tag{5.68}$$

where $N(x^i)$ is the set of all neighboring individuals of x^i arranged in the two-dimensional mesh. There can be different types of neighborhood shape as shown in Figure 5.11 by shaded circles. Neighborhood shape is fixed for a particular implementation. A random individual x^r with $r \in \{1,\ldots,HMS\}$ is selected from the HM, which undergoes improvisation. The improvisation process is carried out by interacting with the neighborhoods of specific individuals. The cHS applies a cellular memory consideration, which can control the diffusion between the individuals in the HM. The best solutions are diffused in the whole cellular HM population, and the diversity is preserved throughout the search. The updating process of the HM actually replaces the worst individual among the neighboring individuals with a new individual but not among the individuals in the whole HM. The process of the cHS is shown in Figure 5.12.

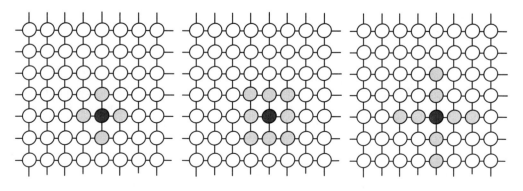

FIGURE 5.11   HM in cellular structure with different types of neighborhood.

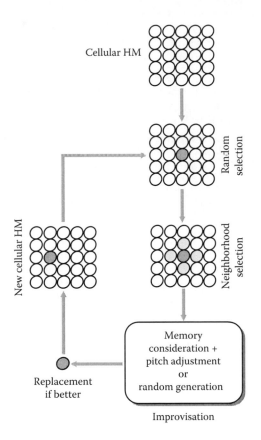

FIGURE 5.12 Process of cHS.

The main steps of cHS are as follows:

Step 1: Initialize cHS parameters and the problem parameters
Step 2: Initialize HM
Step 3: Initialize neighborhood matrix
Step 4: Select a random individual from HM
Step 5: Improvise a new individual (i.e., harmony) by
      Applying cellular memory consideration
      Applying pitch adjustment
         Or
      Random generation
Step 6: Update the HM
      Replace worse individual from the neighbors with new individual
Step 7: If (termination condition not satisfied), Goto Step 4
Step 8: Return solution

The performance of the cHS has been tested on 10 global optimization benchmark problems. The results show that the cHS algorithm outperforms the standard HSA in almost all cases.

5.7.1.15 Design-Driven HS (DDHS)

One of the important procedures in the HSA is the diversification/intensification process by applying pitch adjustment. In standard HS, pitch adjustment is purely a stochastic process specified by the parameter PAR in a random direction. Three strategies are central in the pitch adjustment step: search direction, step size, and the neighborhood in which the adjustment should be carried out. In pursuit of improvement, a number of variants of HS have been reported in the literature, but there is still scope for improvement in efficiency and accuracy by a pitch adjustment procedure of solutions in an intelligent way based on structural data of the design (or decision) variables stored in the HM. More specifically, these data can be used to determine the logical search direction and estimate the appropriate step size of adjustment and logical neighborhood efficiency and accuracy. Based on this idea, Murren and Khandelwal (2014) proposed a DDHS algorithm in order to produce higher-quality steel frame design with high robustness that decreases computational time. The DDHS method is carried out in four steps: determining the search direction and step size from data stored in the HM, mutating the design variables within defined neighborhoods of discrete spaces, performing island-hopping refinement, and storing trial solutions and closing off trial solutions by modifying the HMCR.

In the design data for search direction and step size part, the existing constraint evaluation data are retained after each iteration as member-to-capacity ratio (DCR) values and are used to estimate the search direction deterministically based on the maximum DCR in a group k (denoted as $DCR_{max,k}$) and determine the step size probabilistically. In the mutation step, two types of mutation (namely strength based or drift based) are applied based on the value of $DCR_{max,k}$. In the island-hopping step, member groups, where $DCR_{max,k}$ falls between a specified range, for example, [0.9,1.1], are mutated in a random direction with a small mutation step size. In this step of storing the trial solution and modification of the HMCR, trial solutions are stored where fitness function values are known to prevent unnecessary time-consuming computation of fitness functions. When the solutions start converging to the optimum (approximately when the lowest fittest harmony vector approaches within 8% of the best fitness), random generation of trial solutions is closed off by setting the HMCR to unity. The innovations of the DDHS are the guided mutation schemes, which are different from stochastic mutation schemes. The proposed DDHS method has been verified on a 2-bay 3-story frame, 3-bay 24-story frame, and 3-bay 24-story frame with extended column space design problems.

5.7.1.16   Island-Based HS

Poor performance of population-based meta-heuristic algorithms such as the HSA is mainly caused by decreasing population diversity due to genetic drift (Al-Betar et al., 2013a,b). There have been many methods proposed for improving the diversity. Among the popular nonpanmictic EA models introduced by Corcoran and Wainwright (1994) is the island model. The island model is a structured population mechanism used in meta-heuristic algorithms and EAs to preserve the diversity of the population that counteracts genetic drift and contributes to performance improvement. In the island model, the total population is divided into subpopulations (called islands). Each subpopulation is an independent standard EA (Tomassini, 2005), and each of them runs with its own parameter setting. The cross-island interaction of the population is carried out through a migration process which controls incoming and outgoing individuals across islands at a migration rate and migration frequency. The migration process uses a migration topology which defines the route of incoming and outgoing individuals among islands (Rucinski et al., 2010). The island model further involves consideration of a number of parameters such as the number of islands (I_n), population size of each island (I_s), migration rate (R_m), and migration frequency (F_m).

Al-Betar et al. (2015) introduced the island model embedded into the main framework of the HSA to improve population diversity. The total harmonies of size HMS stored in the HM are distributed into I_n sub-HM or islands of size I_s. The improvisation process and the updating process of the sub-HM are performed separately for each island, which can be done synchronously or asynchronously. The migration process is run according to the migration topology to exchange individuals among islands, determined by a migration rate (R_m) at every predefined number of iterations

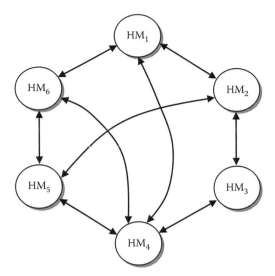

FIGURE 5.13 Island HM and migration topology in the island-based HSA.

and determined by migration frequency (F_m). The migration process follows a migration policy of replacing the worst individuals in the neighboring island by the best individuals from one island. An island HM with the migration topology is shown in Figure 5.13.

The main steps in island-based HS are as follows:

Step 1: Set parameters
        Initialize HM
Step 2: Divide HM into islands HM_i of size I_s, $i = 1, 2, \ldots, I_n$
Step 3: Improvise new harmony
        Each island invokes improvisation separately
        New harmony is generated by applying three operators:
            a. Island memory consideration
            b. Pitch adjustment
            c. Random consideration
Step 4: Update HM_i of each island
Step 5: Perform migration process according to migration topology
        Replace a specified number of the worst individuals by the best individuals from other islands
Step 6: If (termination condition not satisfied), Goto Step 3
Step 7: Return solution

Al-Betar et al. (2015) verified the effectiveness and performance of the proposed island-based HS on 25 unimodal and multimodal test functions with different characteristics selected from the set of test functions of IEEE-CEC2005. The experimental results of the benchmark functions show competitive performances in all cases. Experimentally, higher values of I_n and I_s lead to better results obtained due to better population diffusion. Experimental results also show that the migration frequency (F_m) plays an influential role over the migration rate (R_m).

5.7.1.17 Harmony Memory (HM) Initialization

Geem (2012) carried out extensive simulation on the effects of the initial HM. For enhancing the frequency and reaching the global optimum, he proposed two strategies for initializing the HM:

1. Generate initial harmonies more than the size of the HM
2. Limit the number of identical harmonies in the HM

The first approach is to generate initial harmonies more than HMS. In this approach, more harmonies, that is, m times of HMS, are generated. The harmonies are evaluated against the objective function, and only the good harmonies are selected for the initial HM of size HMS. This approach may increase the chances of identifying the region containing the global optimum.

The second approach is to limit the number of identical harmonies in the HM. Although the new harmony x^{new} is better than the worst harmony x^{worst}, those two harmonies are not swapped if the HM already contains an allowable number (τ) of identical harmonies as follows:

$$(x^{new} \in \text{HM}) \wedge (x^{worst} \notin \text{HM}) \text{ where } (x^{new} \succ x^{worst}) \wedge [n(x^{new} = x^j) < \tau, j = 1, 2, \dots, \text{HMS}] \quad (5.69)$$

The number of identical harmonies n is initially set to zero. The newly generated harmony x^{new} is compared with each harmony x^j stored in the HM. If the function value of x^{new} is identical to that of x^j, it is assumed that x^{new} and x^j are tentatively identical, and then further compare each element of the two harmonies. If every element in x^{new} exactly matches that in x^j, then n is increased by one and n should be less than τ. Geem (2012) verified the performance of the proposed approach on optimization of two-loop water network systems by carrying out extensive simulation and showed that limiting the number of identical harmonies in the HM enhanced the solution quality in terms of global optimum frequency and objective function value. It was also shown that generating more initial harmonies did not affect the solution quality significantly.

Degertekin (2008) also proposed a new HM initialization technique that generated initial harmonies of size $2 \times$ HMS and selected the best HMS harmonies for the HM. The first kind of variants are based on the initialization of the HM or extended HM structure. Cheng (2009) reported a modified HSA where the harmonies are rearranged into several pairs and the better pairs are used to develop several new harmonies. This prevents the HSA for large-scale problems from getting trapped into the local minima.

5.7.1.18 Grouping HS

Falkenauer (1993) showed that the traditional EA approach cannot deal with grouping problems due to huge increase of redundant solutions in the solution space. In order to counteract the redundant solution issue, Falkenauer proposed the grouping GA by modifying the genetic operators. The grouping technique has been applied to other meta-heuristic algorithms. Landa-Torres et al. (2012) utilized the concepts of the grouping GA in HS and proposed the grouping HSA for the access node location problem (ANLP) consisting of N nodes randomly spread over a $K \times K$ grid with different types of T concentrators. In grouping-based HS, the encoding strategy is accomplished by separating each solution vector into parts subject to certain criteria based on the problem restrictions. For example, the solution vector S for the ANLP is encoded into three parts as $S = \{s_x | s_y | s_n\}$. Here, s_x is the assignment part comprising N integer indices from a set $\{1, \dots, M\}$, s_y is the grouping part consisting of M integer indices drawn from the set $\{1, \dots, N\}$, and s_n is the encoding of a solution (i.e., node-type in ANLP) part. It is seen that s_n has the same length as s_y and consists of integer values from the set $\{1, \dots, T\}$. The improvisation operation is applied to the s_x part, and the construction of $\{s_y, s_n\}$ is made based on a deterministic criterion, which renders optimum concentrators for each group and ensures a node model with least cost. To verify the performance and effectiveness of the grouping HS, Monte Carlo simulations in synthetic instances have been carried out, which show that the proposed approach provides a faster convergence rate, less computational complexity, and better statistical performance than alternative algorithms such as grouping GAs. Askarzadeh and Rezazadeh (2011) also reported a grouping-based HSA for modeling of systems.

5.7.1.19 Multiple Pitch Adjustment Rate HS (PAR HS)

In general, single PAR is used in HSA. Geem et al. (2005a) proposed a multipitch adjusting rate (multiple PAR) for a generalized orienteering problem. They proposed three PARs that are the rates of moving to the nearest, second nearest, and third nearest points, respectively. Al-Betar et al. (2010a) also proposed a multipitch adjusting rate strategy for enhancing the performance of HS in solving a course timetabling problem. They proposed eight procedures instead of using one PAR value, each of which is controlled by its PAR value range. Each pitch adjustment procedure is responsible for a particular local change in the new harmony. Furthermore, the acceptance rule for each pitch adjustment procedure is changed to accept the adjustment that leads to a better or equal objective function.

5.7.1.20 Geometric Selective HS

Castelli et al. (2014) introduced a selection process in HS and proposed a new geometric selective HS called geometric selective harmony search (GSHS). GSHS is different from HS and existing selection-based HS variants in the following way: (i) it uses a selection procedure that is typical of GAs, (ii) the process of generating a new harmony makes use of a particular recombination operator that combines the information of two harmonies, and (iii) the algorithm includes a mutation operation that uses the PAR parameter. Specifically, geometric crossover produces an offspring that is not worse than the worst of its parents, and geometric mutation causes a perturbation on the semantics of solutions, whose magnitude is controlled by a parameter. The new approach was applied to 20 benchmark problems of the CEC 2010 suite. In all cases, it outperformed other variants.

5.7.1.21 Adaptive Binary HS

Wang et al. (2010a,b) pointed out that the dysfunction of the pitch adjustment rule degraded the search ability of BHS and redesigned the pitch adjustment operation in the proposed discrete binary harmony search (DBHS) algorithm. Afterward, they extended DBHS to tackle the Pareto-based MOOPs (Wang et al., 2011a,b). Based on the analysis of the drawback of the binary-valued HS, an improved adaptive binary harmony search (ABHS) algorithm is proposed by Wang et al. (2013) with a novel pitch adjustment operator to solve the binary-coded problems more effectively. In ABHS, the HM is an HMS $\times M$ matrix where HMS individuals are represented by binary strings of length M, that is, HM looks like

$$\text{HM} = \begin{bmatrix} h_{1,1} & \cdots & h_{1,M} \\ \vdots & \ddots & \vdots \\ h_{\text{HMS},1} & \cdots & h_{\text{HMS},M} \end{bmatrix}, h_{i,j} \in \{0,1\}, i \in \{1,2,\dots,\text{HMS}\}, j \in \{1,2,\dots,M\} \qquad (5.70)$$

The new improvised harmony vector x_i' can be picked up from the different vectors in the HM for bit selection or randomly selecting the harmony vector from the HM according to the probability of the HMCR. The pitch adjustment is simply done by complementing (i.e., inverting) the bits of the harmony vector with the probability of the PAR. A number of adaptive mechanisms for the HMCR and the PAR have been proposed such as linear increment, linear decrease, nonlinear increment, and random increment. Various adaptive mechanisms are examined and investigated, and a scalable adaptive strategy is developed for ABHS to enhance its search ability and robustness. The experimental results on benchmark functions and 0–1 knapsack problems demonstrate that the proposed ABHS is efficient and effective, and outperforms the binary HS and the global HSA.

There are many other variants of HS published recently, which have not attracted attention from the research community yet such as social HS (Kaveh and Ahangaran, 2012), intelligent tuned HS

(Yadav et al., 2012), and multipopulation-based HS with external archive (Turky and Abdullah, 2014), which has been applied to dynamic optimization problems.

5.7.2 HSA VARIANTS BASED ON HYBRIDIZATION WITH OTHER METHODS

A number of modified variants of HS were developed for enhancing the solution accuracy and the convergence rate. Modifications to the original HS may be classified into several categories: (1) parameters such as the HMCR and the PAR are dynamically adapted by the learning mechanisms, while bw is dynamically adjusted to favor exploration in the early stages and exploitation in the final stages of the search process; (2) new operations are introduced from other heuristic algorithms for improvisation; (3) rules for handling constraints during generation of new harmonies such as after generating a new harmony; (4) different criteria for deciding when to include a new harmony in the HM; and (5) hybrid with other meta-heuristics.

In order to improve exploration or global search ability, exploit local search, increase convergence speed, improve solution quality, and minimize computational cost, the standard HSA has been hybridized with other meta-heuristic algorithms. The hybridization can be categorized into two broad groups: (i) incorporation of some techniques of other meta-heuristic algorithms into HS procedure and (ii) incorporation of some HS techniques into other meta-heuristic algorithms.

In general, different meta-heuristic algorithms have their own advantages and disadvantages. The most common disadvantageous feature of all meta-heuristic algorithms is the premature convergence at local minima. Among other detrimental features are the slow convergence demanding more computational cost, low-quality solution meaning poor exploitation or local search ability, lack of powerful operations, for example, HS does not have any crossover or mutation operations, and a poor representation scheme. In order to minimize the impact of detrimental features, hybridization with other meta-heuristic algorithms is performed that will improve the overall search abilities, solution quality, and performance of these optimization algorithms. In both cases, the original ability and structure of the HSA to be integrated with other meta-heuristics become relatively easy and flexible (Yang, 2009). Alia and Mandava (2011) reported a survey on different hybrid variants of HS in 2011. There have been more hybrid variants reported in the literature in the recent years, which are also included in this section.

5.7.2.1 Hybridizing HS with Other Meta-Heuristic Algorithms

In this type of hybrid algorithms, components of different techniques from other meta-heuristic algorithms are incorporated into HSA to enhance performance and local search. There have been many hybrid algorithms reported by researchers during the past few years such as particle swarm optimization, swarm intelligence (SI), ACO, artificial bee colony algorithm, biogeography-based optimisation, DE, GA, hunting search, simplex, simulated annealing, hill climbing optimization, clonal selection algorithm, sequential quadratic programming, Broydon-Fletcher-Goldfarb-Shanno, K-means, fuzzy C-means, etc. These hybrid HS algorithms are discussed in the following.

5.7.2.1.1 Hybrid of HS and PSO

Omran and Mahdavi (2008) proposed GHS borrowing the concept of the global best particle from PSO, replaced the parameter bw, and introduced random selection of decision variables from the best harmony vector in the HM. GHS and some of its variants are discussed in Section 5.7.1.3. Coelho and Bernert (2009) proposed a hybrid version of the HS by integrating a component from dispersed PSO (Cai et al., 2008) into HS. The hybrid method actually introduced a dynamic PAR using the PSO concept. Geem (2009b) also used the same concept to improve the selection process of harmony vectors from the HM.

In classical PSO, the movement of particles is determined by the velocity in the particular iteration (or generation) and the position is updated using the velocity. Pandi and Panigrahi (2011) proposed a hybrid HSA and PSO approach where this concept is implemented. That is, the position of the particle is adjusted in such a way so that the highly fitted particle moves slowly when compared to the low fit particle. A parameter ω is introduced to preserve the good locations near the global optimum in the search space. The value of ω_i is defined for each particle according to their rank. The velocity $v_i^j(t)$ is updated using ω_i and $v_i^j(t-1)$. Multiplying the velocity of the i-th particle of previous generation $v_i^j(t-1)$ by ω_i is conceptually similar to elitism used in many EAs. The ω_i value is defined as

$$\omega_i = \omega_{\min} + (\omega_{\max} - \omega_{\min}) \frac{rank_i}{Population\_size} \tag{5.71}$$

where $\omega_{\min}$ and $\omega_{\max}$ are arbitrary parameters and chosen as 0.1 and 0.8, respectively. The variable *Population_size* in Equation 5.71 is the size of the HM. The pitch adjustment is carried out as follows:

If rand(0,1) $\leq$ PAR(t), then do pitch adjustment
{
  $$v_i'(t) = \omega_i \cdot v_i^j(t-1) + c_1 r_1 (p_i^l - x_i^l) + c_2 r_2 (p_{gi}^l - x_i^l)$$
  $$v_i'(t) = \text{sign}\left[v_i^j(t)\right] * \min\left(\left|v_i'(t)\right|, v_{\max}^l\right)$$
  $$x_i'(t) = x_i'(t-1) + v_i'(t) \times T$$
Else
  $$x_i' = L_i = \text{rand}(0,1) \times (U_i - L_i) \text{ \{random generation from the range\}}$$
  $$v_i'(t) = r_3 v_{\max}(t)$$

where $v_i'(t)$ is the updated velocity of the current iteration; rand(0,1) is a random number distributed over [0,1]; T is the time, which is unity to convert the velocity into position; x_i^l, p_i^l, and p_{gi}^l are the position, iteration best, and global best position, respectively; $v_{\max}^l(t)$ is the max velocity; c_1 and c_2 are cognitive and social coefficients, respectively, having the value 2.1 as used by the classical PSO algorithm; and r_1, r_2, and r_3 are the random numbers generated between [0, 1]. The rest of the algorithm remains the same. The hybrid PSO and HS approach has been verified on three cases of a dynamic ELD problem, which demonstrates the effectiveness and performance improvement of the algorithm.

Zhao et al. (2010, 2011) proposed a new hybrid approach by combining the exploration capabilities of the dynamic multiswarm particle swarm optimizer (DMS-PSO) and the stochastic exploitation of the HSA. DMS-PSO was introduced by Liang and Suganthan (2005) where the neighborhood topology is dynamic and randomized. DMS-PSO exhibits superior exploratory capabilities on multimodal problems than other PSO variants. The new hybrid approach is known as DMS-PSO-HS. In the DMS-PSO-HS algorithm, the whole DMS-PSO population is divided into a large number of small and dynamic subswarms, which are also individual HS populations. These subswarms are regrouped frequently, and information is exchanged among the particles in the whole swarm. Because of frequent regrouping, DMS-PSO-HS enables the particles having more diverse patterns, learning from them and forming new harmonies. The algorithm has been verified on 14 benchmark optimization functions. DMS-PSO-HS makes good use of information resulting in improved solution quality compared to DMS-PSO and HS.

Wang and Yan (2013) proposed a new approach combining GHS (discussed in Section 5.7.1.21) and PSO where the two control parameters of GHS, namely HMCR and PAR, are dynamically adjusted along with optimization of HS using PSO. The proposed approached is called PSO-CE-GHS. It is similar to SGHS, discussed earlier, except the two parameters HMCR and PAR are coevolved using PSO according to the following rules:

$$\begin{cases} v_{1i}(t+1) = w \times v_{1i}(t) + c_1 \times r_1 \times [pbest_{1i}(t) - HMCR_i(t)] + c_2 \times r_2 \times [gbest_1(t) - HMCR_i(t)] \\ \qquad\qquad HMCR_i(t+1) = HMCR_i(t) + v_{1i}(t+1) \end{cases} \quad (5.72)$$

$$\begin{cases} v_{2i}(t+1) = w \times v_{2i}(t) + c_1 \times r_1 \times [pbest_{2i}(t) - PAR_i(t)] + c_2 \times r_2 \times [gbest_2(t) - PAR_i(t)] \\ \qquad\qquad PAR_i(t+1) = PAR_i(t) + v_{2i}(t+1) \end{cases} \quad (5.73)$$

where $pbest_{1i}$ and $gbest_{1i}$ are the local best and global best values of $HMCR_i$, and $pbest_{2i}$ and $gbest_{2i}$ are the local best and global best values of PAR_i, respectively.

The effectiveness and performance of the PSO-CE-GHS are verified on 14 unimodal, multimodal, and constrained benchmark optimization functions, and the performances were compared with standard HS, IHS, GHS, and SGHS. The experimental results show that the proposed approach is more powerful than existing variants and suitable for constrained optimization problems.

5.7.2.1.2 Hybrid of HS and Swarm Intelligence

Zou et al. (2010) proposed a hybrid of HS and SI of particle swarm. This approach is called novel global HS (NGHS). The proposed hybrid approach introduced two new operations: position updating and genetic mutation. The position update operator actually implements the PSO concept of the global best particle in a swarm. As a consequence, the worst harmony in the HM is replaced with the global best harmony at each iteration. This eventually affects the diversity of HS and leads to a premature convergence problem. To counteract the premature convergence, a simple remedy is the introduction of the second operator of genetic mutation with a small probability, which reinstalls the diversity back. The new hybrid approach and the HS are different in three aspects as follows:

1. HMCR and PAR are excluded from the algorithm, and genetic mutation with small probability (p_m) is introduced.
2. The improvisation of the HSA is modified for each decision variable as follows:
   $x_r = 2 \times (x_i^{best} - x_i^{worst}), \ \forall i = 1, \ldots, N$
   If $(x_r > U_i)$
   $\quad x_r = U_i$
   If $(x_r < L_i)$
   $\quad x_r = L_i$
   $x_i' = x_i^{worst} + r_1 \times (x_r - x_i^{worst})$
   If $(r_2 \le p_m)$
   $\quad x_i' = L_i + r_3 \times (U_i - L_i)$
   where N is the number of decision variables; L_i and U_i are the lower and upper bounds of the decision variables, respectively; r_1, r_2, and r_3 are random numbers uniformly distributed over [0,1]; and x_i^{best} and x_i^{worst} are the global best and worst vectors in the HM, respectively.
3. The worst harmony x_i^{worst} in the HM is replaced with the newly generated harmony vector x_i' even if it were worse than x_i^{worst}.

With these three modifications, the main steps of NGHS become as follows:

Step 1: Initialize HSA parameters
Step 2: Initialize HM

Step 3: Improvise a new harmony

$$x_r = 2 \times (x_i^{best} - x_i^{worst}), \ \forall i = 1, \ldots, N$$

If $(x_r > U_i)$

$$x_r = U_i$$

If $(x_r < L_i)$

$$x_r = L_i$$

$$x_i' = x_i^{worst} + r_1 \times (x_r - x_i^{worst})$$

If $(r_2 \le p_m)$

$$x_i' = L_i + r_3 \times (U_i - L_i)$$

Step 4: Update the HM

Replace x_i^{worst} with x_i'

Step 5: If (termination condition not satisfied), Goto Step 3

Step 6: Return solution

The efficiency of NGHS has been verified on three engineering optimization problems, namely a complex bridge system, overspeed protection system for a gas turbine, and large-scale reliability problem. The experimental results show that NGHS has improved convergence and higher exploration capability.

5.7.2.1.3   Hybrid of HS and ACO

ACO (also known as ant algorithm and ant system) was first proposed by Dorigo et al. (1991). The basic idea of ACO is simple. The ants move on a path leaving pheromone trails as they pass through the path in a random manner. The probability of selecting a path is proportional to the amount of pheromones and inversely proportional to the cost of the path. The amount of pheromones is changed when a certain number of ants (called colony) have crossed the path. A brief discussion on the ACO algorithm is provided in Chapter 8. Two concepts, namely, pheromone and heuristic values (cost of path) are borrowed from ACO and utilized in HS. Amini and Ghaderi (2013) proposed the hybrid AntHS approach combining the ACO and HS algorithm. The pheromone and heuristic values for a solution component $c_i \in \mathbf{C}$ are denoted by $\tau(c_i)$ and $\eta(c_i)$, respectively.

The pheromone values $\tau(c_i)$ are dynamic weight factors assigned to each solution component and updated as the optimization process progresses. The pheromone update rule is defined as

$$\tau(c_i) = (1-\rho) \times \tau(c_i) + w \times \rho \times F_q(x) \quad \text{for } \forall c_i \in x \tag{5.74}$$

where $\rho \in (0,1]$ is the pheromone evaporation rate. The factor ρ can also change dynamically with time defined by $\rho(t) = \rho^t$. Here, w is the quality weight factor; $F_q: x \to R+$ is a quality function, which assigns greater values to the solution with higher objective function values; and $F_q(x)$ is defined as

$$F_q(x) = \frac{\beta}{f(x)} \tag{5.75}$$

where β is a parameter to make the quality function dimensionless and $f(\cdot)$ is the objective function defined for the optimization problem.

The pheromone values $\tau(c)$ associated to the solution components within that solution are updated. This is performed based on the value of the objective function and the quality of the solution. The heuristic values $\eta(c_i)$ are constant weight factors assigned by algorithm to solution components based on existing prejudgments on the quality of solution components. The heuristic values are predominantly problem dependent.

In each iteration, pheromone and heuristic values are used to compute a parameter called probability mass function (PMF) (similar to probability of selecting a path in the original ACO) for a discrete random variable defined on a sample space B. The set B represents a subset of C. The PMF is used every time AntHS samples a solution component c_i either from the set C or the HM. The PMF is defined as

$$\text{PMF}(c_i) = \begin{cases} \dfrac{\tau(c_i) \times \eta(c_i)}{\sum_{c_j \in B} \tau(c_j) \times \eta(c_j)} & \text{if } c_i \in B \subset C \\ 0 & \text{if } c_i \notin B \end{cases} \tag{5.76}$$

where $\sum_{c_j \in B} c_j = 1$.

A new solution x is constructed by selecting solution components c_i, and then the new solution is evaluated against the objective function. If the quality of the new solution is better than the worst solution in the HM, it replaces the worst solution in the HM. Pitch adjustment in AntHS is carried out in the following ways:

1. Assemble a solution x from a set of solution components c_i for $\forall c_i = \{n_j, n_k\} \in x$ near the solution which will undergo pitch adjustment. The set is called the vicinity set $V(c_i)$ of that solution component and defined by

$$V(c_i) = \{c \mid c = \{n_p, n_q\} \in C, c \cap c_i \neq \varnothing\} \tag{5.77}$$

2. Compute the PMF over the vicinity set using Equation 5.76
3. Perform sampling from the vicinity set $V(c_i)$ using the PMF
4. Use the recently sampled solution component in the solution construction procedure

The main steps of the AntHS algorithm are as follows:

Step 1: Initialize HM,
        Initialize pheromone values $\tau(c_i)$ for $\forall c_i \in C$
        Analyze the problem and assign heuristic values $\eta(c_i)$ for $\forall c_i \in C$
ACO part:
Step 2: Update pheromone values $\tau(c_i)$
Step 3: Compute PMF using $\tau(c_i)$ and $\eta(c_i)$
HS part:
Step 4: Improvise solution: sample using PMF
Step 5: Evaluate solution
Step 6: Update HM
Step 7: If (termination condition not satisfied), Goto Step 2
Step 8: Return solution

The proposed AntHS has been applied to finding optimal locations for dampers in structural systems through three benchmark problems such as shear building, two-dimensional truss, and two-dimensional frame. Experiments with the three examples of optimal locations for dampers show that AntHS generally has a better performance than the original HS.

5.7.2.1.4   Hybrid of HS and ABC

The ABC algorithm is a new SI technique inspired by the foraging behavior of honey bees. The ABC algorithm was introduced by Karaboga and Basturk (2007, 2008) and further enhanced by Karaboga and Bahriye (2009). In the ABC algorithm, the colony of bees comprises three groups of bees: employed bees, onlookers, and scouts. An employed bee collects honey from the source, the onlooker bee waits at the dance area to make a decision to choose a food source, and the scout bee carries out random search for a new source. The position of a food source represents a possible solution to the optimization problem. The nectar amount of a food source corresponds to the quality (or fitness) of the associated solution. In the ABC algorithm, the first half of the colony consists of employed bees and the second half consists of onlooker bees. The number of solutions in the population is equal to the number of employed or onlooker bees. The basic idea of the ABC algorithm is simple. A population of food source (i.e., solutions) x_i, $i = 1,2, \ldots ,$ SN is randomly generated in the range of parameters where SN is the number of the food sources, that is, the size of the population:

$$x_{ij} = x_{ij}^{\min} + \text{rand}(0,1)(x_{ij}^{\max} - x_{ij}^{\min}) \tag{5.78}$$

where rand(0,1) is a random number uniformly distributed over the range, $j \in \{1, 2, \ldots , D\}$, D is the number of decision variables of the problem, and $x_{ij}^{\max}$ and $x_{ij}^{\min}$ are the maximum and minimum values of the parameter j, respectively.

A new food source is created by an employed bee in the neighborhood of current food source x_{ij} according to

$$x_{ij}' = x_{ij} + \varphi_{ij}(x_{ij} - x_{kj}), \quad \forall j = 1,\ldots,D \tag{5.79}$$

where $k \in \{1, 2, \ldots , SN\}$, $k = \text{int}(\text{rand} \times SN) + 1$ is the index of the solution x_{kj} chosen randomly from the colony, and φ_{ij} is a random number uniformly distributed within the range of [–1,1].

The population undergoes $C = 1, 2, \ldots ,$ MCN cycles of the search process. As the search approaches the optimum solution, the term $(x_{ij} - x_{kj})$ decreases to zero. The fitness of the new food source x_i' is evaluated by the amount of food. If the fitness value of x_i' is better than x_i, then x_i is replaced by x_i' according to

$$x_i = \begin{cases} x_i' & \text{if } f(x_i') > f(x_i) \\ x_i & \text{otherwise} \end{cases} \tag{5.80}$$

Employed bees complete the search process and share information on food amount with onlooker bees. An onlooker bee evaluates the food source (fitness value) and selects a food source based on the probability defined as

$$p_i = \frac{f(x_i)}{\displaystyle\sum_{i=1}^{SN} f(x_i)} \tag{5.81}$$

where $f(x_i)$ is the fitness value of the food source (solution) x_i, which is proportional to the amount of the food. The number of food sources SN is equal to the number of employed bees.

If a better food source cannot be found in the neighborhood of the current source after a predetermined number of iterations, then the food source is assumed to be abandoned. In that situation, the corresponding employed bee becomes a scout. A new food source is generated by the scout bees using Equation 5.78. The abandoned food source is replaced with a new food source found by a scout. The process is repeated for MCN iterations.

In order to enhance the accuracy and convergence rate of HS, Wu et al. (2012) proposed incorporation of an ABC algorithm into HS to improve the improvisation strategy. The proposed hybrid

HS with ABC is called HHS-ABC. Though the performance of the HSA mostly depends on the improvisation strategy involving the parameters HMCR, PAR, and *bw*, the HM plays the key role by providing the pool of elite solutions. The HHS-ABC algorithm is an enhancement to HS for solving global numerical optimization problems where the HM is considered as food sources that are being explored and exploited by employed, onlooker, and scout bees to improve solution quality. The main steps of the HHS-ABC algorithm are as follows:

Step 1: Set HS and ABC parameters
$$\text{HM} \leftarrow x_i^{\min} + \text{rand}(0,1)(x_i^{\max} - x_i^{\min}), \forall j = 1, \ldots, D$$
$f(x_i) = \text{Evaluate } (x_i)$

Step 2: Improvise HM using ABC

2.1: Employed bees search
For $i = 1$: HMS
$$x_i' = x_i + \phi_i (x_i - x_k)$$

$$x_i = \begin{cases} x_i' & \text{if } f(x_i') > f(x_i) \\ x_i & \text{otherwise} \end{cases}$$

2.2: Compute selection probability p_i
For $i = 1$: HMS

$$p_i = \frac{f(x_i)}{\sum_{i=1}^{SN} f(x_i)}$$

2.3: Onlooker bees search
For $i = 1$: HMS
{

Find an index value k using roulette wheel selection
Onlooker bees search as employed bees (same as Step 2.1)

}

2.4: Scout bees search
If the number of trials is greater than limit
$$\text{HM} \leftarrow x_i^{\min} + \text{rand}(0,1)(x_i^{\max} - x_i^{\min})$$

Step 3: Improvise new harmony x_{new} using HS
If $(r_1 < \text{HMCR})$
{

$x_{\text{new}} = x_k + r_2 * bw$
If $(r_3 < \text{PAR})$
$x_{\text{new}} = x_{\text{best}}$

}
Else
$$x_{\text{new}} = x^{\min} + r_4 * (x^{\max} - x^{\min})$$

Step 4: Update HM
$$x_{\text{worst}} = x_{\text{new}} \text{ if } f(x_{\text{new}}) > f(x_{\text{worst}})$$

Step 5: If (termination condition not satisfied), Goto Step 2

Step 6: Return solution

where r_1, r_2, r_3, and r_4 are random numbers uniformly distributed over [0,1].

Wu et al. (2012) employed chaotic ABC in the HHS-ABC algorithm using logistic function (see Appendix C for different chaotic maps). Two versions of chaotic ABC are used in this hybrid approach: employed bees carry out chaotic search, and onlooker bees carry out chaotic search. Another modification is made to ABC by incorporating the idea of PSO (called PABC) where bees use historical information about the food source and their quality. The new food source is computed using bees' own experience, that is, cognitive knowledge, and the knowledge of the other bees, that is, social knowledge. The bees update the food source according to

$$x'_{ij} = \omega \cdot x_{ij} + c_1 \varphi_{ij}(x_j^{\text{best}} - x_{ij}) + c_2 \varphi_{ij}(x_{kj} - x_{ij}), \quad j = 1,\dots,D \tag{5.82}$$

where ω is the inertia weight, $\{c_1, c_2\}$ are positive constants representing cognitive and social weighs, x^{best} is the best food source, and x_k represents the random neighbor.

The effectiveness and performance of the proposed HHS-ABC algorithm and the three variants have been verified on a number of well-known benchmark global optimization problems. Experimental results show that the proposed algorithms can find better solutions when compared to standard HSA and other heuristic algorithms and are powerful search algorithms for various global optimization problems.

5.7.2.1.5 Hybrid Differential HS

Practical experiences demonstrate that the HSA suffers from slow and/or false convergence over multimodal and rough fitness landscapes when applied to numerical function optimization. A more powerful explorative approach is deemed necessary to overcome such problems. A practical optimization technique should satisfy three demands for any problem. First, the method should find the true global optimum. Second, the algorithm should ensure fast convergence. Third, the algorithm should have a minimum number of control parameters. Considering these three demands, DE is a fast and simple technique. DE is a population-based direct search algorithm developed by Storn and Price (1997), involves only three parameters, and performs extremely well on a wide variety of test problems (Storn, 1999). It has been demonstrated by Zaharie (2002) that DE has better explorative power than other EAs. The main difference between DE and other EAs is that DE uses differences of two randomly selected individuals (parameter vectors) as the source of perturbing the vector population rather than probability function as an evolution strategy. It performs mutation based on the distribution of the solutions in the current population first and then applies a crossover operator to generate offspring. In this way, search directions and possible step sizes depend on the location of the individuals selected to calculate the mutation values. A detailed description of DE and its operators can be found in Siddique and Adeli (2013). The basic algorithm of DE is simple and straightforward, consisting of the following main steps:

Step 1: Initialize population and parameters
Step 2: Evaluate individuals in population
Step 3: Perform mutation using the difference vectors
Step 4: Perform crossover using exponential (two-point modulo) or binomial (uniform) operator
Step 5: Select individuals based on fitness

Chakraborty et al. (2009) proposed a new improvement to HS by introducing the mutation operator borrowed from the DE algorithm and called it differential HS (DHS), which replaces the pitch adjustment operation in classical HS with a mutation strategy from DE. The target vector x'_i is the newly generated vector considering the memory consideration step (with the probability of the HMCR) and randomness (with the probability of $1 - \text{HMCR}$). The target vector

x_i' is mutated using the difference of two randomly selected members from the HM using the following rule:

$$x_i' = x_i' + F \cdot (x_{r1} - x_{r2})$$ (5.83)

where $F \in [0,1]$ is a uniformly distributed random scalar number called the scale factor, and x_{r1} and x_{r2} are the two randomly selected members of the HM. The pitch adjustment operation using the PAR is excluded in DHS. Chakraborty et al. (2009) also found that the mutation strategy DE/rand/1 employed by Storn et al. (2005) produces the best result when hybridized with classical HS. The main steps of DHS are as follows:

Step 1: Initialize HSA parameters
Step 2: Initialize HM
Step 3: Improvise a new harmony

$$x_i' = \begin{cases} x_i' \in \left\{ x_i^1, x_i^2, \ldots, x_i^{HMS} \right\} & \text{if} \left(r \leq \text{HMCR} \right) \\ L_i + \text{rand}(0,1) \times (U_i - L_i) & \text{if} \left(r \leq 1 - \text{HMCR} \right) \end{cases}$$

Mutate x'_i using DE/target-to-best/1 scheme
$x'_i = x'_i + F \cdot (x_{r1} - x_{r2})$
Step 4: Update the HM
Step 5: If (termination condition not satisfied), Goto Step 3
Step 6: Return solution

The performance of the resulting DHS algorithm is tested on six well-known benchmark functions and compared with standard HS, GHS, and a very popular variant of DE based on the performance indices of accuracy, which showed higher computational speed and frequency of hitting the optima. Chakraborty et al. (2009) also provided rigorous mathematical analysis of performance. All the experimental results supported the theoretical performance analysis and demonstrated superior performance of DHS over the others. Poursalehi et al. (2013a,b) also applied the DHS to a reactor core loading pattern optimization problem with the scale factor $F \in [-1,1]$, which again demonstrated near-optimal performance of the algorithm.

Wang and Li (2013) proposed an effective DHS algorithm by combining all the mechanisms of both DE and HS. Three effective mechanisms are incorporated in the improvisation step to enhance the performance of DHS. First, a new individual is generated for each individual in the population using memory consideration with the parameter HMCR, followed by pitch adjustment operation combining differential mutation with the parameter PAR to enhance the exploitation ability of HS. Second, exponential crossover operation is employed between individuals selected from the new and original population to strengthen the exploration ability. Third, a repair mechanism using simple selection rules is included for constraint handling of the problem at hand. The rest of the steps of DHS remain the same. The effectiveness of the approach has been verified on an ELD problem, and the experimental results confirmed improvement.

To handle more nonsmooth and nonconvex optimization problems such as the dynamic economic dispatch (DED) problem, Arul et al. (2013) proposed a chaotic self-adaptive DHS algorithm, where a chaotic self-adaptive differential mutation operator is used instead of the pitch adjustment operator to enhance the search performance and quality of the solution. In chaotic self-adaptive DHS, the scale factor $F \in [0,1]$ in the differential mutation Equation 5.83 is modified with an adaptive mechanism and a chaotic number defined by

$$F = F_L + x_{n+1} \cdot (F_U - F_L)$$ (5.84)

where F_U and F_L are the upper and lower limits of F, respectively, and x_{n+1} is a chaotic sequence.

A chaotic sequence has been incorporated into many heuristic optimization algorithms to enrich search capability and help in avoiding local optimum. Different chaotic maps are discussed in Appendix C. The logistic map, a discrete chaotic system with any dimensionality that can exhibit strange attractors, has been widely used in many research works due to its efficiency for heuristic optimization. The other advantage is that its average computation time is less than that of other chaotic maps. The mathematical expression of the logistic map to generate x_{n+1} is simple and represented by

$$x_{n+1} = 4 \cdot x_n \cdot (1 - x_n) \tag{5.85}$$

where $x_n \in [0,1]$, n denotes the iteration number and x_n is distributed over the interval of $[0,1]$. The first x_{n+1} value is generated using $x_0 = 0.91$ for $n = 0$.

In chaotic self-adaptive DHS, the new harmony vector is generated using Equation 5.83 with scale factor F calculated using Equation 5.84. The main steps of the chaotic self-adaptive DHS are the same as DHS except for the computation of scale factor F. The effectiveness of the proposed algorithm is verified on five different cases of the DED problem.

Dash et al. (2014) proposed a self-adaptive variant of DHS (SADEHS) that uses the current to best mutation scheme of DE in the pitch adjustment operation in the improvisation process. The improvisation of a new harmony is computed using the following mechanism:

For $i = 1$ to N

$$x_i' = \begin{cases} x_i' \in \{x_i^1, x_i^2, \ldots, x_i^{HMS}\} & \text{if} \left(\text{rand}(0,1) \leq HMCR(t)\right) \\ L_i + \text{rand}(0,1) \times (U_i - L_i) & \text{if} \left(\text{rand}(0,1) \leq 1 - HMCR(t)\right) \end{cases}$$

If $(\text{rand}(0,1) < PAR(t))$

$$x_i' = x_i' + bw(t) \times (x_i^{best} - x_i') + bw(t) \times (x_i^b - x_i^c), \text{ where } b,c \in \{1,2,\ldots,HMS\}$$

where x_i^b and x_i^c are randomly selected harmony vector from the HM. At each iteration t, the parameters $HMCR(t)$, $PAR(t)$, and $bw(t)$ are adapted using the following rules:

$$\begin{cases} HMCR(t) = HMCR_{min} + (HMCR_{max} - HMCR_{min}) \times \dfrac{t}{NI} \\ PAR(t) = PAR_{min} + (PAR_{max} - PAR_{min}) \times \dfrac{t}{NI} \\ bw(t) = bw_{min} + (bw_{max} - bw_{min}) \times \dfrac{t}{NI} \end{cases} \tag{5.86}$$

The proposed SADEHS has been applied in combination with extreme machine learning (EML) to train a NN with a single hidden layer, and the performance is compared against other algorithms. The simulation results confirmed superior performance.

Gao et al. (2008, 2009a,b) proposed two modifications to HS by integrating the DE technique to fine-tune the vectors in the HM by applying the usual DE procedure. The second modification is to handle multimodal problems where a HM-updating strategy is proposed such that the new harmony meets the following criteria:

1. New harmony is better than the worst harmony in the HM.
2. There are less than a critical number of similar harmonies in the HM.
3. The fitness of the new harmony is better than the average fitness of the similar harmonies.

Qin and Forbes (2011) proposed a new hybrid version of HS integrating HS and DE by employing self-referential DE mutation operation for pitch adjustment, which also dynamically adapts the properties of the fitness landscape. To confirm improved performance, the proposed approach has been verified on 16 unimodal and multimodal test problems.

5.7.2.1.6   Hybrid of HS and GA

It is well known that HS has poor ability for local search, which leads the algorithm to converge to the local optimum. One of the reasons for such premature convergence is the lack of diversity. Therefore, increasing the diversity of the population can improve local search capability and lead to better-quality solution. There are meta-heuristic algorithms such as GAs, which have powerful techniques for population diversity. For example, arithmetic crossover (AC) operation in the GA can be utilized in HS. The GA is an exploratory search and optimization method that is based on Darwinian evolution theory, where individuals with higher fitness in the population have a higher chance of surviving to the next generation. Each individual in the GA population, represented by means of a string similar to the way genetic information is coded in organisms as chromosomes (Holland, 1975), represents a potential solution to the problem. The GA does not require mathematical descriptions of the optimization problem, but instead it relies on an objective function, in order to assess the fitness of a particular solution (Goldberg, 1989). The GA then iteratively creates new populations by applying mainly three genetic operators, namely selection, crossover, and mutation on the individuals in the old population, leading to the optimum solution. A detailed description of AC operations can be found in Amjady and Nasiri-Rad (2009) and Siddique and Adeli (2013). The end result is a search strategy that is tailored for vast, complex, multimodal search spaces.

Niu et al. (2014a,b) proposed a hybrid HS with AC operation borrowed from the GA. The hybrid approach combining AC with HS is called ACHS. In ACHS, the global best information and AC are used to update the newly generated solution, which contributes to the algorithm's exploitation capability and speeds up the convergence. AC is carried out by performing some arithmetic operation on two individuals to make a new offspring. Michalewic (1992) described three types of AC: simple arithmetic, single arithmetic, and whole AC. In all three types of operations, it works by taking the weighted sum of the two parental alleles x_j and y_j, that is, the first offspring is $o_{j1} = \alpha * x_j + (1 - \alpha) * y_j$ and the second offspring is $o_{j2} = (1 - \alpha) * x_j + \alpha * y_j$, where $j = 1,\ldots, n$ are the alleles of the parental chromosomes and $\alpha \in [0,1]$. In simple arithmetic, a random point at k in the chromosome is chosen, the first k-alleles from both parents to both children are copied, and the rest of the alleles are obtained by arithmetic averaging of parents P1 and P2. In single arithmetic, a random allele k is chosen, all alleles from parents P1 and P2 are copied to children O1 and O2, and the arithmetic average of the k-th allele from parents P1 and P2 is copied to both children. In whole arithmetic, all alleles in children are calculated by the arithmetic average of parents P1 and P2. The three crossover operators are illustrated in Figure 5.14.

Niu et al. (2014a,b) applied an AC operator to recombine only the current global best and newly generated harmony vector rather than all individuals in the population. This makes the implementation of crossover simple and preserves the simple structure of HS without increasing the computation complexity. In ACHS, Niu et al. (2014a,b) applied an opposition-based strategy to create a new harmony vector alongside random generation. The main steps of the ACHS are described as follows:

Step 1: Initialize parameters
Step 2: Initialize HM
Step 3: Improvise new harmony
     If ($r_1 \leq$ HMCR)
     {
     Choose random harmony from HM

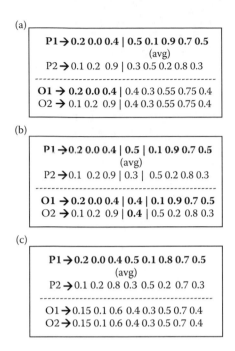

(a)

P1→ 0.2 0.0 0.4 | 0.5 0.1 0.9 0.7 0.5
(avg)
P2 → 0.1 0.2 0.9 | 0.3 0.5 0.2 0.8 0.3
O1 → 0.2 0.0 0.4 | 0.4 0.3 0.55 0.75 0.4
O2 → 0.1 0.2 0.9 | 0.4 0.3 0.55 0.75 0.4

(b)

P1 →0.2 0.0 0.4 | 0.5 | 0.1 0.9 0.7 0.5
(avg)
P2 →0.1 0.2 0.9 | 0.3 | 0.5 0.2 0.8 0.3
O1 →0.2 0.0 0.4 | 0.4 | 0.1 0.9 0.7 0.5
O2 →0.1 0.2 0.9 | 0.4 | 0.5 0.2 0.8 0.3

(c)

P1→0.2 0.0 0.4 0.5 0.1 0.8 0.7 0.5
(avg)
P2 →0.1 0.2 0.8 0.3 0.5 0.2 0.7 0.3
O1→0.15 0.1 0.6 0.4 0.3 0.5 0.7 0.4
O2→0.15 0.1 0.6 0.4 0.3 0.5 0.7 0.4

FIGURE 5.14   AC used in the GA: (a) simple AC, (b) single AC, and (c) whole AC.

$$x_i \in \left\{ x_i^1, x_i^2, \ldots, x_i^{HMS} \right\}$$

If $(r_2 \le PAR)$

    Apply pitch adjustment

    $x_i' = x_i + 2 \times \text{rand}(0,1) \times bw$

}

Else

{

    If $(r_3 \le 0.5)$

        Create new harmony applying random selection

        $x_i' = L_i + \text{rand}(0,1) \times (U_i - L_i)$

    Else

        Create new harmony applying opposition-based strategy

        $x_i' = L_i + U_i - x_i$

}

    Apply arithmetic crossover

$$x_i' = \alpha * x_i^{best} + (1 - \alpha) * x_i'$$

Step 4: Update HM

Step 5: If (termination condition not satisfied), Goto Step 3

Step 6: Return solution

where r_1, r_2, and r_3 are random numbers distributed over [0,1].

    To verify the performance of the proposed ACHS, three groups of simulation experiments are carried out based on eight commonly used benchmark functions. Three other commonly used crossover operators such as uniform crossover (Picek and Golub, 2010), blend crossover (BLX-α) (Eshelman and Schaffer, 1993), and simulated binary crossover (SBX) (Deb and Agrawal, 1995) are also investigated, and the AC shows superior performance over the others when incorporated into HS. To make a comprehensive study on the scalability, ACHS is first tested on the benchmark functions with 100 dimensions and compared with several state-of-the-art methods. The ACHS is also used to solve seven different economic dispatch cases and compared with the results reported in the literature. In all cases, the results confirm the superiority of the ACHS for different optimization problems.

5.7.2.1.7   Hybrid of HS and Hunting Search Algorithm

Kulluk (2013) proposed a novel hybrid algorithm by integrating the HSA and Hunting Search algorithms. The HSA builds the main structure of the search algorithm, and HuS forms the pitch adjustment phase of the proposed hybrid algorithm. HuS is used as a local search technique and applied in the pitch adjustment phase. The HuS algorithm is a recent meta-heuristic algorithm inspired from the hunting behavior of animals hunting in groups, such as lions, wolves, and dolphins (Oftadeh et al., 2010). Animals that hunt in groups have certain strategies of chasing or encircling the prey, cooperate with each other in positioning themselves around the prey, and correct positions while the prey is dynamically changing its position. To simulate the animals' hunting process, a hunting group (HG) is generated. The HuS algorithm uses a leader toward which all other hunters move and correct their positions. The leader in HuS is the hunter that has the best position (i.e., solution) at the current iteration, defined as

$$x^{best}(t) = \max_{j \in 1,...,N} \text{fit}[x_j(t)] \tag{5.87}$$

The best solution depends on the definition of fitness and the problem at hand. The hunter that finds a better point in the search space becomes the leader in the next iteration. The parameters of the HuS are hunting group size (HGS), maximum movement toward the leader (MML), hunting group consideration rate (HGCR), number of epochs (NE), and iteration number (IE) for HuS. NE also indicates the number of reorganizations. The hybrid HSA and HuS algorithm can be described as follows:

Step 1: Set parameters of HSA and HuS
Step 2: Initialize HM and HG
Step 3: Improvise new harmony
    Perform HM consideration

$$x_i' = \begin{cases} x_i' \in \left\{ x_i^1, x_i^2, ..., x_i^{HMS} \right\} & \text{if} \left(r \leq HMCR \right) \\ L_i + \text{rand}(0,1) \times (U_i - L_i) & \text{if} \left(r \leq 1 - HMCR \right) \end{cases}$$

    Perform pitch adjustment
    $x_i' = x_i' + r \times bw$ if $\left(r \leq PAR \right)$
    Perform HuS
    {
    Determine leader
    Move toward leader
    Correct position
    If (IE reached)
      Reorganize HG
    If (IE × NE not reached), repeat HuS
    }
Step 4: Update HM
Step 5: If (Termination condition not satisfied), Goto Step 3
Step 6: Return solution

The hybrid HSA and HuS algorithm has been applied to train feedforward NNs with significant success, which demonstrates the efficiency of the hybrid algorithm.

5.7.2.1.8 Hybrid of HS and Simplex Method

The simplex method was originally proposed by Spendley et al. (1962) and improved by Nelder and Mead (1965). It is also known as the Nelder–Mead simplex method and is good at local search. A simplex is a geometric object determined by an assembly of $n + 1$ points $\{p_0, p_1, \ldots, p_n\} \in R^n$ in n-dimensional space such that

$$\det \begin{bmatrix} p_0 & p_1 & \cdots & p_n \\ 1 & 1 & \cdots & 1 \end{bmatrix} \neq 0 \tag{5.88}$$

The condition in Equation 5.88 ensures that two points in R do not coincide, three points in R^2 are not colinear, four points in R^3 are no coplanar, and so on. The operations of the simplex search method are to rescale the simplex at each iteration based on the landscape and local behavior of the function by using four basic steps: reflection, expansion, contraction, and shrinkage. The simplex search method can successfully improve itself and get closer to the optimum if an initial point is properly chosen through these steps. It starts with an initial p_0, and the remaining points of the initial simplex are generated as follows (Chong and Zak, 2008):

$$p_i = p_0 + \lambda_i e_i, \quad i = 1, 2, \ldots, n \tag{5.89}$$

where e_i are the unit vectors in R^n and λ_i are coefficients selected in such a way that the magnitudes reflect the length scale of the optimization problem.

Jang et al. (2008) proposed the hybrid approach by incorporating the simplex method into the HSA to enhance the local search. In simplex HS, M vectors are randomly generated in the HM. The M vectors are sorted according to the objective function value. For an n-dimensional problem, the best n HM vectors are saved for the next iteration. The simplex search operation is performed on the top $n + 1$ harmony vectors, and the $(n + 1)$th HM is updated with the new vector. The remainder of the HM vectors are adjusted by the HSA. After the process, the harmony vectors are sorted for use in the next iteration. The simplex HS process is shown pictorially in Figure 5.15.

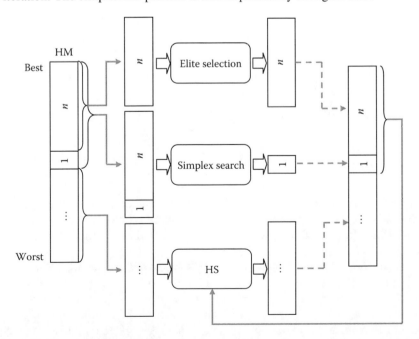

FIGURE 5.15 Hybrid simplex HS method.

The simplex HSA is a simple direct search that has been widely used in optimization. It is very sensitive to the initial point. The proposed method has been tested on two unconstrained and two constrained function minimization problems. Experimental results show that the proposed algorithm not only performs global exploration efficiently but also provides better-quality optimal solution than HS and other algorithms.

5.7.2.1.9   Hybrid of HSA and SA

The mechanism of accepting a solution in SA is based on the thermodynamic process of annealing following a cooling schedule (decreasing temperature) to achieve the energy minimum. This makes SA a suitable for traditional local search method in optimization algorithms, which has found many successful applications. On the other hand, increasing the PAR continuously in the HS is not a good strategy. A very high PAR with a wide bandwidth (bw) may cause the solution to scatter around some potential optimal solution. Taherinejad (2009) introduced the notion of cooling strategy (accepting some bad solutions in the early stage of search) in HS and proposed modification to the dynamic version of PAR(t) and $bw(t)$ (Mahdavi et al. 2007). This modification changed the direction of linear adaptation of PAR(t) from an increasing manner as in Mahdavi et al. (2007) to a decreasing manner. Both PAR(t) and bandwidth $bw(t)$ values are updated in a decreasing manner as defined in Equation 5.86, which has been proved to explore the maximum search space in HS and leads to significant improvement of HS performance.

In SA, a new solution is produced, and the decision on acceptance or rejection is made based on the probability calculated using a defined energy function. Askarzadeh (2013a,b) proposed a new hybrid SA and HSA approach where the concept of production of a new solution is borrowed from HS. The basic mechanism of producing a new solution in the hybrid approach is based on HS rules as follows:

For k = 1:3
If (r_1 > HMCR(t))
   $x'(k)$ = feasible random integer number
Else
  {
     $x'(k) = x(t,k)$
    If [r_2 < PAR(t)]
      $x'(k) = x'(k) + r_w$
  }

where t is the current iteration and $r_w \in \{-1,1\}$ is defined as

$$r_w = \begin{cases} 1 & \text{if } (r_3 > 0.5) \\ -1 & \text{else} \end{cases} \tag{5.90}$$

where $r_1, r_2, r_3 \in [0, 1]$ are random numbers uniformly distributed over [0,1].

The proposed approach has been applied to an optimal design of photovoltaic array and wind hybrid system. Simulation results show that the proposed hybrid method provides better results compared to other well-known methods.

Hosseini et al. (2014) proposed a hybrid HS- and SA-based heuristics (HS-SA algorithm) in order to solve the transportation problem of a consolidation network where a set of vehicles are used to transport goods from suppliers to their corresponding customers via three transportation mechanisms: direct shipment, shipment through cross-dock (indirect shipment), and milk run (traditional way of selling milk from door to door by milkman following a specific route and collecting empty bottles). HSA builds the main structure of the proposed approach to provide the new solutions, and the SA algorithm is incorporated into this structure to enhance the initial solution. The objective is formulated as the minimization problem of the total shipping cost in the network.

5.7.2.1.10   Hybrid of HS and Hill Climbing Search

It is well known that HS is a global search technique and has poor local search ability. Therefore, inclusion of a local search algorithm will exploit the search space better and yield a higher-quality optimum solution. HC search is an iterative local search technique. HC search starts with an arbitrary solution x' to a problem and attempts to find a better solution by exploring the neighborhood $N(x')$ by incrementally changing one element of the solution at each iteration. Al-Betar et al. (2012) proposed a hybrid HS and HC search to improve local exploitation. HS produces a harmony solution x' using its own operators. The HC search then fine-tunes the new harmony solution x' with a probability of the HC rate HCR. $N(x')$ denotes a neighborhood function defined around the solution x' found by HS, which is problem dependent. The main steps of the HS-HC are described in the following:

Step 1: Initialize parameters
Step 2: Initialize HM
Step 3: Improvise new harmony solution x' using HS operators
Step 4: Apply local HC search:
> If $(r_1 < \text{HCR})$
>> While (local optimal solution not found)
>>> {
>>> $x'' = \text{Explore}\left[N(x'')\right]$, where $x'' \in N(x'')$
>>> If $(f(x'') \leq f(x'))$
>>>> $x' = x''$
>>> }
Step 5: Update HM
Step 6: If (termination condition not satisfied), Goto Step 3
Step 7: Return solution

Here r_1 is a random number uniformly distributed over [0,1]. The proposed hybrid approach has been applied on a timetabling problem and tested using 11 data sets. The experimental results were compared against 27 other methods reported in the literature, which have also used the same 11 data sets. The results demonstrated the effectiveness of the proposed hybrid approach.

5.7.2.1.11   Hybrid of HS and Sequential Quadratic Programming

The efficiency and ability of the HSA in finding near-global region solutions within a reasonable time have been evident from many applications. It is also well known that HS is comparatively inefficient in performing local search. To support the exploitation mechanism of HS, Fesanghary et al. (2008) proposed a new framework combining HS with nonlinear programming methods to speed up local search and possible improvement of solution quality and computational efficiency. Plenty of synergism lies in the combination of the HS and local search methods. Fesanghary et al. (2008) used SQP as a local search. The basic idea of SQP is to model nonlinear programming at a given approximate solution x^k by a quadratic programming subproblem and then to use the solution to this subproblem to construct a better approximation x^{k+1}. This process is iterated to create a sequence of approximations that will converge to a solution x^* (Boggs and Tolle, 1996). SQP is applied a few times to improve the quality of the new improvised solution vector through the improvisation process of HS. In order to obtain a high-quality solution with minimum resource requirement, switching between global search and local search is controlled by p_c, which is experimentally set to a low value (e.g., 0.1). Also as a final step after HS met the stopping criterion, SQP is applied to the best vector in the HM measured by the objective function value as a final improvement step. The main steps of the hybrid HS-SQP approach can be described by the following pseudocodes.

```
While (Not stopping criteria met)
    {
    Construct a new vector
    For each new vector
        {
        Compute fitness
        If (rand() < p_c)
          Improve solution vector applying local search (SQP)
        Update HM (if fitness is better)
        }
    }
    For each vector in HM
        {
        Improve solution vector applying local search (SQP)
        Update the best solution
        }
```

The proposed hybrid method is tested on a number of unconstrained and constrained function minimization problems with continuous design variables and structural engineering optimization problems from the literature. The experimental results reveal that the hybrid approach is able to obtain good results in terms of quality and the required number of fitness function evaluations.

5.7.2.1.12   Hybrid HS and the Broydon–Fletcher–Goldfarb–Shanno Method

As mentioned in earlier sections, an efficient local search technique can improve the quality of solution provided by HS and improve convergence. The BFGS is a quasi-Newton method widely used as a local search method for solving an unconstrained nonlinear optimization problem (Yang, 2010). The basic idea is the replacement of Hessian matrix H by an approximate matrix B in terms of an iterative updating formula with rank-one matrices* as its increment. The minimization of a function $f(x)$ with no constraint, the search direction s_k at each iteration is given by

$$B_k s_k = -\nabla f(x_k) \tag{5.91}$$

where B_k is the approximation of the Hessian matrix H at the k-th iteration.

To find an optimal step size β_k, a line search is performed such that the new trial solution x_{k+1} can be computed as

$$x_{k+1} = x_k + \beta_k s_k \tag{5.92}$$

The new B_{k+1} can be estimated as follows:

$$B_{k+1} = B_k + \frac{v_k v_k^T}{v_k^T u_k} - \frac{(B_k u_k)(B_k u_k)^T}{u_k^T B_k u_k} \tag{5.93}$$

where $u_k = x_{k+1} - x_k = \beta_k s_k$ and $v_k = \nabla f(x_{k+1}) - \nabla f(x_k)$.

---

* A rank-one matrix is a matrix which can be written as $r = ab^T$ where a and b are vectors and have at most one nonzero eigenvalue. The eigenvalue can be calculated by $b^T a$.

Using Equations 5.91 through 5.93, the BFGS method can be described as follows:

Initialize a guess of x_0 and approximate B_0 (e.g., $B_0 = I$)
While (termination condition not met)
{
Compute s_k using Equation 5.91
Compute optimal step size β_k using line search
Update x_{k+1} using Equation 5.92
Update B_{k+1} using Equation 5.93
Set $k = k + 1$
}

The BFGS is an efficient local search technique. Karahan et al. (2013) proposed a hybrid HS-BFGS algorithm for the parameter estimation of the nonlinear Muskingum flood-routing model. The BFGS algorithm is used with a low probability for accelerating the HSA. In the proposed technique, an indirect penalty function approach is imposed on the model to prevent negativity of outflows and storages. The proposed algorithm finds the global or near-global minimum regardless of the initial parameter values with fast convergence. The proposed algorithm found the best solution among 12 different methods. The results demonstrate that the proposed algorithm can be applied confidently to estimate optimal parameter values of the nonlinear Muskingum model.

5.7.2.1.13 *Hybrid of HS and K-means and Fuzzy C-means Clustering*

K-means clustering, first proposed by MacQueen (1967), is a method of vector quantization and clustering. Fuzzy *C*-means (FCM), developed by Bezdek (1981), is a method for partitioning a finite collection of *n* elements into *C* fuzzy clusters with respect to some given criterion. These are popular methods for cluster analysis in data mining. A number of hybrid approaches were developed by integrating HS and *k*-means and FCM clustering algorithms. *k*-Means and FCM clustering algorithms are applied as local search methods to improve the performance HS.

Forsati et al. (2008a) and Mahdavi and Abolhassani (2009) proposed hybrid approach by integrating the *k*-means algorithm into HS. The *k*-means algorithm is applied a few times with the best vector stored in the HM as initial cluster centers, and the returned vector is added to the HM if it has a better fitness value than the harmony vectors in the HM. The hybrid method demonstrated significant improvement when applied to web documents clustering. Malaki et al. (2008) proposed a hybrid IHS-FCM clustering algorithm, where FCM is integrated into IHS, firstly, as a local search component to increase the convergence speed and, secondly, as a final clustering step to enhance the partitioning results. The hybrid methods were tested on a ~58,000 element NASA radiator data set, which demonstrated the efficiency of the methods. Alia et al. (2009a–c) proposed a hybrid HS and FCM method for image segmentation problems, where FCM is used as the local search method to determine the appropriate number of clusters and cluster centers. To support the selection mechanism of the number of clusters, Alia et al. (2010) proposed another IHS-FCM method by introducing a new operator, which improved the quality of the segmentation results.

Besides these well-known hybrid algorithms discussed above, there are still many other hybrid approaches reported in the literature. Alia and Mandava (2011) and Siddique and Adeli (2015a,b) already published survey papers reporting a good number of hybrid methods with very brief discussion on those methods.

Lee and Zomaya (2009) proposed a parallel meta-heuristic framework in which HS is the key component, and the three meta-heuristics GA, SA, and Artificial Immune System (Dasgupta, 2006) are auxiliary components of the method to enhance the solutions in the HM as an extra step to speed up the convergence and at the same time to prevent the HS from getting stuck in the local optimum. Ayvaz et al. (2009) reported a hybrid approach combining HS and a spreadsheet "Solver" instead of SQP to improve the results of the HSA in solving continuous engineering optimization problems.

Castelli et al. (2012) reported a hybrid grouping HS and extreme learning machine approach, which was applied to the assessment of success of companies. Yildiz (2008) and Yildiz and Ozturk (2010) proposed a new framework that combined HS with Taguchi's robust design approach (Taguchi, 1990) to improve the performance of HS. The proposed method is based on two stages: (i) Taguchi's method to find appropriate interval levels of design parameters to be used as an initialization step for the HM and (ii) HS to generate optimal multiobjective solutions using refined intervals from the previous stage. This hybridization step is also introduced to reduce the effects of noise factors in the optimization process. Lamberti and Pappalettere (2013) proposed a hybrid approach by integrating the Big Bang–Big Crunch method (discussed in Chapter 10) into HS.

5.7.2.1.14   Hybrid of HS and Clonal Selection Algorithm

Wang et al. (2009) proposed a hybrid approach to improve the convergence speed of HS by integrating the clonal selection algorithm (CSA) into HS. CSAs are a special class of immune algorithms, which are inspired by the clonal selection principle of the human immune system to produce effective methods for search and optimization (Dasgupta, 2006; Wang et al., 2004). All HM vectors are updated using the CSA, where they are considered as individual antibodies and evolve in the population of the CSA. This operation is considered as a fine-tuning mechanism for HS. Even though this approach moderately increases the computational complexity of the standard HS method, it improves the convergence capability of HS.

5.7.2.1.15   Hybrid of HS, ABC, and Biogeography-based Optimization Algorithm

BBO is based on mathematical models developed by MacArthur and Wilson (1967) on biogeography that describe how species migrate from one island to another, how new species arise, and how species become extinct. Since then, biogeography has become a major area of research, which studies the geographical distribution of biological species. The concept of the biogeography can be used to derive a new family of algorithms for optimization called BBO (Simon, 2008). The BBO algorithm has two main operators: habitat modification (Ω) and mutation (M). The former is an $H^n \rightarrow H$ probabilistic operator that adjusts an habitat H_i based on the ecosystem H^n according to its immigration rate λ_i and to the emigration rate μ_j, taking H_j as the source of the modification. The second operator is an $H \rightarrow H$, a probabilistic operator that randomly modifies the SIVs of a habitat H_i according to a probability p_i based on both the λ_i and μ_i parameters (Simon, 2008). Both operators let BBO implement elitism for the best b habitats (b is a user-selected elitism parameter) setting $\lambda_i = 0$ in the habitat modification operator and $p_i = 0$ in the mutation operator. BBO starts initializing the search with a set of random habitats, and the HSI is computed from each of them. The habitat modification (Ω) and mutation (M) operators are subsequently applied on each non-elite habitat (N-b worst habitats), while the stopping criterion is not satisfied. The performance and effectiveness of BBO have been verified on benchmark functions and sensor selection problems, which demonstrate competitive performance compared to other population-based methods. García-Torres et al. (2014) proposed a hybrid approach combining ABC, BBO, and HS for 3-D modeling application.

5.7.2.2   Hybridizing HS Components into Other Meta-Heuristic Algorithms

In this type of hybrid algorithms, components of the HSA are incorporated into other meta-heuristic algorithms to enhance the performance of these algorithms. There have been many such algorithms reported in the literature. Alia and Mandava (2011) also covered some known hybrid algorithms in a survey published in 2011. In this section, some of the recent and important hybrid algorithms are presented.

The PSO algorithm uses the personal best or pbest concept. Since the selection mechanism used in generating a pbest swarm let the new vector to go out of the boundary of the variables, to keep the variables within the feasible space, Li et al. (2007) borrowed the concept of HM and integrated it into the PSO algorithm where all feasible vectors are stored. The approach is applied to the designing of optimal pin connected structures by handling the particles that may fly out of bound. Another similar variant is reported by Kaveh and Talatahari (2009) with a modified PSO version called PSO with passive congregation (PSOPC), in which ACO and HS are integrated together. The PSOPC is used as the global search technique, ACO is used as local search method for updating the positions of the particles, and the HM concept is used to control the variable constraints. The combination improved the performance of PSO.

In general, a new population is created by performing crossover (with crossover rate p_c) and mutation (with mutation rate p_m) operation on selected vectors or individuals (as parents) from the population according to a chosen selection mechanism. Li et al. (2008) proposed a hybrid approach by introducing the HM into the GA to improve the performance. The proposed modification mimics the HS improvisation method, where the newly generated vector is selected from all vectors stored in the HM. Nadi et al. (2010) introduced a new technique for maintaining balance between the exploration and exploitation of the EA. An adaptive parameter is used to control the parameter values of the EA through the search process using HS, which directs the search from the current state to a desired state by determining suitable parameter values.

Linear discriminant analysis (LDA), also known as Fisher's linear discriminant, is a collection of methods used in statistics, pattern recognition, and machine learning to find a linear combination of features which characterizes or separates classes of objects (Duda et al., 2000; Fisher, 1936). The problem with LDA is the Gaussian distribution of each class and the same within-class covariance while having different means. For improving the accuracy of classification using LDA, low interclass covariance and high intraclass covariance are desirable. Moeinzadeh et al. (2009) used HS to improve the accuracy of the LDA classification algorithm. HS is used as a preprocessing technique and used to compute a transformation matrix with the aim of decreasing the interclass covariance and increasing the intraclass covariance. The HM is initialized by representation of the transformation matrix as a float numerical vector, where each component of this vector is an element of $[-1, 1]$.

5.8 APPLICATION OF HSA TO ENGINEERING PROBLEMS

The HSA was proved to be very successful and was able to attract many researchers to develop HS-based solutions for many optimization problems. Early applications of the HSA were in the domain of water distribution and water-related studies since the first publication of the HSA in 2000 (Geem, 2000) and followed by introduction to the wider research community in 2001 (Geem et al., 2001; Geem, 2009b,c). In the following years, the HSA has found many applications to various domains such as engineering design, construction engineering, telecommunication, water networks and system management, medical, and control and robotics applications. An overview of the applications of the HSA was reported in 2009 (Ingram and Zhang, 2009). Surveys on applications of the HSA were published by Manjarres et al. (2013) and Siddique and Adeli (2015a,b), which thoroughly reviews and analyzes different features and the applications of the HSA in engineering. This section presents an overview of applications and outlines the current trends of the HSA.

5.8.1 Function and Constrained Optimization Problems

The effectiveness of the HSA was first investigated by Geem (2000) and Geem et al. (2001) by applying the algorithm to well-known benchmark and constrained optimization problems. Lee and Geem (2005) applied the algorithm to many unimodal, multimodal, and constrained optimization problems. Some of these are as follows:

1. Rosenbrock's function (Lee and Geem, 2005)
2. Goldstein and Price function I and II (Lee and Geem, 2005)
3. Eason and Fenton's gear train inertia function (Lee and Geem, 2005)

4. Wood function (Lee and Geem, 2005)
5. Powell quadratic function (Lee and Geem, 2005)
6. Six-hump Camelback problem (Lee and Geem, 2004)
7. Constrained functions (Lee and Geem, 2005)

Each nonlinear function has only one global optimum, but they have several local optima. Lee and Geem (2004) tested the HSA on a six-hump camelback optimization problem with an MH size of 10, HMCR value of 0.85, and PAR value of 0.45. The HSA achieved a near-optimal solution of $x = [0.08984, -0.71269]$ with the minimum objective function value of -1.0316285 after 4870 searches. The algorithm also found other optimal solutions with different random number seeds.

Lee and Geem (2005) applied the HSA to these benchmark functions and constrained functions to find not only the global optimum but also possible local optima. They applied the standard HSA as well as the modified HSA to optimization problems mentioned above with HMS = 100, HMCR = 0.75, and PAR = 0.6. The standard HSA could find the global optimum, while the modified HSA could find the global as well as the local optimum. Kumar et al. (2014) applied the standard HSA and the adaptive parameter HSA on seven unimodal and eight multimodal benchmark functions. The experiments were carried out with different HM sizes from 5 to 50. An HM size of 5 provided the best result on the majority of the benchmark functions.

5.8.2 STRUCTURAL DESIGN OPTIMIZATION

Structural design of a cylindrical pressure vessel is to minimize the total cost of material, forming, and welding. The four design variables are shell thickness x_1, spherical head thickness x_2, radius of the cylindrical shell x_3, and shell length x_4. Here, x_1 and x_2 are integer multipliers of 0.0622; and x_3 and x_4 are continuous values within the intervals of $40 \le x_3 \le 80$ and $20 \le x_4 \le 60$, respectively. The optimization problem can be formulated as

$$f(x) = 0.6224 x_1 x_3 x_4 + 1.7781 x_2 (x_3)^3 + 3.1611 (x_1)^2 x_4 + 19.84 (x_1)^2 x_3 \tag{5.94}$$

subject to satisfying the design constraints defined as

$$\begin{cases} g_1(x) = 0.0193 x_3 - x_1 \le 0 \\ g_2(x) = 0.00954 x_3 - x_2 \le 0 \\ g_3(x) = 750.0 * 1728.0 - \pi (x_3)^2 x_4 - \dfrac{4}{3} \pi (x_3)^3 \le 0 \\ g_4(x) = x_4 - 240.0 \le 0 \\ g_5(x) = 1.1 - x_1 \le 0 \\ g_6(x) = 0.64 - x_2 \le 0 \end{cases} \tag{5.95}$$

Lee and Geem (2005) applied the HSA and achieved a design with a best solution of $[x_1 = 1.125, x_2 = 0.625, x_3 = 58.2789, x_4 = 43.7549]$ and a minimum cost of \$7198.433 satisfying all the constraints $\{g_1(x), g_2(x), g_3(x), g_4(x), g_5(x), g_6(x)\}$ defined in Equation 5.95.

A second benchmark problem in structural design is the welded beam design. The structure consists of a beam A, and the weld required holding the beam to a member B. The objective is to find the feasible set of dimensions for the design variables $0.125 \le h \le 5$, $l \ge 0.1$, $t \le 10$, and $0.1 \le b \le 5$

that can carry certain load P, which will minimize the total manufacturing cost comprising setup, welding, and material cost. The optimization problem is formulated as

$$f(x) = 1.10471h^2 l + 0.04811 tb(L + l) \tag{5.96}$$

which is subject to a set of design constraints. Here L denotes the overhang length and set to a fixed value of 14. Lee and Geem (2005) applied the HSA to the welded beam design for optimization of the dimensions and achieved a best solution of [$h = 0.2442$, $l = 6.2231$, $T = 8.2915$, $b = 0.2443$] and a minimum cost of \$2.38 after 110,000 iterations. The result is comparable to the known best solution achieved by some other well-known methods.

A third benchmark problem in structural design is the truss or frame structural sizing and configuration optimization. Structural optimization for solving shape and sizing optimization of trusses with multiple frequency constraints is a highly nonlinear problem, which has been used for testing the effectiveness and performance of many optimization algorithms. The optimization problem can be formulated using three types of design variables: sizing variables, geometric variables, and topological variables. Sizing optimization determines the cross-sectional size of structural members. Configuration optimization looks for a set of geometric and sizing variables for a predefined topology. Topology optimization selects the appropriate topology from different structural types, which is a combinatorial optimization problem. The truss optimization problem can be formulated as

$$f(x) = \gamma \sum_{i-1}^{n} L_i A_i \tag{5.97}$$

subject to satisfying the k inequality constraints,

$$g_j^L \le g_j \le g_j^U, \quad j = 1, 2, \dots, k \tag{5.98}$$

Here, L_i is the member length, A_i is the member cross-sectional area, γ is the material density, and g_j^L and g_j^U are the lower and upper bound of the constraint function g_j, respectively. The sizing optimization is to find the optimal values for the cross-sectional areas A that minimizes the structural weight $f(x)$ satisfying k inequality constraints. The configuration optimization is to find the optimal values for the cross-sectional areas A and nodal coordinates R that minimizes the structural weight $f(x)$ satisfying k inequality constraints. A 10-bar cantilever plane truss is shown in Figure 5.16. A nonstructural mass is attached to each free node {(1)–(4)}.

Lee and Geem (2004) applied the HSA to the sizing optimization problem of a 10-bar plane truss with a material density of $\gamma = 0.1$ lb/in^3 and minimum member cross-sectional area of 0.1 in^2 subject to stress limitations of ±25 ksi and displacement limitations ±2.0 in imposed on all nodes in both directions (x and y). There are 10 independent design variables. A single loading condition of $P_1 = 100$ kips and $P_0 = 0$ kips was considered. The HSA was able to find 20 different solutions, and the best solution found was {$A_1 = 30.15$, $A_2 = .102$, $A_3 = 22.71$, $A_4 = 15.27$, $A_5 = .102$, $A_6 = .544$, $A_7 = 7.541$, $A_8 = 21.56$, $A_9 = 21.45$, $A_{10} = .1$} within 20,000 searches, achieving the minimum objective function value of $f(x) = 5057.88$ lb. For a second case considering the single loading condition of $P_1 = 150$ kips and $P_0 = 50$ kips, the HSA was able to find the best solution of {$A_1 = 23.25$, $A_2 = .102$, $A_3 = 25.73$, $A_4 = 14.51$, $A_5 = .1$, $A_6 = 1.97$, $A_7 = 12.21$, $A_8 = 12.61$, $A_9 = 20.36$, $A_{10} = .1$} within 15,000 searches and achieved the minimum objective function value of $f(x) = 4668.81$ lb. In another investigation, Lee and Geem (2005) applied the HSA to the sizing optimization problem of a 10-bar with the same material density, member cross-sectional area, stress limitations, and displacement limitations. The HSA was able to find 20 different solutions and the best objective function value of $f(x) = 4668.81$ lb. Lee and Geem (2004) also applied the HSA to the sizing optimization problem

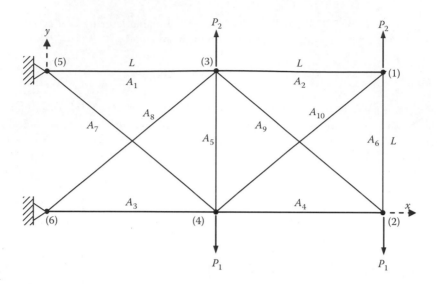

FIGURE 5.16 10-Bar plane truss.

of a 17-bar plane truss problem and achieved an optimal solution at the objective function value of $f(x) = 2580.81$ lb approximately after 20,000 searches.

Lee and Geem (2004, 2005) applied the HSA to the configuration optimization problem of an 18-bar plane truss with the same material density, member cross-sectional area, stress limitations, and displacement limitations, which is a popular optimization benchmark problem due to its simple configuration. The 18 cross-sectional members were divided into four groups. The HSA found an optimal configuration within 25,000 searches with the minimum objective function values of $f(x) = 4515.6$ lb. Lee and Geem (2004) also applied the HSA to a 22-bar, 25-bar, 72-bar space truss, 200-bar planar truss, and 120-bar dome truss. The optimal design obtained using the HSA showed an excellent agreement with the previous mathematical designs reported in the literature. Miguel and Miguel (2012) also applied the HSA to 10-bar and 37-bar plane truss, 52-bar space truss dome structure, and 72-bar space truss problem. In all cases, the HSA achieved competitive results and illustrated the effectiveness and applicability of the HSA.

The optimum design problem of cellular beams turns out to be a discrete nonlinear programming problem when formulated according to the constraints specified in Steel Construction Institute Publication Number 100. The constraints to be considered in the design of a cellular beam include the displacement limitations, overall beam flexural capacity, beam shear capacity, overall beam buckling strength, web post flexure and buckling, Vierendeel bending of upper and lower tees, local buckling of the compression flange, and practical restrictions for cell diameter and the spacing between cells. The design problem also needs to include the sequence number of the universal beam section, diameter and total number of holes, and the space between holes in the beam as design variables. Erdal et al. (2011) applied the HSA to the optimum design of cellular beams and compared the performance to that of PSO. Saka (2007, 2009a,b) applied HSA design of geodesic domes and steel sway frames.

5.8.3 Hydrologic Model Optimization

Flooding is caused by exceeding the capacity limit of water bodies such as rivers or lakes. This causes a lot of sufferings to people and damage to infrastructure, harvest, and the economy. Researchers have been making efforts to model river floods. The Muskingum model is one of the popular hydrologic models comprising two hydrologic parameters developed by McCarthy (1938) based on the research of the Muskingum river basin in Ohio state in the USA. The Muskingum model was later modified by Gill (1978), adding a third parameter to represent flood characteristics. Calibration of the parameters of the Muskingum model is a difficult task. Many researchers formulated the calibration as an optimization problem. The objective is to minimize the difference between the real-world flood amount and a simulated one. Thus, the objective function is defined as the residual sum of squares (SSQ) between the observed and computed hydrologic values (e.g., river overflows):

$$f(x) = \text{SSQ} = \sum_{t=1}^{n} (\text{RO}_{ob}(t) - \text{RO}_{com}(t))^2, \tag{5.99}$$

where RO_{ob} is the observed river overflow and RO_{com} is the computed river overflow, which is computed using a nonlinear Muskingum model defined by

$$\text{RO}_{com} = \left(\frac{1}{1-h}\right)\left(\frac{S_t}{K}\right)^{1/r} - \left(\frac{h}{1-h}\right)I_t \tag{5.100}$$

In the Muskingum model, S_t is the channel storage and I_t is the river inflows, which can be calculated and provided based on the calibration of the three model parameters $h \in [1.5, 2.5]$, $K \in [0.01, 0.20]$, and $r \in [1.5, 2.5]$.

Kim et al. (2001) first applied the HSA to estimate the parameters of the nonlinear Muskingum model. In a later study by Lee and Geem (2005), the HSA was applied to calibrate the three hydrologic model parameters using the HSA parameters as follows: HMS = 20, HMCR = 0.90, and pitch-adjusting rate (PAR) = 0.35. The HSA found the best optimal values for $h = 0.287$, $K = 0.087$, and $r = 1.8661$ within 20,000 searches with the minimum objective function value for SSQ = 36.77. This is a competitive optimal solution compared to many other meta-heuristic algorithms reported in the literature. Ayvaz (2007) used the HSA with FCM clustering for determination of aquifer parameters.

5.8.4 WATER DISTRIBUTION NETWORK (WDN)

The WDN is the most useful network system in daily urban life that brings purified water from the source to the end user. The network comprises various interconnected hydraulic components such as water distribution pipes, pumps, tanks, reservoirs, and water treatment plants. Designing water networks requires several factors to be considered such as water demand, hydraulic criteria, network layout, and size of the network. The water demand can be estimated by the number of end users. Hydraulic criteria are engineering design issues relating to the level of service, maximum and minimum pressures along the pipeline, pressure fluctuation, headloss pipes, and maximum velocity of water flow in the pipelines. Network layout and size (e.g., pipe diameter) of the WDN are issues relating to the efficiency of water supply where capacity decreases along the distance from the source node to the end user. Thus, the network layout and size are subject to optimization. The objective function for the WDN design can be formulated as the minimization of the total costs of pipes and pumps that are functions of diameter sizes and pumping capacities. There are three design constraints:

1. Mass conservation: For every node, the total inflows must be equal to the total outflows.
2. Energy conservation: For every loop, the summation of energy loss due to friction is assumed to be zero.
3. Minimal nodal pressure: For every demand node, the residual pressure should be greater than or equal to the minimum nodal pressure.

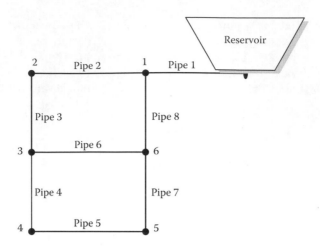

FIGURE 5.17 Alperovits and Shamir's two-loop benchmark WDN.

If any design constraint is not satisfied, a penalty is imposed to the objective function.

The first benchmark problem for the WDN was proposed by Alperovits and Shamir (1977) comprising a two-loop network, one reservoir, six demand nodes, and eight pipes. Alperovits and Shamir's two-loop benchmark WDN is shown in Figure 5.17.

There are 8 decision variables (i.e., number of pipes), and there are 14 candidate diameters for each pipe, which leads to 14^8 design solutions. A number of meta-heuristic algorithms have been applied to optimize the WDN problem. Geem (2006b), Geem (2009b,c), and Geem et al. (2002, 2009) applied the HSA to an optimal design of the WDN. Experimental results were compared with other meta-heuristic algorithms such as the GA (Wu et al., 2001), SA (Cunha and Sousa, 1999), and Shuffled Frog Leaping Algorithm (SFLA) (Eusuff and Lansey, 2003). Results showed that the HSA reached an optimal design solution, that is, a cost of $419,000, within 1,121 evaluations, which is much less than the GA, SA, or SFLA.

Fujiwara and Khang (1990) first proposed the real-world problem of Hanoi Network (located in Vietnam) with 1 reservoir, 31 demand nodes, and 34 pipes, which is now used as a benchmark problem for the WDN. Geem (2009b) applied the HSA to the Hanoi WDN and the results show very competitive performance of the HSA compared to the GA (Reca et al., 2008) and ACO (Zecchin et al., 2006), achieving the optimal design solution of $6.08 million. Ayvaz (2009a,b) applied the HSA to the solution of ground water management models.

5.8.5 WATER PUMP SWITCHING PROBLEM

Water system optimization is a challenging optimization problem for design engineers, which received considerable attention from the research community. The problem is to handle a water supply system consisting of n pipes and n pumping stations with m pumps in series within each station while minimizing the associated energy cost and satisfying a number of constraints such as adequate pressure requirement in the system. A simple water pumping system is illustrated in Figure 5.18. Water is delivered to the consumers' destination point from the water source by applying pumps at different stations.

For example, to supply water to a consumer's destination, the pump j in station i needs to be turned on to add pressure P_{ij}. This will cost an amount of energy E_{ij}. The energy E_{ij} can be calculated using the following equation:

$$E_{ij} = \frac{\gamma Q_0 P_{ij}}{\alpha \eta_{ij}} \qquad (5.101)$$

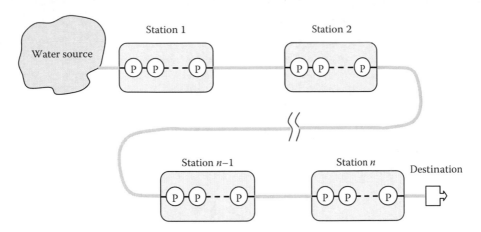

FIGURE 5.18   Schematic diagram of water pump switching problem

where γ is a weight function (e.g., 62.4 lb/ft3), Q_0 is the flow rate (e.g., 19 cfs), P_{ij} is the pressure rise (in psi) across the pump, η_{ij} is the motor pump efficiency, and $\alpha = 238.3$ is a constant. By applying an on–off switch to each pump at each station, the optimization of the water pump switching problem can be formulated as the minimization of the total electric energy, and the objective function is defined as

$$\text{Min } f(E) = \sum_{i=1}^{n} \sum_{j=1}^{m} E_{ij} x_{ij} \tag{5.102}$$

where $x_{ij} \in \{0,1\}$ is a binary variable that denotes on–off scheduling of the pump. The objective function is subject to a discharge pressure constraint, discharge pressure bound, suction pressure constraint, and suction pressure bound. The discharge pressure P_i^D at pumping station i should be equal to the summation of suction pressure P_i^S and all operating pump pressures $\sum_{j=1}^{m} P_{ij} x_{ij}$.

$$P_i^D = P_i^S + \sum_{j=1}^{m} P_{ij} x_{ij}, \quad i = 1,\dots,n \tag{5.103}$$

Discharge pressure P_i^D in any station should be placed at less than the upper bound, that is, $P_i^D \le U(P_i^D)$, $i = 1, \dots , n$. The suction pressure in pumping station $i + 1$ is equal to the difference of discharge pressure P_i^D and pressure loss P_i^L.

$$P_{i+1}^S = P_i^D - P_i^L, \quad i = 1,\dots,n \tag{5.104}$$

The suction pressure should be within the upper and lower bounds, that is, $L(P_i^S) \le P_i^S \le U(P_i^S)$ and initial suction pressure P_1^S is assumed to be zero.

Geem (2005) applied the HSA to a water pump switching system consisting of 10 pipes ($n = 10$) and 10 pump stations ($n = 10$), with 4 pumps ($m = 4$) in series within each station. Thus, the total number of decision variables is 40 ($n \times m = 40$). For known pumping pressure P_{ij}, energy E_{ij} and pressure loss P_i^L (Darcy–Weisbach and Colebrook–White method) were calculated. Geem investigated the problem for 10 different runs with different HMS of (1 ~ 100) using an HMCR value of 0.95. For all cases, maximum improvisations (function evaluations) of 3500 are used. After 3500 improvisations, the HSA found pumping energy solutions for 10 runs. The minimum pumping energy of 11,169.43 units was obtained for HMS = 2. The best HSA solution was far better compared to GA solutions reported by Goldberg and Kuo (1987).

5.8.6 Transmission Network Expansion Planning Problem

The transmission network expansion planning (TNEP) problem deals with the least cost expansion of new lines such that no overloads are produced during the planning perspective. The TNEP problem is generally formulated as a mixed-integer nonlinear optimization problem. Since there are many possible configurations of the network that may satisfy the demand with very similar or equal costs, the TNEP problem constitutes a large combinatorial, nonpolynomial, multidimensional problem that cannot be solved either without employing further simplifications, or heuristic and meta-heuristic approaches. A common approach to tackle the problem is to use the direct current (DC) model formulation for the power flow equations. The classical DC model for the TNEP problem is described by

$$\min_{\theta, g, n} \sum_{\forall (i,k) \in \Omega} c_{i,k} n_{i,k} + \sum_{\forall i \in \Omega_g} c_{gi} p_{gi} \tag{5.105}$$

subject to

$$\begin{cases} -\sum_{k \neq i} f_{i,k} + p_{gi} = d_i \\ f_{i,k} - b_{i,k} \left(n_{ik}^0 + n_{ik} \right) (\theta_i - \theta_j) = 0 \\ |f_{ik}| \leq \left(n_{ik}^0 + n_{ik} \right) \bar{f}_{ik} \\ 0 \leq p_{gi} \leq \bar{p}_{gi} \\ 0 \leq n_{ik} \leq \bar{n}_{ik} \end{cases} \tag{5.106}$$

The cost function in Equation 5.105 is divided into two parts, the first relating to expansion costs and the second to operational costs, which should satisfy the set of equality and inequality constraints in Equation 5.106 comprising the flow balance constraint in each node network, the flow calculation formula, and the capacity constraints. The objective of TNEP is to minimize the network construction and operational cost while satisfying the demand increase, considering technical and economic conditions. TNEP is a basic part of power system planning that determines where, when, and how many new transmission lines should be added to the network. TNEP should satisfy the required adequacy of the lines for delivering safe and reliable electric power to load centers along the planning horizon. A detailed discussion on TNEP can be found in Garver (2007). A number of meta-heuristic algorithms have been applied to the TNEP problem. An SA approach for TNEP is proposed in Romero et al. (1990) for long-term TNEP. A parallel tabu search algorithm for TNEP is discussed in Gallego et al. (2000) and Da Silva et al. (2001). Applications of the GA have been reported in Da Silva et al. (2000).

Verma et al. (2010) applied the HSA to a simplified TNEP problem with and without security constraints. The simplified model of the TNEP considered is defined as follows:

$$\min \sum_{\forall l \in \Omega} c_l n_l \tag{5.107}$$

where c_l is the cost of adding circuits in the l-th right-of-way (or corridor), Ω is the set of all right-of-ways, and n_l is the number of circuits that can be added in the l-th right-of-way, which is subject to the set of equality and inequality constraints defined as follows:

$$
\begin{cases}
\mathbf{S}f^k + g = d, & k = 0,1,\ldots,N_c \\
f_l^k - \gamma_l\left(n_l^0 + n_l\right)(\Delta\theta_l^k) = 0 & \text{for } l \in \{1,2,\ldots,n_l\}, l \neq k \\
f_l^k - \gamma_l\left(n_l^0 + n_l - 1\right)(\Delta\theta_l^k) = 0 & \text{for } l = k \\
\left|f_l^k\right| \leq \left(n_l^0 + n_l\right)\overline{f_l} & \text{for } l \in \{1,2,\ldots,n_l\}, l \neq k \\
\left|f_l^k\right| \leq \left(n_l^0 + n_l - 1\right)\overline{f_l} & \text{for } l = k \\
0 \leq n_l \leq \overline{n_l} &
\end{cases}
\tag{5.108}
$$

where $\mathbf{S}$ is the branch-node incidence transpose matrix of the power system, f^k is the vector with elements f_l^k, γ_l is the susceptance of the circuit that can be added to the l-th right-of-way, n_l^0 is the number of circuits in the base case, $\Delta\theta_l^k$ is the phase angle difference in the l-th right-of-way when the k-th line is out, f_l^k is the total real power flow by the circuit in the l-th right-of-way when the k-th line is out, $\overline{f_l}$ is the maximum allowed real power flow in the circuit in the l-th right-of-way, and $\overline{n_l}$ is the maximum number of circuits that can be added in the l-th right-of-way. It is seen that f_l^k and θ_l^k are unbounded, n_l is an integer, and $n_l > 0$, $(n_l + n_l^0 - 1) \geq 0$ is an integer for $l = k$, $l \in \Omega$, and $k = 0, 1, \ldots, N_c$ where N_c is the number of credible contingencies and $k = 0$ represents the base case without any line outage. The objective is to minimize the total investment cost of the new transmission lines to be constructed, satisfying the constraint on real power flow in the lines of the network for base case and $(N - 1)$ contingency cases. Constraints in Equation 5.108 represent the power balance at each node, the real power flow equations in DC network, the line real power flow, and the restriction on the construction of lines per right-of-way.

Verma et al. tested the TNEP problem on three standard IEEE bus test systems: an IEEE 24-bus system, south Brazilian 46-bus system, and 93-bus Colombian systems. The fitness function $f(x)$ represents the Equation 5.107 where x denotes the set of candidate lines of the solution vector. Each element in x represents the right-of-way in which a candidate line is constructed. The range of each variable defined by X_i indicates the list of available right-of-ways. If two lines are added in a particular right-of-way, then two elements with the same number is included in the vector x. To apply the HSA to the TNEP problem discussed above, the fitness function is defined as follows:

$$
f = \sum_{l \in \Omega} c_l n_l + w_1 \sum_{k=0}^{N_c} \sum_{ol} \left(abs(f_l^k) - \overline{f_l}\right) + w_2\left(n_l - \overline{n_l}\right)
\tag{5.109}
$$

where ol is the set of overloaded lines, and w_1 and w_2 are the weighting factors. The first term in Equation 5.109 indicates the total investment cost of a transmission expansion plan. The second term is added to the objective function for the real power flow constraint violations in the base case, and $(N - 1)$ contingency cases. The third term is added to the objective function if the maximum number of circuits that can be added in the l-th right-of-way exceeds the maximum limit.

The comparison of results is presented for an IEEE 24-bus system and 46-bus south Brazilian system, with the ones obtained with the basic binary GA and bacteria foraging-differential evaluation algorithm (BF-DEA) to confirm the potential of the proposed approach. For the 24-bus system and the 46-bus south Brazilian system, the HSA parameters are set as follows: HMS = 50, HMCR = 0.98, maximum PAR = 0.99, minimum PAR = 0.1, and the number of improvisations (NI) or stopping criteria is set to 2500. The HSA has also been proved to be the best for the case of the 93-bus Colombian systems. The GA and the BF-DEA were implemented to compare the results. The detailed results are shown only for the case of TNEP with security constraints. However, the results for TNEP without security constraints are also obtained. The results are compared with the BF-DEA and the binary GA. In all cases, the HSA confirms the potential of its application to the

TNEP problem. Rastgou and Moshtagh (2014) also applied an improved HSA to TNEP considering adequacy–security.

5.8.7   JOB SHOP SCHEDULING

Mahdavi (2009) demonstrated the application of HSA to NP-complete problems. Attempts were made to apply HSA to solve NP-hard problems such as JSSP, which plays an important role in manufacturing and production systems. A brief introduction to the JSSP is provided in Chapter 8. Considering its application in industry, there are namely three different objective functions used such as makespan, tardiness, and mean flow. Makespan C_{max} is the time interval between the time at which the schedule begins and the time at which the schedule ends. Thus, the makespan of a schedule is defined as

$$F = \max(C_i), \quad i = 1,\ldots,m \tag{5.110}$$

where C_i is the completion time of job J_i.

The tardiness T_i of a job J_i is the nonnegative amount of time by which the completion time exceeds the due date d_i, that is, it is the differences between the completion time C_i and due date d_i for each job and is defined as

$$T_i = \max\left[0,(C_i - d_i)\right] \tag{5.111}$$

The mean flow time $\overline{F}$ is the average flow time of a schedule defined as

$$F = \frac{1}{n}\sum_{i=1}^{n} F_i \tag{5.112}$$

This means that $\overline{F}$ is the average time of processing a single job.

Following the successful applications of the HSA to various domains, the HSA has been applied to solve the classic JSSP and its variants. There are a good number of research papers on the application of the HSA to the JSSP reported in the literature (Gao et al., 2014; Liu and Zhou, 2013; Pan et al., 2011a,b; Wang et al., 2011a,b; Yuan et al., 2013; Zammori et al., 2014).

The flexible JSSP (FJSSP) is a generalization of the classic JSSP, where operations can be processed by any machine from a given set of machines rather than by one specified machine. Like the classic JSSP, the FJSSP is also an NP-hard problem. Various meta-heuristic algorithms have been applied to the FJSSP. Yuan et al. (2013) applied a hybrid HSA to the FJSSP to optimize the makespan. The hybrid HSA incorporates the HSA and local search to strike a balance between global exploration and local exploitation of the search space. The initial HM is generated by an initialization scheme combining heuristic and random strategies that ensures certain qualities and diversity of the harmonies. One harmony vector is generated by a heuristic approach, and the remainder are produced randomly. The heuristically generated harmony vector is a two-part vector formed by heuristics in advance, and thereafter, the two-part vector code is converted to a harmony vector X. The two parts are the machine assignment vector and the operation sequence vector. A conversion technique is developed to map the continuous two-vector codes to a feasible active schedule for the FJSSP. The randomly generated harmony vector $X = [x_1, x_2, \ldots, x_n]$ is created according to the following formula:

$$x(j) = x_{min}(j) + [x_{max}(j) - x_{min}(j)] \times \text{rand}(0,1), \quad j = 1,\ldots,n \tag{5.113}$$

where $x_{min}(j) = -\delta$ and $x_{max}(j) = +\delta$ for each dimension j and rand(0,1) is a random number uniformly distributed over [0,1].

To speed up the local search procedure, the concept of common critical operations represented by a disjunctive graph is introduced; that is, a schedule is represented by a disjunctive graph. The effective neighborhoods are based on the critical path in the disjunctive graph. The local search is performed on the disjunctive graph, which introduces problem-specific knowledge for the local search. The optimization process is executed by providing cooperation between the global search-based HS and the local search. The effectiveness and performance of the proposed approach have been tested on a number of benchmark problems. Thirty independent runs are carried out for each data set. The chosen parameter settings are as follows: HMS = 5, HMCR = 0.95, PAR = 0.3, and $\delta = 1$. The number of iterations was adjusted according to data sets. The computational results demonstrate the effectiveness and efficiency of the proposed algorithm for solving the FJSSP with the makespan criterion, and also results are competitive in terms of solution quality and time requirements.

5.8.8 Timetabling and Rostering Problem

The university timetabling problem (UTP) includes assigning a set of events to a set of given rooms each with particular features and timeslots on a certain period of time, for example, weekly or monthly basis according to a set of constraints (Burke et al., 2004). A timetabling solution must satisfy a set of constraints. There are usually two types of constraints: hard constraints and soft constraints. The solution must satisfy hard constraints in order to be a feasible solution, but it need not satisfy all soft constraints. The quality of solution is ensured by minimizing the number of soft-constraint violations. The most common soft constraints in university timetabling are the preferences of students and lecturers.

There have been many meta-heuristic algorithms applied to UTP such as the hybrid EA with a variable neighborhood search (Abdullah et al., 2007), hybridized GA with local search (Yang and Jat, 2010), hybridized evolutionary approach with nonlinear great deluge (Landa-Silva and Obit, 2009), hybrid ant colony systems (Ayob and Jaradat, 2009), and hybridized electromagnetism-like mechanism, which is a population-based method with great deluge (Turabieh et al., 2009). There has been a number of research works reported in the literature on the application of the HSA to the timetabling problem including UTP (Al-Betar and Khader, 2009; Al-Betar et al., 2008, 2010b,c, 2012). In all cases, the HSA demonstrated the effectiveness and efficiency of its application to the timetabling problem.

The RP is the assignment of a set of people with various skills and work contracts to a set of shift types over a predefined scheduling period. The RP solution (or roster) is subject to hard and soft constraints. Soft constraints generally reflect the preferences of staff and organizational requirements. The basic objective is to find a feasible roster with a good quality that optimizes an objective function defined for specific organization, for example, healthcare center or hospitals. The main goal of solving RPs is to efficiently utilize the available staff by producing a well-balanced duty roster that also satisfies staffs' preferences in general. The roster is evaluated using the objective function formalized in Equation 5.114 that adds up the penalty of soft constraint violations in a feasible roster (Awadallah et al., 2013, 2014; Hadwan et al., 2013):

$$\min f(x) = \sum_{s=1}^{10} c_s g_s(x) \tag{5.114}$$

where c_s refers to the penalty weight for the violation of the soft constraint, and $g_s(x)$ is the total number of violations for the soft constraint c_s in solution roster x. Nurse RPs are highly constrained combinatorial problems which are difficult to be solved to optimality. Various meta-heuristic approaches have been applied to a nurse RP such as the GA (Tsai and Li, 2009), ACO (Gutjahr and Rauner, 2007), tabu search (Burke et al., 1999), SA (Brusco and Jacobs, 1995), variable neighborhood search

Allocation	→	x_1	x_2	x_3	...	x_N
Nurse	→	1	4	13	...	1
Day	→	3	3	2	...	8
Shift	→	0	2	3	...	2
MCF	→	1	0	1	...	1

FIGURE 5.19   Representation of roster.

(Bilgin et al., 2012), and scatter search (Burke et al., 2009). There have been also a number of researches reported in the literature on the application of the HSA to the RP (Awadallah et al., 2013, 2014; Hadwan et al., 2013).

The solution vector of the nurse RP is represented as a vector of allocations $x = (x_1, x_2, \ldots, x_N)$, where each allocation x_j, $j = 1, 2, \ldots N$ is assigned four values [Nurse number, Day number, Shift number, and Memory Consideration Flag (MCF)] as shown in Figure 5.19. The MCF is a binary variable that is set to one when the current allocation is assigned by the memory consideration or zero otherwise. This is an example of a roster, which is evaluated using the objective function defined in Equation 5.114.

With the roster representation shown in Figure 5.19, an initial HM will look like as follows:

$$HM = \begin{bmatrix} (10,3,0,1) & (4,3,1,0) & (13,2,3,1) & \cdots & (1,8,2,1) \\ (30,0,2,1) & (1,0,0,1) & (2,0,5,1) & \cdots & (2,0,5,0) \\ (9,3,0,1) & (5,2,1,1) & (3,3,3,1) & \cdots & (1,7,3,0) \\ \vdots & \vdots & \vdots & \ddots & \vdots \\ (0,0,0,0) & (3,0,3,1) & (3,0,1,1) & \cdots & (13,3,2,1) \end{bmatrix} \quad (5.115)$$

Awadallah et al. (2013, 2014) applied the HSA to benchmark data sets of the nurse rostering problem published by the first International Nurse Rostering Competition INRC2010. The parameters' settings are HMS = 10, HMCR = 0.99, PAR = 0.7, and NI = 100,000. Awadallah et al. (2014) tested different memory consideration selection methods experimentally and found that the tournament memory consideration selection method provided the best rate of convergence and the best results compared to other memory consideration selection methods. The choices of the parameter values are based on the best parameter settings achieved in Awadallah et al. (2013). Hadwan et al. (2013) also applied the HSA to two sets of very different nurse RPs. The first set is a real-world data set obtained from a hospital. The second set is a well-known benchmark data set from the literature and widely used by researchers. Experimental results show that the proposed HSA produces better-quality rosters for all considered instances than the GA and other meta-heuristic algorithms reported in the literature.

5.8.9   TRAINING NN

The most popular and widely used training algorithm for NNs is the backpropagation learning (BP) algorithm (Siddique and Adeli, 2013). Since NNs generate complex error surfaces with multiple local minima, BP tends to become trapped in local minima. To overcome the local minimum problem, many meta-heuristic optimization techniques have been adopted for the training of NNs such as the GA (Siddique and Tokhi, 2001), ACO (Socha and Blum, 2007), PSO (Setteles et al., 2003; Zamani and Sadeghian, 2010), DE (Slowik and Bialko, 2008), ABC algorithm (Karaboga and Ozturk, 2009), central force optimization (Green et al., 2012), and spiral dynamics optimization (Khan and Sahai, 2013). To apply such meta-heuristic optimization algorithms, the training problem of an NN has been formulated as an optimization problem where the training of the NN can be carried out by minimizing the sum squared of the network error function defined by

$$\text{SSE}(w,b) = \min_{w,b}\left\{\sum_{j=1}^{P}\sum_{i=1}^{N}(y_i - \hat{y}_i)^2\right\} \tag{5.116}$$

The minimum of SSE(w,b) leads to optimum values of the weights and biases of the NN where y_i is the target output, $\hat{y}_i$ is the actual output of the network, and P is the number of patterns or data points. The architecture of a feedforward NN is presented in Chapter 2 in Figure 2.10.

Kattan and Abudullah (2011) applied the HSA as a new training technique to a feedforward NN for binary classification. They modified the standard stopping criteria that are based on the number of improvisation steps to the BtW harmony ratio in the current HM. Therefore, they modified the existing improved version of HS (Mahdavi et al., 2007) to suit the new stopping criterion. These modifications would be more suitable for NN training since parameters and termination would depend on the quality of the attained solution as reported by the authors.

Kulluk et al. (2011, 2012) applied five different variants of the HSA with special emphasis on the SGHS algorithm for the supervised training of feedforward NNs designed for classification problems. The SGHS algorithm employs a new improvisation scheme and an adaptive parameter tuning method. A suitable data representation for NNs is adapted for the SGHS algorithm. The SGHS algorithm for training NNs is applied to binary classification problems as well as n-ary classification problems. The performance and effectiveness of HSA and SGHS algorithms have been tested on six data sets obtained from the machine learning repository of University of California at Irvine. These are glass identification, ionosphere, Iris, Thyroid disease, Wine, and Wisconsin Breast Cancer (WBC) data sets. WBC and Iris data sets have a binary classification problem and the rest have an n-ary classification problem. A three-layered feedforward NN architecture with a sigmoid activation function is designed for all data sets. The number of neurons used in the hidden layers are 8, 33, 3, 17, 11, and 8 for glass, ionosphere, iris, thyroid, wine and WBC data sets, respectively. As defined in Equation 5.116, SSE(w,b) was used as the fitness function for the training of the NN. The HSA parameter settings used are HMS is set to 10, HMCR is set to 0.98, and PAR = 0.9. The algorithm ran for 50,000 iterations in each run. Overall training time, sum of squared errors, training and testing accuracies of the SGHS algorithm are compared with the results of the standard BP, standard HS, IHS, modified IHS, and GHS algorithms. All cases of the experimental results show that the SGHS algorithm is well suited to the training of NNs and it is also highly competitive when compared with other methods.

Kulluk (2013) applied a novel hybrid HSA and HuS algorithm to train feedforward neural networks for pattern classification. The HuS algorithm is a recently proposed meta-heuristic algorithm inspired from the hunting behavior of animals (Oftadeh et al., 2010). HuS is used as a local search technique and applied in the pitch adjustment phase. The performance and effectiveness of the hybrid HSA-HuS algorithms have been tested on nine data sets. These are cancer, card, diabetes, gene, glass, heart, horse, soybean, and thyroid data sets. A three-layered feedforward NN architecture with sigmoid activation function is designed for all data sets. A single hidden layer with six neurons is used. Mean squared error MSE(w,b) is used as the fitness function for the training of the NN. Classification error percentage (CEP) is considered as a performance metric for classification accuracy. The algorithm ran for 30 times for all data sets. The standard HSA parameter setting is used: HMS is set to 10, HMCR is set to 0.9, $\text{PAR}_{\min} = 0.01$, and $\text{PAR}_{\max} = 0.99$. The algorithm ran for 50 iterations in each run. The results were compared with the results of the standard HSA, BP, GA, PSO, DE, and ABC algorithms. In all cases, the experimental and statistical results show that the HSA-HuS algorithm can train NNs with lower CEP values in almost all of the nine data sets and proved to be highly competitive compared to other methods.

5.8.10   Clustering Problem

Clustering process involves partitioning data sets into homogeneous subgroups subject to satisfying two objectives: data items within a cluster should be similar to each other and data items within different clusters should be dissimilar based on a similarity metric. The clustering algorithms mainly aim to minimize within-cluster variation (i.e., intracluster distance) and maximize the between-cluster variation (i.e., intercluster distance). The mathematical approach to clustering has been discussed in Chapter 2 in Equations 2.93 through 2.96.

The HSA has been applied for clustering applications such as web page, data, and document clustering (Forsati et al., 2008a,b; Kumar et al. 2012, 2014; Mahdavi and Abolhassani, 2009; Mahdavi et al., 2008). Data or document sets are represented by vectors in feature space or term space with n different features or terms. The HSA is to find the optimal cluster centroids when the number of clusters is known in advance. Let there be K clusters in the data or document set. The HSA has to find K cluster centroids for the d-dimensional data or document set. Each cluster center is considered as a decision variable. Therefore, each row of the HM contains K decision variables that represent one possible solution. In d-dimensional data set with K clusters, the number of decision variables is $K \times d$. The HM is initialized with random values chosen from the given data set. The HM is defined as follows:

$$
\text{HM} = \begin{bmatrix}
c_{11}^{1} & \cdots & c_{1d}^{1} & c_{21}^{1} & \cdots & c_{2d}^{1} & \cdots & c_{k1}^{1} & \cdots & c_{kd}^{1} \\
c_{11}^{2} & \cdots & c_{1d}^{2} & c_{21}^{2} & \cdots & c_{2d}^{2} & \cdots & c_{k1}^{2} & \cdots & c_{kd}^{2} \\
\vdots & \ddots & \vdots & \vdots & \ddots & \vdots & \ddots & \vdots & \ddots & \vdots \\
c_{11}^{\text{HMS}} & \cdots & c_{1d}^{\text{HMS}} & c_{21}^{\text{HMS}} & \cdots & c_{2d}^{\text{HMS}} & \cdots & c_{k1}^{\text{HMS}} & \cdots & c_{kd}^{\text{HMS}}
\end{bmatrix} \tag{5.117}
$$

In general, Euclidean distance metric is used as distance measure and squared error function as an objective function. The clustering algorithm aims to minimize the objective function defined by

$$
f(X,C) = \min \sum_{i=1}^{d} \| x_i - c_k \|^2, \quad k = 1,2,\ldots,K \tag{5.118}
$$

where $\| x_i - c_k \|^2$ is a distance measure between data point x_i and cluster center c_k.

In document clustering, each document is represented by a weight vector w_{ij} of n features such as words, terms, or N grams. This representation is called vector-space model of the i-th document described by

$$
\mathbf{d}_i = [w_{i1}, w_{i2}, \ldots, w_{in}] \tag{5.119}
$$

where w_{ij} is the weight of the i-th document and j-th feature, $i = 1, 2, \ldots, m$ and $j = 1, 2, \ldots, n$. Some researchers use the combination of term frequency t_f and inverse document frequency id_f in defining the weight w_{ij} (Salton, 1989) as follows:

$$
w_{ij} = t_f(i,j) \times id_f(i,j) = t_f(i,j) \times \log \frac{m}{d_f(j)} \tag{5.120}
$$

where $t_f(i,j)$ is the frequency of the feature j in the i-th document $\mathbf{d}_i$, the inverse document frequency $id_f(i,j)$ is the factor which enhances the terms that appear in fewer documents and degrades the terms that appear in many documents, $id_f(i,j)$ is defined in an elegant way in terms of m representing the total number of documents and $d_f(j)$ representing the number of documents with feature j.

The vector-space model can be used for similarity measure between documents $\mathbf{d}_1$ and $\mathbf{d}_2$ for inclusion or exclusion in a cluster. There are two prominent methods of similarly measure widely used in text document clustering: Minkowski distance measure and cosine correlation measure.

1. Minkowski distance measure: Given two documents described by $\mathbf{d}_1 = [w_{11}, w_{12}, ..., w_{1n}]$ and $\mathbf{d}_2 = [w_{21}, w_{22}, ..., w_{2n}]$, the Minkowski distance is defined by

$$D_p(\mathbf{d}_1, \mathbf{d}_2) = \left(\sum_{j=1}^{n} \left| w_{1j} - w_{2j} \right|^p \right)^{1/p} \tag{5.121}$$

   where $D_p(\cdot)$ is converted into Euclidean distance for $p = 2$. If the dimensions of $\mathbf{d}_1$ and $\mathbf{d}_2$ are close to each other, the Minkowski distance measure is more useful.

2. Cosine correlation measure: For two documents $\mathbf{d}_1 = [w_{11}, w_{12}, ..., w_{1n}]$ and $\mathbf{d}_2 = [w_{21}, w_{22}, ..., w_{2n}]$, the cosine correlation measure is defined by

$$\cos(\mathbf{d}_1, \mathbf{d}_2) = \frac{\mathbf{d}_1 \cdot \mathbf{d}_2}{\| \mathbf{d}_1 \| \times \| \mathbf{d}_2 \|} \tag{5.122}$$

   where . denotes the dot product and $\|\cdot\|$ denotes the length of the vector. It is seen that $\cos(\mathbf{d}_1, \mathbf{d}_2)$ becomes 1 if $\mathbf{d}_1$ and $\mathbf{d}_2$ are identical and $\cos(\mathbf{d}_1, \mathbf{d}_2)$ becomes 0 if $\mathbf{d}_1$ and $\mathbf{d}_2$ are dissimilar. If the dimensions of $\mathbf{d}_1$ and $\mathbf{d}_2$ vary largely, the cosine correlation measure is more useful. For higher-dimensional vector spaces, the pairwise adaptive dissimilarity measure proposed by D'hondt et al. (2010) can be used.

Most of the clustering algorithms are sensitive to randomly selected initial cluster centers, which are optimized using the search ability of the HSA. If the new harmony vector is better than the harmony in the HM in terms of objective function, the new harmony is included in the HM. Computation is terminated when the maximum number of improvisations is satisfied.

Forsati et al. (2008a,b, 2013), Mahdavi et al. (2008), and Mahdavi and Abolhassani (2009) applied the HSA to web page document clustering, where each document is represented by an n-dimensional vector in feature or term space. The objective is to find the optimal cluster centroids and each solution of the clustering problem is a vector of centroids. Let $C = [c_1, c_2, ... , c_K]$ be the set of K centroids corresponding to a row in the HM representing a solution of document clustering. The objective function is to determine the locus of the cluster centroids in order to maximize intracluster similarity and minimize intercluster similarity. The centroid of the k-th cluster $c_k = [c_{k1}, c_{k2}, ... , c_{kn}]$ is computed by

$$c_{kj} = \frac{\sum_{i=1}^{m} a_{ki} d_{ij}}{\sum_{i=1}^{m} a_{ki}} \tag{5.123}$$

where $j = 1, 2, ... , n$ is the feature in document d_{ij}, $a_{ki} \in \{0,1\}$ is the weight for the document d_{ij}, and $a_{ki} = 1$ if d_{ij} belongs to cluster k. The fitness value of a clustering corresponding to a row in the HM is determined by the average distance of documents to the cluster centroid (ADDC) and defined by

$$f_{\text{ADDC}} = \frac{\sum_{k=1}^{K} \left((1/m_k) \sum_{i=1}^{m_k} D(\mathbf{c}_k, \mathbf{d}_i) \right)}{K} \tag{5.124}$$

where m_k is the number of documents in cluster k, $m_k = \sum_{i=1}^{m} a_{ki}$, and $D(\mathbf{c}_k, \mathbf{d}_i)$ is the similarity function. The solution is replaced with a row in the HM if the optimized vector has a higher fitness value than the existing solutions in the HM.

The F measure (Banergee et al., 2005) is a widely used method for performance evaluation of clustering algorithms. The F measure combines the precision and recall with equal weight and evaluates whether the clustering can remove the noise pages and generate clusters with high quality. Precision and recall compare each cluster k with each class j in the classification; precision and recall are obtained by

$$\begin{cases} P(k, j) = \dfrac{n_{kj}}{n_k} \\[2mm] R(k, j) = \dfrac{n_{kj}}{n_j} \end{cases} \tag{5.125}$$

where n_{kj} is the number of members of class j in cluster k (i.e., the number of overlapping members), n_k is the number of members of cluster k, and n_j is the number of members in class j. By $P(k,j)$, one measures the accuracy with which cluster k reproduces class j, and by $R(k,j)$, the completeness with which cluster k reproduces class j. Both $P(k,j)$ and $R(k,j)$ take the values between [0,1]. The F measure for a cluster k and class j is given by:

$$F(k, j) = \frac{2[P(k, j) \cdot R(k, j)]}{P(k, j) + R(k, j)} \tag{5.126}$$

The F measure of the whole clustering is then defined by

$$F = \sum_j \frac{n_j}{n} \max_k \{F(k, j)\} \tag{5.127}$$

Forsati et al. (2008a) applied the standard HSA and combined HSA and K-means clustering algorithm to cluster five data sets comprising different web page collections. They used the fitness function defined in Equation 5.124. The HSA parameter settings used are as follows: for each data set the HMS is set 2 times the number of clusters in the data set, the HMCR is set to 0.9, $PAR_{min} = 0.09$, and $PAR_{max} = 0.099$. The algorithms ran for an average of 30 runs for each algorithm and 500 iterations in each run. The experimental results on the five web document sets showed that combined HSA and K-means clustering produces better solutions with high quality in comparison with the HSA or the K-means algorithm alone. Forsati et al. (2008b, 2013) applied a sequential and interleaved hybrid HSA and K-means approach to a clustering problem of five different document data sets. The cosine correlation measure defined in Equation 5.122 is used as the similarity measure in each algorithm. The HSA parameter settings used are as follows: for each data set the HMS is set to 2 times the number of clusters in the data set, with varying HMCR from 0.1 to 0.6, and $PAR_{min} = 0.45$ and $PAR_{max} = 0.9$. The algorithms ran for an average of 20 runs for each algorithm and 1000 iterations in each run for the convergence of algorithms. Forsati et al. (2008b, 2013) used the F measure (Banergee et al., 2005; Larsen and Aone, 1999), a widely used quality/performance measure of clustering algorithms defined in Equation 5.127, in their experimental investigations. In all experiments, results show that the hybrid HSA and K-means algorithms are performing better than the HSA, K-means, and GA-based approaches.

Mahdavi et al. (2008) and Mahdavi and Abolhassani (2009) have also applied the hybrid HSA and K-means algorithm to document clustering problems where they applied the HSA to find the promising areas of the search space and then applied the K-means algorithm as a local search method to fine-tune the clustering. Documents are represented by a vector-space model described in Equation

5.119. A similarity measure is performed by a Minkowski and cosine correlation metric. ADDC and *F*-measure are used as the quality measure of clustering algorithms described in Equations 5.124 and 5.126, respectively. *F*-measure expresses the clustering quality and ADDC examines how much the clustering satisfies the optimization constraints. To verify the effectiveness of the proposed algorithms, three different web page collections were used. A standard parameter setting was used for the HSA. Mahdavi and Abolhassani (2009) used finite Markov chains for convergence analysis. The results are compared with other approaches. In all cases, the proposed hybrid HSA-*K*-means algorithm outperformed the GA, PSO, and Mises-Fisher generative model-based algorithms.

Kumar et al. (2012, 2014) used the modified HSA for data clustering. The modification of the HSA is proposed in the form of variations in the values of the PAR and the HMCR during the improvisation step. Values of the PAR and the HMCR lie between the maximum and minimum range specified for them. The PAR and the HMCR change dynamically (combination of linear and exponential) with generation number. The proposed HSA has been tested on five real-life data sets: wine, haberman, bupa, CMC, and breast cancer data sets. The Euclidean distance metric is used as a fitness function. The results have been compared in terms of means and standard deviation over 10 independent runs in each case. On comparing the results of the proposed technique with the others, it has been found that the modified HSA resulted in better cluster quality metrics than the HSA, GA, and *K*-means. Experimental results demonstrate that the proposed HSA yields better intracluster distance than other techniques for all the data sets except the wine data set.

5.8.11 COMBINED HEAT AND POWER ECONOMIC DISPATCH PROBLEM

The principle of combined heat and power (CHP) economic dispatch, also known as cogeneration, is to recover and make valuable use of the heat, significantly raising the overall efficiency of the conversion process. Economic dispatch must be applied in order to obtain the optimal utilization of CHP units. The primary objective of the economic dispatch problem is to minimize the total cost of generation while satisfying the operational constraints of the available generation resources (Wood and Woolenberg, 1984). Complication arises if one or more units produce both electricity and heat. In this case, both heat and power demands must be met simultaneously.

The CHP economic dispatch problem of a system is to determine the unit heat and power production so that the system production cost is minimized, while the heat–power demands and other constraints are satisfied. Mathematically, the CHP problem is to minimize the following objective function:

$$C = \text{Min}\left\{\sum_{i=1}^{N_p} c_i(p_i) + \sum_{j=1}^{N_c} c_j(h_j p_j) + \sum_{k=1}^{N_h} c_k(h_k)\right\} \tag{5.128}$$

$$\text{subject to} \quad P_D = \left\{\sum_{i=1}^{N_p} p_i + \sum_{j=1}^{N_c} p_j\right\} \tag{5.129}$$

$$H_D = \left\{\sum_{j=1}^{N_c} h_j + \sum_{j=1}^{N_h} h_k\right\} \tag{5.130}$$

$$\begin{cases} p_i^{\min} \le p_i \le p_i^{\max}, & i = 1,\dots,n_p \\ p_j^{\min}(h_j) \le p_j \le p_j^{\max}, & j = 1,\dots,n_c \\ h_j^{\min}(p_j) \le h_j \le h_j^{\max}(p_j), & j = 1,\dots,n_c \\ h_k^{\min} \le h_k \le h_k^{\max}, & k = 1,\dots,n_h \end{cases} \tag{5.131}$$

where C is the total heat and power production cost (\$); c is the unit production cost (\$); p is the unit power generation; h is the unit heat production; H_D and P_D are the system heat and power demands, respectively; i, j, and k are the indices of conventional power units, cogeneration units, and heat-only units, respectively; n_p, n_c, and n_h are the numbers of the kinds of units mentioned above; p_{min} and p_{max} are the unit power capacity limits; and h_{min} and h_{max} are the unit heat capacity limits.

Vasebi et al. (2007) verified the effectiveness of the HSA by applying the algorithm to two examples of the CHP problem. For both examples, the HSA parameters were set as follows: HMS = 6, HMCR = 0.85, and PAR = 0.5. The first CHP problem consists of a conventional power unit, two cogeneration units, and a heat-only unit. The objective was to assess the minimum overall cost of units. The cost functions of the four units are as follows:

$$
\begin{cases}
c_1 = 50p_1 \\
c_2 = 2650 + 14.5p_2 + .034p_2{}^2 + 4.2h_2 + .03h_2{}^2 + .031p_2h_2 \\
c_3 = 1250 + 36p_3 + .0435p_3{}^2 + .6h_3 + .027h_3{}^2 + .011p_3h_3 \\
c_4 = 23.4h_4
\end{cases}
\tag{5.132}
$$

$$
\text{Minimize } C = \sum_{i=1}^{4} c_i, \quad i = 1,\ldots,4
\tag{5.133}
$$

$$
\text{subject to } \begin{cases}
0 \le p_1 \le 150 \text{ MW} \\
0 \le h_4 \le 2695.2 \text{ MWth} \\
P_D = p_1 + p_2 + p_3 \\
H_D = h_2 + h_3 + h_4
\end{cases}
\tag{5.134}
$$

With a system power and heat demand of 200 MW and 115 MW, respectively, the HSA obtained the minimum cost of \$9257.07 after 25,000 evaluations at $p_1 = 0$, $p_2 = 160$, $h_2 = 40$, $p_3 = 40$, $h_3 = 75$, and $h_4 = 0$. Vasebi et al. (2007) verified the effectiveness of the HSA by applying it to a second test example with 5 cost functions. The performance of the HSA is compared with that of the GA, which indicates that the HSA can yield better results in comparison with those obtained using the GA.

Khorram and Jaberipour (2011) suggested solving the CHP economic dispatch problem using a new HSA, where bw is adjusted according to the following formulation:

$$
bw_i = \begin{cases}
\dfrac{x_i^U - x_i^L}{1000} & \text{if } \text{rand}(0,1) < \text{PAR}/2 \\[2ex]
\dfrac{\dfrac{df}{dx_i}(X)}{x_i^U - x_i^L} & \text{if } \text{rand}(0,1) \ge \text{PAR}/2
\end{cases}
\tag{5.135}
$$

$df(X)/dx_i$ is approximated as $\dfrac{df}{dx_i}(X) \approx \dfrac{f(X + 2\varepsilon e_i) - f(X - 2\varepsilon e_i)}{4\varepsilon(x_i^U - x_i^L)}$,

where e_i is the i-th unit vector, and $\varepsilon = 10^{-8}$. Two standard examples were tested to demonstrate the effectiveness and robustness of this algorithm. In all cases, the solutions obtained using the HSA were better than those obtained by other methods.

Javadi et al. (2012) and Huang and Lin (2013) also applied the HSA to CHP economic dispatch problem. In all cases, they found competitive results compared to other meta-heuristic algorithms.

5.8.12 ELD PROBLEM

The ELD problem is to schedule the power-generating units to meet the system load demand at minimum operating cost, that is, minimize the total fuel cost, while satisfying the various system equality and inequality constraints. The ELD problem is multimodal, nondifferentiable, and highly nonlinear and formulated as Min[F], where F is the total fuel cost of N_g number of generators. The ELD problem is described in greater detail in Chapter 2 by Equations 2.76 through 2.79. A detailed mathematical model and description of the ELD problem is also provided in Chatterjee et al. (2012a).

A number of meta-heuristic algorithms have been applied to the ELD problem such as the GA (Kuo, 2008; Walters and Sheble, 1993), PSO (Gaing, 2003; Panigrahi et al., 2008), evolutionary programming (Jayabarathi et al., 2005), DE (Coelho and Mariani, 2007), tabu search algorithm (Khamsawang and Jiriwibhakorn, 2010), self-organizing migrating algorithm (Coelho and Mariani, 2010), and cuckoo search algorithm (Basu and Chowdhury, 2013; Vo et al., 2013).

Coelho and Mariani (2009) applied the HSA to the ELD problem for a power system with 13 thermal generation units having valve-point effects with the load demand P_D of 1800 MW. The HSA parameters are set as follows: the HM size was 15, the HMCR and PAR were set to 0.85 and 0.45, respectively. In the improved HSA, the same parameter setting was used. Numerical results show that the improved HSA has both a better economic cost and lower mean cost than the standard HSA. The best results were obtained for solution vector P_i, $i = 1, \ldots, 13$ with the improved HSA, with a minimum cost of 17960.3661 \$/h. The results were compared with those of other studies reported in the literature. It is found that the best result obtained in this study is comparatively lower than recent studies reported in the literature.

Pandi et al. (2010) applied the HSA to a dynamic ELD problem with thermal generators and renewable energy source such as wind energy. The nonlinear constraints considered are power generation limits, reserve limits, and ramp rate limits. The optimal dispatch in the presence of wind generation is obtained using the modified HSA. The wind generator cost model is developed and used in the dynamic ELD problem. The forecasted wind speed is utilized to calculate the average wind power output and this power is included in the economic dispatch model by means of the negative load approach. Numerical results for the sample test system having five thermal units, one wind unit, and one diesel unit have been presented to demonstrate the performance of the algorithm.

Pandi and Panigrahi (2011) applied the hybrid PSO-HSA to a dynamic ELD problem. The idea here is to adjust the position of the particle such that the highly fitted particle moves slowly when compared to low fit particle. To preserve the good locations near the global optimum in the search space, an ω value is defined for each particle according to their rank. Multiplying the velocity of the i-th particle of previous generation $v_i^j(g-1)$ by ω_i is conceptually similar to the elitism incorporated in many EAs. The ω_i value is defined as

$$\omega_i = \omega_{\min} + (\omega_{\max} - \omega_{\min}) \frac{\text{rank}_i}{\text{Total\_population}} \tag{5.136}$$

where $\omega_{\min}$ and $\omega_{\max}$ are arbitrary parameters and chosen as 0.1 and 0.8, respectively. The pitch adjustment is carried out as follows:

If $U(0,1) \leq \text{PAR}(g)$, then do pitch adjustment

$$v_i'(g) = \omega_i \cdot v_i^j(g-1) + c_1 r_1 \left(p_i^l - x_i^l \right) + c_2 r_2 \left(p_{gi}^l - x_i^l \right)$$

$$v_i'(g) = \text{sign}\left[v_i^j(g) \right] * \min \left(|v_i'(g)|, v_{\max}^l \right)$$

$$x_i'(g) = v_i'(g-1) + v_i'(g) \times T$$

where T is the time, which is unity to convert the velocity into position, and x_i^l, p_i^l, and p_{gi}^l are the position, iteration best, and global best position, respectively.

A different fitness function is used in this study defined as follows:

$$\min F_T = \sum_{t=1}^{T}\sum_{i=1}^{N_g} C_i^t(P_i^t) + \lambda_1\left(\sum_{t=1}^{T}\sum_{i=1}^{N_g} P_i^t - P_D^t\right)^2 + \lambda_r\left(\sum_{t=1}^{T}\sum_{i=1}^{N_g} P_i^t - P_{r\lim}\right)^2 \qquad (5.137)$$

where F_T is the minimized total electricity generation cost, C_i^t is the cost function of i-th generating unit at time t, T is the total time period of dispatch, N_g is the number of power-generating units each loaded to P_i^t in MW, $C_i^t(P_i^t)$ is expressed as the quadratic cost function, P_i^t is the power output of the i-th generating unit, P_D^t is the total load demand at time t, λ_1 and λ_r are penalty parameters, and $P_{r\lim}$ is the ramp range limit defined as

$$P_{r\lim} = \begin{cases} P_i^{t-1} - R_i^L & \text{if } P_i^t < P_i^{t-1} - R_i^L \\ P_i^{t-1} + R_i^U & \text{if } P_i^t > P_i^{t-1} + R_i^U \\ P_i^t & \text{otherwise} \end{cases} \qquad (5.138)$$

where R_i^L and R_i^U are the lower and upper ramp range limits of the i-th generating unit, respectively.

Pandi et al. (2011) applied hybrid SI-based HSA to the ELD problem. The SI-based HSA was verified on four standard test systems with different numbers of generating units, and the results were compared with other meta-heuristic algorithms, which confirmed the robustness and efficiency of HSA.

Jeddi and Vahidinasab (2014) applied the modified HSA, and Niu et al. (2014a,b) applied a hybrid IISA with AC operation to the ELD problem. The experimental results were compared to other meta-heuristic algorithms, which again demonstrate HSA's effectiveness and applicability to the ELD problem.

5.8.13 ECONOMIC AND EMISSION DISPATCH PROBLEM

The objective of EED problem of power generation is to schedule the committed generating unit outputs to meet the load demand at minimum operating cost while satisfying all the environmental constraints of minimum emission levels of releasing toxic gases such as carbon dioxide, sulfur dioxide, and nitrogen oxide, which cause huge pollution to the environment. Thus, the EED problem is a more complicated optimization problem than ELD in power system operation and forms the basis of a benchmark problem for optimization algorithms. Thus, the EED poses a bi-objective optimization problem formulated as Min[F,E] where F is the total fuel cost of N_g number of generators and E is the total emission dispatch from the generators expressed as the sum of all types of emission considered. The EED problem is described in greater detail in Chapter 2 by Equations 2.81 through 2.83. Interested readers are also directed to Wood and Wollenberg (1984) for further details of the problem formulation of EED.

Many meta-heuristic algorithms have been applied to the EED problem by researchers (Güvenç et al., 2012; Jiang et al., 2014; Mondal et al., 2013; Shaw et al., 2012). Sivasubramani and Swarup (2011a,b) applied the HSA to an EED problem, which is formulated as a nonlinear and constrained optimization problem with competing and noncommensurable objectives. The EED problem with two competing objectives [F,E] (fuel cost and emission) is formulated as a constrained MOOP, that is, minimize [F,E] subject to

$$\sum_{i=1}^{N_g} P_i = P_D + P_{\text{Loss}} \qquad (5.139)$$

Here, P_D is the total load demand and P_{Loss} is the transmission loss. The power output P_i of each generating unit should be within the minimum and maximum limits. The inequality constraint is defined by

$$P_i^{\min} \leq P_i \leq P_i^{\max}, \quad i = 1, 2, \dots, N_g \tag{5.140}$$

Here, $P_i^{\min}$ and $P_i^{\max}$ are the minimum and maximum power outputs of the i-th generating unit, respectively. The HSA has been extended to handle the EED problem by nondominated sorting and ranking with a dynamic crowding distance strategy.

To verify the effectiveness and performance of the HSA, a standard IEEE 30-bus system with 6 generators and IEEE 118-bus system with 14 generators have been used. The HSA parameters used are as follows: HMS = 20, harmony memory considering rate HMCR = 0.85, pitch-adjusting rate $PAR_{\min} = 0.2$ and $PAR_{\max} = 2$, bandwidth $bw_{\min} = 0.45$ and $bw_{\max} = 0.9$, and the number of improvisations NI = 1000. For each of the test systems, two cases are considered: power balance constraint without transmission loss P_{Loss} and with transmission loss P_{Loss}. In all cases of the two test systems, the experimental results show that the HSA is able to provide a well-distributed Pareto-optimal solution.

Chatterjee et al. (2012b) proposed an opposition-based HSA to accelerate the algorithm. The proposed HSA employs opposition-based learning for HM initialization and also for the generation jumping. The opposite numbers have been utilized to improve the convergence rate of the HS. The proposed HSA is successfully applied to different EED problems of power systems. The test results demonstrate the effectiveness and robustness of the proposed algorithm.

Niu et al. (2014a,b) applied the HSA with a new pitch adjustment rule to a dynamic EED problem. The pitch adjustment rule is based on the perturbation information and the mean value of the HM. The new HSA is simple and helps in enhancing solution quality and convergence speed. Kherfane et al. (2014) applied the HSA to a power network system with injected renewable energy. To validate the robustness of the proposed approach, the algorithm is tested on the IEEE 30-bus system with six generating units. Three case studies were investigated: minimization of fuel cost, emission of gas, and integration of renewable energy into the network. Comparison of the results with recent global optimization methods shows the superiority of the HSA.

5.8.14 MOOP

It is amply clear from the applications discussed above that the HSA has been very successful in optimizing mono-objective problems. Unfortunately, most real-world engineering optimization problems are multiobjective in nature. Problems requiring simultaneous optimization of more than one objective functions are known as MOOPs. They can be defined as the problems consisting of multiple objectives, which are to be minimized or maximized while maintaining some constraints. The formal definition of MOOP is provided by the Equations 2.121 through 2.123 in Chapter 2. The main difference between mono- and multiobjective HSA is the solution sorting procedure, also known as ranking assignment. Ranking comprises associating a number to each solution and determining an order for the solutions where the best solutions have a lower ranking than worse solutions. There are many methods used for sorting and ranking assignments based on the concepts of Pareto optimality (Ben-Tal, 1980). It is now traditional to compare the performance of any new approach to the Nondominated Sorting Genetic Algorithm-2 (NSGA-II) developed by Deb et al. (2002), currently regarded as one of the most representative algorithms of the state-of-the-art in EAs for multiobjective optimization.

There has been very little reported on the application of the HSA to MOOPs apart from Geem (2009d). Xu et al. (2010) proposed a multiobjective HSA for a Pareto-optimal solution for the design of a reconfigurable mobile robot prototype. The objectives to be minimized are stability of the mobile robot, torque resistance of rear wheels, and mass of the robot. The ranking procedure in a Pareto-optimal solution is not explained by Xu et al.

Ricart et al. (2011) presented two proposals for a general resolution of multiobjective problems using HS considering a representative test bed that is classic in the optimization literature, namely ZDT (Zitzler, Deb, and Thiele) functions. These functions consider the minimization of two objectives. Experimental results show that the proposed algorithms are competitive when compared to NSGA-II.

Sivasubramani and Swarup (2011a) proposed a multiobjective HS approach for solving an OPF problem. OPF is a nonlinear, constrained optimization problem where many competing objectives are present. A formal definition of OPF is provided in Chapter 2. A fast elitist nondominated sorting strategy and a crowding distance procedure are used to find and estimate the Pareto-optimal front. The objective functions to be minimized are the total fuel cost and the real power transmission line losses of the system. Finally, a fuzzy-based mechanism selects a compromised solution from the estimated Pareto set. In Sivasubramani and Swarup (2011b) a multiobjective HSA is applied to an environmental and economic dispatch problem in order to optimize simultaneously two competing objectives: fuel cost and emission. The algorithm utilizes a nondominated sorting and a ranking procedure based on crowding distance values to develop and maintain a well-distributed Pareto-optimal set. The obtained results are compared with NSGA-II, showing a better performance of the proposed multiobjective HS method in the achievement of the Pareto-optimal solutions.

Optimal placement and sizing of the active power filters (APFs) is a nonconvex optimization problem of nonlinear and mixed-integer nature. Shivaie et al. (2014) proposed a new multiobjective HS framework for the APF with satisfactory and acceptable standard levels considering four objective functions, namely total harmonic distortion of voltage, harmonic transmission line loss, motor load loss function, and total APF currents in the optimization. The proposed approach was tested on the IEEE 18-bus and IEEE 30-bus test systems with different scenarios. Experimental results demonstrated the feasibility and effectiveness of the proposed method.

Multiserver location–allocation problem (MSLAP) with facilities modeled as an M/M/m queuing system is the most applied queuing model in analyzing service stations with more than one server. The model requires minimizing the waiting time and the maximum idle time of all facilities. Hajipour et al. (2014) proposed multiobjective HS for MSLAP and verified the performance on various test problems of different sizes. The results showed that MO-HS works better than the other meta-heuristic algorithms, especially for larger-size problems.

There are few other MO-HS that have been reported in the literature. Gao et al. (2014) applied MO-HS to a multiobjective FJSSP. Landa-Torres et al. (2013) applied multiobjective grouping HS to Optimal Distribution of 24-hour Medical Emergency Units.

Besides the above applications of the HSA, there are many applications reported in the literature over the years. These applications seem not very attractive and simply went out of the attention of the research community without further extension or application to other problems. These can be broadly classified into feature classification and selection (Alexandre et al., 2009; Wang et al. 2015), heat exchanger design (Fesanghary, 2009), satellite heat pipe design (Geem and Hwangbo, 2006), music composition (Geem and Choi, 2007), dam scheduling (Geem, 2007a), Sudoku puzzle solving (Geem, 2007b), transportation energy modeling (Ceylan and Ceylan, 2009), medical physics (Panchal, 2009, 2010, Alia et al., 2009a–c), RNA structure prediction (Mohsen et al., 2010), fractional-order controller (Roy et al., 2010), traveling salesman problem and tour planning (Geem, 2005), optimal allocation of shunt Var compensators in power systems (Sirjani et al., 2012), and multipass face milling (Zarei et al., 2009).

5.9   CONCLUSION

The HSA is a recent meta-heuristic optimization algorithm originally developed by Geem (2000) with many interesting features that distinguish it from other meta-heuristic algorithms. The HSA has attracted widespread attention from the research community within a very short time. It is due to the structural simplicity of the algorithm that is relatively easy to implement. The HSA possesses several advantages over the traditional optimization techniques such as the following:

(i) It is a simple meta-heuristic algorithm and does not require for initial setting for decision variables.

(ii) It uses stochastic random searches, so derivative information is not necessary.

(iii) It has few parameters for tuning.

(iv) HS generates a new solution considering all existing solutions stored in the HM.

(v) It considers each decision variable independently.

(vi) It considers continuous decision variable values without loss of generality.

(vii) It does not require the derivative of the objective function.

Due to these features, performance of the HSA has been found to be better than earlier existing meta-heuristic algorithms. The HMCR, PAR, and *bw* are the major and dominating factors in the optimization process in HS (Das et al., 2011). The HSA was further improved by Mahdavi et al. (2007) in the form of the IHSA by dynamically changing parameters. However, in the IHSA, the effect of the HMCR is not considered.

The HSA has already attracted significant interest from the research community, which will help its further development and hopefully will involve both empirical algorithmic improvements and theoretical refinements. This is again evident from the different variants that have already come into being in the literature in the recent past. Some researchers made some valuable source code in MATLAB available in Appendix B of Yang (2010), which will help new researchers apply the HSA algorithm to new applications.

The structural simplicity is an added advantage for the HSA to be combined with other meta-heuristic algorithms. A good number of hybrid algorithms have been reported in the literature combining HS and other meta-heuristic algorithms such as PSO, SI, ACO, ABC, BBO, DE, GA, HuS, Simplex, SA, CSA, SQP, BFGS, *K*-means, FCM, and many others. These hybrid algorithms have found many applications, which again demonstrate the strength and applicability of HS.

There is a recent criticism by Weyland (2010) about the HSA, claiming HS as a special case of evolution strategies. Despite Weyland's smearing assertion that HS is fundamentally misguided, HS is an emerging meta-heuristic optimization algorithm finding its application to wider research domains. Even HS has been well accepted within the wider research community, which is evident from the number of research papers reported in the recent years and the citation of these papers. More theoretical and empirical analyses have been reported in the recent years (Das et al., 2011; Yong et al., 2012). Despite all of these successes so far, there are many open research questions.

REFERENCES

Abdullah, S., Burke, E. K., and McCollum, B. 2007. A hybrid evolutionary approach to the university course timetabling problem, *Proceedings of IEEE Congress on Evolutionary Computation*, September 2007, Singapore, pp. 764–1768.

Alatas, B. 2010. Chaotic harmony search, *Applied Mathematics and Computation*, 216, 2687–2699.

Alba, E. and Dorronsoro, B. 2005. The exploration/exploitation tradeoff in dynamic cellular genetic algorithms, *IEEE Transactions on Evolutionary Computation*, 9(2), 126–142.

Alba, E. and Dorronsoro, B. 2008. *Cellular Genetic Algorithms*, Vol. 42, Berlin, Germany: Springer.

Al-Betar, M. A., Awadallah, Khader, A. T., and Z. A. Abdallkareem 2015. Island-based harmony search for optimization problems, *Expert Systems with Applications*, 42, 2026–2035.

Al-Betar, M. A. and Khader, A. T. 2009. A hybrid harmony search for university course timetabling, *Proceedings of 4th Multidisciplinary Conference on Scheduling: Theory and Applications (MISTA 2009)*, August, Dublin, Ireland, pp. 157–179.

Al-Betar, M. A., Khader, A. T., Awadallah, M. A., Alawan, M. H., and Zaqaibeh, B. 2013a. Cellular harmony search for optimization problems, *Journal of Applied Mathematics*, Hindawi Publishing, Article ID: 139464, 20 pages.

Al-Betar, M. A., Khader, A. T., and Gani, T. A. 2008. A harmony search algorithm for university course timetabling, *The Proceedings of the 7th International Conference on the Practice and Theory of Automated Timetabling*, August 18–22, Montreal, Canada.

Al-Betar, M. A., Khader, A. T., Geem, Z. W., Doush, I. A., and Awadallah, M. A. 2013b. An analysis of selection methods in memory consideration for harmony search, *Applied Mathematics and Computation*, 219(22), 10753–10767.

Al-Betar, M. A., Khader, A. T., and Liao, I. 2010a. A harmony search with multi-pitch adjusting rate for the university course timetabling, in: Z. Geem, ed., *Recent Advances in Harmony Search Algorithm*, Berlin, Heidelberg: Springer-Verlag, pp. 147–161.

Al-Betar, M. A., Khader, A. T., and Nadi, F. 2010b. Selection mechanisms in memory consideration for examination timetabling with harmony search, *Proceedings of the 12th Annual Conference on Genetic and Evolutionary Computation*, ACM, Portland, Oregon, USA, pp. 1203–1210.

Al-Betar, M. A., Khader, A. T., and Thomas, J. J. 2010c. A combination of metaheuristic components based on harmony search for the uncapacitated examination timetabling, *Proceedings of 8th International Conference on Practice Theory of Automated Timetabling*, August 2010, Belfast, Northern Ireland, pp. 57–80.

Al-Betar, M. A., Khader, A. T., and Zaman, M. 2012. University course timetabling using a hybrid harmony search metaheuristic algorithm, *IEEE Transaction on Systems, Man and Cybernetics – Part C: Applications and Reviews*, 42(5), 664–681.

Alexandre, E., Cuadra, L., and Gil-Pita, R. 2009. Sound classification in hearing aids by the harmony search algorithm, in: Zong Woo Geem, ed., *Music Inspired Harmony Search Algorithm: Theory and Applications*, Berlin, Heidelberg: Springer-Verlag, pp. 173–188.

Alia, O. M. and Mandava, R. 2011. The variants of the harmony search algorithm: An overview, *Artificial Intelligence Review*, 36, 49–68.

Alia, O. M., Mandava, R., and Aziz, M. E. 2010. A hybrid harmony search algorithm to MRI brain segmentation, *The 9th IEEE International Conference on Cognitive Informatics, ICCI2010*, Tsinghua University, Beijing, China, pp. 712–719.

Alia, O. M., Mandava, R., Ramachandram, D., and Aziz, M. E. 2009a. A novel image segmentation algorithm based on harmony fuzzy search algorithm, *International Conference of Soft Computing and Pattern Recognition, SOCPAR '09*, December 4–7, 2009, Malacca, Malaysia, pp. 335–340.

Alia, O. M., Mandava, R., Ramachandram, D., and Aziz, M. E. 2009b. Harmony search-based cluster initialization for fuzzy c-means segmentation of MR images, *Proceedings of the 2009 International Technical Conference of IEEE Region 10 (TENCON 2009)*, November 23–26, 2009, Singapore, pp. 1–6.

Alia, O. M., Mandava, R., Ramachandram, D., and Aziz, M. E. 2009c. Dynamic fuzzy clustering using harmony search with application to image segmentation, *IEEE International Symposium on Signal Processing and Information Technology (ISSPIT09)*, December 14–17, 2009, Ajaman, UAE, pp. 538–543.

Alperovits, E. and Shamir, U. 1977. Design of optimal water distribution systems, *Water Resources Research*, 13, 885–900.

Amini, F. and Ghaderi, P. 2013. Hybridization of harmony search and ant colony optimization for optimal locating of structural dampers, *Applied Soft Computing*, 13(5), 2272–2280.

Amjady, N. and Nasiri-Rad, H. 2009. Nonconvex economic dispatch with AC constraints by a new real coded genetic algorithm, *IEEE Transaction on Power Systems*, 24(3), 1489–502.

Arul, R., Ravi, G., and Velusami, S. 2013. Chaotic self-adaptive differential harmony search algorithm based dynamic economic dispatch, *Electrical Power and Energy Systems*, 50, 85–96.

Ashrafi, S. M. and Dariane, A. B. 2013. Performance evaluation of an improved harmony search algorithm for numerical optimization: Melody search (MS), *Engineering Applications of Artificial Intelligence*, 26(4), 1301–1321.

Askarzadeh, A. 2013a. A discrete chaotic harmony search-based simulated annealing algorithm for optimum design of PV/wind hybrid system, *Solar Energy*, 97, 93–101.

Askarzadeh, A. 2013b. Developing a discrete harmony search algorithm for size optimization of wind–photovoltaic hybrid energy system, *Solar Energy*, 98, 190–195.

Askarzadeh, A. and Rezazadeh, A. 2011. A grouping-based global harmony search algorithm for modeling of proton exchange membrane fuel cell, *International Journal of Hydrogen Energy*, 36(8), 5047–5053.

Askarzadeh, A. and Rezazadeh, A. 2012. An innovative global harmony search algorithm for parameter identification of a PEM fuel cell model, *IEEE Transactions on Industrial Electronics*, 59(9), 3473–3480.

Awadallah, M. A., Khader, A. T., Al-Betar, M. A., and Bolaji, A. L. 2013. Global best harmony search with a new pitch adjustment designed for nurse rostering, *Journal of King Saud University – Computer and Information Sciences*, 25(2), 145–162.

Awadallah, M. A., Khader, A. T., Al-Betar, M. A., and Bolaji, A. L. 2014. Harmony search with novel selection methods in harmony memory consideration for nurse rostering problem, *Asia-Pacific Journal of Operational Research*, 31(3), 1450014. (39 pages)

Ayob, M. and Jaradat, G. 2009. Hybrid ant colony systems for course timetabling problems, *Proceedings of the IEEE 2nd Conference on Data Mining and Optimization*, October 27–28, 2009, Selangor, Malaysia, pp. 120–126.

Ayvaz, M.T. 2007. Simultaneous determination of aquifer parameters and zone structures with fuzzy C means clustering and meta-heuristic harmony search algorithm, *Advances in Water Resources*, 30(11), 2326–2338.

Ayvaz, M. T. 2009a. Identification of groundwater parameter structure using harmony search algorithm, in: Zong Woo Geem, ed., *Music Inspired Harmony Search Algorithm: Theory and Applications*, Berlin, Heidelberg: Springer-Verlag, pp. 129–140.

Ayvaz, M. T. 2009b. Application of harmony search algorithm to the solution of groundwater management models, *Advances in Water Resources*, 32(6), 916–924.

Ayvaz, M. T., Kayhan, A. H., Ceylan, H., and Gurarslan, G. 2009. Hybridizing the harmony search algorithm with a spreadsheet 'solver' for solving continuous engineering optimization problems, *Engineering Optimisation*, 41(12), 1119–1144.

Baker, J. E. 1985. Adaptive selection methods for genetic algorithms, in: *Proceedings of the First International Conference on Genetic Algorithms and their applications*, July 24–26, Carnegie-Mellon University, Pittsburgh, PA, pp. 101–111.

Ball, P. 2008. A sound theory? *Nature*, 13 June 2008, doi: 10.1038/news.2008.883.

Banergee, A., Krumpelman, C., Basu, S., Mooney, R., and Ghosh, J. 2005. Model based overlapping clustering, *Proceedings of the International Conference on Knowledge Discovery and Data Mining*, Chicago, IL, pp. 532–537.

Banerjee, A., Mukherjee, V., and Ghoshal, S. P. 2014. An opposition-based harmony search algorithm for engineering optimization problems, *Ain Shams Engineering Journal*, 5(1), 85–101.

Basu, M. and Chowdhury, A. 2013. Cuckoo search algorithm for economic dispatch, *Energy*, 60(1), 99–108.

Ben-Tal, A. 1980. Characterization of Pareto and lexicographic optimal solutions, in: G. Fandel, T. Gal, eds., *Multiple Criteria Decision Making: Theory and Application, Lecture Notes in Economics and Mathematical Systems*, Vol. 17, Berlin: Springer, pp. 1–11.

Bezdek, J. C. 1981. *Pattern Recognition with Fuzzy Objective Function Algorithms*, Norwell, MA: Kluwer Academic Publishers.

Bilgin, B., De Causmaecker, P., Rossie, B., and Berghe, G. V. 2012. Local search neighbourhoods for dealing with a novel nurse rostering model, *Annals of Operations Research*, 194, 33–57.

Boggs, P. T. and Tolle, J. W. 1996. Sequential quadratic programming, *Acta Numerica*, 4, 1–52.

Borrell, B. 2009. Mosquitoes mate in perfect harmony, *Nature*, December 31, 2009, doi: 10.1038/news.2009.1167.

Brusco, M. and Jacobs, L. 1995. Cost analysis of alternative formulations for personnel scheduling in continuously operating organizations, *European Journal of Operational Research*, 86, 249–261.

Burke, E., Curtois, T., Qu, R., and Berghe, G. 2009. A scatter search methodology for the nurse rostering problem, *Journal of the Operational Research Society*, 61, 1667–1679.

Burke, E., De Causmaecker, P., and Berghe, G. V. 1999. A hybrid tabu search algorithm for the nurse rostering problem. in: B. McKay, X. Yao, C. Newton, J.-H. Kim, and T. Furuhashi, eds., *Simulated Evolution and Learning*, Vol. 1585 of LNCS, Berlin/Heidelberg: Springer, pp. 187–194.

Burke, E. K., deWerra, D., and Kingston, J. 2004. Applications to timetabling, in: J. L. Gross and J. Yellen, eds., *Handbook of Graph Theory*, London: CRC Press, pp. 445–474.

Cai, X., Cui, Z., Zeng, J., and Tan, Y. 2008. Dispersed particle swarm optimisation, *Information Processing Letters*, 105(6), 231–235.

Castelli, M., Ortiz-Garcia, E. G., Salcedo-Sanz, S., Segovia-Vargas, M. J., Gil-Lopez, S., Miranda, M., Leiva-Murillo, J. M., and Del Ser, J. 2012. Evaluating the internationalisation success of companies through a hybrid grouping harmony search – Extreme learning machine approach, *IEEE Journal of Selected Topics in Signal Processing*, 6(4), 388–398.

Castelli, M., Silva, S., Manzoni, L., and Vanneschi, L. 2014. Geometric selective harmony search, *Information Sciences*, 279, 468–482.

Ceylan, H. and Ceylan, H. 2009. Harmony search algorithm for transport energy demand modeling, in: Zong Woo Geem, ed., *Music Inspired Harmony Search Algorithm: Theory and Applications*, Berlin, Heidelberg: Springer-Verlag, pp. 163–172.

Chakraborty, P., Roy, G. G., Das, S., Jain, D., and Abraham, A. 2009. An improved harmony search algorithm with differential mutation operator, *Fundamental Informatics*, 95, 1–26.

Chatterjee, A., Ghoshal, S. P., and Mukherjee, V. 2012a. A maiden application of gravitational search algorithm with wavelet mutation for the solution of economic load dispatch problems, *International Journal of Bio-Inspired Computation*, 4(1), 33–46.

Chatterjee, A., Ghoshal, S. P., and Mukherjee, V. 2012b. Solution of combined economic and emission dispatch problems of power systems by an opposition-based harmony search algorithm, *Electrical Power and Systems*, 39, 9–20.

Cheng, Y.-M. 2009. Modified harmony methods for slope stability problems, in: Zong Woo Geem, ed., *Music Inspired Harmony Search Algorithm: Theory and Applications*, Berlin, Heidelberg: Springer-Verlag, pp. 141–162.

Chong, E.K. P. and Zak, S. H. 2008. *An Introduction to Optimisation*, New York: John Wiley & Sons Inc.

Cobos, C., Estupinan, D., and Perez, J. 2011. GHS+LEM: Global-best harmony search using learnable evolution models, *Applied Mathematics and Computation*, 218, 2558–2578.

Coelho, L. d. S. and Bernert, D. L. A. 2009. An improved harmony search algorithm for synchronization of discrete-time chaotic systems, *Chaos, Solitons and Fractals*, 41, 2526–2532.

Coelho, L.d. S. and Mariani, V. C. 2007. Improved differential evolution algorithms for handling economic dispatch optimization with generator constraints, *Energy Conversions Management*, 48(5), 1631–1639.

Coelho, L. d. S. and Mariani, V. C. 2009. An improved harmony search algorithm for power economic load dispatch, *Energy Conversion and Management*, 50(10), 2522–2526.

Coelho, L. d. S. and Mariani, V. C. 2010. An efficient cultural self-organizing migrating strategy for economic dispatch optimization with valve-point effect, *Energy Conversions Management*, 51(12), 2580–2587.

Contreras, J., Amaya, I., and Correra, R. 2014. An improved variant of the conventional harmony search, *Applied Mathematics and Computation*, 227, 821–830.

Corcoran, A. L. and Wainwright, R. L. 1994. A parallel island model genetic algorithm for the multiprocessor scheduling problem, *Proceedings of the 1994 ACM Symposium on Applied Computing*, March 6–8, 1994, Phoenix, Arizona, USA, pp. 483–487.

Cunha, M. C. and Sousa, J. 1999. Water distribution network design optimisation: Simulated annealing approach, *ASCE Journal of Water Resources Planning and Management*, 125, 215–221.

Das, S., Mukhopadhyay, A., Roy, A., Abraham, A., and Panigrahi, B. K. 2011. Exploratory power of the harmony search algorithm: Analysis and improvements for global numerical optimization, *IEEE Transactions on Systems, Man, and Cybernetics – Part B: Cybernetics*, 41(1), 89–106.

Dasgupta, D. 2006. Advances in artificial immune systems, *IEEE Computational Intelligence Magazine*, 1(4), 40–49.

Dash, R., Dash, P. K., and Bisoi, R. 2014. Self-adaptive differential harmony search based optimisation extreme learning machine for financial time series prediction, *Swarm and Evolutionary Computation*, 19, 25–42.

Da Silva, E. L., Gil, H. A., and Areiza, J. M. 2000. Transmission network expansion planning under an improved genetic algorithm, *IEEE Transaction on Power Systems*, 15(3), 1168–1175.

Da Silva, E. L., Ortiz, M. A., Oliveira, G. C., and Binatos, S. 2001. Transmission network expansion planning under a Tabu search approach, *IEEE Transaction on Power Systems*, 16(1), 62–68.

Deb, K. and Agrawal, R. B. 1995. Simulated binary crossover for continuous search space, *Complex Systems*, 9, 115–148.

Deb, K., Pratap, A., Agarwal, S., and Meyarivan, T. 2002. A fast and elitist multiobjective genetic algorithm: NSGA-II, *IEEE Transaction on Evolutionary Computing*, 6(2), 182–97.

Degertekin, S. 2008. Optimum design of steel frames using harmony search algorithm, *Structural Multidisciplinary Optimisation*, 36(4), 393–401.

Degertekin, S. O. 2012. Improved harmony search algorithms for sizing optimisation of truss structures, *Computer and Structures*, 92-93, 229–241.

D'hondt, J., Vertommen, J., Verhaegen, P. A., Cattrysse, D., and Duflou, J. R. 2010. Pairwise-adaptive dissimilarity measure for document clustering, *Information Sciences*, 180(2), 2341–2358.

Dorigo, M., Maniezzo, V., and Colorni, A. 1991. The ant system, *IEEE Transaction on Systems, Man and Cybernetics – Part B*, 26(1), 29–41.

Draa, A., Meshoul, S., and Talbi, H. 2010. Batouche a quantum-inspired differential evolution algorithm for solving the N-queens problem, *The International Arab Journal of Information Technology*, 7(1), 21–27.

Duan, H. and Li, J. 2014. Gaussian harmony search: A novel method for Loney's solenoid problem, *IEEE Transactions on Magnetics*, 50(3), 7026405.

Duda, R. O., Hart, P. E., and Stork, D. H. 2000. *Pattern Classification*, 2nd ed., New York: Wiley Interscience.

El-Abd, M. 2013. An improved global-best harmony search algorithm, *Applied Mathematics and Computation*, 222, 94–106.

Elaydi, S. N. 1999. *Discrete Chaos*, Boca Raton, FL: Chapman & Hall, CRC Press.

Erdal, F., Doğan, E., and Saka, M. P. 2011. Optimum design of cellular beams using harmony search and particle swarm optimizers, *Journal of Constructional Steel Research*, 67, 237–247.

Ergezer, M. and Simon, D. 2014. Mathematical and experimental analyses of oppositional algorithms, *IEEE Transactions on Cybernetics*, 44(11), 2178–2189.

Eshelman, L. J. and Schaffer, J. D. 1993. Real-coded genetic algorithms and interval schemata, in: L. D. Whitley, ed., *Foundations of Genetic Algorithms*, Vol. 2. San Mateo, CA: Morgan Kaufmann, pp. 187–202.

Eusuff, M. and Lansey, K. E. 2003. Optimisation of water distribution network design using the shuffled frog leaping algorithm, *ASCE Journal of Water Resources Planning and Management*, 129, 210–225.

Falkenauer, E. 1993. The grouping genetic algorithms—widening the scope of the Gas, *Belgian Journal of Operations Research, Statistics and Computer Science*, 33, 79–102.

Fesanghary, M. 2009. Harmony search applications in mechanical, chemical and electrical engineering, in: Zong Woo Geem, ed., *Music Inspired Harmony Search Algorithm: Theory and Applications*, Berlin, Heidelberg: Springer-Verlag, pp. 71–86.

Fesanghary, M., Mahdavi, M., Minary-Jolandan, M., and Alizadeh, Y. 2008. Hybridizing harmony search algorithm with sequential quadratic programming for engineering optimization problems, *Computer Methods in Applied Mechanics and Engineering*, 197(33–40), 3080–3091.

Feynman, R. P. 1942. The principle of least action in quantum mechanics, PhD dissertation, Princeton University, in: L. M. Brown, ed., *Feynman's Thesis: A New Approach to Quantum Theory*, Singapore: World Scientific, 2005.

Fisher, R. A. 1936. The use of multiple measurements in taxonomic problems, *Annals of Eugenics*, 7(2), 179–188.

Forsati, R., Mahdavi, M., Kangavari, M., and Safarkhani, B. 2008a. Web page clustering using harmony search optimization, *Canadian Conference on Electrical and Computer Engineering 2008 (CCECE 2008)*, May 4–7, 2008, Niagara Falls, Canada, pp. 1601–1604.

Forsati, R., Mahdavi, M., Shamsfard, M., and Meybodi, M. R. 2013. Efficient stochastic algorithms for document clustering, *Information Sciences*, 220, 269–291.

Forsati, R., Meybodi, M. R., Mahdavi, M., and Neiat, A. G. 2008b. Hybridization of k-means and harmony search methods for web page clustering, *Proceedings of the 2008 IEEE/WIC/ACM International Conference on Web Intelligence and Intelligent Agent Technology*, December 9–12, 2008, Sydney, Australia, pp. 329–335.

Fujiwara, O. and Khang, D. B. 1990. A two-phase decomposition method for optimal design of looped water distribution networks, *Water Resources Research*, 26, 539–549.

Gaing, Z-L. 2003. Particle swarm optimization to solving the economic dispatch considering the generator constraints, *IEEE Transactions on Power Systems*, 8(3), 1187–1195.

Gallego, R. A., Romero, R., and Monticellia, A. J. 2000. Tabu search algorithm for network synthesis, *IEEE Transaction on Power Systems*, 15(2), 490–495.

Gao, K. Z., Suganthan, P. N., Pan, Q. K., Chua, T. J., Cai, T. X., and Chong, C. S. 2014. Pareto-based grouping discrete harmony search algorithm for multi-objective flexible job shop scheduling, *Information Sciences*, 289, 76–90.

Gao, X. Z., Wang, X., and Ovaska, S. J. 2008. Modified harmony search methods for uni-modal and multi-modal optimization, in: F. Xhafa, F. Herrera, A. Abraham, M. Köppen, and J. M. Benítez, eds., *Proceedings of the 2008 8th International Conference on Hybrid Intelligent Systems (HIS 2008)*, September 10–12, 2008, Barcelona, Spain, pp. 65–72.

Gao, X. Z., Wang, X., and Ovaska, S. J. 2009a. Harmony search methods for multi-modal and constrained optimization, in: Zong Woo Geem, ed., *Music Inspired Harmony Search Algorithm: Theory and Applications*, Berlin, Heidelberg: Springer-Verlag, pp. 39–52.

Gao, X. Z., Wang, X., and Ovaska, S. J. 2009b. Uni-modal and multi-modal optimization using modified harmony search methods, *International Journal of Innovative Computing and Information Control*, 5(10A), 2985–2996.

García-Torres, J. M., Damas, S., Cordón, O., and Santamaría, J. 2014. A case study of innovative population-based algorithms in 3D modeling: Artificial bee colony, biogeography-based optimization, harmony search, *Expert Systems with Applications*, 41(4 Part 2), 1750–1762.

Garver, L. L. 2007. Transmission network estimation using linear programming, *IEEE Transaction on Power Apparatus and Systems*, 89(7), 1688–1697.

Geem, Z. 2006a. Improved harmony search from ensemble of music players, in: B. Gabrys, R. J. Howlett, L. Jain, eds., *Knowledge-Based Intelligent Information and Engineering Systems*, Vol. 4251, Lecture Notes on Artificial Intelligence, Heidelberg: Springer, pp. 86–93.

Geem, Z. 2007a. Optimal scheduling of multiple dam system using harmony search algorithm, in: *Computational and Ambient Intelligence*, Berlin: Springer, pp. 316–323.

Geem, Z. 2007b. Harmony search algorithm for solving sudoku, in: B. Apolloni, R. J. Howlett, L. Jain, eds., *Knowledge-Based Intelligent Information and Engineering Systems*, Lecture Notes in Computer Science, Vol. 4692, Berlin, Heidelberg: Springer, pp. 371–378.

Geem, Z. W. 2000. Optimal design of water distribution networks using harmony search, PhD thesis, Department of Civil and Environmental Engineering, Korea University.

Geem, Z. W. 2005. Harmony search in water pump switching problem, in: L. Wang, K. Chen, and Y. S. Ong, eds., *Advances in Natural Computation, ICNC 2005, LNCS 3612*, Berlin, Heidelberg: Springer-Verlag, pp. 751–760.

Geem, Z. W. 2006b. Optimal cost design of water distribution networks using harmony search, *Engineering Optimisation*, 38(3), 259–280.

Geem, Z. W. 2008. Novel derivative of harmony search algorithm for discrete design variables, *Applied Mathematics and Computations*, 199(1), 223–230.

Geem, Z. W. 2009a. *Harmony Search Algorithms for Structural Design Optimization*, Berlin, Heidelberg: Springer Verlag.

Geem, Z. W. 2009b. Particle-swarm harmony search for water network design, *Engineering Optimisation*, 41(4), 297–311.

Geem, Z. W. 2009c. Harmony search optimisation to the pump-included water distribution network design, *Civil Engineering and Environmental Systems*, 26(3), 211–221.

Geem, Z. W. 2009d. Multiobjective optimization of time-cost trade-off using harmony search, *Journal of Construction Engineering and Management*, 136(6), 711–716.

Geem, Z. W. 2010a. *Recent Advances in Harmony Search Algorithm, Studies in Computational Intelligence*, Berlin, Heidelberg: Springer Verlag.

Geem, Z. W. ed. 2010b. State-of-the-art in the structure of harmony search algorithm, in: *Recent Advances in Harmony Search Algorithm*, Berlin: Springer, pp. 1–10.

Geem, Z. W. 2012. Effects of initial memory and identical harmony in global optimization using harmony search algorithm, *Applied Mathematics and Computation*, 218, 11337–11343.

Geem, Z. W. and Choi, J. 2007. Music composition using harmony search algorithm, in: M. Giacobini, ed., *Evo Workshops 2007, LNCS*, Vol. 4448, Heidelberg: Springer, pp. 593–600.

Geem, Z. W. and Hwangbo, H. 2006. Application of harmony search to multi-objective optimization for satellite heat pipe design, *Proceedings of US-Korea Conference on Science, Technology, & Entrepreneurship (UKC 2006)*, Teaneck, NJ, USA, Citeseer, pp. 1–3.

Geem, Z. W., Kim, J. H., and Loganathan, G. V. 2001. A new heuristic optimization algorithm: Harmony search, *Simulation*, 76(2), 60–68.

Geem, Z. W., Kim, J. H., and Loganathan, G. V. 2002. Harmony search optimisation: Application to pipe network design, *International Journal of Modelling and Simulation*, 22(2), 125–33.

Geem, Z. W., Lee, K. S., and Park, Y. 2005a. Application of harmony search to vehicle routing, *American Journal of Applied Science*, 2(12), 1552–1557.

Geem, Z.W., Tseng, C. L., and Park, Y. 2005b. Harmony search for generalized orienteering problem: Best touring in China, in: L. Wang, K. Chen, Y. Ong, eds., *Advances in Natural Computation*, Vol. 3412, Berlin: Springer, pp. 741–750.

Geem, Z. W., Tseng, C.-L., and Williams, J. C. 2009. Harmony search algorithms for water and environmental systems, in: Zong Woo Geem, ed., *Music Inspired Harmony Search Algorithm: Theory and Applications*, Berlin, Heidelberg: Springer-Verlag, pp. 113–128.

Geem, Z. W. and Williams, J. C. 2007. Harmony search and ecological optimization, *International Journal of Energy and Environment*, 1(2), 150–154.

Geem, Z. W. and Williams, J. C. 2008. Ecological optimization using harmony search, *Proceedings of the American Conference on Applied Mathematics, World Scientific and Engineering Academy and Society (WSEAS)*, 2008, Cambridge, MA, pp. 148–152.

Gill, M. A. 1978. Flood routing by the Muskingum method, *Journal of Hydrology*, 36, 353–363.

Goldberg, D. E. 1989. *Genetic Algorithms in Search, Optimization, and Machine Learning*, Reading, MA: Addison Wesley Publishing Company.

Goldberg, D. E. and Kuo, C. H. 1987. Genetic algorithms in pipeline optimization, *Journal of Computing in Civil Engineering, ASCE*, 1(2), 128–141.

Greblicki, J. and Kotowski, J. 2009. Analysis of the properties of the harmony search algorithm carried out on the one dimensional binary knapsack problem, in: R. Moreno-Díaz, F. Pichler, A. Quesada-Arencibia, eds., *12th International Conference on Computer Aided Systems Theory, EUROCAST 2009*, Berlin, Heidelberg: Springer Verlag, pp. 697–704.

Green, R. C., Wang, L., and Alam, M. 2012. Training neural networks using central force optimization and particle swarm optimization: Insights and comparisons, *Expert Systems with Applications*, 39, 555–563.

Gutjahr, W. and Rauner, M. 2007. An ACO algorithm for a dynamic regional nurse-scheduling problem in Austria, *Computers and Operations Research*, 34, 642–666.

Güvenç, U., Sönmez, Y., Duman, S., and Yörükeren, N. 2012. Combined economic and emission dispatch solution using gravitational search algorithm, *Scientia Iranica*, 19(6), 1754–1762.

Gutjahr, W. and Rauner, M. 2007. An ACO algorithm for a dynamic regional nurse-scheduling problem in Austria, Computers and Operations Research, 34, 642–666.

Güvenç, U., Sönmez, Y., Duman, S., and Yörükeren, N. 2012. Combined economic and emission dispatch solution using gravitational search algorithm, Scientia Iranica, 19(6), 1754–1762.

Hadwan, M., Ayob, M., Sabar, N. R., and Qu, R. 2013. A harmony search algorithm for nurse rostering problems, *Information Sciences*, 233, 126–140.

Hajipour, V., Rahmati, S. H. A., Pasandideh, S. H. R., and Niaki, S. T. A. 2014. A multi-objective harmony search algorithm to optimize multi-server location–allocation problem in congested systems, *Computers & Industrial Engineering*, 72, 187–197.

Han, K. H. and Kim, J. H. 2004. Quantum-inspired evolutionary algorithms with a new termination criterion, Hε gate, and two phase scheme, *IEEE Transactions on Evolutionary Computation*, 8(2), 156–169.

Hasancebi, O., Erdal, F., and Saka, M. P. 2009. An adaptive harmony search method for structural optimization, *Journal of Structural Engineering*, 136(4), 419–431.

Hénon, M. 1976. A two-dimensional mapping with a strange attractor, *Communications in Mathematical Physics*, 50(1), 69–77.

Holland, J. H. 1975. *Adaptation in Natural and Artificial Systems*, Ann Arbor, MI: University Michigan Press.

Hosseini, S. D., Shirazi, M. A., and Karimi, B. 2014. Cross-docking and milk run logistics in a consolidation network: A hybrid of harmony search and simulated annealing approach, *Journal of Manufacturing Systems*, 33, 567–577.

Huang, S.-H. and Lin, P.-C. 2013. A harmony-genetic based heuristic approach toward economic dispatching combined heat and power, *International Journal of Electrical Power & Energy Systems*, 53, 482–487.

Ingram, G. and Zhang, T. 2009. Overview of applications and developments in the harmony search algorithm, in: Zong Woo Geem, ed., *Music Inspired Harmony Search Algorithm: Theory and Applications*, Berlin, Heidelberg: Springer-Verlag, pp. 15–37.

Jaberipour, M. and Khorram, E. 2010. Two improved harmony search algorithms for solving engineering optimisation problems, *Communications in Nonlinear Science and Numerical Simulation*, 15, 3316–3331.

Jaeger, G. 2006. *Quantum Information: An Overview*, Berlin: Springer, 2006.

Jang, W. S., Kang, H. I., and Lee, B. H. 2008. Hybrid simplex-harmony search method for optimization problems, *IEEE World Congress on Computational Intelligence, IEEE Congress on Evolutionary Computation (CEC 2008)*, June 1–6, 2008, Hong Kong, pp. 4157–4164.

Javadi, M. S., Nezhad, A. E., and Sabramooz, S. 2012. Economic heat and power dispatch in modern power system: Harmony search algorithm versus analytical solution, *Scientia Iranica*, 19(6), 1820–1828.

Jayabarathi, T., Jayaprakash, K., Jeyakumar, D., and Raghunathan, T. 2005. Evolutionary programming techniques for different kinds of economic dispatch problems, *Electrical Power Systems Resources*, 73(2), 169–176.

Jeddi, B. and Vahidinasab, V. 2014. A modified harmony search method for environmental/economic load dispatch of real-world power systems, *Energy Conversion and Management*, 78, 661–675.

Jiang, S., Ji, Z., and Shen, Y. 2014. A novel hybrid particle swarm optimization and gravitational search algorithm for solving economic emission load dispatch problems with various practical constraints, *International Journal of Electrical Power & Energy Systems*, 55, 628–644.

Karaboga, D. and Bahriye, A. 2009. A comparative study of artificial bee colony algorithm, *Applied Mathematics and Computing*, 214, 108–132.

Karaboga, D. and Basturk, B. 2007. A powerful and efficient algorithm for numerical function optimization: Artificial bee colony algorithm, *Journal of Global Optimisation*, 3, 9459–9471.

Karaboga, D. and Basturk, B. 2008. On the performance of artificial bee colony algorithm, *Applied Soft Computing*, 8, 687–697.

Karaboga, D. and Ozturk, C. 2009. Neural networks training by artificial bee colony algorithm on pattern classification, *Neural Network World*, 19, 279–292.

Karahan, H., Gurarslan, G., and Geem, Z. W. 2013. Parameter estimation of the nonlinear muskingum flood-routing model using a hybrid harmony search algorithm, *Journal of Hydrologic Engineering*, 18(3), 352–360.

Kattan, A. and Abdullah, R. 2011. Training of feed-forward neural networks for pattern-classification applications using music inspired algorithm, *International Journal of Computer Science and Information Security*, 9, 44–57.

Kattan, A. and Abdullah, R. 2013. A dynamic self-adaptive harmony search algorithm for continuous optimisation problems, *Applied Mathematics and Computation*, 219, 8542–8567.

Kaveh, A. and Ahangaran, M. 2012. Social harmony search algorithm for continuous optimization, *IJST – Transactions of Civil Engineering*, 36, 121–137.

Kaveh, A. and Talatahari, S. 2009. Particle swarm optimizer, ant colony strategy and harmony search scheme hybridized for optimization of truss structures, *Computers & Structures*, 87(5–6), 267–283.

Kennedy, J. and Eberhart, R. 1995. Particle swarm optimization, *Proceedings of the IEEE International Conference on Neural Networks*, 27 Nov–1 Dec, 1995, Perth, WA, pp. 1942–1948.

Khamsawang, S. and Jiriwibhakorn, S. 2010. DSPSO–TSA for economic dispatch problem with nonsmooth and noncontinuous cost functions. *Energy Conversions Management*, 51(2), 365–375.

Khan, K. and Sahai, A. 2013. Spiral dynamics optimization-based algorithm for human health improvement, *GESJ: Computer Science and Telecommunications*, 37(1), 31–38.

Kherfane, N., Kherfane, R. L., Younes, M., and Khodja, F. 2014. Economic and emission dispatch with renewable energy using HSA, *Energy Procedia*, 50, 970–979.

Khorram, E. and Jaberipour, M. 2011. Harmony search algorithm for solving combined heat and power economic dispatch problems, *Energy Conversion and Management*, 52, 1550–1554.

Kim, J. H., Geem, Z. W., Kim, E. S. 2001. Parameter estimation of the nonlinear Muskingum model using harmony search, *Journal of the American Water Resources Association*, 37(5), 1131–1138.

Krumhansl, C. L. 1990. *Cognitive Foundations of Musical Pitch*, New York: Oxford University Press.

Kulluk, S. 2013. A novel hybrid algorithm combining hunting search with harmony search algorithm for training neural networks, *Journal of the Operational Research Society*, 64, 748–761.

Kulluk, S., Ozbakir, L., and Baykasoglu, A. 2011. Self-adaptive global best harmony search for training neural networks, *Procedia Computer Science*, 3, 282–286.

Kulluk, S., Ozbakir, L., and Baykasoglu, A. 2012. Training neural networks with harmony search algorithms for classification problems, *Engineering Applications of Artificial Intelligence*, 25(1), 11–19.

Kumar, V., Chhabra, J. K., and Kumar, D. 2012. Effect of harmony search parameters' variation in clustering, *Procedia Technology*, 6, 265–274.

Kumar, V., Chhabra, J. K., and Kumar, D. 2014. Parameter adaptive harmony search algorithm for unimodal and multimodal optimisation problems, *Journal of Computer Science*, 5, 144–155.

Kuo, C.C. 2008. A novel string structure for economic dispatch problems with practical constraints, *Energy Conversion Management*, 49(12), 3571–3577.

Lamberti, L. and Pappalettere, C. 2013. Truss weight minimization using hybrid harmony search and Big Bang–Big Crunch algorithms, *Metaheuristic, Applications in Structures and Infrastructures*, 2013, 207–240.

Landa-Silva, D. and Obit, J. H. 2009. Evolutionary non-linear great deluge for university course timetabling, in: E. Corchado, X. Wu, E. Oja, E. Hristozov, and T. Jedlovcnik, eds., *Hybrid Artificial Intelligence Systems*, Lecture Notes in Computer Science (Lecture Notes in Artificial Intelligence, Vol. 5572), Berlin, Germany: Springer-Verlag, pp. 269–276.

Landa-Torres, I., Gil-Lopez, S., Salcedo-Sanz, S., Del Ser, J., and Portilla-Figueras, J. A. 2012. A novel grouping harmony search algorithm for the multiple-type access node location problem, *Expert Systems with Applications*, 39(5), 5262–5270.

Landa-Torres, I., Manjarres, D., Salcedo-Sanz, S., Del Ser, J., and Gil-Lopez, S. 2013. A multi-objective grouping harmony search algorithm for the optimal distribution of 24-hour medical emergency units, *Expert Systems with Applications*, 40(6), 2343–2349.

Larsen, B. and Aone, C. 1999. Fast and effective text mining using linear-time document clustering, *Proceedings of the Fifth ACM SIGKDD International Conference on Knowledge Discovery and Data Mining*, August 15–18, 1999, San Diego, CA, USA, pp. 16–22.

Layeb, A. 2013. A hybrid quantum inspired harmony search algorithm for 0–1 optimization problems, *Journal of Computational and Applied Mathematics*, 253, 14–25.

Layeb, A. and Saidouni, D. E. 2008. A new quantum evolutionary local search algorithm for MAX3-SAT problem, in: *Proceedings of the 3rd International Workshop on Hybrid Artificial Intelligence Systems*,

HAIS 2008, Burgos, Spain, September 24–26, 2008, Corchado, Emilio, Abraham, and Ajith, eds., *Lecture Notes in Artificial Intelligence*, Vol. 5271, Berlin, Heidelberg: Springer-Verlag, pp. 172–179.

Lee, K. S. and Geem, Z. W. 2004. A new structural optimization method based on the harmony search algorithm, *Computers and Structures*, 82(9–10), 781–798.

Lee K. S. and Geem Z. W. 2005. A new meta-heuristic algorithm for continuous engineering optimization: Harmony search theory and practice, *Computer Methods in Applied Mechanics and Engineering*, 194(36–38), 3902–3933.

Lee, Y. C. and Zomaya, A. Y. 2009. Interweaving heterogeneous metaheuristics using harmony search, *IEEE International Symposium on Parallel & Distributed Processing, IPDPS 2009*, May 25–29, 2009, Rome, Italy, pp. 1–8.

Li, J. and Duan, H. 2014. Novel biological visual attention mechanism via Gaussian harmony search, *Optik*, 125, 2313–2319.

Li, M. J., Ng, M. K., Cheung, Y. M., and Huang, J. Z. 2008. Agglomerative fuzzy k-means clustering algorithm with selection of number of clusters, *IEEE Transaction on Knowledge and Data Engineering*, 20(11), 1519–1534.

Li Q., Mitianoudis N., and Stathaki T. 2007. Spatial kernel k-harmonic means clustering for multi-spectral image segmentation, *Image Process IET*, 1(2), 156–167.

Liang, J. J., and Suganthan, P. N. 2005. Dynamic multi-swarm particle swarm optimizer, *Proceedings of IEEE International Swarm Intelligence Symposium*, June 8–10, 2005, Pasadena, California, pp. 124–129.

Liu, L. and Zhou, H. 2013. Hybridization of harmony search with variable neighborhood search for restrictive single-machine earliness/tardiness problem, *Information Sciences*, 226, 68–92.

MacArthur, R. and Wilson, E. 1967. *Theory of Biogeography*, Princeton, NJ: Princeton University Press.

MacQueen, J. B. 1967. Some methods for classification and analysis of multivariate observations, in: L. M. LeCam, J. Neyman, eds., *Proceedings of 5th Berkeley Symposium on Mathematical Statistics and Probability*, Berkeley: University of California Press, pp. 281–297.

Mahdavi, M. and Abolhassani, H. 2009. Harmony k-means algorithm for document clustering, *Data Mining and Knowledge Discovery*, 18(3), 370–391.

Mahdavi, M., Chehreghani, M. H., Abolhassani, H., and Forsati, R. 2008. Novel meta-heuristic algorithms for clustering web documents, *Applied Mathematics and Computation*, 201(1-2), 441–451.

Mahdavi, M., Fesanghary, M., and Damangir, E. 2007. An improved harmony search algorithm for solving optimization problems, *Applied Mathematics and Computation*, 188(2), 1567–1579.

Malaki, M., Pourbaghery, J. A., and Abolhassani, H. 2008. A combinatory approach to fuzzy clustering with harmony search and its applications to space shuttle data, *SCIS & ISIS 2008*, Nagoya, Japan.

Manjarres, D., Landa-Torres, I., Gil-Lopez, S., Del Ser, J., Bilbao, M. N., Salcedo-Sanz, S., and Geem, Z. W. 2013. A survey on applications of the harmony search algorithm, *Engineering Applications of Artificial Intelligence*, 26(8), 1818–1831.

May, R. M. 1976. Simple mathematical models with very complicated dynamics, *Nature*, 261, 459.

McCarthy, G. T. 1938. The unit hydrograph and flood routing, *Proceedings of the Conference of the North Atlantic Division*, U.S. Army Corps of Engineers, New London.

Michalewic, Z. 1992. *Genetic Algorithms + Data Structures = Evolution Programs*, Berlin, Heidelberg: Springer Verlag.

Michalski, R. S. 2000. Learnable evolution model: Evolutionary processes guided by machine learning, *Machine Learning*, 38(1), 9–40.

Miguel, L. F. F. and Miguel, L. F. F. 2012. Shape and size optimisation of truss structures considering dynamic constraints through modern metaheuristic algorithms, *Expert Systems with Applications*, 39, 9458–9467.

Moeinzadeh, H., Asgarian, E., Zanjani, M., Rezaee, A., and Seidi, M. 2009. Combination of harmony search and linear discriminate analysis to improve classification. *Third Asia International Conference on Modelling and Simulation*, AMS '09, May 25–29, 2009, Bandung, Bali, Indonesia, pp. 131–135.

Mohsen, A. M., Khader, A. T., and Ramachandram, D. 2010. An optimization algorithm based on harmony search for RNA secondary structure prediction, *Studies in Computational Intelligence*, 270, 163–174.

Mondal, S., Bhattacharya, A., and Dey, S. H. 2013. Multi-objective economic emission load dispatch solution using gravitational search algorithm and considering wind power penetration, *International Journal of Electrical Power & Energy Systems*, 44(1), 282–292.

Mukhopadhyay, A., Roy, A., Das, S., and Abraham, A. 2008. Population-variance and explorative power of harmony search: An analysis. *Second National Conference on Mathematical Techniques Emerging Paradigms for Electronics and IT Industries (MATEIT 2008)*, New Delhi, India.

Murren, P. and Khandelwal, K. 2014. Design-driven harmony search (DDHS) in steel frame optimisation, *Engineering Structures*, 59, 798–808.

Nadi, F., Khader, A. T., and Al-Betar, M. A. 2010. Adaptive genetic algorithm using harmony search, *Proceedings of the 12th Annual Conference on Genetic and Evolutionary Computation, ACM*, Portland, Oregon, USA, pp. 819–820.

Nelder, J. A. and Mead, R. 1965. A simplex method for function minimisation, *Computer Journal*, 7(4), 308–313.

Nettl, B. 1974. Thoughts on improvisation: A comparative approach, *The Musical Quarterly, Oxford Journals*, 60(1), 1–19.

Neumann, J. V. 1966. *Theory of Self-Reproducing Automata*, Urbana, IL: University of Illinois Press.

Niu, Q., Zhang, H., Li, K., and Irwin, G. W. 2014a. An efficient harmony search with new pitch adjustment for dynamic economic dispatch, *Energy*, 65, 25–43.

Niu, Q., Zhang, H., Wang, X., Li, K., and Irwin, G. W. 2014b. A hybrid harmony search with arithmetic crossover operation for economic dispatch, *International Journal of Electrical Power & Energy Systems*, 62, 237–257.

Oftadeh, R., Mahjoob, M. J., and Shariatpanahi, M. 2010. A novel meta-heuristic optimization algorithm inspired by group hunting of animals: Hunting search, *Computers and Mathematics with Applications*, 60(7), 2087–2098.

Omran, M. G. H. and Mahdavi, M. 2008. Global-best Harmony Search, *Applied Mathematics and Computation*, 198, 643–656.

Pan, Q. K., Suganthan, P. N., and Fatih Tasgetiren, M. 2010. A self-adaptive global-best harmony search algorithm for continuous optimization problems, *Applied Mathematics and Computation*, 216(3), 830–848.

Pan, Q.-K., Suganthan, P. N., Liang, J. J., and Tasgetiren, M. F. 2011a. A local-best harmony search algorithm with dynamic sub-harmony memories for lot-streaming flow shop scheduling problem, *Expert Systems with Applications*, 38(4), 3252–3259.

Pan, Q.-K., Wang, L., and Gao, L. 2011b. A chaotic harmony search algorithm for the flow shop scheduling with limited buffers, *Applied Soft Computing*, 11, 5270–5280.

Panchal, A. 2009. Harmony search in therapeutic medical physics, in: Zong Woo Geem, ed., *Music Inspired Harmony Search Algorithm: Theory and Applications*, Berlin, Heidelberg: Springer-Verlag, pp. 189–204.

Pandi, V. R. and Panigrahi, B. K. 2011. Dynamic economic load dispatch using hybrid swarm intelligence based harmony search algorithm, *Expert Systems with Applications*, 38(7), 8509–8514.

Pandi, V. R., Panigrahi, B. K., Bansal, R. C., Das, S., and Mohapatra, A. 2011. Economic load dispatch using hybrid swarm intelligence based harmony search algorithm, *Electric Power Components and Systems*, 39(8), 751–767.

Pandi, V. R., Panigrahi, B. K., Das, S., and Cui, Z. 2010. Dynamic economic load dispatch with wind energy using modified harmony search, *International Journal of Bio-Inspired Computation*, 2(3/4), 282–289.

Panigrahi, B. K., Pandi, V. R., and Das, S. 2008. Adaptive particle swarm optimization approach for static and dynamic economic load dispatch, *Energy Conversion Management*, 49(6), 1407–1415.

Panigrahi, B. K., Pandi, V. R., Das, S., and Abraham, A. 2010. Population variance harmony search algorithm to solve optimal power flow with non-smooth cost function, in: Z. W. Geem, ed., *Recent Advances in Harmony Search Algorithm. Studies in Computational Intelligence*, Vol. 270, Berlin, Germany: Springer-Verlag, pp. 65–75.

Parncutt, R. 1989. *Harmony: A Psycho-Acoustical Approach*, New York: *Springer-Verlag*.

Peitgen, H., Jurgens, H., and Saupe, D. 1992. *Chaos and Fractals*, Berlin, Germany: Springer-Verlag.

Picek, S. and Golub, M. 2010. Comparison of a crossover in binary-coded genetic algorithms, *WSEAS Transaction on Computers*, 9(9), 1064–73.

Poursalehi, N., Zolfaghari, A., and Minuchehr, A. 2013a. Differential harmony search algorithm to optimise PWRs loading pattern, *Nuclear Engineering and Design*, 257, 161–174.

Poursalehi, N., Zolfaghari, A., Minucherhr, A., and Valavi, K. 2013b. Self-adaptive global-best harmony search algorithm applied to reactor fuel management optimisation, *Annals of Nuclear Energy*, 62, 86–102.

Qin, A. K. and Forbes, F. 2011. Harmony search with differential mutation based pitch adjustment, *GECCO'11*, July 12–16, Dublin, Ireland, pp. 545–552.

Rahnamayan, S., Tizhoosh, H. R., and Salama, M. M. A. 2008. Opposition-based differential evolution, *IEEE Transaction on Evolutionary Computation*, 12(1), 64–79.

Rastgou, A. and Moshtagh, J. 2014. Improved harmony search algorithm for transmission expansion planning with adequacy–security considerations in the deregulated power system, *International Journal of Electrical Power & Energy Systems*, 60, 153–164.

Reca, J., Martinez, J., Gil, C. et al. 2008. Application of several meta-heuristic techniques to optimisation of real looped water distribution networks, *Journal of Water Resources Management*, 22, 1367–1379.

Ricart, J., Huttemann, G., Lima, J., and Baran, B. 2011. Multiobjective harmony search algorithm proposals, *Electronic Notes in Theoretical Computer Science*, 281, 51–67.

Romero, R., Gallego, R. A., and Monticellia, A. J. 1990. Transmission network expansion planning by simulated annealing, *IEEE Transaction on Power Systems*, 11(1), 364–369.

Roy, G. G., Chakraborty, P., and Das, S. 2010. Designing fractional-order piλdμ controller using differential harmony search algorithm, *International Journal of Bio-Inspired Computation*, 2(5), 303–309.

Rucinski, M., Izzo, D., and Biscani, F. 2010. On the impact of the migration topology on the island model, *Parallel Computing*, 36(10), 555–571.

Saka, M. 2007. Optimum geometry design of geodesic domes using harmony search algorithm, *Advances in Structural Engineering*, 10(6), 595–606.

Saka, M. and Hasancebi, O. 2009. Adaptive harmony search algorithm for design code optimization of steel structures, in: Z. Geem, ed., *Harmony Search Algorithms for Structural Design Optimization*, Berlin, Heidelberg: Springer-Verlag, pp. 79–120.

Saka, M. P. 2009a. Optimum design of steel skeleton structures, in: Zong Woo Geem, ed., *Music Inspired Harmony Search Algorithm: Theory and Applications*, Berlin, Heidelberg: Springer-Verlag, pp. 87–112.

Saka, M. P. 2009b. Optimum design of steel sway frames to BS5950 using harmony search algorithm, *Journal of Constructional Steel Research*, 65(1), 36–43.

Salton, G. 1989. *Automatic Text Processing*, Reading, MA: Addison-Wesley.

Schuster, H. G. 1988. *Deterministic Chao: An Introduction*, 2nd ed., Weinheim, Germany: Physika Verlag.

Setteles, M., Rodebaugh, B., and Soule, T. 2003. Comparison of genetic algorithm and particle swarm optimizer when evolving a recurrent neural network, in: *Genetic and Evolutionary Computation, GECCO'2003*, Vol. 2723, Berlin, Heidelberg: Springer, pp. 148–149.

Shaw, B., Mukherjee, V., and Ghoshal, S. P. 2012. A novel opposition-based gravitational search algorithm for combined economic and emission dispatch problems of power systems, *Electrical Power and Energy Systems*, 35(1), 21–33.

Shi, Y., Liu, H., Gao, L., and Zhang, G. 2011. Cellular particle swarm optimization, *Information Sciences*, 181(20), 4460–4493.

Shivaie, M., Salemnia, A., and Ameli, M. T. 2014. A multi-objective approach to optimal placement and sizing of multiple active power filters using a music-inspired algorithm, *Applied Soft Computing*, 22, 189–204.

Shokri, M., Tizhoosh, H. R., and Kamel, M. 2006. Opposition-based Q(λ) algorithm, *Proceedings of IEEE International Joint Conference on Neural Networks: IEEE World Congress on Computational Intelligence*, Vancouver, Canada, pp. 646–653.

Siddique, N. and Adeli, H. 2013. *Computational Intelligence: Synergies of Fuzzy Logic, Neural Networks and Evolutionary Computing*, Chichester, West Sussex: John Wiley & Sons.

Siddique, N. and Adeli, H. 2015a. Hybrid harmony search algorithms, *International Journal on Artificial Intelligence Tools,* Vol. 24(6), 1530001 (16 pages).

Siddique, N. and Adeli, H. 2015b. Applications of harmony search algorithms in engineering, *International Journal on Artificial Intelligence Tools*, Vol. 24(6), 1530002 (15 pages).

Siddique, N. H. and Tokhi, M. O. 2001. Training neural networks: Backpropagation vs genetic algorithms, *Proceedings of the IEEE International Joint Conference on Neural Networks* (IJCNN-2001), Washington DC, July 15–19, 2001, pp. 2673–2678.

Simon, D. 2008. Biogeography-based optimization, *IEEE Transaction on Evolutionary Computing*, 12(6), 712–713.

Sirjani, R., Mohamed, A., and Shareef, H. 2012. Optimal allocation of shunt Var compensators in power systems using a novel global harmony search algorithm, *International Journal of Electrical Power & Energy Systems*, 43(1), 562–572.

Sivasubramani, S. and Swarup, K. S. 2011a. Multi-objective harmony search algorithm for optimal power flow problem, *Electrical Power and Energy Systems*, 33, 745–752.

Sivasubramani, S. and Swarup, K. S. 2011b. Environmental/economic dispatch using multi-objective harmony search algorithm, *Electric Power Systems Research*, 81(9), 1778–1785.

Slowik, A. and Bialko, M. 2008. Training of artificial neural networks using differential evolution algorithm, *Proceedings of the IEEE Conference on Human System Interaction*, Sracow, Poland, pp. 60–65.

Socha, K. and Blum, C. 2007. An ant colony optimization algorithm for continuous optimization: Application to feed-forward neural network training, *Neural Computing and Applications*, 16, 235–247.

Spendley, W., Hext, G. R., and Himsworth, F. R. 1962. Sequential application of simplex designs in optimization and evolutionary operation, *Technometrics*, 4, 441–461.

Storn, R. 1999. System design by constraint adaptation and differential evolution, *IEEE Transaction on Evolutionary Computation*, 3(1), 22–34.

Storn, R. and Price, K. V. 1997. Differential evolution—A simple and efficient heuristic for global optimiza-tion over continuous spaces, *Journal of Global Optimization*, 11(4), 341–359.

Storn, R., Price, K. V., and Lampinen, J. 2005. *Differential Evolution—A Practical Approach to Global Optimization*, Berlin: Springer.

Taguchi, G. 1990. *Introduction to Quality Engineering*, Tokyo: Asian Productivity Organization.

Taherinejad, N. 2009. Highly reliable harmony search algorithm, *European Conference on Circuit Theory and Design (ECCTD 2009)*, August 23–27, 2009, Antalya, Turkey, pp. 818–822.

Tizhoosh, H. R. 2005a. Reinforcement learning based on actions and opposite actions, *Proceedings of ICGST International Conference on Artificial Intelligence and Machine Learning*, 19–21 December, 2005, Cairo, Egypt, pp. 94–98.

Tizhoosh, H. R. 2005b. Opposition-based learning: A new scheme for machine intelligence, *Proceedings of International Conference on Computational Intelligence for Modelling Control and Automation, CIMCA'2005*, Vol. I, Vienna, Austria, pp. 695–701.

Tizhoosh, H. R. 2006. Opposition-based reinforcement learning, *Journal of Advanced Computational Intelligence and Intelligent Informatics*, 10(5), 578–585.

Tomassini, M. 2005. *Spatially Structured Evolutionary Algorithms: Artificial Evolution in Space and Time (Natural Computing Series)*, New York: Springer-Verlag.

Tsai, C. and Li, S. 2009. A two-stage modeling with genetic algorithms for the nurse scheduling problem, *Expert Systems with Applications*, 36, 9506–9512.

Turabieh, H., Abdullah, S., and McCollum, B. 2009. Electromagnetism-like mechanism with force decay rate great deluge for the course timetabling problem, *Rough Sets and Knowledge Technology*, July 14–16, 2009, Gold Coast, Australia, pp. 497–504.

Turky, M. and Abdullah, S. 2014. A multi-population harmony search algorithm with external archive for dynamic optimization problems, *Information Sciences*, 272, 84–95.

Upadhyay, P., Kar, R., Mandal, D., Ghoshal, S. P., and Mukherjee, V. 2014. A novel design method for optimal IIR system identification using opposition-based harmony search algorithm, *Journal of the Franklin Institute*, 351, 2454–2488.

Vasehi, A., Fesanghary, M., and Bathaee, S. M. T. 2007. Combined heat and power economic dispatch by har-mony search algorithm, *International Journal of Electrical Power & Energy Systems*, 29(10), 713–719.

Ventresca, M. and Tizhoosh, H. R. 2006. Improving the convergence of backpropagation by opposite transfer functions, *Proceedings of IEEE International Joint Conference on Neural Networks*, Vancouver, BC, pp. 4777–4784.

Verma, A., Panigrahi, B. K., and Bijwe, P. R. 2010. Harmony search algorithm for transmission network expansion planning, *IET Generation, Transmission & Distribution*, 4(6), 663–673.

Vo, D. N., Schegner, P., and Ongsakul, W. 2013. Cuckoo search algorithm for non-convex economic dispatch, *IET Generation, Transmission & Distribution*, 7(6), 645–654.

Walters, D. C. and Sheble, G. B. 1993. Genetic algorithm solution of economic dispatch with valve point load-ing, *IEEE Transaction on Power Systems*, 8(3), 1325–1332.

Wang, C. M. and Huang, Y. F. 2010. Self-adaptive harmony search algorithm for optimization, *Expert Systems with Applications*, 37(4), 2826–2837.

Wang, L. and Li, L.-P. 2013. An effective differential harmony search algorithm for the solving non-convex economic load dispatch problems, *Electrical Power and Energy Systems*, 44, 832–843.

Wang, L., Mao, Y. F., Niu, Q., and Fei, M. R. 2011a. a multi-objective binary harmony search algorithm, *2nd International Conference on Swarm Intelligence*, Chongqing, China, pp. 74–81.

Wang, L., Pan, Q.-K., and Tasgetiren, M. F. 2010a. Minimizing the total flow time in a flow shop with block-ing by using hybrid harmony search algorithms, *Expert Systems with Applications*, 37(12), 7929–7936.

Wang, L., Pan, Q.-K., and Tasgetiren, M. F. 2011b. A hybrid harmony search algorithm for the blocking per-mutation flow shop scheduling problem, *Computers & Industrial Engineering*, 61(1), 76–83.

Wang, L., Xu, Y., Mao, Y., and Fei, M. 2010b. A discrete harmony search algorithm, *International Conference on Life System Modeling and Simulation, LSMS 2010 and International Conference on Intelligent Computing for Sustainable Energy and Environment*, ICSEE 2010, Wuxi, China, September 17–20, 2010, pp. 37–43.

Wang, L., Yang, R., Xu, Y., Niu, Q., Pardalos, P.M., and Fei, M. 2013. An improved adaptive binary harmony search algorithm, *Information Sciences*, 232, 58–87.

Wang, X., Gao, X. Z., and Ovaska, S. J. 2004. Artificial immune optimization methods and applications – A survey, *Proceedings of the IEEE International Conference on Systems, Man and Cybernetics*, Vol. 4, The Hague, The Netherlands, pp. 3415–3420.

Wang, X., Gao, X. Z., and Ovaska, S. J. 2009. Fusion of clonal selection algorithm and harmony search method in optimisation of fuzzy classification systems, *International Journal of Bioinspired Computing*, 1(1), 80–88.

Wang, X. and Yan, X. 2013. Global best harmony search algorithm with control parameters co-evolution based on PSO and its application to constrained optimal problems, *Applied Mathematics and Computation*, 219, 10059–10072.

Wang, Y., Liu, Y., Feng, L., and Zhu, X. 2015. Novel feature selection method based on harmony search for rmail classification, *Knowledge-Based Systems*, 73, 311–323.

Weyland, D. 2010. A rigorous analysis of the harmony search algorithm—How the research community can be misled by a 'novel' methodology, *International Journal of Applied Metaheuristic Computing*, 1(2), 50–60.

Wood, A. J. and Woolenberg, B. F. 1984. *Power Generation, Operation and Control*, New York: John Wiley & Sons.

Wu, B., Qian, C., Ni, W., and Fan, S. 2012. Hybrid harmony search and artificial bee colony algorithm for global optimization problems, *Computers and Mathematics with Applications*, 64(8), 2621–2634.

Wu, Z.Y., Boulos, P. F., Orr, C. H. et al. 2001. Using genetic algorithms to rehabilitate distribution systems, *Journal of the American Water Works Association*, 93, 74–85.

Xiang, W.-L., An, M.-Q., Li, Y.-Z., He, R.-C., and Zhang, J.-F. 2014. An improved global-best harmony search algorithm for faster optimization, *Expert Systems with Applications*, 41, 5788–5803.

Xu, H., Gao, X. Z., Wang, T., and Xue, K. 2010. Harmony search optimization algorithm: Application to a reconfigurable mobile robot prototype, in: Z. W. Geem, ed., *Recent Advances in Harmony Search Algorithm*, *Studies in Computational Intelligence*, Berlin, Heidelberg: Springer-Verlag, Vol. 270, pp. 11–22.

Yadav, P., Kumar, R., Panda, S. K., and Chang, C. S. 2012. An intelligent tuned harmony search algorithm for optimization, *Information Sciences*, 196, 47–72.

Yang, S. and Jat, S. N. 2010. Genetic algorithms with guided and local search strategies for university course timetabling, *IEEE Transaction on System, Man, Cybernetics – Part C: Applications. Reviews*, 41(1), 93–106.

Yang, X.-S. 2008. *Nature-Inspired Meta-Heuristic Algorithms*, UK: Luniver Press.

Yang, X.-S. 2009. Harmony search as a metaheuristic algorithm, in: Zong Woo Geem, ed., *Music Inspired Harmony Search Algorithm: Theory and Applications*, Berlin, Heidelberg: Springer-Verlag, pp. 1–14.

Yang, X.-S. 2010. *Engineering Optimisation – An Introduction with Metaheuristic Applications*, Hoboken, NJ: John Wiley & Sons.

Yanofsky, N. S. and Mannucci, M. A. 2008. *Quantum Computing for Computer Scientists*, New York: Cambridge University Press.

Yildiz, A. and Ozturk, F. 2010. Hybrid Taguchi-harmony search approach for shape optimization, in: Z. Geem, ed., *Recent Advances in Harmony Search Algorithm*, Berlin: Springer, pp. 89–98.

Yildiz, A. R. 2008. Hybrid Taguchi-harmony search algorithm for solving engineering optimization problems, *International Journal of Industrial Engineering: Theory, Applications and Practice*, 15(3), 286–293.

Yong, L., Liu, S., Zhang, J., and Feng, Q. 2012. Theoretical and empirical analyses of an improved harmony search algorithm based on differential mutation operator, *Journal of Applied Mathematics*, 2012, Article ID 147950, 20 pages.

Yousefi, M., Enayatifar, R., Darus, A. N., and Abdullah, A. H. 2013. Optimization of plate-fin heat exchangers by an improved harmony search algorithm, *Applied Thermal Engineering*, 50(1), 877–885.

Yuan, X., Zhao, J., Yang, Y., and Wang, Y. 2014. Hybrid parallel chaos optimization algorithm with harmony search algorithm, *Applied Soft Computing*, 17, 12–22.

Yuan, Y., Xu, H., and Yang, J. 2013. A hybrid harmony search algorithm for the flexible job shop scheduling problem, *Applied Soft Computing*, 13, 3259–3272.

Zaharie, D. 2002. Critical values for the control parameters of differential evolution algorithms, in: R. Matousek, P. Osmera, eds., *Proceedings of Mendel 2002, 8-th International Conference on Soft Computing*, June 2002, Brno, Czech Republic, pp. 62–67.

Zamani, M. and Sadeghian, A. 2010. A variation of particle swarm optimization for training of artificial neural networks, in: Al-Dahoud Ali, ed., *Computational Intelligence and Modern Heuristics*, INTECH, February 2010, ISBN: 978-953-7619-28-2.

Zammori, F., Braglia, M., and Castellano, D. 2014. Harmony search algorithm for single-machine scheduling problem with planned maintenance, *Computers & Industrial Engineering*, in press, Accepted Manuscript, Available online 13 August 2014.

Zarei, O., Fesanghary, M., Farshi, B., Saffar, R., and Razfar, M. 2009. Optimization of multi-pass face-milling via harmony search algorithm, *Journal of Materials Processing Technology*, 209(5), 2386–2392.

Zecchin, A. C., Simpson, A. R., Maier, H. R. et al. 2006. Application of two ant colony optimisation algorithms to water distribution system optimisation, *Mathematical and Computer Modelling*, 44, 451–468.

Zhao, S., Suganthan, P. N., and Das, S. 2010. Dynamic multi-swarm particle swarm optimizer with sub-regional harmony search, *Proceedings of the IEEE Congress on Evoluionary Computation*, 18–23 July 2010, CCIB, Barcelona, Spain, pp. 1–8.

Zhao, S.-Z., Suganthan, P. N., Pan, Q.-K., and Tasgetiren, M. F. 2011. Dynamic multi-swarm particle swarm optimizer with harmony search, *Expert Systems with Applications*, 38, 3735–3742.

Zheng, W. M. 1994. Kneading plane of the circle map, *Chaos, Solitons and Fractals*, 4, 1221.

Zou, D., Gao, L., Wu, J., Li, S., and Li, Y. 2010. A novel global harmony search algorithm for reliability problems, *Computers and Industrial Engineering*, 58(2), 307–316.

6 Water Drop Algorithm

6.1 INTRODUCTION

In nature, water has the general tendency of flowing toward the ocean. By virtue of this natural propensity, water drops (WDs) very often end up in the rivers, lakes, and the seas. As WDs move, they affect the environment in their way of flowing. Moreover, the environment itself has substantial effects on the paths that the WDs follow. Consider a hypothetical river in which water is flowing and moving from a high terrain to a lower terrain and finally joins a lake or sea. The paths that the river follows are not straight but rather full of twists and turns. The water drops have no sensing means to be able to find the destination (lake or river). The gravitational force from the center of earth is the source of motion of the water drops of the river. If there were no obstacles and barriers, the water drops would have followed a straight path toward the destination, which is the shortest path from the source to the destination. However, due to different kinds of obstacles, the real path becomes different from the ideal path with lots of bends and turns in a river path. In contrast, the natural forces always try to change the real path of the water drops to be straight or close to the ideal path. This continuous interplay of efforts changes the path of the river as time passes by. One feature of a WD is the velocity with which it flows, which enables the WD to transfer an amount of soil from one place to another place in the front. This soil is usually transferred from fast flowing parts to the slow flowing parts of the path. As the fast flowing parts get deeper by removal of soil, they can hold more volume of water and thus may attract more water. The removed soil which is carried in the WDs is unloaded in the river beds where the water flow is slower.

Thus, the water drops change the amount of soil on the paths the water drops traversed. This variation depends on the velocity and the soil carried by the water drops, and it can increase or decrease to attract or obstruct more water drops. The idea of WDA arose by observing flowing water drops in the river with certain velocity and transporting an amount of soil from one place to another. WDA is formulated by taking these two properties into account. Shah-Hosseini (2007, 2008a,b) proposed a novel nature-inspired optimization algorithm, called intelligent water drop (IWD), which imitates the dynamics of river systems and the behavior of water drops such as the variation of velocity, the change of sediment (or soil) in the river bed, the change of direction of the flow and so on. The IWD algorithm can be used for maximization or minimization problems. The solutions are incrementally constructed by the IWD algorithm. Therefore, the IWD algorithm is a population-based constructive optimization algorithm, which has been applied to many optimization problems such as well-known benchmark problems.

This chapter will present river systems, mechanisms of sediment transportation, the natural water drops and mechanisms of flowing water. The chapter will then present the principles of WDA and its parameters, variants of WDA and finally present few applications to engineering problems.

6.2 RIVER SYSTEMS

The complex river systems are an integral part of any landscape and are created as water flows from higher to lower elevations. There is an inherent supply of potential energy in the river systems created by the change in elevation between the beginning and ending points of the river or within any discrete stream reach. This potential energy is expressed in a variety of ways as the river moves through and shapes the landscape, developing a complex fluvial network, with a variety of channel and valley forms and associated aquatic and riparian habitats. Excess energy is dissipated in many ways: contact with vegetation along the banks, in turbulence at steps and riffles in the river profiles, in erosion at meander bends, in irregularities, or roughness of the channel bed and banks, and in sediment, ice, and debris transport (Allan, 1995; Kondolf et al., 2002).

Natural river systems are amazing and consist of many twists and turns along their paths. Both naturalists and scientists are trying to reveal the mysteries behind these twists. Can these phenomena be explained by physics or described by mathematics? If that is so, can we use the mechanisms that happen in river systems and develop intelligent algorithms? The WDA is a step in the direction of modelling natural rivers and implementing them in the form of algorithms.

6.2.1 Sediment Production, Transport, and Storage in the Working River

In ideal form, WDs in the river would follow a straight path to reach their destination, which should be the shortest path from the source to the destination. In real form, a river changes its course many times, making the path full of bends and turns. Finally, the river falls into a sea or lake. A path from the source to the destination of a river is shown in Figure 6.1. On the other hand, the river always tries to change the environment in order to make its path from the source to the destination an optimal path satisfying the environmental constraints. It is the energy of the water that enables the flowing river to find an optimal path.

The watershed through which a river flows or drains dictates the amounts and types of sediments that will be transported and/or stored. Within the watershed, there are locations where sediment is produced, transported, or stored. These zones are often referred to as source (production), transfer (transport), and response (storage or deposition) as shown in Figure 6.2.

Source streams: It refers to Zone 1 shown in Figure 6.2. Primarily, it is the zone where nonalluvial sediments (colluvial material) enter into the stream system, from landslides and mass wasting failures, and are transported with debris during large and infrequent flow events.

Transfer streams: It refers to Zone 2 shown in Figure 6.2. It is the zone that is geomorphically resilient with high sediment transport capacity. These streams are able to convey limited increases

FIGURE 6.1   Natural river system.

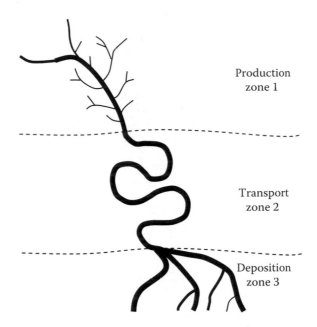

FIGURE 6.2 Watershed sediment production, transport, and deposition locations: Zone 1—sediment production zone; Zone 2—sediment transfer zone; and Zone 3—sediment deposition zone.

in sediment loads and will change little in response to reduction in sediment supply. Generally, the sediment volume supplied to transport reaches is balanced by the sediment exported from the reach.

Response streams: It refers to Zone 3 shown in Figure 6.2, with storage reaches in which significant geomorphic adjustment occurs in response to changes in sediment supply. Zones of transition from transport to response or storage reaches are locations where changes in sediment supply may result in both pronounced and persistent channel instability.

Sediment production is influenced by many factors, including soil type, vegetation type, coverage, land use, climate, and weathering or erosion rates. Once the sediment enters the fluvial system, it will be transported and/or stored within the system including the flood plains (Robert, 2003). The movement of sediment is influenced not only by the zone of the watershed of the river but also by the local conditions of the river. The sediment transport capacity refers to the amount and size of sediments that the river has the ability, or energy to transport. The key components that control the sediment transport capacity are the velocity and depth of the water moving through the channel. Velocity and depth are controlled by the channel slope and dimensions, discharge (volume of flow), and roughness of the channel. Changes in any of these parameters will result in a change in the sediment transport capacity of the river.

The specific characteristics of the sediment load are another key factor influencing channel form and process. The load is the total amount of sediment being transported. There are three types of sediment load in the river: dissolved, suspended, and bed load, as shown in Figure 6.3. The dissolved load is made up of the solutes that are generally derived from chemical weathering of bedrock and soils. Fine sands, clay, and silt are typically transported as suspended load. The suspended load is held up in the water column by turbulence. The bed load is made up of sands, gravel, cobbles, and boulders. Bed load is transported by rolling, sliding, and bouncing along the bed of the channel (Allan, 1995). While dissolved and suspended loads are important components of the total sediment load, in most river systems, the bed load is what influences the channel morphology and stability (Kondolf et al., 2002).

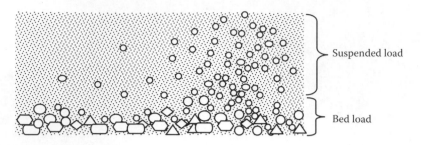

FIGURE 6.3   Sediment load in river.

By comparing the sediment transport capacity with the sediment load, some general assumptions can be made as to whether a river will erode more sediment, deposit extra sediment, or be in balance with the amount of erosion and deposition happening.

1. *if(capacity > load)* → *erosion*: The symbol "→" here means implies. If the capacity is greater than the load, erosion would happen. This is due to the river having the excess energy needed to transport more sediment than is currently being transported.
2. *if(capacity < load)* → *deposition*: If the capacity is less than the load, deposition would happen. The amount of excess energy needed to move the extra sediments is not available in the system, so the sediments are deposited in the river beds.
3. *if(capacity = load)* → *no erosion/deposition*: If the capacity equals the load, no net change in erosion or deposition would happen.

River systems or reaches are considered in equilibrium when there is a balance between the amount of sediment load being supplied to the system and the capacity of the system to carry that sediment load. Another way to view this concept is by using Lane's balance diagram (Allan, 1995), which demonstrates how the channel may respond to a change in various parameters, such as sediment load, channel geometry, channel slope, erosion resistance, and discharges (hydrologic load). For example, by increasing the amount of sediment load, the scale will tip toward sediment deposition. To bring the scale back in balance, a change in the channel geometry, slope, and/or hydrologic load would be needed.

There are natural fluctuations in the balance of any of these inputs such as flood events, valley wall slope failures increasing sediment loads, beaver dams or debris jams causing changes in channel geometry, etc. Changes caused by humans in this balance also occur in the watershed, along the floodplains, and in the channel. The degree and type of adjustment will depend on whether it is a source, transfer, or response reach; the sediment transport capacity; and the type and magnitude of change that is introduced to the system.

Turbulence and velocity are the two closely related issues causing the erosion, transportation, and deposition in a river system. Some researchers termed it as work of a river (Morisawa, 1968). Work is measured by energy. The total work of a stream is measured by the amount of material it erodes, transports, and deposits. Erosion is carried on by rivers through the process of corrosion (chemical process resulting from the reaction of water and rocks on the surface of the land), corrosion (mechanical wearing away of the land), and cavitations (erosion of land under very high velocity of water).

Potential energy is converted into kinetic energy by the down flow of the river. Most of the energy is lost through friction on channel walls and the bed. Only a small part of the energy of a river is left for transportation of sediment load. Rivers carry materials as debris load, in suspension and along the bed of the channel and in solution. The bed load refers to the part of the sediment load consisting of grains finer than those on the bed of the channel. The bed load moves more slowly than the water flow in a stream. The particles may move individually along the bottom, or they may travel in groups. Once in motion, larger grains move faster than smaller ones and often move by rolling or sliding. The suspended load is composed of finer particles than the bed load. These particles have settling velocities, which are less than the buoyant velocity. Little or no energy is required to transport the suspended load. The distribution of suspended load varies with depth below the surface of the stream.

6.3 NATURAL WDs

In nature, flowing water in rivers consisting of small particles of WDs form huge moving swarms. The flowing river creates its own path to reach its destination influenced by many factors discussed in the previous section. The flowing river is then a part of the environment that has been dramatically changed by the swarm of WDs and will also keep changing. The environment itself has substantial effects on the paths that the WDs flow. For example, the parts of the environment having hard soils resist the water flow more than the parts with soft soils. In fact, a natural river is the result of a competition between WDs in a swarm and the environment that resists the flow of WDs. When resisted, water flow changes the course of its path to its destination. It is also attracted by the earth's gravitational force, which pulls everything toward the center of the earth in a straight line. The earth's gravitational force makes the river to reach its ultimate destination. Therefore, with no obstacles and barriers, the WDs should follow a straight path toward its destination, which is ideally the earth's center. This gravitational force creates acceleration such that WDs gain speed as they come near to the earth's center. However, in reality, due to different obstacles and constraints in the way of this ideal path, the real path is so different from the ideal path such that lots of twists and turns in a river path are seen, and the destination is not the earth's center but a lake, sea, or even a bigger river. It is often observed that the constructed path seems to be an optimal one in terms of the distance from the destination and the constraints of the environment.

An important feature of a WD flowing in a river is its velocity. It is assumed that each WD of a river can also carry an amount of soil. When an imaginary natural WD is flowing from one point of a river to the next point in the front, three obvious changes occur during this transition:

- Velocity of the WD is increased.
- Soil in the WD is increased.
- Soil of the river bed between the two points is decreased.

In fact, an amount of soil of the river's bed is removed by the WDs, and this removed soil is added to the soil of the WDs. Moreover, the speed of the WD is increased during the transition. This velocity of WDs plays an important role in removing soil from the beds of rivers. The following properties are assumed for flowing WDs:

1. *High-speed WDs gather more soil than slower WDs*: The WDs with higher speed remove more soil from the river's bed than WDs with slower speed. The soil removal is thus related to the velocities of WDs. It has been mentioned earlier that when WDs flow over a part of a river's bed, they gain speed. But the increase in velocity depends on the amount of soil of that part.
2. *Increase in velocity of WDs is higher on a path with light soil than a path with heavy soil*: The velocity of the WDs is changed such that, on a path with a little amount of soil, the velocity of the WDs is increased more than a path with a considerable amount of soil. Therefore, a path with a little soil lets the flowing WDs gather more soil and gain more speed, whereas the path with heavy soil resists more against the flow of WDs such that it lets the flowing WDs gather less soil and gain less speed.

3. *WDs prefer a path with less soil than a path with more soil*: The WDs prefer an easier
   path to a harder path when it has to choose between several branches of a river that
   exist in the path from the source to the destination. The easiness or hardness of a path is
   denoted by the amount of soil on that path. A path with more soil is considered a hard
   path, whereas a path with less soil is considered an easy path. The selection of a branch
   is performed using a probabilistic function, which is inversely proportional to the amount
   of soil.

A number of terminologies and definitions need to be explained before discussing the WDA.
The environment in which WDs are flowing is assumed to be discrete and comprised of N nodes.
WDs move from one node to another node with a velocity denoted as vel_{WD}. The nodes joining
together form a path. While moving along a path, WDs carry some amount of soil, denoted as
$soil_{WD}$.

The properties of $soil_{WD}$ and vel_{WD} do change as the WDs flow through its environment. WDs are
assumed to be a representation of a problem in an environment that flow from a source to a desired
destination seeking an optimal path. There exist numerous paths from a given source to a desired
destination. Even the location of the destination may be unknown. In the case of an unknown destination, the solution to the problem is sought by finding the best (very often the shortest) path from
the source to the destination. However, there are cases in which the destination is known. In such
cases, the solution is obtained by finding the optimum path to the destination in terms of cost or any
other measure for the given problem.

A WD moves in discrete finite-length steps in its environment from its current location i at a
velocity of vel_{WD} to a next location j and increases its velocity by an amount of Δvel_{WD}. It is seen that
Δvel_{WD} is nonlinearly proportional to the inverse of the amount of soil denoted as $soil(i,j)$ in WD
and can be written as

$$\Delta vel_{WD} \overset{NL}{\propto} \frac{1}{soil(i,j)} \tag{6.1}$$

The symbol NL denotes the nonlinear proportionality. Shah-Hosseini (2009a,b) formulated the
change of velocity of the WD $vel_{WD}(t)$ by the increase in the amount of soil $soil(i,j)$ between two
locations i and j:

$$vel_{WD}(t+1) = vel_{WD}(t) + \frac{a_v}{b_v + c_v \cdot soil^{2\alpha}(i,j)} \tag{6.2}$$

$$\Delta vel_{WD}(t) = vel_{WD}(t+1) - vel_{WD}(t) = \frac{a_v}{b_v + c_v \cdot soil^{2\alpha}(i,j)} \tag{6.3}$$

Here, a_v, b_v, c_v, and α are positive parameters arbitrarily chosen by the users. The WD's soil is
increased by extracting some more soil from the environment, that is, river bed, on its way from
the location i to location j. The amount of soil added to the WDs while flowing from location i to
location j is denoted by the change of soil as $\Delta soil(i,j)$. It is seen that $\Delta soil(i,j)$ is inversely (and
nonlinearly) proportional to the time needed for the WDs to flow from their current location to
the next. The time taken from location i to location j is denoted by $T(i,j)$. We define $\Delta soil(i,j)$ as
follows:

$$\Delta soil(i,j) \overset{NL}{\propto} \frac{1}{T(i,j)} \tag{6.4}$$

Shah-Hosseini (2009a,b) provided an elegant formulation for Equation 6.4 in terms of travel time $T(i,j)$ and the static arbitrary parameters a_v, b_v, and c_v. The change of soil is thus given by

$$\Delta soil(i,j) = \frac{a_s}{b_s + c_s \cdot T^{2\theta}(i,j)} \qquad (6.5)$$

where θ is a positive parameter arbitrarily chosen by the user.

The duration of time for the WD is calculated by the simple laws of physics for linear motion. Thus, the time $T(i,j)$ taken for the WD to move from location i to j is proportional to the velocity vel_{WD} of the WD, which can be expressed more specifically as follows:

$$T(i,j) \overset{L}{\propto} \frac{1}{vel_{WD}} \qquad (6.6)$$

where $\overset{L}{\propto}$ denotes linear proportionality. The time $T(i,j)$ taken for the WD to travel from location i to j with velocity vel_{WD} can be formulated as

$$T(i,j) = \frac{HUD(i,j)}{vel_{WD}} \qquad (6.7)$$

Here, $HUD(.)$ is a local heuristic function defined for a given problem to measure the degree of undesirability of WD to move between nodes i and j. The undesirability measure by the function $HUD(.)$ is illustrated in Figure 6.4. The local heuristic function $HUD(.)$ is problem dependent and should be defined for each problem.

One important mechanism that each WD must contain is how to select its next node. A WD prefers a path that contains less amount of soil compared to the other paths. This preference is implemented by assigning a probability to each path from the current node to all valid nodes that do not violate constraints of the given problem. Let a WD be at the node i; then, the probability $p_{WD}(i,j)$ of going from node i to node j is calculated by Equations 6.12.

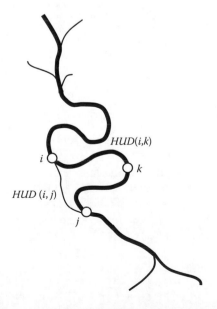

FIGURE 6.4   Undesirability measure by heuristic functions $HUD(i,j)$ and $HUD(i,k)$.

Some soil is extracted from the traversed distance between locations i and j. The increased amount of soil obtained in this path denoted by $\Delta soil(i,j)$ is proportional to the amount of soil extracted by the WD flowing along this path:

$$soil(i, j) \overset{L}{\propto} \Delta soil(i, j) \tag{6.8}$$

One such formulation is given for the WD algorithm such that $soil(i,j)$ is incremented by the amount of soil taken by the WD from its way from location i to location j:

$$soil(i, j) = \rho_0 \cdot soil(i, j) - \rho_n \cdot \Delta soil(i, j) \tag{6.9}$$

where ρ_0 and ρ_n are positive values and $\rho_0, \rho_1 \in [0,1]$. Shah-Hosseini (2007) defined $\rho_0 = 1 - \rho_n$ in his experiments. The soil content in the WD denoted by $soil_{WD}$ is increased by the amount of $\Delta soil(i,j)$ and is given by the following equation:

$$soil_{WD} = soil_{WD} + \Delta soil(i, j) \tag{6.10}$$

Another mechanism that exists in the behavior of a WD is that it prefers the paths with light soils on its beds to the paths with heavier soils on its beds. To implement this behavior for the choice of path, a uniform random distribution is used among the soils of the available paths such that the probability of the WD to flow from location i to j, denoted by $p_{WD}(i,j)$, is proportional to the amount of soils on the available paths:

$$p_{WD}(i, j) \overset{L}{\propto} soil(i, j) \tag{6.11}$$

The lighter the soil of the path between locations i and j, the higher chance this path has for being selected by the WD located at i. The formulation in Equation 6.11 has been used for the calculation of the probability of choice of location j expressed by

$$p_{WD}(i, j) = \frac{f(soil(i,j))}{\displaystyle\sum_{k \notin vc(WD)} f(soil(i,k))} \tag{6.12}$$

where $f(soil(i,j))$ is defined as

$$f(soil(i, j)) = \frac{1}{\varepsilon_s + g(soil(i,j))} \tag{6.13}$$

The constant parameter ε_s is a small positive number to prevent singularity of the function $f(soil(i,j))$ or possible division by zero. Shah-Hosseini (2008) suggested a small value such as $\varepsilon_s = 0.001$. The set $vc(WD)$ is the set of nodes that the WD should not visit to satisfy the constraints of the problem. The function $g(soil(i,j))$ is used to carry the $soil(i,j)$ of the path joining nodes i and j toward positive values and is expressed by

$$g(soil(i, j)) = \begin{cases} soil(i, j) & \text{if } \min_{l \notin vc(WD)}(soil(i,l)) \geq 0 \\ soil(i, j) - \min_{l \notin vc(WD)}(soil(i,l)) & \text{otherwise} \end{cases} \tag{6.14}$$

where the function min(.) returns the minimum value of its arguments. To decide on the next location of the WD, a uniform random distribution is used to compare with the probability defined in Equation 6.12.

6.4   WDs ALGORITHM

The basic idea of the WDA is based on the observation of flow of water in rivers. The water in rivers is seen as a collection of WDs that flow from a point in a high terrain to a point in a low terrain with a certain velocity depending on the zones shown in Figure 6.1. The WDs also carry some amount of soil in them while flowing along a path and transport an amount of soil from one place to another. By taking these two properties of WDs into account, WDA has been developed (Shah-Hosseini, 2007, 2008, 2009a,b). For an optimization problem with N decision variables, the problem is represented by graph $G = \langle V, E \rangle$, where $V = \{v_i | i = 1, \ldots, N\}$ is the set of nodes, $E = \{e_{i,j} | e_{i,j} \in V \times V, i \neq j,$ $(i,j) = 1, \ldots, N\}$ is the set of edges, $|V| = N$, and N is the number of decision variables to be optimized. The number of WDs is the user's choice. Initially, the same amount of soil is present on each edge (or path). Each WD is initialized with zero amount of soil. Each WD starts its journey from the first node, travels through all nodes, and completes at the N-th node. While going from one node to another, the WD chooses any one of the edges depending on the probability. The probability of the edge again depends on the amount of soil on the edge, which is calculated based on Equations 6.12 through 6.14.

The WDA employs a population of WDs to find the optimal solutions to a given problem (Shah-Hosseini, 2009b). The graph $G = \langle V, E \rangle$ is the environment for the WDs and the WD flows on the edges of the graph. Each WD begins constructing its solution gradually by traveling between the nodes of the graph along the edges until the WD finally completes its solution denoted by T_{WD}. Each solution T_{WD} is represented by the edges that the WD has visited. An iteration of the WDA is completed when all WDs complete their solutions. After each iteration, the iteration-best solution T_{ibest} is found. The iteration-best solution T_{ibest} is the best solution based on a quality function among all solutions obtained by the WDs in the current iteration. One uses T_{ibest} to update the total-best solution $T_{total-best}$. The total-best solution $T_{total-best}$ is the best solution since the beginning of the WDA, which has been found in all iterations.

For a given problem, an objective or quality function is defined to measure the fitness of solutions. Consider the quality function of a problem to be denoted by $f(.)$. Then, the quality of a solution T_{WD} found by the WD is given by $f(T_{WD})$. Therefore, the iteration-best solution T_{ibest} is given by

$$T_{ibest} = \arg\max_{\forall WDs} f(T_{WD}) \tag{6.15}$$

where arg(.) returns its argument.

At the end of each iteration of the algorithm, the total-best solution $T_{total-best}$ is updated by the current iteration-best solution T_{ibest} as follows:

$$T_{total-best} = \begin{cases} T_{total-best} & \text{if } f(T_{total-best}) \geq f(T_{ibest}) \\ T_{ibest} & \text{otherwise} \end{cases} \tag{6.16}$$

At the end of each iteration of the WDA, the amount of soil on the edges of the iteration-best solution T_{ibest} is reduced based on the fitness (quality) of the solution. One such mechanism (Shah-Hosseini, 2007) is used to update the $soil(i,j)$ of each edge (i,j) of the iteration-best solution T_{ibest}:

$$soil(i,j) = \rho_s \cdot soil(i,j) - \rho_{WD} \cdot \frac{1}{(N_{ibest} - 1)} \cdot soil_{ibest}^{WD}, \quad \forall (i,j) \in T_{ibest} \tag{6.17}$$

where $soil_{ibest}^{WD}$ represents the soil of the iteration-best WD. The iteration-best WD is the WD that has constructed the iteration-best solution T_{ibest} at the current iteration. By N_{ibest}, we denote the number of nodes in the solution T_{ibest} and by ρ_{WD} the global soil updating parameter, which should be chosen from [0,1]; ρ_s is often set as $(1 + \rho_{WD})$. Then, Equation 6.17 becomes

$$soil(i,j) = (1+\rho_{WD}) \cdot soil(i,j) - \rho_{WD} \cdot \frac{1}{(N_{ibest}-1)} \cdot soil_{ibest}^{WD}, \quad \forall (i,j) \in T_{ibest} \qquad (6.18)$$

The first term on the right-hand side of Equation 6.18 is the amount of soil that remains from the previous iteration, whereas the second term represents the quality of the solution. Then, the algorithm begins a new iteration with new WDs but with the same soils on the paths of the graph, and the whole process is repeated. The WDA stops when it reaches the maximum number of iterations $iter_{max}$ or the total-best solution $T_{total\text{-}best}$ achieves the expected quality demanded for the given problem.

The WDA was first proposed for solving a TSP by Shah-Hosseini (2007). Later on, the WDA was generalized for a number of benchmark problems in Shah-Hosseini (2008, 2009a,b). The WDA can be described in the following ten steps:

Step 1: Initialization of static parameters:
- The representative graph $G = \langle V, E \rangle$ of the problem is given to the algorithm, which contains N nodes, that is, $|V| = N$.
- The quality of the total-best solution $T_{total\text{-}best}$ is initially set to the worst value: $f(T_{total\text{-}best}) = -\infty$.
- An iteration count $iter$ is used with an initial value of 0.
- The maximum number of iterations $iter_{max}$ is specified by the user, and the algorithm stops when it reaches $iter_{max}$.
- The number of WDs N_{WD} is set to a positive integer value. The value of N_{WD} should be at least equal to two. However, N_{WD} is usually set to the number of nodes N of the graph.
- The velocity parameters a_v, b_v, and c_v are initially set to 1, 0.01, and 1, respectively.
- The soil updating parameters a_s, b_s, and c_s are initially set to 1, 0.01, and 1, respectively.
- The local soil updating parameter ρ_n is set to 0.9.
- The global soil updating parameter ρ_{WD} is set to 0.9.
- The initial soil on each edge of the graph is denoted by the constant *InitSoil* such that the soil of the edge between every two nodes i and j is set by $soil(i,j) = InitSoil$ and the initial value is set to 10,000.
- The initial velocity of each WD is set to *InitVel*. Usually, it is set to 200.

Step 2: Initialization of dynamic parameters:
- Every WD has a list of visited nodes $V(WD)$, which is initially empty, that is, $V(WD) = \{\emptyset\}$.
- Each WD's velocity is set to *InitVel*.
- All WDs are set to have zero amount of soil.

Step 3: Spread the WDs randomly on the nodes of the graph as their first visited nodes.

Step 4: Update the visited node list of each WD to include the nodes just visited.

Step 5: Repeat Steps 5.1–5.4 for those WDs with partial solutions.

5.1 For the WD residing in node i, choose the next node j, which does not violate any constraints of the problem and is not in the visited node list $V(WD)$ of the WD, using the probability $p_i^{WD}(j)$, defined as follows:

$$p_i^{WD}(j) = \frac{f(soil(i,j))}{\displaystyle\sum_{k \notin vc(WD)}} \qquad (6.19)$$

such that

$$f(soil(i,j)) = \frac{1}{\varepsilon_s + g(soil(i,j))}$$ (6.20)

and

$$g(soil(i,j)) = \begin{cases} soil(i,j) & \text{if } \min_{l \notin vc(WD)}(soil(i,j)) \geq 0 \\ soil(i,j) - \min_{l \notin vc(WD)}(soil(i,j)) & \text{otherwise} \end{cases}$$ (6.21)

Then, add the newly visited node j to the list $V(WD)$.

5.2 For each WD moving from node i to node j, the velocity $vel_{WD}(t)$ is updated by using the following equation:

$$vel_{WD}(t+1) = vel_{WD}(t) + \frac{a_v}{b_v + c_v \cdot soil^2(i,j)}$$ (6.22)

where $vel_{WD}(t+1)$ is the updated velocity of the WD.

5.3 For the WD moving on the path from node i to j, compute the change of soil $\Delta soil(i,j)$ that the WD loads from the path by

$$\Delta soil(i,j) = \frac{a_s}{b_s + c_s \cdot T^2(i,j;vel_{WD}(t+1))}$$ (6.23)

such that $T(i,j; vel_{WD}(t+1)) = (HUD(j)/vel_{WD}(t+1))$ where the heuristic undesirability function $HUD(j)$ is defined appropriately for a given problem. The local heuristic function $HUD(i,j)$ measures the undesirability of a WD to move from node i to node j. For small deficit values, the heuristic measure $HUD(i,j)$ would be small, whereas, for large deficits, $HUD(i,j)$ also becomes large. For the TSP problem, $HUD(i,j)$ is defined as $HUD_{TSP}(i,j) = \|c(i) - c(j)\|$, where $\|\cdot\|$ is the Euclidean norm and $\{c(i), c(j)\}$ represents two nodes.

5.4 Update the soil $soil(i,j)$ of the path from node i to j traversed by that WD, and also update the soil that the WD carries $soil_{WD}$ by

$$soil(i,j) = (1-\rho_n) \cdot soil(i,j) - \rho_n \cdot \Delta soil(i,j)$$ (6.24)

$$soil_{WD} = soil_{WD} + \Delta soil(i,j)$$ (6.25)

Step 6: Find the iteration-best solution T_{ibest} from all the solutions T_{WD} found by the WDs using

$$T_{ibest} = \arg\max_{\forall T^{WDs}} f(T_{WD})$$ (6.26)

where function $f(\cdot)$ gives the quality of the solution.

Step 7: Update the soils on the paths that form the current iteration-best solution T_{ibest} by

$$soil(i,j) = (1+\rho_{WD}) \cdot soil(i,j) - \rho_{WD} \cdot \frac{1}{(N_{ibest}-1)} \cdot soil_{ibest}^{WD}, \quad \forall(i,j) \in T_{ibest} \qquad (6.27)$$

where N_{ibest} is the number of nodes in solution T_{ibest}.
Step 8: Update the global (total)-best solution $T_{\text{total-best}}$ by comparing the current iteration-best solution T_{ibest} as follows:

$$T_{\text{total-best}} = \begin{cases} T_{\text{total-best}} & \text{if } q(T_{\text{total-best}}) \geq q(T_{ibest}) \\ T_{ibest} & \text{otherwise} \end{cases} \qquad (6.28)$$

where $q(\cdot)$ is the quality function defined for the specific problem.
Step 9: Increment the iteration number by $iter = iter + 1$. Then, go to Step 2 if $iter < iter_{\max}$.
Step 10: The algorithm stops at this step and returns the total-best solution $T_{\text{total-best}}$.

The ten steps of the WDA discussed above are illustrated as a flow diagram in Figure 6.5.

6.5   PARAMETERS OF WDA

Once the graph $G = \langle V, E \rangle$ with $|V| = N$ of the problem at hand is provided, the WDA needs to initialize two kinds of parameters, namely:

* Static parameters
* Dynamic parameters

Static parameters are those parameters that remain constant during the whole iterations of the WDA. Dynamic parameters are those that are changed and reinitialized after each iteration of the WDA.
The static parameters are as follows:

* The maximum number of iterations $iter_{\max}$ is specified by the user.
* The number of WDs N_{WD} is set to a positive integer value.
* Velocity updating parameters a_v, b_v, and c_v are defined.
* Soil updating parameters a_s, b_s, and c_s are defined.
* Local soil updating parameter ρ_n is defined within $0 < \rho_n < 1$.
* Global soil updating parameter ρ_{WD} is defined within $0 < \rho_{WD} < 1$.
* Quality of the total-best solution $T_{\text{total-best}}$ is initially set to the worst value.

In general, static parameter values are determined experimentally. Shah-Hosseini has investigated these parameters in many experiments (Dariane and Sarani, 2013; Shah-Hosseini, 2007, 2008, 2009a,b). The values of the parameters' range are presented in Table 6.1.
The dynamic parameters are as follows.

* Every WD has a list of visited nodes $V(WD)$, which is initially empty, that is, $V(WD) = \{\varnothing\}$.
* Initial amount of soil of WDs is set to $soil_{WD} = 0$.
* Initial velocity of WDs is set to $vel_{WD} = InitVel$.
* Initial soil amount for each edge is randomly determined in the interval [MinSoil, MaxSoil].
* Initial velocity of every WD is randomly determined in the interval [MinVel, MaxVel].

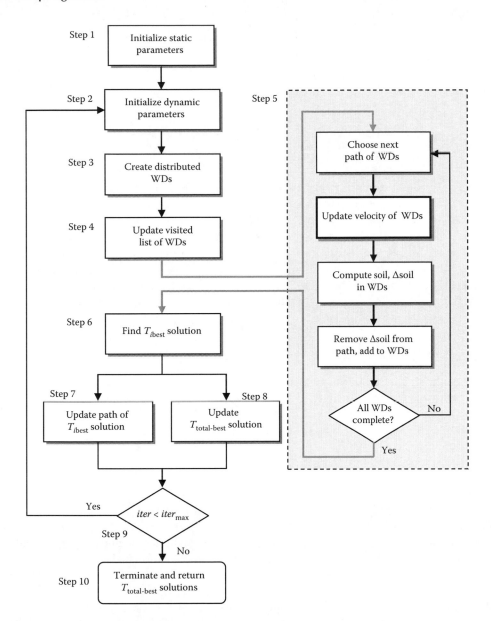

FIGURE 6.5 Flow diagram of the WDA.

TABLE 6.1
Static Parameters

Parameters	Values
$iter_{max}$	User's choice
N_{WD}	30
a_s, a_v	1
b_s, b_v	0.01
c_s, c_v	1
ρ_n, ρ_{WD}	0.95
$T_{total\text{-}best}$	Set to the worst value

TABLE 6.2
Dynamic Parameters

Parameters	Values
$V_c(WD)$	$V_c(WD) = \{\emptyset\}$
vel_{WD}	$InitVel$
$soil_{WD}$	$soil_{WD} = 0$
$HUD(i,j)$	Problem dependent

In general, dynamic parameter values vary from problem to problem and should be determined experimentally for specific problems. Shah-Hosseini and other researchers have investigated these parameters in many applications over the years. Some initial values of the parameters are presented in Table 6.2. More details will be discussed in Section 6.7.

In the original WDA (Shah-Hosseini, 2007), all the edges are set with the same amount of initial soil and all the WDs have the same initial velocity. As mentioned earlier, it is necessary to define the meta-heuristic function $HUD(i,j)$ used in the WDA. $HUD(i,j)$ should be defined separately for any particular problem. In general, for small deficit values, $HUD(i,j)$ would be small, whereas, for large deficits, $HUD(i,j)$ becomes large.

6.6   CONVERGENCE ANALYSIS

It is important to demonstrate for any algorithm that it converges and it is able to find the optimal solution at least once within its lifetime for the number of iterations, which is sufficiently large. For any WD in the proposed algorithm, the next node j of the WDA is found using the probability $p_{WD}(i,j)$ defined in Equation 6.12. It is expected that it will choose the node j at some stage during the iteration of the WDA if $p_{WD}(i,j) > 0$. As stated earlier, a problem is presented as the graph $G = \langle V, E \rangle$. Any solution S of the given problem is composed of a number of nodes $\{v_p, v_q, \dots, v_r\}$, $v_{k=p,q,\dots,r} \in V$ selected by the WDA during an iteration. As a result, if it is shown that the chance of selecting any node v_k, $v_k \in V$ in the graph $G = \langle V, E \rangle$ of the problem is greater than zero, then the chance of finding any feasible solution from the set of all solutions of the problem is nonzero. As a consequence, if it is proved that there is positive chance for any feasible solution to be found by the WDA within a finite number of iterations, it will be guaranteed that the optimal solution is found. Because, once a WDA finds an optimal solution, that solution becomes the iteration-best solution in the algorithm, and thus the total-best solution is updated to the newly found optimal solution.

Let the graph $G = \langle V, E \rangle$ represent the graph of a given problem. This graph is a fully connected graph with $|V| = N$ nodes. Also, let N_{WD} represent the number of WDs in the algorithm. In the soil updating of the algorithm, two cases are considered.

Case 1: Includes only those terms that increase soil to an edge of $G = \langle V, E \rangle$, and the highest possible value of soil that an edge can hold is computed.

Case 2: Includes only those terms that decrease the soil to an edge of $G = \langle V, E \rangle$, and the lowest possible value of soil for an edge is computed.

Case 1: The initial soil of an edge (i,j) is denoted by IS_0. The edge (i,j) is supposed to contain the maximum possible value of soil and is called $edge_{max}$. For Equation 6.9, the first term on the right-hand side, that is, $\rho_0 soil(i,j)$, is the only term with positive sign. To consider the extreme case, it is assumed that in one iteration of the algorithm, this term is applied just once to the edge because the parameter ρ_0 is supposed to have its value between [0,1]. For Equation 6.17, the first term on the right-hand side, that is, $\rho_s soil(i,j)$, has positive sign. In the extreme case, this term is applied once in one iteration of the algorithm. As a result, by replacing $soil(i,j)$ with IS_0 in the mentioned terms, the amount of soil of $edge_{max}$ will be $((\rho_s\rho_0)IS_0)$ after one iteration.

Let m denote the number of iterations that the algorithm has executed so far. Therefore, the soil of $edge_{max}$, $soil(edge_{max})$, will have the soil $((\rho_s\rho_0)^m IS_0)$ at the end of m iterations of the algorithm:

$$soil(edge_{max}) = ((\rho_s\rho_0)^m IS_0) \tag{6.29}$$

Case 2: In this case, the lowest amount of soil of an edge (i,j) is estimated. Let $edge_{min}$ denote the minimum soil of the edge (i,j). Here, only the negative terms of Equations 6.9 and 6.17 are considered. From Equation 6.9, the term $-\rho_n \cdot \Delta soil(i,j)$ is supposed to be applied N_{WD} times to the $edge_{min}$ in one iteration, which is the extreme case for making the soil as lowest as possible. The extreme high value for $\Delta soil(i,j)$ is obtained from Equation 6.5 by setting the time in the denominator to zero, which yields the positive value (a_s/b_s). Therefore, the most negative value in one iteration that can come from Equation 6.9 is the value $(-\rho_n N_{WD}(a_s/b_s))$.

From Equation 6.17, the term $\rho_{WD} \cdot (1/(N_{ibest} - 1)) \cdot soil_{ibest}^{WD}$ is the negative term. The highest value of $soil_{ibest}^{WD}$ can be $((N_{WD} - 1)(a_s/b_s))$, and the most negative value for the term will be $(\rho_{WD}(a_s/b_s))$. As a result, in one iteration of the algorithm, $edge_{min}$ has the amount of soil that is greater than or equal to the value $((\rho_{WD} - \rho_n N_{WD})(a_s/b_s))$. Similar to Case 1, for m number of iterations, the soil of $edge_{min}$, that is, $soil(edge_{min})$, will be

$$soil(edge_{min}) = \left(m(\rho_{IWD} - \rho_n N_{WD})\left(\frac{a_s}{b_s} \right) \right) \tag{6.30}$$

The $soil(edge_{min})$ and $soil(edge_{max})$ are the extreme lower and upper bounds of the soil of edges in the graph $G = \langle V, E \rangle$ of the given problem, respectively. Therefore, the soil of any edge after m iterations of the WDA remains within the interval of $[soil(edge_{min}), soil(edge_{max})]$.

Consider that the WDA is at the stage of choosing the next node j from node i. The value $g(soil(i,j))$ of $edge(i,j)$ calculated using Equation 6.14 positively moves $soil(i,j)$ by the amount of the lowest negative soil value of any edge, $\min_{l \notin vc(WD)}(soil(i, j))$. To consider the worst case, let this lowest negative soil value be $soil(edge_{min})$ and the $soil(i,j)$ be equal to $soil(edge_{max})$. As a result, the value of $g(soil(i,j))$ becomes $(soil(edge_{max}) - soil(edge_{min}))$ with the assumption that $soil(edge_{min})$ is negative, which is the worst case. The probability of selecting node j is calculated using Equation 6.12. Therefore, $f(soil(i,j))$ needs to be computed using Equation 6.13, which yields

$$f(soil(i, j)) = \frac{1}{\varepsilon_s + (soil(edge_{max}) - soil(edge_{min}))} \tag{6.31}$$

The denominator of formula (6.12) becomes its largest possible value when it is assumed that each $soil(i,k)$ in Equation 6.12 is zero. Consequently, the probability of the WD going from node i to node j, $p_{WD}(i,j)$, will be bigger than p_{lowest} such that

$$p_{WD}(i, j) > p_{lowest} = \frac{\varepsilon_s}{(N_{WD} - 1)(\varepsilon_s + soil(edge_{max}) - soil(edge_{min}))} \tag{6.32}$$

The value of p_{lowest} is greater than zero.

With some assumptions on the relations between parameters of the algorithm, p_{lowest} can become even bigger. For example, if it is assumed that $(\rho_s\rho_0) < 1$, then $soil(edge_{max})$ in Equation 6.29 becomes zero as m increases. If $\rho_n = \rho_{WD}/N_{WD}$, then $soil(edge_{min})$ will be zero. These two assumptions yield that $soil(edge_{max}) - soil(edge_{min}) = 0$. Therefore, $p_{lowest} = 1/(N_{WD} - 1)$, which is again above zero, and it is the biggest value that p_{lowest} can get in the worst case.

The probability of finding any feasible solution by the WDA in m iterations will be $(p_{\text{lowest}})^{(N_{WD}-1)}$. Since there are N_{WD} WDs, it follows that the probability $p(s;m)$ of finding any feasible solution s by the WDs in m iterations is

$$p(s;m) = N_{WD}(p_{\text{lowest}})^{(N_{WD}-1)} \tag{6.33}$$

The probability of finding any feasible solution s at the end of m iterations of the algorithm will be

$$p(s;M) = 1 - \prod_{m=1}^{M}(1 - p(s;m)) \tag{6.34}$$

Since $0 < p(s;m) \leq 1$, by making M large, the term tends toward zero. That is,

$$\lim_{M \to \infty} \prod_{m=1}^{M}(1 - p(s;m)) = 0 \tag{6.35}$$

Therefore,

$$\lim_{M \to \infty} p(s;M) = 1 \tag{6.36}$$

This fact indicates that any solution s of the given problem can be found at least once by at least one WD of the algorithm if the number of iterations M is sufficiently large.

6.7   VARIANTS OF WDA

The WDA has been applied to a variety of application domains with significant success. In order to apply the algorithm to specific problems; improve the quality of the solution and convergence, ease of application, and parameter settings; and finally to reduce computational cost, there have been some variants reported in the literature. Some of the variants and modifications of the WDA are discussed in the following sections.

6.7.1   IMPROVED WDA

Niu et al. (2012) proposed an improved WDA to solve the JSSP. Five new schemes were introduced to increase the diversity of the search space and enhance the performance of the original WDA, as follows:

1. Consider diverse soil and velocity initializations, whereby the initial amount of soil for each edge and the initial velocity of every WD are randomly initialized.
2. Consider conditional probability computation, whereby the selection probability considers the processing time (operation) of the next node in order to increase the diversity of the search space by introducing some degree of randomness.
3. Consider bounded local soil update, where the soil update is bounded by the lower and upper values that control the convergence rate.
4. Consider elite global soil update, where all the soil values of all edges included in the elite group are updated.
5. Combined local search, whereby a local search that integrates both breadth and depth search to improve the quality of the produced solution is introduced in the improved WDA.

The optimization objective is the makespan of the schedule. Experimental results show that the improved WDA is able to find better solutions for the standard benchmark instances than the existing algorithms. The new approach has made contribution in two aspects. Firstly, this is the first application of the improved WDA to the JSSP. This work can inspire further studies of applying the improved WDA to other scheduling problems, such as open shop scheduling and flow shop scheduling. Secondly, further improvements to the original improved WDA are made by employing five schemes to increase the diversity of the solution space as well as the solution quality. Niu et al. (2013) have also customized the improved WDA for a multiobjective JSSP. The experimental evaluation shows that the customized improved WDA can identify the Pareto nondominance schedules efficiently.

6.7.2 Modified WDA

The standard WDA introduced by Shah-Hosseini utilizes the fitness proportionate selection (FPS) method and determines the probability of going to node j from node i by

$$p_j = \frac{f_j}{\sum_{i=1}^{N} f_i} \tag{6.37}$$

where p_j is the probability of selecting node j, f_j is the fitness function value, and N is the number of candidate nodes from node i. The fitness function value of candidate node j is inversely proportional to the absolute soil value between nodes (i,j) defined by

$$f_j = \frac{1}{soil(i,j)} \tag{6.38}$$

The FPS has three distinct shortcomings:

1. Inability to accommodate negative soil value between nodes
2. Inability to select nodes having similar fitness values, that is, indistinguishable fitness
3. Inability to select nodes with very high fitness values, that is, dominating nodes

To overcome the limitations of the FPS method, the modified WDA uses a ranking-based selection method. Fitness of all nodes are computed according to their soil values and ranked according to higher fitness values. The selection probabilities are computed using the ranked fitness. Alijla et al. (2014) proposed two types of ranking-based selection probabilities: linear ranking selection and exponential ranking selection. The performance of this selection method depends on both the mapping function and the parameter called selection pressure (SP). The SP parameter refers to the tendency of selecting the best node (Al-Betar, 2012, 2013).

The linear ranking selection is defined by the following function:

$$p_j = \frac{1}{N}\left(SP - 2(SP-1)\cdot\frac{r_{f_j}-1}{N-1}\right) \tag{6.39}$$

Here, r_{f_j} is the rank of the fitness f_j of node j according to their soil values, $j \in \{1, \ldots, N\}$. The parameter SP, $1 \le SP \le 2$, is used to control the gradient of the linear selection function. A larger SP value leads to higher selection pressure.

The exponential ranking selection is defined by the following function:

$$p_j = SP^{r_{fj}} \cdot \frac{1-SP}{1-SP^N} \tag{6.40}$$

Here, the probabilities of the ranked nodes are exponentially weighted, and the parameter SP, $0 < SP < 1$, is used to control the slope. A smaller SP value leads to higher selection pressure.

To evaluate the effectiveness of the proposed ranking selection methods, a series of experimental studies using three combinatorial optimization problems, that is, rough set feature subset selection, multiple knapsack, and TSPs, was conducted and compared against the standard WDA with the FPS method. The proposed ranking-based WDA can successfully avoid local minima though being computationally intensive.

6.7.3   ADAPTIVE WDA

The velocity and soil update parameters $\{a_v, b_v, c_v\}$ and $\{a_s, b_s, c_s\}$, in general, are fixed over the whole iterations of the WDA. Sur et al. (2013) proposed an adaptive WDA where these parameters are adapted over the iterations. The velocity update parameters are defined as follows:

$$\begin{cases} a_v(t) = a_v(t-1) + \lambda_v \cdot \text{rand}() \\ b_v(t) = b_v(t-1) + \lambda_v \cdot \text{rand}() \\ c_v(t) = c_v(t-1) + \lambda_v \cdot \text{rand}() \end{cases} \tag{6.41}$$

where rand() is a random number distributed over the interval of [−1,+1]. There is a possibility that the velocity may increase or decrease depending on the variation between the current and previous iteration. The constant λ_v is important for defining the step size and the range it can cover between the initial value and the final value of the respective parameters. In a similarly way, the soil update parameters are defined as follows:

$$\begin{cases} a_s(t) = a_s(t-1) + \lambda_s \cdot \text{rand}() \\ b_s(t) = b_s(t-1) + \lambda_s \cdot \text{rand}() \\ c_s(t) = c_s(t-1) + \lambda_s \cdot \text{rand}() \end{cases} \tag{6.42}$$

The soil update parameters $\{a_s, b_s, c_s\}$ will also introduce the probability of increasing or decreasing the soil content of the WDs. All the steps of the adaptive WDA remain the same as the standard WDA except for the computation of the velocity vel_{WD} and change of soil $\Delta soil_{WD}$. The adaptive WDA was applied to vehicle guidance in a road graph network optimization problem, which demonstrates superior performance compared to the standard WDA.

6.7.4   WDA CONTINUOUS OPTIMIZATION ALGORITHM

Nagalakshmi et al. (2011) introduced a local search for the WDA and called the new approach WD-Continuous Optimization (WD-CO). To improve the quality of the solution, a mutation operation is performed to the solutions found by the WDA. In the WD-CO approach, each solution undergoes a mutation. That is, an $edge(i,j)$ is randomly selected based on the probability defined in Equation 6.19 from the edges of a solution. The mutation is carried out by randomly choosing an edge and replacing it with another edge, that is, the $edge(i,j)$ is replaced with another edge connecting nodes i and j. If the new edge improves the quality of the solution, then the new edge is accepted;

otherwise, the previous solution is retained. This process is repeated for a number of times for each solution. The main steps are described below:

Step 1: Generate new WDs
Step 2: WDA on the first node of the graph
Step 3: Select edge
Step 4: Update local soil
Step 5: If (Not end of graph), go to Step 3
Step 6: Perform mutation on randomly selected edge
Step 7: Update global soil
Step 8: If (Termination condition not met), go to Step 1
Step 9: Return solution

Once the global soil updating (Step 7) is finished, one iteration of the WD-CO algorithm has been completed and a new iteration begins with a new set of WDs. The process continues until the maximum number of iterations is performed. The effectiveness and performance of the WD-CO algorithm have been applied to a number of well-known benchmark functions taken from Yao and Liu (1996).

It is also possible to solve continuous optimization problems by the WDA. In a continuous optimization problem, a number of continuous variables (parameters) are needed such that a function is minimized or maximized. Shah-Hosseini (2012) applied WD-CO by encoding real continuous variables into binary strings like binary chromosomes in the standard GA. The WDA then tries to optimize the given function in the binary representation. Finally, the best solution is reported as the final solution.

There are also many other WDAs reported in the literature, but they are mere modifications for specific applications rather than variants. Therefore, these methods are not discussed further. Some of these modifications are mentioned in the application section.

6.8 APPLICATIONS TO ENGINEERING PROBLEMS

The WDA has been applied to a variety of application domains and has been a great success in engineering and technological problems. In addition, the IWD algorithm has also been applied to different optimization problems in different fields of study, and it provides better or at least comparable performance in comparison with other well-known optimization methods, such as ACO and PSO. Initially, the WDA has been applied to various well-known benchmark functions to verify its performance as an optimization method. The applications can be categorized into the following domains:

1. Traveling salesman problem (TSP) (Shah-Hosseini, 2007, 2009a,b)
2. n-queen problem (Shah-Hosseini, 2008, 2009a)
3. Multi-dimensional knapsack problem (Shah-Hosseini, 2008, 2009a,b)
4. Vehicle routing problem (VRP) (Kamkar et al., 2010)
5. Economic load dispatch problem (Rayapudi, 2011)
6. Combined economic and emission dispatch problem (Nagalakshmi et al., 2011)
7. Reactive power dispatch problem (Lenin and Kalavathi, 2012)
8. Vehicle guidance in road graph networks (Sur et al., 2013)
9. Path planning (Duan et al., 2008, 2009)
10. Trajectory planning (Duan et al., 2009)
11. Feature selection (Alijla et al., 2013; Hendrawan and Murase, 2011; Shah-Hosseini, 2012)
12. Automatic multilevel thresholding (ATM) (Shah-Hosseini, 2009b)
13. Data aggregation and routing in wireless networks (Hoang et al., 2012)

14. Mobile ad hoc networks (Khaleel and Ahmed, 2012; Sensarma and Majumder, 2013)
15. Job shop scheduling (Niu et al., 2012, 2013)
16. Web service selection (Palanikkumar et al., 2012)
17. Max-Clique problem (Al-Taani and Nemrawi, 2012)
18. Reservoir operation problem (Dariane and Sarani, 2013)
19. Data clustering (Shah-Hosseini, 2013)
20. Steiner tree problem (Noferesti and Shah-Hosseini, 2012)
21. Other applications

6.8.1 Traveling Salesman Problem

The TSP is a combinatorial optimization problem and is generally considered as a benchmark problem for testing many optimization algorithms (Flood, 1955). In the TSP, a map of n cities is given to the salesman. The salesman is allowed to visit every city only once, starting from an initial city and returning to the starting city on completion of the tour through all cities. The goal of the TSP is to find a tour with the minimum total length among all such possible tours for the given map. In order to apply an optimization algorithm to the TSP, the TSP is represented by a graph $G = (V, E)$, where the set of nodes V denotes the n cities and the edge set E denotes the paths between cities. Here, the graph of the TSP is considered a complete graph. That is, every city has a direct connectivity to another city. It is also assumed that the connectivity between two cities is undirected. A solution to the TSP represented by the graph $G = (V, E)$ is an ordered set of n distinct cities. For such a TSP with n cities, there are $(n - 1)!/2$ feasible solutions in which the global optimum is sought. The successful application of an optimization algorithm to the TSP can lead to solutions to many other combinatorial optimization problems such as scheduling, routing, placement of goods in the warehouse, placement of machines in workshops, and printed circuit design (Dariane and Sarani, 2013). A solution to the TSP can be represented by the ordered set $T^k = \langle c_i^k, \ldots, c_{i+1}^k \rangle$, $i \in \{1, \ldots, n\}$, $k = 1, \ldots, (n - 1)!/2$. Here, $i + 1$ is meant to be the starting city again after visiting the last city. The tour leads to the tour length T_L^k, defined by

$$T_L^k(c) = \sum_{i=1}^{n} d(c_i^k, c_{i+1}^k) \tag{6.43}$$

Here, $d(\cdot)$ is the Euclidean distance between two cities. The objective is to find the optimum tour $T_L^*(c^*)$ such that, for every other feasible tour T^k, it satisfies

$$\forall T_L : \ T_L^*(c^*) \le T_L^k(c) \tag{6.44}$$

Shah-Hosseini (2007, 2009a,b) first applied the WDA to solve the TSP. In order to apply the WDA to the TSP, the problem is represented by a complete undirected graph (V, E) with $V = \{c_1, c_2, \ldots, c_n\}$, where the WDs travel between nodes V meaning physical positions of the cities. A WD starts its tour from a random node, travels through all nodes, and returns to the first node. The WDA maintains a list of visited cities $V_c(WD)$. The next city c_j is chosen from the nodes that are not in $V_c(WD)$, that is, $c_j \notin V_c(WD)$ depending on the local heuristic function $HUD(i,j)$ for the TSP defined (Shah-Hosseini, 2009a,b) as

$$HUD_{TSP}(i, j) = \| c_i - c_j \| \tag{6.45}$$

where $\|\cdot\|$ is the Euclidean norm. The time taken to travel from city c_i to city c_j is proportional to the heuristic function $HUD_{TSP}(i,j)$.

Shah-Hosseini (2009a,b) proposed a modification of the WDA for the TSP. To escape the local optimum solution after a specific number of iterations, the amounts of soil on all paths are reinitialized such that the paths of the global (total)-best solution $T_{\text{total-best}}$ are given less soil than the other paths (Shah-Hosseini, 2009b). This is implemented using a simple mechanism as follows:

$$soil(i,j) = \begin{cases} \alpha_I \Gamma_I Initsoil(i,j) & \text{for every } (soil(i,l)) \in T_{\text{total-best}} \\ Initsoil(i,j) & \text{otherwise} \end{cases} \qquad (6.46)$$

where α_I is a small positive number. Shah-Hosseini (2009b) used $\alpha_I = 0.1$. Γ_I denotes a random number within the interval of [0,1]. Shah-Hosseini (2009a,b) carried out experiments using the proposed modified WDA for the TSP with 10–100 cities and parameter settings as given in Tables 6.1 and 6.2, and local and global soil updating parameters $\rho_n = 0.9$ and $\rho_{WD} = 0.9$, respectively. The TSP problems are selected from the library TSPLIB95 for these experiments. For a 10-city problem, the average number of iterations required was 10.4, and for 20 cities 39.6 iterations.

A modified WDA for the TSP was reported by Kesavamoorthy et al. (2011). They suggested that after a few number of iterations, the soils of all paths of the graph of the given TSP should be reinitialized with the initial soil except for the paths of the total-best solution which are given less soil than the initial soil. The modified WDA finds better tours and escapes local optima. Msallam and Hamdan (2011) also developed an improved WDA based on an adaptive scheme, which can avoid premature convergence. The performance of the improved WDA is verified, and the performance is compared with that of the original WDA and other meta-heuristic algorithms in solving the TSP. The results show clearly that the proposed algorithm has better performance than the original WDA.

6.8.2 THE N-QUEEN PROBLEM

The n-queen puzzle is the problem of putting n chess queens on an $n \times n$ chessboard such that no two queens attack each other (Watkins, 2004). Thus, a solution requires that no two queens be placed on the same row, the same column, or on the diagonal. The n-queen problem is the generalization of the 8-queen puzzle with 8 queens on an 8×8 chessboard. The 8-queen puzzle was originally proposed by the chess player Max Bezzel in 1848. It has been of general interest to many mathematicians for long. Mathematicians like Gauss and Cantor have worked on the 8-queen puzzle and generalized it to an n-queen puzzle. The n-queen puzzle leads to a huge search space of n^{2n}. For $n = 8$, the search space is 64^8. The 8-queen puzzle has 92 distinct solutions. If the solutions that differ only by rotations and reflections of the board are counted as one, the 8-queen puzzle has 12 unique solutions (Shah-Hosseini, 2009b). A solution to the 8-queen puzzle is depicted in Figure 6.6.

The researchers follow a general strategy to reduce the size of the search space. The n queens are placed one by one such that the first queen is placed on any row of the first column. The second queen is placed on any row of the second column except for the row of the first queen. Following the strategy, the n-th queen is placed on any row except for those where queens were placed earlier in the n-th column. This incremental strategy reduced the search space to $n!$.

Shah-Hosseini (2009a,b) first applied the WDA to the n-queen problem. The n-queen problem is to be transformed into a TSP by applying the incremental strategy and considering every row as a city. The first row chosen by the first queen is considered the first city of the tour. The second row chosen by the second queen is the second city of the tour and so on; the n-th row chosen by the n-th queen is the n-th city of the tour for the TSP. The constraint of no two queens being on the same row in the n-queen problem is seen as the constraint of no city visited twice in the TSP. With this supposition, the n-queen problem is represented by a complete undirected graph (Shah-Hosseini,

FIGURE 6.6   A solution to the 8-queen puzzle on the chessboard.

2009a,b). The local undesirability heuristic function $HUD(i,j)$ for the n-queen problem is defined (Shah-Hosseini, 2009a,b) as follows:

$$HUD_{NQ}(i, j) = (1+r)\left| |i-j| - \frac{n}{2} \right|$$ (6.47)

where $|\cdot|$ denotes the absolute value, $r \in [0,1]$ is a random number, and n is the number of cities (i.e., rows or columns). Here, $HUD_{NQ}(i,j)$ is a heuristic function for the WDA to go from the current row i (i.e., city c_i) to the next row j (i.e., city c_j). The quality of a solution function is measured by

$$f(T) = -\sum_{i=1}^{n-1} \sum_{j=i+1}^{n} \text{attack}(c_i, c_j)$$ (6.48)

Attack(c_i,c_j) is the attack of two queens in cities c_i and c_j defined by

$$\text{attack}(c_i, c_j) = \begin{cases} 1 & \text{if } c_i \text{ and } c_j \text{ attack each other} \\ 0 & \text{else} \end{cases}$$ (6.49)

The optimal solution (T^*) to the n-queen problem has the quality value of $f(T^*) = 0$. It is expected that the total-best solution reaches the quality value of 0. While doing experiment with different parameter settings, it is found that the WDA gets trapped into local optima where only two queens attack each other. A considerable number of iterations were required to overcome the local optima. Shah-Hosseini (2009a,b) proposed a simple local search algorithm, called n-queen local search (NQLS), which is activated when the iteration-best solution of the WDA has a two-queen attack. Shah-Hosseini (2009a,b) investigated experimentally with 50 WDs for 10 different n-queen puzzles by increasing n by 10–100. It is found that the number of iterations does not necessarily depend on n. For example, a 90-queen problem takes more iterations (average 2951 iterations) than a 100-queen problem (average 1294 iterations).

6.8.3 MULTIDIMENSIONAL KNAPSACK PROBLEM

The problem is to select a subset of items $i \subseteq I$ with each item i yielding the profit b_i and resource (capacity) requirement r_i such that all the selected items fit in the knapsack of limited capacity and the sum of profits of the selected items is maximized. This problem is known as the knapsack problem (Keller et al., 2004). The knapsack problem is then extended to multiple knapsacks with multiple constraints. Thus, the inclusion of an item i in the m knapsacks is denoted by setting the variable y_i to one; otherwise, y_i is set to zero. Let the variable r_{ij} represent the resource requirement of an item i with respect to the resource constraint (knapsack) j having the capacity a_j. This problem is called the MKP. The MKP with m knapsacks and n items wants to maximize the total profit of including a subset of the n items in the m knapsacks without exceeding the capacities of the knapsacks. The MKP is then formulated as a maximization of profit defined in specific terms:

$$\max \sum_{i=1}^{n} y_i b_i \tag{6.50}$$

subject to the following constraints:

$$\sum_{i=1}^{n} r_{ij} y_i \le a_j \tag{6.51}$$

where $y_i = \{0,1\}$, b_i is the profits, and r_{ij} are the resource requirements of having nonnegative values for $i = 1, 2, \dots, n$ and $j = 1, 2, \dots, m$. The MKP is an NP-hard combinatorial optimization problem. The successful application of an optimization algorithm to the MKP can lead to solutions to many other combinatorial optimization problems such as cutting stock problems, processor allocation in distributed systems, cargo loading, and capital budgeting and economics (Shah-Hosseini, 2009b).

Shah-Hosseini (2008, 2009a,b) first applied the WDA to solve the MKP. In order to apply the WDA, the MKP is represented by a graph $\langle V, E \rangle$ where the node set V is the set of items and the edge set E denotes paths between items. A feasible solution is a set $V' \subseteq V$, which satisfies the constraints in Equation 6.51. An optimal solution $V^* \subset V'$ comprising n items must ensure the maximum profit defined by Equation 6.50. The local undesirability heuristic function $HUD(i,j)$ for the MKP is defined (Shah-Hosseini, 2009a,b) as follows:

$$HUD_{MKP}(j) = \frac{1}{mb_j} \sum_{k=1}^{m} r_{jk} \tag{6.52}$$

where r_{jk} is the resource requirement for item j from knapsack k. The proposed WDA was tested with a set of MKPs available in the OR-Library. For each test, the WDA is run 10 times and averaged. The qualities of optimal solutions are known for the eight MKPs. The WDA was able to find the global solution for the first 6 MKPs with 2 constraints and 28 items. The rest 2 MKPs were able to reach a near-optimal solution with 2 constraints and 105 items. The best run of the WDA to the MKP converged to an optimal solution of 119,377 in 12 iterations, whereas the worst run required 60 iterations.

6.8.4 VEHICLE ROUTING PROBLEM (VRP)

The VRP originated from the problem where a company wants to deliver goods to a number of customers at their delivery point using a limited number of vehicles. More specifically, there are m

identical vehicles located at a depot, which are to deliver discrete quantities of goods to n customers located at different points. Each customer has a demand for goods, and each vehicle has a capacity. A vehicle can make only one tour, starting at the depot, visiting a subset of customers (i.e., delivery points), and returning to the depot. Therefore, a solution is a set of tours for a subset of vehicles such that all customers are served only once, respecting capacity constraints and the time window. The goal is to minimize cost and additionally to reduce the number of vehicles used. Since the cost is associated with the traveled distance, the VRP needs to find the minimum distance traveled by a number of vehicles in order to meet customers' demand. Thus, the VRP is a combinatorial optimization and integer programming problem (Dantzig and Ramser, 1959), and determining the optimal solution is an NP-complete problem. The VRP has been an important problem in the field of distribution and logistics since the early 1960s (Clark and Wright, 1964).

Mathematically, the VRP is described as finding the minimum distance of the combined routes of m vehicles that must service n customers. The problem can be presented as a weighted graph $G = \langle V, E, d \rangle$ where the delivery points for the customers are the nodes represented by $V = \{v_0, v_1, \ldots, v_n\}$, and the traveled distances comprise edges represented by $E = \{v_i, v_j\}$ with $i \neq j$. Each vehicle starts its route from node v_0 (meaning a central depot), and each of the other nodes $\{v_1, \ldots, v_n\}$ represents the n customers' service points. The distances associated with each edge are represented by the variable $d_{ij} = dist(v_i, v_j)$, which is the Euclidean distance between the two nodes. Each customer is assigned a nonnegative demand q_i, and each vehicle is given a capacity constraint Q. The problem is solved under the following constraints:

1. Each customer v_i is visited only once by a single vehicle.
2. Each vehicle must start and end its route at the depot v_0.
3. Total demand serviced by each vehicle cannot exceed the capacity Q.

In practice, heuristic and deterministic methods have been developed for solving the VRP that yield acceptable good solutions. Kamkar et al. (2010) were the first to apply the WDA to the VRP. They used the number of WDs equal to the number of vehicles, that is, $N_{WD} = m$. The velocity and soil updating parameters are set to $a_v = a_s = 1000$, $b_v = b_s = 0.01$, and $c_v = c_s = 1$. The initial soil of each edge and initial velocity are set to $Initsoil = 1000$ and $Initvel = 100$, respectively. At each iteration, the WDA builds a solution for the VRP, moving to the next node according to the selection rule based on the amount of soil and the length of the edge. A list of visited nodes is updated each time a node is visited. A route is determined in one of two ways: sequential or parallel. In sequential route determination, the WDA starts the route for the first vehicle until its capacity is complete and then continues with other vehicles. In parallel route determination, all vehicles start at the same time. At each iteration, only one node is chosen as per the selection rule followed by an extension of the best route. The selection of the next node is based on the probability defined in Equation 6.12. The heuristic undesirability function $HUD(i,j)$ for the VRP is defined by

$$HUD_{VRP}(i,j) = \| c_i - c_j \| \tag{6.53}$$

where $\{c_i, c_j\}$ represents the two-dimensional positional vectors and $\|\cdot\|$ is the Euclidean norm. When two nodes are close to each other, $HUD(i,j)$ becomes small, which reduces the travel time. The performance of the WDA was tested on a set of 14 benchmark instances available from the OR-Library. The performance of the WDA is competitive compared to other methods in terms of solution quality.

6.8.5 ECONOMIC LOAD DISPATCH PROBLEM

ELD is a method of determining the most efficient, low-cost, and reliable operation of a power system by dispatching available electricity generation resources to supply load on the system. The objective of the ELD problem is to find the optimal combination of power generations that minimizes the total cost generation while satisfying different equality and inequality constraints. Thus, the optimization problem is formulated in the following.

Let a power plant have some N generating units each loaded to P_i MW. The units should be loaded in such a way that the total fuel cost C_T of generators is the minimum that satisfies the operating constraints. Therefore, the objective function of the classical ELD problem is written as

$$C_T = \text{Min} \sum_{i=1}^{N} C_i(P_i) \tag{6.54}$$

where C_i is the fuel input–power output cost function of the i-th generator. The search space of a constrained optimization problem consists of feasible and unfeasible points. Feasible points satisfy all the constraints, while unfeasible points violates at least one constraint. Therefore, a solution to such constrained optimization problems must be feasible. A constrained optimization method usually employs a concept of penalty function which handles the constraints, that is, penalizes the unfeasible solutions. The objective function is then modified using the penalty function as follows:

$$C_T = \begin{cases} \text{Min} \sum_{i=1}^{N} C_i(P_i) & \text{if } P_i \in F \\ \text{Min} \sum_{i=1}^{N} C_i(P_i) + \lambda(P_i) & \text{else} \end{cases} \tag{6.55}$$

where F is the set of feasible solutions, and $\lambda(P_i)$ is zero when all constraints are satisfied.

Rayapudi (2011) applied the WDA to solve the ELD problem with the objective function defined in Equation 6.55. Two networks having 6 and 20 generators are simulated, and the test results are compared with popular methods reported in the literature. For the 6-unit system, the velocity and soil updating parameters are set to $a_v = a_s = 1$, $b_v = b_s = 0.01$, and $c_v = c_s = 1$. The local and global soil update parameters, ρ_n and ρ_{WD}, are set to 0.8. The initial soil is $InitSoil = 10,000$, and initial velocity is $InitVel = 200$. The WDA reached the optimal solution with 6 WDs within 100 iterations. For 20-unit systems with the same parameter setting, the WDA reached the optimal solution with 20 WDs within 200 iterations. It has been demonstrated that the WDA is easy to implement and capable of finding feasible, near global optimal solutions with less computational effort and shows good convergence properties.

6.8.6 COMBINED ECONOMIC AND EMISSION DISPATCH PROBLEM

The combined economic and emission dispatch (CEED) problem is one of the most important optimization problems in power system operation and forms the basis of many application programs. The main objective of CEED is to determine the optimum share of each generating unit that satisfies the required load conditions minimizing the operational cost and satisfying the emission cost. The CEED problem has been discussed in Chapter 7. Further details of problem formulation of CEED can be found in Wood and Wollenberg (1994). A number of optimization algorithms have been applied to solve the CEED problem such as evolutionary programming, genetic algorithms, PSO, and spiral dynamic optimization algorithms. Nagalakshmi et al. (2011) first applied the WDA to solve the CEED problem with 3- and 6-unit generator systems. The WD-CO algorithm was run with 6 WDs, 32-bit precision, with Minsoil and Maxsoil set to 2000 and 10,000, respectively. The number of WDs was equal to three. Each solution undergoes a mutation-based local search (Shah-Hosseini, 2012), where an edge is randomly selected and replaced with another. If the new edge improves the solution, then it is accepted; otherwise, the previous solution is retained. The WDA reached an optimal demand of 300 MW for the CEED problem within 10 iterations.

6.8.7 REACTIVE POWER DISPATCH PROBLEM

A power system comprises generators, synchronous condensers, capacitors, static compensators, and tap changing transformers. The electric power loads vary from hour to hour. The change in load causes variation in the reactive power requirement. The reactive power depends on voltage so that the variation of load causes the variation of voltage. Therefore, the important operating task of a power system is to maintain the voltage within the allowable range for high-quality consumer service. In other words, it is to minimize the system's real power loss. This objective can be achieved by utilizing the generator bus voltage regulators, transformer tap setting, and the output of the reactive power sources so as to minimize the loss and maximize the voltage stability margin of the system. This problem is known as the RPD problem, which is a nonlinear optimization problem with a number of equality and inequality constraints.

The objective of the RPD problem is to minimize or maximize one or more objective functions while satisfying a number of constraints such as load flow, generator bus voltages, load bus voltages, switchable reactive power compensations, reactive power generation, transformer tap setting, and transmission line flow. In general, the objective functions are the minimization of real power loss and maximization of the static voltage stability margin (SVSM) subject to satisfying a set of equality and inequality constraints defined in the following.

Minimization of real power loss: The objective function can be formulated as the minimization of the active power losses in the transmission lines defined by

$$\text{Min } P_{\text{loss}} = \sum_{k=1}^{NTL} g_k \left[V_i^2 + V_j^2 - 2V_i V_j \cos\theta_{ij} \right) \right] \tag{6.56}$$

where *NTL* is the number of transmission lines; g_k is the conductance of the k-th line; and V_i and V_j are the voltage magnitudes of the i-th and j-th busses, respectively; and θ_{ij} is the voltage angle difference between the i-th and j-th busses.

Maximization of SVSM: SVSM is defined as the largest load change for which the power system may sustain. It is the most accepted index for voltage collapse.

The equality constraints are the typical load flows defined by

$$P_i - V_i \sum_{j=1}^{NN_B} V_j (G_{ij} \cos\theta_{ij} + B_{ij} \sin\theta_{ij}) = 0, \quad i \in N_B - 1 \tag{6.57}$$

$$Q_i - V_i \sum_{j=1}^{N_g} V_j (G_{ij} \sin\theta_{ij} - B_{ij} \cos\theta_{ij}) = 0, \quad i \in N_{PQ} \tag{6.58}$$

The inequality constraints represent the system operating constraints. These are the generator bus voltage limits (V_{gi}) defined by $V_{gi}^{\min} \le V_{gi} \le V_{gi}^{\max}$, $i \in N_B$; generator reactive power capability limit (Q_{ci}) defined by $Q_{gi}^{\min} \le Q_{gi} \le Q_{gi}^{\max}$, $i \in N_g$; capacitor reactive power generation limit (Q_{ci}) defined by $Q_{ci}^{\min} \le Q_{ci} \le Q_{ci}^{\max}$, $i \in N_c$; transformer tap setting limit (t_k) defined by $t_k^{\min} \le t_k \le t_k^{\max}$, $k \in N_T$; and transmission line flow limit (S_l) defined by $S_l \le S_l^{\max}$, $l \in N_l$.

Lenin and Kalavathi (2012) applied the WDA to an RPD problem and demonstrated the effectiveness and robustness on an IEEE-30 bus system (Lee et al., 1985). The WDA parameters used for simulation are as follows: number of WDs $N_{wd} = 30$; velocity updating parameters $a_v = 1$, $b_v = 0.01$, and $c_v = 1$; soil updating parameters $a_s = 1$, $b_s = 0.01$, and $c_s = 1$; local soil updating parameter $\rho_n = 0.9$; and global soil updating parameter $\rho_{WD} = 0.9$. The initial soil and initial velocity were set to $InitSoil = 10,000$ and $InitVel = 200$, respectively. The WDA shows improvement in real power loss within 100 iterations with 30 WDs.

6.8.8 Vehicle Guidance in Road Graph Networks

Vehicle guidance is an interesting optimization problem for determination of the route for vehicles through underutilized paths in a road graph network. There is a usual tendency of drivers to follow the same path due to preference of the shortest distance to the destination. This very often causes high traffic congestion and increase in waiting time. Sur et al. (2013) first applied the multiobjective WDA to optimize road graph networks for vehicle guidance. The two objective functions considered are minimization of the distance function defined as $f_d = \min\left(\sum_{k=1}^n d_k\right)$ and minimization of the average waiting time defined as $f_{awt} = \min\left(\sum_{k=1}^n awt_k\right)$ for the path $k \in \{i,j\}$ for all $\{i,j\}$, $i \neq j$. The adaptive mechanism was used for parameters $\{a_v, b_v, c_v\}$ and $\{a_s, b_s, c_s\}$. The fitness of the multiobjective problem is a nonweighted multiparameter function comprising f_d and f_{awt} and defined as

$$f = f_d + f_{awt} \tag{6.59}$$

Three different simulation setups were used for experimentation with 20 WDs for 500 iterations. The results demonstrated better performance compared to other meta-heuristic methods.

6.8.9 Path Planning

Path planning is to generate a path between an initial location and the destination location that has an optimal or near-optimal performance under specific constraints. The flight path planning in a large mission area is typical of a large-scale optimization problem. Incorporation of threat sources is an inherent part and key in air robot optimal path planning. In order to simplify the air robot path planning problem, the robot's task space is divided into a two-dimensional mesh, which forms a two-dimensional network diagram connecting the starting point and the goal point. A simple air robot task space discretization is shown in Figure 6.7. Thus, the air robot optimal path planning problem can be transformed into a general path planning optimization problem.

There are some threatening areas in the task region, as shown in Figure 6.7. The flight region is divided by $(m - 1)$ vertical lines and $(2n + 1)$ horizontal lines, which are labeled with $L_1, L_2, \ldots, L_m$.

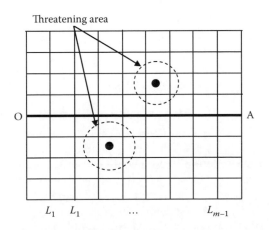

FIGURE 6.7 Typical air robot flight area.

The $(m-1)$ vertical lines and the $(2n+1)$ horizontal lines cross-constitute $(m-1)*(2n+1)$ nodes. These nodes are named as

$$L_1(x_1,y_1), L_2(x_2,y_1), \ldots, L_{m-1}(x_{m-1},y_1), \ldots, L_1(x_1,y_{2n+1}), \ldots, L_{m-1}(x_{m-1},y_{2n+1}) \qquad (6.60)$$

where $L_i(x_i, y_i)$ is the i-th node in the vertical line L_i. In this way, the path from a starting node to a target node can be described by

$$Path = \{O, L_1(x_1,y_{k1}), L_2(x_2,y_{k2}), \ldots, L_{m-1}(x_{m-1},y_{k(m-1)}), A\} \qquad (6.61)$$

where $k = 1, 2, \ldots, 2n+1$. Here, O is the starting node, and A is the target node in the path.

The objective function can be simplified to contain only two terms (Jin et al., 2002) when horizontal path optimization is to be considered:

$$J = L_k + \delta \sum_{i=1}^{m-1} \frac{1}{d_{i\,min}} \qquad (6.62)$$

where L_k is the flight distance, $d_{i\,min}$ is the distance from the node to the nearest threat, and δ is the threat avoiding coefficient. The bigger the δ is, the safer the air robot flight would be. Suppose that there are q threats, each of which is described by a circle with the center point (x_j, y_j) and the radius r_j; then, the distance between the $node(x_i, y_{ki})$ and the threat j can be defined as follows:

$$d = \sqrt{(x_i - x_j)^2 + (y_{ki} - y_j)^2} \qquad (6.63)$$

Thus, the distance between the $node(x_i, y_{ki})$ and the nearest threat can be defined as follows:

$$d_{i\,min} = \left\{ \left(\sqrt{(x_i - x_j)^2 + (y_{ki} - y_j)^2} - r_1 \right), \ldots, \left(\sqrt{(x_i - x_q)^2 + (y_{ki} - y_q)^2} - r_q \right) \right\} \qquad (6.64)$$

Air robot path planning has to find out the safe flight path to the destination and avoid the hostile threats. There are some similarities between the TSP and air robot path planning (Flood, 1955). Considering the common features, Duan et al. (2008) applied the WDA to air robot path planning.

In order to investigate the feasibility and effectiveness of the WDA to air robot path planning, a series of experiments have been conducted under complicated environments. The basic WDA and the improved WDA were applied to a path planning problem. In all experiments, the same set of parameter values were used: $a_s = 1000$, $b_s = 0.1$, $c_s = 1$, $a_v = 1000$, $b_v = 0.1$, $c_v = 1$, $InitSoil = 1000$, $InitVel = 100$, and $\rho = 0.05$. It is found that the WDA can find a feasible and optimal path for the air robot with 60 WDs within 10 iterations.

6.8.10 TRAJECTORY PLANNING

Trajectory planning is the moving point from a point A to point B while avoiding collisions over time. It is also referred to as motion planning. It is a rather complicated global optimum problem for unmanned combat aerial vehicle (UCAV) mission planning. Trajectory planning for a UCAV is a complex problem in dimension optimization, mainly concerned with optimizing the flight route taking into consideration that there are various constraints under complicated field environments (Wang et al., 2012). The objective function of the trajectory planning problem for the UCAV can be formulated as the same objective function defined in Equation 6.62. Duan et al. (2009) applied the WDA to a UCAV trajectory planning problem. An improved WDA approach was used for solving the smooth trajectory planning problems in various combating environments.

In order to investigate the feasibility and effectiveness of the WDA to single UCAV trajectory planning, a series of experiments have been conducted under complicated environments. The basic WDA and the improved WDA were applied to the trajectory planning problem. In all experiments, the same set of parameter values were used: $a_s = 1000$, $b_s = 0.1$, $c_s = 1$, $a_v = 1000$, $b_v = 0.1$, $c_v = 1$, $InitSoil = 1000$, $InitVel = 100$, and $\rho = 0.05$. It is found that the WDA can find a feasible and optimal path for the UCAV with 60 WDs within 10 iterations. Series experimental comparison results show that the proposed WDA is more effective and feasible in the single UCAV smooth trajectory planning than the basic WDA.

6.8.11 FS

Removing useless, irrelevant, and redundant features can reduce the dimension of the data set while improving the learning and classification accuracy. The problem is then how to select the minimum subset of features that can preserve the meaning of the original features without affecting the knowledge represented by the entire set of features. If there is a data set with N features, then FS can be seen as a search process over a search space of 2^N possible subsets of features. Therefore, FS is a combinatorial optimization problem.

Textural features of cultured Sunagoke moss can be used to predict the water content in the moss. It is required to input a set of features to the neural network-based prediction system that can assess the water content. The accuracy of the prediction system depends on the optimal set of features. Hendrawan and Murase (2011) applied the WDA for features selection from textural features of Sunagoke moss. The total number of textural features extracted is 120. The WDA was employed for selecting an optimal set of features. In all experiments, the same set of parameter values were used: $a_s = 1$, $b_s = 0.1$, $c_s = 1$, $a_v = 1$, $b_v = 0.1$, $c_v = 1$, $InitSoil = 10,000$, and $InitVel = 4$. The local soil updating parameter ρ_n and global soil updating parameter were chosen from the interval of [0,1]. It is found that the WDA can find an optimal set of features for the prediction system with 70 WDs within 500 iterations.

Random search (RS) theory is a mathematical tool for FS that helps in analyzing the minimal knowledge representation required for a data set. RS has been used with many meta-heuristic algorithms for FS. Alijla et al. (2013) adopted the WDA for FS with RS. Specifically, the WDA is used to search for a subset of features based on RS dependency as an evaluation function. The resulting system, called intelligent water drops for rough set feature selection (IWDRSFS), is evaluated with six benchmark data sets. The performance of IWDRSFS is analyzed and compared with other methods reported in the literature. It demonstrates that the IWDRSFS is able to provide competitive and comparable results.

6.8.12 AUTOMATIC MULTILEVEL THRESHOLDING

Image segmentation is widely used in many computer applications such as computer vision, image processing, and object recognition. For such applications, an image is segmented into subimages based on the similarities and differences between pixels in an image. Multilevel thresholding is a technique used for image segmentation. It is assumed that each object has a distinct continuous area of the image histogram. By separating the histogram appropriately, the objects of the image can be segmented correctly.

A number of thresholds $\{t_1, t_2, \ldots, t_M\}$ in the histogram of an image $f(x, y)$ are used to separate the pixels of the objects in the image. The segmented image $T(f(x, y))$ is created as per the thresholds:

$$T(f(x,y)) = \begin{cases} g_0 & \text{if } f(x,y) < t_1 \\ g_1 & \text{if } t_1 \leq f(x,y) < t_2 \\ \vdots & \vdots \\ g_M & \text{if } f(x,y) \geq t_M \end{cases} \tag{6.65}$$

The image is divided into $M + 1$ regions determined by the M threshold values $\{t_1, t_2, \ldots, t_M\}$. Here, g_i is the gray level assigned to all pixels of the region i, which eventually represents object i. The value of g_i may be chosen to be the mean value of gray levels of the region's pixels, or a maximum range of gray levels 255 can be used to distribute the gray levels of regions equally. Specifically, $g_i = i \cdot [255/M]$ such that the function [.] returns the integer value of its argument. In general, the numbers of thresholds are provided to the algorithm. In automatic multilevel thresholding (AMT), the problem becomes harder because the search space increases with the number of thresholds.

Shah-Hosseini (2009b) applied the WDA to an AMT problem where he transformed the problem into a TSP-like problem so that the number of nodes can be assumed to be 256. Each WD starts traveling from Node 0 and traverses all 255 nodes. The WDA was tested with three gray-level images using 50 WDs. The most important features of the images have been preserved, and the number of gray levels has been reduced.

6.8.13 Data Aggregation and Routing in Wireless Networks

Wireless sensor networks (WSNs) are made up of numerous sensors with wireless communication (Hasan et al., 2012). The sensors are mostly ad hoc by nature, having a processor with limited computational capability, memory storage, and energy resource that allow seamless data communication between sensors distributed over a region. WSNs are employed for a wide range of applications such as surveillance, medical healthcare systems, environmental monitoring, and target tracking. Finite energy sources like batteries or ultra-capacitors are used to power these sensor nodes and are hardly replaced or recharged during lifetime. Energy-efficient use can only ensure long-time operation of the WSN. Therefore, energy-efficient strategies of routing data within the network are important for operation of the WSN. There are three types of sensor nodes defined in a WSN: source nodes, relay nodes, and a base station (BS). Each node can send data to neighboring nodes within its limited communication range. Whenever a source node needs to send data to the BS, it has to find a route formed by a set of relay nodes to transfer data to the BS. The relay nodes receive data from different sources or at the junction of different routes. The aggregation node gathers data, combines all of the data received, and then forwards the concise data to the next relay node or the BS, as shown in Figure 6.8.

In order to optimize the total energy consumption in the WSN, the number of data relayed in the network needs to be reduced, which means the number of hops on the routes should be minimized. Furthermore, the aggregation nodes need to be as close as possible to the source nodes in order to perform data aggregation at an early stage of transmission. The WSN now can be thought of as a directed graph defined by $G = \langle V, E \rangle$, where V is the set of sensor nodes with $|V| = n$, and $E = (v_i, v_j), i \neq j, \{v_i\} \in V, i,j = 1, 2, \ldots, n, d(v_i, v_j) \leq R$ denotes the set of edges connected to nodes in their mutual neighborhood, $d(v_i, v_j)$ is the Euclidean distance between nodes v_i and v_j, and R is the

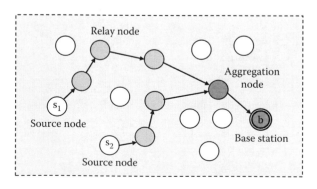

FIGURE 6.8 Data aggregation scheme in the WSN.

communication range of every node in V. The set $S = \{s_1, s_2, \ldots, s_m\}$, $S \subset V$ is the set of source nodes that collect data from the field of the WSN. The node b denotes the BS, which is the destination node for the data to reach. The optimization problem is to find a subset $G_{agg} = \langle V_{agg}, E_{agg} \rangle \subset G$, $V_{agg} \subset V$, which is the set of all the source nodes and selected relay nodes and $E_{agg} \subset E$, which is the set of edges that connect the nodes in V_{agg} and the BS in order to minimize the number of hops, defined as

$$\min \sum_{s \in S, v \in V} (h_{sv} + h_{vb}) \tag{6.66}$$

Here, h_{sv} and h_{vb} are the number of hop counts from source node s to aggregation node v and from the aggregation node v to the BS node b, respectively, such that $V_{agg} \supseteq S$. That means the objective of the optimization problem can be formulated as the minimization of the number of edges from source nodes to the destination node in the network, which is often called a data aggregation tree. The total energy consumption in the WSN can be minimized by constructing an optimal data aggregation tree. A data aggregation tree is constructed during a run. The total energy consumption of the WSN is defined by

$$E_{Tx} = E_{elec} \times k + \varepsilon_{amp} \times k \times d^2 \tag{6.67}$$

$$E_{Rx} = E_{elec} \times k \tag{6.68}$$

where E_{Tx} is the amount of energy consumed by a node to transmit data over a distance d, E_{Rx} is the energy consumption of the node, E_{elec} is the energy required to operate the transmitter or receiver electronic circuitry, ε_{amp} is the energy consumption for the transmitter amplifier, and k is the data length.

Hoang et al. (2012) applied the WDA to construct the optimal data aggregation tree in WSN. The computational experimental results show that the WDA is able to find an optimal data aggregation tree with a smaller number of hops that ensures the shortest path to the BS. Initially, the same amount of soil is assigned to every edge. An additional hop count term $h_d(j)$ is included into the definition of $f(soil(i,j))$ given by

$$f(soil(i, j)) = \frac{1}{\varepsilon_s + g(soil(i, j))} h_d^{-\beta}(j) \tag{6.69}$$

where $h_d(j)$ is the hop count from node j to destination and β is a parameter for determining the relative influence of $h_d(j)$. The value for β was set to 20. The heuristic undesirability function $HUD(i,j)$ for the optimal data aggregation problem is defined by

$$HUD_{ODAP}(i, j) = \sum_{k \in S} h_s(k, j) + h_d(j) \tag{6.70}$$

where $h_s(k, j)$ is the hop count from source node k to node j and $h_d(j)$ is the hop count from node j to the destination node. The edge soil update equation defined in Equation 6.24 is modified in the WSN application in order to eliminate an edge that moves the edge away from the BS. This is defined as

$$soil(i, j) = (1 - \rho_n) \cdot soil(i, j) - \rho_n (1 + h_d(i) - h_d(j)) \cdot \Delta soil(i, j) \tag{6.71}$$

The term $1 + h_d(i) - h_d(j)$ controls the edge moving away from the BS.

The initial settings of the parameters for all experiments are as follows: $a_s = 1$, $b_s = 0.1$, $c_s = 1$, $a_v = 1$, $b_v = 0.1$, $c_v = 1$, *InitSoil* = 10,000, and *InitVel* = 200. The local soil updating parameter ρ_n and global soil updating parameter ρ_{WD} are 0.9 and 0.9, respectively. The simulations were carried out for 500 runs. It is found that the WDA can find an optimal set of data aggregation nodes using 300 WDs (i.e., sensors).

6.8.14 MOBILE AD HOC NETWORKS (MANET)

MANETs work with no presetup infrastructure. Khaleel and Ahmed (2012) and Sensarma and Majumder (2013) suggested application of the WDA for optimization of the routing protocols for MANETs. The approach helps in conveying simple packets containing basic WD characteristics between adjacent nodes without knowledge of the entire network topology and the link status between nodes. These algorithms feature a small traffic overhead, adaptation to the dynamic topology, and support for redundant routing. The main feature of this new algorithm is that it calculates all possible relay node sets (RNSs) and then selects the set with the minimum number of hops using WDA. The algorithm demonstrates a good performance compared with other routing protocols.

6.8.15 JSSP

In JSSP, there are a set of jobs $J = \{J_i | i = 1, \ldots, n\}$ to be performed by a set of machines $M = \{M_j | j = 1, \ldots, m\}$. Each job has a sequence of operations $O = \{O_k | k = 1, \ldots, l\}$ to be completed. The n jobs have to be processed on m machines. The constraints are that splitting a job is not allowed, interruption of an operation is not permitted, each machine can perform an operation at a time, and each operation is performed only once on a machine and to a unique machine. JSSP has to find a feasible schedule of all the operations on the given machines with optimized objectives. Thus, JSSP is a typical NP-hard problem. Considering the application, there can be different objectives such as makespan, tardiness, and mean flow. Makespan C_{max} is the time interval between the time at which the schedule begins and the time at which the schedule ends. Thus, the makespan of a schedule is equal to $\max(C_i)$, $i = 1, \ldots, m$. The tardiness T_i of a job J_i is the nonnegative amount of time by which the completion time exceeds the due date d_i, $T_i = \max[0, (C_i - d_i)]$, that is, it is the difference between the completion time C_i and due date d_i for each job. The mean flow time $\bar{F}$ is the average flow time of a schedule defined as $\bar{F} = (1/n)\sum_{i=1}^{n} F_i$. This means that $\bar{F}$ is the average time of processing a single job.

Niu et al. (2012) proposed an improved WDA for JSSP. Niu et al. (2013) then applied WDA to a multiobjective JSSP. They conducted the experiments on the benchmark data set for JSSP available from the OR-Library. The makespan (C_{max}), tardiness (T_i), and the mean flow time ($\bar{F}$) are considered in the experiments in this research. They applied the original WDA with a standard setting for the parameters for all experiments as follows: $a_s = 1$, $b_s = 0.01$, $c_s = 1$, $a_v = 1$, $b_v = 0.01$, $c_v = 1$, *InitSoil* = 10,000, and *InitVel* = 200. The local soil updating parameter ρ_n and global soil updating parameter ρ_{WD} were 0.9 and 0.9, respectively. A set of optimal solutions were achieved with 50 WDs within 100 iterations. Three objectives were considered for the computation of Pareto optimal solutions using 10 good solutions.

6.8.16 WEB SERVICE SELECTION

A service is a sequence of activities provided by a service provider or system to a client as a solution to a problem. A client actually selects a set of services from a larger pool of services. Nowadays, web services, typically web applications, are very popular for almost everything in daily life among Internet users. Web services comprise a variety of software applications and components distributed over the Internet using some standard protocols, which have been the main key player in achieving service-oriented computing in recent years. With the increasing number of web services, there are many alternatives of meeting users' needs in web service. Therefore, quality of service (QoS) is a decisive factor for selecting the most suitable service from a set of services with similar functionality. For measuring the QoS, Palanikkumar et al. (2012) applied the WDA to web service selection and composition. They compared the performance of the WDA in identifying the best web service composition and compared it to that of the PSO algorithm. Based on the experimental results, the performance of the WDA proved to supersede that of PSO. It is shown that the WDA exhibits higher accuracy, feasibility, and effectiveness than the PSO algorithm.

6.8.17 MAX-CLIQUE PROBLEM

The maximum-clique problem (MCP) is a classical graph theory problem, which is considered as an NP-hard optimization problem. A clique C of a graph G is a subset of V such that $G(C)$ is complete. The graph G is an arbitrary directed graph defined by $G = \langle V, E \rangle$, where $V = \{v_1, \ldots, v_n\}$ is the set of vertices, $E \subseteq V \times V$, $E = \{(v_i, v_j)\}$, $i \neq j$, $\{v_i\} \in V$, $i,j = 1,2, \ldots, n$ is the set of edges in G. A positive weight w_i is associated with each vertex $v_i \in V$. $A_G = [a_{ij}]_{n \times n}$ is an $n \times n$ adjacency matrix of graph G where $a_{ij} = 1$ if $(v_i, v_j) \in E$ is an edge and $a_{ij} = 0$ if $(v_i, v_j) \notin E$. For a subset $S \subseteq V$ the weight of S is defined as $W(S) = \sum_{v_i \in S} w_i$. The subgraph induced by S is defined as $G(S) = (S, E \cap S \times S)$. Thus, $G(C)$ is complete if all its vertices are pairwise adjacent. The maximum clique then requires $G(C)$ to be of maximum weight (Pardalos and Xue, 1994). The MCP has many formulations, but the simplest is defined as follows:

$$\max \sum_{v_i \in C} w_i x_i \quad \text{with} \quad x_i \in \{0,1\} \tag{6.72}$$

The MCP has many applications in various domains including project selection, classification, fault tolerance, coding, computer vision, economics, information retrieval, and signal transmission. The application of the WDA requires the problem at hand to be represented in the form of a graph. In this case, the WDA can easily be applied to the MCP without any further effort. Al-Taani and Nemrawi (2012) proposed a WDA to solve the MCP, calling it WD-Clique. They used available benchmark graph to test the algorithm. The initial soil and velocity are set to *InitSoil* = 10,000 and *InitVel* = 20, respectively. The maximum number of WDs and iterations was 200 and 100, respectively. It is shown that the performance of the WD-Clique algorithm mainly depends on its variables, the number of drops, and the maximum number of iterations based on the conducted experiment. It is also observed that the execution time spent by the algorithm increases, being directly proportional to the increased value of the variables. The results of the WD-Clique algorithm were compared with Ant Clique, and genetic algorithms, which indicated that the WD Clique results are the best.

6.8.18 RESERVOIR OPERATION PROBLEM

The reservoir operation problem (ROP) is considered as an NP-hard optimization problem. Finding an exact solution within an acceptable finite time is difficult. Therefore, the ROP has been a test problem for many meta-heuristic algorithms. Dariane and Sarani (2013) were the first to apply the WDA to an ROP. Data from the Dez reservoir are used to evaluate the algorithm. The objective function is defined in the form of a loss function as the minimization of the sum of the squared water deficit.

$$\text{Min } Z = \sum_{i=1}^{T} (\text{Deficit}_t)^2 \tag{6.73}$$

$$\text{Deficit}_t = \begin{cases} (TD_t - R_t) & \text{if } R_t < TD_t \\ 0 & \text{otherwise} \end{cases} \tag{6.74}$$

where Z is the loss function. The objective function is one sided and only calculates the damages related to deficits. The problem is subject to the following constraints.
For all

$$t \begin{cases} S_{t+1} = S_t + Q_t - R_t - E_t \\ S_{\min} \leq S_t \leq S_{\max} \\ S_1 = S_{T+1} \\ R_t \geq R_{\min} \end{cases} \tag{6.75}$$

where S_t, Q_t, R_t, E_t, and TD_t are reservoir storage, inflow, release, evaporation, and target demand at time t, respectively. The meta-heuristic undesirability function $HUD(i,j)$ is defined by

$$HUD_t^{ROP}(i, j) = (R_t(i, j) - TD_t) \tag{6.76}$$

The WDA was applied to the ROP with standard parameter settings. The results obtained using the WDA validate those found by earlier researchers.

6.8.19 DATA CLUSTERING

Clustering is the task of assigning a set of objects into homogeneous groups (called clusters) so that the objects in the same cluster are more similar to each other than to those in other clusters based on a similarity metric. In general, clustering algorithms aim to minimize within-cluster variation (i.e., intracluster distance) and maximize the between-cluster variation (i.e., intercluster distance). Some mathematical description of clustering has been discussed in Chapter 2 in Equations 2.93 through 2.96. The K-means clustering algorithm is a partitioning clustering algorithm and also known as the generalized Lloyd algorithm. In K-means clustering, the number of clusters K is given, K cluster centers are randomly chosen from the data set or randomly generated, and the algorithm partitions the data set into K clusters by minimizing an objective function.

Shah-Hosseini (2013) applied a modified WDA to solve a K-means clustering problem. In the proposed approach, data vectors $X = \{x_1, x_2, \ldots, x_M\}$ is to be partitioned into K clusters. A graph with $M + 1$ nodes and $K + M$ edges is constructed to be used by the WDA. The WDA–K-means algorithm is described by the following steps:

Step 1: Select edge and update local soil.
Step 2: Calculate cluster center.
Step 3: Apply local chaotic search.
Step 4: Calculate cluster center.
Step 5: Update global soil.
Step 6: If (not termination condition), go to Step 1.
Step 7: Return solution.

The WDA begins a trip starting from node i and ends at visiting node $M + 1$. A directed $edge(i, i + 1)(k)$ connecting node i to node $i + 1$ is devoted to cluster k. Since there are K clusters, there will be directed K edges. The cluster center calculation is performed as follows.

When all WDs have completed their solutions $T^{WD} = \left\{t_1^{WD}, t_2^{WD}, \ldots, t_M^{WD}\right\}$ with M elements, its corresponding cluster centers m_k^{WD} with $k = 1, 2, \ldots, K$ are computed. Here m_k^{WD} is defined as

$$m_k^{WD} = \frac{1}{\left|J_k^{WD}\right|} \sum_{j \in J_k^{WD}} x_j \tag{6.77}$$

where $J_j^{WD} = \left\{j \mid t_j^{WD} = k\right\}$.

Several well-known data sets examined the modified algorithm for clustering. Comparing its performance with the K-means algorithm, the results of the proposed algorithm are superior.

6.8.20 STEINER TREE PROBLEM

The Steiner tree problem (also known as the motorway problem or the minimum Steiner tree problem) is to find the shortest interconnect for a given set of points. It is named after Swiss mathematician Jacob Steiner who introduced the problem in the 1830s. It is basically a minimum spanning tree problem. The problem is to construct a graph $G = (V, E)$ of the shortest length by connecting the vertices (or nodes) of a given set of vertices V where the length is the sum of the lengths of all edges E. In the Steiner tree problem, extra intermediate vertices and edges can be added to the graph G to reduce the length of the spanning tree. These vertices are known as Steiner points or vertices. The resulting graph $G = (V', E')$ is the Steiner tree. The Steiner tree consists of two prominent cases: the Euclidean Steiner tree and rectilinear Steiner tree (Hougardy and Promel, 1999). The problem is to find the shortest tree connecting the two given nodes. It is a combinatorial optimization problem, which is considered as an NP-hard problem. There have been a lot of works done to find an algorithm or heuristic to solve the Steiner tree problem. The GA (Kapsalis et al., 1993), PSO (Ma and Liu, 2010), and ACO (Tashakori et al., 2004) have been applied successfully to the Steiner tree problem.

Noferesti and Shah-Hosseini (2012) proposed a hybrid algorithm combining the modified WDA and learning automata to solve the Steiner tree problem. The candidate Steiner points are numbered from 1 to m, where m is the number of nonterminal nodes. A graph $G = (V, E)$ is constructed where V denotes the candidate Steiner nodes, and E denotes directed edges between nodes. The number of WDs is set equal to the number of nonterminal nodes. Each WD constructs a solution by selecting some of the candidate Steiner nodes. The probability of selecting the next node is based on Equation 6.19, the velocity is updated using Equation 6.22, the amount of soil is calculated using Equations 6.24 and 6.25, soil on the path of the best-iteration WD is updated using Equation 6.27, and the total-best solution is updated using Equation 6.28. The algorithm is repeated for the prescribed number of times, and the global best solution is chosen as the minimum Steiner tree. Noferesti and Shah-Hosseini extended the WDA for solving the Steiner tree problem, whereby, at the end of each iteration, prescribed numbers of the best WDs (called elitist WDs) will perform global soil updating. In addressing the premature convergence problem, when all elitist WDs produce the same tree, the soil amount on all edges of the graph is set to the initial value. Global soil updating is performed for the best resulting tree up to the current iteration. Furthermore, the performance of the extended WDA for solving the Steiner tree problem is improved by hybridizing it with Krylov learning automata for parameter adaptation. The extended WDA has been verified on the B-problem set and the C-problem set of the OR-Library. Experimental results on the C-problem set showed that the hybrid algorithm achieved better results.

6.8.21 OTHER APPLICATIONS

There have been many other applications of the WDA dispersedly reported in the literature. Agarwal et al. (2012) proposed the use of WDA to optimize the code coverage by generating automated test sequences and determine the software quality. Aldeeb (2014a,b) applied WDA to the time tabling problem.

6.9   CONCLUSION

The WDA is a very recent nature-inspired meta-heuristic optimization algorithm mimicking the features of natural WDs and has been very successful in applications. The salient features of the WDA can be stated as follows:

1. It offers good-quality solutions based on average values.
2. WDA has fast convergence in comparison to other methods.
3. It works well in the dynamic environment.

It is efficient to address the problem with respect to its global search and fast convergence ability. It appears that due to the nature of the algorithm, the problem needs to be transformed into a graph for the WDA to be applied. This seems a disadvantage for the WDA for application to a wide range of problem domains, for example function optimization. But WDA can be applied without much effort and computation to problems that can be represented by a graph. The representation of a problem into a graph-like structure demands extra computation for WDA.

The WDA requires several parameters to be chosen and tuned for optimal performance and convergence of the algorithm (Palanikkumar et al., 2012). A good number of local search methods and heuristics are employed to get the best performance of the WDA. There is still open space and scope for further development of the algorithm.

REFERENCES

Agarwal, K., Goyal, M., and Srivastava, P. R. 2012. Code coverage using intelligent water drop (IWD), *International Journal of Bio-Inspired Computation*, 4(6), 392–402.

Allan, J. D. 1995. Channels and flow, the transport of materials, in: *Stream Ecology, Structure and Function of Running Waters*, London: Chapman & Hall Publishing.

Al-Betar, M. A., Doush, I. A., Khader, A. T., and Awadallah, M. A. 2012. Novel selection schemes for harmony search, *Applied Mathematics and Computation*, 218(10), 6095–6117.

Al-Betar, M. A., Khader, A. T., Geem, Z. W., Doush, I. A., and Awadallah, M. A. 2013. An analysis of selection methods in memory consideration for harmony search, *Applied Mathematics and Computation*, 219(22), 10753–10767.

Aldeeb, B. A., Al-Betar, M. A., and Norita, M. N. 2014a. Intelligent water drops algorithm for university examination timetabling, *International Parallel Conferences on Researches in Industrial and Applied Sciences*, Dubai, UAE, pp. 18–30.

Al Deeb, B. A. M., Norwawi, N. M., and Al-Betar, M. A. 2014b. A survey on intelligent water drop algorithm, *International Journal of Computers and technology*, 13(10), 5075–5084.

Al-Taani, A. and Nemrawi, M. K. 2012. Solving the maximum clique problem using intelligent water drops algorithm, *The International Conference on Computing, Networking and Digital Technologies*, Bahrain, pp. 142–151.

Alijla, B. O., Lim, C. P., Khader, A. T., and Al-Betar, M. A. 2013. Intelligent water drops algorithm for rough set feature selection, in: A. Selamat et al. eds., *Intelligent Information and Database Systems, Lecture Notes in Computer Science*, Vol. 7803, Berlin: Springer Verlag, pp. 356–365.

Alijla, B. O., Wong, L.-P., Lim, C. P., Khader, A. T., and Al-Betar, M. A. 2014. A modified intelligent water drops algorithm and its application to optimization problems, *Expert Systems with Applications*, 41, 6555–6569.

Clark, G. and Wright, J. W. 1964. Scheduling of vehicles from a central depot to a number of delivery points, *Operations Research*, 12, 568–581.

Dantzig, G. B. and Ramser, J. H. 1959. The truck dispatching problem, *Management Science*, 6(1), 80–91.

Dariane, A. B. and Sarani, S. 2013. Application of intelligent water drop algorithm in reservoir operation, *Water Resources Management*, 27, 4827–4843.

Duan, H., Liu, S., and Lei, X. 2008. Air robot path planning based on Intelligent Water Drops optimization, *IEEE International Joint Conference on Neural Networks (IJCNN2008), IEEE World Congress on Computational Intelligence*, Hong Kong, China, pp. 1397–1401.

Duan, H., Liu, S., and Wu, J. 2009. Novel intelligent water drops optimization approach to single UCAV smooth trajectory planning, *Aerospace Science and Technology*, 13, 442–449.

Flood, M. M. 1955. The travelling salesman problem, *Journal of Operations Research Society of America*, 4, 61–75.

Hasan, S. F., Siddique, N., and Chakraborty, S. 2012. *Intelligent Transport Systems: 802.11-Based Roadside-to-Vehicle Communications*, New York: Springer.

Hendrawan, Y. and Murase, H. 2011. Neural-intelligent water drops algorithm to select relevant textural features for developing precision irrigation system using machine vision, *Computers and Electronics in Agriculture*, 77(2), 214–228.

Hoang, D. C., Kumar, R., and Panda, S. K. 2012. Optimal data aggregation tree in wireless sensor networks based on intelligent water drops algorithm, *IET Wireless Sensor Systems*, 2(3), 282–292.

Hougardy, S. and Promel, H. J. 1999. A 1.598 approximation algorithm for the Steiner problem in graphs, *The Tenth Annual ACM-SIAM Symposium on Discrete Algorithms*, Baltimore, USA, pp. 448–453.

Jin, F. H., Hong, B. R., and Gao, Q. J. 2002. Path planning for free-flying space Robot using ant algorithm, *Robot*, 24(6), 526–529.

Kamkar, I., Akbarzadeh-T, M.-R., and Yaghoobi, M. 2010. Intelligent water drops a new optimization algorithm for solving the vehicle routing problem, *Proceedings of the 2010 IEEE Conference on System, Man and Cybernetics*, pp. 4142–4146.

Kapsalis, A., Rayward-smith, V. J., and Smith, J. D. 1993. Solving the graphical Steiner tree problem using genetic algorithms, *Journal of the Operational Research Society*, 44, 397–406.

Keller, H., Pferschy, U., and Pisinger, D. 2004. *Knapsack Problems*, New York: Springer.

Kesavamoorthy, R., Shunmugam, D. A., and Mariappan, L. T. 2011. Solving traveling salesman problem by modified intelligent water drop algorithm, *International Conference on Emerging Technology Trends (ICETT), Proceedings published by International Journal of Computer Applications*, Number 2, Article 5, pp. 18–23.

Khaleel, T. A. and Ahmed, M. Y. 2012. Using intelligent water drops algorithm for optimisation routing protocol in mobile ad-hoc networks, *International Journal of Reasoning-Based Intelligent Systems*, 4(4), 227–234.

Kondolf, G. M., Smeltzer, M., and Kimball, L. 2002. *Freshwater Gravel Mining and Dredging Issues*, Prepared for Washington Department of Fish & Wildlife, Department of Ecology, and Department of Transportation.

Lenin, K. and Kalavathi, M. S. 2012. An intelligent water drop algorithm for solving optimal reactive power dispatch problem, *International Journal on Electrical Engineering and Informatics*, 4(3), 450–462.

Lee, K. Y., Paru, Y. M., and Oritz, J. L. 1985. A united approach to optimal real and reactive power dispatch, *IEEE Transactions on Power Apparatus and Systems*, PAS-104, 1147–1153.

Ma, X. and Liu, Q. 2010. A particle swarm optimization for Steiner tree problem, *Sixth International Conference on Natural Computation*, ICNC 2010, pp. 2561–2565.

Morisawa, M. 1968. *Streams: Their Dynamics and Morphology*, New York: McGraw-Hill.

Msallam, M. M. and Hamdan, M. 2011. Improved intelligent water drops algorithm using adaptive schema, *International Journal of Bio-Inspired Computation*, 3(2), 103–111.

Nagalakshmi, P., Harish, Y., Kranthi, K. R., and Chakravarthi, J. 2011. Combined economic and emission dispatch using intelligent water drops-continuous optimization algorithm, *2011 International Conference on Recent Advancements in Electrical, Electronics and Control Engineering*, pp. 168–173.

Noferesti, S. and Shah-Hosseini, H. 2012. A hybrid algorithm for solving Steiner tree problems, *International Journal of Computer Applications*, 41(5), 14–20.

Niu, S. H., Onq, S. K., and Nee, A. Y. C. 2012. An improved intelligent water drops algorithm for achieving optimal job shop scheduling solutions, *International Journal of Production Research*, 50(15), 4192–4205.

Niu, S. H., Ong, S. K., and Nee, A. Y. C. 2013. An improved intelligent water drops algorithm for solving multi-objective job shop scheduling, *Engineering Applications of Artificial Intelligence*, 26, 2431–2442.

Palanikkumar, D., Gowsalya, E., Rithu, B., and Anbuselven, P. 2012. An intelligent water drops algorithm based service selection and composition in service oriented architecture, *Journal of Theoretical and Applied Information Technology*, 39(1), 45–51.

Pardalos, P. M. and Xue, J. 1994. The maximum clique problem, *Journal of Global Optimisation*, 4, 301–328.

Rayapudi, S. R. 2011. An intelligent water drop algorithm for solving economic load dispatch problem, *International Journal of Electrical and Electronics Engineering*, 5(1), 43–49.

Robert, A. 2003. *River Process: An Introduction to Fluvial Dynamics*, London, UK: Routledge.

Sensarma, D. and Majumder, K. 2013. IWDRA: An intelligent water drop based QoS-aware routing algorithm for MANETs, *Proceedings of the International Conference on Frontiers of Intelligent Computing: Theory and Applications (FICTA)*, Odisha, India, pp. 329–336.

Shah-Hosseini, H. 2007. Problem solving by intelligent water drops, *Proceedings of IEEE Congress on Evolutionary Computation (CEC 2007), Swissotel The Stamford*, Singapore, September, pp. 3226–3231.

Shah-Hosseini, H. 2008. Intelligent water drops algorithm – A new optimisation method for solving the multiple knapsack problem, *International Journal of Intelligent Computing and Cybernetics*, 1(2), 193–212.

Shah-Hosseini, H. 2009a. Intelligent water drops algorithm – A nature-inspired Swarm-based optimisation algorithm, *International Journal of Bio-Inspired Computation*, 1(1–2), 71–79.

Shah-Hosseini, H. 2009b. Optimization with the nature-inspired intelligent water drops algorithm, in: Wellington Pinheiro dos Santos ed., *Evolutionary Computation*, Vienna, Austria: In-Tech, p. 572.

Shah-Hosseini, H. 2012. An approach to continuous optimisation by the intelligent water drops algorithm, *Procedia Social and Behavioural Science*, 32, 224–229.

Shah-Hosseini, H. 2013. Improving K-means clustering algorithm with the intelligent water drops (IWD) algorithm, *International Journal of Data Mining, Modeling and Management*, 5(4), 301–317.

Sur, C., Sharma, S., and Shukla, A. 2013. Multi-objective adaptive intelligent water drops algorithm for optimization & vehicle guidance in road graph network, *2013 International Conference on Informatics, Electronics & Vision (ICIEV)*, May 17–18, 2013, Dhaka, pp. 1–6.

Tashakori, M., Adibi, P., Jahanian, A., and Norallah, A. 2004. Solving dynamic Steiner tree problem by using ant colony system, *9th Annual Conference of Computer Society of Iran*, Tehran, Iran.

Wang, G., Guo, L., Duan, H., and Liu, L. 2012. Path planning for uninhabited combat aerial vehicle using hybrid meta-heuristic DE/BBO algorithm, *Advanced Science, Engineering and Medicine*, 4, 550–564.

Watkins, J. 2004. *Across the Board: The Mathematics of Chessboard Problems*, Princeton, NJ: Princeton University Press.

Wood, A. J. and Wollenberg, B. F. 1994. *Power Generation, Operation and Control*, New York: John Wiley.

Yao, X. and Liu, Y. 1996. Fast evolutionary programming, in: L. J. Fogel, P. J. Angeline, and T. Back eds., *Proceedings of the 5th Annual Conference on Evolutionary Programming*, San Diego, CA: MIT Press, pp. 451–460.

7 Spiral Dynamics Algorithms

7.1 INTRODUCTION

Most recently, a new multipoint meta-heuristics search method has emerged for 2-D continuous optimization problems based on the analogy of spiral phenomena in nature, called 2-D spiral optimization first proposed by Tamura and Yasuda in 2010. Focused spiral phenomena are approximated to logarithmic spirals, which frequently appear in nature, such as whirling currents, low-pressure fronts, nautilus shells, and arms of spiral galaxies. A remarkable point about logarithmic spirals is that their discrete processes generating spirals can realize effective behavior in meta-heuristics.

The spiral optimization is a multipoint search for continuous optimization problems. The spiral optimization model is composed of plural logarithmic spiral models and their common center. The spiral model has two specific setting parameters: the convergence rate and the rotation rate whose values characterize its trajectory. The common center is defined as the best point in all search points. The search points moving toward the common center with logarithmic spiral trajectories can find better solutions and update the common center.

This chapter will investigate the spiral phenomena in nature and will attempt to explain and describe the features mathematically. The best way to explain is to describe them in terms of a set of parameters so that the curves representing spirals or certain features of spirals can be manipulated in a convenient and befitting manner to fit certain real-life applications. With this pursuit in mind, different forms of parametric curves, spiral models, 2-D spiral models, and *n*-dimensional spiral models are discussed. As a final attempt, spiral dynamics (SpD)-based optimization algorithms will be discussed along with its different variants and parameter sets reported in the literature. An important aspect of the optimization algorithm is its stability, which needs careful analysis for deployment in various domains such as its effectiveness as an optimization technique and applications to engineering problems.

7.2 SPIRAL PHENOMENA IN NATURE

The spirals are mysterious and resonate with the human spirit. They are complex, intriguing, and beautiful. The spiral pattern is found extensively in nature—encoded into plants, animals, humans, the earth, and galaxies around us. Mathematics can explain the complex algorithms that make up a spiral pattern, but it cannot explain the lure and fascination of the spiral to the human. Mathematicians from ancient times to modern days have shown interest in spirals and have been making efforts to explain the complexities, sequences, equations, and algorithms that make up a spiral pattern. The spiral phenomena are approximated to logarithmic spirals, which are seen frequently in the nature, such as a whirling current, a low-pressure front, a nautilus shell, and arms of spiral galaxies.

Frequently encountered examples such as spiral galaxies, tropical cyclone, tornado, millipede, vine tendrils, and Romanesco broccoli are shown in Figure 7.1. A remarkable point about logarithmic spirals is that their discrete processes generating spirals can realize effective search behavior in meta-heuristics.

7.3 PARAMETRIC REPRESENTATION OF CURVES

Spiral phenomena in nature can be approximated to curves and describe mathematically. Mathematically, a spiral is a curve that emanates from a central point, getting progressively

(a)   (b)   (c)

(d)   (e)   (f)

FIGURE 7.1   Spiral shapes in nature. (a) Galaxies, (b) tropical cyclone, (c) tornado, (d) millipede, (e) vine tendrils, and (f) Romanesco broccoli. (Courtesy of Google.)

farther away as it revolves around the point (Bromstein and Semendyayev, 1998). If $F(x,y) = 0$ is an equation in the coordinate system S, which can be satisfied by the set of ordered pairs $M = \{(a,b) \in \Re \,|\, F(a,b) = 0\}$ of real numbers, then the set $K = \{P(a,b)|(a,b) \in M\}$ of all points P in the plane having the coordinates a and b in S is called a curve defined by the equation $F(x,y) = 0$ in S. If S is a Cartesian coordinate system, then K is the graph of the function $y = f(x)$. The equation of a curve depends not only on the curve itself but also on the coordinate system. A curve K is called algebraic of order n if there is a Cartesian coordinate system S and a polynomial $F(x,y)$ in x and y of total degree n, such that $F(x,y) = 0$ is the equation of K in S. If S is a polar coordinate system, then one restricts K as a function of $\zeta = f(\varphi)$. If $x = x(t)$ and $y = y(t)$ are two functions defined within the same interval and if S is a Cartesian coordinate system, then the curve $K = \{P(x(t), y(t))|t \in I\}$ is called a parametric representation of the two functions.

7.3.1 PARAMETRIC REPRESENTATION OF SPIRALS

There are plenty of examples of spirals in the nature. Scientists, mathematicians, physicists, and biologists are trying to reveal the mysteries of spirals in the nature, explain the complexities, and describe them mathematically (Costa and Martinez, 1995). As the first attempt, different forms of spiral are represented by parametric curves, an initiative which started many years ago. Some of them are discussed in this section.

Archimedean spiral is named after the Greek mathematician Archimedes. An Archimedean spiral is shown in Figure 7.2. This spiral was studied by Conon, and later by Archimedes about 225 BC. Archimedes was able to work out the lengths of various tangents to the spiral. The curvatures are traced out by a point that moves at a constant velocity v along a line (rod) that is rotating about the origin at a constant angular velocity ω. The Archimedean spiral can also be used to describe the group of spirals given by the general equation:

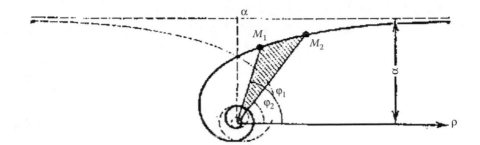

FIGURE 7.2 Archimedean spiral.

$$\rho = a\varphi^x \qquad (7.1)$$

where $a = v/\omega$ and $-\infty < \varphi < \infty$. Two symmetrical branches (with respect to y-axis) ranging from $-\infty < \varphi \leq 0$ and $0 \leq \varphi < \infty$ form the spiral and can be distinguished (as shown in Figure 7.2). The two branches together form a cusp at the origin. Every ray going out of the origin intersects each of the two branches at a constant of $2\pi a$. The arch length s is defined by

$$s(\varphi) = \frac{1}{2}a\left(\varphi\sqrt{[\varphi^2 + 1]} + \text{arcsinh}(\varphi)\right) \qquad (7.2)$$

$$s(\varphi) = \frac{1}{2}a\left[\varphi\sqrt{[\varphi^2 + 1]} + \ln\left(\varphi + \sqrt{[\varphi^2 + 1]}\right)\right] \qquad (7.3)$$

where $\lim(2s/a\varphi^2) = 1$, which is useful for approximating the arch length s for large values of φ. $s(\varphi)$ can be expressed by a series expansion as follows:

$$s(\varphi) = a\left\{\varphi + \frac{1}{2}\sum_{k=3}^{\infty}\left[P_{n-3}(0) + \frac{n+1}{n}P_{n-1}(0)\right]\varphi^k\right\} \qquad (7.4)$$

$$s(\varphi) = a\left(\varphi + \frac{1}{6}\varphi^3 - \frac{1}{40}\varphi^5 + \frac{1}{112}\varphi^7 - \frac{5}{1152}\varphi^9 + \cdots\right) \qquad (7.5)$$

where $P_n(\cdot)$ is a Legendre polynomial. The radius of the curvature r is defined by

$$r(\varphi) = \frac{a(\varphi^3 + 1)^{3/2}}{(\varphi^2 + 2)} \qquad (7.6)$$

The term Archimedean spiral is now used to describe a group of spirals. These are

- Archimedes or arithmetic spiral
- Hyperbolic spiral
- Parabolic or Fermat spiral
- Lituus spiral

These Archimedean spirals are discussed in further details in the following sections.

7.3.1.1 Archimedes or Arithmetic Spiral

Setting $x = 1$ in Equation 7.1, the Archimedes or arithmetic spiral is obtained. The mathematical description of Archimedes spiral in polar coordinates is expressed as

$$\rho = a\varphi \tag{7.7}$$

where $a = v/\omega$ and $-\infty < \varphi < \infty$. Two symmetrical branches (with respect to y-axis) ranging from $-\infty < \varphi \leq 0$ (Figure 7.3b) and $0 \leq \varphi < \infty$ (Figure 7.3a) form the spiral and can be distinguished. The two branches, shown separately in Figure 7.3, together form a cusp at the origin. The radius increases by a fixed amount for each 2π turn. Equation 7.7 can be translated onto Cartesian coordinate as follows:

$$\begin{cases} x = \rho\cos(\varphi) = a\varphi\cos(\varphi) \\ y = \rho\sin(\varphi) = a\varphi\sin(\varphi) \end{cases} \tag{7.8}$$

Archimedes' spiral can be used for compass and straightedge division of an angle into n parts (including angle trisection) and can also be used for circle squaring (Steinhaus, 1999).

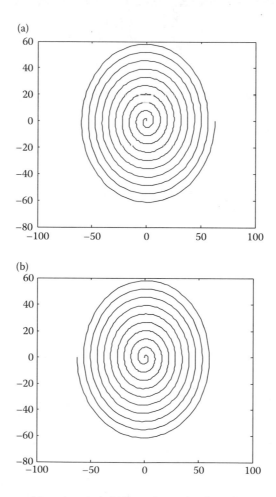

FIGURE 7.3   Archimedes or arithmetic spiral. (a) Branch ranging from $0 \leq \varphi < \infty$. (b) Branch ranging from $-\infty < \varphi \leq 0$.

7.3.1.2 Hyperbolic Spirals

For $x = 1$ in Equation 7.1, it is the Archimedean or arithmetic spiral discussed in the above section. For $x = -1$, it is hyperbolic spiral. That is, a plane transcendental curve in polar coordinates is described by the equation

$$\rho = \frac{a}{\varphi} \tag{7.9}$$

where $a = v/\omega$ and $-\infty < \varphi < \infty$. It consists of two branches, which are symmetrical (with respect to y-axis) ranging from $-\infty < \varphi \leq 0$ (Figure 7.4a) and $0 \leq \varphi < \infty$ (Figure 7.4b). The two branches shown separately in Figure 7.4, together form a cusp at the origin. Equation 7.9 can be translated onto Cartesian coordinate as follows:

$$\begin{cases} x = \rho\cos(\varphi) = \dfrac{a}{\varphi}\cos(\varphi) \\[2mm] y = \rho\sin(\varphi) = \dfrac{a}{\varphi}\sin(\varphi) \end{cases} \tag{7.10}$$

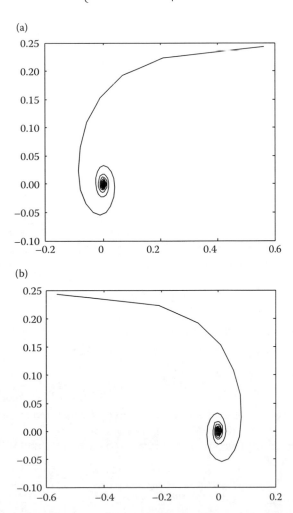

FIGURE 7.4 Hyperbolic spiral. (a) Branch ranging from $-\infty < \varphi \leq 0$. (b) Branch ranging from $0 \leq \varphi < \infty$.

The pole is an asymptotic point. The asymptote is the straight line parallel to the polar axis at a distance α from it. The arc length between two points, say $M_1(\rho_1,\phi_1)$ and $M_2(\rho_2,\phi_2)$, is defined as

$$l = \alpha\left[-\frac{\sqrt{1+\phi^2}}{\phi} + \ln\left(\phi + \sqrt{1+\phi^2}\right)\right]_{\phi_1}^{\phi_2} \tag{7.11}$$

The area of the sector bounded by an arc of the hyperbolic spiral and the two radius vectors ρ_1 and ρ_2 corresponding to the angles ϕ_1 and ϕ_2 is

$$S = \frac{\alpha^2(\rho_1 - \rho_2)}{2} \tag{7.12}$$

A hyperbolic spiral and an Archimedean spiral may be obtained from each other by inversion with respect to the pole O of the hyperbolic spiral. Using the representation of the hyperbolic spiral in polar coordinates, the curvature can be found by

$$k = \frac{r^2 + 2r_\theta^2 - rr_{\theta\theta}}{(r^2 + r_\theta^2)^{3/2}} \tag{7.13}$$

$$r_\theta = \frac{dr}{d\theta} = \frac{-a}{\theta^2} \tag{7.14}$$

$$r_{\theta\theta} = \frac{d^2r}{d\theta^2} = \frac{2a}{\theta^3} \tag{7.15}$$

Then the curvature at θ reduces to

$$k(\theta) = \frac{\theta^4}{\alpha(\theta^2 + 1)^{3/2}} \tag{7.16}$$

The curvature tends to infinity as θ tends to infinity. For values of $\theta = [0,1]$, the curvature increases exponentially, and for values of $\theta > 1$, the curvature increases at an approximately linear rate with respect to the angle. The tangential angle of the hyperbolic curve is

$$\phi(\theta) = -\tan^{-1}\theta \tag{7.17}$$

7.3.1.3 Farmat's or Parabolic Spiral

For $x = 1/2$, Equation 7.1 becomes a Fermat's spiral. That is, Fermat's spiral is defined by

$$\rho = a\phi^{1/2} = a\sqrt{\phi} \tag{7.18}$$

where $a = v/\omega$ and $-\infty < \phi < \infty$. It consists of two branches, which are symmetrical (with respect to y-axis) ranging from $-\infty < \phi \leq 0$ (Figure 7.5a) and $0 \leq \phi < \infty$ (Figure 7.5b). The two branches, shown separately in Figure 7.5, together form a cusp at the origin. Equation 7.18 can be translated onto Cartesian coordinate as follows:

$$\begin{cases} x = \rho\cos(\phi) = a\sqrt{\phi}\cos(\phi) \\ y = \rho\sin(\phi) = a\sqrt{\phi}\sin(\phi) \end{cases} \tag{7.19}$$

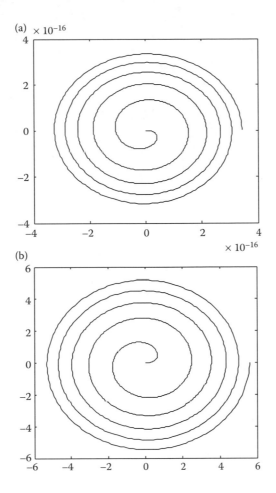

FIGURE 7.5 Farmat's or parabolic spiral. (a) Branch ranging from $-\infty < \varphi \leq 0$. (b) Branch ranging from $0 \leq \varphi < \infty$.

7.3.1.4 Lituus Spiral

The lituus curve was named for the ancient Roman lituus. The spiral curve was first described by English mathematician Roger Cotes in a collection of papers entitled *Harmonia Mensurarum*, which was published by the Scottish mathematician Colin Maclaurin in 1722 after the death of Roger Cotes. Lituus is Latin for an augur's staff and was later used for a *crook*, such as a bishop's crosier. The lituus is the locus of a point that moves in such a manner that the area of a circular sector remains constant. Mathematically, a lituus is a spiral in which the angle is inversely proportional to the square of the radius as expressed in polar coordinates. Equation 7.1 becomes a lituus spiral by setting $x = -1/2$. That is, lituus spiral has the polar equation as follows:

$$\rho = a\varphi^{-1/2} = \frac{a}{\sqrt{\varphi}} \tag{7.20}$$

where $a = v/\omega$ and $-\infty < \varphi < \infty$. It consists of two branches, which are symmetrical (with respect to y-axis) ranging from $-\infty < \varphi \leq 0$ (Figure 7.6a) and $0 \leq \varphi < \infty$ (Figure 7.6b). The two branches together form a cusp at the origin. Equation 7.20 can be translated onto Cartesian coordinate as follows:

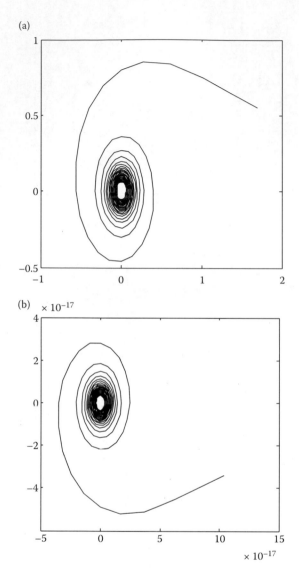

FIGURE 7.6   Lituus spiral. (a) Branch ranging from $-\infty < \varphi \le 0$. (b) Branch ranging from $0 \le \varphi < \infty$.

$$\begin{cases} x = \rho \cos(\varphi) = \dfrac{a}{\sqrt{\varphi}} \cos(\varphi) \\[3mm] y = \rho \sin(\varphi) = \dfrac{a}{\sqrt{\varphi}} \sin(\varphi) \end{cases} \tag{7.21}$$

As with Fermat's spiral, the lituus has two branches but typically only the positive branch is shown (Figure 7.6a and b).

7.3.1.5   Clothoid Spirals

Italian mathematician Ernesto Cesàro, at the beginning of the twentieth century, gave the name "clothoid." Clothoid was first described by Euler and then further studied by Alfred Cornu and

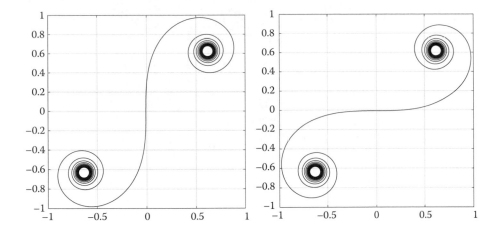

FIGURE 7.7 Euler–Cornu–Cesaro clothoid.

Ernesto Cesàro. An Euler–Cornu–Cesàro clothoid is shown in Figure 7.7. These curves have a point of origin with property that the curvature (the radius of curvature) of $1/r$ at every point M on the curve is proportional to the length of the arc on the curve, that is, $1/r = l/a^2$, where l is the length of the arc and $1/a^2$ is the factor of proportionality. The first major application was found in optics, and it has many applications in engineering. Details of treatments of clothoids can be found in Gray (1997) and Venema (2005).

7.4 LOGARITHMIC SPIRALS

Spiral phenomena are often found in nature such as nautilus shell (shown in Figure 7.8) or spider webs (shown in Figure 7.9). Such spirals can be approximated mathematically by parametric curves called logarithmic spiral. The logarithmic spiral was first investigated by Rene Descartes and later extensively studied by Jacob Bernoulli. Bernoulli called it *Spira mirabilis* meaning the miraculous spiral (Darling, 2004). Logarithmic spirals are also known as exponential spiral, geometric spiral, golden spiral, equiangular spiral, Bernoulli's spiral (named after Jakob Bernoulli), the Pheidian

FIGURE 7.8 Logarithmic spiral of nautilus shell. (From Google image.)

FIGURE 7.9   Logarithmic spiral of spider web. (From Google image.)

spiral, Fibonacci spiral, or growth spiral. The logarithmic spiral is defined by the following equation on the 2-D polar coordinate system (ρ, φ) as

$$\rho = ae^{-b\phi} \tag{7.22}$$

where $a = v/\omega$ and $-\infty < \varphi < \infty$, a and b are arbitrary real positive constants, and e is the base of natural logarithms. In case of $b > 0$, Equation 7.22 describes curves like natural nautilus shell (shown in Figure 7.8), which converges at the origin as φ increases. Equation 7.22 can be translated onto Cartesian coordinate as follows:

$$\begin{cases} x = \rho\cos(\varphi) = ae^{-b\phi}\cos(\varphi) \\ y = \rho\sin(\varphi) = ae^{-b\phi}\sin(\varphi) \end{cases} \tag{7.23}$$

Logarithmic spiral based on Equation 7.23 is shown in Figure 7.10. Logarithmic spiral can be distinguished from Archimedean spiral by the fact that the distance between the arms

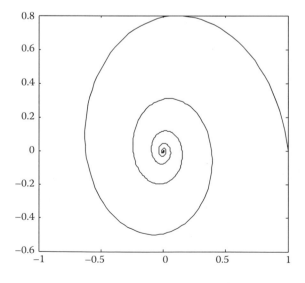

FIGURE 7.10   Logarithmic spiral.

of a logarithmic spiral increases in a geometric sequence while in Archimedean spiral the distance is constant.

A symbolic feature of logarithmic spirals is that the distance between a point on the trajectory and a center decreases (or increases) in equal ratio as the angular increases at the rate of an equal amount. This feature can be utilized for a meta-heuristic search strategy.

7.5 SPIRAL MODELS

Different spirals and their mathematical descriptions were reviewed briefly in Section 7.4. On the basis of these spirals, different models have been developed and reported in the literature (Tamura and Yasuda, 2011a,b, 2013).

7.5.1 2-D Spiral Models

In this section, a 2-D spiral model is described. The model will provide an apparent picture for developing an n-dimensional spiral model, which is discussed in the next section (Tamura and Yasuda, 2011b). A point x in 2-D orthogonal coordinates $x_1 x_2$ rotating anticlockwise around the origin by an angle θ, shown in Figure 7.11, can be described by x' as

$$x' = R^{(2)}(\theta)x \qquad (7.24)$$

The rotation matrix $R^{(2)}(\theta)$ is defined as

$$R^{(2)}(\theta) = \begin{bmatrix} \cos\theta & -\sin\theta \\ \sin\theta & \cos\theta \end{bmatrix} \qquad (7.25)$$

Using the rotation matrix $R^{(2)}(\theta)$, a discrete logarithmic spiral model can be formulated, which generates a point converging at the origin on the $x_1 x_2$ plane initiating from an arbitrary initial point x_0 while drawing a discrete logarithmic spiral.

$$\begin{bmatrix} x_1(k+1) \\ x_2(k+1) \end{bmatrix} = rR^{(2)}(\theta) \begin{bmatrix} x_1(k) \\ x_2(k) \end{bmatrix} \qquad (7.26)$$

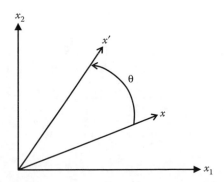

FIGURE 7.11   Rotation of a point in 2-D space.

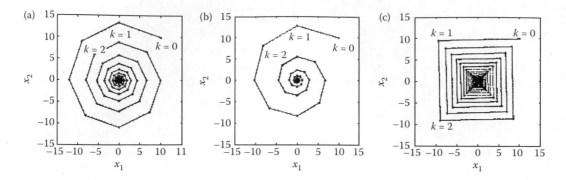

FIGURE 7.12   2-D spiral model described by Equation 4.27. (a) Case 1, (b) Case 2, and (c) Case 3.

Equation 7.26 is written in matrix notation form as

$$\mathbf{x}(k+1) = \mathbf{S}_2(r,\theta)x(k) \tag{7.27}$$

where $\mathbf{x}(0) = x_0$ and $k = 0, 1, \ldots$. $\mathbf{S}_2(r,\theta)$ is a stable matrix from the range of r and eigenvalues $\lambda_i \in C$, $i = 1, 2$ such that $|\lambda_i| = 1$ of rotation matrix $R^{(2)}(\theta)$. The model by Equation 7.27 has two parameters r and θ. $0 < r < 1$ is the convergence rate of distance between a point and the origin at each k and $0 \le \theta < 2\pi$ is a rotation angle around the origin at each k. Different 2-D spiral models described by Equation 7.27 are illustrated in Figure 7.12a–c for different parameter values of r and θ that are shown in Table 7.1.

Spiral model in Equation 7.27 does not have enough flexibility for applications due to its center at the origin. The model is thus enhanced by translating the origin of Equation 7.27 toward an arbitrary point x^* and defined as follows:

$$\mathbf{x}(k+1) = \mathbf{S}_2(r,\theta)x(k) - (\mathbf{S}_2(r,\theta) - I_2)x^* \tag{7.28}$$

where $\mathbf{I}_2 \in R^{2\times2}$ is the identity matrix. The convergence of the trajectory at x^* can be shown by transforming Equation 7.28 into

$$e(k+1) = \mathbf{S}_2(r,\theta)e(k) \tag{7.29}$$

where the error function $e(k)$ is defined as

$$e(k) = x(k) - x^* \tag{7.30}$$

It is appropriate to select x^* as the best solution obtained in the search process and satisfy the strategy of diversification to intensification. Therefore, a multipoint search model should be adopted.

TABLE 7.1
Values for Parameters r and θ

Case 1	$r = 0.95$	$\theta = \pi/4$
Case 2	$r = 0.90$	$\theta = \pi/4$
Case 3	$r = 0.95$	$\theta = \pi/2$

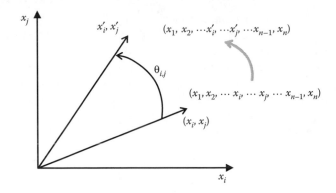

FIGURE 7.13 Rotation on $x_i x_j$-plane in n-dimensional space.

For a single-point search model, x^* does not work as the initial point becomes the best solution and center when evaluating the initial point (Tamura and Yasuda, 2011a).

7.5.2 N-DIMENSIONAL SPIRAL MODELS

Rotation of a point in n-dimensional orthogonal coordinates can be represented by composing 2 D rotations defined in n-dimensional space (Takada, 1999). A 2-D rotation in n-dimensional space means the operation of only two elements on the $x_i x_j$-plane of a point rotating anticlockwise about the origin by $\theta_{i,j}$ and let other elements unchanged except x_i and x_j. This is shown in Figure 7.13.

Rotations in the plane of a pair of coordinate axes (x_i, x_j), $i, j = 1, ..., N$, can be written as a block matrix (Takada, 1999) defined as follows:

$$R_{i,j}^{(n)}(\theta_{i,j}) = \begin{bmatrix} 1 & \cdots & 0 & 0 & 0 & \cdots & 0 & 0 & 0 & \cdots & 0 \\ \vdots & \ddots & \vdots & \vdots & \vdots & \ddots & \vdots & \vdots & \vdots & \ddots & \vdots \\ 0 & \cdots & 1 & 0 & 0 & \cdots & 0 & 0 & 0 & \cdots & 0 \\ 0 & \cdots & 0 & \cos\theta_{i,j} & 0 & \cdots & 0 & -\sin\theta_{i,j} & 0 & \cdots & 0 \\ 0 & \cdots & 0 & 0 & 1 & \cdots & 0 & 0 & 0 & \cdots & 0 \\ \vdots & \ddots & \vdots & \vdots & \vdots & \ddots & \vdots & \vdots & \vdots & \ddots & \vdots \\ 0 & \cdots & 0 & 0 & 0 & \cdots & 1 & 0 & 0 & \cdots & 0 \\ 0 & \cdots & 0 & \sin\theta_{i,j} & 0 & \cdots & 0 & \cos\theta_{i,j} & 0 & \cdots & 0 \\ 0 & \cdots & 0 & 0 & 0 & \cdots & 0 & 0 & 1 & \cdots & 0 \\ \vdots & \ddots & \vdots & \vdots & \vdots & \ddots & \vdots & \vdots & \vdots & \ddots & \vdots \\ 0 & \cdots & 0 & 0 & 0 & \cdots & 0 & 0 & 0 & \cdots & 1 \end{bmatrix} \quad (7.31)$$

From this definition, there are at most $n(n-1)$ kinds of rotation matrices, that is, in case of all permutations $^nP_2$ on selecting two axes from n axes. In case of all combinations $^nC_2$, there are possible $n(n-1)/2$ rotation matrices. Various composition rotation matrices can be generated by multiplying the rotation matrices represented by Equation 7.31 with each other. Thus, the distinct rotation matrices $R_{(i,j)}^{(n)}(\theta_{i,j})$ may be concatenated in some order to produce a rotation matrix such as

$$M^{(n)}(\theta) = \prod_{i<j} R_{i,j}^{(n)}(\theta_{i,j}) \quad (7.32)$$

with $n(n-1)/2$ degrees of freedom parameterized by $\theta_{i,j}$. However, since the matrices $R_{(i,j)}^{(n)}(\theta_{i,j})$ do not commute, different orderings give different results and it is difficult to intuitively understand the global rotation. In fact, as is the case for 3-D Euler angles, one may even repeat some matrices (with distinct parameters) and omit others, and still not miss any degrees of freedom.

There exist many rotation matrices represented by Equation 7.31. In general, an arbitrary n-dimensional rotation matrix can be defined by composing them in space. A general n-dimensional spiral model using composition of the basic rotation matrices of $R_{(i,j)}^{(n)}(\theta_{i,j})$ is being proposed in this section. The following n-dimensional spiral model can be deduced based on all combinations of two axes.

$$x(k+1) = rR^{(n)}(\theta_{1,2}, \theta_{1,3}, \ldots, \theta_{n-1,n})x(k) \qquad (7.33)$$

where $0 \le \theta_{i,j} < 2\pi$ are rotation angles for each plane around the origin at every $k = 0, 1, \ldots$ $0 < r < 1$ is the convergence rate of distance between a point and the origin at each k. $rR^{(n)}$ is a stable matrix from the range of r and eigenvalues $\lambda_i \in C$, $i = 1, 2, \ldots, n$, of the rotation matrix $R^{(n)}$ such that $|\lambda_i| = 1$. The rotation matrix $R^{(n)}(\theta_{1,2}, \theta_{1,3}, \ldots, \theta_{n-1,n})$ is defined as

$$\begin{aligned} R^{(n)}(\theta_{1,2}, \theta_{1,3}, \ldots, \theta_{n-1,n}) &= R_{n-1,n}^{(n)}(\theta_{n-1,n}) \\ &\times R_{n-2,n}^{(n)}(\theta_{n-2,n})R_{n-2,n-1}^{(n)}(\theta_{n-2,n-1}) \times \cdots \times R_{2,n}^{(n)}(\theta_{2,n}) \\ &\times \cdots \times R_{2,3}^{(n)}(\theta_{2,3})R_{1,n}^{(n)}(\theta_{1,n}) \times \cdots \times R_{1,3}^{(n)}(\theta_{1,3})R_{1,2}^{(n)}(\theta_{1,2}) \end{aligned} \qquad (7.34)$$

The rotation matrix $R^{(n)}(\theta_{1,2}, \theta_{1,3}, \ldots, \theta_{n-1,n})$ defined in Equation 7.34 can be rewritten in short form as

$$R^{(n)}(\theta_{1,2}, \theta_{1,3}, \ldots, \theta_{n-1,n}) = \prod_{i=1}^{n-1}\left(\prod_{j=1}^{i} R_{n-i,n+1-j}^{(n)}(\theta_{n-i,n+1-j})\right) \qquad (7.35)$$

For n-dimensional spiral model, $\theta_{i,j}$ can be set individually on each plane, but as n grows larger this becomes difficult to adjust $\theta_{i,j}$. A simple n-dimensional spiral model based on Equations 7.33 and 7.35, which has the same rotation angles $\theta = \theta_{i,j}$ on each plane in Equations 7.33 and 7.35, can be derived:

$$x(k+1) = rR^{(n)}(\theta)x(k) = S_n(r, \theta)x(k) \qquad (7.36)$$

where $\theta = [\theta_{1,2}, \theta_{1,3}, \ldots, \theta_{n-1,n}]^T \in R^{n(n-1)/2}$. $S_n(r,\theta) = rR_n(\theta)$ is called the spiral matrix, $0 < \theta \le 2\pi \in R$ is rotation rate, and $0 < r < 1 \in R$ is convergence rate. The spiral model in Equation 7.36 generates $x(k) \in R^n$ and lets the distance between the point and the origin to asymptotically converge toward zero and rotates the point around the origin by θ in the sense of Equation 7.35. The convergence $\lim_{k \to \infty} x(k) = 0$ that equals the origin of the spiral model in Equation 7.36 is the unique asymptotic stable equilibrium point. Figure 7.14 shows the dynamic behavior of the spiral model in Equation 7.36 with parameter values $r = 0.95$, $\theta = \pi/4$, and $\theta = \pi/2$ (Cases 1 and 3 in Table 7.1) in three-dimension.

The dynamic behavior of the spiral point generated by the spiral model in Equation 7.36 is shown in Figure 7.14a and b. It is seen clearly from the model that it demonstrates the diversification in

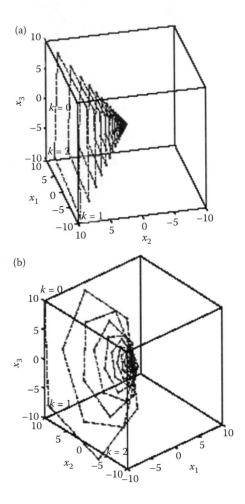

FIGURE 7.14 Trajectories of spiral in 3-D. (a) 3-D trajectory of the spiral with $r = 0.95$ and $\theta = \pi/4$ (Case 1 in Table 7.1). (b) 3-D trajectory of the spiral with $r = 0.95$ and $\theta = \pi/2$ (Case 3 in Table 7.1).

the early phase and intensification around the center in the later phase. The model in Equation 7.36 can further be taken as a search model. It is useful to have the asymptotic stable equilibrium point not only at the origin but also at an arbitrary point x^* for realization of various search strategies. To have the center of the spiral at an arbitrary point x^*, which will be useful for various applications, the spiral model in Equation 7.36 can be extended as follows:

$$x(k+1) = S_n(r,\theta)x(k) - (S_n(r,\theta) - I_n(k))x^* \tag{7.37}$$

where $I_n \in R^{n \times n}$ is the identity matrix. The spiral model in Equation 7.37 generates $x(k)$, which converges at x^* from any $x(0)$ creating the logarithmic spiral trajectory as time step k increases. The convergence of the trajectory to x^* can be proven by introducing an error function defined by $e(k) = x(k) - x^*$ so that Equation 7.37 can be transformed into an error function $e(k + 1) = Sn(r,\theta)e(k)$. Therefore, $\lim_{k \to \infty} x(k) = x^*$, which means x^* is the unique asymptotic stable equilibrium point. Thus, Equation 7.37 is the general n-dimensional logarithmic spiral model. The model has many parameters $\theta_{i,j}$ causing difficulty to set and adjust $\theta_{i,j}$ as n becomes larger.

7.6 SpD-BASED OPTIMIZATION ALGORITHM

In meta-heuristics search algorithms, a well-known effective strategy is that the search point dynamics should have diversification in the early phase and intensification in the later phase of the search (Aiyoshi and Yasuda, 2011). There are two strategies followed in meta-heuristics:

1. Diversification of search by exploring wider region for better solutions
2. Intensification of search by exploiting around a good solution for improving solutions

Diversification helps exploring wider search space. It continuously converges with a smaller radius providing dynamic step size, which helps approaching regions of high fitness values that lead to better solutions located at the center. Intensification is deployed at later stage for better exploitation for local search within the region identified during diversification stage. Intensification works well for problems having multipeak structure where optimum solution exists around the good solutions. The distance between a point in a path trajectory and the center point is varied constantly if the radius of the trajectory is changing at constant rate, thus making the radius an important converging parameter for the algorithm. An effective optimization algorithm maintains a good balance between diversification and intensification, that is, exploration and exploitation. This strategy is realized in a natural way in the SpD model described in Equation 7.37 by setting the center at an arbitrary point.

A common feature of logarithmic spiral is that it can realize an effective search strategy in meta-heuristics and construct a new optimization algorithm based on a discrete model of logarithmic spiral. A conclusion can be drawn from the discussions in Sections 7.4 and 7.5 that a logarithmic spiral explores a diverse region in the first half and intensively exploits around a region in the second half and makes it a powerful search and optimization technique. This technique is relatively new. There is not much work reported in the literature. The original work on the algorithm was first reported in Tamura and Yasuda (2011a,b). Two such techniques reported by Tamura and Yasuda are discussed in the following sections.

7.6.1 2-D SPIRAL DYNAMICS OPTIMIZATION ALGORITHM

The search behavior in meta-heuristic algorithms is generated by interaction, which enables achieving an effective and global search. Extending the spiral model in Equation 7.28 to a multipoint search model with a proper interaction is thus a valid approach to constructing better meta-heuristics. The multipoint search model using Equation 7.28 is formulated as follows:

$$x_i(k+1) = S_2(r,\theta)x_i(k) - (S_2(r,\theta) - I_2)x^* \tag{7.38}$$

where $i = 1,2, \ldots, m$. x^* is set as the common center of the best solution among all search points during the search. Namely, x^* becomes an interaction. Thus, applying the multipoint and using the interaction will contribute the following features to the heuristic search algorithm.

1. Interaction contributes to realization of intensification that exploits the search around a good solution.
2. Multipoint search model contributes to enhancing both diversification and intensification by exploring the search space.

In a minimization problem, the algorithm based on the multipoint search model in Equation 7.38 is described here. Search points drawing spiral trajectories toward common center x^* that is the best solution, which can naturally realize the strategy from diversification to intensification, can be expected to search for a better solution.

The spiral dynamics optimization (SpDO) algorithm based on 2-D spiral model can be described by the following steps:

Step 1: Select the number of search points, that is, $m \geq 2$
Set parameters $0 \leq \theta \leq 2\pi$, $0 < r < 1$ of $S_2(r, \theta)$
Set maximum iteration k_{max} and set $k = 0$
Step 2: Randomly initialize points $x_i(0) \in R^{(2)}$, $i = 1, 2, \ldots, m$ in the feasible region
Set $x^* = x_{i_g}(0)$, $i_g = \arg \min f[x_i(0)]$
Step 3: Compute $x_i(k+1) = S_2(r, \theta)x_i(k) - (S_2(r, \theta) - I_2)x^*$
Step 4: Compute $x^* = x_{i_g}(k+1)$, $i_g = \arg \min f[x_i(k+1)]$
Step 5: If $(k < k_{max})$, set $k = k + 1$ and Goto Step 3
Step 6: Return solution

where $f(\cdot)$ is the fitness function of the problem at hand. There have been some good applications of 2-D SpDO algorithm reported in the literature (Tamura and Yasuda, 2011a,b, 2013), some of which are discussed in Section 7.9.

7.6.2 N-Dimensional SpDO Algorithm

Using the same principle as the 2-D SpDO algorithm, the n-dimensional SpDO algorithm is developed for multipoint search model using m points in Equation 7.37 as follows:

$$x_i(k+1) = S_n(r, \theta)x_i(k) - (S_n(r, \theta) - I_n(k))x^* \qquad (7.39)$$

where $i = 1, 2, \ldots, m$, with the common center x^* set as the best solution among all search points during the search.

The SpDO algorithm can search for a better solution in n-dimensional space by using search points that draw spiral trajectories toward common center x^* set as the best solution, which naturally realizes the strategy from diversification to intensification.

The algorithm of the n-dimensional SpDO algorithm based on the spiral model in Equation 7.37 can be described by the following steps:

Step 1: Select the number of search points, that is, $m \geq 2$
Select parameters $0 \leq \theta \leq 2\pi$, $0 < r < 1$ of $S_n(r, \theta)$
Set maximum iteration k_{max} and $k = 0$
Step 2: Randomly initialize points $x_i(0) \in R^{(n)}$, $i = 1, 2, \ldots, m$ in the feasible region
Set $x^* = x_{i_g}(0)$, $i_g = \arg \min f[x_i(0)]$
Step 3: Compute $x_i(k+1) = S_n(r, \theta)x_i(k) - (S_n(r, \theta) - I_2)x^*$
Step 4: Compute $x^* = x_{i_g}(k+1)$, $i_g = \arg \min f[x_i(k+1)]$
Step 5: If $(k < k_{max})$, set $k = k + 1$ and Goto Step 3
Step 6: Return solution

There have been some good applications of n-dimensional SpDO algorithm reported in the literature (Tamura and Yasuda, 2011b, 2013), some of which are discussed in Section 7.9.

7.6.3 Parameters of SpDO Algorithm

In general, the SpDO algorithm has fewer parameters compared to other meta-heuristic algorithms. These are shown in Table 7.2 with brief descriptions.

TABLE 7.2

Parameters of SpDO Algorithm

Parameters	Description
θ	Search point angular displacement on $x_i - x_j$ plane
r	Spiral radius
$x_i(k)$	Position of i-th point of the k-th iteration, $i = 1, 2, ..., m$ and $k = 1, 2, 3, ..., k_{max}$
R^n	Composition of rotational matrix on combination of all axes leading to an $n \times n$ matrix
m	Total number of search points
k_{max}	Maximum number of iterations

The parameters θ and r have an independent role. $\theta > 0$ is the rotation rate to rotate its state around the origin at each k. Since θ exists only in the rotation matrix R^n, the domain of θ is restricted to $0 \leq \theta < 2\pi$. $0 < r < 1$ is a convergence rate of distance between a point and the origin at each iteration k. The value of r close to 1 shows improved performance. The effectiveness of r depends on k_{max}. Different types of spirals are presented in Table 7.1 for varying values of the parameters r and θ. The value for k_{max} is chosen between 100 and 1000. Larger value for k_{max} shows improved performance.

7.7 VARIANTS OF SpDO ALGORITHM

There have been many successful implementations of the SpDO algorithm since the first publication of the algorithm by Tamura and Yasuda in 2011 (Tamura and Yasuda, 2011a,b). Despite many successes, the SpDO algorithm suffers from some well-known problems such as convergence speed, trapping into local optima, accuracy, and computational time. The accuracy of the optimum solution mainly depends on how the algorithm explores and exploits the search space. A number of modifications of parameters have been introduced to improve exploration, exploitation of the search space, and solution quality. The exact number of variants is unknown though but there are very limited number of publications reported, which are scattered mostly in unknown publications. Some of the important variants of the algorithm are discussed in this section.

7.7.1 ADAPTIVE SpDO ALGORITHM

The dynamic step size of SpDO algorithm depends on a unique constant spiral radius throughout the search process regardless of the fitness value at particular point within the solution space, which leads to low accuracy of solution and slow convergence. In order to improve the search efficiency, step size is varied dynamically based on the fitness value of the current search point by adapting the spiral radius leading to higher accuracy and convergence speed. Varying spiral radius within a specific range, for example [0,1], can produce better variation of step size. In other words, varying step sizes from very small to very large can provide higher chances of finding accurate solutions. Using this strategy, Nasir et al. (2012c) proposed four adaptive SpDO algorithms:

1. Linear adaptive SpDO
2. Quadratic adaptive SpDO
3. Exponential adaptive SpDO
4. Fuzzy adaptive SpDO

7.7.1.1 Linear Adaptive SpDO

A simple linear function is defined to establish a relationship between the spiral radius and absolute fitness value of a point within search space (Nasir et al., 2012c). The linear adaptive spiral radius r_{la} is defined by

$$r_{la} = \frac{r_l - r_u}{1 + (c_1/|\,f(x_i(k))\,|)} + r_u \qquad (7.40)$$

where r_u and r_l are the upper and lower bounds of the spiral radius, respectively, r_u and r_l should be chosen from the range of [0,1]. c_1 is a positive constant value, which defines the rate of change of fitness value and spiral radius. A small value for c_1 tends to select the upper bound value of the radius, while a large value for c_1 tends to select the lower bound value of the radius. The selection of the spiral radius between r_u and r_l has a dynamic and effective influence on the change of step size of a search point. $|f(x_i(k))|$ is the absolute fitness of a search point.

Adapting only the spiral radius may not allow the adaptive SpDO algorithm to explore the search space thoroughly. An adapting rotation angle (or angular displacement) along with adapting spiral radius will explore the search space better. On the basis of this strategy, Nasir et al. (2013b) proposed an extension to the linear adaptive SpDO algorithm where an adaptive mechanism is implemented in the spiral radius and rotation angle. This strategy determines different values of radius and angle of rotation for each search point according to the current fitness level and the best fitness level. In this case, the adaptive spiral radius r_a is defined by

$$r_a = \frac{r_l - r_u}{1 + (c_1/|\,f(x_i(k)) - \min f(x_i(k))\,|)} + r_u \qquad (7.41)$$

where c_1 is a positive constant value, which defines the rate of change of fitness deviation value and spiral radius, $f(x_i(k))$ is the fitness value of a search point, and $\min f(x_i(k))$ is the best global fitness value in the current iteration. The upper and lower bounds of the spiral radius $r_u, r_l \in [0,1]$ ensure that a point in the search space converges toward the current best location.

The adaptive rotation angle is defined by

$$\theta_a = \frac{\theta_l - \theta_u}{1 + (c_a/|\,f(x_i(k)) - \min f(x_i(k))\,|)} + \theta_u \qquad (7.42)$$

where θ_a is the adaptive rotation angle, c_a is a positive constant value, which defined the rate of change of fitness deviation value and spiral angle, θ_u and θ_l are the upper and lower bounds of the rotation angle, respectively, which should be chosen from the range of $[0,2\pi]$. $\theta_u, \theta_l \in [0,2\pi]$ ensures that a point in the search space converges toward the current best location.

This linear adaptive SpDO algorithm is the same as of the standard SpDO algorithms described in Sections 7.6.1 and 7.6.2 except Step 3. Once the adaptive spiral radius r_a and the adaptive rotation angle θ_a are defined, Step 3 is modified by replacing the spiral radius r and rotation angle θ with the adaptive spiral radius r_a and rotation angle θ_a. A drawback of this approach is that the total computation time may increase due to the increase in the size of the rotation square matrix when the dimension of the problem increases. The adaptive SpDO algorithm has been tested on well-known benchmark problems such as Rastrigin function, Sphere function, Griewank function, and Ackley function by Nasir et al. (2013b), which demonstrated better performance and faster convergence against standard SpDO algorithm. Nasir et al. (2013b) also tested the adaptive SpDO algorithm on optimization of model parameters of a flexible robot manipulator.

Nasir et al. (2016) proposed another version of linear adaptive SpDO, which is slightly different from Equation 7.41 and r_{la} is defined as

$$r_{la} = \frac{r_l - r_u}{1 + (c_1/c_2 \mid f(x_i(k)) - \min f(x_i(k)) \mid)} + r_u \tag{7.43}$$

where c_1 and c_2 are positive constant values. The constant c_2 is used to control the value of r_{la} within an interval of not very small and not very big by scaling the term $\mid f(x_i(k)) - \min f(x_i(k)) \mid$. Using the scaled term $c_1/c_2 \mid f(x_i(k)) - \min f(x_i(k)) \mid$, a rule of thumb can be formulated such that search point step size will vary only linearly within the interval of $[r_l, r_u]$. Two rules of thumb are described as follows:

1. If $\mid f(x_i(k)) - \min f(x_i(k)) \mid$ is small, then the term $c_1/c_2 \mid f(x_i(k)) - \min f(x_i(k)) \mid$ is big. When this rule is true, r_{la} becomes big or approaches to the upper bound value of spiral radius r_u.
2. If $\mid f(x_i(k)) - \min f(x_i(k)) \mid$ is big, then the term $c_1/c_2 \mid f(x_i(k)) - \min f(x_i(k)) \mid$ is small. When this rule is true, r_{la} becomes small or approaches to the lower bound value of spiral radius r_l.

The performance of the linear adaptive SpDO approach was verified on eight different unimodal and multimodal benchmark problems and on a flexible manipulator system.

7.7.1.2 Quadratic Adaptive SpDO

In quadratic adaptive SpDO, Equation 7.40 is modified with the quadratic absolute fitness value (Nasir et al., 2012c). The quadratic adaptive spiral radius r_{qa} is defined by

$$r_{qa} = \frac{r_l - r_u}{1 + (c_1 \times \mid f(x_i(k)) \mid^2 + \mid f(x_i(k)) \mid)} + r_u \tag{7.44}$$

where c_2 is a constant value to be tuned using heuristic method and $\mid f(x_i(k)) \mid^2 + \mid f(x_i(k)) \mid$ is the quadratic term of the fitness function. The term $\mid f(x_i(k)) \mid^2 + \mid f(x_i(k)) \mid$ produces a steeper slope of the spiral radius that ensures a higher acceleration toward the best location.

7.7.1.3 Exponential Adaptive SpDO

In exponential adaptive SpDO, Equation 7.40 is modified with an exponential absolute fitness value to change the spiral radius and the step size of a search point dynamically (Nasir et al., 2012c). The exponential adaptive spiral radius r_{ea} is defined by

$$r_{ea} = \frac{r_l - r_u}{1 + (c_1/\exp(c_2 \mid f(x_i(k)) \mid))} + r_u \tag{7.45}$$

where $\exp(c_2 \mid f(x_i(k)) \mid)$ is the exponential term of the absolute fitness function of a search point. The constant c_2 is a parameter tuned mainly using heuristic approach. The exponential term $\exp(c_2 \mid f(x_i(k)) \mid)$ provides a sharper slope of the spiral radius versus fitness value compared to linear or quadratic term. This results in higher acceleration of the search point toward global best solution.

7.7.1.4 Fuzzy Adaptive SpDO

In fuzzy adaptive SpDO, the relationship between the spiral radius and absolute fitness value of a point within search space is modified by a fuzzy function or mapping (Nasir et al., 2012c). The fuzzy adaptive spiral radius r_{fa} is defined by

$$r_{fa} = F(\mid f(x_i(k)) \mid) \tag{7.46}$$

where $|f(x_i(k))|$ is the absolute fitness of a search point and $F(\cdot)$ is the fuzzy mapping of the input and output. The absolute fitness $|f(x_i(k))|$ is the input and spiral radius r_{fa} is the output of the fuzzy mapping and they are defined within the range of [0,1]. The input $|f(x_i(k))|$ and the output r_{fa} are represented by Gaussian membership functions. A Mamdani-type fuzzy inference system is used. The general rule for the Mamdani-type fuzzy system is defined as

$$\text{If } \left| f(x_i(k)) \right| \text{ is } A, \text{ then } r_{fa} \text{ is } B \tag{7.47}$$

where A and B are the Gaussian membership functions for $|f(x_i(k))|$ and r_{fa}, respectively, defined within the universe of discourse. A detailed discussion of the fuzzy system is provided in Siddique and Adeli (2013).

The other three adaptive SpDO algorithms, that is, quadratic, exponential, and fuzzy adaptive SpDO, discussed above are the same as of the standard SpDO algorithms described in Sections 7.6.1 and 7.6.2 except Step 3. Once the adaptive spiral radii r_{qa}, r_{ea}, and r_{fa} are defined, Step 3 is modified by replacing the spiral radius r with the adaptive spiral radius r_{qa}, r_{ea}, and r_{fa}. The adaptive SpDO algorithms have been tested on well-known unimodal and multimodal benchmark problems such as Sphere function, Ackley function, Rastrigin function, and Griewank function by Nasir et al. (2012c), which demonstrated better performance in terms of convergence speed and accuracy against standard SpDO algorithm. The fuzzy adaptive SpDO algorithm shows higher computation time due to fuzzy inference rule processing, which results in slower convergence speed than other adaptive SpDO algorithms.

7.7.2 HYBRID SpDO ALGORITHMS

The use of constant spiral radius and rotation angle in the SpDO algorithm limits the movements of the search points within the search space. Adaptive step size of spiral radius improves the exploration of the search space. In general, the search mechanism of SpDO is good at exploration of larger search space but it is not good at exploitation, which very often leads to local optima and poor accuracy of solutions. In order to improve the solution quality, an efficient local search technique is to be incorporated into the SpDO algorithm.

Bacterial foraging algorithm (BFA) (Passino, 2002) is the adaptation of foraging strategy of *Escherichia coli* bacteria to survive in their lives. The algorithm comprises three basic phases, namely, chemotaxis, reproduction, and elimination-dispersal. In the chemotaxis phase, each bacterium tumbles one step randomly around its current location looking for higher nutrient location. The bacterium then swims a few more steps in the direction of that tumble toward the position with higher nutrient than the previous position and this endeavor continues until the end of bacterium life. This tactic is called chemotactic strategy of BFA. The chemotaxis feature is incorporated into BFA using a very crucial parameter, namely, bacteria step size, which enables BFA to effectively move from one position to another. The parameter predominantly determines the performance of BFA. A large step size may be defined to expedite bacteria movement or to accelerate the algorithm's convergence speed. In the case of large step size, BFA suffers from oscillation when approaching the global optimum. However, it may cause the algorithm to get trapped into local optima if the global optimum point is located in remote area and hence decrease the accuracy. Therefore, in this case, a small step size should be defined to increase the accuracy of the solution, but it may lead to slower convergence since more steps are required to reach the global optimum point.

7.7.2.1 Hybrid Spiral-Dynamics Bacterial-Chemotaxis Algorithm

The tumble and swim actions in chemotaxis phase of BFA are good schemes to find a global optimum point in search space. The BFA algorithm has faster convergence speed to feasible solutions in the defined search space but exhibits oscillations toward the end of the search operation, whereas the SpDO algorithm has faster computation time and higher accuracy than the BFA. The SpDO algorithm has also robust stability due to the spiral steps when searching toward the optimum point. On the basis of the advantages of the both approaches, Nasir et al. (2012a) first proposed the hybridization between the SpD algorithm and bacterial chemotaxis from BFA for global optimization and developed the hybrid spiral-dynamics bacterial-chemotaxis (HSDBC) optimization algorithm. The HSDBC incorporates the BFA chemotaxis part into the SpDO algorithm, which reduces the computational time and retains the strength and performance of the SpDO algorithm. Thus, the HSDBC algorithm combines the strengths of BFA and SpDO algorithm into a faster, stable, and accurate global optimization algorithm. The main steps of HSDBC algorithm are as follows:

Step 1: Select the number of search points (i.e., bacteria), $m \geq 2$
      Set N_s (max number of swim for chemotaxis)
      Select parameters of SpDO $0 \leq \theta < 2\theta, 0 < r < 1$
      Set max iteration k_{max}, $k = 0$, $s = 0$
Step 2: Initialization
      Set initial points $x_i(0) \in R^n$, $i = 1, 2, \ldots, m$
      Set center $x^* = x_{i_g}(0)$, $i_g = \arg \min_i f[x_i(0)]$
Step 3: Apply bacteria chemotaxis
      3.1: Update $x_i(k + 1) = S_n(r, \theta)x_i(k) - [S_n(r, \theta) - I_n]x^*$
      3.2: Bacteria swim
      Check number of swim for bacteria i
         If $s < N_s$, then check fitness
         Else set $i = i + 1$ and Goto Step 3.1
      Check fitness
         If $f[x_i(k + 1)] < f[x_i(k)]$, then update x_i
         Else set $s = N_s$ and Goto Step 3.1
      Update $x_i(k + 1) = S_n(r, \theta)x_i(k) - [S_n(r, \theta) - I_n]x^*$
Step 4: Update $x^* = x_{i_g}(k + 1)$, $i_g = \arg \min_i f[x_i(k + 1)]$
Step 5: If $(k < k_{max})$, set $k = k + 1$ and Goto Step 3
Step 6: Return solution

The performance of the HSDBC algorithm was verified on a number of unimodal, for example, sphere and Ackely function, and multimodal, for example, Rastrigin and Griewank function. HSDBC algorithm was also applied to optimize controller parameters for a flexible manipulator system.

Nasir and Tokhi (2015) later investigated the performance of HSDBC algorithm on eight well-known benchmark functions such as Rastrigin, Sphere, Griewank, Goldstein and Price, Ackley, Dixon and Price, and Rosenbrock. They termed the algorithm as HSDBC-S and provided performance in comparison with HSDBC-R algorithm (discussed later in this section). HSDBC-S algorithm was also applied to real-world problems like control design for flexible manipulator system (Nasir and Tokhi, 2015). Almeshal et al. (2013) extended the HSDBC algorithm by combining the linear adaptive SpDO algorithm and chemotactic behavior of BFA. HSDBC algorithm was applied to fuzzy logic controller design for a two-wheeled robotic vehicle, which showed very fast convergence compared to standard SpDO algorithm.

7.7.2.2 Hybrid Spiral-Dynamics Random-Chemotaxis Algorithm

The SpDO algorithm has very good strategy for global search, but its capability is limited in local search. Incorporating chemotactic strategy into SpDO algorithm balances the exploration and exploitation parts of the algorithms. After the bacteria move one-step ahead using spiral model, they perform random tumble and swim actions without referring to the global best position to represent a local search strategy. Nasir and Tokhi (2013, 2015) proposed the hybrid spiral-dynamics random-chemo-taxis (denoted as HSDBC-R) algorithm by incorporating chemotaxis phase of BFA into SpDO algorithm where bacteria swim in random direction instead of swimming in spiral direction. The hybrid approach preserves the strengths of global search of SpDO and local search of BFA. An additional step is incorporated into the chemotaxis phase of the HSDBC-R approach to ensure that the best solution and best fitness are always preserved when the bacteria move from one location to another while performing the tumble and swim actions. The main steps of HSDBC-R algorithm are as follows:

Step 1: Initialize the number of search points (i.e., bacteria), $m \geq 2$
 Select parameters of SpDO $0 \leq \theta < 2\theta$, $0 < r < 1$, and max iteration k_{max}
 Set N_s (max number of swim for chemotaxis)
 Set $k = 0$, $s = 0$
Step 2: Randomly place $x_i(k)$ in the search space
 Compute $f[x_i(k)]$
 Set $x_i^{best} = x_i(k)$ and $J_i^{best} = J_i(k)$
 Set initial points $x_i(0) \in R^n$, $i = 1, 2, \ldots, m$
 Set center $x^* = x_{i_g}(0)$ as the center of spiral, $i_g = \arg \min_i f[x_i(0)]$
Step 3: Move spirally toward global best position
 Update $x_i(k + 1) = S_n(r_{tumble}, \theta_{tumble})x_i(k) - [S_n(r_{tumble}, \theta_{tumble}) - I_n]x^*$
Step 4: Evaluate fitness
 If $f[x_i(k + 1)] < f[x_i(k)]$
 Set $x_i^{best} = x_i(k+1)$ and $J_i^{best} = J_i(k+1)$
 Else Goto Step 6
Step 5: Swim randomly around the local best position
 $x_i(k + 1) = x_i(k + 1) +$ step size $\times$ random direction
 Step 5.1: Compute $f[x_i(k + 1)]$
 Step 5.2: If $s < N_s$
 Set $s = s + 1$ and Goto Step 4
Step 6: Set $x_i(k+1) = x_i^{best}$ and $f_i(k+1) = f_i^{best}$
Step 7: If $i < m$
 Update $x^* = x_{i_g}(k+1)$, $i_g = \arg \min_i f[x_i(k+1)]$
 Else set $i = i + 1$ and Goto Step 3
Step 8: If $(k < k_{max})$, set $k = k + 1$ and Goto Step 3
Step 9: Return solution

The performance of the HSDBC-R approach was initially verified on an eighth-order autore-gressive with exogenous input (ARX) model of a flexible manipulator system, which demonstrates superior performance over SpDO algorithm and BFA (Nasir and Tokhi, 2013). Later, Nasir and Tokhi (2015) verified the performance on eight well-known benchmark functions such as Rastrigin, Sphere, Griewank, Goldstein and Price, Ackley, Dixon and Price, and Rosenbrock. Nasir and Tokhi (2015) also applied HSDBC-R to controller design.

7.7.2.3 Hybrid Spiral-Dynamics Bacterial-Foraging Algorithm

HSDBC algorithm proposed by Nasir et al. (2012a) employed bacterial chemotaxis strategy into SpDO algorithm, which resulted in faster convergence of the search space. The standard BFA has faster convergence due to chemotaxis approach and better exploitation strategy at local search due to tumble and swim actions. BFA has good diversity, which enables it to avoid trapping in local optimum. Undesirably, BFA suffers from oscillation problem toward the end of the search process resulting in low accuracy of solutions. HSDBC is good at exploration using dynamic spiral step, which provides better stability and accuracy when approaching the optimum point. The problem with HSDBC is that it has higher possibility of getting trapped into local optima when dealing with multimodal problems. In order to handle the oscillation problem of BFA, retain stability and accuracy of HSDBC, and avoid local optima that arise in HSDBC, Nasir et al. (2012b) proposed to hybridize the desirable features of the BFA and HSDBC algorithm into hybrid spiral-dynamics bacterial-foraging (HSDBF) algorithm. In HSDBF algorithm, an n-dimensional spiral model is utilized for a uniform initial distribution of bacteria throughout the search space and spiral adaptive chemotaxis strategy is incorporated instead of random approach as in the BFA. The HSDBF is a new optimization algorithm comprising simple structure and reduced computational cost. The performance of the HSDBF algorithm was verified on unimodal Sphere function, a number of multimodal benchmark functions, for example, Rastrigin, Shubert function, and optimizing control parameters of a collocated proportional-derivative (PD)-type controller.

Nasir et al. (2013a) proposed three synergies of SpDO algorithm and BFA, namely, hybrid bacteria chemotaxis-spiral dynamics (HBCSD), hybrid spiral-bacteria foraging (HSBF), and hybrid chemotaxis-spiral (HCS) algorithms. The three synergies are discussed in the sequel.

7.7.2.3.1   HBCSD Algorithm

Nasir et al. (2013a) proposed the HBCSD algorithm representing the synergy between chemotaxis strategy of bacteria through randomly tumble and swim actions of BFA and linear adaptive SpDO algorithm. In order to balance between exploration and exploitation in the search process, chemotaxis strategy through random tumble and swim actions is applied at the initial search phase and linear adaptive SpDO algorithm is used through linear adaptive spiral radius r_{la} and angular displacement θ_{la} toward the final phase of the search process. Chemotaxis strategy is used to ensure exploration of the search space by adopting large value for step size in bacteria motion. Linear adaptive SpDO algorithm is used to ensure exploitation to complete the search operation. The algorithm is divided into two parts: exploration and exploitation. In exploration stage, selection of suitable step size in bacterial movement and the number of chemotaxis are the crucial parameters and must be defined to provide the bacteria enough lifetime to search optimum fitness. The final locations of bacteria are the optimum points that are stored and considered as initial search points for the exploitation part. In exploitation stage, efficient exploitation strategy is used by applying linear adaptive SpDO algorithm (linear adaptive spiral radius and angular displacement discussed earlier in this section). The spiral search enables finding optimum point in the remote area of the search space and helps speeding convergence. The main steps of HBCSD algorithm are summarized as follows:

Part 1: Exploration stage
      Step 1: Randomly place bacteria in search space
      Step 2: Bacteria tumble randomly
      Step 3: Check fitness: If $f[x_i(k+1)] > f[x_i(k)]$, Goto Step 7
      Step 4: Swim on a similar direction as tumble with large step size
      Step 5: Compute fitness: $f[x_i(k+1)]$
      Step 6: Check max swim: If $(s < N_s)$, $s = s + 1$ and Goto Step 3
      Step 7: Check max bacteria: If $(i < m)$, $i = i + 1$ and Goto Step 2
Part 2: Exploitation stage
      Step 8: Set optimum locations found in Part 1 as initial locations in SpDO
      Step 9: Compute fitness: $f[x_i(k)]$
      Step 10: Set $x^* = x_{i_g}(0)$ as center of spiral
      Step 11: Update $x_i(k+1) = S_n(r_{la}, \theta_{la})x_i(k) - [S_n(r_{la}, \theta_{la}) - I_n]x^*$
      Step 12: Check fitness: $f[x_i(k+1)] < f[x_i(k)]$
      Step 13: Check max bacteria: If $(i < m)$, $i = i + 1$ and Goto Step 11
      Step 14: Set $x^* = x_{i_g}(k+1)$ as center of spiral
      Step 15: If $(k < k_{max})$, set $k = k + 1$ and Goto Step 11
      Step 16: Return solution

A comprehensive study of HBCSD algorithm was carried out by Nasir and Tokhi (2014) with applications to modeling of flexible manipulator and twin rotor systems.

7.7.2.3.2 HSBF Algorithm

Once the initialization of the bacteria position and associated variables is done, the fitness of all the bacteria is computed and the best bacterium found at this stage is considered x^* as the center point of the spiral model. The best bacterium guides the move of bacteria over the entire search space in a spiral toward the global best position. In order to improve the position of the bacteria in the search space, spiral dynamics model is employed with the chemotaxis phase of BFA. This approach of HSBF algorithm was proposed by Nasir et al. (2012a). After completing the tumble and swim action, the bacterium is directed toward the global best position spirally. Bacteria are updated using spiral model with an adaptive mechanism for bacteria step size establishing a relationship between step size and fitness. Bacteria fitness is calculated, best bacterium is determined, and is used as x^* in the chemotaxis phase. It then starts a new reproduction phase. The process continues with the bacteria guided based on local best position using chemotaxis strategy and global best position using spiral dynamics model until the end of search process. The main steps of HSBF algorithm are summarized as follows:

Step 1: Initialize variables
Step 2: Compute fitness $f[x_i(p)]$
   Set the best bacteria as x^*
Step 3: Compute fitness $f[x_i(p)]$ for $i = 1, 2, \ldots, S$
Step 4: Apply tumble action
Step 5: Check fitness: If $f[x_i(p + 1)] > f[x_i(p)]$, Goto Step 8
Step 6: Apply swim action
Step 7: Check: If $sw(i) < N_s$
   $sw(i) = sw(i + 1)$ and Goto Step 5
Step 8: Update $x_i(p + 1) = S_n(r_{la}, \theta_{la})x_i(p) - [S_n(r_{la}, \theta_{la}) - I_n]x^*$
Step 9: Check: If $i < S$
   $i = i + 1$ and Goto Step 3
Step 10: Compute fitness $f[x_i(p)]$
   Set the best bacteria as x^* for chemotaxis phase
Step 11: Check: If $j < N_c$
   $j = j + 1$ and Goto Step 3
Step 12: Do reproduction of bacteria
Step 13: Check: If $k < N_{reprod}$ (total number of reproduction)
   $k = k + 1$ and Goto Step 3
Step 14: Do elimination and dispersal
Step 15: Check: If $l < N_{ed}$ (total number of elimination and dispersal)
   $l = l + 1$ and Goto Step 3
Step 16: Return solution

The performance of the HSBF algorithm was verified on optimization of eighth-order ARX model parameters.

7.7.2.3.3 HCS Algorithm

HCS algorithm is a combination of chemotaxis and spiral dynamics algorithm proposed by Nasir et al. (2012a). HCS algorithm is a simplified version of HSBF algorithm, where the spiral dynamics model is incorporated into adaptive BFA to improve the bacteria position in the search area. The simplified version reduces algorithmic complexity as well as computational time by removing reproduction, elimination, and dispersal elements of search process. This combination retains the advantages and strengths possessed by its predecessor algorithms. The main steps of HCS algorithm are as follows:

Step 1: Initialize variables
Step 2: Compute fitness $f[x_i(k)]$
         Set the best bacteria as $x*$
Step 3: Compute fitness $f[x_i(k)]$ for $i = 1, 2, ..., S$
Step 4: Apply tumble action
Step 5: Check fitness: If $f[x_i(k + 1)] > f[x_i(k)]$, Goto Step 8
Step 6: Apply swim action
Step 7: Check: If $sw(i) < N_s$
         $sw(i) = sw(i+1)$ and Goto Step 5
Step 8: Update $x_i(k + 1) = S_n(r_{la}, \theta_{la})x_i(k) - [S_n(r_{la}, \theta_{la}) - I_n]x*$
Step 9: Check: If $i < S$
         $i = i+1$ and Goto Step 3
Step 10: Compute fitness $f[x_i(p)]$
         Set the best bacteria as $x*$ for chemotaxis
Step 11: Check: If $j < N_c$
         $j = j+1$ and Goto Step 3
Step 12: Return solution

The performance of HCS algorithm was verified on parameter optimization problem of an eighth-order ARX model.

7.8 STABILITY OF SPIRAL MODELS

The dynamics of spiral has been modeled by Equations 7.28 and 7.37 for two-dimension and n-dimension, respectively. It is important to confirm that the center $x*(k)$ of a dynamic spiral model is stable, that is, the center $x*(k)$ is asymptotically stable.

The SpDO algorithm based on the model in Equation 7.39 eventually comprises m spiral models with the common center $x*(k)$ that is being updated each iteration in the search process. The common center $x*(k)$ must be a unique stable equilibrium point, where the search algorithm should be converging to over the search process. If $x*(k)$ is the global equilibrium point, then $x_i(k)$, $i = 1, 2, ..., m$, must converge to $x*(k)$ over the iterations.

$$\lim_{k \to \infty}[x_i(k)] = x^*(k), \quad i = 1, 2, ..., m \tag{7.48}$$

Let u be an unknown equilibrium point, then the following hold according to Equation 7.39:

$$u = S_n(r, \theta)u - (S_n(r, \theta) - I_n)x^*(k) \tag{7.49}$$

Manipulating Equation 7.49 yields

$$(S_n(r, \theta) - I_n)(u - x^*(k)) = 0 \tag{7.50}$$

where $S_n(r, \theta) - I_n$ is a nonsingular matrix.

$$(u - x^*(k)) = 0 \tag{7.51}$$

The unique solution to Equation 7.51 yields

$$u = x^*(k) \tag{7.52}$$

That is, $x^*(k)$ is the unique and dynamic equilibrium point of the dynamic model of the spiral in Equation 7.39. From the above derivation for the unique equilibrium point, it can be proved that there exists at least one point from $x_i(k)$, $i = 1, 2, \ldots, m$, that stays at $x^*(k)$ in the search space. Thus, the asymptotic stability can be proved from

$$\lim_{k \to \infty} \| x_i(k) - x^*(k) \| = 0, \quad i = 1, 2, \ldots, m \tag{7.53}$$

where $\|\cdot\|$ is the Euclidean norm.

7.9   APPLICATIONS TO ENGINEERING PROBLEMS

To investigate and verify the search performance and identify the parameter properties of SpDO algorithm, SpDO algorithm has been applied, firstly, to a number of standard benchmark functions and then to some industrial and real-world problems. The functions such as (i) Rosenbrock function, (ii) 2^n minima, (iii) Rastrigin function, (iv) Schwefel function, (v) Griewank function, (vi) Sphere function, and (vii) Ackley function are benchmark functions commonly used to verify optimization problems. These functions are defined in the following.

The Rosenbrock function is defined as

$$f(x) = \sum_{i=1}^{n-1} \left[100 \left(x_{i+1} - x_i^2 \right)^2 + (x_i)^2 \right] \tag{7.54}$$

The function has a maximum value of zero at $x_i = 26.123$.

The 2^n minima function is defined by

$$f(x) = \sum_{i=1}^{n} \left(x_i^4 - 16x_i^2 + 5x_i \right), \quad -5 \le x_i \le 5 \tag{7.55}$$

The Rastrigin function is defined as

$$f(x) = \sum_{i=1}^{n} \left(x_i^2 - 10 \cos 2\pi x_i + 10 \right) \tag{7.56}$$

The Rastrigin function is a multimodal function, which has a global minimum at $x_i = [0,0]$ with a fitness value of $f(x) = 0$.

The Schwefel problem is highly multimodal and is defined by

$$f(x) = \sum_{i=1}^{n} \left(\sum_{j=1}^{i} x_j \right)^2, \quad -5 \le x_i \le 5 \tag{7.57}$$

Griewank function is continuous, highly multimodal function and is defined by

$$f(x) = \frac{1}{4000} \sum_{i=1}^{n} x_i^2 + \prod_{i=1}^{n} \cos\left(\frac{x_i}{\sqrt{i}}\right), \quad -50 \leq x_i \leq 50 \tag{7.58}$$

The Griewank function has a global minimum at $x_i = [0,0]$ with a fitness value of $f(x) = 0$. Sphere function is a unimodal function defined as

$$f(x) = \sum_{i=1}^{n} x_i^2, \quad -5.12 \leq x_i \leq 5.12 \tag{7.59}$$

The Sphere function has a global minimum at $x_i = [0,0,0]$ with a fitness value of $f(x) = 0$. Ackley function is a unimodal function defined as

$$f(x) = -20 \exp\left(-0.2\sqrt{\frac{1}{n}\sum_{i=1}^{n} x_i^2}\right) - \exp\left(\frac{1}{n}\sum_{i=1}^{n} \cos(2\pi x_i)\right) + 20 + e \tag{7.60}$$

The Ackley function has a global minimum at $x_i = [0,0,0]$ with a fitness value of $f(x) = 0$.

The effectiveness of SpDO algorithm was first investigated by Tamura and Yasuda (2011a). They applied the SpDO algorithm to 2-D continuous optimization benchmark problems such as Rosenbrock, 2^n minima, and Rastrigin functions defined in Equations 7.54 through 7.56. The number of search points was set to 5 and the maximum interaction number was set to 100. The results demonstrate that SpDO algorithms are powerful optimization tool and yield better performance, which are comparable to that of PSO.

Tamura and Yasuda (2011b) then extended the 2-D model to n-dimensional model based on rotation matrices and applied SpDO algorithm to a set of standard benchmark functions such as 2^n minima, Rastrigin, Schwefel, and Griewank functions defined in Equations 7.55 through 7.58 for dimensions of $n = 3$, $n = 30$, and $n = 100$ whose extrema are known to verify the search performance and parameter properties of these functions. The number of search points was set to 20 and number of iterations was $k_{max} = 100$ and $k_{max} = 1000$. They confirmed that the parameter rotational angle $\theta = \pi/2$ was superior to $\theta = \pi/4$ and the effectiveness of spiral radius r depends on the number of maximum iteration.

Nasir et al. (2012c) considered 2-D Rastrigin function within the range of $-5.12 \leq x_i \leq 5.12$ for 50 search points with an angular displacement of $\theta = \pi/4$ and spiral radius of $r = 0.96$. The Rastrigin function is a multimodal function, which has a global minimum at $x_i = [0,0]$ with a fitness value of $f(x) = 0$. Nasir et al. (2012a) applied the HSDBC algorithm to Rastrigin function. The function achieved its minimum within 120 iterations, which showed faster convergence over SpDO and BFA.

Nasir et al. (2012c) considered 2-D Griewank function within the range of $-600 \leq x_i \leq 600$, which has a global minimum at $x_i = [0,0]$ with a fitness value of $f(x) = 0$. They applied adaptive SpDO algorithm, which results in faster convergence. Nasir et al. (2012a) applied HSDBC algorithm to Griewank function for 50 search points with an angular displacement of $\theta = \pi/4$ and spiral radius of $r = 0.96$. The function achieved its minimum within 120 iterations, which shows improved performance over SpDO and BFA.

Nasir et al. (2012c) applied SpDO algorithm to 3-D Sphere function defined in Equation 7.59, which has a global minimum at $x_i = [0,0,0]$ with a fitness value of $f(x) = 0$. They used 30 search points with an angular displacement of $\theta = \pi/4$ and spiral radius of $r = 0.96$. The function achieved its minimum within 80 iterations. Nasir et al. (2012b) applied the HSDBC algorithm to Sphere function within the range of $-5.12 \leq x_i \leq 5.12$ with 200 chemotaxis, which shows improved performance over SpDO and BFA.

Nasir et al. (2012c) applied the SpDO algorithm to 3-D Ackley function defined in Equation 7.60 within the range of $-32.768 \leq x_i \leq 32.768$ for 30 search points with an angular displacement of $\theta = \pi/4$ and spiral radius of $r = 0.96$. The Ackley function has a global minimum at $x_i = [0,0,0]$ with a fitness value of $f(x) = 0$. The function achieved its minimum within 200 iterations. Nasir et al. (2012a) applied the HSDBC algorithm to Ackley function within the range of $-32.768 \leq x_i \leq 32.768$ using 30 search points. The function achieved its minimum within 200 iterations, which shows improved performance over SpDO algorithm and BFA.

7.9.1 Modeling and Control Design

Controller design for flexible manipulator system is complicated and challenging task due to the fact that the manipulator system is highly nonlinear and complex. Nasir et al. (2012b,c) applied the hybrid of spiral dynamic and bacterial chemotaxis (SpD-BC) algorithm to optimize the gains of a PD controller design for the flexible-link manipulator. The algorithm used 30 search points with an angular displacement of $\theta = \pi/4$ and spiral radius of $r = 0.96$ to optimize the controller parameters. The function achieved its minimum within 200 iterations. The hybrid SpD-BC approach shows better performance than that of the root locus technique (Mohamed and Ahmad, 2008). Nasir et al. (2012a) also applied the HSDBF algorithm for global optimization with applications to control design. The hybrid HSDBF algorithm shows better performance.

Modeling flexible robot manipulator is also complicated task due to its inherent nonlinearity and modes of vibration to be considered in the model description. Nasir and Tokhi (2013) applied a hybrid SpDO algorithm with random-chemotaxis to modeling flexible robot manipulator. The combined bacterial chemotactic strategy with spiral dynamics effectively preserves the strength of both algorithm and improves their performances. The proposed algorithm is used as an optimization tool to estimate the parameters of high-order dynamic model of a flexible manipulator system. The results show that the proposed algorithm has faster convergence speed and higher accuracy compared to individual algorithms.

SpDO algorithm has been applied to modeling of a single-input multi-output flexible manipulator system. From a control point of view, the hub-angle and end-point acceleration are of interest for control (Siddique, 2014). An eighth-order ARX model is considered for end-point acceleration model of the flexible manipulator. Nasir and Tokhi (2014) and Nasir et al. (2013b) applied HBCSD algorithm, which is a combination of chemotaxis strategy in BFA used for fast exploration strategy due to its large step size and linear adaptive SpDO algorithm used for exploitation strategy due to its dynamic step size. Employing the chemotaxis and SpDO strategies at the initial and final stages, respectively, balances the exploration and exploitation. To make the search operation more efficient, spiral radius and angular displacement are adaptively defined on the basis of linear relationship with respect to fitness of each bacterium. The advantages of the combination of the two approaches are avoidance of local optima and faster convergence speed. The root mean square error with gain function is used as the objective function to be minimized. The objective function is defined as

$$
f(t) = \begin{cases} \left(\dfrac{1}{K} \sqrt{\displaystyle\sum_{i=1}^{K} e(i)^2} \right) \cdot w & \text{if poles} < 1 \\[4ex] \left(\dfrac{1}{K} \sqrt{\displaystyle\sum_{i=1}^{K} e(i)^2} \right) \cdot w + \lambda_{\text{gain}} & \text{else} \end{cases} \tag{7.61}
$$

Here K is the total number of sampled data, $e(i) = y(i) - \hat{y}(i)$ is the difference between actual output $y(i)$ and predicted output $\hat{y}(i)$, w is a gain factor, and poles are the roots of the denominator polynomial in ARX model. If there is any pole outside the stable region, the solution is penalized by adding a large constant λ_{gain} to the error function. The proposed algorithm is used to optimize parameters of a linear parametric model of a flexible robot manipulator system. The number of search points and number of bacteria were set to 50. The proposed hybrid algorithm achieved good results within 1000 iterations. Nasir et al. (2013a) also applied an adaptive SpDO algorithm to the eighth-order ARX model of the flexible manipulator system using the adaptive spiral radius and angle for 5000 iterations. The adaptive SpDO algorithm achieved the best fitness of 15.24, which is far better than the standard SpDO algorithm.

The SpDO algorithm has also been applied to optimize the parameters of a fuzzy controller developed for a five degree-of-freedom two-wheeled robotic vehicle with a moveable payload and that can steer on irregular and inclined surface. The control system of the vehicle consists of five hybrid fuzzy logic controllers with a total of 15 parameters. The control objective is chosen to minimize the MSE of the fuzzy controller measured for each five control loops of the vehicle defined as

$$
\text{MSE} = \min \left\{ \frac{1}{N} \sum_{i=1}^{N} (e_i)^2 \right\}
\tag{7.62}
$$

MSE leads to optimum controller parameters. Almeshal et al. (2013) applied HSDBC algorithm to the controller, which resulted in very fast convergence compared to standard SpDO algorithm. Due to the coupling of the vehicle model and complexity, the stability of the control system is critical subject to changes of the parameter values. Therefore, a constrained optimization approach is to be applied by defining a feasible solution space within the stability region of the robotic vehicle model. The HSDBC algorithm obtained a minimum MSE value of 0.3682 within 150 iterations that clearly improved the control performance.

7.9.2 TRAINING OF NEURAL NETWORK

Apart from many successful applications discussed earlier, the SpDO algorithm has also been applied to training of neural networks (NNs). An NN consists of a set of interconnected neurons (or processing elements) structured in layers. In general, feed-forward NN consists of one input layer, one or more hidden layer with nonlinear activation function, and one output layer neuron with linear activation function. Biases can be set to nonzero or zero. A generic architecture of a feed-forward NN is provided in Chapter 2 (Figure 2.10). The problem is to find the weights of the connectivity of the network for the given architecture to produce the correct output for the function for each corresponding inputs. The weights training of NN can be carried out by minimizing (optimizing) the mean squared of the network error function defined by

$$
\text{MSE}(w) = \min_{w} \left\{ \frac{1}{N} \sum_{i=1}^{N} (y_i - \hat{y}_i)^2 \right\}
\tag{7.63}
$$

Minimum of MSE(w) leads to optimum behavior of the NN where y_i is the target output and $\hat{y}_i$ is the actual output of the network. An NN can be trained using backpropagation learning algorithm with the possible risk of being trapped in local minima without reaching the global optimum (Siddique and Adeli, 2013; Siddique and Tokhi, 2001). Khan and Sahai (2013) collected data on human health risk at workplace that represent a checklist of nine human health dimensions. They presented the data for the calculation of health risk index in the form of an NN and proposed a neuro-swarm spiral dynamics search (NSSS) algorithm for the training of the NN.

The idea underlying NSSS is to interpret the weight matrices of the NNs as solutions and to update the weights by means of an iterative SpDO-based optimal finding for nine search points. Also exploitation of the SpDO algorithm was increased by using subspirals within the neural network structure, which ensured faster convergence within the maximum number of iteration, which was set to 2000.

7.9.3 COMBINED ECONOMIC AND EMISSION DISPATCH

CEED problem is one of the most important optimization problems in power system operation and forms the basis of many application programs. The main objective of economic load dispatch of electric power generation is to schedule the committed generating unit outputs to meet the load demand at minimum operating cost while satisfying all unit and system constraints. One of those constraints that are always taken into account is the environmental constraints. That is minimization of pollution emission (NO_x, CO_2, SO_x, toxic metals, etc.) in case of power plants. Thus, the CEED poses a bi-objective optimization problem formulated as Min[F,E]. F is the total fuel cost of generators $i = 1, 2, ..., n_g$ defined as

$$\operatorname{Min} F = \left[\sum_{i=1}^{n_g} F_i(P_{G_i}) \right] \tag{7.64}$$

where $F_i(P_G) = a_i P_{G_i}^2 + b_i P_{G_i} + c_i (/h)$, $i = 1, 2, ..., n_g$. P_{G_i} is the real power output, n_g is the number of generators, and a_i, b_i, and c_i are the cost coefficients. The total emission dispatch E from the generators is expressed as the sum of all types of emission considered, such as NO_x, SO_2, CO_2, particles, and thermal emissions:

$$\operatorname{Min} E = \left[\sum_{i=1}^{n_g} E_i(P_{G_i}) \right] \tag{7.65}$$

Here, $E_i(P_{G_i}) = 10^{-2} \left(\alpha_i P_{G_i}^2 + \beta_i P_{G_i} + \delta_i \right)$ (Ton/h), $i = 1, 2, ..., n_g$, α_i, β_i, and δ_i are emission coefficients. The bi-objective CEED problem can then be converted into a single objective optimization problem defined by the following equation (Sayah et al., 2014):

$$\operatorname{Min} C_T = [h \cdot F + (1-h) \cdot P_f \cdot E] \tag{7.66}$$

where C_T is the total operating cost, $h = [0,1]$ is the weighting factor, and P_f is the price penalty factor defined as $P_f = (F(P_{Gi\max})/E(P_{Gi\max}))$/Ton, $i = 1, 2, ..., n_g$, which blends the emission cost with the normal fuel costs. The objective of the CEED problem is to minimize the total operating cost C_T. Benasla et al. (2014) applied SpDO algorithm to three test systems of CEED with 3, 6, and 40 generators. Results of the SpDO algorithm-based optimization of CEED confirm effective high-quality solution. Detailed analysis of experimental results of CEED is provided in Benasla et al. (2014).

7.9.4 CLUSTERING APPLICATIONS

Finding a high-performance search method for mining of huge data known as *big data* is of great current interest (Al-Mubaid and Moazzam, 2012). A requisite for mining of big data is an effective clustering algorithm. While a number of clustering and classification algorithms have been developed in recent years (Cabrerizo et al., 2012; Hsu, 2013a,b; Kodogiannis et al., 2013; Li et al., 2013), the search for effective algorithms continues.

Since the most of traditional search methods are unable to satisfy the current needs of data mining, finding a high-performance search method for data mining has gradually become a critical issue. Tsai et al. (2014) applied a distributed SpDO (dSpDO) algorithm to solve the clustering problem. Clustering is the task of assigning a set of objects into homogeneous groups (called clusters) so that the objects in the same cluster are more similar to each other than to those in other clusters based on a similarity metric. In general, clustering algorithms aim to minimize within-cluster variation (i.e., intracluster distance) and maximize the between-cluster variation (i.e., intercluster distance). Some mathematical approach to clustering has been discussed in Chapter 2 in Equations 2.93 through 2.96.

Unlike the standard SpDO algorithm, which rotates the points around the elitist center iteratively, the proposed dSpDO algorithm splits the population into several subpopulations so as to increase the diversity of search to further improve the clustering result. The *k*-means (Bello-Orgaz et al., 2012) and oscillation methods are also used to enhance the efficiency of dSpDO. The *k*-means dSpDO algorithm is applied to the clustering problem. The results show that the proposed algorithm is quite promising.

7.9.5 Heat Sink Design

The design of optimal heat sinks for modern electronics has been a major problem in electronics industry. The aim is to enhance the thermal efficiency of the heat sink that maximizes heat transfer from the chip to surrounding media using dissipating structures such as pin-fins, plate-fins, or microchannels. A popular model of heat sink is a rectangular microchannel heat sink based on entropy generation minimization criterion. Cruz et al. (2015) applied SpDO algorithm to optimize the design of the heat sink under different scenarios, materials (e.g., silicon, aluminum, and copper), working fluids (e.g., air and ammonia gas), and volume flow rates and compared the performance with unified PSO (UPSO) and SA. The best configuration was an aluminum heat sink with ammonia gas as working fluid. They observed that SpDO algorithm required less iterations than UPSO and SA.

7.10 CONCLUSIONS

Spiral phenomena are very often found in the nature. The phenomena can be described mathematically using parametric curves. Many of the spirals are described especially by logarithmic spiral (Kennelly, 1916). Interestingly enough, the logarithmic spiral can realize an effective meta-heuristic algorithm. On the basis of the logarithmic spirals. The SpDO algorithm has been proposed firstly by Tamura and Yasuda and then followed by many others. The algorithm is simple and needs few parameters to set. The SpDO algorithm has been applied to well-known benchmarks problems and many other engineering problems with great success. A number of variants of the SpDO algorithm have also been reported in the recent years, and among them are the adaptive SpDO algorithms and hybridization with BFA, which show better accuracy, faster convergence, and cost-effective computation. The SpDO algorithm is new but very emerging meta-heuristic approach. Yet, a lot of theoretical studies are to be undertaken by researchers for it to be robust meta-heuristic optimization algorithm.

REFERENCES

Aiyoshi, E. and Yasuda, K. 2011. *Metaheuristics and Their Applications*, Tokyo, Japan: Ohmsa.
Almeshal, A. M., Goher, K. M., Nasir, A. N. K., Tokhi, M. O., and Agouri, S. A. 2013. Fuzzy logic optimised control of a novel structure two-wheeled robotic vehicle using HSDBC, SDA and BFA: A comparative study, *2013 18th International Conference on Methods and Models in Automation and Robotics (MMAR)*, August 26–29, Miedzyzdroje, Poland, pp. 656–661.

Al-Mubaid, H. and Moazzam, D. 2012. A model for mining material properties for radiation shielding, *Integrated Computer-Aided Engineering*, 19(2), 151–164.

Bello-Orgaz, G., Menendez, H., and Camacho, D. 2012. Adaptive K-means algorithm for overlapped graph clustering, *International Journal of Neural Systems*, 22(5), 1250018.

Benasla, L., Belmadani, A., and Rahli, M. 2014. Spiral optimization algorithm for solving combined economic and emission dispatch, *Electrical Power and Energy Systems*, 62, 163–174.

Bromstein, I. N. and Semendyayev, K. A. 1998. *Handbook of Mathematics*, Berlin, Germany: Springer Verlag.

Cabrerizo, M., Ayala, M., Goryawala, M., Jayakar, P., and Adjouadi, M. 2012. A new parametric feature descriptor for the classification of epileptic and control EEG records in pediatric population, *International Journal of Neural Systems*, 22(2), 1250001–1250016.

Costa, A. F. and Martinez, E. 1995. On hyperbolic right-angled polygons, *Geometricae Dedicata*, 58(3), 313–326.

Cruz, J., Amaya, I., and Correa, R. 2015. Optimal rectangular microchannel design using simulated annealing, unified particle swarm and spiral algorithm in presence of spreading resistance, *Applied Thermal Engineering*, 84, 126–137.

Darling, D. 2004. *The Universal Book of Mathematics: From Abracadabra to Zenos's Paradoxes*, Hoboken, NJ: John Wiley & Sons.

Gray, A. 1997. *Modern Differential Geometry of Curves and Surfaces with Mathematica, Section 3.7: Clothoids*, 2nd ed., Boca Raton, FL: CRC Press, pp. 64–66.

Hsu, W. Y. 2013a. Application of quantum-behaved particle swarm optimization to motor imagery EEG classification, *International Journal of Neural Systems*, 23(6), 1350026.

Hsu, W. Y. 2013b. Single-trial motor imagery classification using asymmetry ratio, phase relation and wavelet-based fractal features, and their selected combination, *International Journal of Neural Systems*, 23(2), 1350007.

Kennelly, A. E. 1916. *The Application of Hyperbolic Functions to Electrical Engineering Problems*, New York, NY: McGraw-Hill.

Khan, K. and Sahai, A. 2013. Spiral dynamics optimization-based algorithm for human health improvement, *GESJ: Computer Science and Telecommunications*, 37(1), 31–38.

Kodogiannis, V. S., Amina, M., and Petrounias, I. 2013. A clustering-based fuzzy-wavelet neural network model for short-term load forecasting, *International Journal of Neural Systems*, 23(5), 1350024.

Li, D., Xu, L., Goodman, E., Xu, Y., and Wu, Y. 2013. Integrating a statistical background-foreground extraction algorithm and SVM classifier for pedestrian detection and tracking, *Integrated Computer-Aided Engineering*, 20(3), 201–216.

Mohamed, Z. and Ahmad, M. A. 2008. Hybrid input shaping and feedback control scheme of a flexible robot manipulator, *Proceedings of the 17th World Congress of the International Federation of Automatic Control*, July 6–11, Seoul, South Korea, pp. 11714–11719.

Nasir, A. N. K., Raja Ismail, R. M. T., and Tokhi, M. O. 2016. Adaptive spiral-dynamics metaheuristic algorithm for global optimisation with application to modelling of a flexible system, *Applied Mathematical Modelling*, 40(9–10), 5442–5461.

Nasir, A. N. K. and Tokhi, M. O. 2013. A novel hybrid spiral-dynamics random-chemotaxis optimization algorithm with application to modelling of a flexible robot manipulator, *Proceedings of 2013 16th International Conference on Climbing and Walking Robots (CLAWAR2013)*, July 14–17, Sydney, Australia, pp. 667–674.

Nasir, A. N. K. and Tokhi, M. O. 2014. A novel hybrid bacteria-chemotaxis spiral-dynamic algorithm with application to modelling of flexible systems, *Engineering Applications of Artificial Intelligence*, 33, 31–46.

Nasir, A. N. K. and Tokhi, M. O. 2015. Novel metaheuristic hybrid spiral-dynamic bacteria-chemotaxis algorithm for global optimisation, *Applied Soft Computing*, 27, 357–375.

Nasir, A. N. K., Tokhi, M. O., AbdGhani, N. M., and Ahmad, M. A. 2012a. A novel hybrid spiral dynamics bacterial-chemotaxis algorithm for global optimization with application to controller design, *Proceedings of 2012 UKACC International Conference on Control*, September 3–5, Cardiff, UK, pp. 753–758.

Nasir, A. N. K., Tokhi, M. O., AbdGhani, N. M., and Ahmad, M. A. 2012b. A novel hybrid spiral dynamics bacterial-foraging algorithm for global optimization with application to control design, *Proceedings of 12th Annual Workshop on Computational Intelligence (UKCI2012)*, September 5–7, Edinburgh, UK, pp. 1–7.

Nasir, A. N. K., Tokhi, M. O., AbdGhani, N. M., and Raja Ismail, R. M. T. 2012c. Novel adaptive spiral dynamics algorithms for global optimization, *Proceedings of the 11th IEEE International Conference on Cybernetic Intelligent System 2012*, August 23–24, Limerick, Ireland, pp. 99–104.

Nasir, A. N. K., Tokhi, M. O., and AbdGhani, N. M. 2013a. Novel hybrid bacteria foraging and spiral dynamics algorithms, *Proceedings of 13th Annual Workshop on Computational Intelligence (UKCI2013)*, September 9–11, Surrey, UK, pp. 199–205.

Nasir, A. N. K., Tokhi, M. O., Sayidmarie, O., and Raja Ismail, R. M. T. 2013b. A novel adaptive spiral dynamic algorithm for global optimisation, *Proceedings of 13th Annual Workshop on Computational Intelligence (UKCI2013)*, September 9–11, Surrey, UK, pp. 334–341.

Passino, K. M. 2002. Biomimicry of bacterial foraging for distributed optimization and control, *IEEE Control Systems Magazine*, 22, 52–67.

Sayah, S., Hamouda, A., and Bekrar, A. 2014. Efficient hybrid optimization approach for emission constrained economic dispatch with non-smooth cost curves, *Electric Power Energy Systems*, 56, 127–129.

Siddique, N. 2014. *Intelligent Control: Hybrid Approach Using Fuzzy Logic, Neural Networks and Genetic Algorithms*, Springer-Verlag, Heidelberg, New York, London.

Siddique, N. and Adeli, H. 2013. *Computational Intelligence: Synergies of Fuzzy Logic, Neural Networks and Evolutionary Computing*, Chichester, UK: John Wiley & Sons.

Siddique, N. H. and Tokhi, M. O. 2001. Training neural networks: Backpropagation vs genetic algorithms, *Proceedings of the IEEE International Joint Conference on Neural Network*, July 15–19, Washington, DC, pp. 2673–2678.

Steinhaus, H. 1999. *Mathematical Snapshots, 3rd ed.* New York: Dover Publications.

Takada, I. 1999. On rotations and orthogonal projections in n-dimensional Euclidean space, *Journal of Graphic Science of Japan*, 33(1), 33–43.

Tamura, K. and Yasuda, K. 2011a. Primary study of spiral dynamics inspired optimisation, *IEEJ Transactions on Electrical and Electronic Engineering*, 6(S1), 98–100.

Tamura, K. and Yasuda, K. 2011b. Spiral dynamics inspired optimisation, *Journal of Advanced Computational Intelligence and Intelligent Informatics*, 15(8), 1116–1122.

Tamura, K. and Yasuda, K. 2013. The spiral optimisation and its stability analysis, *2013 IEEE Congress on Evolutionary Computation*, June 20–23, Cancun, Mexico, pp. 1075–1082.

Tsai, C.-W., Huang, B.-C., and Chiang, M.-C. 2014. A novel spiral optimization for clustering, *Mobile, Ubiquitous, and Intelligent Computing Lecture Notes in Electrical Engineering*, 274, 621–628.

Venema, G. 2005. *Foundations of Geometry*, Upper Saddle River, NJ: Prentice-Hall.

8 Simulated Annealing

8.1 INTRODUCTION

SA is motivated by analogy to physical annealing process in metals. The principle of the annealing process is to heat the solid-state metal to a high temperature so that the atoms in the metal are in random state and then cooling it down very slowly according to a specific temperature schedule. If the heating temperature is sufficiently high to ensure random state and the cooling process is slow enough to ensure thermal equilibrium, then the atoms will place themselves in a pattern that corresponds to the global energy minimum of a perfect crystal. The patterns of atoms define the state of the metal in the annealing process. Such states are also referred to configurations. The annealing process is also termed as thermal system. Metropolis et al. (1953) proposed a strategy to move to a new state in which the probability that a proposed move to be accepted or rejected is determined by the Metropolis criterion. The criterion mimics the way thermodynamic systems go from one energy level to another. He thought of this after simulating a heat bath on certain chemicals. The method requires that a system of particles exhibits energy levels in a manner that maximizes the thermodynamic entropy at a given temperature value. Also, the average energy level must be proportional to the temperature. The structural properties of solid depend on the rate of cooling. If the liquid is cooled slowly enough, large crystals are formed. However, if the liquid is cooled quickly (quenched), the crystals contain imperfections. Metropolis's method simulated the material as a system of particles. The algorithm simulates the cooling process by gradually lowering the temperature of the system until it converges to a steady (or frozen) state. Kirkpatrick et al. (1983) used the idea of the Metropolis method and applied it to optimization problems. The idea is to use SA process to search for feasible solutions and converge to an optimal solution. This method is called SA. SA is a popular local search algorithm (meta-heuristic) capable of escaping local minima due to its easy-to-implement and convergence properties. There have been many survey articles, book chapters, and books devoted to SA's theoretical development and domains of applications (Aarts and Korst, 1989; Aarts and Lenstra, 1997; Eglese, 1990; Fleicher, 1995; Henderson et al., 2003; Koulamas et al., 1994; Romeo and Sangiovanni-Vincentelli, 1991; van Laarhoven and Aarts, 1987).

This chapter presents an introductory description on statistical thermodynamics and process of annealing. The chapter then presents the notion of optimization based on SA and different annealing or cooling schedules used in SA and neighborhoods of solutions. There are many variants of SA algorithm such as improved and modified variants and hybrids of SA, which are presented next followed by convergence analysis of SA algorithms. The chapter then finally presents some recent applications of SA of selected benchmark and engineering problems.

8.2 PRINCIPLES OF STATISTICAL THERMODYNAMICS

In statistical thermodynamics, predictions are made on the behavior of macroscopic system, for example, annealing in metals, on the basis of laws governing the component atoms. Annealing process at a given temperature approaches a state of equilibrium spontaneously, which is characterized by a mean value of energy depending on the temperature. The annealing process is also seen as a system, which changes from state to state due to temperature change. It is possible to find the smaller values of the mean energy of the annealing by simulating the transition to equilibrium and decreasing the temperature (Cerny, 1985). Let x be the current state of a thermal system and $E(x)$ is the energy of the system at temperature T. The system randomly changes its state from one possible state to another toward the equilibrium state. The probability that the system will be in state x is expressed by the Boltzmann–Gibbs distribution (Kirkpatrick et al., 1983):

$$p(x) = K \cdot \exp(-E(x)/T) \tag{8.1}$$

where K is a constant. If the system is assumed to have a discrete number of possible states, then the mean energy $\bar{E}(x)$ of the system in equilibrium is given by

$$\bar{E}(x) = \left[\frac{\sum_c E_c(x)\exp((-E_c(x))/T)}{\sum_c \exp((-E_c(x))/T)} \right] \tag{8.2}$$

E_c denotes the energy in a particular configuration of x. The numerical calculation of $\bar{E}(x)$ is very cumbersome if the number of configuration is high. Metropolis et al. (1953) first proposed an algorithm for calculation of $\bar{E}(x)$. This algorithm of Metropolis established a connection between statistical thermodynamics and multivariate combinatorial optimization, which was first noted by Pincus (1970). It was Kirkpatric et al. (1983) who proposed that this algorithm could form the basis of a general-purpose optimization technique that can be applied to solve combinatorial problems. This approach is a variant of the well-known heuristic search in which a subset of the feasible solutions is explored by repeatedly moving from the current solution to a neighboring solution. For a minimization problem, the local search employs a descent strategy, that is, the search always moves in the direction of improvement. Such solutions are dependent on the initial starting point, which is chosen randomly.

8.3 ANNEALING PROCESS

In condensed matter physics, annealing is known as a thermal process where low energy of a solid in bath is obtained. This process is carried out in two steps:

1. Increase the temperature of the heat bath to a maximum value at which the solid will start melting
2. Lower the temperature of the heat bath slowly until it solidifies again in the ground state (organized state with the minimum of energy) of the solid

In liquid state, particles of the solid are in random arrangement and have the highest level of energy. In ground state, the particles are highly organized in lattice structure and reach the minimum of energy. The ground state can be obtained only if the solid is heated to a sufficient high temperature and the cooling process is significantly slow; otherwise, the solid will be into a metastable state where the particles are not yet organized and the minimum of energy has not been reached. This process of heating and cooling of material (e.g., glass or metal) is known as annealing, which increases strength and improves material properties. It involves heating a material to above its critical temperature, that is, melting point, maintaining a suitable temperature, and then cooling. Annealing can induce ductility, soften material, relieve internal stresses, refine the structure by making it homogeneous, and improve strain hardening properties. If solid material is heated to its melting point and then cooled back to its solid state, the structural properties of the material depend on the rate of cooling. For example, large crystals are formed by slow cooling. If cooling is done in a faster rate, crystals are formed imperfectly. The laws of thermodynamics state that at temperature T, the probability of an increase in energy of magnitude ΔE is given by (Boltzmann, 1866)

$$p(\Delta E) = \exp\left(-\frac{\Delta E}{k_B T}\right) \tag{8.3}$$

where $\Delta E = E_j - E_i$, E_i is the current energy state and E_j is the new energy state, and k_B is a physical constant known as the Boltzmann constant. If the energy difference has decreased, that is, $\Delta E \leq 0$, the annealing process, that is, the state of the metal, moves to new state j. If the energy difference has increased, that is, $\Delta E > 0$, the annealing process can still move to new state according to the probability given by Equation 8.3. This rule of acceptance of the new state based on energy decrease or on the probability $p(\Delta E)$ is known as the Metropolis criterion or acceptance rule. After accepting a new state, the temperature is decreased. The process is repeated for a number of times. The algorithm based on this rule is known as the Metropolis algorithm (Metropolis et al., 1953).

8.4 SA ALGORITHM

The basic idea of SA algorithm came from statistical thermodynamics, a branch of physics, where a thermal equilibrium is sought to ensure global minimum energy that gives the best properties of metal. This is basically a search procedure whereby the temperature is lowered slowly following a cooling schedule to achieve the energy minimum. Thus, it makes SA a variant of traditional local search or optimization algorithm (usually based on the Metropolis algorithm).

Kirkpatrick et al. (1983) and later Cerny (1985) showed that the Metropolis algorithm can be applied to optimization problems by mapping the physical cooling process onto combinatorial optimization problem. In SA algorithm, system states are equivalent to feasible solutions, energy is equivalent to value of the cost (or fitness) function for a solution, ΔE is considered as the difference of computational cost between solutions i and j, change in states is meant the solution in the neighborhood of the current solution, temperature is comparable to control parameter, and the frozen state is equivalent to the equilibrium state when a heuristic solution is found by SA algorithm. Suppose it is a minimization problem over a set of feasible solution space S and a cost function $f(\cdot): S \to R$ defined for all $x \in S$. In general, an optimal solution can be found by an exhaustive computation of the cost function $f(x)$ for all $x \in S$ and selecting the minimum. But practically, this is nearly impossible because the solution space S is really big. An alternative is to search a small set of feasible solution space $S' \subset S$. A neighborhood structure is defined and search is performed on the neighborhood of the current solution $x \in S$. If an improvement in the cost function is found, the current solution is replaced by the improved solution and the process is repeated. If there is no improvement found to the cost function in the neighborhood, the current solution is considered as an approximation to the optimum. The local search process can be described as follows:

Step 1: Select an initial solution $x_0 \in S$
        Calculate fitness $f(x_0)$
Step 2: Select $x_i \in N(x_0)$ from the neighborhoods
Step 3: Calculate fitness $f(x_i)$
Step 4: If $f(x_i) < f(x_0)$
        $x_0 = x_i$
Step 5: If (termination condition not met), Goto Step 2
Step 6: Return solution

where $N(\cdot)$ is the neighborhood function that generates a new solution in the neighborhood of x_0.

On the basis of the concept of the local search, an outline of the SA algorithm can be formulated in the following five steps:

Step 1: The algorithm starts with an initial solution, which is considered as the current solution. A new solution is generated randomly in the neighborhood of the current solution. The acceptance of the new solution over the current solution is determined by a probability distribution with a scale proportional to the current temperature.

Step 2: The new solution is evaluated. It determines whether the new solution is better or worse than the current solution. If the new solution is better, then the current solution is replaced with the new solution. If the new solution is worse, the algorithm may still accept the new solution based on a probability threshold value.

Step 3: The algorithm systematically lowers the temperature and the threshold value and stores the best solution found so far.

Step 4: Reannealing raises the temperature after a certain number of new points have been accepted, and starts the search again at the higher temperature. This avoids the algorithm getting stuck at local minima.

Step 5: The algorithm stops when the objective function reaches a desired minimum value, or when any other stopping criteria are met.

On the basis of the five steps outlined above, a standard SA procedure begins by generating an initial solution at random. Let x_i be the current solution of an optimization problem and x'_i be the new solution after small random change to the current solution in the neighborhood of x_i at temperature T. The objective function value $f(x_i)$ is also termed as energy denoted by $E(x_i)$. The energy $E(x'_i)$ of the new solution is computed. The change in energy $\Delta E = f(x'_i) - f(x_i) = E(x'_i) - E(x_i)$ is defined as the change in the objective function value. On the basis of the value of ΔE, a decision is made on the solution x'_i to be accepted or rejected. If $\Delta E \leq 0$, then x'_i is accepted and x'_i is used as the starting point for the next iteration. If $\Delta E > 0$, then the probability $p(x)$ is the chance of acceptance of x'_i. Otherwise, a new solution is generated. The probability $p(x)$ of accepting a new solution is given by

$$p(x) = \begin{cases} 1 & \text{if } f(x'_i) < f(x_i) \\ \exp\left(-\dfrac{f(x'_i) - f(x_i)}{T}\right) & \text{otherwise} \end{cases} \tag{8.4}$$

The calculation of this probability in Equation 8.4 relies on the parameter T, which is referred to as temperature, since it plays a similar role as the temperature in the physical annealing process. Thus, at the beginning of SA, most worsening solutions may be accepted, but at the end, only improving ones are likely to be taken. This helps the procedure avoid a local minimum in an optimization process. The algorithm is terminated when the objective function $f(x)$ reaches a desired minimum value or after a prespecified run time. The standard SA algorithm is described by the following steps:

Step 1: Randomly initialize x_i within the solution space
    Initialize temperature T, set iteration $k = 1$
Step 2: Calculate fitness $f(x_i)$
Step 3: Select $x'_i = N(x_i)$
Step 4: Calculate fitness $f(x'_i)$
Step 5: Calculate $\Delta E = f(x'_i) - f(x_i) = E(x'_i) - E(x_i)$
    Calculate $p(x)$ according to Equation 8.3
    If $f(x'_i) < f(x_i)$
     $x_i = x'_i$
    Else If [rand(0, 1) $< p(x)$]
     $x_i = x'_i$
    $f(x_i) = f(x'_i)$
Step 6: Decrease $T = g(T, k)$ using a cooling scheme
Step 7: If (termination condition not met), set $k = k + 1$ and Goto Step 3
Step 8: Return solution

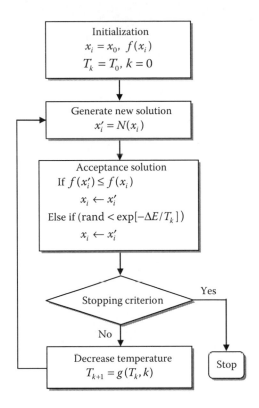

FIGURE 8.1 Flow chart of a typical SA algorithm.

where $g(T,k)$ is cooling function and usually a decreasing function that provides the next value for the temperature and rand[0,1) is a random number without 0. $N(x)$ is a neighborhood function, which generates a new solution in the neighborhood of a solution x.

Annealing is the technique of closely controlling the temperature when cooling a material to ensure that it is brought to an optimal state. The temperature T is the control parameter in SA. It determines the probability of accepting a worse solution at any step and is used to limit the extent of the search in a given dimension. Associated with the temperature parameter T is a cooling or annealing schedule function $g(T,k)$, which determines systematically the decrease of the temperature as the algorithm proceeds. As the temperature decreases, the algorithm reduces the extent of its search to converge to a minimum. The process of SA algorithm is illustrated in a flow diagram in Figure 8.1. There is a number of cooling (or annealing) schedules or functions used in SA. These are discussed in detail in Section 8.5. A general problem in SA is that it can be stuck at local minima that may exist in the solution space. Reannealing is a technique that raises the temperature after a certain number of new solutions have been accepted, and starts the search again at the higher temperature and helps avoiding in getting caught at local minima.

8.5 COOLING (ANNEALING) SCHEDULE

It can be seen from Equation 8.4 that the probability of accepting solution mainly depends on the temperature. Therefore, the cooling (or annealing) schedule is critical to the performance of the SA algorithm. For a given random process, cooling at a too fast rate will likely freeze in a nonglobal minimum. Cooling at a too slow rate, while reaching the desired minimum, is a waste of computational resources and time. The challenge then becomes deriving the fastest cooling schedule that will guarantee convergence to the global minimum. In locating the minimum, one must start at an initial

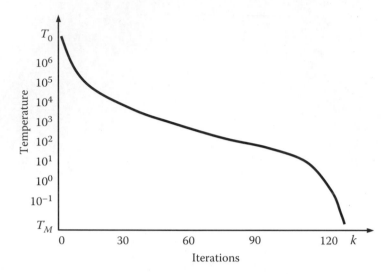

FIGURE 8.2 Generic annealing schedule.

temperature, which is a state in a D-dimensional space, evaluate the function at that state, generate the next state by reducing or changing the temperature, and stop the SA algorithm using a criterion.

The cooling schedule is the rate by which the temperature is decreased as the annealing algorithm proceeds. The slower the rate of decrease, the better the chances are of finding an optimal solution, but the longer the run time. The schedule of a combinatorial optimization problem using SA algorithm is a decreasing sequence of temperatures. A generic annealing schedule is a curve dictating the decrease of temperature with iterations as shown in Figure 8.2. This is a free parameter and a critical factor contributing to the performance of the annealing algorithm. The cooling schedule depends on the initial temperature T_0, the final temperature T_M, decrement in temperature in each step, and the number of iterations at each temperature. The cooling schedule should be selected in such a way so that the computation is minimized and at the same time the algorithm converges to a global optimum. It is practical to perform small number of iterations at high temperature as the rate of acceptance is high anyway.

In general, cooling schedules are almost always heuristic and depend on the problem domain. A good cooling schedule is one that strikes a balance between moderate execution time and SA's asymptotic behavior. There are several theoretical and empirical cooling schedules being suggested in the literature that have been applied to various applications by researchers. These schedules can be broadly categorized into three classes:

- Monotonic schedules
- Geometric schedules
- Adaptive schedules

By monotonic it is meant a monotonic function that is decreasing over time or iteration. If monotonic schedule is adopted, the end of the search has less chance of getting escaped from the local minima. In geometric schedule, annealing procedure involves first "melting" the system at a high temperature, then lowering the temperature by a constant factor. Adaptive schedules use an adjustment mechanism for the rate of decrease of temperature from available information during the execution of the SA algorithm. Some of the widely used cooling schedules found in the contemporary literature are discussed in the following. Interested readers are also directed to Koulamas et al. (1994) and Rose et al. (1990) for further annealing schedules and analyses.

A cooling schedule has the following parameters: (i) an initial value of temperature T_0, (ii) a decrement function for temperature T, (iii) the number of iterations to be performed at each temperature, and (iv) a termination criterion, which may be a temperature T_M or the maximum number of iterations.

8.5.1   MONOTONIC SCHEDULES (OR SIMPLE TIME SCHEDULE)

In monotonic schedules, temperature can decrease in a linear or nonlinear fashion. Figure 8.3 shows a linear and nonlinear temperature schedule. Linear schedule is widely used since it was introduced by Kirkpatrick et al. (1983). It is defined as

$$T(k) = -\eta k + T_0 \tag{8.5}$$

where $T(k)$ is the temperature at iteration k, T_0 is the initial temperature, and η is a decreasing factor. The decreasing factor η is eventually the slope describing the decrease of temperature from T_0 to the final temperature T_M. The value of η decides how fast the temperature will fall from T_0 to T_M with increasing iterations.

There are other monotonic cooling schedules as well, for example,

$$\frac{1}{T(k+1)} = \frac{1}{T(k)} + c \tag{8.6}$$

$$T(k+1) = \frac{T(k)}{1 + cT(k)} \tag{8.7}$$

where c is a constant and the value of c can be chosen arbitrarily.

Hoffmann and Salamon (1990) set up equations to derive an optimal schedule that minimizes the mean energy at a particular time (i.e., iteration). This optimum schedule did not follow a set of the Boltzmann distributions but rather a turnpike solution. The turnpike solutions can be used as efficient initial approximations in iterative search methods. This schedule is an approximation of temperature defined by

$$T(k) \sim \frac{D-1}{\log(k)} \tag{8.8}$$

where $D > 1$ is the dimension of the search space and $k > 1$ is the iteration number.

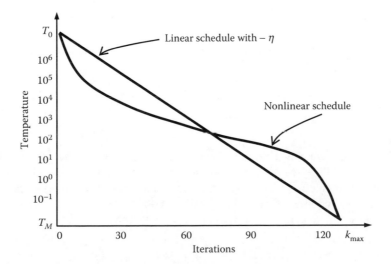

FIGURE 8.3   Linear and nonlinear annealing schedule.

Golden and Skiscim (1986) used a different formulation for linear cooling as defined by

$$T(k) = T_0 \frac{c-k}{c} \tag{8.9}$$

where c is a constant to be chosen arbitrarily, k is the current iteration, and the cooling starts with an initial temperature T_0.

Sekihara et al. (1992) used a temperature decrement rule defined as follows:

$$T(k) = \begin{cases} \dfrac{T_0}{(1+k)} & \text{if } k \le k_{\text{lim}} \\ \alpha \cdot T(k-1) & \text{if } k > k_{\text{lim}} \end{cases} \tag{8.10}$$

where the decrease of temperature will be very slow for large k due to the term $T_0/(1+k)$. Therefore, when k starts growing large, that is, $k > k_{\text{lim}}$, it is switched to faster decrement according to $\alpha \times T(k-1)$ at k_{lim}. Sekihara et al. (1992) used $\alpha = 0.95$ in their experiment. k_{lim} is a parameter to be determined empirically intended to improve the convergence and consequently the efficiency of the algorithm.

Kirkpatrick et al. (1983) suggested a proportional cooling schedule where the temperature at k-th iteration is determined by

$$T(k) = \alpha T(k) - 1 \tag{8.11}$$

where $0 < \alpha < 1$ is a parameter called cooling ratio. If proportional cooling is used, the parameter α should be determined in advance. The value of α can be determined according to Potts and Van Wassenhove (1991) from the initial temperature T_0, the total number of iterations M, and the temperature at final iteration T_M as follows:

$$\alpha = \left(\frac{T_M}{T_0}\right)^{1/(M-1)} \tag{8.12}$$

The values of the parameters T_0, T_M, and M should be determined in advance for this scheme empirically.

Johnson's cooling schedule is widely used in many applications (Johnson et al., 1989, 1991). The schedule is described by

$$T(k) = \frac{T_0}{(1+\beta T_0)} \tag{8.13}$$

where $T(k)$ is the temperature at iteration k and β is a coefficient for the initial temperature T_0. Johnson's cooling schedule is simple and linear and can be applied to a wide range of applications, which saves a lot of computation time. When the search space is not big and does not involve many search points, Johnson's cooling schedule is a suitable choice.

Lundy's cooling schedule is the classic cooling schedule defined by

$$T(k+1) = \exp\left(-\frac{c}{T(k)}\right) \tag{8.14}$$

where $T(k)$ is the temperature at iteration k and c is an arbitrary constant. In general, a large number of iterations are done at a lower temperature and a small number of iterations are carried out at a higher temperature. In order to use an average number of iterations at both temperatures, Lundy's average cooling schedule is used, which is defined by

$$T(k+1) = 1 - \frac{c}{T(k)} \tag{8.15}$$

The average cooling schedule in Equation 8.15 reduces the average number of iterations.

Lundy and Mees (1986) proposed another cooling schedule defined by

$$T(k+1) = \frac{T(k)}{1 + \beta T(k)} \tag{8.16}$$

where $T(k)$ is the temperature at iteration k and $\beta > 0$. This cooling scheme becomes slower than proportional cooling scheme at later stages in SA. Connoly (1990) determined the value of β from the initial temperature T_0, the total number of iterations M, and the temperature at final iteration T_M using the following formulation:

$$\beta = \frac{T_0 - T_M}{(M-1)T_0 T_M} \tag{8.17}$$

Van Laarhoven et al. (1992) used an adaptive value for β in their application where β is based on the standard deviation $\sigma(k)$ of the objective function values of the solutions generated at k-th iteration defined as follows:

$$\beta(k) = \ln(1+\delta) \cdot 3\sigma(k) \tag{8.18}$$

where δ is an empirical distance parameter.

Geman and Geman (1984) introduced a logarithmic cooling scheme, which has some theoretical importance due to asymptotically very slow temperature decrease and is defined by

$$T(k) = \frac{c}{\ln(d+k)} \tag{8.19}$$

where d is usually set equal to 1. c is a constant. It has been proven that c is greater than or equal to the largest energy barrier in the problem. In other words, c should be greater than the depth of local minimum. Johnson et al. (1989) used an alternative formulation for logarithmic cooling defined by

$$T(k) = \frac{c}{1 + \log(k)} \tag{8.20}$$

where c is a constant. Their experiments showed that the algorithm almost converges to an optimal solution for a long enough time. This schedule will lead the system to the global minimum state in the limit of infinite iterations (Hajek, 1988). However, logarithmic cooling is asymptotically very slow in decreasing the temperature. It is impractical outright as it simply undoes the exponential Boltzmann acceptance function and causes a random search.

Van Laarhoven (1992) proposed a cooling schedule, which decreases according to

$$T(k+1) = \frac{T(k)}{1+T(k)((\ln(1+\delta))/3\sigma(k))} = \frac{T(k)}{1+T(k)((\ln(1+\delta))/3\sqrt{\text{Var}[T(k)]})} \tag{8.21}$$

where $\sigma(k)$ is the standard deviation of the previously visited solutions and defined as $\sigma(k) = \sqrt{\text{Var}[T(k)]}$. δ is an empirical distance parameter.

A similar decrement rule was provided by Lundy and Mees (1986),

$$T(k+1) = \frac{T(k)}{1+T(k)(\gamma/E_u)} \tag{8.22}$$

where γ is a small positive value and E_u is the upper bound to the deviation of the occurring energy values to the optimum. The cooling schedule proposed by Lundy and Mees (1986) is a good simplification of the cooling schedule in Equation 8.21.

Another related cooling schedule was provided by Otten and van Ginneken (1984), where the temperature variance is considered as follows

$$T(k+1) = T(k) - \frac{1}{Z(k)} \times \frac{T^3(k)}{\text{Var}[T(k)]} \tag{8.23}$$

where $Z(k)$ is defined as

$$Z(k) = T(k)\frac{C_{\max} + T(k)\ln(1+\delta)}{\text{Var}[T(k)]\ln(1+\delta)} \tag{8.24}$$

$C_{\max}$ is an estimation of the maximum value of the cost function and δ is an empirical distance parameter and usually set to a small positive value.

8.5.2 GEOMETRIC (OR EXPONENTIAL) COOLING SCHEDULE

The disadvantage of a logarithmic schedule is that the time required to reach the next temperature is exponential. This is impractical in terms of computation time. A common solution to this would be to use geometrical schedule (Ortner et al., 2007). The annealing procedure involves first "melting" the system at a high temperature, then lowering the temperature by a factor β^k, taking enough steps at each temperature to keep the system close to equilibrium, until the system approaches the ground state. These result in a geometrical cooling schedule defined (Nourani and Andresen, 1998) as

$$T(k) = T_0\beta^k \tag{8.25}$$

where $T(k)$ is the temperature at iteration k and T_0 is the initial temperature and β is a small positive value within the range of $0 < \beta < 1$ that tunes the cooling speed.

8.5.3 ADAPTIVE COOLING

In standard SA, search begins with high temperature allowing higher chances of overcoming local minima. As the search continues, temperature decreases monotonically allowing lower chances of uphill move. If the local minima are encountered towards the end of search at relatively low temperature, the SA will be stuck at local minima. The adaptive cooling schedule takes into consideration the statistical behavior of the search trajectory and adjusts the temperature dynamically at each iteration. The simplest mechanism is to decrease the temperature by a factor γ and γ is varied in an adaptive way. Fodorean et al. (2012) used the cooling method described by

$$T(k+1) = \gamma T(k) \tag{8.26}$$

where $T(k)$ is the temperature at iteration k. To avoid getting trapped at a local minimum, the rate of reduction in temperature should be slow. This is ensured by choosing γ between [0.90,0.99]. This cooling is similar to proportional cooling when the factor γ is fixed.

The adaptive cooling schedule is based on the energy scale of the system and is defined by

$$T(k+1) = T(k) - \frac{cT_0^2}{\sigma_E} \tag{8.27}$$

where T_0 is the initial temperature, σ_E/T_0^2 is the rate of change of energy, and c is a small positive constant.

Azizi and Zolfaghari (2004) proposed an adaptive cooling schedule that adjusts the temperature dynamically based on the profile of the search path. Such adjustment can be in any direction including possible reheating. The single function defined in Equation 8.28 can maintain the cooling and heating up when necessary.

$$T(k) = T_{min} + \lambda \ln(1 + r(k)) \tag{8.28}$$

T_{min} is the minimum temperature SA can have, λ is a coefficient that controls the rate of temperature increase, and $r(k)$ is the number of consecutive upward moves at iteration k. The initial value of $r(k)$ is 0, which leads to initial temperature $T(0) = T_0 = T_{min}$. T_{min} can take any value greater than zero. The initial temperature T_0 has twofold use. It prevents the probability function in Equation 8.3 from division by zero, and secondly, it determines the initial value for temperature. λ is the rate of temperature change. Higher value of λ causes faster changes of the temperature. When large value is assigned to λ, less time is spent in the exploitation in the current neighborhood. When small value is assigned to λ, more time is spent for search in the current neighborhood. The study by Azizi and Zolfaghari (2004) suggests that the best results are obtained when T_0 and λ are set to 1. Depending on the value of computational cost (or energy) $\Delta E(k)$, $r(k)$ is updated according to Equation 8.29, which leads to a good chance of downhill moves at the beginning of the search that ensures less need to a high temperature to push the search out of local minima.

$$r(k) = \begin{cases} r(k-1)+1 & \text{if } \Delta E(k) > 0 \\ r(k-1) & \text{if } \Delta E(k) = 0 \\ 0 & \text{if } \Delta E(k) < 0 \end{cases} \tag{8.29}$$

Another widely used adaptive cooling schedule was introduced by Huang et al. (1986) incorporating the variance of temperature

$$T(k+1) = T(k)\exp\left(-\frac{\lambda T(k)}{\sqrt{\text{Var}[T(k)]}}\right) \tag{8.30}$$

with $0 < \lambda \le 1$. A similar approach to Equation 8.30 was introduced by Triki et al. (2005)

$$T(k+1) = T(k)\left(1 - T(k)\frac{\Delta T(k)}{\text{Var}[T(k)]}\right) \tag{8.31}$$

Triki et al. (2005) demonstrated that these adaptive cooling schedules, Equations 8.30 and 8.31, are equivalent to each other.

A number of researchers defined adaptive cooling as a parallel approach called ensemble-based annealing (Frost and Salamon, 1993; Hoffmann et al., 1991; Ruppeiner et al., 1991). In order to reduce the fluctuation of energy values after temperature decrease, the ensemble-based scheduling estimates the ensemble average of the energy. If $\Delta E(k + 1)$ is greater than $\Delta E(k) + \gamma \text{ Var}[\Delta E(k + 1)]$, then it is assumed that the energy value fluctuates around the expectation value of $\Delta E(k + 1)$. By using a suitable arbitrary value of γ, the rate of temperature decrease can be controlled. These adaptive cooling schemes require the measurement of at least the variance and sometimes mean value of the fluctuating energy values over some time (Schneider and Puchta, 2010).

Beside the cooling schemes discussed in Sections 8.5.1–8.5.3, a number of researchers investigated different cooling schemes, initial temperature, number of iterations at each temperature, rate of decrease of temperature, final temperature, stopping criterion, and related performance for SA in the 1990s. Romeo and Sangiovanni-Vincentelli (1991) performed theoretical investigations into the performance of SA in terms of initial temperature, the number of loops for each temperature, the rate of decrease of temperature, and the stopping criteria. Their theoretical investigations were unable to explain the mysteries of the success of SA. These investigations suggest that the effectiveness of cooling schedules can only be compared through experimentation. This also led them to guess that the neighborhood and the corresponding topology may be responsible for the performance of the SA algorithm. Strenski and Kirkpatrick (1991) investigated a number of small problems experimentally. They reported that optimal cooling schedules are not monotone decreasing in temperature. They also found that geometric and linear cooling rates provide a better result than logarithmic scheduling. Further, they revealed that there was not a significant difference in performance between the linear and exponential cooling schedules. With regard to the starting temperature, they found that excessively high initial temperatures do not greatly improve the optimization of the algorithm when a geometric cooling factor is used. Cohn and Fielding (1999) conducted detailed analysis of various cooling schedules to assess the performance of SA and provided examples of when very high, fixed temperature, or even fast cooling schedules can be applied. Fielding (2000) conducted a series of studies for fixed temperature cooling schedules for a number of benchmark problems that demonstrate a fixed temperature cooling schedule can yield superior results in locating optimal and near-optimal solutions. Orosz and Jacobson (2002) also reported finite-time performance measures for SA with fixed temperature cooling schedules.

8.5.4   INITIAL TEMPERATURE

Kirkpatrick et al. (1983) suggested that the value of initial temperature T_0 should be set large enough to make the initial probability of accepting transitions be close to 1. Strenski and Kirkpatrick (1991) reported that excessively high initial temperatures do not greatly improve the optimization of the algorithm when a geometric cooling is used. They also suggested that the initial temperature be chosen such that the fraction of accepted uphill transitions (when $\Delta \ge 0$) in a trial run is equal to a given acceptance probability p_0. If the initial temperature is too low at the beginning of the standard SA, the algorithm can get trapped at local optimum leading to a poor solution. If the initial temperature is too high, the algorithm will require too many iterations leading to time-consuming computation by accepting too many hill-climbing moves. On the basis of this observation, Varanelli and Cohoon (1999) proposed a method for determining an initial temperature for two-staged SA algorithm using traditional cooling schedules. The initial temperature T_0 can be evaluated by the probability p_0 defined according to Equation 8.4.

$$p_0 = \exp\left(-\frac{\Delta E}{T_0}\right) \tag{8.32}$$

From Equation 8.32, T_0 can be estimated as

$$T_0 = -\frac{\Delta E}{\log(1 - p_0)} \tag{8.33}$$

The initial temperature should ideally be set to very high for all trials at that temperature so that it increases the value of the objective function to be accepted.

8.5.5 FINAL TEMPERATURE

Theoretically, the temperature should be allowed to decrease to zero before stopping condition is met. However, there is no need to decrease the temperature to zero. In a method suggested by Van Laarhoven et al. (1992), SA is stopped when temperature drops down to a preselected final temperature (T_M). As temperature T approaches zero, probability of accepting any uphill moves will be indistinguishable from zero. It is likely that the chances of an escape from the current local optimum will be negligible. Thus, the criterion for stopping can be formulated in terms of a minimum value of the temperature or in terms of freezing the system at current solution. Lundy and Mees (1986) suggested the following formulation:

$$T \leq \frac{\varepsilon}{\ln[(|S|-1)/P]} \tag{8.34}$$

where S is the solution space, $|S|$ is the cardinality of the solution space, ε is a small allowable range, and the solution should be within ε of the optimum solution with probability of P.

8.5.6 STOPPING CONDITION

SA can be terminated when the iteration count reaches a predetermined maximum count M or the total number of trials reaches a predetermined limit K_M. In a method suggested by Van Laarhoven et al. (1992), SA is stopped when temperature drops down to a preselected final temperature (T_M). On the other hand, in a method suggested by Johnson et al. (1989), a counter is incremented by one when an iteration is completed with the fraction (or percentage) of accepted moves less than a predetermined limit and the counter is reset to 0 when a new solution is found. In Johnson et al.'s method, SA is terminated when the counter reaches a predetermined limit K_M. It is difficult to select an annealing schedule that gives the best performance of an SA algorithm since various schemes for defining the schedule should be selected and values of parameters included in the schemes should be determined at the same time. Moreover, some of these parameter values must be determined simultaneously because of correlation among them. Geman and Geman (1984) provided a sufficiency proof for a lower bound on a schedule described by $1/\log(t)$, where t is an artificial time measure for the annealing schedule.

8.6 NEIGHBORHOODS

A neighborhood of a point $x \in R^n$ is the set defined by $\{x' \in R^n \mid \|x' - x\| < \varepsilon\}$, where ε is some small positive number. A neighborhood is also called a ball with radius ε and center x. A general supposition for SA algorithm is that the number of iterations at each temperature depends on the structure and size of the neighborhood function and solution space and may vary from temperature to temperature, which have direct influence and impact of the performance of the SA algorithm and the quality of the solution. For example, it is important to perform fewer iteration at high temperature to minimize computation and more iteration at low temperature to ensure full exploitation for local optimum. Therefore, suitable neighborhood function definition is essential to influence the efficiency of SA (Moscato, 1993). While defining neighborhood structure, smooth topology with shallow local minima is preferred to bumpy topology with many deep local minima (Eglese, 1990; Fleischer and Jacobson, 1999). This is also supported by Hajek (1988) where it is shown that the convergence to global optima depends on the depth of the local minima.

Another important factor to consider for neighborhood functions is the size of neighborhood. Henderson et al. (2003) reported in a practical guideline for the implementation of SA algorithm that no theoretical results are available for determining the size of neighborhood. Cheh et al. (1991) and Goldstein and Waterman (1988) supported the view that small neighborhoods are seemingly the best, while Ogbu and Smith (1990) provided evidence that larger neighborhoods result in better performance (Figure 8.4).

Most research on SA has concentrated on the *update* and *accept* function and various algorithmic parameters. Limited research has been reported on the *generate* function, which describes the neighborhood to create a new solution $x' \in N(x)$ with higher probability. However, the *generate* function decides on the neighborhood structure and size of a local search algorithm regardless of whether it is a deterministic one or a stochastic one. The neighborhood $N(x)$ of a configuration x is defined by

$$N(x) = \{x' \mid x' \in S, g_{xy}(T_M) > 0\} \tag{8.35}$$

where $g_{xy}(T_M)$ is the probability of generating configuration y from configuration x at temperature T_M.

Research on SA hitherto only assumed that $g_{xy}(T_M) = 1/(\mid N(x)\mid)$, where $\mid N(x)\mid$ is the size of neighborhood (shown in Figure 8.4), that is, number of configurations to be considered in $N(x)$ for all $x \in S$, where S is the search space.

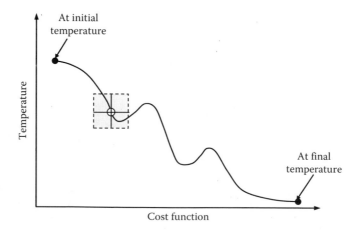

FIGURE 8.4   Neighborhood.

Moreover, $|N(x)|$ is fixed during the search. Goldstein and Waterman (1988) and Cheh et al. (1991) carried out some experiments on different neighborhood sizes in SA, but the sizes are all fixed during the run in all cases.

The interesting feature of SA is that it is able to perform exploration of the search space at the high-temperature stages to locate promising regions and exploitation of a subspace at low temperatures to find a good near-optimal solution into the same algorithm. The fixed size neighborhood clearly does not conform to the basic strategy of the SA. Fast SA (Szu and Hartley, 1987a,b) can be regarded as an example of SA with a dynamic neighborhood size, but it is used only in the continuous case. The application of dynamic neighborhood size in combinatorial optimization has not been well-studied. Yao (1992) has studied SA with extended and dynamic neighborhood size, which can adjust itself in the different search stages.

8.7 VARIANTS OF SA

Metropolis et al. (1953) first proposed an algorithm to simulate the behavior of physical systems in the presence of a heat bath. Thirty years later, Kirkpatrick et al. (1983) applied the Metropolis algorithm to combinatorial optimization problems and named it SA. Bohachevsky et al. (1986) applied the SA algorithm to solve continuous optimization problems followed by a number of applications of SA to optimization problems by many researchers. Since then, a number of researchers have devoted to improvement of the SA algorithm (Corana et al., 1987; Dekkers and Aarts, 1991; Ingber, 1989, 1993, 1996; Szu and Hartley, 1987a,b). There are many variants and improvements made to standard SA. There are mainly three approaches applied to improve SA, which in turn yielded many variants of SA: (i) variants based on cooling schedule, (ii) variants based on neighborhood selection, and (iii) variants based on learning mechanism. In the sequel, some of these variants are discussed.

8.7.1 BOLTZMANN ANNEALING

The standard SA needs a lot of time to converge to the optimal solution. It is well-known that the standard SA requires a cooling schedule in which the temperature T must satisfy the following equation:

$$T(k) = \frac{T_0}{\ln(1+k)} \qquad (8.36)$$

where $T(k)$ is the temperature at iteration k. Equation 8.36 characterizes the simplest cooling scheme for SA, which is also called the Boltzmann annealing.

8.7.2 FAST ANNEALING

The Boltzmann cooling in Equation 8.36 characterizes the simplest cooling scheme for SA. As k becomes very large over the iterations while $T(k)$ approaching to zero, the Boltzmann cooling scheme becomes slow. A faster cooling scheme can be achieved by using a Cauchy distribution. The Cauchy distribution is a probability obeying function, which is inversely linear in time (Szu and Hartley, 1987b). The Cauchy distribution defined by Szu and Hartley (1987a,b) is as follows:

$$g(k) = \frac{T(k)}{(T(k)^2 + |\Delta x|^2)^{((D+1)/2)}} \qquad (8.37)$$

where D denotes the dimension of x, Δx denotes an increment of the state vector x, and $|\Delta x|$ denotes the Euclidean norm. The Cauchy distribution has some definite advantages over the Boltzmann form (Szu and Hartley, 1987b), where the Cauchy distribution has a fatter tail than the Gaussian form of the Boltzmann distribution, permitting easier access to test local minima in the search for global minimum. The Cauchy distribution of the form $g(k)$ statistically finds a global minimum. The method of fast annealing is thus seen to have an annealing schedule exponentially faster than the method of the Boltzmann annealing. This method has been tested on a variety of problems (Szu and Hartley, 1987b). Fast SA offers a big improvement over standard SA due to the adoption of the Cauchy distribution.

8.7.3 Very Fast Simulated Reannealing

The Boltzmann cooling in Equation 8.36 characterizes the simplest cooling scheme for SA. As k becomes very large over the iterations while $T(k)$ approaching to zero, the Boltzmann cooling scheme becomes slow. A very fast annealing with exponential cooling schedule (Ingber, 1989, 1993; Ingber and Rosen, 1992) was proposed for modification of the fast SA proposed by Szu and Hartley (1987a) and also for further improvement of the performance of the original SA. The variant was first called very fast simulated reannealing (VFSA) by Ingber (1989). In VFSA algorithm, the neighborhood function is the product of all the neighborhood in every parameter dimension. If the optimization problem consists of D variables, it will have D dimensions. Ingber suggests a cooling schedule for VFSA in which the temperature T decreases exponentially in annealing time. That means for the dimension i, the cooling schedule must satisfy the following equation:

$$T^i(k) = T_0^i \exp\left[-c^i k^{1/D}\right]$$ (8.38)

where $T(k)$ is the temperature at iteration k, and c^i is an empirical parameter. From Equation 8.38, it is obvious that the cooling schedule in VFSA is faster than the Boltzmann annealing characterized by Equation 8.36.

8.7.4 Adaptive SA

The variant VFSA was later renamed by Ingber (1993) as adaptive SA (ASA). ASA is now the most used variant of SA, which found many applications. In a D-dimensional search space, the individual variables have different finite ranges as well as time-dependent sensitivities, which are demonstrated and observed by the curvature of the fitness landscape. There is no mechanism for considering these differences in each variable dimension in the SA algorithm. These are among several considerations that lead to an ASA. Another issue in many stochastic optimization algorithms is the long periods of poor improvement toward the global optimum. Such poor improvements are mainly caused by various parameter settings. In SA, cooling schedule is mainly responsible for such behavior. The convergence speed is limited by the characteristics of probability density functions, which are employed with the purpose of generating new candidate points. If the Boltzmann annealing is chosen, the temperature decrease should be at a maximum rate of $T(k) = T_0/\ln(k)$. In the case of fast annealing, the cooling schedule becomes $T(k) = T_0/k$, resulting in a faster cooling. Therefore, Ingber (1993) proposed the modification given by Equation 8.38 that has an even better default scheme, where c^i is an arbitrary parameter defined by user and D is the number of variables of the optimization problem (dimension of the domain). To have better control over the empirical parameter c^i in Equation 8.38, it can now be defined as

$$c^i = m^i \exp\left(-\frac{n^i}{D}\right)$$ (8.39)

where m^i and n^i are considered free parameters to help tune ASA for specific problems.

Another adaptive feature of ASA is its ability to perform quenching[*] in a methodical fashion. Ingber (1996) also proposed to take advantage of simulated quenching defined by

$$T^i(k) = T_0^i \exp\left[-c^i k^{Q^i/D}\right] \tag{8.40}$$

Q^i is the quenching parameter that facilitates obtaining a gain in speed but it does not assure global optimum. The effectiveness of the ASA has been verified through application to several relevant areas (Pachter and Wang, 2008).

Oliveira et al. (2009) proposed a fuzzy ASA, where the progress of the SA was sampled during run time and updated the selected parameters that govern the dynamics of SA using a fuzzy controller. The adaptive mechanism helps SA escaping from local minima. This can be accomplished when the parameters to be tuned and their influence on the dynamics of SA are known. The effectiveness and performance of the fuzzy ASA were verified on digital filter design, global optimization problems, and solving nonlinear systems (Oliveira and Petraglia, 2011, 2013; Oliveira et al., 2009). To carry out the optimization process, a cost or error function is synthesized and a global minimization process is executed. At the end, the point that minimizes globally the particular cost function at hand determines the optimal filter or the system.

8.7.5 DISCRETE SA

In SA, the current solution $x(k)$ at iteration k corresponds to the objective function value $f[x(k)]$. The probability of the next solution $x(k+1)$ depends on both the difference between the corresponding fitness values of $x'(k)$ (a randomly generated solution around $x(k)$) and $x(k)$, i.e., $\Delta E = f[x'(k)] - f[x(k)]$) and the temperature T (Askarzadeh, 2013). The next solution is then defined by

$$x(k+1) = \begin{cases} x'(k) & \text{if } \exp(-\Delta E/T) > r \\ x(k) & \text{otherwise} \end{cases} \tag{8.41}$$

where $r \in [0,1]$ is an uniform random number. According to Equation 8.40, the probability term $p = \exp(-\Delta E/T)$ is always greater than 1 for $\forall \Delta E \leq 0$. That means $x(k+1) = x'(k)$ for all $\Delta E \leq 0$. The probability p depends on ΔE and T. The search for new solution continues until maximum iteration is reached. In discrete SA, $x'(k)$ and T are updated at every iteration according to the definition below:

$$x'(k) = x(k) + w \tag{8.42}$$

$$T(k+1) = \eta \cdot T(k) \tag{8.43}$$

where w is a vector randomly distributed over $[-w_f, w_f]$ and η is the step size. The algorithm starts at an initial temperature T_0.

8.7.6 COUPLED SA

In general, SA is a method where a single solution is initially generated randomly and then from the neighborhoods. The solution is then tested for acceptance based on the probability defined in

---

[*] Quenching is a process converse of annealing in which the temperature of the heat bath is instantaneously lowered.

Equation 8.3. Thus, optimizing certain fitness function requires several runs with different initial conditions. This eventually slows down the convergence. To increase the convergence speed, Xavier-de-Souza and Suykens (2010) proposed a new approach to SA where a set of distributed SA algorithms are run separately. The individual SA algorithms are coupled with each other by an acceptance probability function. The coupling term is a scalar function according to

$$0 \le p_\Theta(\gamma, x_i, y_i) \le 1 \tag{8.44}$$

where $x_i \in \Theta$ is the current state and $y_i \in \Omega$ is the probing state ($i = 1, \dots, m$, m is the number of elements in Θ). Ω is the set of all possible states and $\Theta \subset \Omega$. The coupling term γ is a function of energy of the elements in Θ defined as

$$\gamma = \left[E(x_1), E(x_2), \dots, E(x_m) \right] \tag{8.45}$$

The primary objective of the coupling in coupled SA is to assimilate more information on deciding to accept solutions that comprise the energy of many current states or solutions. In addition, coupling can be used to steer the overall optimization process toward the global optimum. Moreover, a particular coupled SA instance method is distinguished by the form of its coupling term and acceptance probability. The main steps of the coupled SA are described as follows:

Step 1: Initialization
    Assign random initial solutions to Θ
    Compute the costs $E(x_i)$, $\forall\, x_i \subset \Theta$ and evaluate the coupling term γ
    Set initial temperatures $T_k = T_0$ and $T_k^{ac} = T_0^{ac}$ and $k = 0$
Step 2: Generate solution
    Generate solution $y_i = x_i + \varepsilon_i$, $y_l \in \Theta$, $\forall x_l \in \Theta$, $\varepsilon_i \in g(\varepsilon_i, T_k)$
    and $g(\varepsilon_i, T_k)$ is the given random distribution
    Assess the costs for all probing solutions: $E(y_i)$, $\forall i = 1, \dots m$.
Step 3: Accept solution according to

$$x_i = \begin{cases} y_i & \text{if } E(y_i) \le E(x_i) \\ y_i & \text{if } p_\Theta(\gamma, x_i, y_i) > r, \quad \text{for} \quad i = 1, \dots, m \\ x_i & \text{otherwise} \end{cases}$$

    Evaluate $\gamma = [E(x_i)]$ and Goto Step 2 for N inner iterations
Step 4: Decrease temperature according to $U(T_k, k)$ and $V(T_k^{ac}, k)$
    Set $k = k + 1$
Step 5: If (termination criterion not met), Goto Step 2
Step 6: Return solution

where $r \in [0, 1]$ is a uniformly distributed random number.

Xavier-de-Souza and Suykens (2010) presented three coupled SA instance methods and compare them with the uncoupled case or multistart SA. This approach leads to much better optimization efficiency, because it reduces the sensitivity of the algorithm to initialization parameters while guiding the optimization process to quasi-optimal runs. Results of extensive experiments show that the addition of the coupling and the variance control leads to considerable improvements with respect to the uncoupled case and a more recently proposed distributed version of SA.

8.7.7 Modified SA

In general, the SA algorithm accepts all new solutions that improve the fitness value. The SA algorithm also accepts worse solutions with decreasing the fitness value base on the current acceptance probability. This acceptance probability even depends on the current wrong distance and temperature. As the temperature decreases according to the selected annealing schedule, the algorithm reduces the extent of its search to converge to a minimum.

Hedayat (2014) proposed a modified SA algorithm, which introduces a single evolutionary-based objective where each new solution is accepted if it has a positive evolution. Thus, the acceptance probability in the modified SA is defined as

$$p_{ac} = \exp\left[\frac{\text{Evolution}}{T}\right] \tag{8.46}$$

In this case, two evolutions are proposed by Hedayat (2014) as follows:

$$\begin{cases} \text{Evolution}_A = f(\text{current solution}) - f(\text{previous accepted solution}) \\ \text{Evolution}_B = f(\text{current solution}) - f(\text{best solution}) \end{cases} \tag{8.47}$$

where $f(\cdot)$ is the fitness of the solution. The proposed modified SA has been applied to optimization problem of refueling program for a research reactor. The experimental results show high searching abilities and quality global solution by the modified SA.

8.7.8 Corana SA

Corana et al. (1987) suggested a variant of the SA algorithm, called CSA, using an adaptive move along the coordinate directions (It is to be noted that the coupled SA in section 8.7.6 is also called CSA). Each new candidate solution point is obtained by changing only one coordinate of the current point x^k. The new point is generated using

$$x' = x^k + r_i^k \lambda_i^k e_i \tag{8.48}$$

where r_i^k is a random value uniformly distributed over [−1,1], λ_i^k is a component of the step vector λ^k, and e_i is the unit vector for coordinate i. After N iterations, step vector λ^k is updated. The r_i^k is computed using the equation $r_i^k = 2u - 1$, where u is a random number uniformly distributed over [0,1]. The step vector components λ_i^k is adjusted as follows:

$$\lambda_i^k = \begin{cases} \lambda_i^{k*}\left[1 + V_i^*\left(\dfrac{z - 0.6}{0.4}\right)\right] & 0.6 < z \\[2ex] \lambda_i^k & 0.4 \leq z \leq 0.6 \\[2ex] \dfrac{\lambda_i^k}{\left[1 + V_i^*\left(\dfrac{0.4 - z}{0.4}\right)\right]} & z < 0.4 \end{cases} \tag{8.49}$$

where z is the percentage of points accepted per coordinate i, that is, $z = $ (No. accepted points per coordinate i)/N and V_i is a fixed value. The idea in CSA is to accept 50% of the generated points. Goffe et al. (1994) suggested that the number of the iterations with the same control parameter value should be constant during the process. Pereira and Fernández (2004) tested the performance of the CSA algorithm on a number of benchmark functions. Experimental results demonstrated the effectiveness of the CSA algorithm.

8.7.9 ORTHOGONAL SA

Designing an efficient mechanism of generating a new solution plays an important role in SA algorithms. Ho et al. (2004a,b, 2006) proposed an orthogonal SA (OSA) algorithm. OSA algorithm is mainly an efficient approach to generating a new good candidate solution based on an intelligent generation mechanism (IGM), which helps SA performing an efficient search. The concept of IGM is simple. Let $X = [x_1, x_2, \ldots, x_n]^T$ be the current solution. IGM generates two intermediate solutions $X_1 = \left[x_1^1, x_2^1, \ldots, x_n^1 \right]^T$ and $X_2 = \left[x_1^2, x_2^2, \ldots, x_n^2 \right]^T$ by perturbing X according to the following rule:

$$\begin{cases} x_i^1 = x_i + \bar{x}_i \\ x_i^2 = x_i - \bar{x}_i \end{cases}, \quad i = 1, 2, \ldots, n \tag{8.50}$$

where the values of $\bar{x}_i$ are generated using the Cauchy–Lorentz distribution (Szu and Hartley, 1987a,b). The Cauchy–Lorentz distribution is a continuous probability distribution defined by the probability distribution function (PDF):

$$f_{PDF}(x) = \frac{1}{\pi} \left(\frac{\gamma}{(x - x_0)^2 + \gamma^2} \right) \tag{8.51}$$

where x_0 is the mean of the distribution and γ is the scale parameter specifying the half width of the PDF at half the maximum height. If the mean is zero (i.e., $x_0 = 0$), and the scale is 1 (i.e., $\gamma = 1$), then the result is a standard Cauchy distribution. The advantage of the IGM is its efficiency in combining the good values from solutions X, X_1, and X_2 to generate a new good candidate solution X^{new}. The total number of variables n is divided into N nonoverlapping groups with sizes l_j, $j = 1, 2, \ldots, N$, such that $\sum_{j=1}^{N} l_j = n$. The value of N is problem-dependent. A larger value of N means a weaker interaction between parameter groups but it ensures an efficient IGM. A smaller value of N means a stronger interaction between parameter groups but it results in a more accurate estimated main effect. A trade-off would be to minimize the interactions among parameter groups while maximizing the value of N. The main steps in OSA (Ho et al., 2004a) are described as follows:

Step 1: Initialize SA parameters, and an initial solution X
          Compute objective function value $f(X)$
Step 2: Generate a new candidate solution X^{new} applying IGM on X
Step 3: Accept new solution X^{new} with probability $p(X^{new})$
          Repeat Steps 2 and 3 until max_count
Step 4: Decrease temperature
Step 5: If (termination condition not met), Goto Step 2
Step 6: Return solution

The performance and effectiveness of OSA have been verified on a number of control design and floor planning problems by Ho et al. (2004a,b, 2006). The experimental results demonstrated the efficiency of the proposed OSA algorithm, which are competitive to other meta-heuristic algorithms.

8.7.10 Chaotic SA

In general, random initialization using Gaussian or other distributions is used in SA algorithm. Random sequences with uniform distribution have been applied to simulation of complex phenomena, sampling, numerical analysis, decision making, and heuristic optimization algorithms (Kroese et al., 2011). Chaotic sequence has been found to be useful in many application domains such as high-performance circuits and devices and shows promising performance over random sequence. Chaos is typically a mathematical property of a dynamical system, which exhibits dynamic, unstable, pseudo-random, ergodic, and nonperiodic behavior. Recently, chaotic sequences have been adopted instead of random sequences and it is very interesting that somewhat good results have been achieved in many heuristic algorithms and applications (Ju and Hong, 2013; Yang and Chen, 2002). The behavior of chaotic system is sensitive to the initial value and can be controlled using a set of parameters (Elaydi, 1999; May, 1976).

Mingjun and Huanwen (2004) introduced two chaotic maps to SA and proposed two new chaotic SA algorithms. The first chaotic map is produced using one-dimensional logistic map defined by

$$z_{k+1} = f(\mu, z_k) = \mu z_k (1 - z_k) \quad \text{with} \quad k = 0, 1, \ldots \tag{8.52}$$

where $z_k \in [0,1]$ is the value of the variable z at the k-th iteration. μ is the so-called bifurcation parameter of the system. The logistic map has special characteristic of ergodic and stochastic property and sensitivity dependence on initial conditions of chaos.

The second chaotic map is derived from a chaotic neuron defined by

$$z_{k+1} = \eta z_k - 2 \tanh(\gamma z_k) \exp\left(-3 z_k^2\right) \quad \text{with} \quad k = 0, 1, \ldots \tag{8.53}$$

where z is the internal state of the neuron, η is a damping factor of nerve membrane, and $0 \le \eta \le 1$. The term $2 \tanh(\gamma z_k) \exp\left(-3 z_k^2\right)$ is a nonlinear self-feedback.

The main steps of the chaotic SA are as follows:

Step 1: Generate chaotic variables z_k with a given initial value z_0
   Generate an initial solution $x_0 = \{x_0^1, x_0^2, \cdots, x_0^n\}$.
Step 2: Set temperature to $T = T_{max}$
   Best solution to $x^* = x_0$ and the best value to $f^* = f(x^*)$
Step 3: Generate new solution using $x'_k = x_k + \alpha \times (x_{max} - x_{min}) \times z_k$
   Generate z_k using Equation 8.52 or 8.53
Step 4: Evaluate $\Delta E^* = f(x'_k) - f^*$ and $\Delta E = f(x'_k) - f(x_k)$
Step 5: If $\Delta E^* \le 0$
   Update $x^* = x'_k$ and $f^* = f(x^*)$
Step 6: If $\Delta E \le 0$
   Update $x_{k+1} = x'_k$
Step 7: If $\Delta E > 0$, then update current state with new state with probability

$$P = \exp\left(\frac{-\Delta E}{T}\right)$$

Step 8: Decrease temperature by $T = \delta \times T$
   Set $k = k + 1$
Step 9: If (max iteration not reached), Goto Step 3
Step 10: Return solution x^*

where x_{max} and x_{min} are the maximum and minimum value of x, respectively, and α is defined as $\alpha = \alpha \times e^{-\beta}$.

Mingjun and Huanwen (2004) verified the performance of the proposed chaotic SA on six well-known benchmark functions, for example, Goldstein-Price, Branin, Hurtman, Rastrigin, and Shubert functions. The simulation results confirmed the efficiency of the proposed chaotic SA.

Hopfield and Tank (1985) showed that the classical combinatorial optimization problem of n-city TSP can be represented using a NN of $n \times n$ neurons. Using the NN representation of TSP, Chen and Aihara (1995) proved mathematically and demonstrated experimentally by applying a transiently chaotic NN that a chaotic SA has better search efficiency for solving the combinatorial optimization problems. They showed that chaotic SA leads to good solutions for TSP compared to Hopfield–Tank approach. He (2002) proposed a chaotic SA by adding decaying chaotic noise to each neuron of the discrete-time continuous Hopfield NN. The proposed chaotic SA has been applied to TSP, which resulted in high-quality robust solution. Wang et al. (2004) proposed a combination of stochastic SA and chaotic SA by adding a decaying stochastic noise in the transiently chaotic NN of Chen and Aihara (1995). The new method is called stochastic chaotic SA, which was applied to TSP and channel assignment problem of cellular mobile communications. The experimental results demonstrated remarkable improvements.

8.7.11 QUANTUM ANNEALING

Kadowaki and Nishimori (1998) investigated the possibility of making use of quantum tunneling processes for state transitions, which is termed as quantum annealing (QA). Particularly, it is aimed to find out how quantum tunneling processes lead to the global minimum in comparison to temperature-driven processes such as the conventional SA. It has been found experimentally that QA converges faster and finds better local optimums for Ising spin models. It has a parameter inducing quantum fluctuation, so the search space is controlled in a way different from that in SA. A number of QA has been reported in the literature (Das and Chakrabarti, 2005; Kadowaki, and Nishimori, 1998; Santoro et al., 2002; Sato et al., 2013).

8.8 HYBRID SA

SA algorithm is usually a "generate solution and test" algorithm and good at local search. The performance of SA algorithm very often depends on the initial solution and the neighborhood search. Attempts have been made to provide good initial solution using population-based meta-heuristics such as GA, DE, HSA, PSO, and many other algorithms. Thus, a bunch of hybrid SA algorithms are reported in the literature (Pereira and Fernández, 2004). Some of these important hybrid approaches are discussed in this section.

8.8.1 HYBRID SA AND GA

GA is a population-based search algorithm and good at global exploration to ensure a global optimum solution. On the other hand, SA is good at local search, which can be utilized for the exploitation of the local search space around a global solution found by GA. The combination of the GA and SA will make an efficient search strategy such that the final global best solution found via GA can be used as the initial solution for SA. Incorporation of GA into SA has been proposed by researchers in early 1990s by Lin et al. (1993) and Yip and Pao (1995). The basic technique of the hybrid GASA approach can be described by the following steps:

Step 1: Generate an initial population of solutions $\{x_i\}$ where $i = 1, 2, \ldots, N$, N is the size of population or number of solutions, $x_i = \left\{x_i^1, x_i^2, \ldots, x_i^n\right\}$ and n is the number of variables

Step 2: Fitness of each solution x_i is evaluated

Step 3: A subset of solutions $x_k \subset x_i$, $k = 1, 2, \ldots, N_1$, $N_1 < N$, is selected according to probabilities of proportional fitness

Step 4: A new set of solutions is generated by applying genetic operators, for example, crossover and mutation operation

$\quad\quad x'_k \leftarrow$ crossover (x_k) with crossover rate ρ_c

$\quad\quad x'_k \leftarrow$ mutation (x'_k) with mutation rate ρ_m

Step 5: $f(x'_k) \leftarrow$ evaluate (x'_k)

Step 6: Apply SA operator for selection of x'_{k+1} from the parents and offsprings

$\quad\quad$ if $f(x'_k) < f(x'_{k+1})$

$\quad\quad\quad x' = x'_k$

$\quad\quad$ else if $\rho < \exp[-\Delta E/T]$

$\quad\quad\quad x' = x'_{k+1}$

Step 7: Decrease temperature T

Step 8: If (termination condition not met), Goto Step 3

Step 9: Return solution

where $\Delta E = f(x'_k) - f(x'_{k+1})$ and $\rho_c, \rho_m \in [0,1]$ are random numbers. The crossover and mutation are genetic operators of GA. A detailed discussion on the basic mechanism of GA and the genetic operators can be found in Siddique and Adeli (2013).

Orkcu (2013) also proposed similar hybrid GASA approach where the acceptance test mechanism from SA is incorporated into GA. GASA was applied to subset selection in multiple linear regression models, which demonstrates the efficiency of GASA as an alternative to traditional subset selection methods. Kalantari and Abadeh (2013) proposed a new hybrid evolutionary-annealing approach by combining GA and SA for solving JSSP. It is tested on a set of 23 standard instances and compared with 3 other heuristic algorithms reported in the literature. The computational results show that the approach GASA for JSSP could produce the best-known solution on 78.26% of all instances tested. Zameer et al. (2014) proposed a hybrid GASA scheme where a batch composition preserving crossover GA is applied to nuclear power plant core loading pattern optimization problem. When GA becomes stagnant at a search point, the optimization process is switched to inner SA layer. At the beginning of the search process, switching is less frequent, but toward the final phases of the search, SA is found to be frequent in order to perform local optimization in a refined manner. The hybrid approach is found to outperform GA and SA. Shokouhifar and Jalali (2015) also applied hybrid GASA to simplification of complex analog symbolic expressions where GA provides a final global best solution x_{gbest}. SA generates a new solution x_{new} in the neighborhood of the global best solution at every iteration. The global best solution x_{gbest} is replaced with new solution x_{new} or accepted according to the SA scheme. The methodology was tested on three analog circuits, which shows promising results.

8.8.2 HYBRID HARMONY SEARCH-BASED SA

In SA, a new solution is produced around the current solution and the new solution is accepted or rejected based on the probability defined by $p = \exp(-\Delta F/T)$. Askarzadeh (2013) proposed the discrete HSA (DHSA)-based SA where a new solution $x'(k)$ is produced using DHSA rule (Siddique and Adeli, 2015a–c). A detailed description on HSA and DHSA is provided in Chapter 5. Two key parameters of HSA are $HMCR \in [0,1]$ (harmony memory consideration rate) and $PAR \in [0,1]$ (pitch adjustment rate), which are used to produce the new solution and accept the new solution. The main steps used in DHSA-SA are given by the pseudocode in the following:

```
for k = 1 to 3
    if r₁ < HMCR
        x'(k) = a feasible random integer number
    else
        x'(k) = x(t, k)
    if r₂ < PAR
        x'(k) = x'(k) + rₓ
    endif
    endif
endfor
```

where t is the iteration and r_w is a parameter defined using the random number r_3 as follows:

$$r_w = \begin{cases} 1 & r_3 < 0.5 \\ -1 & \text{otherwise} \end{cases} \tag{8.54}$$

where $r_1, r_2, r_3 \in [0,1]$ are random numbers uniformly distributed over [0,1].

Askarzadeh (2013) also further proposed the discrete chaotic harmony search-based SA (DCHS-SA). Discrete chaotic HSA is also discussed in Chapter 5. Chaotic variables are very useful due to their regularity without repetition and the dynamic properties of chaos variables. The main steps used in DCHS-SA are shown in the following pseudocode:

```
for k = 1 to 3
    if r₁ < HMCR
        x'(k) = a feasible random integer number by chaotic sequence
    else
        x'(k) = x(t, k)
        if r₂ < PAR
            x'(k) = x'(k) + rₓ
        endif
    endif
endfor
```

where r_w is defined using Equation 8.54 and $r_1, r_2, r_3 \in [0,1]$ are random numbers defined earlier.

The proposed approaches have been applied to find the optimum design of a PV/wind hybrid system. The experimental results demonstrated the superior performance of the DHS-SA and DCHS-SA algorithms.

A novel model for the transportation problem of a consolidation network was introduced by Hosseini et al. (2014), where direct shipment, cross-docking, and shipment through milk run trips are used simultaneously in order to send the loads from the suppliers to the customers. Due to the complexity of the model, Hosseini et al. (2014) proposed a hybrid of HSA and SA-based heuristics algorithm in order to minimize the total transportation cost in the network. The effectiveness and performance of the proposed approach were verified on four problem sets, which show competitive performance.

Cross-docking refers to a process where the product is received in a facility, occasionally married with other products going to the same destination, then shipped at the earliest opportunity, without going into long-term storage. The cross-docking of a product through a distribution network is recognized as one of the basic distribution strategies. The objective of this problem is to minimize the total shipping cost in the network, so it tries to reduce the number of required vehicles using an efficient vehicle routing strategy in the algorithm (Chopra, 2003).

The milk run logistics comes from the traditional system for selling milk in which the milkman used to walk to the doors of the customers' houses with his dray in a specified route and delivered the milk in bottles to his customers and finally took back the empty bottles. This system has been used in various industries, and the automobile manufacturing companies around the world have been the most important clients of this system (Dua et al., 2007).

8.8.3　HYBRID PSO-SA

Sadati et al. (2009) proposed a combination of PSO and SA. PSO was developed by Kennedy and Eberhart (1995, 2001) based on the social behavior of swarms such as bird flocking and fish schooling in nature. Each particle i, referred to an individual in the swarm representing a candidate solution to the optimization problem, is moving with velocity v_i^d in the d multidimensional search space, adjusting its position x_i^d in search space according to its own experience and that of neighboring particles. Particles make use of the best positions encountered and the best position of its neighbors to position themselves toward an optimum solution. The performance of each particle is measured according to a predefined fitness function related to the problem being solved. The velocity of the i-th particle of the swarm in a d-dimensional space is represented by

$$v_i^d(k+1) = w \cdot v_i^d(k) + c_1 \cdot \rho_1 \cdot \left[x_{pbest}^d(k) - x_i^d(k) \right] + c_2 \cdot \rho_2 \cdot \left[x_{gbest}^d - x_i^d(k) \right] \qquad (8.55)$$

where $\rho_1, \rho_2 \in [0,1]$ are random variables, $c_1, c_2 \in [0,2]$ are learning rates, $x_{pbest}^d(k)$ is the best position that the i-th particle (or agent) has ever encountered until iteration k, that is, personal best position until k-th iteration. x_{gbest}^d is the global best position. If v_i^d exceeds the upper limit, it is limited to $v_{i\max}^d$. w is a weighting function defined by

$$w = w_{\max} - \frac{w_{\max} - w_{\min}}{k_{\max}} k \qquad (8.56)$$

where $k_{\max}$ represents the maximum number of iteration.

The solution is obtained from the updated position of the particle based on the velocity updates as follows:

$$x_i^d(k+1) = x_i^d(k) + v_i^d(k+1) \times \Delta t \qquad (8.57)$$

where time Δt is unity to convert velocity into position.

In the process of SA, the new solution is generated randomly around an original one using the formulation below:

$$x'_i = x_i + r_1 \cdot \text{rand}(\cdot) \qquad (8.58)$$

where r_1 is an arbitrary constant and rand($\cdot$) is random number within the range of [0,1].

The PSO-SA algorithm starts with the initialization of a group of random particles. The search process starts with the initialized group of particles. Two mechanisms are used for selecting a new solution in the PSO-SA algorithm: the new solution x_i from the best swarm and the new solution is taken from SA. The main steps of the PSO-SA procedure are as follows:

Step 1: Initialize randomly N particles x_i, $i = 1, 2, \ldots, N$
Step 2: Evaluate particles $f(x_i) \leftarrow$ evaluate (x_i)
Step 3: Obtain global best position x_{gbest} using SA
Step 4: For each particle x_i, compare it with its personal best position x_{pbest} and replace
        if $f(x_i) > f(x_{pbest})$
            $x_{pbest} \leftarrow x_i$
Step 5: For each particle x_i, compare it with its global best position x_{gbest} and replace
        if $f(x_i) > f(x_{gbest})$
            $x_{gbest} \leftarrow x_i$
Step 6: Update each particles velocity according to Equation 8.55
        Update position according to Equation 8.57
Step 7: If (termination condition not met), Goto Step 3
Step 8: Return solution

The proposed hybrid PSO-SA approach is applied to under-voltage load shedding scheme for IEEE 14 and 18 bus test systems. An optimal load shedding scheme was achieved for long-term voltage stability in minimum runs, which demonstrates its efficiency and global property for online application of the proposed approach.

The most common features and pitfalls of SA algorithm, standard PSO, and hybrid PSO-SA algorithms are the immature convergence and long convergence time. An appropriate combination of parameters for some applications such as the robust support vector regression (SVR) model may improve the quality of the algorithm. Moreover, due to the decrease in the diversity in the later part of the search process, the PSO-SA algorithm can still be stuck in local optimum. To enhance performance, a hybrid PSO-SA algorithm based on the cat mapping is used. Cat mapping function has good ergodic uniformity in the interval of [0,1] and does not easily fall into a minor cycle (Li et al., 2013). Two-dimensional cat mapping function is described by

$$\begin{cases} x_{n+1} = (x_n + y_n) \bmod 1 \\ y_{n+1} = (x_n + 2y_n) \bmod 1 \end{cases} \tag{8.59}$$

where $(x \bmod 1) = x - [x]$.

Geng et al. (2015) proposed the combination of chaotic SA with PSO and called it CSAPSO algorithm. While the PSO algorithm evolves a certain number of generations, the algorithm applies the cat mapping to implement global disturbance of the poorer individuals. Geng et al. (2015) used cat mapping function to implement the chaos distribution. Any variable of this kind of chaotic space can travel ergodically over the whole space of interest to determine the improved solution eventually. The CSAPSO algorithm also determines a suitable combination of parameters and enhances the algorithm's ability to move out of the local optimum that enables reducing the convergence time. The basic operations of the CSAPSO are as follows:

Step 1: Initialize particle swarm using PSO
Step 2: Adopt SA for convergence of the particles into the lowest energy state
Step 3: Obtain improved particles from SA
Step 4: Employ cat mapping to implement chaos disturbance for particle swarm,
Step 5: Send back new modified individual to PSO for the next generation
Step 6: If (termination condition not met), Goto Step 2
Step 7: Return solution

The proposed CSAPSO algorithm has been applied to transform the three hyperparameters of an SVR model from the solution space to the chaotic space.

8.8.4 Hybrid Ant Colony Optimization and SA

ACO (also known as ant algorithm and ant system) was first proposed by Dorigo et al. (1991, 1996). ACO has features suitable for applying to problems that can be represented by a graph. A graph $G = \langle V, E \rangle$ is defined by a set of vertices V and a set of edges E. The basic idea of ACO is simple—the ants cross the graph and leave a pheromone trail as they cross the graph in a random manner. The probability of selecting a path (defined by a sequence of consecutive edges of the graph) is proportional to the amount of pheromone and inversely proportional to the length of the path. The ants are divided into colonies. The amount of pheromone is changed when the colony has crossed the path. The objective of ACO is to find the shortest path in the graph $G = \langle V, E \rangle$. The basic steps of ACO are described in the following:

1. Initialize the parameters of the ACO: number of colonies, number of ants N_A, initial amount of pheromone $\tau_{ij}(0)$, and length of edge d_{ij}.
2. Compute the probability of selecting the path $p(i,j)$ from node i to node j for ant k:

$$p_{ij}^k(t) = \begin{cases} \dfrac{[\tau_{ij}(t)]^\alpha (1/d_{ij})^\beta}{\displaystyle\sum_{l \in M_i^k} [\tau_{il}(t)]^\alpha (1/d_{il})^\beta} & \text{if } j \in M_i^k \\ \\ 0 & \text{if } j \notin M_i^k \end{cases} \tag{8.60}$$

where M_i^k is the set of possible and acceptable nodes for ant k, α is a parameter associated with the amount of pheromone, and β is a parameter related to length of edge.

3. The amount of pheromone dropped by ant k is given by

$$\Delta\tau_{ij}^k(t) = \begin{cases} 1/L^k(t) & \text{if ant } k \text{ visits path } p(i,j) \\ 0 & \text{otherwise} \end{cases} \tag{8.61}$$

where $L^k(t)$ is the length of the path marked dropping pheromone by ant k.

4. The amount of pheromone accumulated by N_A number of ants is given by

$$\tau_{ij}^k(t+1) = (1-\rho)\tau_{ij}(t) + \sum_{k=1}^{N_A} \Delta\tau_{ij}^k(t) \tag{8.62}$$

where $\rho \in (0,1]$ is the factor for pheromone evaporation. The factor ρ can also change dynamically with time defined by $\rho(t) = \rho^t$.

There are many variants of ACO such as ACO with elite strategy (Dorigo et al., 1996), ACO with ranking (Bullheimer et al., 1999), ant colony system (Dorigo and Gamberdella, 1997), and ACO with local search (Aarts and Lenstra, 1997).

In the hybrid ACO and SA algorithm, ACO is used to supply a good initial solution for SA runs. That is, ACO is used as an exploration mechanism and SA is used as an exploitation mechanism. Sitarz (2009) proposed a hybrid ACO-SA algorithm for multicriteria dynamic programming. Zhang et al. (2012) also applied a hybrid ACO-SA algorithm to path planning problem for the navigation of a mobile robot, where the path planning problem is represented by a quad-tree representation. There are still scopes to develop new hybrid methods combining the ACO variants with SA.

8.8.5 Hybrid ACO, GA, and SA

Hoseini and Shayesteh (2013) proposed a hybrid approach combining GA, ACO, and SA where GA controls the parameters of ACO and moving direction of the ants. That is, GA has the responsibility of modifying the characteristics of the ants. ACO is the core of the hybrid algorithm, which finds the near-optimum solution. At this stage, the algorithm needs to run several times in order to reach an optimum or near-optimum solution. In the SA phase, a local search is carried out on the randomly selected solutions obtained from ACO. Local search is performed frequently at the last stage to yield better solutions and optimize pheromone trains. The hybrid approach is applied to contrast enhancement of images. The results demonstrated higher contrast and preserved the natural look of the image.

8.8.6 Hybrid DE and SA

DE was developed by Kenneth Price in an attempt to solve the Chebyshev polynomial fitting problem. This was done by modifying genetic annealing originally developed by Price (1994). DE is a population-based direct search algorithm for global optimization (Storn, 1995, 1999; Storn and Price, 1997), where it utilizes the distance and difference direction information from the current population to guide the next search. Differences of two randomly selected individuals (solution vectors) are used as the source of perturbing the population rather than probability function as an evolution strategy. The mutation operation is performed based on the distribution of the solutions in the current population first and then applies crossover operator to generate offspring. DE has been proved to be an efficient, reasonably fast, and robust optimization algorithm for combinatorial optimization problems and real-world applications (Storn et al., 2005). The detail of DE and its variants can be found in Siddique and Adeli (2013) and Das et al. (2009). The basic algorithm of DE is simple and straightforward and consists of four steps.

> Step 1: Initialization of population
> Step 2: Mutation with difference vectors
> Step 3: Crossover operation
> Step 4: Selection

Considering the demands of global optimization technique, DE is a fast and, simple technique, involves only three parameters, and performs extremely well on a wide variety of test problems (Storn et al., 2005). The standard DE algorithm has the general problem of being trapped at local optima because of its updating strategy and intrinsic differential property characterized by mutation operator. Yan et al. (2006) proposed a novel SA-DE algorithm to improve the performance of the standard DE by incorporating SA updating method into DE. This helps DE escaping the local optima as well as striking the balance between exploration and exploitation. The other advantage of the SA updating method is the ability to protect the promising individuals and improve the diversity of the population. The main steps of the SA-DE are as follows:

> Step 1: Initialize population and parameters
> > Set acceptance probability p_a to p_0
> > Set annealing speed S_A
> > Set generation $g = 0$
> Step 2: Generate child solution according to recombination rule for each parent
> > Apply crossover operation with crossover rate p_c

Step 3: For each pair of parent and child do
       If $[f(\text{child}) < f(\text{parent})]$
           Replace parent by child
       Else
           Replace parent by child with probability p_a
Step 4: Annealing
       Update the probability $p_a = [p_0 / \log(10 + g * S_A)]$
Step 5: If $(g \leq g_{\max})$, set $g = g + 1$ and Goto Step 2
Step 6: Return solution

The new updating mechanism introduces a deceasing p_a with increasing generations. The advantage of the new acceptance probability is that the promising solutions have a higher probability of surviving to next generation and the diversity of population is maintained. This ensures exploration of the search space at the beginning of the SA-DE search. Toward the end of the search process, the updating strategy evolves to a greedy property with low probability of acceptance that ensures SA-DE focuses on exploitation. To verify the performance of the hybrid algorithm, SA-DE has been applied to eight benchmark functions. In all cases, the experimental results indicate that SA-DE outperforms DE in the global search ability.

8.8.7 Hybrid Artificial Immune System and SA

Artificial immune system (AIS)-based algorithms are recent development in the meta-heuristic algorithms inspired from the biological immune system. The function of the immune system is to fight against infection and disease in biological species. Toxins or bacteria that cause disease or infection in human body are referred to as antigens. The immune system produces antibody when invaded by antigens. The interesting feature of the immune system is that it produces specific antibodies aiming to fight specific type of antigens, which are not effective for other types of antigens. In order to help the body to generate specific antibodies, weaker antigens are injected into body, for example, smallpox. This process is referred to as vaccination. Vaccine usually consists of only a portion of the pathogen's structure and it is injected into the human body to induce the production of appropriate antibodies against a specific disease. In this process of vaccination, the antigens are considered as problems and generation of antibodies is seen as solutions. High-quality antibodies that can eradicate the complete antigens correspond to an optimal solution. Therefore, vaccination process is seen equivalent to optimization process of a meta-heuristic algorithm.

On the basis of the simple idea, an AIS-based algorithm has been proposed by Farmer et al. (1986). The AIS-based heuristic algorithms comprises two major operators in the immune mechanism:

1. *The vaccination process:* Vaccination (actually a heuristic algorithm) is the process of modifying the existing solutions toward optimal solutions.
2. *The immune selection:* It is the selection or acceptance mechanism of a solution. If the modified solution is better (meaning objective function value) than the previous solution, the new solution is accepted. Otherwise, an acceptance probability is used to accept the solution.

The main steps of the AIS-based algorithm are as follows:

Step 1: Randomly generate a solution x
       Set iteration $k = 1$
Step 2: Execute vaccination
       $x(k) = \text{vaccination}(x)$

Step 3: Immune selection
        If $[f(x(k)) < f(x)]$
            $x = x(k)$
        Else If $(r_1 < p_a)$
            $x = x(k)$
Step 4: If (termination condition not met), set $k = k + 1$ and Goto Step 2
Step 5: Return solution

where $f(\cdot)$ is the fitness function value, r_1 is a random number uniformly distributed over [0,1], and p_a is the predetermined acceptance probability.

Due to simplicity of implementation, AISs have been successfully applied to a wide range of theoretical and real-world applications (Hart and Timmis, 2008). In fact, immune-inspired algorithms provide a new way to make use of the particular characteristics of specific problems to improve the solution quality. Zhang and Wu (2010) proposed a hybrid immune system-based SA, called ISA. The main steps of the ISA are as follows:

Step 1: Generate a new solution x' from $x(k)$ at k-th iteration
        Compute $\Delta E = f(x') - f[x(k)]$
Step 2: If $\Delta E \leq 0$
            $x(k + 1) = x'$
        If $\Delta E > 0$
        If $(r_2 \leq p_a)$
            $x(k + 1) = x'$
Step 3: If (x' is accepted), apply immune algorithm onto x'
        $x'' = $ immune algorithm(x')
        If $f(x'') \leq f(x')$
            $x(k + 1) = x''$
        Else
            $x(k + 1) = x'$
Step 4: If (x' not accepted)
        Set $x(k + 1) = x(k)$
Step 5: If (termination condition not met), Goto Step 1
Step 6: Return solution

where $f(\cdot)$ is the fitness function value, r_2 is a random number uniformly distributed over [0,1], and p_a is the acceptance probability defined by $p_a = \exp[-\Delta E/T]$.

Zhang and Wu (2010) applied the proposed ISA algorithm on a number of job-shop scheduling instances. Experimental results for different sized instances show that the proposed algorithm is effective and converges to satisfactory solutions. Yu et al. (2012) and Zhao et al. (2013) also reported hybrid of SA and AIS together with DE to improve the solution quality further.

8.8.8 Noising Method with SA

Charon and Hudry (1993) suggest a simple descent algorithm called the noising method. The algorithm first perturbs the solution space by adding random noise to the problem's objective function values. The noise is gradually reduced to zero during the execution of the algorithm allowing the original problem structure to reappear. The basic principle of noising method is the elementary transformations. Any operation on a solution $s \in S$ that changes s into $s' \in S$ is called a transformation where S is the solution space. The transformation starts with changing one feature of s without changing the global structure. A set of elementary transformations Γ is defined. A transformation $\varphi \in \Gamma$ is applied on $s \in S$ to generate a neighbor $s' \in \Gamma(s)$ such that $s' \in S$. If $f(s') < f(s)$, then the current solution s is replaced with new solution s'; otherwise, s is replaced with s' with a probability $p = \exp[-\Delta E/T]$. The noising method is hybridized with SA. The process is repeated until an optimal solution s^* is obtained $\forall s' \in T(s^*)$ such that $f(s^*) \le f(s')$. The hybrid approach has been applied to graph partitioning, linear ordering problem, TSP, and coding theory problem. The experimental results show that the approach is competitive over SA. Charon and Hudry (1993) provided computational results, but did not provide any proof of the convergence to solutions. Charon and Hudry (2001) also show how the noising method is a generalization of SA.

8.9   CONVERGENCE ANALYSIS

A number of SA variants are discussed in the previous section. Each of these algorithms has very specific performance in terms of, firstly, quality of solutions and, secondly, computation time. From these two issues, the SA algorithm raises many open questions: convergence of the solutions to the optimal set $S^* \subset S$ and the rate of convergence to S^*. The important thing among these is how to minimize the computation time while achieving a good quality solution.

The convergence of SA can be analyzed in two ways: considering the SA modeled as a sequence of homogeneous Markov chains or considering SA modeled as a single inhomogeneous Markov chain. The homogeneous Markov chain approach assumes that each temperature $T(k)$ is held constant for a number of iterations m (where $m = 1, 2, \ldots, M_k$ iterations take place at fixed temperature) such that the stochastic transition matrix P_k can reach its steady state (or stationary) distribution π_k. Let $P_k^m(x, x')$ be the probability of moving from solution $x \in S^*$ to solution $x' \subset S^*$ in m iterations at temperature $T(k)$. If the Markov chain $P_k^m(x, x')$ is irreducible and aperiodic with finitely many solutions, then $\lim_{m \to \infty} P_k^m(x, x') = \pi_k(x')$ exists $\forall x, x' \in S^*$ and for all iterations k. $\pi_k(x')$ is the unique strictly positive solution of

$$\pi_k(x') = \sum_{x \in S} \pi_k(x) P_k(x, x'), \quad \forall x \in S \tag{8.63}$$

and

$$\sum_{x \in S} \pi_k(x) = 1 \tag{8.64}$$

The issue of convergence of SA has been investigated by many researchers, and a fair amount of research has been carried out for finding the necessary and sufficient conditions for convergence of SA. A detailed account of research has been reported by Aarts and Korst (1989) and Henderson et al. (2003).

The convergence approach for SA based on inhomogeneous Markov chain theory need not reach a stationary distribution. That means the number of iterations at temperature $T(k)$ can be of finite length but an infinite sequence for temperature must be examined, that is, infinite length of k should be considered. In other words, the cooling of temperature T should be sufficiently slow. This means that the SA algorithm can be described as a sequence of homogeneous Markov chains of finite length generated at descending values of temperature. This process can be described by combining the homogeneous Markov chains into one single inhomogeneous Markov chain of finite length (Aarts and Korst, 1989).

For a finite sequence of $T(k)$, Hajek's conditions must be satisfied (Hajek, 1988), which is defined as follows:

$$T(k) = \frac{d}{\log(k+2)}, \quad k = 1, 2, \ldots \tag{8.65}$$

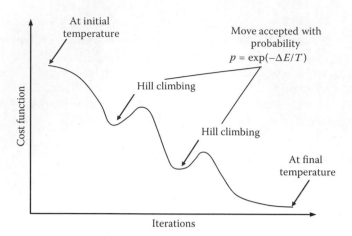

FIGURE 8.5   Convergence of SA.

where d is a constant. The asymptotic convergence of SA, using the transition probability and acceptance probability defined earlier, is guaranteed if and only if

$$\lim_{k \to \infty} T(k) = 0 \qquad (8.66)$$

$$\sum_{k=1}^{\infty} \exp\left[\frac{-d^*}{T(k)}\right] = \infty \qquad (8.67)$$

where d^* is the smallest number such that every $x, x' \in S$ reachable from x to x' and reachable from x' to x at the same height. Further, Hajek's condition says that SA converges if and only if $d \geq d^*$. The constant d^* is a measure of the difficulty for x to escape from a local minimum and nonoptimal state to S^*. Only $d^* > 0$ is of interest. If x is stuck at a local minimum with depth of d^*, SA makes an infinite number of trials to escape from the local minimum. The probability of success at each trial is of order $\exp[-d^*/T(k)]$, and Equation 8.67 says it will succeed escaping local minimum (Figure 8.5). Further theoretical discussion on convergence analysis can be found in Aarts and Korst (1989) and Henderson et al. (2003).

8.10   APPLICATION TO ENGINEERING PROBLEMS

Since its introduction in 1983, the SA algorithm has been proved to be very successful and was able to attract many researchers from different disciplines to develop SA-based solutions for a wide range of application domains over the last three decades. Some general aspects related to the application of SA are discussed so as to give some guidelines, which should be considered while implementing SA. Early applications of SA were in the domain of operation research and more theoretical problems in the 1980s such as TSP (Aarts and Van Laarhoven, 1988; Aarts et al., 1988; Cerny, 1985; Kirkpatrick, 1984), pattern recognition (Geman and Geman, 1984; Hinton et al., 1984), graph partitioning problems (Kirkpatrick, 1984), matching problems (Sasaki and Hajek, 1988), and linear arrangement problems (Nahar et al., 1985). In the following years, SA has found many applications to more practical and industry-related problems in the 1990s such as engineering design, VLSI design (Jespen and Gelatt, 1983; Kirkpatrick et al., 1983; Vecchi and Kirkpatrick, 1983), and JSSP (Van Laarhoven et al., 1992).

A survey of SA was reported in 2006 (Suman and Kumar, 2006). A selected set of applications are presented in this section so as to illustrate some characteristic features of the application of SA and useful implementation techniques.

Implementation of SA Algorithms requires three distinct steps: (i) a description of the problem, (ii) a transition mechanism for transforming a current solution into a subsequent (or new) solution, and (iii) a cooling schedule. The cooling schedule has been discussed in Section 8.5.

8.10.1 TRAVELLING SALESMAN PROBLEM

In TSP, it is required to find a tour of minimum distance traversing through a given set of cities or nodes. The distance measure is assumed to be the Euclidean distance between two cities. TSP is modeled as an undirected weighted graph $G = (V,E)$, where the set of vertices V denotes the n cities and the edge set E denotes the edges between cities, such that cities are the graph's vertices, paths are the graph's edges, and a path's distance is the edges' length. A proper mathematical description of TSP is provided in Chapter 2. The goal of TSP is to find a tour with the minimum total length among all such possible tours for the given graph. If there are n cities, that is, $|V| = n$, any tour can be represented as a permutation of the numbers 1 to n. Physically, this forms a circuit connecting cities 1 through n and finally n and the first city. The permutations of cities form the solution space. A TSP is called a symmetric when the distance from city i to city j (denoted as C_{ij}) is the same as the distance from city j to city i (denoted as C_{ji}), that is, $C_{ij} = C_{ji}$. The symmetric TSP is formulated as follows:

$$F = \min \sum_{i=1}^{n} \sum_{j=1}^{n} C_{ij} x_{ij} \qquad (8.68)$$

where $x_{ij} \in \{0,1\}$.

$$x_{ij} = \begin{cases} 1 & \text{if travelled from city } i \text{ to } j \\ 0 & \text{otherwise} \end{cases} \qquad (8.69)$$

where $\sum_{j=1}^{n} x_{ij} = 1$ for $i = 1, 2, \ldots, n$, $\sum_{i \in P} \sum_{j \in P, i \neq j} x_{ij} \leq |P| - 1$, $\forall \ P \subset \{1, 2, \ldots, n\}$, $2 \leq |P| \leq n - 2$. If the cities can be represented as points in the plain such that the distance between two cities i and j is equal to the Euclidean distance between two points, then the corresponding TSP is called the Euclidean TSP.

An early application of SA to TSP was reported in 1980s by Cerny (1985), where the algorithm generates random permutations of the cities for the salesman. Each permutation of the cities on the travelling salesman trip is considered to be the configuration of the statistical system. The corresponding length of the trip is then called the energy of the system. The probability of accepting a solution was computed by the Boltzmann–Gibbs distribution, which again depends on the length of the corresponding route. Three case studies with 100 cities showed that the SA algorithm can provide solution very close to the optimal solution. There are a good number of applications of SA to TSP reported in the 1980s (Aarts and Van Laarhoven, 1988; Aarts et al., 1988; Kirkpatrick, 1984)

The seminal work of Hopfield and Tank (1985) proposed solving the TSP using Hopfield NN (HNN).

Geng et al. (2011) considered a symmetric TSP and applied an ASA to it. They used three effective mutation strategies such as vertex insertion mutation, block insertion mutation, and block reverse mutation with different frequencies. To achieve faster convergence, a local search technique called greedy search was considered. The mechanism of greedy search is simple as follows:

If $(f(C_{ij}^{new}) < f(C_{ij}^{old}))$
$\quad C_{ij}^{old} \leftarrow C_{ij}^{new}$
Else $C_{ij}^{new} \leftarrow$ neighbour (C_{ij}^{old})

To validate the effectiveness of the proposed ASA, it was tested on 60 TSP benchmark instances with different conditions. The probability of acceptance of the best solution is calculated as

$$p = e^{(-\Delta E/T) \times (10.0 \times N/L_{Opt})} \tag{8.70}$$

where $\Delta E = f(C^{best}) - f(C^{old})$, N is the number of cities, L_{Opt} is the optimal tour length, and $f(C^{best}) = \min \{f[C^{new(1)}, C^{new(2)}, \dots, C^{new(m)}]\}$. The experimental results show that the proposed ASA with greedy search achieves a reasonable trade-off among computation time, solution quality, and complexity of implementation.

Chen and Chien (2011) applied a hybrid approach of SA, which is a combination of GA, SA, ACO, and PSO to TSP problem. It is assumed that there are n cities in the TSP. Initially, the system is divided into g groups $\{G_0, G_1, \dots, G_g\}$. Each group has N ants. Each ant will randomly choose a city as its starting city. Ant colony system is used to generate an initial population for GA. SA mutation is performed to obtain better solutions. For every C cycle (a predefined value), PSO technique is applied to exchange the pheromone information between groups. The initial pheromone level between any two cities is set to τ_0. A cycle consists of one execution of the ant colony system and some execution of the GA with SA mutation. The main steps of proposed method are as follows:

Step 1: Construct travelling sequence using local pheromone update rule for k-th ant of i-th group
Step 2: Evaluate distance of the route completed by ants in each group
Step 3: Perform global pheromone update
Step 4: Encode travelling sequences of cities searched by ant colony into chromosome
Step 5: Perform GA operators, that is, selection and crossover
Step 6: Perform SA mutation
Step 7: Evaluate fitness of offspring for each group
Step 8: Perform pheromone information exchange between groups for every C cycles
Step 9: If (termination condition not met), Goto Step 2
Step 10: Return solution

To verify the effectiveness and performance of the proposed approach, it was tested on 25 data sets obtained from TSP library. The experimental results show that both the average solution and the percentage deviation of the average solution to the best-known solution of the proposed method are better than the other methods reported in the literature. It appears that the proposed approach has no significant advantages in terms of computational effort due to the integration of GA, SA, ACO, and PSO.

8.10.2 Job-Shop Scheduling Problem

Scheduling is concerned with the allocation of limited resources to tasks with the objective of optimizing certain cost function. In JSSP, there is a set of jobs to be performed by a set of machines. Each job has a sequence of operations to be completed by a machine. The constraints are that splitting of a job is not allowed, interruption of an operation is not permitted, each machine can perform

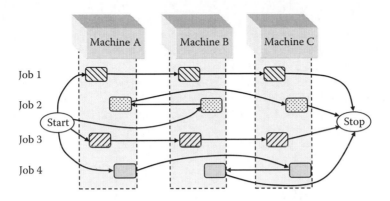

FIGURE 8.6 JSSP with four jobs and three machines.

an operation at a time, and each operation is performed only once on a machine and to a unique machine. Considering the application, there can be different objectives for JSSP such as makespan, tardiness, and mean flow. If C_i denotes the completion time of job i, the minimization objective of JSSP can be makespan defined as $C_{max} = \max\{C_i\}$, $i = 1, \ldots , n$, and maximum lateness defined as $L_{max} = \max\{C_i - d_i\}$. The tardiness T_i of a job is the nonnegative amount of time by which the completion time exceeds the due date d_i defied as $T_i = \max[0,(C_i - d_i)]$, that is, it is the difference between the completion time C_i and due date d_i for each job. The number of tardy jobs defined as $U = \sum_{i=1}^{n} U_i$, where $U_i = 1$, if $C_i > d_i$, otherwise $U_i = 0$. The TWT is defined as $\text{TWT} = \sum_{i=1}^{n} w_i T_i$, where w_i is a preset weight for each job that represents priority or importance of the job. The mean flow time $\bar{F}$ is the average flow time of a schedule defined as $\bar{F} = (1/n)\sum_{i=1}^{n} F_i$. This means that $\bar{F}$ is the average time of processing a single job. A detailed discussion on JSSP is also presented in Chapter 5. A JSSP instance can be depicted using a directed graph as shown in Figure 8.6, where the nodes denote the operations to be performed by a particular machine and the arcs joining the nodes define the required order of operations. A solution s is represented by the arcs connecting the operations performed by each machine in an order determined by s. Such a feasible solution is illustrated in Figure 8.7 by bold arcs. A solution s is feasible if there is no deadlock present in s. A deadlock is identified by a cycle in a graph.

The major contributions on the application of SA to JSSP were made in the 1990s. The first attempt was possibly by Van Laarhoven et al. (1992), which dates back to early 1992. There are also a good number of applications of SA to JSSP reported in the late 1990s (Ponnambalam et al. 1999; Wang and Wu, 2000).

Kolonko (1999) proposed a new approach using a small population of SA that runs in a GA framework. A population of K individuals is generated where each individual is an independent SA run with a starting solution and the best schedule found during the run. The starting values for SA runs are chosen at random. The SA uses an adaptive temperature control that allows reheating.

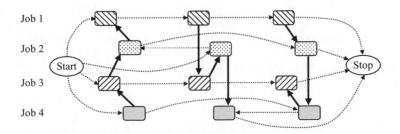

FIGURE 8.7 Feasible solution of JSSP represented by a graph.

The fitness of individuals is the length of its best schedule. A pair of individuals from the population is selected randomly and the crossover operation is performed on the two individuals to create a new generation of population. GA uses a time-oriented crossover operation on schedules. The resulting schedules are considered as starting schedules for SA runs and the best solutions found during runs are the new offspring. A final selection of K individuals is carried out on the offspring population. The new approach can be considered as parallel SA of K individuals with crossover operation or GA with SA local search. The proposed approach was verified on several benchmark problems from the test library, for example, JSP library.

Kalantari and Abadeh (2013) proposed a combination of GA and SA (GASA) algorithm for solving JSSP. The approach is compared with some algorithms in the literature. It is tested on a set of 23 standard instances and compared with 3 other approaches. The computational results show that the approach, GASA-JOSH, could produce the best-known solution on 78.26% of all instances tested.

Steinhoefel et al. (1999) presented two SA-based algorithms for classical JSSP where the objective is to minimize the makespan. Disjunctive graphs have been used to represent the problem. They proposed two cooling schedules as follows:

$$T(t+1) = (1-p_2) \cdot T(t) \tag{8.71}$$

$$T(t+1) = \frac{T(0)}{1 + \varphi(p_3) \cdot T(0)} \tag{8.72}$$

where $T(0) = -(\Delta E^{\max})/(\ln(1 - p_1))$ is the starting temperature, $\varphi(p_3)$ is defined as $\varphi(p_3) = (\ln(1 + p_3))/(E^{\max} - E^{\min}) < 1$, and $p_1, p_2, p_3 > 0$ are small positive numbers. The makespan of a feasible solution can be found by determining the length of the longest path from start to end of schedule. Computational experiments carried out on small to large benchmark problems have shown that within short series of consecutive trials stable results equal to or close to optimal solutions are repeatedly achieved.

Damodaran and Vélez-Gallego (2012) applied SA algorithm to JSSP with the objective function as the minimization of the makespan on a group of identical batch processing machines. A batch processing machine can process several jobs simultaneously such that all the jobs in a batch start and complete their processing at the same time. The SA algorithm was tested on the benchmark cases where each job has an arbitrary processing time, nonidentical size, and nonzero ready time. The batch processing time is equal to the largest processing time among those jobs in the batch. Computational experiments showed that the SA approach is comparable to greedy randomized adaptive search procedure (GRASP) approach with respect to solution quality, and less computationally costly.

In manufacturing process, each job has a fixed processing route, which traverses all the machines in a predetermined order. The duration of each operation is fixed and known. Usually, a preset due date d_i and a preset weight w_i are given for each job. Due dates are the preferred latest finishing time of each job, and the completion of the job after this specific time will result in losses. Minimizing tardiness is becoming more important in manufacturing sector due to completion time related losses and performance, which has further impact on market reputation. TWT as an objective in JSSP is not very common partly because this objective function is difficult and complex. Weights reflect the importance level of the orders from different customers, larger values suggesting higher strategic importance. The TWT is described as mixed-integer linear disjunctive programming model defined as

$$\text{TWT} = \sum_{i=1}^{n} w_i T_i \tag{8.73}$$

where w_i is the predefined weight for each job. The quality of SA search mainly depends on two factors: definition of a neighborhood function and method of selecting the next solution from the neighborhood. A small neighborhood is easy to perform but demands exhaustive search. Zhang and Wu (2011) applied SA to JSSP with the objective of minimizing TWT using a block-based neighborhood structure where a block is a sequence of operations in a critical path with a pre-defined minimum and maximum number of operations processed by the same machine. Such block structure considerably promotes the search ability of the SA and helps converging to high-quality solutions. The performance of the proposed block-based SA was tested on 10 problem sizes and 3 due date levels. The initial acceptance probability was set to 0.6, the cooling function was defined as $T(t + 1) = 0.95 \times T(t)$. The SA is comparatively insensitive to initial solution and usually converges fast to satisfactory solutions. According to the computational results, the new neighborhood considerably promotes the searching capability of SA and helps it converge to high-quality solutions.

8.10.3 Training NN

A brief introduction to NN has been provided in Chapter 2. Detailed description of NNs and their training algorithms can be found in Siddique and Adeli (2013). The problem of training an NN can be formulated as an optimization problem, where a set of parameters representing the connection weights, biases, and the connectivity of the internal structure of the NN is optimized subject to some optimality criteria. NNs have found diverse applications in many domains of engineering, identification, modeling, and control due to its ability to approximate any arbitrary nonlinear functions. This characteristic feature can be exploited to determine the control function of optimal control problems. The optimal control problem is transformed into a nonlinear function approximation problem, which can be represented by a feedforward NN. If $u(x)$ is a scaled control vector, an approximating function $F(x,\Phi)$ with a fixed number of real-valued parameters $\Phi \in R$ can be used to estimate $\hat{u}(x)$:

$$\hat{u}(x) = F(x,\Phi) \tag{8.74}$$

where x is the inputs to NN and $F(\cdot)$ is the NN model to be defined with a set of parameters Φ. Φ eventually represents the vector of learnable parameters, namely, the weights and biases of the NN. Thus, finding the specific $F(\cdot)$ (i.e., NN) becomes a problem of finding the optimal set of parameters Φ that provides the best possible approximation for $\hat{u}(x)$. The optimal control problem reduces to a nonlinear programming problem where the decision variables are the weights and biases of the chosen NN. The objective function for the optimal control problem can be formulated as

$$\max_{\Phi} J = \Gamma[\Phi, x] \tag{8.75}$$

The problem in Equation 8.75 can be solved using any nonlinear optimization technique. The performance index J also depends on the architecture of the NN. A three-layer NN architecture will be sufficient for such an optimal control problem. If there are H number of neurons in the hidden layer and m number of neurons in the output layer, then there are $2H$ connection parameters $(w_{ji}; b_j)$ between input and hidden layers, and $m \cdot (H + 1)$ connection parameters $(w_{kj}; b_k)$ between hidden and output layers. The total number of parameters in $\Phi = \{w_{ji}, w_{kj}, b_j, b_k\}$ are $2H + m \cdot (H + 1)$, which have to be trained using an algorithm. The training algorithm eventually has to find the optimal values of the parameter set. Sarkar and Modak (2003) applied SA to determine the optimal values of the parameter set $\Phi = \{w_{ji}, w_{kj}, b_j, b_k\}$ for its simplicity and the ability to produce global optimal solutions for complex problems, which outweigh its relatively large computational requirements. The initialization of parameters Φ was done randomly within a suitable lower and upper bounds and the initial temperature was set relatively high. The effectiveness of SA as optimization approach was verified on three case studies: optimal production of secreted protein in a fed-batch bioreactor, foreign protein production by recombinant bacteria, and continuous steer tank reactor with complex reactions. The experimental results show good approximation of the control profiles for all cases.

There have been a number of other applications of SA to optimization and/or training of NN reported in the literature. The application of SA is mainly to find the optimal design of the layered feedforward NN, where the NN represents a variety of problems. Chen and Aihara (1995) proposed a transiently chaotic NN, which gradually approaches through transient chaos to a dynamical structure or model that converges to a stable equilibrium point. The optimization process of the transient chaotic NN is similar to SA and can be regarded as a chaotic SA. The effectiveness of the chaotic SA has been verified on TSP. Sexton et al. (1999) reported the application of global search technique for training of NN and carried out a comparative analysis of these techniques such as backpropagation algorithm, SA, and GA. It was revealed from the simulation experiments that SA outperformed backpropagation algorithm but solution quality was not superior to GA. Richards et al. (2008) developed an NN model for prediction of the annual radial growth of Scottish pine using three competition indices. SA was applied to obtain an optimized predictive NN model.

Response surface methodology (RSM) is a collection of mathematical and statistical techniques to explore the relationship between several explanatory variables and response variables. RSM is also used for modeling and analysis of problems in which a response of interest is influenced by several variables where the objective is to optimize the response. There are interests from researchers of developing polynomial estimation model. The problem of polynomial estimation of RSM is that it may not perform well in providing good representation of the objective function. Abbasi and Mahlooji (2012) used NN as a means to improve the estimation of the RSM, which in turn also reduces the computation. SA was used to maximize the estimated objective function. The approach was then verified on three different examples of different complexities and the results were encouraging.

Ensemble learning is an approach where multiple models aggregate their individual decisions or their learning algorithms, or different data to improve the prediction performance. Ensemble learning has been very effective in a broad set of machine learning problems such as feature selection, small data sets, local learning, and concept drift theory among others. A large number of models are necessary in an ensemble to improve the accuracy, which will necessitate an adequate selection of subset of models from a large pool. Soares et al. (2013) applied SA and GA for selecting the best subset of NN models to be aggregated, improving generalization ability and reliability of the whole system. Experiments were carried out on two widely known and used data sets, namely Friedman and Boston Housing. Experiments have shown that SA-NN and GA-NN have similar good performance when compared to single model system.

Singh and Dixit (2013) reported a study on the optimization of stochastic Hopfield* NN (HNN) and hybrid self-organizing map.† HNN was used for storage and recalling of fingerprint images. It was revealed by Singh and Dixit that the capacity and recall efficiency of pattern storage in Hopfield NNs can be substantially enhanced and the probability of error in the recall phase can be reduced when SA is employed for training. Experimental results show an enhancement over the traditional limits on capacity and recall efficiency of HNNs.

8.10.4 Clustering Problem

Clustering process involves partitioning data sets into homogeneous subgroups subject to satisfying two objectives: data items within a cluster should be similar to each other and data items

---

* A Hopfield network is a form of recurrent neural network developed by John Hopfield in 1982. Hopfield nets can be considered as content-addressable memory system for storing and recalling patterns.

† Self-organizing map (SOM) or self-organizing feature map (SOFM) is a type of NN developed by T. V. Kohonen in the 1990s. SOM can preserve topological properties or features of the input space (called a map).

within different clusters should be dissimilar based on a similarity metric. The main objectives of any clustering algorithms are to minimize within-cluster variation (i.e., intracluster distance) and maximize the between-cluster variation (i.e., intercluster distance). Some mathematical description of the basic clustering techniques has been presented in Chapter 2 in Equations 2.93 through 2.96.

Early application of SA to clustering problems dates back to 1990s where simple SA algorithms were applied to clustering problems. Most of the popular clustering approaches are center-based clustering techniques, widely known as K-means clustering method. Such methods start with K number of initial centers and try to minimize the distances of the data sample x_i from the K centers $c_j, j = 1, 2, \ldots, K$. The problem of finding the centers is eventually defined as an optimization of a performance metric or function. Mostly, the performance function is defined as the mean squared error, which has the general problem of converging to local optimum. There have been many clustering methods proposed in the literature, most of which suffer from local minima problem. In order to overcome the local optima problem, Gungor and Unler (2007) proposed K-harmonic means (KHM) clustering combined with SA. The performance function for the KHM clustering is defined as

$$\text{KHM}(x,c) = \sum_{i=1}^{N} \left(\frac{K}{\sum_{j=1}^{K} \left(1 / \left(\left\| x_i - c_j \right\|^q \right) \right)} \right) \tag{8.76}$$

The KHM, defined by the function in Equation 8.76, has the property that if any one element of $x_i, i = 1, 2, \ldots, N$, is small, the harmonic average will move closer to the small value. It behaves like a minimum function but also gives some weight to all the other values as well. The KHM is used for updating the cluster centers $c_j, j = 1, 2, \ldots, K$. The parameter q is associated with the distance calculation. It was found that KHM works better with values of $q > 2$. The basic idea of the SA-KHM clustering is to use SA to generate the nonlocal move for the cluster centers and select the best solution based on KHM. The steps of the SA-KHM scheme are given as follows:

Step1: Randomly choose K initial centroids
     Set $T = T_{\max}$
Step 2: Set cooling scheme
Step 3: Select a new neighboring solution using one of the two strategies for clustering
     (i) *Random swap*: Replace a randomly chosen cluster center c_j with a randomly chosen
         point x_i
     (ii) *Logical swap*: Compute the distortion for each cluster D_i and average distortion D_{avg}
         of the clustering process
     Compute a utility index according to $U_i = (D_i / D_{\text{avg}})$
     Shift centers with $U_i < 1$ toward centers with $U_i > 1$
Step 4: Compute fitness
     If $(\Delta E < 0)$
       $x_i \leftarrow x_i^{\text{new}}$
     Else If $(p = e^{-\Delta E / T_n} > \text{rand}(0,1))$
       $x_i \leftarrow x_i^{\text{new}}$
Step 5: If $(k < k_n)$, Set $k = k + 1$, Goto step 3
     $k = k + 1$
Step 6: If (termination condition not met), Goto step 3
Step 7: Return solution

Gungor and Unler (2007) implemented the SA-KHM clustering algorithm and tested on several well-known data sets and the results have been compared with KM and KHM. The SA-KHM clustering algorithm outperforms to KM, and KHM from the point of performance value in most cases, and KHM also shows capability of smoothing the hard nature of KM algorithm.

Most of the existing clustering techniques are based on single objective, which reflects a single measure of goodness of a partitioning (Handl and Knowles, 2007). Rather than optimizing a single objective, it may be better to optimize multiple objective measures for cluster quality. Saha and Bandyopadhyay (2009) proposed a multiobjective SA-based clustering approach. Two objectives are used: total compactness based on the Euclidean distance and total symmetry in the cluster. Symmetry is a natural phenomenon, which can be capitalized to measure the performance of data clustering. The symmetrical (reflected) point of $\bar{x}$ with respect to a particular center c is defined as $\bar{x}^* = 2 \times c - \bar{x}$ (Bandyopadhyay and Saha, 2007). On the basis of this point symmetry, a new definition of point symmetry-based distance $d_{ps}(\bar{x}, \bar{c})$ can be measured. If there are $K_{near} > 1$ unique nearest neighbors of $\bar{x}^*$ at the Euclidean distances of d_i, $i = 1, 2, \ldots, K_{near}$, respectively, then $d_{ps}(\bar{x}, \bar{c})$ can be defined as

$$d_{ps}(\bar{x}, \bar{c}) = d_{sym}(\bar{x}, \bar{c}) \times d(\bar{x}, \bar{c}) = \frac{\sum\limits_{i=1}^{K_{near}} d_i}{K_{near}} \times d(\bar{x}, \bar{c}) \qquad (8.77)$$

where $d_{sym}(\bar{x}, \bar{c})$ is a symmetry measure of $\bar{x}$ with respect to $\bar{c}$ and $d(\bar{x}, \bar{c})$ is the Euclidean distance between the point $\bar{x}$ and $\bar{c}$. The $d_{sym}(\bar{x}, \bar{c})$ quantifies the amount of symmetry of a point with respect to a particular cluster center. It means if $\bar{x}^*$ exists in the data set, then $d_{sym}(\bar{x}, \bar{c})$ will be small. In other words, $d_{sym}(\bar{x}, \bar{c})$ will be large if the point is not at all symmetrical with the cluster center $\bar{c}$. On the basis of the idea of $d_{sym}(\bar{x}, \bar{c})$ and the $d(\bar{x}, \bar{c})$, the two objective functions are defined as

$$\begin{cases} S_K = \min \sum\limits_{k=1}^{K} \sum\limits_{j=1}^{N_k} d_{ps}(\bar{x}_j^k, \bar{c}^k) \\ D_K = \min \sum\limits_{k=1}^{K} \sum\limits_{j=1}^{N_k} d(\bar{x}_j^k, \bar{c}^k) \end{cases} \qquad (8.78)$$

where K is the number of clusters, N_k is the number of points in k-th cluster, $\bar{x}_j^k$ denotes the j-th point in k-th cluster, $d(\bar{x}_j^k, \bar{c}^k)$ denotes the Euclidean distance between the j-th point in k-th cluster, and $\bar{c}^k$ is the k-th cluster center. S_K is the total symmetry of all points in the cluster and a low value of S_K means a good symmetry of a cluster. D_K is the total Euclidean distance of all points in a cluster and a low value of D_K represents the compactness of a cluster. The two objective functions are optimized using SA simultaneously. The approach was verified on artificial and real-life data sets and proved to be well suited to partitioning of clusters having hyperspherical shape or point symmetric structure.

Torun and Tohumoglu (2011) proposed a cooperative SA and subtractive clustering (SC)-based fuzzy classifier. The SA is used to optimize the SC parameters, feature subspace, and output threshold value of fuzzy-based classifier. The notion of the SC method is that it considers each data point as a potential cluster center. It then calculates the likelihood of each data point to be considered for cluster center based on the density of surrounding data points. The density measurement at a data point x_i is defined as

$$D_i = \sum\limits_{j=1}^{m} \exp\left(-\frac{\|x_i - x_j\|^2}{(r_a/2)^2} \right) \qquad (8.79)$$

where m is the total number of data points, $x_i = [x_{i1}, x_{i2}, \dots, x_{in}]$ and $x_j = [x_{j1}, x_{j2}, \dots, x_{jn}]$ are data points $x_i, x_j \in X$ for $X = [x_1, x_2, \dots, x_m]^T$, which describe m observations for n attributes, and $r_a \in [0, \infty)$ is the neighborhood range of the cluster. The density value of i-th data point will be large if it has many neighboring data points and the distance between the data points and its location is small. The first cluster center is chosen as c_{l1}, which has the largest density value D_{cl1}. For the second cluster center, the effect of the first cluster center is subtracted in the determination of the new density values as follows:

$$D_i = D_i - D_{cl1}{}^* \exp\left[-\frac{\|x_i - x_{cl1}\|^2}{(r_b/2)^2} \right] \tag{8.80}$$

where $r_b = \eta \cdot r_a$ with $r_b \in [0, \infty)$ is neighborhood that has measurable reduction in density measurement and $\eta > 1$ is a positive constant.

According to Equation 8.80, the data points which are near the first cluster center c_{l1} will have reduced the density measurement significantly. The data point c_{l2}, which is chosen corresponding to the larger density value according to Equation 8.80, is selected for second cluster center. The selection of next cluster center process is carried out iteratively, until the stopping criteria is achieved. In order to demonstrate the effectiveness of the proposed approach, it was applied to four different classifier systems. The performance and rule base complexity of proposed classifiers are compared with each other and also some classifier tools on some well-known classification tasks. The results show that the proposed classifiers have a satisfactory performance in comparison with other well-known methods.

Che (2012) proposed a two-phase method to solve the two mathematical models for clustering and selection of suppliers. In Phase 1, a model is developed for clustering of suppliers according to customers' demand attributes. It integrates k-means and an SA algorithm with the Taguchi method to help developing the model. In Phase 2, the cluster information is used to determine appropriate supplier combinations. At this stage, it uses an analytic hierarchy process for the second model to weight every factor and then uses an SA algorithm with the Taguchi method to solve the second model.

Capacitated clustering problem (CCP) is an interesting benchmark optimization problem, which has been used for verifying performance of many optimization algorithms. The problem is to partition a given set of objects. Each object has an associated weight (or demand) and must belong to exactly one cluster. Each cluster has a given capacity, which must not exceed the total weight of objects in the cluster. A pictorial representation of the CCP is given in Figure 8.8. The cost of an object is measured by the dissimilarity (or distance) between any two objects. For a given cluster, a center is that object of the cluster from which the sum of the dissimilarities to all other objects in the cluster is minimized. This sum is called the scatter of the cluster. The objective is to minimize the total scatter of objects from the center of the cluster to which they have been allocated and secondly to find a set of centers which minimizes the total scatter of all clusters. Osman and Christofides (1994) applied the hybrid SA and Tabu search to CCP and verified the performance of the hybrid approach by extensive simulation on randomly generated test problems of sizes varying from 50 to 100 customers. The simulation results have confirmed the effectiveness of the approach.

8.10.5 Vertex Covering Problem

In the mathematical discipline of graph theory, a vertex cover (sometimes node cover) of a graph is a set of vertices such that each edge of the graph is incident to at least one vertex of the set. The problem of finding a minimum vertex cover is a classical optimization problem in computer science and is a typical example of an NP-hard optimization problem that has an approximation algorithm. Its decision version, the vertex cover problem, was one of Karp's 21 NP-complete

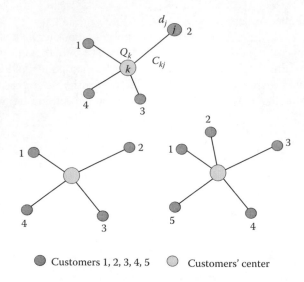

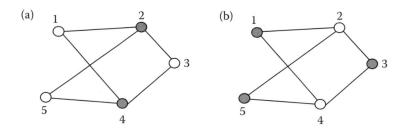

FIGURE 8.8   Capacitated clustering problem.

problems and is therefore a classical NP-complete problem in computational complexity theory. Furthermore, the vertex cover problem is fixed-parameter tractable and a central problem in parameterized complexity theory.

A vertex cover of an undirected graph $G = (V,E)$ is a subset $S \subseteq V$ such that if edge $\{u,v\}$ is an edge of G, then $u \in S$, $v \in V$, or both. The set S is said to "cover" the edges of G. Figure 8.9 shows examples of vertex covers in two graphs where the set S is shaded in gray.

An undirected graph $G = (V,E)$ can be described using an adjacency matrix where the elements of the matrix is represented by a_{ij} defined as

$$a_{ij} = \begin{cases} 1 & \text{if } \{i, j\} \in E \\ 0 & \text{otherwise} \end{cases} \tag{8.81}$$

where $\{i,j\} = 1, 2, \ldots, n$, $i \neq j$ and $a_{ii} = 0$.

The cover for vertex $v_i \in V$ can be determined by

$$v_i = \begin{cases} 1 & \text{if } v_i \text{ is in cover} \\ 0 & \text{otherwise} \end{cases} \tag{8.82}$$

FIGURE 8.9   Vertex covering problem: (a) $S = \{2,4\}$ and (b) $S = \{1,3,5\}$.

Using Equation 8.82, the number of vertices in the cover can be expressed by

$$F_1 = \sum_{i=1}^{n} v_i \tag{8.83}$$

If an edge $\{i,j\} \in E$ is not in a cover, then v_i and v_j will be zero. This information helps defining the constrained condition given as

$$F_2 = \sum_{i=1}^{n} \sum_{j=1}^{n} a_{ij} \overline{v_i \vee v_j} \tag{8.84}$$

Now the objective function for the minimum vertex cover problem can be defined as

$$F = A \cdot \sum_{i=1}^{n} v_i + B \cdot \sum_{i=1}^{n} \sum_{j=1}^{n} a_{ij} \overline{v_i \vee v_j} \tag{8.85}$$

where A and B are coefficients.

Xu and Ma (2006) proposed an efficient SA algorithm for the minimum vertex cover problem. An initial solution s is generated randomly. Then a neighboring solution s' is generated using SA. Both the solutions are evaluated using the objective function defined in Equation 8.85 and the change in cost function ΔF is calculated. ΔF is the same as change in energy ΔE used in previous equations. The probability of acceptance is calculated as follows:

$$p = \begin{cases} e^{-\frac{\Delta F(1-\mathrm{Deg}(v_i))}{T}} & \text{if } v_i = 1 \\ e^{-\frac{\Delta F(1+\mathrm{Deg}(v_i))}{T}} & \text{if } v_i = 0 \end{cases} \tag{8.86}$$

where $\mathrm{Deg}(v_i)$ is defined as

$$\mathrm{Deg}(v_i) = \frac{\mathrm{Degree}(i)}{E_{\mathrm{Total}}} \tag{8.87}$$

Degree(i) is the degree of vertex i meaning the number of edges linking to vertex i. E_{Total} is the total number of edges of the graph G. Simulations are performed on several benchmark graphs, which show that the proposed algorithm has higher convergence rate to optimal solutions.

8.10.6 FLOW SHOP SEQUENCING PROBLEM

Flow shop sequencing problem (FSSP) is a special case of job-shop scheduling where there is strict order of all operations to be performed on all jobs. Recently, several heuristic approaches based on iterative improvement procedures were applied to the FSSP as the computation power of available computers was rapidly improved. Ishibuchi et al. (1995) proposes two SA algorithms for the n-job and m-machine FSSP with an objective of minimizing the makespan. The advantage of the proposed approach is that the performance of the algorithm is less sensitive to the choice of a cooling schedule than that of the standard simulated SA. A second advantage is that one of the algorithms is easily executed in parallel.

8.10.7   MULTIOBJECTIVE OPTIMIZATION

SA has been applied to a variety of optimization problems discussed in the previous sections. Recently, SA has been applied to multiobjective optimization problems. Problems requiring simultaneous optimization of more than one objective functions are known as MOOPs. They can be defined as the problems consisting of multiple objectives, which are to be minimized or maximized while maintaining some constraints. A formal definition of MOOP is provided in Chapter 2 as well. The interested readers are referred to Deb (2008) for further reading. MOOP requires optimization of n objectives satisfying J inequality and K equality constraints. There is no unique perfect solution that can satisfy all objectives, inequalities, and equalities simultaneously. Moreover, the objectives can interact or conflict with each other. As a result, trade-offs or good compromises among the objectives are sought in MOOPs. Pareto optimal set (Ben-Tal, 1980) is an old notion of optimality for generating such trade-off solutions in MOOPs. SA has some distinct advantages for multiobjective framework due to its inherent simplicity and capability of producing a Pareto optimal set of solutions in a single run with comparatively smaller computational cost. A comprehensive review of multiobjective SA algorithms and their applications to various domains are provided in Suman and Kumar (2006). Brief account of multiobjective SA algorithms and their applications in the recent years are also reported by many others (Bandyopadhyay et al., 2008).

8.11   CONCLUSION

The advantages of SA algorithm reside in its high efficiency, simplification, and the generality. The high efficiency does mean that the SA algorithm can obtain a solution in a relatively short time with better approximation of solution. The simplification of SA means its quality of the solution does not depend on the initial solution and the SA algorithm can freely choose any initial solution. The SA algorithm can be applied to deal with a variety of combination of optimization problems, which characterizes its generality. Due to these advantages, SA has been successful in many engineering and industrial applications. Rigorous physical analogy, theoretical underpinning, and wide range of applications of SA are now well documented in the literature. Although a lot of research on the theoretical analysis, variants, hybridization, and applications to new domains has been carried out in the last three decades, but there are still many open issues where new research can be focused.

The performance of any SA-based application depends on the appropriate choice of the annealing schedule, which can save computational time as well as can improve the quality of solution. However, it has not been explored properly though few attempts have been made in the past. This is an open issue where a genuine effort of research is required for an optimal annealing schedule for a problem. More focused attention is necessary for selection of the initial temperature, and the number of iterations to be performed at each temperature.

Another issue with search space of SA is that it explores the search space one by one and in a sequential manner. It is very likely that the algorithm will be revisiting a subset of solutions multiple times demanding extra computation time, which do not yield any improvement to the quality of solution. This is where a hybridization of SA with other heuristic algorithms and parallelization can play a decisive role. Since the SA algorithm is good at local search, a global search algorithm would be a better choice for hybridization with SA.

The selection of an initial solution vector and the next move require knowledge about the lower and upper bound of the solution vector, specifically the neighborhood of all solution vectors. In real-life optimization problem, the neighborhood of all solution vectors may not be known a priori. SA can be extended to search the neighborhood of the solution vector in a manner it has been extended to estimate the algorithmic parameters. SA generates the news solution vector randomly. It is inefficient for searching for the optimal solution of large parameter optimization problems. Dedicated research is required for improving performance of this type of problems.

Implementation issues for SA can follow one of two directions: problem-specific choices such as neighborhood, objective function, and constraints, and generic choices such as generation and acceptance probability functions, and the cooling schedule. The major disadvantage of SA is that it often requires extensive computer time. In particular, the algorithm requires subtle adjustment of parameters in the annealing schedule such as the size of the temperature steps during annealing, the temperature range, the number of restarts and redirection of the search, and so on due to the intrinsic characteristics of its stochastic mechanics. Implementation modifications generally strive to retain the asymptotic convergence character of SA, but at reduced computer run time. Clever implementation techniques can help improving the computation time, which is an open issue in SA. The generation probability function is typically not temperature-dependent. An IGM to modify the neighborhood and its probability distribution accommodating search intensification or diversification would be an open research topic.

Lots of research has been done in SA over the last three decades. There are lots of valuable resources and MATLAB source codes available, which will help researchers implementing new SA algorithms and applications.

Reviews of the theory and many successful applications of SA published in the 1980s and 1990s can be found in Van Laarhoven and Aarts (1987), Rutenbar (1989), Connolly (1990), Eglese (1990), Romeo and Sangiovanni-Vincentelli (1991), Van Laarhoven et al. (1992), and Dowsland (1995). Among other surveys on SA carried out recently are by Henderson et al. (2003), Suman and Kumar (2006), and Siddique and Adeli (2016).

REFERENCES

Aarts, E. and Korst, J. 1989. *Simulated Annealing and Boltzmann Machines: A Stochastic Approach to Combinatorial Optimization and Neural Computing*, Chichester, UK: John Wiley & Sons.

Aarts, E. H. L., Korst, J. H. M., and Van Laarhoven, P. J. M. 1988. A quantitative analysis of simulated annealing algorithm: A case study for travelling salesman problem, *Journal of Statistical Physics*, 50, 189–206.

Aarts, E. H. L. and Lenstra, J. K. 1997. *Local Search in Combinatorial Optimisation*, Chichester, UK: John Wiley & Sons.

Aarts, E. H. L. and Van Laarhoven, P. J. M. 1988. Simulated annealing: An introduction, *Statistica Neerlandica*, 43, 31–52.

Abbasi, B. and Mahlooji, H. 2012. Improving response surface methodology by using artificial neural networks and simulated annealing, *Expert Systems with Applications*, 39, 3461–3468.

Askarzadeh, A. 2013. A discrete chaotic harmony search-based annealing algorithm for optimum design of PV/wind hybrid system, *Solar Energy*, 97, 93–101.

Azizi, N. and Zolfaghari, S. 2004. Adaptive temperature control for simulated annealing: A comparative study, *Computers & Operation Research*, 31, 2439–2451.

Bandyopadhyay, S. and Saha, S. 2007. GAPS: A clustering method using a new point symmetry based distance measure, *Pattern Recognition*, 40, 3430–3451.

Bandyopadhyay, S., Saha, S., Maulik, U., and Deb, K. 2008. A simulated annealing-based multiobjective optimization algorithm: AMOSA, *IEEE Transaction on Evolutionary Computation*, 12(3), 269–283.

Ben-Tal, A. 1980. Characterization of Pareto and lexicographic optimal solutions, in: G. Fandel and T. Gal, eds., *Multiple Criteria Decision Making: Theory and Application, Lecture Notes in Economics and Mathematical Systems*, vol. 17, Berlin: Springer, pp 1–11.

Bohachevsky, I. O., Johnson, M. E., and Stein, M. L. 1986. Generalized simulated annealing for function optimization, *Technometrics*, 28(3), 209–217.

Boltzmann, L. 1866. Über die Mechanische Bedeutung des Zweiten Hauptsatzes der Wärmetheorie, *Wiener Berichte*, 53, 195–220.

Bullheimer, B., Hart, R. F., and Strauss, C. 1999. A new rank-based version of the ant system, *Central European Journal for Operations Research*, 7(1), 25–38.

Cerny, V. 1985. Thermo-dynamical approach to the travelling salesman problem: An efficient simulation algorithm, *Journal of Optimal Theory and Application*, 45(1), 41–51.

Charon, I. and Hudry, O. 1993. The noising method—A new method for combinatorial optimization, *Operations Research Letters*, 14, 133–137.

Charon, I. and Hudry, O. 2001. The noising methods—A generalization of some metaheuristics, *European Journal of Operational Research*, 135, 86–101.

Che, Z. H. 2012. Clustering and selecting suppliers based on simulated annealing algorithms, *Computers and Mathematics with Applications*, 63, 228–238.

Cheh, K. M., Goldberg, J. B., and Askin, R. G. 1991. A note on the effect of neighborhood structure in simulated annealing algorithm, *Computers & Operations Research*, 18, 537–547.

Chen, L. and Aihara, K. 1995. Chaotic simulated annealing by a neural network model with transient chaos, *Neural Networks*, 8(6), 915–930.

Chen, S.-M. and Chien, C.-Y. 2011. Solving the travelling salesman problem based on the genetic simulated annealing ant colony system with particle swarm optimisation techniques, *Expert Systems with Applications*, 38, 14439–14450.

Chopra, S. 2003. Designing the distribution network in a supply chain, *Transport Research E*, 39(2), 123–140.

Cohn, H. and Fielding, M. 1999. Simulated annealing: Searching for an optimal temperature schedule, *SIAM Journal on Optimization*, 9, 779–802.

Connolly, D. 1990. An improved annealing scheme for QAP, *European Journal of Operation Research*, 46, 93–100.

Corana, A., Marchesi, M., Martini, C., and Ridella, S. 1987. Minimizing multimodal functions of continuous variables with the simulated annealing algorithm, *ACM Transactions on Mathematical Software*, 13(3), 262–280.

Damodaran, P. and Vélez-Gallego, M. C. 2012. A simulated annealing algorithm to minimize makespan of parallel batch processing machines with unequal job ready times, *Expert Systems with Applications*, 39, 1451–1458.

Das, S., Abraham, A., Chakraborty, U. K., and Konar, A. 2009. Differential evolution using a neighborhood based mutation operator, *IEEE Transaction on Evolutionary Computation*, 13(3), 526–553.

Das, A. and Chakrabarti, B. K. 2005. *Quantum Annealing and Related Optimization Methods*, Lecture Note in Physics, Vol. 679, Heidelberg, Germany: Springer.

Deb, K. 2008. *Multi-objective Optimisation using Evolutionary Algorithms*, 2nd ed. Chichester: John Wiley and Sons Ltd.

Dekkers, A. and Aarts, E. 1991. Global optimization and simulated annealing, *Mathematical Programming*, 50, 367–393.

Dorigo, M. and Gamberdella, L.M. 1997. Ant colonies for travelling salesman problem, *BioSystems*, 43, 73–81.

Dorigo, M., Maniezzo, V., and Colorni, A. 1996. Ant system: Optimization by a colony of cooperating agents, *IEEE Transactions on System, Man and Cybernetics: Part B*, 26(1), 29–41.

Dorigo, M., Maniezzo, V., and Colorni, A. 1991. *The Ant System: An Autocatalytic Optimising Process*, Technical Report 91-016, Department of Electronics, Polytechnic of Milan.

Dowsland, K. A. 1995. Simulated annealing, in: C. R. Reeves, ed., *Modern Heuristic Techniques for Combinatorial Problems*, London, UK: McGraw-Hill Book Company, Chapter 2.

Dua, T., Wang, F. K., and Lu, P.-Y. 2007. A real-time vehicle dispatching system for consolidating milk runs, *Transport Research E*, 43, 565–577.

Eglese, R. W. 1990. Simulated annealing: A tool for operational research, *European Journal of Operation Research*, 46, 271–281.

Elaydi, S. N. 1999. *Discrete Chaos*, Boca Raton, FL: Chapman & Hall/CRC Press.

Farmer, J. D., Packard, N., and Perelson, A. 1986. The immune system, adaptation and machine learning, *Physica D*, 2, 187–204.

Fielding, M. 2000. Simulated annealing with an optimal fixed temperature, *SIAM Journal of Optimization*, 11, 289–307.

Fleicher, M. A. 1995. Simulated annealing: Past, present and future, in: C. Alexopoulos, K. Kang, W. R. Lilegdon, and D. Goldsman, eds., *Proceedings of the 1995 Winter Simulation Conference*, Piscataway, NJ: IEEE Press, pp. 155–161.

Fleischer, M. A. and Jacobson, S. H. 1999. Information theory and the finite-time behavior of the simulated annealing algorithm: Experimental results. *INFORMS Journal on Computing*, 11, 35–43.

Fodorean, D., Idoumghar, L., N'diaye, A., Bouquain, D., and Miraoui, A. 2012. Simulated annealing algorithm for the optimisation of an electrical machine, *IET Electric Power Applications*, 6(9), 735–742.

Frost, R. and Salamon, P. 1993. *Ensemble-Based Simulated Annealing*, Technical Report, 2nd ed., San Diego Supercomputer Center.

Geman, S. and Geman, D. 1984. Stochastic relaxation, Gibbs distributions and Bayesian restoration of images, *IEEE Transaction on Pattern Analysis and Machine Intelligence, PAMI-6*, 6, 721–41.

Geng, X., Chen, Z., Yang, W., Shi, D., and Zhao, K. 2011. Solving the travelling salesman problem based on an adaptive simulated annealing algorithm with greedy search, *Applied Soft Computing*, 11, 3680–3689.

Geng, J., Li, M.-W., Dong, Z.-H., and Liao, Y.-S. 2015. Port throughput forecasting by MARS-RSVR with chaotic simulated annealing particle swarm optimisation algorithm, *Neurocomputing*, 147, 239–250.

Goffe, W. L., Ferrier, G. D., and Rogers, J. 1994. Global optimization of statistical functions with simulated annealing, *Journal of Econometrics*, 60, 65–99.

Golden, B. L. and Skiscim, C. C. 1986. Using simulated annealing to solve routing and location problems, *Naval Research Logistics Quarterly*, 33, 261–279.

Goldstein, L. and Waterman, M. 1988. Neighborhood size in the simulated annealing algorithm, *American Journal of Mathematical and Management Science*, 8, 409–423.

Gungor, Z. and Unler, A. 2007. K-harmonic means data clustering with simulated annealing heuristic, *Applied Mathematics and Computation*, 184, 199–209.

Hajek, B. 1988. Cooling schedules for optimal annealing, *Mathematics of Operations Research*, 13(2), 311–329.

Handl, J. and Knowles, J. 2007. An evolutionary approach to multi-objective clustering, *IEEE Transaction on Evolutionary Computation*, 11(1), 56–76.

Hart, E. and Timmis, J. 2008. Application areas of AIS: The past, the present and the future, *Applied Soft Computing*, 8(1), 191–201.

He, Y. 2002. Chaotic simulated annealing with decaying chaotic noise, *IEEE Transactions on Neural Networks*, 13(6), 1526–1531.

Hedayat, A. 2014. Developing a practical optimization of the refuelling program for ordinary research reactors using a modified simulated annealing, *Progress in Nuclear Energy*, 76, 191–205.

Henderson, D., Jacobson, S. H., and Johnson, A. W. 2003. The theory and practice of simulated annealing, in: F. Glover and G. A. Kochenberger, eds., *Handbook of Metaheuristics, International Series in Operation Research and Management Sciences*, Vol. 57, New York, NY: Kluwer Academic Publishers, Chapter 10.

Hinton, G. E., Sejnowski, T. J., and Ackley, D. H. 1984. *Boltzmann Machines: Constraint Satisfaction Networks that Learn*, Report No. CMU-CS-84-119, Carnegie-Mellon University, Pittsburgh.

Ho, S.-Y., Ho, S.-J., and Lin, Y.-K. 2004a. An orthogonal simulated annealing for floor planning problems, *IEEE Transaction on VLSI Systems*, 12, 874–876.

Ho, S.-J., Ho, S.-Y., and Shu, L.-S. 2004b. OSA: Orthogonal simulated annealing algorithm and its application to designing mixed H_2/H_∞ optimal controllers, *IEEE Transaction on Systems, Man and Cybernetics—Part A: Systems and Humans*, 34(5), 588–600.

Ho, S.-J., Shu, L.-S., and Ho, S.-Y. 2006. Optimising fuzzy neural networks for tuning PID controllers using an orthogonal simulated annealing algorithm (OSA), *IEEE Transactions on Fuzzy Systems*, 14(3), 421–433.

Hoffmann, K. H. and Salamon, P. 1990. The optimal simulated annealing schedule for a simple model, *Journal of Physics A: Mathematical and General*, 23(15), 3511–3523.

Hoffmann, K.-H., Wuertz, D., De Groot, C., and Hanf, M. 1991. Concepts in optimising simulated annealing schedules: An adaptive approach for parallel and vector machines, in: M. Grauer and D. Pressmar, eds., *Parallel and Distributed Optimisation*, Heidelberg, Germany: Springer, pp. 154–175.

Hopfield, J. J. and Tank, D. W. 1985. Neural computation of decisions in optimisation problems, *Biological Cybernetics*, 52, 141–152.

Hoseini, P. and Shayesteh, M. G. 2013. Efficient contrast enhancement of images using hybrid ant colony optimisation, *Genetic Algorithm and Simulated Annealing, Digital Signal Processing*, 23, 879–893.

Hosseini, S. D., Shirazi, M. A., and Karimi, B. 2014. Cross-docking and milk run logistics in a consolidation network: A hybrid of harmony search and simulated annealing approach, *Journal of Manufacturing Systems*, 33, 567–577.

Huang, M. D., Romeo, F., and Sangiovanni-Vincentelli, A. L. 1986. An efficient general cooling schedule for simulated annealing, *Proc. of the IEEE International Conference on Computer Aided Design*, Santa Clara, CA, pp. 381–384.

Ingber, L. 1989. Very fast simulated re-annealing, *Mathematical and Computer Modelling*, 12(8), 967–973.

Ingber, L. 1993. Simulated annealing: Practice versus theory, *Mathematical Computer Modelling*, 18(11), 29–57.

Ingber, L. 1996. Adaptive simulated annealing (ASA): Lessons learned, *Control and Cybernetics*, 25(1), 33–54.

Ingber, L. and Rosen, B. 1992. Genetic algorithms and very fast simulated re-annealing: A comparison, *Mathematical Computer Modelling*, 16(11), 87–100.

Ishibuchi, H., Misaki, S., and Tanaka, H. 1995. Modified simulated annealing algorithms for the flow shop sequencing problem, *European Journal of Operational Research*, 81, 388–398.

Jespen, D. W. and Gelatt, C. D. 1983. Macro placement by Monte Carlo annealing, *Proceedings of the International Conference on Computer Design*, Port Chester, NY, pp. 495–498.

Johnson, D. S., Aragon, C. R., McGeoch, L. A., and Schevon, C. 1989. Optimization by simulated annealing: An experimental evaluation—Part I: Graph partitioning, *Operations Research*, 37, 865–892.

Johnson, D. S., Aragon, C. R., McGeoch, L. A. M., and Schevon, C. 1991. Optimization by simulated annealing: An experimental evaluation—Part II: Graph coloring and number partitioning, *Operations Research*, 39(3), 378–406.

Ju, F-Y. and Hong, W-C. 2013. Application of seasonal SVR with chaotic gravitational search algorithm in electricity forecasting, *Applied Mathematical Modelling*, 37, 9643–9651.

Kadowaki, T. and Nishimori, H. 1998. Quantum annealing in the transverse Ising model, *Physics Review E*, 58, 5355–5363.

Kalantari, S. and Abadeh, M. S. 2013. GASA-JOSH: A hybrid evolutionary-annealing approach for job-shop scheduling problem, *Bulletin of Electrical Engineering and Informatics*, 2(2), 132–140.

Kennedy, J. and Eberhart, R. 1995. Particle swarm optimization, *Proceedings of the IEEE International Conference on Neural Networks*, Nov 27–Dec 1, Perth, WA, pp. 1942–1948.

Kennedy, J. and Eberhart, R. 2001. *Swarm Intelligence*, San Francisco, CA: Morgan Kaufmann Publishers.

Kirkpatrick, S. 1984. Optimization by simulated annealing: Quantitative studies, *Journal of Statistical Physics*, 34(5–6), 975–986.

Kirkpatrick, S., Gelatt, C. D., and Vecchi, M. P. 1983. Optimisation by simulated annealing, *Science*, 220, 671–680.

Kolonko, M. 1999. Some new results on simulated annealing applied to the job shop scheduling problem, *European Journal of Operational Research*, 113(1), 123–136.

Koulamas, C., Anthony, S. R., and Jaen, R. 1994. A survey of simulated annealing applications to operations-research problems, *Omega—International Journal of Management Science*, 22, 41–56.

Kroese, D. P., Taimre, T., and Botev, Z. I. 2011. *Handbook of Monte Carlo Methods*, New York, NY: John Wiley & Sons, Chapter 1: Uniform random number generation.

Li, M. W., Kang, H. G., Zhou, P. H., and Hong, W. C. 2013. Hybrid optimization algorithm based on chaos, cloud and particle swarm optimization algorithm, *Journal of Systems Engineering and Electronics*, 24(2), 324–334.

Lin, F. T., Kao, C. Y., and Hsu, C. C. 1993. Applying genetic approach to simulated annealing in solving some NP-hard problems, *IEEE Transaction on Systems, Man and Cybernetics*, 23(6), 1752–1767.

Lundy, M. and Mees, A. 1986. Convergence of an annealing algorithm, *Mathematical Programming*, 34, 111–124.

May, R. M. 1976. Simple mathematical models with very complicated dynamics, *Nature*, 261, 459.

Metropolis, N., Rosenbluth, A., Rosenbluth, M., Teller, A., and Teller, E. J. 1953. Equation of state calculation by fast computing machines, *Journal of Chemical Physics*, 21, 1087–1091.

Mingjun, J. and Huanwen, T. 2004. Application of chaos in simulated annealing, *Chaos, Solitons and Fractals*, 21, 933–941.

Moscato, P. 1993. An introduction to population approaches for optimisation and hierarchical objective functions: A discussion on the role of Tabu search, *Annals of Operations Research*, 41, 85–121.

Nahar, S., Sahni, S., and Shragowitz, E. 1985. Experiments with simulated annealing, *Proceedings of the 22nd Design Automation Conference*, Las Vegas, New Mexico, pp. 748–752.

Nourani, Y. and Andresen, B. 1998. A comparison of simulated cooling strategies, *Journal of Physics A: Mathematical and General*, 31, 8373–8385.

Ogbu, F. A. and Smith, D. K. 1990. The application of the simulated annealing algorithm to the solution of the $N/M/C_{max}$ flowshop problem, *Computers and Operations Research*, 17, 243–253.

Oliveira, H. A., Jr. and Petraglia, A. 2011. Global optimization using dimensional jumping and fuzzy adaptive simulated annealing, *Applied Soft Computing*, 11, 4175–4182.

Oliveira, H. A., Jr. and Petraglia, A. 2013. Solving nonlinear systems of functional equation with fuzzy adaptive simulated annealing, *Applied Soft Computing*, 13, 4349–4357.

Oliveira, H. A., Jr., Petraglia, A., and Petraglia, M. R. 2009. Frequency domain FIR filter design using fuzzy adaptive simulated annealing, *Circuits, Systems and Signal Processing*, 28(6), 899–911.

Orkcu, H. H. 2013. Subset selection in multiple linear regression models: A hybrid of genetic and simulated algorithms, *Applied Mathematics and Computation*, 219, 11018–11028.

Orosz, J. E. and Jacobson, S. H. 2002. Finite-time performance analysis of static simulated annealing algorithms, *Computational Optimization and Applications*, 21, 21–53.

Ortner, M., Descombes, X., and Zerubia, J. 2007. *An Adaptive Annealing Cooling Schedule for Object Detection in Images*, Research Report 6336, October 2007, INRIA, Sophia Antipolis Cedex, France.

Osman, I. and Christofides, N. 1994. Capacitated clustering problems by hybrid simulated annealing and Tabu search, *International Transaction on Operation Research*, 1(3), 317–336.

Otten, R. H. J. M. and van Ginneken, L. P. P. P. 1984. Floor plan design using simulated annealing, *Proc. IEEE International Conference on Computer Aided Design*, Santa Clara, CA, pp. 96–98.

Pachter, R. and Wang, Z. 2008. Adaptive simulated annealing and its application to protein folding, in: C. A. Floudas and P. M. Pardalos, eds., *Encyclopedia of Optimization*, New York, NY: Springer-Verlag, pp. 21–26.

Pereira, A. I. P. N. and Fernández, E. M. G. P. 2004. A study of simulated annealing variants, *XXVIII Congreso Nacional de Estadstica e Investigación Operativa (SEIO'04)*, 25–29 October, 1–16.

Pincus, M. 1970. A Monte Carlo method for the approximate solution of certain types of constraint optimisation problems, *Operation Research*, 18, 125–1228.

Ponnambalam, S. G., Jawahar, N., and Aravindan, P. 1999. A simulated annealing algorithm for job shop scheduling, *Production Planning & Control*, 10(8), 767–777.

Potts, C. N. and Van Wassenhove, L. N. 1991. Single machine tardiness sequencing heuristics, *IIE Transactions*, 23, 346–354.

Price, K. 1994. Genetic annealing, *Dr. Dobb's Journal*, October, 127–132.

Richards, M., McDonald, A. J. S., and Aitkenhead, M. J. 2008. Optimisation of competition indices using simulated annealing and artificial neural networks, *Ecological Modelling*, 214, 375–384.

Romeo, F. and Sangiovanni-Vincentelli, A. 1991. A theoretical framework for simulated annealing, *Algorithmica*, 6, 302–345.

Rose, J., Klebsch, W., and Wolf, J. 1990. Temperature measurement and equilibrium dynamics of simulated annealing placement, *IEEE Transactions on Computer Aided Design*, 9, 253–259.

Ruppeiner, G., Peddersen, J., and Salamon, P. 1991. Ensemble approach to simulated annealing, *Journal of Physics I*, 1(4), 455–470.

Rutenbar, R. A. 1989. Simulated annealing algorithms: An overview, *IEEE Proceedings*, 5, 19–26.

Sadati, N., Amraee, T., and Ranjbar, A. M. 2009. A global particle swarm-based simulated annealing optimisation technique for under-voltage load shedding problem, *Applied Soft Computing*, 9, 652–657.

Saha, S. and Bandyopadhyay, S. 2009. A new multi-objective simulated annealing based clustering technique using symmetry, *Pattern Recognition Letters*, 30, 1392–1403.

Santoro, G. E., Martoňák, R., Tosatti, E., and Car, R. 2002. Theory of quantum annealing of an Ising spin glass, *Science*, 295, 2427–2430.

Sarkar, D. and Modak, J. M. 2003. ANNSA: A hybrid artificial neural network/simulated annealing algorithm for optimal control problems, *Chemical Engineering Science*, 58, 3131–3142.

Sasaki, G. H. and Hajek, B. 1988. The time complexity of maximum matching by simulated annealing, *Journal of Association for Computing Machinery* 35, 387–403.

Sato, I., Tanaka, S., Kurihara, K., Miyashita, S., and Nakagawa, H. 2013. Quantum annealing for Dirichlet process mixture models with applications to network clustering, *Neurocomputing*, 121, 523–531.

Schneider, J. J. and Puchta, M. 2010. Investigation of acceptance simulated annealing—A simplified approach to adaptive cooling schemes, *Physica A*, 389, 5822–5831.

Sekihara, K., Haneishi, H., and Ohyama, N. 1992. Details of simulated annealing algorithm to estimate parameters of multiple current dipoles using biomagnetic data, *IEEE Transactions on Medical Imaging*, 11(2), 293–299.

Sexton, R. S., Dorsey, R.E., and Johnson, J.D. 1999. Optimisation of neural networks: A comparative analysis of the genetic algorithm and simulated annealing, *European Journal of Operational Research*, 114, 589–601.

Shokouhifar, M. and Jalali, A. 2015. An evolutionary-based methodology for symbolic simplification of analog circuits using genetic algorithm and simulated annealing, *Expert Systems with Applications*, 42, 1189–1201.

Siddique, N. and Adeli, H. 2013. *Computational Intelligence: Synergies of Fuzzy Logic, Neural Networks and Evolutionary Computing*, Chichester, UK: John Wiley & Sons.

Siddique, N. and Adeli, H. 2015a. Harmony search algorithm and its variants, *International Journal of Pattern Recognition and Artificial Intelligence*, 29(8), 1539001, 22 pp.

Siddique, N. and Adeli, H. 2015b. Hybrid harmony search algorithms, *International Journal on Artificial Intelligence Tools*, 24(6), 1530001, 16 pp.

Siddique, N. H. and Adeli, H. 2015c. Applications of harmony search algorithms in engineering, *International Journal on Artificial Intelligence Tools*, 24(6), 1530002, 15 pp.

Siddique, N. H. and Adeli, H. 2016. Simulated annealing, its variants and engineering applications, *International Journal on Artificial Intelligence Tools*, 25(6), (25 pages)

Singh, M. P. and Dixit, R. S. 2013. Optimization of stochastic networks using simulated annealing for the storage and recalling of compressed images using SOM, *Engineering Applications of Artificial Intelligence*, 26, 2383–2396.

Sitarz, S. 2009. Ant algorithms and simulated annealing for multi-criteria dynamic programming, *Computers & Operation Research*, 36, 433–441.

Soares, S., Antunes, C. H., and Araujo, R. 2013. Comparison of a genetic algorithm and simulated annealing for automatic neural network ensemble development, *Neurocomputing*, 121, 498–511.

Steinhoefel, K., Albrecht, A., and Wong, C. K. 1999. Two simulated annealing-based heuristics for the job shop scheduling problem, *European Journal of Operational Research*, 118, 524–548.

Storn, R. 1995. Constrained optimization, *Dr. Dobb's Journal*, May, 119–123.

Storn, R. 1999. System design by constraint adaptation and differential evolution, *IEEE Transaction on Evolutionary Computation*, 3(1), 22–34.

Storn, R. and Price, K. 1997. Differential evolution—A simple and efficient heuristic for global optimisation over continuous space, *Journal of Global Optimisation*, 11(4), 431–459.

Storn, R., Price, K., and Lampinen, J. 2005. *Differential Evolution—A Practical Approach to Global Optimisation*, Berlin, Germany: Springer Verlag.

Strenski, P. N. and Kirkpatrick, S. 1991. Analysis of finite length annealing schedules, *Algorithmica*, 6, 346–366.

Suman, B. and Kumar, P. 2006. A survey of simulated annealing as a tool for single and multi-objective optimisation, *Journal of the Operational Research Society*, 57, 1143–1160.

Szu, H. and Hartley, R. 1987a. Fast simulated annealing, *Physics Letters A*, 122(3–4), 157–162.

Szu, H. H. and Hartley, R. L. 1987b. Non-convex optimization by fast simulated annealing, *Proceedings of the IEEE*, 75(11), 1538–1540.

Torun, Y. and Tohumoglu, G. 2011. Designing simulated annealing and subtractive clustering based fuzzy classifier, *Applied Soft Computing*, 11, 2193–2201.

Triki, E., Collete, Y., and Siarry, P. 2005. A theoretical study on the behavior of simulated annealing leading to a new cooling schedule, *European Journal of Operation Research*, 166, 77–92.

Van Laarhoven, P. J. M. and Aarts, E. H. L. 1987. *Simulated Annealing: Theory and Applications*, Dordrecht, Netherlands: Kluwer Academic Publishers.

Van Laarhoven, P. J., Aarts, E. H. L., and Lenstra, J. K. 1992. Job shop scheduling by simulated annealing, *Operations Research*, 40, 113–125.

Varanelli, J. M. and Cohoon, J. P. 1999. A fast method for generalized starting temperature determination in homogeneous two-stage simulated annealing systems, *Computers and Operations Research*, 26, 481–503.

Vecchi, M. P. and Kirkpatrick, S. 1983. Global wiring by simulated annealing, *IEEE Transactions on Computer Aided Design*, CAD-2, 215–222.

Wang, L., Li, S., Tian, F., and Fu, X. 2004. A noisy chaotic neural network for solving combinatorial optimisation problems: Stochastic simulated annealing, *IEEE Transactions on Systems, Man and Cybernetics—Part B: Cybernetics*, 34(5), 2119–2125.

Wang, T.-Y. and Wu, K.-B. 2000. A revised simulated annealing algorithm for obtaining the minimum total tardiness in job shop scheduling problems, *International Journal of Systems Science*, 31(4), 537–542.

Xavier-de-Souza, S. and Suykens, J. A. K. 2010. Coupled simulated annealing, *IEEE Transactions on Systems, Man and Cybernetics—Part B: Cybernetics*, 40(2), 320–335.

Xu, X. and Ma, J. 2006. An efficient simulated annealing algorithm for the minimum vertex cover problem, *Neurocomputing*, 69(7–9), 913–916.

Yan, J.-Y., Ling, Q., and Sun, D.-M. 2006. A differential evolution with simulated annealing updating method, *Proceedings of the Fifth International Conference on Machine Learning and Cybernetics*, August 13–16, Dalian, China, pp. 2103–2106.

Yang, L. and Chen, T. 2002. Application of chaos in genetic algorithms, *Communication in Theoretical Physics*, 38, 168–172.

Yao, X. 1992. Dynamic neighbourhood size in simulated annealing. In *Proceedings of International Joint Conference on Neural Networks (IJCNN'92)*, November 3–6, 1992, Beijing, China, Vol. 1, pp. 411–416.

Yip, P. P. C. and Pao, Y.-H. 1995. Combinatorial optimisation with use of guided evolutionary simulated annealing, *IEEE Transaction on Neural Networks*, 6(2), 290–295.

Yu, C., Chen, J., Huang, Q., Wang, S., and Zhao, X. 2012. A new hybrid differential evolution algorithm with simulated annealing and adaptive Gaussian immune, *2012 8th International Conference on Natural Computation (ICNC 2012)*, May 29–31, Chongqing, China, pp. 600–607.

Zameer, A., Mirza, S. M., and Mirza, N. M. 2014. Core loading pattern optimisation of a typical two-loop 300 MWe PWR using simulated annealing (SA), novel crossover genetic algorithms (GA) and hybrid GA(SA) schemes, *Annals of Nuclear Energy*, 65, 122–131.

Zhang, Q., Ma, J., and Liu, Q. 2012. Path planning based quad-tree representation for mobile robot using hybrid-simulated annealing and ant colony optimization algorithm, *Proceedings of the 10th World Congress on Intelligent Control and Automation*, July 6–8, Beijing, China, pp. 2537–2542.

Zhang, R. and Wu, C. 2010. A hybrid immune simulated annealing algorithm for job shop scheduling problem, *Applied Soft Computing*, 10, 79–89.

Zhang, R. and Wu, C. 2011. A simulated annealing algorithm based on block properties for the job shop scheduling problem with total weighted tardiness objective, *Computers & Operations Research*, 38, 854–867.

Zhao, X., Lin, W., Yu, C., Chen, J., and Wang, S. 2013. A new hybrid differential evolution with simulated annealing and self-adaptive immune operation, *Computers and Mathematics with Applications*, 66, 1948–1960.

9 Chemical Reaction Optimization

9.1 INTRODUCTION

Chemistry is a branch of natural science that studies the properties, composition, and structure of matter and how it reacts and changes under certain conditions when it comes in contact with other kinds of matter. Early chemistry was dominated by alchemy and phlogiston theories, which dominated the subject for centuries. It was Antoine Laurent Lavoisier (1743–1794) who disproved the phlogiston theory and thus emancipated chemistry from alchemy in the eighteenth century. He introduced modern chemistry by discovering the law of conservation of mass. The law of conservation of mass states that mass can neither be created nor be destroyed. It changes from one form to another. John Dalton (1766–1844) proclaimed the existence of elementary atoms and postulated that groups of atoms disassociate and then rejoin in new arrangements during chemical reactions. Atomic structure constrains different atoms to form groups in fixed ways as molecules. A molecule comprising several atoms is characterized by the atom type, bond between atoms, angle, and torsion (twisting of structure), which is termed molecular structure. Chemical reactions are explained in terms of atomic structure of matter and of energy changes that occur during reaction. In a chemical reaction, chemical bonds are broken into molecules by absorption of energy and new bonds are formed producing different molecules with release of energy. A chemical reaction comprises different types of unimolecular and multimolecular elementary reactions, each of which releases or absorbs different levels of energy. Tobern Olof Bergman (1735–1784) used diagrams and symbols to explain chemical reactions. Thus, chemistry took its present scientific form in which chemical reactions are represented by chemical equations. The law of conservation holds during any chemical reaction. Thus, the same collection of atoms is present after a reaction as before the reaction. The changes that occur during a reaction involve just the rearrangement of atoms.

In chemistry, chemical reactions are written in the form of an equation using symbols representing chemical elements and molecules. In the equation, the reactants are written on the left-hand side and the products are written on the right-hand side separated by an equal sign or a single or double arrow. The arrow signifies the direction of the reaction. For example, a chemical reaction is represented by a chemical equation shown in Figure 8.1. The chemical equation is also called chemical formulas. The hydrogen gas (H_2) can react (burn) with oxygen gas (O_2) to form water (H_2O). The plus (+) sign is read as "reacts with," and the arrow ($\rightarrow$) means "produces." The numbers in front of the formulas are called coefficients where the number 1 is usually omitted.

According to Dalton, atoms are neither created nor destroyed in a reaction; a chemical equation must have an equal number of atoms for each element on each side of the equation, when the equation is said to be balanced. The chemical reaction presented in Figure 9.1 is presented as a balanced equation in Figure 9.2 where the arrow is replaced with an equal sign.

The chemical reaction shown in Figures 9.1 and 9.2 can be seen as a process that leads to the transformation of one set of chemical substances to another. It is usually characterized by a chemical change yielding one or more products with different properties, which are different from the chemical reactants taking part in the reaction. Historically, the term chemical reaction has been applied only to changes to substances. Each chemical reaction displays the same unique characteristics: transformation, conservation of mass, fixed atomic composition, and energy effects.

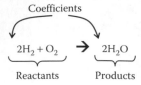

FIGURE 9.1   Chemical reaction equation.

$$\underbrace{2H_2 + O_2}_{\text{Reactants}} = \underbrace{2H_2O}_{\text{Product}}$$

FIGURE 9.2   Balanced equation.

Transformation: In a chemical reaction, transformation of substances occurs, where the substances that were present initially convert to new substances that were not present initially.

Conservation of mass: Although substances transform into different substances, the total mass does not change. The total mass is conserved, which means that the loss of mass as the reactants vanish is equal to the gain of mass as the new substances form.

Fixed atomic composition: The substances formed in chemical reactions as products have the same compositions every time irrespective of the reaction. For example, if water is formed as a product for whatever the reaction is, water will have the same molecular composition of H_2O.

Energy effect: In all but a very few chemical reactions, energy is either absorbed or released. Energy is absorbed when chemical bonds are broken down, and energy is released when new bonds are formed. The amount of energy mainly depends on the mass of the product formed.

Every chemical reaction stabilizes at an equilibrium, a state determined by the minimum free energy (also called Gibbs' free energy), which depends on parameters such as temperature, pressure, and the chemical reactants involved. Gibbs' energy (also referred to as G) is also the chemical potential that is minimized when a system reaches equilibrium at constant pressure and temperature (Gibbs, 1873). In general, the change in Gibbs' free energy ΔG must be zero when equilibrium is reached. Chemical reactions, which are determined by the laws of thermodynamics, can take place when energy is released. The Gibbs free energy G of the reaction comprises two different thermodynamic quantities: enthalpy and entropy (Atkins and De Paula, 2006). Entropy, denoted as S, is a thermodynamic quantity representing the unavailability of a system's thermal energy for conversion into mechanical work. Entropy is often interpreted as the degree of disorder or randomness in the system. Enthalpy, denoted as H, is also a thermodynamic quantity equivalent to the total heat content of a system. It is equal to the internal energy of the system plus the product of pressure and volume. The total enthalpy H in a system cannot be measured directly. Therefore, a change in enthalpy ΔH is measured. The change in Gibbs' free energy ΔG is defined by

$$\Delta G = \Delta H - T \cdot \Delta S \tag{9.1}$$

where ΔS is the change in entropy and T is the temperature.

According to energy consideration in chemical reactions, chemical reactions can be classified into three types: exoergic (or exothermic) when energy is released during the reaction, endoergic (or endothermic) when energy is absorbed from the surroundings, and aergic (or athermic) when the reaction takes place without a change in energy. In exothermic reactions, heat is transferred to the environment (i.e., ΔH is negative). Precipitation and crystallization are chemical reactions that occur at low temperatures in which ordered solids are formed from disordered gaseous or liquid phases. These reactions are exothermic reactions. In endothermic reactions, heat is consumed from the environment (i.e., ΔH is positive). In contrast, in gaseous reactions, products are formed by increasing the entropy of the system, resulting in endothermic reactions. Since entropy increases with temperature, many endothermic reactions take place at high temperatures. Changes in temperature can sometimes reverse the sign of the enthalpy of a reaction. At some stage of the reaction, it reaches equilibrium point (i.e., at the final stage of the reaction) when the products have no (or minimum) free energy such that no further reaction can take place. Every chemical reaction seeks to achieve a minimum free energy. That means chemical reactions tend to release energy, and thus, products generally have less energy than reactants. In terms of stability, the lower the energy of the substance, the more stable it is.

Chemical reactions do not happen instantly. Some reactions are slow, and some are fast to reach chemical equilibrium. The speed at which a reaction takes place (also known as reaction kinetics) depends on various parameters such as reactant concentrations, surface area available for contact between the reactants, pressure, activation energy, temperature, catalyst, and the presence of electromagnetic radiation (e.g., ultraviolet light). The temperature dependence of the rate constant k usually follows the Arrhenius equation:

$$k = k_0 e^{-E_a/k_B T} \qquad (9.2)$$

where E_a is the activation energy, k_B is the Boltzmann constant, and k_0 is the initial rate.

Molecules take part in chemical reactions, which can be of different types such as unimolecular and multimolecular elementary reactions. A molecule is distinct with its structure comprising several atoms, that is, two molecules even with the same set of atoms are different from each other owing to the difference in their chemical bonds. Molecules undergo structural changes during chemical reactions. The molecules must acquire the necessary energy to be activated. In general, all chemical reactions require the introduction of energy in order to make the reactions happen. Very often but not always, the energy is supplied in the form of heat. There are instances where the reaction takes place in the presence of light, in which the requisite energy is supplied in the form of light. Molecules store energy in chemical bonds. Breaking the bonds releases energy to the environment. Forming the bonds requires energy supplied from outside. There are two types of energy that a molecule can possess:

1. Potential energy denoted as PE
2. Kinetic energy denoted as KE

PE is the energy that a molecule contains in its structure. In a chemical reaction (e.g., exothermic reaction), chemical bonds break and new bonds are formed, that is, during the reaction, molecules transform from a structure of higher PE into a structure of lower PE with release of energy. Molecules need to collide for the chemical reaction to happen. Collision between molecules provides the KE needed to break the bonds. Sometimes, there is not enough KE for collisions. In such a case, energy is provided as heat. Temperature is a measure of the average KE of molecules. Raising the temperature increases the KE required for breaking the bonds.

If $x = \{x^1, x^2, \ldots, x^n\}$ is a molecule with x^k, $k = 1, \ldots, n$ atoms of certain structure, then the molecule tends to change from x to a new molecule x' by changing its structure during a chemical reaction only if the following condition holds:

$$PE(x) \geq PE(x') \qquad (9.3)$$

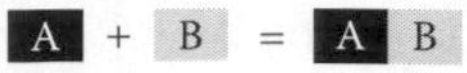

FIGURE 9.3   Pictorial representation of a synthesis reaction.

A reaction converts a molecule with higher PE to a molecule with lower PE. If condition (9.3) does not hold, the reactant molecules need to move to a higher-energy state for the reaction to happen. KE allows the molecules to move to a higher-energy state such that the reaction can happen with their change to new molecules, which satisfies the following condition:

$$PE(x) + KE(x) \geq PE(x') \tag{9.4}$$

A molecule with higher KE has a higher possibility of having a new structure with a higher PE. A reactant (generally a molecule or a compound) with high energy is unstable and tends to have a reaction when it comes in contact with another molecule or compound. The reactants in the high-energy states undergo a sequence of reaction phases and dissipation of energy when the conditions of reactions (9.3) and (9.4) hold. At the final stage, the reaction produces products with low energy and a stable state. The final state is considered as an optimal stable state of a chemical reaction. There are many types of unimolecular and multimolecular elementary chemical reactions. Some basic types of reactions (Brown et al., 2008; Kotz et al., 2008) are discussed in the following. The symbol x for a molecule is replaced with symbols such as A, B, C, and D representing reactants and products. Reactants and products are molecules or compounds (group of molecules). Examples are given from chemistry using proper chemical equations and notations.

1. *Synthesis (combination)*: A synthesis reaction is the simplest chemical reaction that involves two or more reactants and forms a product, often as the only product. A synthesis reaction can be expressed symbolically as $A + B \rightarrow AB$ and is shown pictorially in Figure 9.3. For example, the synthesis of water (H_2O) from hydrogen gas (H_2) and oxygen gas (O_2) is represented by the reaction equation: $2H_2 + O_2 = 2H_2O$.
2. *Polymerization*: Polymerization reactions are not unlike synthesis reactions in that simpler reactants combine to form more complex products. Polymerization is restricted to chemical reactions in which the product is composed of many hundreds or thousands of simpler reactants.
3. *Decomposition*: A decomposition reaction is the reaction in which complex reactants are broken down into simpler products. A decomposition reaction can be expressed symbolically as $AB \rightarrow A + B$ and is shown pictorially in Figure 9.4. For example, the decomposition of ammonia gas ($2NH_3$) into nitrogen (N_2) and hydrogen (H_2) is represented by the reaction equation: $2NH_3 = N_2 + 3H_2$.
4. Combustion: Combustion is the reaction in which the reaction takes place between oxygen gas (O_2) and another chemical reactant to produce carbon dioxide (CO_2) and water (H_2O). The reaction is represented by the following equation: $O_2 + (hydrocarbon) = CO_2 + H_2O$.
5. *Displacement or substitution*: Displacement or substitution reaction is a type of reaction where part of one reactant is replaced by another reactant. Single- and double-replacement reactions are the most common reactions in chemical process.
   Single-displacement or single-replacement reaction: A substitution or single-displacement reaction is characterized by displacement of one element from a compound by another

FIGURE 9.4   Pictorial representation of a decomposition reaction.

(a) A + B C = A C + B

(b) A B + C D = A C + B D

FIGURE 9.5 Pictorial representation of displacement reactions: (a) single-displacement reaction and (b) double-displacement reaction.

element. In this reaction, two or more reactants participate in the chemical reaction and an element of a compound is replaced by another element to produce two or more products. A single-displacement reaction can be expressed symbolically as $A + BC \rightarrow AC + B$ and is shown pictorially in Figure 9.5a. For example, chlorine (Cl_2) displaces bromine (Br_2) from sodium bromide (2NaBr), and this single-displacement reaction is represented by the following equation: $Cl_2 + 2NaBr = 2NaCl + Br_2$.

Double-displacement or double-replacement reaction: This type of reaction is also known as metathesis. In this reaction, two or more reactants participate in the chemical reaction and elements of compounds are interchanged between one another to produce two new products. A double-displacement reaction can be expressed symbolically as $AB + CD \rightarrow AC + BD$ and is shown pictorially in Figure 9.5b. For example, two elements in sodium chloride (2NaCl) displace the two elements in sulfuric acid (H_2SO_4) by producing sodium sulfate (Na_2SO_4) and hydrochloric acid (2HCl), which is represented by $2NaCl + H_2SO_4 = Na_2SO_4 + 2HCl$.

6. *Elimination*: Elimination reaction is a type of organic reaction in which two atoms (or constituents) are eliminated from a molecule. An elimination reaction can be expressed symbolically as $ABCD \rightarrow AD + BC$. Symbolically, it looks like a decomposition reaction. For example, monochlorodifluoromethane ($CHClF_2$) undergoes reaction when heated strongly; a hydrogen (H) atom and a chlorine (Cl) atom are eliminated as molecular hydrogen chloride (HCl). The reaction is represented by the following reaction equation: $2CHClF_2 = C_2F_4 + 2HCl$.

7. *Addition*: Addition reactions are reactions in which atoms are added to a molecule. If the added atoms are hydrogen atoms, the reaction is called hydrogenation reaction. Many vegetable oils are hydrogenated by addition reaction. For example, addition of hydrogen (H_2) to oleic acid ($C_{18}H_{34}O_2$) is represented by the following reaction equation: $C_{13}H_{34}O_2 + H_2 = C_{13}H_{36}O_2$. The product is a solid and used for food preparation.

8. *Redox or oxidation–reduction reaction*: Redox is short for *red*uction–*ox*idation (REDOX) reactions. Any chemical reaction in which the oxidation numbers (oxidation states) of the atoms are changed is an oxidation–reduction reaction. Such reactions transfer electrons from one reactant to another. The chemical agent that gains electrons is called oxidizing agent. The agent that loses electrons is called oxidized agent. For example, the reaction between hydrogen and fluorine is the oxidation–reduction reaction represented by the following equation: $H_2 + F_2 \rightarrow 2HF$.

9. *Acid–base*: Acid–base reactions are when two reactants form salts and water. The combination of hydrochloric acid (HCl) and sodium hydroxide (NaOH) produces sodium chloride (NaCl), the common table salt, and water (H_2O). The word salt is a general term which applies to the products of all such acid–base reactions. This acid–base reaction is represented by the following equation: $HCl + NaOH = NaCl + H_2O$.

10. *Chain reaction*: A chain reaction is a series of reactions in which the products become reagents in each step of the reaction series. Polymerization of chemicals happens through chain reactions. For example, formation (or synthesis) of hydrogen bromide given by the chemical reaction $H_2 + Br_2 \rightarrow 2HBr$ is believed to involve a series of reactions as follows:

$Br_2 \rightarrow 2Br$ $\qquad$ (begins with endoergic reaction)

$Br + Br \rightarrow Br_2$ $\qquad$ (some bromine (Br) atoms recombine in exoergic reaction)

$Br + H_2 \rightarrow HBr + H$ $\qquad$ (Br atom can react with H atom)

$$\left\{\begin{array}{l} H + Br_2 \rightarrow HBr + Br \\ \qquad or \\ H + HBr \rightarrow H_2 + Br \end{array}\right.$$ $\qquad$ (H atom can react with Br or HBr already formed)

HBr is formed, and the chain reaction is propagated by the following two reactions: $Br + H_2 \rightarrow HBr + H$ and $H + Br_2 \rightarrow HBr + Br$.

The basic types of chemical reactions discussed above can be considered as mechanisms of natural selection and variation that can be applied on a population of molecules taking part in chemical reactions in a predefined manner to create new molecules, compounds, and products. The reactants take part in a sequence of chemical reactions and move toward a chemical equilibrium and reach a steady state when no further reactions can occur. The mechanisms of chemical reactions are very similar to the mechanisms of selection and variation used in evolutionary algorithms (Holland, 1975; Siddique and Adeli, 2013), which lead to the new concept of search and optimization algorithms.

9.2 MECHANISMS OF CHEMICAL REACTION

By the mechanism of chemical reaction, we mean the detailed process of transformation of chemical reactants into other products. This process involves the interaction between atoms, molecules, ions, electrons, and free radicals (Brückner, 2004). The interaction may take place in gases, liquids, or solids. Such reaction mechanisms are important for choosing conditions that may maximize the desired quality and amount of product and minimize waste and undesired product.

The most important part of the reaction mechanism is the energy requirements, where energy is consumed for a reaction to happen. The actual course that any reaction should follow is the one that requires least activation energy. The minimum amount of energy required for a chemical reaction to occur is known as the activation energy denoted as E_a, discussed in the earlier section. For a reaction to occur, molecules must collide. The collision frequency describes how many times a particular molecule collides with others per unit of time. Collision theory is an aspect of kinetic molecular theory and was proposed by Max Trautz (1916), which explains how chemical reactions occur and why reaction rates differ for different reactions. In order for molecules to react, a physical chemist named Svante Arrhenius explained in 1889 (Arrhenius, 1889) that colliding molecules must possess enough kinetic energy. Energetic collisions are collisions between molecules with enough kinetic energy to cause the reaction to occur. Not all collisions are energetic collisions because they do not provide the necessary amount of E_a, so not all collisions lead to reactions and product formation. The higher the E_a of a reaction, the smaller is the amount of energetic collisions present, and the slower is the reaction. In contrast, the lower the E_a of a reaction, the greater is the amount of energetic collisions present, and the faster is the reaction. The molecular orientation of the energetic collision has to be right to generate the activated complex in the transition state that leads to the product.

9.3 CHEMICAL REACTION OPTIMIZATION

Based on the theories of chemical reactions discussed in the preceding sections, Lam and Li (2009) first reported the chemical reaction-inspired meta-heuristic algorithm for optimization in their technical report in 2009. Shortly after that, they developed the meta-heuristic optimization algorithm based on the nature of chemical reactions and coined the term CRO (Lam and Li, 2010a,b).

In CRO, molecules are seen as the population of solutions of an optimization problem and the molecules undergo a process of modification through a sequence of elementary chemical reactions.

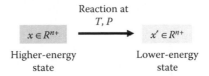

FIGURE 9.6   Change in molecular structure by chemical reaction.

A chemical reaction starts with some unstable molecules with high energy; molecules interact with each other and are transformed into stable products with minimum energy. The reaction process will continue until it reaches the equilibrium state, producing a final product, which is to be considered the optimal solution of the process. This chemical reaction process can be seen as an optimization process. This concept is comparable to any optimization algorithms used in evolutionary computing.

If the problem defines the feasible solution set as a set of n-dimensional positive real numbers R^{n+}, then any vector $x_i \in R^{n+}$ with $x_i = \{x_i^1, x_i^2, \ldots, x_i^n\}$, $i = 1, \ldots, N$, is a valid molecule and x_i^k, $k = 1, \ldots, n$, is thought of as an atom; that is, a molecule $x \in R^{n+}$ is a mathematical representation of a solution of the problem at hand. The representation of $x \in R^{n+}$ can be in the form of a number, an array (similar to a chromosome in the genetic algorithm), a matrix, or a graph (similar to the tree structure in genetic programming). The molecular structure depends on the problem to be solved. In terms of chemical reaction, a change in a molecule $x \in R^{n+}$ with a higher-energy state can only occur by means of a chemical reaction, which changes the energy state of the molecule and results in a new molecule $x' \in R^{n+}$ with a lower-energy state. A hypothetical model of such a reaction that changes the energy state at temperature T and pressure P is shown in Figure 9.6.

As discussed in an earlier section, PE quantifies the energy in the structure of a molecule. The chemical reaction changes the energy state of the molecule and results in a new molecule $x' \in R^{n+}$ with a lower energy state. Therefore, minimization of energy is considered as the objective in CRO and hence is defined as the objective function when evaluating a solution. PE is the energy responsible for stable structure whereas KE is the energy needed for the movement of molecules and collision between molecules such that the reaction can happen. KE of the molecule helps it escape from the local minimum if the solution is stuck at a local minimum. If x represents the molecule with a certain structure (i.e., configuration of atoms), then an objective function, $f(x)$, is defined as equal to PE:

$$PE(x) = f(x) \tag{9.5}$$

A chemical reaction comprises different types of unimolecular and multimolecular elementary reactions, each of which releases different levels of energy of the involved molecule(s). Successful completion of an elementary reaction results in a structural change of a molecule, releasing or absorbing energy and producing a new molecule. Four types of elementary reactions normally are considered in CRO proposed by Lam and Li (2010a).

There are two unimolecular reactions used in CRO:
1.   On-wall ineffective collision
2.   Decomposition
and two multimolecular reactions:
3.   Intermolecular ineffective collision
4.   Synthesis

On-wall ineffective collision: A single molecule is involved in this reaction operation where the molecule does not collide with another molecule; rather, a molecule x (with its inherent molecular structure) hits the wall and bounces back. This results in a change in the molecular structure and change in PE and KE. The new molecule is described by

$$x' = x + \Delta \tag{9.6}$$

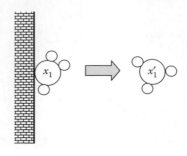

FIGURE 9.7   Illustration of on-wall ineffective collision operator.

where Δ is a perturbation of the molecule due to structural change caused by the collision on the wall.

The perturbation Δ can be modeled as the probability distribution over a finite interval. There are many probability distribution functions available in the literature, for example, Gaussian, Cauchy, lognormal, exponential, and many more. Each of the distribution functions has its own advantages and disadvantages. The on-wall ineffective collision of molecule x is illustrated in Figure 9.7. Due to the change in molecular structure, PE and KE also change from $PE(x)$ to $PE(x')$ and from $KE(x)$ to $KE(x')$, respectively, which must satisfy the energy equation described in Equation 9.4. Otherwise, a change in the molecular structure x will not happen except release (or loss) of some amount of $KE(x)$. The amount of $KE(x)$ is stored in an energy buffer when the reaction is complete. According to the law of conservation of energy, energy cannot be created or destroyed. Adding or removing of energy to and from the system is also not allowed according to this theory. Therefore, a central energy *buffer* is used to store the energy released from reactions. The amount of released energy $KE(x')$ is modeled using a random number $\rho_1 \in [KE_{LossRate}, 1]$, where $KE_{LossRate}$ is considered as a parameter of the chemical reaction and represents the maximum percentage of KE lost in environment at a time. $KE(x')$ and energy *buffer* update can be described by

$$KE(x') = (PE(x) - PE(x') + KE(x)) \times \rho_1 \tag{9.7}$$

$$buffer = buffer + (PE(x) - PE(x') + KE(x)) \times (1 - \rho_1) \tag{9.8}$$

This central energy *buffer* can also be used when additional energy is required for a reaction to occur.

Decomposition: It is a unimolecular operation and does not collide with another molecule. A molecule x hits the wall and breaks into two or more molecules, for example, x_1 and x_2. The decomposition of molecule x is illustrated in Figure 9.8. Due to the change in molecular structure, PE and KE also change from $PE(x)$ to $\{PE(x_1), PE(x_2)\}$ and from $KE(x)$ to $\{KE(x_1), KE(x_2)\}$, respectively, which must satisfy the energy equation described by

$$PE(x) + KE(x) \geq PE(x_1') + PE(x_2') \tag{9.9}$$

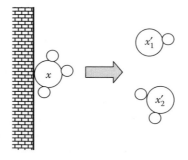

FIGURE 9.8   Illustration of decomposition operator.

$$KE(x_1') = [PE(x) + KE(x) - PE(x_1') - PE(x_2')] \times \rho \qquad (9.10)$$

$$KE(x_2') = [PE(x) + KE(x) - PE(x_1') - PE(x_2')] \times (1 - \rho) \qquad (9.11)$$

where ρ is a random number uniformly distributed over the interval [0, 1].

Sometimes, x may not have enough energy (both $PE(x)$ and $KE(x)$), meaning that Equation (9.9) does not hold for the reaction to happen and decompose the molecule x into x_1 and x_2. It can only happen when $KE(x)$ is large enough. Therefore, for the reaction to happen and decomposition to occur, an amount of energy from the central *buffer* is utilized to help the change taking place. The process is described by the following modified conditions:

$$PE(x) + KE(x) + buffer \geq PE(x_1') + PE(x_2') \qquad (9.12)$$

$$KE(x_1') = [\{PE(x) + KE(x) - PE(x_1') - PE(x_2')\} + buffer] \times \rho_1 \times \rho_2 \qquad (9.13)$$

$$KE(x_2') = [\{PE(x) + KE(x) - PE(x_1') - PE(x_2')\} + buffer] \times \rho_3 \times \rho_4 \qquad (9.14)$$

where ρ_1, ρ_2, ρ_3, and ρ_4 are random numbers uniformly distributed over [0,1]. To ensure a small amount of $KE(x'_1)$ and $KE(x'_2)$ from the central energy *buffer*, multiplication of two random numbers $\{\rho_1 \times \rho_2\}$ and $\{\rho_3 \times \rho_4\}$ is used in Equations 9.13 and 9.14. The central energy *buffer* is updated as follows:

$$buffer = buffer + [PE(x) + KE(x) - PE(x_1') - PE(x_2')] - KE(x_1') - KE(x_2') \qquad (9.15)$$

If Equations (9.9) and (9.12) do not hold, the decomposition will not take place and the molecule retains its structure and energy.

Intermolecular ineffective collision: It is a multimolecular operation. In an intermolecular ineffective collision, two molecules x_1 and x_2 collide with each other and then bounce away. The collision generates two new molecules x'_1 and x'_2 by changing their structures in the neighborhood of the old molecular structures, causing a change in $PE = \{PE(x_1), PE(x_2)\}$ and $KE = \{KE(x_1), KE(x_2)\}$. The intermolecular operation between two molecules x_1 and x_2 is illustrated in Figure 9.9. This elementary reaction involves more than one molecule without any KE drawn from the central energy buffer. The PE changes from $\{PE(x_1), PE(x_2)\}$ to $\{PE(x'_1), PE(x'_2)\}$, and the KE changes from $\{KE(x_1), KE(x_2)\}$ to $\{KE(x'_1), KE(x'_2)\}$. The intermolecular ineffective collision must satisfy the following energy equations:

$$PE(x_1) + PE(x_2) + KE(x_1) + KE(x_2) \geq PE(x_1') + PE(x_2') \qquad (9.16)$$

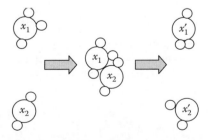

FIGURE 9.9   Illustration of intermolecular ineffective collision operator.

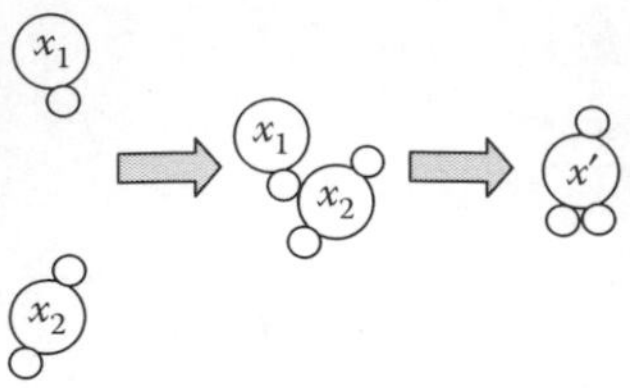

FIGURE 9.10 Illustration of synthesis operator.

$$KE(x_1') = [PE(x_1) + PE(x_2) + KE(x_1) + KE(x_2) - PE(x_1') - PE(x_2')] \times \rho \qquad (9.17)$$

$$KE(x_2') = [PE(x_1) + PE(x_2) + KE(x_1) + KE(x_2) - PE(x_1') - PE(x_2')] \times (1 - \rho) \qquad (9.18)$$

where ρ is a random number uniformly distributed over [0,1]. If the condition in Equation 9.16 does not hold, the reaction will not take place and the molecules retain their structures and energy.

Synthesis: It is a multimolecular operation. In synthesis, two or more molecules x_1 and x_2 collide with each other and combine to form a single molecule x'. The synthesis operation between two molecules x_1 and x_2 is illustrated in Figure 9.10. Synthesis causes a massive change in energy. The *PE* changes from $\{PE(x_1), PE(x_2)\}$ to $PE(x')$, and the *KE* changes from $\{KE(x_1), KE(x_2)\}$ to $KE(x')$. The modification by synthesis must satisfy the following energy equations:

$$PE(x_1) + PE(x_2) + KE(x_1) + KE(x_2) \geq PE(x') \qquad (9.19)$$

$$KE(x') = [PE(x_1) + PE(x_2) + KE(x_1) + KE(x_2) - PE(x')] \qquad (9.20)$$

If the condition in Equation 9.19 does not hold, the reaction will not take place and the molecules retain their structures and energy.

In CRO, a population of reactant molecules represented by $x_i = [x_i^1, x_i^2, \ldots, x_i^n]$, $i = 1, \ldots, N$ with a high-energy level and unstable states (i.e., molecules of different structures) undergo a sequence of elementary chemical reactions. The chemical reactions are mechanisms of transformation through different energy levels to generate products (i.e., molecules of new structures). The final products of low-energy and stable states satisfying the respective energy equations (9.7)–(9.20) are the optimal solutions of the problem. The process of the chemical reaction is seen as an optimization process. This process is implemented in the CRO algorithm with some additional parameters such as MoleColl, KE_{LossRate}, and *buffer*. The main steps of the CRO algorithm can be described as follows:

Step 1: Initialize a population of molecules $x_i = [x_i^1, x_i^2, \ldots, x_i^n]$, $i = 1, \ldots, N$
   Initialize MoleColl, and set initial *KE*
Step 2: Compute $PE_i = f(x_i)$ for all $i = 1, \ldots, N$
Step 3: If ($\rho <$ MoleColl)
   Select molecule $\{x_i, x_j\}$
   If (synthesis criteria hold)
     Create x' by synthesis operation on (x_i, x_j)
     Remove x_i, x_j from population
     Add x' to population
   Else
     Create $\{x'_i, x'_j\}$ by intermolecular ineffective collision of (x_i, x_j)
Step 4: If ($\rho >$ MoleColl)
   Select molecule x_i
   If (decomposition criteria hold)

Create $\{x'_1, x'_2\}$ by decompose operation on $(x_i, buffer)$
Remove x_i from population
Add x'_1, x'_2 to population
Update *buffer* using (9.15)
Else
Create x'_i by on-wall ineffective collision operation on $(x_i, buffer)$
Update *buffer* using (9.8)
Step 5: Check for new minimum point for solution
Step 6: If (termination condition not met), Goto Step 2
Step 7: Return solution

where ρ is a random number and MoleColl is a parameter of the CRO algorithm, which decides on the fraction of all elementary reactions corresponding to intermolecular reactions.

Figure 9.11 illustrates the process of the CRO algorithm. CRO can be applied to tackle problems in both the discrete and continuous domains. The main operators of the CRO algorithm are the decomposition and synthesis operations as mechanisms for generating new solutions for exploring the search space, and the on-wall ineffective and intermolecular ineffective collision operations as the mechanisms for generating solutions from the neighborhood structure for exploiting the search space. The decomposition and synthesis operators act as diversification, and ineffective and intermolecular ineffective collision operators act as intensification for the algorithm. Detailed algorithms for on-wall ineffective collision, decomposition, intermolecular ineffective collision, and synthesis operators are provided in Lam and Li (2012).

9.4   FEATURES OF CRO

The total energy (TE) plays a key role in searching the solution space. TE is constant during the search process. TE is given by

$$\text{TE} = \sum_{i=1}^{N} PE(x_i) + KE(x_i) = \sum_{i=1}^{N} f(x_i) + KE_0 \times N \tag{9.21}$$

where N is the population size.

PE depends on the fitness values $f(x_i)$ defined in Equation 9.5, and *KE* depends on the initial energy KE_0 and population size N. By setting appropriate values for the initial energy KE_0 and population size, TE can be controlled. A lower TE will increase the convergence rate at the risk of getting stuck in a local minimum. If the fitness function $f(\cdot)$ has a very uneven landscape, large KE_0 is suitable. Suppose a solution x changes to x' due to reaction. The change will happen with probability 1, only when $\Delta PE = f(x) - f(x') \geq 0$. If $\Delta PE < 0$, the change can only occur with large KE that compensates ΔPE. In an intermolecular ineffective collision, two or more molecules take part, which means that their KE is shared among molecules. The fitness value $f(x')$ is larger in an intermolecular ineffective collision than in an ineffective collision. In an ineffective collision, KE is transferred to buffer.

$$\text{TE} = \sum_{i=1}^{N} PE(x_i) + KE(x_i) + buffer \tag{9.22}$$

When a series of on-wall ineffective collisions occur, the molecules tend to have lower KE and result in subsequent solutions x with lower $f(x)$. The local minima can then be reached.

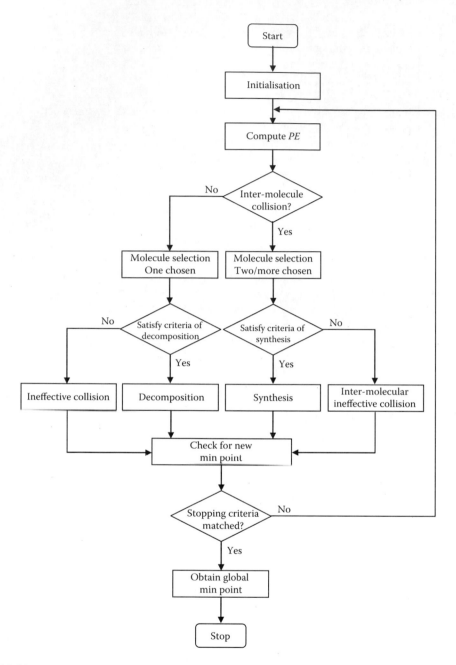

FIGURE 9.11   Flow diagram of CRO algorithm.

Close observation of the life cycle of a molecule reveals that the CRO algorithm searches a region of the solution space for a certain period of time and then jumps out to another region. The algorithm exploits the search space for the local minimum for a number of iterations (say t_k) and then the decomposition mechanism is used to move to a new region for continuing the search. This condition can be used as the decomposition criterion. Once the decomposition condition is reached, the molecule splits into two and searches a new region of the solution space. The search process continues as along as the excessive energy from buffer can be recycled. If a molecule lacks the necessary KE to transform to a new molecule, a synthesis takes place to form a new molecule with higher energy. It allows the CRO algorithm to explore a new region of solution space. Thus, the CRO algorithm can explore every possible region of the solution space and eventually find the global minimum. An interesting feature of CRO is that several CROs can be carried out simultaneously on the same problem without synchronization. Characteristic features and convergence analysis of the CRO algorithm are provided in Lam et al. (2013a) and Astudillo et al. (2015).

9.5 PARAMETERS OF CRO

There are mainly four parameters in CRO. These are population size, KE loss rate, fraction of unimolecular reaction, and initial KE. Population size, denoted as $PopSize$, is the initial number of solutions generated randomly in the solution space. KE loss rate, denoted as $KE_{LossRate}$, is the upper limit of percentage of KE lost to the environment during on-wall ineffective collisions; and fraction of unimolecular reaction, denoted as MoleColl, is the fraction of molecules that undergo unimolecular or intermolecular reactions. If MoleColl is less than a random number ρ, it will result in a unimolecular collision. Otherwise, an intermolecular collision will take place. It is to be noted that a unimolecular collision will always take place when there remains only one molecule in the population. Initial KE, denoted as KE_0, is the initial value assigned to KE in the initialization stage.

9.5.1 CRO OPERATORS

CRO is a variable population-based meta-heuristic where the total number of solutions kept by the algorithm may change from time to time. Decomposition increases and synthesis decreases the number of molecules in the population. Decomposition breaks a molecule into multiple molecules, whereas synthesis combines multiple molecules into one. Four types of operators are applied to the population of solutions.

Decomposition operator: This operator breaks a molecule into two or more molecules. Firstly, the solution x is copied onto x_1 and x_2. Then, half of the variables (i.e., $n/2$ variables) of the original solution x are perturbed by adding random variation to create new solutions, where n is the number of decision variables (or dimension) of the problem. The decomposition operator is implemented using the following pseudocode:

```
Copy x to x₁ and x₂
For k = 1, ... ,n do
  {
  If (k ≤ n/2)
    x₁(k)= x(k) + random[0,1]
  Else
    x₂(k)= x(k) + random[0,1]
  }
```

Synthesis operator: This operator combines two molecules, that is, two solutions x_1 and x_2, by applying "probabilistic selection" to implement synthesis into a new solution x'. Each component of x' in the same position is chosen either from x_1 or from x_2 randomly. The synthesis operator is implemented using the following pseudocode:

```
For k = 1, ... ,n do
  {
  ρ = rand[0,1]
  If (ρ > 0.5)
    x′(k) = x₁(k)
  Else
    x′(k) = x₂(k)
  }
```

The effect of synthesis operation is similar to the recombination (or crossover) operation used in many other evolutionary algorithms.[*] The molecules then can undergo a series of on-wall and intermolecular ineffective collisions. In any one of these events, they search around their own neighborhoods for (local) minimum solutions.

Ineffective collision: Some small change occurs in the molecular attributes during this operation, and the molecule obtains new structure x' in the neighborhood of x, which is expressed as

$$x' = N(x) \tag{9.23}$$

where $N(\cdot)$ is the neighborhood operator.

Yu et al. (2011) used a neighborhood operator to generate a new solution x' from a solution x. It is done by perturbing one element of x chosen randomly. The perturbation is done by adding a Gaussian perturbation $\rho(m,\sigma)$ to the randomly chosen element, where m is the mean and σ is the standard deviation. It is seen that σ is a user-defined parameter. The neighborhood operator is implemented using the following pseudocode:

```
k = random integer (1,n)
Find the k-th element m in x, that is, m = x(k)
x'(k) = x(k) + ρ(m,σ)
Replace x(k) with x'(k) in x
```

Intermolecular ineffective collision: Two molecules x_1 and x_2 collide with each other and then bounce away. Small changes create new molecules x'_1 and x'_2 in the neighborhood of x_1 and x_2 by this collision. The changes are expressed as

$$\begin{cases} x'_1 = N(x_1) \\ x'_2 = N(x_2) \end{cases} \tag{9.24}$$

where $N(\cdot)$ is the neighborhood operator.

In CRO, decomposition and synthesis are exploration mechanisms and give the effect of global search, while ineffective collision and intermolecular ineffective collision are exploitation mechanisms and give the effect of local search. These two operators are local search mechanisms for improving the quality of solutions.

9.6 VARIANTS OF CRO

The CRO algorithm is a very recent addition to the meta-heuristic algorithm family. Researchers made simple modifications while applying the algorithm to different application domains, which later became known as variants of CRO algorithms such as real-coded CRO (RCCRO), opposition-based CRO, and orthogonal CRO. These variants are discussed briefly in this section.

9.6.1 REAL-CODED CRO (RCCRO)

The current version of CRO (Lam and Li, 2010a,b, 2012) is designed for discrete optimization problems. An extension of the CRO algorithm is proposed for continuous problems by Lam et al. (2012a), which has become known as real-coded CRO. Three modifications are introduced: solution representation, neighborhood operator, and boundary constraint handling.

---

[*] Recombination or crossover operation in the EA generally selects two solutions randomly from the population pool and combine them for producing one or two offspring by choosing individual elements from each of them. There are a number of crossover operations reported in the literature. A detailed account of discussion on EA crossover operations can be found in Siddique and Adeli (2013).

A molecule $x_i = [x_i^1, x_i^2, \ldots, x_i^n]$, $i = 1, \ldots, N$ is a mathematical representation of the solution of a problem. In RCCRO, each variable x^k, $k = 1, 2, \ldots, n$ is implemented using a floating-point number. To deal with continuity, continuous search ability is incorporated into the neighborhood search operator $N(\cdot)$. If the problem does not impose any constraints on relations between solution variables, x^k can be treated independently and the perturbation is defined as

$$x_i'^k = N(x_i^k) = x_i^k + \delta_i^k \tag{9.25}$$

where δ_i^k is a perturbation in the k-th dimension. It is seen that δ_i^k can be modeled as a probabilistic perturbation, for example, Gaussian, Cauchy, and Lévy. Gaussian distribution can be used for δ_i^k (Lam et al., 2012a):

$$\delta_i = N(\mu, \sigma^2) \tag{9.26}$$

where $N(\mu, \sigma^2)$ is the Gaussian distribution with mean μ and variance σ^2.

The perturbation depends mainly on the starting point from the solution x_i, direction from the mean μ, and step size based on the spread σ. The starting point of δ_i is the solution itself, that is, x_i. If there is no specific information available for the direction, then setting $\mu = 0$ will generate perturbation from x_i in all directions. The step size is the spread of the Gaussian distribution. Considering the three components, δ_i becomes $\delta_i = N(0, \sigma^2)$. The perturbation of x_i can now be defined as

$$x_i' = x_i + N(0, \sigma^2) \tag{9.27}$$

In general, σ is fixed in RCCRO during the execution of the algorithm. Too large or too small a value for σ will make the algorithm inefficient. Lam et al. (2012a), therefore, proposed an adaptive scheme for σ, where the initial value of σ is set equal to solution space $(U - L)$ and decreased gradually by a fixed factor θ, that is, $\sigma = \sigma \times \theta$. It is found by experiment that the logarithmic decrease can avoid overly aggressive refinements.

The perturbation of a solution x_i, defined by Equation 9.25 or 9.27, may go out of bound defined by the lower (L_i) and upper (U_i) bounds. The boundary constraints can be handled by bringing the solution x_i back within boundary (Lam et al., 2012a). A simple technique such as a reflecting scheme, where x_i reflects back by the same amount from the boundary, can be used. x_i' is defined by

$$x_i' = \begin{cases} 2 \times L_i - x_i & \text{if } x_i < L_i \\ 2 \times U_i - x_i & \text{if } x_i > U_i \end{cases} \tag{9.28}$$

Lam et al. (2012a) also proposed another hybrid scheme, which combines the absorbing and reflecting approach; x_i' is then defined by

$$x_i' = \begin{cases} L_i & \text{if } (\rho \leq 0.5) \text{ and } (x_i < L_i) \\ U_i & \text{if } (\rho \leq 0.5) \text{ and } (x_i > U_i) \\ 2 \times L_i - x_i & \text{if } (\rho > 0.5) \text{ and } (x_i < L_i) \\ 2 \times U_i - x_i & \text{if } (\rho > 0.5) \text{ and } (x_i > U_i) \end{cases} \tag{9.29}$$

where $\rho \in [0, 1]$ is a random number.

The effectiveness and performance of the RCCRO have been verified by Lam et al. (2012a) by applying it to a number of unimodal, high-dimensional multimodal, and low-dimensional multimodal benchmark functions. Bhattacharjee et al. (2014a) applied RCCRO to a short-term hydrothermal scheduling problem. The objective is to minimize the total power generation cost by scheduling the generation of different hydrothermal power plants for certain intervals of time such that it satisfies various constraints such as the hydro, thermal plant, and power system network.

9.6.2 Opposition-Based CRO

In order to improve the computational efficiency and the convergence rate of different optimization techniques, opposition-based learning (OBL) was proposed by Tizhoosh (2005a,b, 2006). OBL has been proposed to improve the candidate solution by considering the current population as well as its opposite population at the same time. OBL has been applied successfully to the learning process by many researchers into different soft computing paradigms. The opposition-based gravitational search algorithm has been discussed in Chapter 2 (Shaw et al., 2012, 2014), and the opposition-based harmony search algorithm has been discussed in Chapter 5 (Chatterjee et al., 2012; Upadhyay et al., 2014). Bhattacharjee et al. (2014b) used opposite and quasi-opposite numbers in one-dimensional space to the CRO algorithm. If x be any real number between $[a,b]$, its opposite number $\breve{x}$ is defined by

$$\breve{x} = a + b - x \tag{9.30}$$

If x be any real number between $[a,b]$, its quasi-opposite number $\breve{x}_q$ is defined as

$$\breve{x}_q = \text{rand}(c, \breve{x}) \tag{9.31}$$

where c is the center of the interval $[a,b]$ and estimated as the mean of the interval $[a,b]$, that is, $c = (a + b)/2$. Here, $\breve{x}_q$ is a random number uniformly distributed over $[c, \breve{x}]$. Similarly, the reflected quasi-opposite number is defined by

$$\breve{x}_{qr} = \text{rand}(c, x) \tag{9.32}$$

where $\breve{x}_{qr}$ is a random number uniformly distributed over $[c,x]$.

One-dimensional representation of x can easily be extended to higher dimensions. A population of molecules is generated, and then a quasi-opposite molecular matrix QOM is formed from the molecular set using a jumping rate $J_r \in [0,1]$ as follows:

```
If (rand < J_r){
   For i = 1 to N
      For i = 1 to n
         QOM(i,j) = rand(c,x̆ )
}
```

The main steps of the opposition-based real-coded CRO (ORC-CRO) are as follows:

Step 1: Generate quasi-opposite molecular matrix QOM
Step 2: Compute fitness PE
Step 3: If (decomposition criterion satisfied)
            Perform decomposition
            Else perform on-wall ineffective collision

Step 4: If (synthesis criterion satisfied)
        Perform decomposition
      Else perform inter-molecular ineffective collision
Step 5: If (Termination condition not met), Goto Step 2
Step 6: Return solution

The effectiveness and performance of the ORC-CRO have been verified on a short-term hydro-thermal scheduling problem. Three test systems have been chosen for the verification of the algorithm. Test System 1 comprises 4 hydro plants coupled hydraulically and an equivalent thermal plant. Test System 2 consists of 4 cascaded hydro plants and three composite thermal plants. Test System 3 is a more practical representation of a hydrothermal system comprising 4 hydro plants and 10 thermal plants. ORC-CRO has good exploration and exploitation ability, which helps the algorithm in reaching an optimal solution within a very small number of iterations.

9.6.3 Orthogonal CRO

CRO shows random behavior and traverses through the solution space in a random manner, which eventually limits the search scope and slows down the convergence speed of the algorithm (Lam et al., 2013a). A meta-heuristic algorithm needs to explore promising regions. This even becomes more difficult when there are a large number of decision variables to be optimized within a limited number of experiments or trials. The concept behind the orthogonal experimental design (OED) is an approach based on an orthogonal array to find out the best combination of different factors from a small number of experiment samples (Montgomery, 2008). The orthogonal array has been used in other heuristic algorithms such as simulated annealing (Ho et al., 2004), PSO (Zhan et al., 2011), and the genetic algorithm (Hu et al., 2006). Based on the OED concept, Li et al. (2015b) proposed an orthogonal CRO algorithm where they introduced a quantization orthogonal crossover (QOX) operator. In the QOX operator, the decision variables are quantized into levels, the variables are divided into groups, and each group is treated as a factor and then from there creates an individual, that is, a molecule. The QOX operator is used for synthesis operation in the CRO algorithm as follows:

If ($r <$ MoleColl)
    Select two molecules x_1, x_2 from population
    If (synthesis criteria hold)
        Create x' using orthogonal crossover QOX(x_1, x_2)

The effectiveness and performance of the orthogonal CRO have been verified on 23 well-known unimodal, high-dimensional multimodal, and low-dimensional multimodal benchmark functions by Li et al. (2015b). In all cases, the effectiveness of the proposed approach in solving optimization problems has been demonstrated. The results also show the less efficiency for low-dimensional functions.

9.6.4 Adaptive Collision CRO

In standard CRO, intermolecular ineffective collision operation ultimately results in two on-wall ineffective collisions occurring at the same time. In principle, these two operators are performing similar operations. In other words, there is an overlap between functionalities of intermolecular and on-wall ineffective collision operators, which may lead to waste of valuable computation time to some extent. To reduce this functional overlap, an intermolecular adaptive collision scheme is introduced to CRO by Yu et al. (2014, 2015). The adaptive collision comprises a new intermolecular ineffective collision operator and an adaptive collision scheme. The new approach is called adaptive collision CRO.

    In the intermolecular ineffective collision, generally two on-wall ineffective collisions take place at the same time. Therefore, an interoperator is introduced to ineffective collisions such that it makes a difference between these two on-wall ineffective collisions. Two molecules $\{x_i, x_j\}$ are randomly selected, and their fitness values, that is, $PE(x_i)$ and $PE(x_j)$, are calculated. Let $PE(x_i)$ be greater than $PE(x_j)$. Based on the PE values, two approaches are deployed to modify the molecules instead of the neighborhood operation usually applied for on-wall ineffective collision (Yu et al., 2014, 2015).

$$\begin{cases} x_i'^k = (x_i^k - x_j^k) \times r^k + x_i^k \\ x_j'^k = (x_i^k - x_j^k) \times r^k + x_j^k \end{cases} \tag{9.33}$$

where $x_i'^k$ and $x_j'^k$ are the new molecules, x_i^k and x_j^k are old molecules in the k-th dimension, and r^k is a random number for each dimension over the interval [0,1]. The mechanism expressed by Equation (9.33) ensures that the new molecules $x_j'^k$ and $x_j'^k$ are not similar.

In canonical CRO, collision rate (CollRate) is a user-defined parameter, which is fixed during execution and usually chosen empirically by users (Xu et al., 2011a; Lam et al., 2012a). But experience with CRO suggests that the ratio of occurrence of on-wall ineffective collision to that of intermolecular ineffective collision is more critical to the performance. In the adaptive collision scheme, collision rate (CollRate) is defined as the ratio of occurrence of on-wall ineffective collision to that of intermolecular ineffective collision. CollRate is defined by the following sigmoid function (Yu et al., 2014, 2015), which controls the ratio of the collision adaptively:

$$\text{CollRate} = \frac{1}{1 + \exp[-6 \times (count/\text{FE}_{max})]} \tag{9.34}$$

where *count* is the number of successful intermolecular ineffective collisions and FE_{max} is the maximum number of allowable function evaluations. The value of *count* is incremented when a successful intermolecular ineffective collision occurs, and it is decremented when an on-wall collision occurs.

The performance of the adaptive CRO has been verified experimentally on 16 different benchmark functions and compared with canonical CRO. The simulation results demonstrate the expected superior performance of adaptive CRO over the canonical CRO.

9.6.5 Elitist CRO

In the standard CRO algorithm, the molecules taking part in the reactions are selected randomly, which slows down the convergence rate, though it contributes to the diversity of the population. Duan and Gan (2015b) proposed an elitist CRO (ECRO) algorithm incorporating elitist strategies into the framework of CRO, which include elitist selection, evolution, and crossover. Two attributes of the molecule are defined: affinity and concentration. The affinity represents the quality of a solution, while the concentration reflects the proportion of molecules with similar structures in the current population.

The elitist selection strategy speeds up the convergence of ECRO while still retaining its performance. The efficiency of the ECRO has been verified on a contour-based target recognition problem. The experimental results show that the ECRO has improved searching ability.

9.6.6 Artificial Chemical Reaction Optimization (ACRO)* Algorithm

Alatas (2011) proposed a chemistry-inspired computational method of optimization, which is different from CRO proposed by Lam and Li (2010a) where the operators introduced to ACRO are based on different types of chemical reactions. The advantage of ACRO is that it does not use an extra function for determination of the quality of reactants. The ACRO has been discussed in Chapter 10 in a greater detail.

---

* Alatas (2011) developed the ACRO algorithm independently from Lam and Li (2010a,b). Lam and Li published their idea of CRO as a technical report back in 2009 (Lam and Li, 2009) and subsequently published the complete CRO algorithm in 2010 (Lam and Li, 2010a,b). It is claimed by Alatas that the ACRO approach is significantly different from the CRO of Lam and Li.

9.7   HYBRID CRO

Some researchers attempted to improve the operators by introducing techniques from other meta-heuristic algorithms such as DE and PSO and proposed hybrid CRO. The hybrid variants are discussed briefly in this section.

9.7.1   HYBRID CRO AND DE

Roy et al. (2014) suggested a new improvement to the standard CRO algorithm by introducing the mutation and crossover operators borrowed from the DE algorithm (Storn et al., 2005) and called it hybrid DE-CRO, which implements on-wall ineffective collision operation, decompose operation, intermolecular ineffective collision operation, and synthesis operation using a mutation and crossover strategy. A brief discussion on DE is provided in Chapter 8. A detailed discussion on DE can be found in Siddique and Adeli (2013).

The on-wall ineffective collision operation in CRO is implemented using the mutation operation borrowed from DE. A new molecule is generated using the mutation operation as follows:

$$x'_{ij} = x_{ij} + F \times (x_{mj} - x_{nj}) \tag{9.35}$$

where x'_{ij} is the new j-th component of the i-th molecule, $\{x_{ij}, x_{mj}, x_{nj}\}$ are the j-th components of three different molecules chosen randomly from the current population, and F is a positive control parameter.

The decompose operation is implemented using the crossover operation borrowed from DE. To perform the crossover, one molecule x_m is selected randomly from the population and another molecule x_n is generated randomly. Two new molecules x'_m and x'_n are created applying the crossover operation between x_m and x_n.

The intermolecular ineffective collision operation is implemented using the crossover operation from DE. Two new molecules are created by performing the crossover operation on randomly selected two molecules x_m and x_n from the population.

Molecules are modified using the synthesis collision operation implemented by applying the conventional crossover operation from the genetic algorithm.

The effectiveness and performance of the DE-CRO algorithm have been verified on four test systems of a conventional static ELD problem with 6, 10, 13, and 40 thermal generators, one test system of a restructured ELD problem, and one dynamic ELD problem (Roy et al., 2014). In all cases, the solution obtained through hybridization of CRO with DE has better quality in terms of simulation results and computation time. Dutta et al. (2015) applied the DE-CRO approach to a unified power flow control (UPFC) problem to find the optimal placement and parameter setting for achieving optimal performance of a power system network. Dutta et al. used the same operators developed in Roy et al. (2014). The IEEE 14-bus and IEEE 30-bus power test systems are used for the UPFC implementation. Both single and multiobjective optimizations are considered for experimentation. The performance of the DE-CRO algorithm is compared with other meta-heuristic algorithms reported in the literature, which again demonstrates the competitiveness of the DE-CRO approach.

9.7.2   HYBRID CRO AND PSO

A good balance between exploration and exploitation is a characteristic of an efficient optimization algorithm. Though the performance of the CRO algorithm seems promising, there is still some kind of insufficiency at exploration carried out by using decomposition and synthesis operations. The advantage with PSO is that it can be used either for good exploration or exploitation. Considering the advantages of the compensatory property of CRO and PSO, Nguyen et al. (2014) proposed a hybrid algorithm, combining the explorative and exploitative features of PSO and CRO, denoted as HP-CRO. Due to low efficiency, the decomposition and synthesis operations are eliminated from

HP-CRO and a PSO-based update operation is performed instead. HP-CRO creates new molecules using neighboring operations of CRO and using mechanisms of PSO. These molecules can be considered as molecules of CRO or particles of PSO. PSO and CRO use the same population generated initially. The basic operations involved in the PSO algorithm are described by

$$v_i^d(k+1) = w \cdot v_i^d(k) + c_1 \rho_1 (x_{bi}^d - x_i^d) + c_2 \rho_2 (x_g^d - x_i^d) \tag{9.36}$$

$$x_i^d(k+1) = x_i^d(k) + v_i^d(k+1) \times T \tag{9.37}$$

where v_i^d is the velocity of i-th particle; x_i^d, x_{bi}^d, and x_g^d are the position, iteration best, global best position in the d-th dimension, respectively; w is the inertia weight; c_1 and c_2 are cognitive and social coefficients, respectively; ρ_1 and ρ_2 are the random numbers generated between [0, 1]; and T is the time, which is unity to convert the velocity into position. The computation of molecules (or positions of particles) is easy and straightforward. If a PSO-update criterion is satisfied, then molecules are updated using the PSO algorithm, that is, Equations (9.36) and (9.37); otherwise, the intermolecular ineffective collision operation and the on-wall ineffective collision operation are performed. Thus, the HP-CRO algorithm repeats the PSO update, intermolecular ineffective collision operation, and on-wall ineffective collision operation until the termination condition is satisfied. In order to verify the effectiveness and performance of the HP-CRO algorithm, it has been verified on 23 well-known unimodal, high-dimensional multimodal, and low-dimensional multimodal benchmark functions by Nguyen et al. (2014). In all cases, the effectiveness of the proposed approach in solving optimization problems was validated by experimental results.

Zhang and Duan (2014) proposed another version of the hybrid PSO-CRO approach, called PCRO, for solving an image matching problem. In each iteration, the best molecule is saved. In this PCRO, once the ineffective collisions (i.e., on-wall ineffective collision and intermolecular ineffective collision) are performed, the molecule is updated by the distance between the original molecule and the molecule with the current best solution. The PSO update mechanism is simplified as follows:

$$v_i^d(t+1) = \frac{c_1 \rho_1 (x_{bi}^d - x_i^d)}{T} \tag{9.38}$$

$$x_i^d(t+1) = x_i^d(t) + v_i^d(t+1) \times T \tag{9.39}$$

where v_i^d, x_i^d, and x_{bi}^d are the velocity, position, and iteration best position in the d-th dimension of the i-th particle, respectively; c_1 is the cognitive coefficient; ρ_1 is a random number generated between [0, 1]; and T is the time, which is considered unity for conversion of position into velocity in Equation 9.38 and velocity into position in Equation 9.39. Simulation results from an image matching problem demonstrated the effectiveness of the approach. Li et al. (2015a) proposed a hybrid algorithm by combining the features of PSO and CRO, called HP-CRO, for MOOPs. The proposed algorithm balances the operators of CRO and PSO and is capable of exploring the search space and avoiding premature convergence.

9.7.3 HYBRID CRO AND LIN–KERNIGHAN LOCAL SEARCH

In CRO, decomposition and synthesis operators contribute to exploration of the search space, while on-wall ineffective collision and intermolecular collision operators contribute to exploitation of the search space. A heuristic algorithm needs to strike a balance between the exploration and exploitation of search space for obtaining a good-quality solution. Strengthening the exploitation ability of on-wall ineffective collision can improve solution quality. Sun et al. (2011) proposed the hybrid CRO approach, which considers the trade-off between the exploration abilities using decomposition and synthesis operations and the exploitation abilities using the Lin–Kernighan (LK) local search algorithm (Lin and Kernighan, 1973).

The LK local search algorithm is a well-known heuristic approach used for solving the Euclidean TSP. The main concept of the LK algorithm is the swapping of pairs of subtours for creating new tours. It is a generalization of two-opt and three-opt, that is, switching of two or three paths to make the tour shorter (Lin and Kernighan, 1973). The LK local search builds an exchange of variable size λ by sequentially deleting and adding edges to the current tour while maintaining tour feasibility. The LK algorithm is adaptive where each step decides how many paths between cities need to be switched for finding a shorter tour. The four elementary operations of the CRO-LK algorithm are implemented using LK local search, order crossover (OX), and distance preserving crossover (DPX) operators to improve computation and performance. The on-wall ineffective collision is implemented using the LK local search algorithm, where the LK local search uses the concept of λ optimal. The intermolecular ineffective collision and decomposition operations are implemented using extended modified OX operator introduced by Davis (1985). The synthesis operation is implemented using the DPX operator introduced by Freisleben and Merz (1996).

The effectiveness and performance of the hybrid CRO-LK algorithm were verified on a set of 40 Euclidean sample problems with sizes ranging from 51 to 783 nodes. Optimal solutions have been confirmed for 75% of the problems, which are competitive over other meta-heuristic algorithms.

9.8 APPLICATION OF CRO

CRO is the first of its kind meta-heuristic optimization algorithm that employs the principles of chemical reactions from theories of chemistry. Since its inception in 2009, CRO has been successfully exploited for solving a broad range of problems in many areas of computing, engineering, and optimization. The applications of CRO can be broadly categorized into different domains such as quadratic assignment problem (QAP), RCPS problem, energy systems, for example, OPF problem and ELD problem, hydro-thermal scheduling, training NN, fuzzy rule learning problem, communications and networking problems, for example, peer-to-peer (P2P) streaming, network coding optimization problem (NCOP), and cognitive radio spectrum allocation problem (CRSAP), sensor networks deployment problem, computing grid scheduling problem, bioinformatics, stock portfolio selection problem in finance. These applications are reported with a brief discussion of the problems in the following sections.

9.8.1 QUADRATIC ASSIGNMENT PROBLEM

QAP is a minimization of the total cost for assigning facilities to locations. The facilities can be of different types to be assigned to different locations. The number of facilities and the number of locations are equal. That means each facility is to be assigned to a unique location. The cost is defined as (flow × distance). The distance between the pair of locations and the flow between two locations are known. The total cost is computed by summing the cost of every possible pair of facilities and locations. If there are n facilities to be assigned to n locations, the QAP, which minimizes the total distances and flows between facilities, is mathematically defined by

$$\min \sum_{i,j=1}^{n} \sum_{k,l=1}^{n} f_{ij} d_{kl} x_{ik} x_{jl} \tag{9.40}$$

where f_{ij} is the flow of material from facility i to j and d_{kl} is the distance from location k to l. The constraints subject to satisfaction are

$$x_{ij} \in \{0,1\}, 1 \le i, j \le n \tag{9.41}$$

$$\sum_{j=1}^{n} x_{ij} = 1, 1 \leq i \leq n \tag{9.42}$$

$$\sum_{i=1}^{n} x_{ij} = 1, 1 \leq j \leq n \tag{9.43}$$

where x_{ij} is defined as

$$x_{ij} = \begin{cases} 1 & \text{if facility } i \text{ assigned to location } j \\ 0 & \text{otherwise} \end{cases} \tag{9.44}$$

The constraints in Equations 9.41 and 9.42 assure that each facility i is assigned to exactly one location. The constraints in Equations 9.41 and 9.43 assure that each location j has exactly one facility which is assigned to it.

The term quadratic comes from the formulation of the QAP as an integer optimization problem with a quadratic objective function.

Every possible solution of QAP can be represented in the form of a permutation of n elements. That means a molecule is a permutation of n elements. The positions and the values of the permutation correspond to the location and facility, respectively, as shown in Figure 9.12 for an instance of QAP with $n = 5$.

If the size of instances is larger than 20, the QAP becomes intractable. QAP is one of the most difficult NP-hard combinatorial optimization problems and has many applications to the real world, which can be transformed to other well-known combinatorial optimization problems such as the TSP, maximum clique problem, and graph-partitioning problem that leads to many other real-life applications. Therefore, solutions to QAP using meta-heuristic methods become very important.

The on-wall ineffective collision and intermolecular ineffective collision are implemented using neighborhood structure. In permutation representation, neighborhood structure is a simple swap between positions of two facilities as illustrated in Figure 9.13.

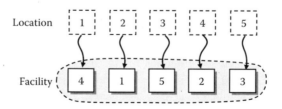

Solution in form of permutation

FIGURE 9.12   Representation of a solution in QAP.

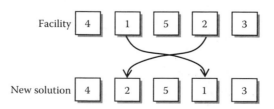

FIGURE 9.13   Neighborhood structure by swapping positions of two facilities.

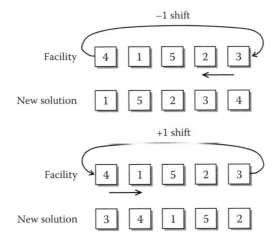

FIGURE 9.14 Decompose operation applying right or left rotate of facilities.

The decomposition operation is implemented using the circular shift operator to generate new solutions. A random integer value is generated between $[-n,n]$, which decides the number of circular shifts (rotate), that is, left shift for a negative number or right shift for a positive number as illustrated in Figure 9.14.

The synthesis operation is implemented using a distance-preserving crossover operator (Merz and Freisleben, 1997) to generate new solutions, where a minimum structure is obtained from two existing molecules.

Lam and Li (2010b) applied CRO using the elementary reaction operators designed for QAP and verified the performance and effectiveness of the CRO algorithm on 23 problem instances. The simulation results show improved performance in terms of the mean and maximum costs compared to three meta-heuristic algorithms, namely, the fast ant algorithm, improved SA, and tabu search. Xu et al. (2010b) also reported good performance of parallel CRO for QAP.

9.8.2 TRAVELING SALESMAN PROBLEM

In TSP, the connectivity of n cities is represented by a directed graph $G = (V, E)$, where V is the set of cities and E is the set of edges (i.e., path between cities). The objective is to find a minimum length tour among all such possible tours for the given graph with n cities. A solution to the TSP is an ordered set of n distinct cities. There are $(n - 1)!/2$ feasible solutions for a TSP with n cities, where the global optimum(s) is sought. In fact, many real-world problems can be formulated as an instance of the TSP, which can lead to solutions to many other combinatorial optimization problems. A detailed discussion and mathematical treatment of TSP is provided in Chapter 2. TSP can be considered as a special case of QAP. The solution representation in TSP is similar to that of QAP. The elementary operations, that is, on-wall ineffective collision, intermolecular ineffective collision, decomposition, and synthesis, are implemented in the same way as in QAP. The on-wall ineffective collision and intermolecular ineffective collision for TSP can be implemented using the neighborhood structure as shown in QAP. The decomposition operation is implemented using the circular shift operator. The synthesis operation is implemented using the distance-preserving crossover operator. Sun et al. (2011) applied the hybrid CRO-LK approach to TSP, where the exploitation abilities are enhanced using the LK local search algorithm.

9.8.3 RESOURCE-CONSTRAINED PROJECT SCHEDULING PROBLEM

An RCPS problem considers resources of limited availability and activities (or tasks) of known durations and resource requests, linked by precedence relations. The objective is to find a schedule of minimal duration by assigning a start time to each activity such that the precedence relations and the resource availabilities are satisfied. RCPS is a generalization of job shop scheduling.

RCPS comprises a set of activities $T = \{t_0, t_1, t_2, \ldots, t_n, t_{n+1}\}$; by convention, t_0 represents the start of schedule and t_{n+1} the end of schedule, R a set of finite resources, each $r \in R$ with its capacity limit, each activity putting some demand on resources, $C: R \rightarrow N$ a capacity function, $D: T \rightarrow N$ a duration function, $S: T \rightarrow N$ an assignment start time, d a deadline, and $U: T \times R \rightarrow N$ a utilization function. A partial ordering P on the activities T is also given, specifying that some activities must precede others. The goal is to minimize the makespan without violating the precedence constraints or over-utilizing the resources.

If the activity $t_1 \in T$ precedes $t_2 \in T$ in the partial ordering P, then the precedence constraints are defined as

$$S(t_1) + D(t_1) \leq S(t_2) \tag{9.45}$$

For any time τ and running time $running(\tau) = \{t | S(t) \leq \tau \leq S(t) + D(t)\}$, the resource constraints are defined as

$$\sum_{t \in running(\tau)} U(t,r) \leq C(r), \forall \tau, \forall r \in R \tag{9.46}$$

For all tasks, $t \in T$, $S(t) \geq 0$, and $S(t) + D(t) < d$.

RCPS problem is one of the classical NP-hard optimization problems. Job shop, flow shop, and open shop problems are special cases of RCPS. A number of comprehensive surveys on algorithms for solving RCPS problems have been reported in Hartmann and Kolisch (2000) and Kolisch and Hartmann (2006). Lam and Li (2010b) applied CRO to solve the RCPS problem. A solution of the RCPS problem can be represented as a permutation, that is, as an activity list. For example, a project with 3 activities with Activity 0 as the start and Activity 4 as the end of schedule is shown in Figure 9.15.

Activity 1 has higher priority than Activities 3 and 2. It can be seen from Figure 9.12 that the solution structure of RCPS is the same as that of QAP, that is, it represents a permutation of n activities. The four operations of CRO, that is, on-wall ineffective collision, intermolecular ineffective collision, decomposition, and synthesis, can easily be applied to RCPS. Lam and Li (2010b) tested the performance of the CRO algorithm to 600 benchmark instances set with 120 activities from the digital library. Simulation results show the best makespan for 116 instances.

9.8.4 Economic Load Dispatch Problem

ELD is a method of determining the most efficient, low-cost, and reliable operation of a power system by dispatching available electricity generation resources to supply load on the system. The objective of the ELD problem is to find an optimal combination of power generation that minimizes the total cost of generation while satisfying different equality and inequality constraints. A detailed mathematical treatment of the ELD problem has been provided in Chapter 6. Bhattacharjee et al. (2014a–c) applied an RCCRO algorithm to an ELD problem considering different constraints such as power balance, ramp rate limits, and prohibited operating zone constraints. Roy et al. (2014) applied the CRO algorithm for optimization of the different load dispatch problems.

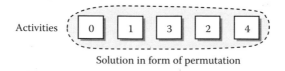

Solution in form of permutation

FIGURE 9.15 Representation of a solution in RCPS.

9.8.5 OPTIMAL POWER FLOW PROBLEM

The OPF problem has become the heart of modern economic power systems and markets. The problem is inherently complex in terms of economy, reliable power supply, and computational effort. Efficient market equilibrium requires multipart nonlinear pricing to be economical. As the power flow is alternating current, it induces additional nonlinearities. The problem poses nonconvexities, which include both binary variables and continuous functions, making the optimization problem computationally difficult to solve (Carpentier, 1979). Moreover, the power system must be able to withstand the loss of any generator or transmission unit, and the system operator must make binary decisions to start up and shut down generation and transmission units in response to system events. There are mainly two objectives which the electrical power system must achieve besides the consideration of the operational constraints: minimization of total generating cost and minimization of active transmission losses. The OPF problem is a large-scale highly nonlinear control optimization problem, which seeks the most favorable settings of a given power system that minimizes the total fuel cost, active power loss, and bus voltage deviation and enhances voltage stability while at the same time satisfying a number of equality and inequality constraints. A detailed mathematical treatment of the OPF problem is provided in Chapter 2. The CRO algorithm has been applied to an OPF problem by a number of researchers (Sun et al., 2012; Dutta et al., 2015).

9.8.6 TRAINING NEURAL NETWORKS

In general, a feed-forward NN consists of one input layer, one or more hidden layers with nonlinear activation function, and one output layer neuron with a linear activation function. Biases can be set to nonzero or zero. The problem is to find the weights of the connectivity of the network for the given architecture to produce the correct output for the function for each corresponding input. The weight (w) and bias (b) training of the NN can be carried out by minimizing or optimizing the mean squared of the network error function, denoted as $MSE(w, b)$. While the most popular training algorithm for an NN such as the backpropagation algorithm requires gradient information and is also prone to local minima, meta-heuristic algorithms are becoming popular such as the genetic algorithm (Siddique and Tokhi, 2001), central-force optimization (Green et al., 2012), harmony search algorithm (Kattan and Abdullah, 2011), hybrid PSO, and gravitational search algorithm (Mirjalili et al., 2012). To apply such meta-heuristic optimization algorithms, the training problem of an NN has been formulated as an optimization problem where the training of the NN can be carried out by minimizing the sum squared of the network error function, that is, $MSE(w, b)$. Details of these procedures and objective functions are discussed in Chapters 2 and 5.

Nayak et al. (2015) applied the CRO algorithm to a higher-order NN with a single hidden layer called pi–sigma neural network (PSNN). The PSNN has the advantage of fast learning capability, avoiding exponential increase in the number of weights and processing units. Therefore, many researchers prefer to use the PSNN for data classification. The effectiveness and performance of the CRO-based training of the PSNN have been tested with various benchmark data sets from the UCI machine learning repository. The experimental results have been compared with the resulting performance of the PSNN, GA-PSNN, and PSO-PSNN. In all cases, the simulation result shows that the CRO-PSNN performs better than the other evolutionary techniques and has better classification accuracy for almost all the data sets. Nayak et al. (2013) also used the CRO algorithm to train a NN for forecasting the stock index. Yu et al. (2011) applied the CRO algorithm, instead of the backpropagation algorithm, to the training of NNs. They used the Iris classification data set, Wisconsin breast cancer classification data set, and Pima Indians diabetes data set.

9.8.7 FUZZY RULES LEARNING

Performance of any fuzzy system depends mainly on the fuzzy rules. There is no exact method for learning fuzzy rules. Generally, fuzzy rules are constructed by trial and error or with the help of experts in the domain. Many researchers use evolutionary algorithms for learning the fuzzy rules (Siddique and Adeli, 2013). Lam et al. (2012b) proposed cooperative rules (COR) to solve the fuzzy rule learning in fuzzy rule-based systems. The COR method consists of two phases: (i) search space construction and (ii) selection of the most cooperative fuzzy rule set. The learning process is then formulated as a combinatorial optimization problem. The proposed CRO-based COR method is verified on two fuzzy modeling benchmark problems, namely, three-dimensional surface function and electrical low-voltage line length estimation. Simulation results demonstrate that the CRO-based COR method is highly competitive and outperforms many other existing optimization methods.

9.8.8 COMMUNICATIONS AND NETWORKING PROBLEMS

The CRO algorithm has found applications in communications and networking problems recently. Among them are the population transition problem in P2P streaming (Lam et al., 2010), CRSAP (Lam and Li, 2010a; Lam et al., 2013b), channel assignment problem (CAP) in wireless mesh networks (Lam and Li, 2010b), NCOP (Pan et al., 2011), and Bus Sensor Deployment Problem (BSDP) (Yu et al., 2012).

9.8.8.1 Peer-to-Peer Streaming

P2P technology provides a new platform for live video streaming with less geographical constraint at a cost-effective price. In P2P streaming, there is a source to provide streaming data for peers to receive data. The networks are heterogeneous in general, which is the main cause of transmission delays from source to peers. Based on the amount of delays, peers are grouped into colonies so that the peers with shorter transmission delays and its colony members can be used as the source of data for other peers. The peers have the flexibility to switch from colony to colony, which is controlled by assigning a population transition probability (PTP). By assigning the appropriate PTP to colonies, streaming data from source to peers can be maximized. Lam et al. (2010) applied the CRO algorithm for maximization of the PTP and showed that CRO performs better than nonevolutionary meta-heuristic algorithms.

9.8.8.2 Cognitive Radio Spectrum Allocation Problem

For proper management, control of agencies, and smooth communication of wireless signals, the frequency spectrum is divided into licensed bands and unlicensed bands and regulated by government agency. Licensed bands are allocated to different agencies, service providers, communication networks, and military for authorized users (primary users), while unlicensed bands are allocated to public domains for unlicensed users (secondary users) for a variety of purposes (Hasan et al., 2012). As a result of this policy, unlicensed bands are used by a majority of users causing overcrowding, while licensed bands are underutilized. The capacity of communication networks can be dramatically increased if the unlicensed users (secondary users) are allowed to use the licensed bands, provided that their activities do not affect the primary users. In order to maximize system utility, spectrum allocation is used to assign channels to users subject to hardware constraints. This problem is known as the CRSAP. A solution of the CRSAP is a channel assignment matrix A, whose entities $a_{ij} \in \{0,1\}$ are indicators specifying whether particular channels are assigned to certain secondary users. The CRO algorithm has been applied to the CRSAP by Lam and Li (2010a) and Lam et al. (2013b), which demonstrated dramatic improvement over other existing approaches.

9.8.8.3 Channel Assignment Problem

A wireless mesh network consists of N routers. Each $i \in N$ is equipped with certain radio interfaces R_i. There are $K = \{1, 2, \ldots, m\}$ radio channels. In order to communicate with other routers, a router assigns channels. A router can be assigned to a maximum number of channels, and it can transmit signals within a certain transmission range. A communication link can be established between two routers that interfere on the same channel and within the transmission range of each other. Based on this configuration of routers, a conflict graph $G_c = (V_c, E_c)$ can be constructed, where V_c is the set of routers and E_c is the set of edges representing the communication link between two routers within their respective transmission range. The channel assignment is defined as a function $f: V_c \rightarrow K$, that is, assigning each link to exactly one channel in the channel set K. The objective of the CAP in wireless mesh networks is to minimize the overall network interference subject to the interface constraint. This problem is NP-hard and combinatorial in nature. A conflict graph is generated from the problem at hand, and the solution is represented in the form of an integer vector of length $|V_c|$. Each element of the vector represents a channel from the set of $K = \{1, 2, \ldots, m\}$ radio channels. For example, the solution vector [1 3 2 1] means that there are 4 routers (i.e., the graph has 4 nodes) with Channel 1 assigned to Router 1, Channel 3 assigned to Router 2, Channel 2 assigned to Router 3, and Channel 1 assigned to Router 4. Now, the CAP looks like the same as the QAP, and all the elementary operators of CRO can be applied to the CAP. Lam and Li (2010b) applied CRO to test problems of the CAP to improve the existing solutions. The simulation results show that CRO outperforms tabu search.

9.8.8.4   Network Coding Optimization Problem

In computer networks, routers are deployed to exchange data between the source and destinations. The routers do not process data. They only receive and forward the data. It is found that network coding can enhance network throughput without any change to the topology of the network if routers have processing or coding ability (Li et al., 2003). Adding in coding to all possible routers will simply increase computational cost and reduce overall efficiency. Therefore, finding the minimum number of coding links in the network that ensures the maximum data transfer rate is an optimization problem. This problem is known as the NCOP. It is an NP-hard problem and combinatorial in nature. The NCOP can be represented by a directed acyclic graph $G = (V, E)$, where V is the set of routers and E is the set of edges representing a communication link between routers within their respective data transmission rate. Two routers can be connected by multiple links, and some links are allowed not to transmit data based on the coding scheme. The objective is to find the minimum number of links to achieve a target transmission rate, and how many links and which links are to be coded. Network coding happens at merging nodes (routers) where data from multiple links come together. Pan et al. (2011) used linear coding by applying exclusive OR operation. A solution (i.e., a molecule) for the NCOP in CRO is a vector of integers of length $|E|$, where $|E|$ is the number of links of the network. Each element of the vector specifies how a certain link receives or routes data from incoming links. For example, the solution vector [1 2 2 1] represents a network of 4 links. Pan et al. (2011) applied the CRO algorithm to the NCOP and showed that it outperformed existing algorithms.

9.8.8.5   Bus Sensor Deployment Problems

To provide sufficient coverage of a large area for surveillance or observation, a large number of sensors are deployed, which demand larger processing and significant computational cost. An alternative is to use embedded sensors on moving objects such as vehicles or animals. The idea is to install sensors on buses (or vehicles of a city transport system), which cover a much larger area with a limited number of sensors (Hasan et al., 2012). Thus, the selection of bus routes on which sensors are deployed is also important so as to minimize the number of sensors required for a satisfactory coverage of the area. This problem is formulated as an optimization problem called bus sensor deployment problem (BSDP). The monitoring area is divided into grids. There are n bus routes. Each bus route R_i, $i = 1, 2, \ldots, n$, goes through multiple bus routes, meaning that the bus route covers the grids. A solution for the BSDP is represented by a binary vector $S = [s_1, s_2, s_3, \ldots, s_n]$, where $s_i \in \{0, 1\}$, $i = 1, 2, \ldots, n$. $s_i = 1$ means a sensor is installed on bus route R_i and $s_i = 0$ means no sensor is installed on bus route R_i. A coverage graph G_{S_f} is created on collection of all s_i sensor information, where $S_f = [1, 1, 1, \ldots, 1]$ means sensors are installed on all bus routes. The number of 1s in S_f gives the total number of covered grids in G_{S_f} denoted as t_{S_f}, that is, t_{S_f} is the total coverable grids. The objective function for the BSDP can be defined as

$$\min\left[\left(1 - \frac{t_{S'}}{t_{S_f}}\right) \times \alpha + \frac{\sum_{i=1}^{n} s_i}{n} \times (1 - \alpha)\right] \tag{9.47}$$

where $t_{S'}$ is the total number of grids covered for solution S'. The term $t_{S'}/t_{S_f}$ defines the percentage of uncovered grids over the total coverable grids, and the term $\left(\sum_{i=1}^{n} s_i\right)/n$ defines the percentage of sensor deployment. Here, α is an arbitrary parameter, which is chosen as 0.5, meaning that equal weight is given to both terms.

Yu et al. (2012) applied the CRO algorithm to the BSDP. The elementary operations such as on-wall ineffective collision, intermolecular ineffective collision, decomposition, and synthesis of CRO are straightforward to apply to the solution vector. They reported the simulation carried out on the Island of Hong Kong, in which the optimization results outperformed the simple GA.

9.8.9 MULTIOBJECTIVE OPTIMIZATION PROBLEMS

Problems requiring simultaneous optimization of more than one objective function are known as MOOPs. Formally, it can be defined as

$$\begin{cases} \text{Minimize/maximize } f(x) \\ \text{Subject to } g_j(x) \geq 0, j = 1, 2, 3, \ldots, J \\ h_k(x) = 0, k = 1, 2, 3, \ldots, K \end{cases} \tag{9.48}$$

where $f(x) = \{f_1(x), f_2(x), \ldots, f_n(x)\}$ is a vector of objective functions, n is the number of objectives, $x = \{x_1, x_2, \ldots, x_p\}$ is a vector of decision variables, and p is the number of decision variables. Here, the problem optimizes n objectives and satisfies J inequality and K equality constraints. This type of problem has no unique perfect solution. The objectives can interact or conflict with each other. Most MOOPs do not provide a single solution; rather, they offer a set of solutions. Such solutions are the "trade-offs" or good compromises among the objectives. In order to generate these trade-off solutions, an old notion of optimality called the Pareto-optimum set (Ben-Tal, 1980) is normally adopted.

CRO has been applied to MOOPs (Bechikh et al., 2014; Chaabani et al., 2014; Li et al., 2015b). Duan and Gan (2015a) used orthogonal multiobjective CRO to the optimal design of a brushless direct-current motor (BLDCM). The simulation results demonstrate that CRO has superior performance when compared with other existing optimization algorithms. Li et al. (2015b) proposed a hybrid algorithm by combining the features of PSO and CRO for an MOOP. The HP-CRO creates new molecules not only for CRO operations but also by mechanisms of PSO. The proposed algorithm balances the operators of CRO and PSO and is capable of exploring the search space and avoiding premature convergence.

9.8.10 OTHER APPLICATIONS

The CRO algorithm has received considerable attention, and it has been successfully applied in various real-world applications such as computing grid scheduling problem (Xu et al., 2010a, 2011a), stock portfolio selection problem (Xu et al., 2011b), and short adjacent repeat identification problem in bioinformatics (Xu et al., 2012). Scheduling tasks on heterogeneous computing platforms with the objective of minimizing makespan has become an important problem in a variety of applications, which involves making decisions about the execution order of tasks and task-to-processor mapping. Xu et al. (2014) applied the hybrid CRO algorithm to a task scheduling problem. Zhang and Duan (2014) applied a hybrid particle CRO algorithm for biological image matching based on lateral inhibition. Duan and Gan (2015b) used the CRO algorithm for contour-based target recognition in aerial images. Li et al. (2015a) showed the application of CRO algorithms to global numerical optimization problems. Finding an optimization model for a monopolistic firm serving an environmentally conscious market, a trade-off between investment in cleaner technology and the penalty of reduced net profit has to be found, which is formulated as a nonlinear programming model with maximization of quadratic profit function. Choudhary et al. (2015) applied the CRO algorithm to solve such a nonlinear programming problem.

9.9 CONCLUSION

The CRO algorithm is based on the principles of chemical reactions. A chemical reaction is a natural process of transforming unstable molecules to stable molecules through formation and destruction of chemical bonds. The process continues until it reaches the minimum free energy. This simple concept is utilized for optimization where molecules that take part in the reactions are representation of solutions of the problem at hand. The population of solutions undergo a series of reactions and reach a stable state, which is termed as the optimal solution.

Though a very recent addition to the meta-heuristic family, CRO has received considerable attention, and it has been successfully applied to various real-world problems since its inception in 2009. CRO nevertheless appears to hold considerable promise. CRO has already attracted significant interest from the research community, which will help its further development and hopefully will involve both empirical algorithmic improvements and theoretical refinements. Some researchers made some valuable source code in MATLAB.

There are so far a good number of variants reported in the literature. CRO still needs further research in order to develop new methods for exploring and exploiting the search space. Any application of CRO should be explored and compared with other population-based meta-heuristic methods.

Another important issue is the convergence of the CRO algorithm. CRO is very likely to stick in local minima. There have been many hybrid variants published so far that help CRO to overcome this situation. There has been little theoretical analysis done so far apart from the convergence analysis by Lam et al. (2013a). In addition, premature avoidance, estimation of convergence rate, searching behaviors explanation, accelerating convergence, and parameter selection are important issues to be addressed.

REFERENCES

Alatas, B. 2011. ACROA: Artificial chemical reaction optimization algorithm for global optimization, *Expert Systems with Applications*, 38, 13170–13180.

Arrhenius, S. A. 1889. Über die Dissociationswärme und den Einfluß der Temperatur auf den Dissociationsgrad der Elektrolyte, *Zeitschrift der Physikalischen Chemie*, 4, 96–116. (in German)

Astudillo, L., Melin, P., and Castillo, O. 2015. Introduction to an optimization algorithm based on the chemical reactions, *Information Sciences*, 291, 85–95.

Atkins, P. W. and De Paula, J. 2006. *Physical Chemistry*, 4th ed., Weinheim: Wiley-VCH.

Bechikh, S., Chaabani, A., and Said, L. B. 2014. An efficient chemical reaction optimization algorithm for multi-objective optimization, *IEEE Transaction on Cybernetics*, 45, 2051–2064, DOI: 10.1109/TCYB.2014.2363878.

Ben-Tal, A. 1980. Characterization of Pareto and lexicographic optimal solutions, in: G. Fandel and T. Gal, eds., *Multiple Criteria Decision Making: Theory and Application, Lecture Notes in Economics and Mathematical Systems*, Vol. 17, Berlin: Springer, pp. 1–11.

Bhattacharjee, K., Bhattacharya, A., and Dey, S. H. 2014a. Real coded chemical reaction based optimization for short-term hydrothermal scheduling, *Applied Soft Computing*, 24, 962–976.

Bhattacharjee, K., Bhattacharya, A., and Dey, S. H. 2014b. Opposition real coded chemical reaction based optimization for short-term hydrothermal scheduling, *Electrical Power and Energy Systems*, 63, 145–157.

Bhattacharjee, K., Bhattacharya, A., and Dey, S. H. 2014c. Chemical reaction optimisation for different economic dispatch problems, *IET Generation, Transmission & Distribution*, 8(3), 530–541.

Brown, T. E., LeMay, H. E., Bursten, B., and Murphy, C. 2008. *Chemistry: The Central Science*, Upper Saddle River, NJ: Prentice Hall.

Brückner, R. 2004. *Reaktionsmechanismen*, 3rd ed., München: Spektrum Akademischer Verlag. (In German.)

Carpentier, J. 1979. Optimal power flows, *International Journal of Electrical Power & Energy Systems*, 1(1), 3–15.

Chaabani, A., Bechikh, S., and Said, L. B. 2014. An indicator-based chemical reaction optimization algorithm for multi-objective search, *Proceedings of the 2014 Conference Companion on Genetic and Evolutionary Computation (GECCO'14)*, ACM, New York, pp. 85–86.

Chatterjee, A., Ghoshal, S. P., and Mukherjee, V. 2012. Solution of combined economic and emission dispatch problems of power systems by an opposition-based harmony search algorithm, *International Journal of Electrical Power & Energy Systems*, 39(1), 9–20.

Choudhary, A., Suman, R., Dixit, V., Tiwari, M. K., Fernandes, K. J., and Chang, P.-C. 2015. An optimization model for a monopolistic firm serving an environmentally conscious market: Use of chemical reaction optimization algorithm, *International Journal of Production Economics*, 164, 409–420.

Davis, L. 1985. Applying adaptive algorithms to epistatic domains, in: A. K. Joshi, ed., *Proceedings of the 9th International Joint Conference on Artificial Intelligence*, Vol. 1, Los Angeles: Morgan Kaufmann, pp. 162–164.

Duan, H. and Gan, L. 2015a. Orthogonal multi-objective chemical reaction optimization approach for the brushless DC motor design, *IEEE Transaction on Magnetics*, 51(1), 7000207.

Duan, H. and Gan, L. 2015b. Elitist chemical reaction optimization for contour-based target recognition in aerial images, *IEEE Transaction on Geosciences and Remote Sensing*, 53(5), 2845–2859.

Dutta, S., Roy, P. K., and Nandi, D. 2015. Optimal location of UPFC controller in transmission network using hybrid chemical reaction optimization algorithm, *Electrical Power and Energy Systems*, 64, 194–211.

Freisleben, B. and Merz, P. 1996. New genetic local search operators for the traveling salesman problem, in: H.-M. Voigt, W. Ebeling, I. Rechenberg, H.-P. Schwefel, eds., *Proceedings of the 4th Conference on Parallel Problem Solving from Nature, Lecture Notes in Computer Science*, Vol. 1141, Berlin: Springer, pp. 890–900.

Gibbs, J.W. 1873. A Method of geometrical representation of the thermodynamic properties of substances by means of surfaces, *Transactions of the Connecticut Academy of Arts and Sciences*, 2, 382–404.

Green, R. C., Wang, L., and Alam, M. 2012. Training neural networks using central force optimization and particle swarm optimization: Insights and comparisons, *Expert Systems with Applications*, 39, 555–563.

Hartmann, S. and Kolisch, R. 2000. Experimental evaluation of state-of-the-art heuristics for the resource-constrained project scheduling problem, *European Journal of Operations Research*, 127(2), 394–407.

Hasan, S. F., Siddique, N., and Chakraborty, S. 2012. *Intelligent Transport Systems: 802.11-based Roadside-to-Vehicle Communications*, New York: Springer Verlag.

Ho, S.-J., Ho, S.-Y., and Shu, L.-S. 2004. OSA: Orthogonal simulated annealing algorithm and its application to designing mixed H_2/H_∞ optimal controllers, *IEEE Transaction on Systems, Man and Cybernetics – Part A: Systems and Humans*, 34(5), 588–600.

Holland, J. 1975. *Adaptation in Natural and Artificial Systems*, Ann Arbor, MI: University of Michigan Press.

Hu, X., Zhang, J., and Zhong, J. 2006. An enhanced genetic algorithm with orthogonal design, In *IEEE Congress on Evolutionary Computation*, CEC 2006, pp. 3174–3181.

Kattan, A. and Abdullah, R. 2011. Training of feed-forward neural networks for pattern-classification applications using music inspired algorithm, *International Journal of Computer Science and Information Security*, 9, 44–57.

Kolisch, R. and Hartmann, S. 2006. Experimental investigation of heuristics for resource-constrained project scheduling: An update, *European Journal of Operations Research*, 174(1), 23–37.

Kotz, J. C., Treichel, P. M., and Townsend, J. 2008. *Chemistry and Chemical Reactivity*, Belmont, CA: Brooks Cole.

Lam, A. Y. S. and Li, V. O. K. 2009. *Chemical-Reaction-Inspired Meta-Heuristic for Optimization*, Technical Report TR-2009-003.

Lam, A. Y. S. and Li, V. O. K. 2010a. Chemical reaction optimization for cognitive radio spectrum allocation, *Proceedings of the IEEE Global Communications Conference (GLOBECOM 2010)*, Miami, FL, USA, pp. 1–5.

Lam, A. Y. S. and Li, V. O. K. 2010b. Chemical-reaction-inspired meta-heuristic for optimization, *IEEE Transaction on Evolutionary Computation*, 14(3), 381–399.

Lam, A. Y. S. and Li, V. O. K. 2012. Chemical reaction optimization: A tutorial, *Memetic Computing*, 4(1), 3–17.

Lam, A. Y. S., Xu, J., and Li, V. O. K. 2010. Chemical reaction optimization for population transition in peer-to-peer live streaming, *Proceedings of the IEEE Congress on Evolutionary Computation (CEC-2010)*, July, 2010, Barcelona, Spain, pp. 1–8.

Lam, A. Y. S., Li, V. O. K., and Wei, Z. 2012a. Chemical reaction optimization for the fuzzy rule learning problem, *WCCI 2012 IEEE World Congress on Computational Intelligence*, June, 10-15, 2012, Brisbane, Australia, pp. 1–8.

Lam, A. Y. S., Li, V. O. K., and Yu, J. J. Q. 2012b. Real-coded chemical reaction optimization, *IEEE Transaction on Evolutionary Computation*, 16(3), 339–353.

Lam, A. Y. S., Li, V. O. K., and Xu, J. 2013a. On the convergence of chemical reaction optimization for combinatorial optimization, *IEEE Transactions on Evolutionary Computation*, 17(5), 605–620.

Lam, A. Y. S., Li, V. O. K., and Yu, J. J. Q. 2013b. Power-controlled cognitive radio spectrum allocation with chemical reaction optimization, *IEEE Transaction on Wireless Communications*, 12, 3180–3190.

Li, S.-Y. R., Yeung, R. W., and Cai, N. 2003. Linear network coding, *IEEE Transaction on Information Theory*, 49, 371–381.

Li, Z., Nguyen, T. T., Chen, S. M., and Truong, T. K. 2015a. A hybrid algorithm based on particle swarm and chemical reaction optimization for multi-object problems, *Applied Soft Computing*, 35, 525–540.

Li, Z.-Y., Li, Z., Nguyen, T. T., and Chen, S.-M. 2015b. Orthogonal chemical reaction optimization algorithm for global numerical optimization problems, *Expert Systems with Applications*, 42, 3242–3252.

Lin, S. and Kernighan, B. W. 1973. An effective heuristic algorithm for the traveling-salesman problem, *Operations Research*, 21(2), 498–516.

Merz, P. and Frcisleben, B. 1997. A genetic local search approach to the quadratic assignment problem, in: T. Bäck, ed., *Proc. 7th Int. Conf. Genetic Algorithms*, San Francisco, CA: Morgan Kaufmann, pp. 465–472.

Mirjalili, S., Hashim, S. Z. M., and Sardroudi, H. M. 2012. Training feed forward neural networks using hybrid particle swarm optimisation and gravitational search algorithm, *Applied Mathematics and Computation*, 218, 11125–11137.

Montgomery, D. C. 2008. *Design and Analysis of Experiments*, Hoboken, NJ: John Wiley & Sons.

Nayak, S. C., Misra, B. B., and Behera, H. S. 2013. Hybridising chemical reaction optimisation and artificial neural networks for stock future index forecasting, *2013 1st International Conference on Emerging Trends and Application in Computer Science (ICETACS)*, September 13–14, 2013, Shillong, India, pp. 130–134.

Nayak, J., Naik, B., and Behera, H. S. 2015. A novel Chemical Reaction Optimization based Higher order Neural Network (CRO-HONN) for nonlinear classification, *Ain Shams Engineering Journal*, 6(3), 1069–1091.

Nguyen, T. T., Li, Z.-Y., Zhang, S.-W., and Truong, T. K. 2014. A hybrid algorithm based on particle swarm and chemical reaction optimization, *Expert Systems with Applications*, 41, 2134–2143.

Pan, B., Lam, A. Y., and Li, V. O. 2011. Network coding optimization based on chemical reaction optimization, *2011 IEEE Global Telecommunications Conference (GLOBECOM 2011)*, December 5–9, Houston, Texas, pp. 1–5.

Roy, P. K., Bhui, S., and Paul, C. 2014. Solution of economic load dispatch using hybrid chemical reaction optimization approach, *Applied Soft Computing*, 24, 109–125.

Shaw, B., Mukherjee, V., and Ghoshal, S. P. 2012. A novel opposition-based gravitational search algorithm for combined economic and emission dispatch problems of power systems, *Electrical Power and Energy Systems*, 35(1), 21–33.

Shaw, B., Mukherjee, V., and Ghoshal, S. P. 2014. Solution of reactive power dispatch of power systems by an opposition-based gravitational search algorithm, *International Journal of Electrical Power & Energy Systems*, 55, 29–40.

Siddique, N. and Adeli, H. 2013. *Computational Intelligence: Synergies of Fuzzy Logic, Neural Networks and Evolutionary Computing*, Chichester, West Sussex, UK: John Wiley.

Siddique, N. H. and Tokhi, M. O. 2001. Training neural networks: Backpropagation vs genetic algorithms, *Proceedings of the IEEE International Joint Conference on Neural Networks (IJCNN-2001)*, Washington DC, 15–19 July 2001, pp. 2673–2678.

Storn, R., Price, K. V., and Lampinen, J. 2005. *Differential Evolution – A Practical Approach to Global Optimization*, Berlin: Springer.

Sun, J., Wang, Y., Li, J., and Gao, K. 2011. Hybrid algorithm based on chemical reaction optimization and Lin-Kernighan local search for the traveling salesman problem, *2011 Seventh International Conference on Natural Computation (ICNC)*, July 26–28, Shanghai, Vol. 3, pp. 1518–1521.

Sun, Y., Lam, A. Y. S., Li, V. O. K., Xu, J., and Yu, J. J. Q. 2012. Chemical reaction optimization for the optimal power flow problem, *Proceedings of the IEEE Congress on Evolutionary Computation (CEC)*, June 10–15, Brisbane, Australia, pp. 1–8.

Tizhoosh, H. R. 2005a. Reinforcement learning based on actions and opposite actions, *Proceedings of ICGST International Conference on Artificial Intelligence and Machine Learning*, December 19–21, Cairo, Egypt, pp. 94–98.

Tizhoosh, H. R. 2005b. Opposition-based learning: A new scheme for machine intelligence, *Proceedings of International Conference on Computational Intelligence for Modelling Control and Automation, CIMCA'2005*, Vienna, Austria, 28–30 November 2005, I, 695–701.

Tizhoosh, H. R. 2006. Opposition-based reinforcement learning, *Journal of Advanced Computational Intelligence and Intelligent Informatics*, 10(5), 578–585.

Trautz, M. 1916. Das Gesetz der Reaktionsgeschwindigkeit und der Gleichgewichte in Gasen. Bestätigung der Additivität von Cv-3/2R, Neue Bestimmung der Integrationskonstanten und der Moleküldurchmesser, *Zeitschrift für Anorganische und Allgemeine Chemie*, 96(1), 1–28. (In German.)

Upadhyay, P., Kar, R., Mandal, D., Ghoshal, S. P., and Mukherjee, V. 2014. A novel design method for optimal IIR system identification using opposition-based harmony search algorithm, *Journal of the Franklin Institute*, 351, 2454–2488.

Xu, J., Lam, A., and Li, V. 2010a. Chemical reaction optimization for the grid scheduling problem, *IEEE International Conference on Communications (ICC)*, May 23–27, Cape Town, South Africa, pp. 1–5.

Xu, J., Lam, A. Y., and Li, V. O. 2010b. Parallel chemical reaction optimization for the quadratic assignment problem, *Proceedings of The 2010 World Congress in Computer Science, Computer Engineering, and Applied Computing (Worldcomp 2010)*, July 12–15, Las Vegas, NV, pp. 125–131.

Xu, J., Lam, A., and Li, V. K. 2011a. Chemical reaction optimization for task scheduling in grid computing, *IEEE Transactions on Parallel and Distributed Systems*, 22(10), 1624–1631.

Xu, J., Lam, A. Y. S., and Li, V. O. K. 2011b. Stock portfolio selection using chemical reaction optimization, *Proceedings of the International Conference on Operations Research and Financial Engineering*, Paris, France, June 2011, pp. 458–463.

Xu, J., Lam, A. Y. S., Li, V. O. K., Li, Q., and Fan, X. 2012. Short adjacent repeat identification based on chemical reaction optimization, *Proceedings of the IEEE Congress on Evolutionary Computation (CEC)*, June 10–15, Brisbane, Australia, pp. 1–8.

Xu, Y., Li, K., He, L., Zhang, L., and Li, K. 2014. A hybrid chemical reaction optimization scheme for task scheduling on heterogeneous computing systems, *IEEE Transactions on Parallel and Distributed Systems*, 26, 3208–3222, DOI: 10.1109/TPDS.2014.2385698.

Yu, J. J. Q., Lam, A. Y. S., and Li, V. O. K. 2011. Evolutionary artificial neural network based on chemical reaction optimization, *Proceedings of the IEEE Congress on Evolutionary Computation (CEC)*, June 5–8, New Orleans, USA, pp. 2083–2090.

Yu, J. J. Q., Li, V. O. K., and Lam, A. Y. S. 2012. Sensor deployment for air pollution monitoring using public transportation system, *WCCI 2012 IEEE World Congress on Computational Intelligence, Proceedings of the IEEE Congress on Evolutionary Computation (CEC-2012)*, 10–15 June 2012, Brisbane, Australia, pp. 1–7.

Yu, J. J. Q., Li, V. O. K., and Lam, A. Y. S. 2014. An inter-molecular adaptive collision scheme for chemical reaction optimization, *Proceedings of the IEEE Congress on Evolutionary Computation (CEC)*, July 6–11, Beijing, China, pp. 1998–2004.

Yu, J. J. Q., Lam, A. Y. S., and Li, V. O. K. 2015. Adaptive chemical reaction optimization for global numerical optimization, *2015 IEEE Congress on Evolutionary Computation*, May 25-28, Sendai, Japan, pp. 3192–3199.

Zhan, Z.-H., Zhang, J., Li, Y., and Shi, Y.-H. 2011. Orthogonal learning particle swarm optimization, *IEEE Transactions on Evolutionary Computation*, 15, 832–847.

Zhang, Z. and Duan, H. 2014. A hybrid particle chemical reaction optimization for biological image matching based on lateral inhibition, *Optik*, 125, 5757–5763.

10 Miscellaneous Algorithms

10.1 INTRODUCTION

Modeling and simulating natural phenomena for solving complex problems has been an interesting research area for several decades (De Castro, 2007). There have been many other physics- and chemistry-based search and optimization algorithms dispersedly reported in the literature (Gendreau and Potvin, 2010). Some of these methods and algorithms are relatively new, which have not attracted much attention from the research community and some of them somehow have gone unnoticed. Some of these methods are really powerful tools but unfortunately have remained unknown to a wider research community. This chapter explores different avenues of the literature such as journals, conferences, symposiums, and books in pursuit of these algorithms and makes an exhaustive effort to assimilate those algorithms with pseudocodes, examples, and applications based on an updated literature survey. These algorithms are broadly categorized into physics- and chemistry-based algorithms. A good, but brief, review on physics-based algorithms is reported in Biswas et al. (2013) and in Siddique and Adeli (2016).

The physics-based algorithms are

1. Big Bang–Big Crunch (BB–BC) algorithm
2. Black hole algorithm
3. Galaxy-based search (GbS)
4. Artificial physics optimization (APO)
5. Integrated radiation search (IRS)
6. Space gravitation optimization (SGO)
7. Gravitational interactions optimization (GIO)
8. Charged system search (CSS)
9. Hysteretic optimization (HO)
10. Colliding bodies optimization (CBO)
11. Ray optimization (RO)
12. Extremal optimization (EO)
13. Particle collision algorithm (PCA)
14. River formation dynamics (RFD)
15. Water cycle algorithm (WCA)

The chemistry-based algorithms are

1. Artificial chemical process algorithm (ACPA)
2. Artificial chemical reaction optimization algorithm (ACROA)
3. Chemical reaction algorithm (CRA)
4. Gases Brownian motion optimization (GBMO)

In the following sections, these algorithms will be presented in a concise form describing the basic idea, working mechanisms, variants, and their applications.

10.2 BIG BANG–BIG CRUNCH (BB–BC) ALGORITHM

10.2.1 INSPIRATION AND ALGORITHM

According to Einstein's theory of relativity, space, time, and matter are connected together. There are three fundamental questions in physics. How did the universe begin? Does time have a beginning and an end? Does space have edges? The question of whether the universe had a beginning in time was posed by German philosopher Immanuel Kant in 1781 (see Kant, 1781). Much later, Edwin Hubble made a landmark discovery in 1929 that the universe is rather expanding (Hubble, 1929). Russian physicist Alexander Friedmann made the exact prediction some years before in 1922. Extrapolating the expansion of the universe backwards in time using the general theory of relativity yields an infinite density and temperature at a finite time in the past. This is known as singularity theory or Big Bang singularity. The Big Bang theory was finally proved mathematically by Stephen Hawking and Roger Penrose in the 1970s (Hawking and Penrose, 1970). This Big Bang singularity theory helps in breaking down the general relativity and refers to the birth of the universe and explains the subsequent evolution of the universe. Einstein's theory of relativity replied to these very old questions with three startling assertions: the universe is expanding from a Big Bang, black holes do distort space and time, and the dark energy could be pulling the space apart sending the galaxies beyond the edge of the visible universe (Hawking, 1998; White and Diaz, 2004). Such theories are also supported by recent studies (Bars et al., 2012; Gibbons and Hawking, 1993; Hawking, 1992, 1993, 1998) that the Big Bang was a bounce, a transition from contraction to expansion, involving a brief effective antigravity phase between the Big Crunch and the Big Bang.

Most scientists believe that all the matter, energy, and space in the universe were once squeezed into an infinitesimally small volume. There was a cataclysmic explosion, which created the universe. The space, time, energy, and matter all came into being at an infinitely dense, infinitely hot gravitational singularity, and began expanding at once. The explosion became known as the Big Bang, which is considered to be the theory behind the birth of the universe. The Big Bang was neither big (the universe was incomparably smaller than the size of a proton), nor a bang (it was more of a snap or a sudden inflation). It is also believed that since the Big Bang, the universe is expanding. However, scientists also believe that this expansion will not continue forever. All matters will collapse into the biggest black hole pulling everything within it, which is referred as Big Crunch (Peebles et al., 1994). On the basis of this notion of expansion phenomenon of Big Bang and shrinking phenomenon of Big Crunch, an optimization algorithm has been formulated by Erol and Eksin (2006) called BB–BC algorithm. The Algorithm is mainly based on the theories of evolution of the universe. There are two phases in the BB–BC algorithm: Big Bang phase and Big Crunch phase. In the Big Bang phase, energy dissipation produces disorder and randomness, which is represented by the creation of a random population of individuals with respect to center of mass within the limitation of the search space. In the Big Crunch phase, randomly distributed particles are brought into an order under the gravitational attraction and all masses collapse into one, that is, the center of mass. The center of mass has heavy gravitational force, which attracts all other masses. The center of mass is computed and resembles the black hole of the system. Thus, the Big Bang phase explores the solution space, and the Big Crunch phase performs necessary exploitation as well as convergence.

Initially, candidate solutions are generated randomly that are distributed all over the search space (hypothetical universe) in a uniform manner. The fitness value is calculated for each individual. The fitness function is problem dependent and defined for a specific problem at hand. The Big Crunch is a convergence (or contraction) operator, which attracts all candidate solutions to the center of mass. It takes the current position of each candidate solution in the population and its associated fitness values to compute a point termed as the center of mass. Here, the term mass refers to the inverse of the fitness function value. In other words, the center of mass is the weighted average of the positions of the candidate solutions with respect to the inverse of the fitness function values. The center of mass denoted as x_c is calculated according to

$$x_c = \frac{\sum_{i=1}^{N} (1/f_i) x_i}{\sum_{i=1}^{N} (1/f_i)}$$

(10.1)

where x_i is the i-th candidate solution in an n-dimensional search space, f_i is a fitness function value of x_i, and N is the population size in the Big Bang phase. Alternatively, the best fit individual can also be used as the center of mass. The convergence operator in the Big Crunch phase is a novel one-step operator where a center of mass is sought using all individuals in the population.

An iteration of the BB–BC algorithm is complete once the Big Crunch phase is performed. The algorithm then creates a new generation of individuals to be used in the Big Bang phase of the optimization process. This can be done in various ways, the simplest one being jumping to the beginning and creating a new population. In that case, the algorithm will be no different than random search method. An optimization algorithm must converge to an optimal solution. The convergence can be ensured if the previous knowledge contained in computing the center of mass is utilized. This can be accomplished by generating new off-springs around the center of mass using a normal distribution operation in every direction where the standard deviation of this normal distribution function decreases with increasing number of iterations. Alternatively, a small decreasing random number can be added or subtracted from the center of mass. Such operations can be performed using the following equations.

$$x(t+1) = x_c(t) + \sigma(t) \tag{10.2}$$

where $x(t+1)$ is the new candidate solution around the center of mass and σ is the standard deviation of normal distribution defined as

$$\sigma(t) = \frac{\rho \cdot \beta \cdot (x_{max} - x_{min})}{t} \tag{10.3}$$

where ρ is a random number with normal distribution, β is a parameter limiting the size of the search space, x_{max} and x_{min} are the upper and lower limit of the candidate solutions, and t is the iteration number. The new solution $x(t+1)$ should be bounded with upper and lower limits by x_{max} and x_{min} values.

Camp (2007) introduced elitist strategy to generate new solutions. In this approach, new solutions are generated between center of mass $x_c(t)$ and global best solution x^{best} according to

$$x(t+1) = \lambda \cdot x_c(t) + (1 - \lambda) \cdot x^{best} + \frac{\rho \cdot \beta \cdot (x_{max} - x_{min})}{t} \tag{10.4}$$

where parameter λ controls the influence of global best solution x^{best}, that is, controls the weighted average of $x_c(t)$ and x^{best}. This approach also helps in improving performance of the original BB–BC algorithm.

After the second Big Bang explosion, the center of mass is recalculated. These successive Big Bang and Big Crunch steps are carried out repeatedly until a stopping criterion is met. The two parameters to be supplied to normal random point generator are the center of mass $x_c(t)$ of the previous step and the standard deviation $\sigma(t)$. The main steps of the BB–BC algorithm are as follows:

Step 1: Create a population of individuals x_i, $i = 1, ..., N$ randomly
Step 2: Calculate the fitness $f(x_i)$ of all individuals
Step 3: Find the center of mass $x_c(t)$ using Equation 10.1
Step 4: Generate new individuals $x_i(t + 1)$ around the center of mass using Equation 10.2 or any other mechanism
Step 5: If (termination condition not met), Goto Step 2
Step 6: Return solution

The performance of the BB–BC method has been verified on six benchmark test problems, which demonstrates superiority over an improved and enhanced GA and outperforms the classical GA for many benchmark test functions. The big advantage of BB–BC algorithm is that it has low computational cost and high convergence speed (Jackson et al., 2006).

Since BB–BC algorithm generates population around the center of mass at every iteration, the algorithm is not capable of exploring the search space very well. Therefore, there is a high probability that the algorithm may not yield the global solution. Kaveh and Talatahari (2009a, 2010a) proposed a hybridization of BB–BC with PSO (BB–BC-PSO) algorithm that enables exploring the search space for global optimal solution. To generate new candidate solutions, the hybrid BB–BC-PSO uses the center of mass, the best position of each individual, and the global best solution. The performance of the BB–BC-PSO was verified on size optimization of space truss problem.

The BB–BC algorithm suffers botching all individuals into a local optimum. If a candidate solution with the best fitness value converges to a local optimum at the very beginning of the optimization process, then all the remaining candidates follow that best solution, causing the algorithm trapped into the local optimum, which eventually causes slow convergence. Moreover, new candidate solutions are generated around the center of mass, which may confine in a small region of the search space. The main cause behind this situation is the nonuniform distribution of the initial population over the search space. Alatas (2011a) suggests a uniform distribution of the initial population to help providing quality solutions and speeding up convergence. Uniform population method (Karci, 2007) has been employed to generate uniformly distributed random initial candidate solutions in the Big Bang phase. All vectors in the search space are created by the linear combination of its elements. This part of the BB–BC method is called uniform Big Bang (UBB).

In order to improve the convergence, Alatas (2011a) suggests a chaotic map (cm) to be introduced in the Big Crunch phase. Small chaotic values in the final iterations of the algorithm may help in fine-tuning the solution by exploiting the local regions. This will improve the convergence speed of the algorithm. In the chaotic Big Crunch phase, the next position of each individual is updated using a chaotic sequence defined by

$$x_i(t+1) = x_c(t) \pm \frac{\alpha(t)(x_{\max} - x_{\min})}{t} \qquad (10.5)$$

where $\alpha(t)$ with $0 < \alpha(t) < 1$ is a cm defined by $\alpha(t) = c \cdot f[\alpha(t-1)]$, where c is a constant.

Chaos has been utilized to rapidly shrink those points to a single representative point via a center of mass in the Big Crunch phase. This part of the BB–BC method is called chaotic Big Crunch (CBC). Thus, the new approach proposed by Alatas (2011a) is called UBB–CBC algorithm. The performance of the UBB–CBC optimization algorithm has been verified on four benchmark optimization functions, which confirms superiority of UBB–CBC over the standard BB–BC optimization.

10.2.2 APPLICATIONS

Kaveh and Talatahari (2009a) employ a hybrid BB–BC optimization algorithm for optimal design of truss structures and compare it to other optimization methods including GA, PSO, ACO, and HSA (Siddique and Adeli 2015a–c). Kaveh and Talatahari (2010a) apply BB–BC algorithm to optimal design of geometrically nonlinear Schwedler and ribbed domes.

Parameter estimation and system identification, which is used in the emerging area of structural health monitoring, has been an active area of research in recent years (Sun et al., 2015). Tang et al. (2010) used the BB–BC algorithm for the parameter estimation of structural systems. The parameter estimation is formulated as a multimodal optimization problem with high dimension.

The performance of the method is investigated under different conditions such as limited output data, noise-polluted signals, and no prior knowledge of mass, damping, or stiffness. Alatas (2011a) proposed different methods to improve the convergence properties of the BB–BC algorithm. Sliding mode control (SMC) is one of the popular approaches for the control of nonlinear systems (Wang and Adeli, 2012). A variation of SMC is the adaptive fuzzy SMC approach based on the integration of SMC and fuzzy logic. Aliasghary et al. (2011) use the BB–BC method in fuzzy SMC of an inverted pendulum. Hatamlou et al. (2011) used the BB–BC algorithm for clustering of data. Zandi et al. (2012) applied the BB–BC algorithm to solve the reactive power dispatch problem. Other applications of the BB BC algorithm include function optimization (Alatas, 2011a; Erol and Eksin, 2006), optimal design of structures (Camp, 2007), automatic target tracking (Genc and Hocaoglu, 2008; Genc et al., 2010), fuzzy control design (Kumbasar et al., 2008a,b, 2011), smart homes (Prudhvi, 2011), and course timetabling (Jaradat and Ayob, 2010).

10.3 BLACK HOLE ALGORITHM

10.3.1 Inspiration and Algorithm

The universe comprises innumerable suns, solar systems, stars, and galaxies. The suns and solar systems are bounded by the outermost stellar rings of the Milky Way. There are nebular patches in gaseous form belonging to the stellar system as stars that have not formed yet. Laplace proposed this model in 1796 and showed how a solar system or star develops from individual nebular mass. The suns and stars have been developed by contraction and cooling from swirling masses of dark matters. The stars will remain stable as long as they have enough fuels in them. At some stage, the stars will run out of fuel and start cooling off and contracting. The stars that are massive (some million times larger than our sun) and compact would have a very strong gravitational field that will drag everything that comes near them.

The astonishing idea of the black hole was first announced in 1783 by John Michell, a tutor at Cambridge at that time, in a paper to the *Philosophical Transactions of the Royal Society of London* (Michell, 1784). He imagined that a star with massive and compact mass can have a strong gravitational field, which attracts everything into itself and swallow them (Ruffini and Wheeler, 1971). The gravitation is so strong that even light cannot not escape from it. Any light emitting from the surface of such star is dragged back by the star's strong gravitational attraction. In 1795, Pierre-Simon Laplace (independently of Michell's work) came to a similar conclusion (Gillispie, 2000). Since light is composed of particles, it must be attracted in the same manner as all other bodies in the universe. Hence, light emitted from the star went back to it, leaving the star invisible. Such body would be invisible to the outside world and appears to be black. These black voids in space from where even light is not emitted due to strong gravity are known as black holes.

Both Michell and Laplace used Newton's gravitational theory to analyze the escape or the capture of light of an existing massive object. The speed of light was well established at that time within the scientific community. The obvious question was if light travels at such high speed, then how gravity could slow down light, and make it fall back. In 1915, Einstein published his famous and revolutionary general theory of relativity (Einstein, 1915, 1916), where he showed that space and time were not separate and independent entities rather they were different directions in a single object called space-time and the space-time is also not flat, but is curved by the matter and energy in it. The general theory of relativity gave the explanation that even the speed of light is influenced by space-time. Thus, the idea of black hole was confirmed in 1916. Karl Schwarzschild then solved Einstein's equations for the case of a black hole (Schwarzschild, 1916), which he envisioned as a spherical volume of warped space surrounding a concentrated mass and completely invisible to the outside world. In the 1920s, Chandrasekhar worked out the size of a star that can support itself against its own gravity (Chandrasekhar, 1931a,b).

In 1939, Robert Oppenheimer and Hartland Snyder published the first detailed treatment of a gravitational collapse, that is, when all thermonuclear sources of energy are exhausted, a sufficiently heavy star will collapse (Oppenheimer and Snyder, 1939). Using Einstein's theory of gravitation, it led to the idea that such an object might be formed by the collapse of a massive star that leaves behind a small and dense remnant core. If the core's mass is more than about three times the mass of the Sun, the equations showed that the force of gravity overwhelms all other forces and produces a black hole. The term black hole was itself coined in 1968 by Princeton physicist John Wheeler, who worked out further details of a black hole's properties. The most common black holes are probably formed by the collapse of massive stars. Larger black holes are thought to be formed by the sudden collapse or gradual accumulation of the mass of millions or billions of stars. Many galaxies are believed to have super massive black holes at their centers. There is a black hole at the center of our Milky Way galaxy, known as Sagittarius A*, has a mass 4 million times that of the solar mass. Most astronomers believe that the black holes in the centers of galaxies grew by swallowing stars and gas, emitting light in the process. But there is not enough light coming from the black holes in active galaxies to explain their growth (White and Diaz, 2004).

The gravity of black hole is very strong because matter has been condensed into a small space. A black hole has a boundary. The sphere-shaped boundary of a black hole in space is known as the event horizon. It is where the gravity is just so strong that drags light back and prevents it escaping and nothing can get away from it. Anything that crosses the boundary of the black hole will be swallowed by it and vanish. The radius of the event horizon is termed as the Schwarzschild radius. At this radius, the escape speed is equal to the speed of light, and once light passes through, even it cannot escape. Nothing can escape from within the event horizon because nothing can go faster than light. The Schwarzschild radius R is calculated by the following equation (Schwarzschild, 1916):

$$R \equiv \frac{2 \cdot G \cdot M}{c^2} \tag{10.6}$$

where G is the gravitational constant, M is the mass of the black hole, and c is the speed of light. If anything moves close to the event horizon or crosses the Schwarzschild radius, then it will be absorbed into the black hole and will disappear forever. The curiosity of general people in black holes revived once again in the 1980s when Stephen Hawking published his book *A Brief History of Time: From the Big Bang to Black Holes* (Hawking, 1998). The black holes are large objects in the universe, the construction of which is based on the concept of space-time (Chadrasekhar, 1998). The recent studies on the black hole in the 1970s and 1980s are presented in Davies (1978) and Hawking (1976, 1992, 1993). The very recent observations on black holes are reported in Abbott et al. (2016).

Inspired by the phenomenon of the black hole, Hatamlou (2013) proposed a new meta-heuristic population-based algorithm, called BH algorithm. BH algorithm starts with a randomly generated initial population of candidate solutions in an n-dimensional space, and an objective function is defined for the optimization problem at hand. After initialization, objective function values are calculated for all individuals in the population. In BH algorithm, candidate solutions are considered as stars. The best candidate solution is selected to be the black hole and the rest form the normal stars. At each iteration, the black hole (i.e., the best solution) starts attracting other stars (i.e., candidate solutions) around it. If a star gets too close to the black hole, it will be swallowed by the black hole and is gone from the universe forever, that is., from the population. In such a case, a new star (i.e., candidate solution) is randomly generated and placed in the search space, and the algorithm starts a new search iteration. In order to move the population of solutions toward optimality, two operations are implemented in the BH algorithm: (1) absorption of stars and (2) sucking of stars.

1. *Absorption of stars*: Absorption of stars is the movement operation of the stars toward the black hole under the gravitational force. If the new location of the star is in the vicinity of the black hole, then it is absorbed by the black hole. Stars move toward the black hole based on their current location and a random number. The new location of stars is calculated according to

$$x_i(t+1) = x_i(t) + \rho \times [x_{\mathrm{BH}} - x_i(t)], \quad i = 1, 2, \ldots, N \tag{10.7}$$

where $x_i(t+1)$ and $x_i(t)$ are the locations of the i-th star at iteration $(t+1)$ and t, respectively, x_{BH} is the location of the black hole in the search space, $\rho \in [0,1]$ is a random number, and N is the number of stars in the search space. A star may reach a new location with lower cost (meaning Euclidean distance between old location and new location) than the black hole. In this case, the black hole moves to the location of that star and vice versa.

2. *Sucking of stars*: There is a probability for a star crossing the event horizon of the black hole, which is the point of no return. Every star that crosses the event horizon of the black hole will be sucked by the black hole. The radius of the event horizon in the BH algorithm is calculated using the following equation:

$$R = \frac{f(x_{\mathrm{BH}})}{\displaystyle\sum_{i=1}^{N} f(x_i)} \tag{10.8}$$

where $f(x_{\mathrm{BH}})$ is the fitness value of the black hole and $f(x_i)$ is the fitness value of the i-th star.

When the distance D_i between a star and the black hole is less than event horizon R, then the star collapses into black hole. The distance between a star and black hole is the Euclidean distance calculated according to

$$D_i = \sqrt{\left(x_{\mathrm{BH}}^1 - x_i^1\right)^2 + \cdots + \left(x_{\mathrm{BH}}^n - x_i^n\right)^2} \tag{10.9}$$

where x_{BH}^k and x_i^k, $k = 1, 2, \ldots, n$, are the components of the black hole and the star, respectively. Every time a star is sucked, a new star is randomly created in order to keep a constant number of stars in the search space. The next iteration takes place after all the stars have been moved. Then the BH algorithm will continue with the black hole in the new location and the stars start moving toward this new location.

On the basis of the above working principle, the main steps of the BH algorithm are summarized as follows:

Step 1: Initialize a population of stars x_i, $i = 1, \ldots, N$ with random locations
Step 2: Compute the fitness of each star $f(x_i)$
Step 3: Determine black hole

$$x_{\mathrm{BH}} \leftarrow \max[f(x_i)]$$

Step 4: Update the location of stars using Equation 10.7

$$x_i(t+1) = x_i(t) + \rho \times [x_{\mathrm{BH}} - x_i(t)], \quad i = 1, 2, \ldots, N$$

Step 5: If a star reaches new location with lower cost than x_{BH}
        Exchange their locations

Step 6: Calculate radius of event horizon R using Equation 10.8

Calculate Euclidean distance D_i between x_i and x_{BH} using Equation 10.9

If $D_i < R$

Generate a random star x^*

$x_i = x^*$

Step 7: If (termination condition not met), Goto Step 2

Step 8: Return solution

The termination condition can be maximum number of iterations, satisfactory accuracy of solution, or sufficient good value of fitness. In BH algorithm, the evolution of the population is done by moving all the candidate solutions toward the best candidate solution termed as the black hole in each iteration and replacing those candidate solutions that enter within the range of the black hole by newly generated candidate solutions in the search space. A brief comparative analysis of BH algorithm in comparison with other meta-heuristic algorithms is carried out by Biswas et al. (2013).

It appears that the terminology "black hole" has been used for the first time in 2008 for solving optimization problems by Zhang et al. (2008). They used the black hole concept to standard PSO algorithm, called random black hole PSO (RBH-PSO), where a particle close to the current best particle is generated randomly and considered it as the black hole denoted as x_{gBH}. A threshold value $p \in [0,1]$ is assigned to it. In each dimension of every particle in the swarm, a random value $l \in [0,1]$ is generated. The positions of the particles are updated using the following:

$$x(t+1) = \begin{cases} x_{gBH} + s & \text{if } l < p \\ x(t) & \text{if } l \geq p \end{cases} \tag{10.10}$$

where s is a random number uniformly distributed over $[-r,r]$ and r is the radius of the region around x_{gBH} from where the black hole is generated. Though the idea of black hole is used in the RBH-PSO approach, it is conceptually different from BH algorithm and merely a variant of PSO algorithm.

Nemati et al. (2013) proposed an extension of BH algorithm by incorporating the charge of the black hole, where the gravitational force of the black hole is used for the global search and the electrical force induced by the charge of the black hole is used for local search. The performance of the proposed algorithm was verified on a set of benchmark problems, which demonstrated better performances over PSO and RBH-PSO. Nemati and Momeni (2014) proposed another extension of BH algorithm by incorporating Hawking's radiation called BH algorithm with fuzzy Hawking radiation. Hawking showed that black holes emit small amounts of thermal radiation (Hawking, 1992), an effect which has become known as Hawking radiation. Hawking radiation is used as a means of mutation mechanism. That is, the individuals went into mutation depending on the fitness value, which is termed as fuzzy Hawking radiation. There was little description provided for the clear understanding of the mechanism of the proposed algorithm, although experimental results based on benchmark problems show improved performance.

The problem with the BH optimization algorithm is that it converges rapidly at the starting stage of the search process. As it goes closer to the global optimal solution, the convergence speed slows down and often gets trapped into the local optimum over a certain number of iterations. To speed up the convergence and also to enhance exploitation at this stage, local search algorithm is to be employed. Chandrasekar and Krishnamoorthi (2014) proposed a hybridized approach using black hole optimization (BHO) as global search method and heuristic search algorithm as local search method. The hybrid approach is called BHOHS. The BH algorithm is used to produce an initial solution for the clustering problem and then the heuristic search algorithm is applied as a tool for local search to improve the initial solution. The performance of BHOHS is compared with K-means, PSO algorithm, and BHO algorithm, which shows competitive performance.

10.3.2 APPLICATIONS

BH algorithm so far has found some interesting applications in data clustering (Chandrasekar and Krishnamoorthi, 2014; Hatamlou, 2013), optimal reactive power dispatch problem (Lenin et al., 2014), and optimal power flow problem (Hasan and El-Hawary, 2014). Some of these applications are discussed very briefly.

To evaluate the performance of the BH algorithm, Hatamlou (2013) applied it to solve the clustering problem, which is considered as an NP-hard problem. A detailed description on the data clustering problem is provided in Chapter 2. Six benchmark data sets with a variety of complexity are chosen for the performance evaluation of the proposed method. The data sets are Iris, Wine, Glass, Wisconsin Breast Cancer, Vowel, and Contraceptive Method Choice (CMC), which are available in the repository of the machine learning databases at UCI. The performance of the BH algorithm is compared against well-known and the most recent algorithms reported in the literature such as K-means, PSO, GSA, and the BB–BC algorithm. The experimental results show that the proposed BH algorithm outperforms other traditional heuristic algorithms. Chandrasekar and Krishnamoorthi (2014) applied BHOHS to data clustering problem. They also applied the BHOHS approach to different data sets such as Iris, Wine, and Glass data set available from UCI machine learning databases. The performance was compared to that of K-means, PSO, and standard BH algorithm. In all runs, it is found that the quality of clusters based on the sum of intracluster distance, the convergence and statistical performance of BHOHS were better compared to those of other methods.

Lenin et al. (2014) applied the BH algorithm to solve the optimal reactive power dispatch problem. A detailed description on the optimal reactive power dispatch problem is provided in Chapter 2. BH algorithm has been tested on standard IEEE 30-bus system with various generator constraints. The performance of the BH algorithm has been demonstrated through the reduction of real power loss and the voltage stability. Hasan and El-Hawary (2014) applied BH algorithm to optimal power flow problem. A detailed description on the optimal power flow problem is provided in Chapter 2. BH algorithm has been tested on standard IEEE 30-bus system and IEEE 57-bus system with various generator constraints and different objective functions for verifying efficiency and applicability, for example, minimizing the fuel cost, enhancing the voltage profile, and improving the voltage stability under both normal and contingency conditions. In all cases, simulation results demonstrated competitive performance of the BH algorithm. The BH algorithm has two significant advantages over other approaches due to its structural simplicity: (1) it is easy to be implemented and (2) it is free of parameter tuning.

10.4 GALAXY-BASED SEARCH

10.4.1 INSPIRATION AND ALGORITHM

The nebular hypothesis was developed by Immanuel Kant in 1755 and published his book *Universal Natural History and Theory of the Heavens*. He argued that nebulae (i.e., gaseous clouds) slowly rotate, gradually collapse, and flatten due to gravity. In this process of the evolution, the solar system and eventually stars are formed (Woolfson, 1993). Galaxies are structures formed by the mutual gravitation of matter into bound systems of stars, stellar remnants, interstellar gas and dust, and dark matter within the universe (Hawking, 1998; Sandage, 1975). It is believed that 27% of the mass–energy of the universe consists of dark matter. Galaxies range from dwarfs with just a few thousand stars to one hundred trillion stars. They are all orbiting their galaxys' own centers of mass. Our galaxy has hundreds of billions of stars, enough gas and dust to make billions of more stars, and

FIGURE 10.1 Spiral shape galaxy.

at least 10 times as much dark matter as all the stars and gas put together, which are all held together by gravity. The modern picture of the universe dates back to 1924, when Edwin Hubble showed that there are many other galaxies in the universe. To date, approximately 170 billion galaxies are known in the observable universe. The space between galaxies is filled with interstellar gas and dust and dark matter. The majority of galaxies are gravitationally organized into associations known as galaxy groups, clusters, and superclusters.

Galaxies can be categorized according to their visual morphology into three groups: elliptical, spiral, and irregular (Sandage, 1975). Elliptical shape galaxies are early type galaxies and exhibit a larger range of mass and less angular variation in brightness. Spiral galaxies are disk-like galaxies first described by Hubble (1936). The characteristic beauty of this type of galaxy is its spiral structure. The spiral arms are sites of ongoing star formation and are brighter than the surrounding disk. More than two-thirds of the known galaxies are spiral shaped. A slightly more extensive description of galaxy types based on their appearance is given by the Hubble sequence (Ma et al., 2000). According to the Hubble classification system, spiral galaxies have three parameters: (i) size of the bulge in galactic center relative to disk, (ii) degree of tightness of winding of the spiral arm, and (iii) degree of disintegration of the arm into stars. Figure 10.1 shows a spiral galaxy.

Hosseini (2011a) proposed GbSA inspired by the spiral arm of spiral galaxies to search its surrounding. GbSA uses a spiral-like movement in each dimension of the search space with the help of chaotic steps and constant rotation around the initial solution. Gradually, the arm of the galaxy opens and covers the search space in order to find a better solution. After generating an initial solution, the spiral chaotic movement helps to find a local optimum. Solutions are adjusted with this spiral movement during local search as well. The GbSA has two components: (i) spiral chaotic move and (ii) local search.

10.4.1.1 Spiral Chaotic Move

This component actually mimics the spiral arm nature of galaxies. It searches the space by applying spiral movement around the current solution. A chaotic variable is used to make a move toward a new solution around the current best solution. There are many chaotic maps used in this kind of chaotic search such as logistic map, Chebyshev map, and iterative cm with infinite collapse (ICMIC). A chaotic sequence can be generated using the following logistic map:

$$z_n = \lambda z_{n-1}(1 - z_n), \quad n = 0,1,2,\ldots \tag{10.11}$$

where the initial value $z_0 \in [0,1]$ should be chosen from the interval $[0,1]$, λ is a control parameter, and z_n is the variable at discrete time n. The function in Equation 10.11, known as the logistic map, shows chaotic dynamics, for example, for $\lambda = 4$ and $z_0 \in \{0, 0.25, 0.5, 0.75, 1\}$. Hosseini (2011a) discarded the first 2000 iterations of the logistic map as it contains the transient motion in the sequence that leads to the chaotic attractor. Using the cm, new solutions are generated in the neighborhood of the current solution.

Let $x = \{x^1, x^2, \ldots, x^n\}$ be a randomly generated solution in n-dimensional space. New solutions $\{x_l^{new}, x_u^{new}\}$ are derived in the neighborhood of x in the spiral chaotic move, that is, using the arm length r and angle of rotation θ of the spiral galaxy. Initially, the arm length r is set to small value, for example, 0.001. r is incremented by Δr at each iteration. The initial angle of rotation θ is calculated using $\theta = (-1 + 2 \times z_n) \cdot \pi$. θ is incremented by $\Delta\theta$ at each iteration. $\Delta\theta$ is considered as a parameter and is initially set to small value, for example, 0.01. If the solutions $\{x_l^{new}, x_u^{new}\}$ generated by spiral chaotic move are proved to be better than the current best solution x, x is updated and local search procedure is initiated. The process is repeated for a maximum number of iterations. The spiral chaotic move consists of the following steps:

Step 1: Consider x as the current best solution provided by local search
   Initialize r, Δr, $\Delta\theta$, and $\theta = (-1 + 2 \times z_n) \cdot \pi$

Step 2: Create new solutions using chaotic function z_n

$$x_u^{new} = x + r \times \cos(\theta) \times z_n$$

$$x_l^{new} = x - r \times \cos(\theta) \times z_n$$

If $f(x_u^{new}) \geq f(x)$

$$x = x_u^{new}$$

Else If $f(x_l^{new}) \geq f(x)$

$$x = x_l^{new}$$

Step 3: Increment arm length and angle of rotation

$$r = r + \Delta r$$

$$\theta = \theta + \Delta\theta$$

If $(\theta > \pi)$ Then $\theta = -\pi$

Step 4: If (max iteration not reached), Goto Step 2

Step 5: Return solution

The spiral chaotic move provides a good exploration mechanism for the search space, ensuring the algorithm to move toward the global optimal solution. Chaotic map is used in this phase so that the solutions do not return to the same solutions again. Chaotic map also ensures high diversity so that solution does not get trapped in the local optimum, which is also useful in dealing with multimodal problems. The spiral chaotic move is generally applied for a small number of iterations, for example, Hosseini (2011a,b) used 100 iterations for this operation.

10.4.1.2 Local Search

Local search is carried out on the current best solution to exploit the search space and to find a local optimum. Hill climbing is a good local search method for finding a local optimum (a solution that cannot be improved by considering a neighboring configuration) but cannot guarantee to find the best possible solution (the global optimum) out of all possible solutions within the search space. Local search used in GbSA is a modified hill-climbing search algorithm, which ensures the exploitation of search space (Aarts and Lenstra, 1997). In the modified hill-climbing method, two solutions are created in the neighborhood of the current best solution provided by the global search methods such as spiral chaotic move. If they are no better than the current best solution x, then the current best solution is returned. Otherwise, a local search exploits the space around the current best solution with small step sizes and gradually increases the step sizes for faster exploration and returns the locally best solution. The local search method consists of the following steps:

Step 1: Consider the current best solution x provided by global search method
        Set $\alpha = 1$, $\Delta\alpha = 0.01$, and define chaos()

Step 2: Generate two solutions x_u and x_l in the neighborhood of x

$$x_l = x - \alpha \times \text{chaos}()$$

$$x_u = x + \alpha \times \text{chaos}()$$

Step 3: If $(f(x_l) < f(x) \wedge f(x_u) < f(x))$
        Goto Step 6

Step 4: If $(f(x_u) > f(x))$

        $x = x_u; x_l = x_u;$

        Else If $(f(x_l) > f(x))$

        $x = x_l; x_u = x_l;$

        Increase α

        $\alpha = \alpha + \Delta\alpha \times \text{chaos}()$

Step 5: If (max iteration not reached), Goto Step 2

Step 6: Return solution

α is a dynamic parameter and chosen arbitrarily within the range of {0.5,1}. α is incremented by a constant value of $\Delta\alpha$ and a chaotic sequence at each iteration. The initial value of $\Delta\alpha$ is set to small value within the range of {0.01,0.05}. The function chaos() returns a chaotic number within the interval of [0,1], which is generated by the logistic map defined in Equation 10.11. The maximum number of iterations for local search is chosen within the range of {500,1500}.

The GbSA performs the spiral-like movement in each dimension of the search space with the help of chaotic steps and constant rotation around the initial solution followed by the local search algorithm for fine-tuning of the solutions obtained by the spiral chaotic move. The main steps of the GbSA are as follows:

Step 1: Initialize a random solution x

Step 2: Perform local search on x
        $x = \text{local search }(x)$

Step 3: Perform spiral chaotic move on x
        $x' = \text{sprial chaotic move }(x)$

Step 4: If $\left[f(x') > f(x) \right]$
        $x = x'$
        $x = \text{local search }(x)$

Step 4: If (termination condition not met), Goto Step 3

Step 5: Return solution

The termination condition for GbSA can be the maximum iteration. Hosseini (2011a) used 50,000 iterations. A second criterion for termination could be the error difference between two successive iterations.

10.4.2 Applications

Dimensionality reduction is one of the main applications of principle component analysis (PCA). The notion of PCA is to transform a high-dimensional data vector into a low dimensional vector, that is, an m-dimensional data vector x is to be represented by a p-dimensional vector $\hat{x}$ such that $p < m$. The reduction is performed by a linear transformation, which is optimal in the sense of mean squared error. The dominant p eigenvectors of the data matrix are sought. Hosseini (2011a) applied GbSA to PCA. The initial solution of the GbSA-based PCA is randomly initialized from a uniform distribution within the interval of $[-1,1]$, and it is considered as current solution $S = [\mathbf{q}_1, \mathbf{q}_2, ..., \mathbf{q}_p]$. The objective function is defined as

$$\min e = \sum_{j=1}^{p} \sum_{i=1}^{n} \mathbf{e}_{i,j}^T \mathbf{e}_{i,j} \tag{10.12}$$

where $\mathbf{e}_{i,j} = \mathbf{x}_i - \sum_{k=1}^{j} \left(\mathbf{q}_k^T \mathbf{x} \right) \mathbf{q}_k$ and $p > 1$. The GbSA has to find a solution $S = [\mathbf{q}_1, \mathbf{q}_2, ..., \mathbf{q}_p]$ iteratively, which minimizes the approximation error e in Equation 10.12. The value of p is chosen by the user. Two synthetic data matrices are used to verify the performance of the GbSA in finding and estimating the dominant principal component vectors. The experimental results reveal that GbSA-based PCA is able to approximate the principal components.

Hosseini (2011b) applied GbSA to image segmentation of gray-level images. In image segmentation, the whole image is segmented into several regions based on similarities and dissimilarities present between the pixels of the input image. Each region should contain an object of the image. The image is divided into several subimages such that each subimage represents an object of the scene. There are different techniques used for segmentation. Multilevel threshold is a widely used technique in image segmentation where the number of thresholds is given in advance. The optimal thresholds are often found by maximizing or minimizing a criterion. One of the commonly used criteria for threshold is Otsu's criterion for multilevel threshold of gray-level images (Otsu, 1979). The images are assumed to be composed of only two regions: object and background. The best threshold is the one that maximizes the between-classes variance of the two regions. GbSA has been applied to find the optimal threshold. The experimental results reveal that GbSA outperforms other methods and demonstrate that GbSA is a promising approach.

Zerigat et al. (2013, 2014) applied GbSA for solving the CEED problem under some equality and inequality constraints. CEED problem is an important optimization problem in power system operation and forms the basis of many applications in engineering. The main objective of CEED is to determine the optimum share of each generating unit that satisfies the required load conditions minimizing the operational cost and the emission cost. CEED problem has been discussed in Chapter 7. Further details of the problem formulation of CEED can be found in Wood and Wollenberg (1994). CEED problem is generally a bi-objective problem. This bi-objective problem is converted into a single objective function using a price penalty factor. The equality constraints are the active power flow balance equations, and the inequality constraints are the minimum and maximum power output of each unit. The performance of the GbSA has been tested on IEEE 30-bus 6-generator system and 9-bus 3-generator system. The performance of GbSA was also compared with that of other methods such as λ-iteration, recursive, PSO, DE, and simplified recursive method. In all cases, GbSA outperforms all other methods in terms of minimizing fuel cost and emission.

10.5 ARTIFICIAL PHYSICS OPTIMIZATION

10.5.1 Inspiration and Algorithm

Motivated by natural physical forces, William Spear, Diana Spear, and Rodney Heil presented physicomimetics or artificial physics (AP) framework (Spears and Spears, 2003; Spears et al., 2005a,b).

According to physicomimetics, the natural physical force acting on each individual mass is calculated based on the "masses" and distances between them using the following formula:

$$F = G \frac{m_i \cdot m_j}{r^p} \tag{10.13}$$

where $F < F_{max}$ is the magnitude of force between two masses m_i and m_j, G is the gravitational constant, and r is the distance between the two masses. The force F is repulsive if $r < R$ and attractive if $r > R$. R is a distance that decides when the force F will be repulsive or attractive. The value of the parameter p can be chosen by the user, which ranges from −5.0 to 5.0. In the real universe, $p = 2$. But in AP-space, it is free to assign an arbitrary value to p for entirely different force variation, for example, $p = -1$. In GSA (gravitational search algorithm) presented in Chapter 2, p is set to $p = 2$.

In AP, each entity or agent is seen as a physical individual with a mass m in two or three dimensions. Each particle i has position x and velocity v. The motion space is two or three dimensional. The continuous movement of each particle can be approximated by many discrete-time perturbations. At each time step Δt, the perturbation of each particle depends on the current velocity, that is, $v = \Delta x/\Delta t$. The change in velocity Δv is controlled by the force acting on the particle, that is, $\Delta v = (F/m) \times \Delta t$, where F is the force acting on the particle of mass m. The system acts like a molecular dynamics simulation. The virtual force drives each individual particle with a positive definite motion toward a heavier mass. The bigger mass determines the higher magnitude of attraction force. The maximum force F_{max} is imposed to restrict the maximum allowable force that can be applied on a particle, which in turn restricts the maximum allowable velocity v_{max} of a particle.

In APO, particles are seen as solutions sampled from the feasible region of the problem space. Particles move toward higher fitness regions and cluster to the optimal region over time. Heavier mass represents higher fitness value and attracts other masses of lower fitness values. The individual with the best fitness attracts all other individuals with lower fitness values. The individuals with lower fitness values repel each other. That means, the individual with the best fitness has the biggest mass and moves with lower velocity than others. Thus, the attractive–repulsive rule can be treated as the search strategy in optimization algorithm, which ultimately leads the population to search the better fitness region of the problem. In an initial state, individuals are randomly generated within the feasible region. In APO, mass is defined as the fitness function for the optimization problem in question. A suitable definition of mass of the individuals is necessary.

The mass of the particle is the value of the objective function, usually a user-defined function, to be minimized or maximized, that is, $m_i = g[f(x_i)]$, where $g[\cdot] \geq 0$ is bounded and monotonically decreasing or increasing depending on the minimization or maximization problem, respectively. The definition of mass should ensure that the value of mass should be within the interval of [0,1], that is, $m_i \in (0,1]$ and defined as

$$m_i = \exp\left[\frac{f(x_{best}) - f(x_i)}{f(x_{worst}) - f(x_{best})}\right] \tag{10.14}$$

where x_{best} and x_{worst} are defined as

$$x_{best} = \arg[\min\{f(x) \mid x \in \{x_1, x_2, \ldots, x_m\}\}] \tag{10.15}$$

$$x_{worst} = \arg[\max\{f(x) \mid x \in \{x_1, x_2, \ldots, x_m\}\}] \tag{10.16}$$

According to the definition in Equation 10.14, the mass of the best individual should be 1 while all other masses have values smaller than 1. This is a requirement for the definition of the mass and can be taken as an example for suitable definition of a mass function. There may be other definitions possible. For example, $g_k(x) = \tanh(x) = (e^{-x} - e^{-x})/(e^{-x} + e^{-x})$ such that for any value of $x \in [-\infty, +\infty]$, $g_k(x)$ can be mapped onto the interval $(0,1)$ (Xie et al., 2009a).

Once the mass is computed using an appropriate mass definition, the force acting on each individual by all other masses has to be determined. Considering the value of $p = -1$ in Equation 10.13, the force between masses m_i and m_j can be defined as

$$F_{ij}^k = \begin{cases} Gm_i m_j \left(x_j^k - x_i^k \right) & \text{if } f(x_j) < f(x_i) \\ -Gm_i m_j \left(x_j^k - x_i^k \right) & \text{if } f(x_j) \geq f(x_i) \end{cases} \tag{10.17}$$

where F_{ij}^k is the k-th component of the force on mass m_i by mass m_j, $\left\{ x_i^k, x_j^k \right\}$ are the k-th dimension of masses m_i and m_j, respectively. Xie et al. (2009b) modified the force definition by including the direction of force. The direction of force is determined by comparing two individuals' fitness values. The distance between x_i^k and x_j^k is defined as $r_{ij}^k = x_j^k - x_i^k$. By including the sign of r_{ij}^k in Equation 10.17, the force function is rewritten as

$$F_{ij}^k = \begin{cases} \mathrm{Sgn}(r_{ij}^k) \cdot G \cdot m_i m_j \left(x_j^k - x_i^k \right) & \text{if } f(x_j) < f(x_i) \\ \mathrm{Sgn}(r_{ij}^k) \cdot G \cdot m_i m_j \left(x_j^k - x_i^k \right) & \text{if } f(x_j) \geq f(x_i) \end{cases} \tag{10.18}$$

where $\mathrm{Sgn}(\cdot)$ defines the sign of the function.

$$\mathrm{Sgn}(r_{ij}^k) = \begin{cases} 1 & \text{if } r > 0 \\ -1 & \text{if } r < 0 \end{cases} \tag{10.19}$$

Some researchers consider that gravitational constant G also varies with the distance variable r_{ij}^k as shown in Figure 10.2 (Xie et al., 2009b). Therefore, G is expressed as a function of r_{ij}^k as follows:

$$G(r_{ij}^k) = \begin{cases} G_0 \left| r_{ij}^k \right|^h & \text{if } r_{ij}^k \leq 1 \\ G_0 \left| r_{ij}^k \right|^q & \text{if } r_{ij}^k > 1 \end{cases} \tag{10.20}$$

where G_0 is the gravitational constant at initialization, $h > 2$, and $0 < q < 2$.

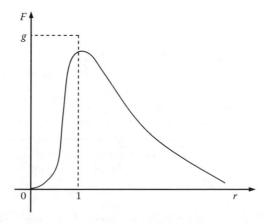

FIGURE 10.2   Force as a function of r.

Finally, the force F_{ij}^k can be expressed as

$$F_{ij}^k = \begin{cases} \operatorname{Sgn}(r_{ij}^k) \cdot G(r_{ij}^k) \cdot m_i m_j \left(x_j^k - x_i^k \right) & \text{if } f(x_j) < f(x_i) \\ \operatorname{Sgn}(r_{ij}^k) \cdot G(r_{ij}^k) \cdot m_i m_j \left(x_j^k - x_i^k \right) & \text{if } f(x_j) \geq f(x_i) \end{cases} \qquad (10.21)$$

The k-th component of the total force F_i^k on mass m_i by all other masses is given by

$$F_i^k = \sum_{j=1}^{N} F_{ij}^k, \quad \forall i \neq \text{best} \qquad (10.22)$$

It is to be noted that the best individual is not attracted or repelled by all other individuals. Therefore, best is excluded in the summation term in Equation 10.22. The force acting on individuals make the masses move within the search space and change their velocity. The velocity of the masses is updated using the following equation:

$$v_i^k(t+1) = w v_i^k(t) + \lambda \frac{F_i^k}{m_i} \cdot \Delta t \qquad (10.23)$$

where F_i^k / m_i is the acceleration term, Δt is the time to achieve the final velocity, and Δt is considered unity. λ is a random number uniformly distributed over [0,1] and w is a user-defined parameter chosen between $0 < w < 1$. Overall exploration of the algorithm is controlled by the weight parameter w. The random parameter λ is actually for putting limitation to convergence and it also serves for exploration.

Once the velocity is known, the position is updated according to the following equation:

$$x_i^k(t+1) = x_i^k(t) + v_i^k(t+1) \cdot \Delta t \qquad (10.24)$$

where Δt is the time to move from old position to new position and Δt is considered unity.

The movement and position are restricted within a feasible domain, $v_i^k \in [v_{\min}, v_{\max}]$ and $x_i^k \in [x_{\min}, x_{\max}]$. The parameters $F_{\max}$ and $v_{\max}$ restrict the maximum force exerted on a particle and the velocity of the particle, respectively.

In APO algorithm, a population of individuals is randomly created in the n-dimensional decision space. The individuals represent the solutions of the problem, but they are thought of as particles in the decision space with a mass, position, and velocity. All the particles are initialized randomly to nonzero values. The value of the objective function $f(\cdot)$ of each individual is calculated. The position vector of the particle with the highest objective function value is denoted as x_{best} and the position vector of the particle with the lowest objective function value is denoted as x_{worst}. Velocities of all individuals are initialized to be random values and are restricted in decision space. The masses of all individuals are calculated using the objective function defined for the problem at hand. The masses change by their objective function values as the algorithm evolves. The change in mass results in the change of total force. The force eventually causes the mass to update the velocity for each particle to move in next iterations. Newton's second law makes the individuals to move until the total force decreased to zero and achieve the lowest potential energy of the system, which is the desired state of optimality. This simple mechanism of the method is called APO algorithm. APO algorithm is a stochastic algorithm based on physicomimetics framework (Xie and Zeng, 2009a,b).

APO algorithm comprises three main procedures: initialization, calculation of force, and motion. The main steps of the APO algorithm are the following:

Step 1: Generate a population in n-dimensional space, initialize position and velocity
   Set parameters G, λ, and w
Step 2: Calculate fitness $f(x_i)$ of individuals, determine x_{best} and x_{worst}
   Calculate mass m_i using Equation 10.14
Step 3: Calculate total force F_i using Equation 10.22
Step 4: Update velocity v_i using Equation 10.23
Step 5: Update position x_i using Equation 10.24
Step 6: If (termination condition not met), Goto Step 2
Step 7: Return solution

Velocity threshold v_{max} is an important parameter of APO algorithm. It affects the performance by restricting the movement step size and the direction of velocity information. Because of the complex optimization problems, the proper v_{max} setting may provide a reasonable solution within an allowed generation. There are only a few researchers who have paid attention to this issue. Xie et al. (2011a) investigated two selection principles of velocity threshold, namely, fixed v_{max} and adaptive v_{max}, with large dimension.

The APO algorithm has been modified firstly to suit different applications and to enhance performance and convergence speed. As a result, a number of variants are reported in the literature. These variants are discussed very briefly in the following sections.

10.5.2 Vector Model APO

Driven by the virtual force, the individuals in APO algorithm move toward the bigger masses, which is an analogy of the particles flying toward the better fitness region. To analyze the algorithm, a vector model of APO (VM-APO) algorithm is proposed by Xie et al. (2009c). On the basis of the vector model, APO algorithm can perform well in diversity if some conditions can be satisfied. VM-APO algorithm works on the nonlinear optimization problems with bounded variables of the form described by Equation 10.25. The VM-APO algorithm comprises three main procedures: initialization, force calculation, and motion. After the initialization of a random population of size N and random initialization of velocities in n-dimensional decision space, the mass $m_i(t)$ is computed according to Equation 10.14. The force vector acting on individual i via individual j is calculated by

$$F_{ij} = \begin{bmatrix} F_{ij}^1 \\ F_{ij}^2 \\ \dots \\ F_{ij}^n \end{bmatrix} = \begin{cases} Gm_im_j \parallel X_j - X_i \parallel r_{ij} & \text{if } j \in N_i \\ -Gm_im_j \parallel X_j - X_i \parallel r_{ij} & \text{if } j \in M_i \end{cases} \quad (10.25)$$

where $\parallel X_j - X_i \parallel = \sqrt{\sum_{j=1}^{n} \left(x_j^k - x_i^k\right)^2}$, $\{x_i^k, x_j^k\}$ are the k-th dimension of the masses, $\{N_i, M_i\}$ are two sets related to individual i defined by $N_i = \{j | f(X_j) < f(X_i), \forall j \in S\}$ and $M_i = \{j | f(X_j) \geq f(X_i), \forall j \in S\}$, $X_i = \{x_i^1, \dots, x_i^k, \dots, x_i^n\}$, and S is the set of a population with N individuals. r_{ij} is the relative direction of individual j toward individual i and it is defined as

$$r_{ij} = \begin{bmatrix} r_{ij}^1 \\ r_{ij}^2 \\ \dots \\ r_{ij}^n \end{bmatrix} \quad (10.26)$$

The component direction r_{ij} in the k-th dimension is defined as

$$r_{ij}^k = \begin{cases} 1 & \text{if } X_j^k > X_i^k \\ 0 & \text{if } X_j^k = X_i^k \\ -1 & \text{if } X_j^k < X_i^k \end{cases} \tag{10.27}$$

Once the force is computed, the velocity is updated using the following equation, while position is updated according to Equation 10.24.

$$v_i(t+1) = wv_i(t) + \lambda \sum_{\substack{j=1 \\ j \neq i}}^{N} \frac{F_{ij}}{m_i} \cdot \Delta t \tag{10.28}$$

where $\lambda \in [0,1]$ is a random number and $w \in [0,1]$ is a parameter.

Xie et al. (2009c) verified the performance of the VM-APO algorithm on a number of benchmark optimization problems, which demonstrates competitive performance over other methods.

10.5.3 Hybrid Vector APO

VM-APO algorithm works on the nonlinear optimization problems with bounded variables. The problem with VM-APO is that the positions of the individuals may go beyond the lower and upper limits. The usual procedure for retrieving the positions is as follows:

$$\begin{cases} \text{If } x_i^k < x_{min}^k, & \text{then } x_i^k = x_{min}^k \\ \text{If } x_i^k > x_{max}^k, & \text{then } x_i^k = x_{max}^k \end{cases} \tag{10.29}$$

The simple way to deal with such individuals is to change the direction of its motion vector. The direction is very useful information for search algorithms. In order to keep the direction of its motion vector and keep the individuals with the decision space, Yang et al. (2010) proposed a shrinkage coefficient δ to be introduced to APO. In addition, one-dimensional search method is used to locate the individual with a better fitness along its velocity vector at iteration $(t + 1)$ than that at t. The new approach is called hybrid vector APO (HV-APO) with one-dimensional search (HV-APO-ODS) method. HV-APO-ODS improves the local search capability of VM-APO algorithm. When an individual i goes beyond the lower or upper limits in the kth dimension, the shrinkage coefficient δ_k can bring the individual back into the decision space.

$$\begin{cases} x_{min}^k + x_i^k(t) = \delta_k \left[x_i^k(t+1) - x_i^k(t) \right] \\ x_{max}^k - x_i^k(t) = \delta_k \left[x_i^k(t+1) - x_i^k(t) \right] \end{cases} \tag{10.30}$$

The shrinkage coefficient δ_k can be calculated from Equation 10.31 as follows:

$$\delta_k = \begin{cases} \dfrac{x_{min}^k + x_i^k(t)}{x_i^k(t+1) - x_i^k(t)} & \text{if } x_i^k(t+1) < x_{min}^k \\ \dfrac{x_{max}^k - x_i^k(t)}{x_i^k(t+1) - x_i^k(t)} & \text{if } x_i^k(t+1) > x_{max}^k \end{cases} \tag{10.31}$$

The position vector of individual at time $(t+1)$ is updated using the shrinkage coefficient δ as follows:

$$x_i(t+1) = x_i(t) + \delta \cdot v_i(t+1) \cdot \Delta t \tag{10.32}$$

where δ is the minimum shrinkage coefficient defined as $\delta = \min\{\delta_k\}$, $k = 1, 2, \ldots, n$. δ guarantees the new position in Equation 10.32 to be within the decision space $[x_{\min}^k, x_{\max}^k]$, but it cannot ensure that $f[x_i(t+1)]$ would be better than $f[x_i(t)]$. To improve this fitness condition, a new variable $\beta \in [0,\delta]$ is introduced into position vector calculation as

$$x_i(t+1) = x_i(t) + \beta \cdot v_i(t+1) \cdot \Delta t \tag{10.33}$$

The introduction of $\beta \in [0,\delta]$ into Equation 10.33 improves the search ability and it is able to locate the individual i on the position where the fitness is minimum along the direction of its velocity vector such that

$$\min_{\beta \in [0,\delta]} f[x_i(t+1)] = \min_{\beta \in [0,\delta]} f[x_i(t) + \beta \cdot v_i(t+1) \cdot \Delta t] \tag{10.34}$$

where β represents only a length of coefficient such that $\min[x_i(t+1)]$ can be considered as a one-dimensional search for the optimization problem. Yang et al. (2010) verified the HV-APO-ODS algorithm on a number of benchmark optimization problems such as Griewank, Rastrigin, Rosenbrock, Ackley, Schaffer's F7, and Penalized functions, which shows improved performance over VM-APO algorithm.

Xie et al. (2011a) proposed a new approach of HV-APO with multidimensional search (HV-APO-MDS) method to improve the local search capability. An n order diagonal matrix of shrinkage coefficient is introduced in HV-APO-MDS. This ensures that each individual is within the decision space. Multidimensional search method is merged into the VM-APO to improve the local exploitation capability. The simulation results confirm that the performance of the HV-APO-MDS search method is effective. Xie and Zeng (2011) introduced an n order diagonal matrix of shrinkage coefficient δ to HV-APO, which actually ensures each individual within the decision space. HV-APO-MDS approach is merged with the VM-APO to ensure that the moving of the whole population is limited in the feasible region.

10.5.4 Extended APO Algorithm

In standard APO, the personal best position of each individual is not included and therefore not stored. Also to improve the convergence property, an extended version of APO is proposed by Xie and Zeng (2009c), where individual particle's fitness history, for example, best position, is stored in every iteration and utilized in the velocity updating. This new version of APO is called extended APO (EAPO). The personal best position of individual i is updated using the following equation:

$$xp_i^k(t+1) = \begin{cases} xp_i^k(t) & \text{if } f\left[x_i^k(t+1)\right] > f(xp_i^k(t)) \\ x_i^k(t+1) & \text{if } f\left[x_i^k(t+1)\right] \leq f(xp_i^k(t)) \end{cases} \tag{10.35}$$

The personal best individual is then included in the mass function calculation as follows:

$$m_i = \exp\left[\frac{f(x_{\text{best}}) - f(xp_i)}{f(x_{\text{worst}}) - f(x_{\text{best}})}\right], \quad \forall i \tag{10.36}$$

Once the mass is computed using Equation 10.36, the component forces acting on each individual are calculated.

$$F_{ij}^k = \begin{cases} Gm_i m_j \left(xp_j^k - x_i^k \right) & \text{if } f(xp_j) \leq f(x_i) \\ -Gm_i m_j \left(xp_j^k - x_i^k \right) & \text{if } f(xp_j) > f(x_i) \end{cases} \quad \forall i \neq \text{best} \tag{10.37}$$

where F_{ij}^k is the k-th component of the force on mass m_i by mass m_j, x_i^k is the mass, and xp_j^k is the personal best individual in the k-th dimension.

The velocity update in EAPO is performed using the following equation, while position is updated using Equation 10.24:

$$v_i^k(t+1) = wv_i^k(t) + r_1 c_1 \left[xp_i^k(t) - x_i^k(t) \right] + r_2 c_2 \left[x_{\text{best}}^k(t) - x_i^k(t) \right] \tag{10.38}$$

where $r_1 = \lambda_1$, $c_1 = Gm_i$, $r_2 = \lambda_{\text{best}}$, and $c_2 = Gm_{\text{best}}$.

The detailed derivation of the velocity update equation is provided in Xie and Zeng (2009c). This is an elegant derivation of standard APO into EAPO where the velocity update equation is comparable with that of PSO, and EAPO can be considered as a special case of PSO. The proposed EAPO algorithm has been verified on a number of benchmark functions such as Sphere, Schaffer's F6, Griewank, Rastrigin, and Rosenbrock. The performances were compared with those of modified EM, DE, EA, and PSO algorithms. The simulation results show that EAPO is effective and competitive compared to other methods. Convergence and performance analysis of EAPO was also carried out by Xie et al. (2011b).

10.5.5 Local APO Algorithm

In standard APO algorithm, a population of individuals is randomly initialized within the feasible region where each individual represents a solution. Each individual has a position and velocity in the search space. Each individual except the optimal one in population is attracted or repelled by all other individuals to move to a new position driven by the virtual force. While calculating the virtual force, the relation (attraction and repulsion) of each individual of the population to other individuals is considered as communication between individuals. If the individuals of the population are denoted as nodes and communication between them are regarded as connections by an edge, then APO can be seen as a full-connected graph. This approach of the APO is named as global APO (GAPO) algorithm. In GAPO, the current best individual in the population can only attract other individuals intensely and cannot be influenced by any other individuals. The best individual is selected directly into the next iteration (i.e., generation). The problem with GAPO is that if the best individual is not updated for many successive generations, it keeps attracting other individuals with low-fitness values. As a result, these low-fitness-valued individuals gradually move toward the best and form a cluster in its vicinity. This will result in decreasing diversity leading to premature convergence, which again causes the algorithm trapped into a local optimum.

To overcome the problem of premature convergence, the population is divided into subpopulations representing different regions. The subpopulations then explore different regions improving the diversity of the population and overcome the problem of getting trapped in local optima. Mo and Zeng (2009) proposed this new variant of APO called local APO (LAPO) algorithm and applied it under some topologies. The idea is to represent the population in GAPO by a connected graph $\tilde{G}$ defined by

$$\tilde{G} = (V, E) \tag{10.39}$$

where V is the set of nodes and E is the set of edges defined by the following equations, respectively,

$$V = \{X_i \mid X_i \in \Omega \subset R^n, i = 1, \ldots, N\} \tag{10.40}$$

$$E = \{\langle X_i, X_j \rangle \mid X_i, X_j \in V, r(X_i, X_j), \forall i, j = 1, \ldots, N\} \tag{10.41}$$

where $r(X_i, X_j)$ is the relation between X_i and X_j.

It is to find a suitable topology, which is a subgraph $G_{\text{LAPO}} \subset \tilde{G}$, and the underlying mechanism that influences the performances of LAPO algorithm so that GAPO algorithm can be replaced with LAPO algorithm under some neighborhood structures. The neighborhood N_i of an individual i is defined by

$$N_i = \{X_j \mid \langle X_i, X_j \rangle \in E, \forall i, j = 1, \ldots, N\} \tag{10.42}$$

There are many problems that contain cliffs leading to complexity of solutions. LAPO algorithm with its variable topography is a way to deal with such difficult problems. The masses, total force, velocity, and position are calculated using the same equations provided by GAPO. Mo and Zeng (2009) applied LAPO algorithm under some ring and star topologies and the performances of LAPO algorithm are verified on six classical benchmark functions such as Rosenbrock, Griewank, Ackley, Penalized, Sphere, and Schaffer's F6 by adjusting the gravitation constant G. Simulation results show that LAPO algorithm is valid under some neighborhood structures. The gravitation constant G has a great influence on the performance of LAPO algorithm with different topologies. The simulation results indicate that the LAPO algorithm outperforms standard APO algorithm as long as the parameter G is selected properly under particular topologies.

10.5.6   APO with Feasibility-Based Rule

APO with a feasibility-based rule (APO-FBR) is proposed by Yin et al. (2010) to solve constrained optimization problems. The constraint violation value of an infeasible individual is calculated using the following equation:

$$\text{viol}(x) = \sum_{i=0}^{m} \max[0, g_i(x)] + \sum_{j=0}^{l} \max[0, \text{abs}(h_j(x))] \tag{10.43}$$

where $g_i(x)$ is the inequality constraint and $h_j(x)$ is the equality constraint. If the individual is in a feasible region, set $\text{viol}(x) = 0$, otherwise $\text{viol}(x) \neq 0$.

For constrained optimization problems, three feasibility rules are applied: (1) any feasible solution is preferred to any infeasible solution—that means, any feasible individuals have bigger mass than any infeasible individuals; (2) between two feasible solutions, the one having better objective function value is preferred—that means, the individual having higher fitness value has the bigger mass; and (3) between two infeasible solutions, the one having smaller constraint violation is preferred—that means, the individual having smaller constraint violation has bigger mass.

To satisfy the feasibility rules, the mass function is redefined in APO-FBR algorithm as follows:

$$m_i = 0.5 + \exp\left[\frac{f(x_{\text{best}}) - f(x_i)}{f(x_{\text{worst}}) - f(x_{\text{best}})}\right] \cdot \text{sign}[\text{viol}(x_i)] + \frac{0.5}{1 + \text{viol}(x_i)} \cdot (1 - \text{sign}[\text{viol}(x_i)]) \tag{10.44}$$

where $\text{viol}(x_i)$ is violation and $\text{sign}(\cdot)$ is defined as

$$\text{sign} = \begin{cases} 1 & \text{if } \text{viol}(x_i) = 0 \\ 0 & \text{if } \text{viol}(x_i) > 0 \end{cases} \tag{10.45}$$

The component force is computed according to three force rules: (1) among two feasible individuals, the individual having the higher fitness value attracts the worse individual and the individual having the lower fitness value repels the better individual; (2) among two infeasible individuals, the individual having smaller constraint violation attracts the bigger one and the individual having bigger constraint violation repels the smaller one; and (3) feasible individuals attract infeasible individuals and feasible individuals repel feasible individuals.

For Case 1, the force F_{ij}^k is the same as standard APO. For Case 2, the force F_{ij}^k is computed as

$$F_{ij}^k = \begin{cases} Gm_i m_j \left(x_j^k - x_i^k\right) & \text{if } \text{viol}(x_j) < \text{viol}(x_i) \\ -Gm_i m_j \left(x_j^k - x_i^k\right) & \text{if } \text{viol}(x_j) \geq \text{viol}(x_i) \end{cases} \quad \forall i \neq j \wedge i \neq \text{best} \tag{10.46}$$

For Case 3, the force F_{ij}^k is computed as

$$F_{ij}^k = \begin{cases} Gm_i m_j \left(x_j^k - x_i^k\right) & \text{if } j \in \text{FI} \wedge i \in \text{IFI} \\ -Gm_i m_j \left(x_j^k - x_i^k\right) & \text{if } i \in \text{FI} \wedge j \in \text{IFI} \end{cases} \quad \forall i \neq j \wedge i \neq \text{best} \tag{10.47}$$

where FI denotes feasible individual and IFI denotes infeasible individual.

The total force F_i^k is calculated using Equation 10.22. The main steps of the APO-FBR algorithm are the following:

Step 1: Generate m individuals, initialize position and velocity
    Set parameters: gravitational constant G, λ, and w
Step 2: Compute individuals' violation value using Equation 10.43
    Compute the feasible individuals' fitness values
    Update global best and worst position x_{best} and x_{worst}
Step 3: Calculate mass m_i using Equation 10.44
Step 4: Calculate component force F_{ij}^k based on Cases 1–3 using Equations 10.46 and 10.47
    Calculate total force F_i^k using Equation 10.22
Step 5: Update velocity using Equation 10.23
    Update position using Equation 10.24
Step 6: If (termination condition not met), Goto Step 3
Step 7: Return solution

The effectiveness and performance of APO-FBR algorithm are verified on well-known benchmark functions with different population size for constrained optimization and compared the performances with those of FAD-ELM algorithm, DE, and homomorphous maps reported in the literature. The experimental results show that APO-FBR has better global search ability.

10.5.7 APPLICATIONS

Though very recent, APO algorithm has found a significant number of applications in various domains. Firstly, the APO framework has been applied to the distributed control of swarms of robots, such as robots formation (Spears et al., 2004, 2005b, 2006). The basic idea was based on robots formations whose mission is to form a hexagonal lattice, thus creating an effective antenna (Spears and Spears, 2003). The hexagon can be created via overlapping circles of radius R. To map this into a force law, each robot repels other robots that are closer than R, while attracting robots that are farther than R in distance. APO has been found very effective in those applications.

APO algorithm and its different variants have been applied to a number of optimization benchmark functions (Xie et al., 2012). Xie et al. (2009a) tested the performance of APO on eight benchmark functions, namely, Tablet, Quadric, Rosenbrock, Griewank, Rastrigin, Schaffer's F7, Schaffer's F6, and Sphere. Xie et al. (2009b) applied APO algorithm to 13 global optimization problems and compared the performance with that of EM (electromagnetism) algorithm. Experimental results show competitive performance of APO. Xie and Zeng (2009c) applied EAPO, a variant comparable with PSO, to a number of benchmark functions, namely, Sphere, Schaffer's F6, Griewank, Rastrigin, and Rosenbrock. EAPO also demonstrates competitive performance over PSO.

Yang et al. (2010) applied HV-APO algorithm to benchmark functions such as Griewank, Rastrigin, Rosenbrock, Ackley, Schaffer's F7, Penalized Function 1, and Penalized Function 2. The simulation results demonstrated that the performance of the variant of APO algorithm is competitive over other meta-heuristic algorithms. Xie et al. (2011a) applied the new variant of HV-APO-MDS to these benchmark functions for improving the local search capability. The simulation results show that HV-APO-MDS outperforms HV-APO-ODS and VM-APO algorithms. Xie and Zeng (2011) conducted a series of experiments by applying HV-APO algorithm to well-known Michalewicz' benchmark functions (Michalewicz and Dcholenauer, 1996). APO-FBR has been applied to benchmark functions by Yin et al. (2010) to solve constrained optimization problems. Simulation results and comparisons show that APO-FBR algorithm is effective for constrained optimization problems.

Xie et al. (2010a,b) carried out theoretical convergence analysis and parameter selection of APO. They also verified the performance of APO algorithm on a number of benchmark functions such as Tablet, Quadric, Rosenbrock, Griewank, Rastrigin, Schaffer's F7. Tablet and Quadric functions are unimodal functions. Rosenbrock is generally viewed as a unimodal function. It can also be treated as a multimodal with two local optima. Griewank, Rastrigin, and Schaffer's F7 functions are well-known multimodal nonlinear functions, which have wide search spaces and lots of local optima and make algorithms easily fall into a local optimum. The simulation results demonstrate the convergent property of APO algorithm, and convergent condition helps in selecting appropriate parameters (Xie and Zeng, 2010).

Yan and Lian-chao (2010) applied APO algorithm to constraint multiobjective optimization problems. They modified the original mass function and force rules of APO algorithm so that the algorithm fits into the constraint multiobjective optimization problem. The approach further presents a method of decreasing the virtual force, which decreased the feasibility of the solutions moving from a feasible region to an infeasible region. The effectiveness of the algorithm was verified on a benchmark problem known as Binh(2). Binh(2) has a convex and continuous Pareto optimal region. The simulation results showed that most obtained solutions were on the true Pareto optimal fronts. APO algorithm has been very successful in benchmark function optimization problems but as yet its application to real-world engineering applications has not been reported in the literature.

10.6 SPACE GRAVITATIONAL OPTIMIZATION

10.6.1 Inspiration and Algorithm

Albert Einstein determined that the laws of physics are the same for all nonaccelerating observers, and that the speed of light in a vacuum is independent of the motion of all observers. This theory is known as the special theory of relativity published in 1905 (Einstein, 1905). The special theory of relativity introduced a new framework and the concept of space and time in physics. For the next 10 years Einstein worked for the development of the theory of general relativity and presented his work to the Prussian Academy of Science in 1915 of what are now known as the Einstein field equations

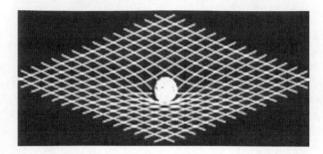

FIGURE 10.3 Curvature of the space-time. (From Internet.)

(Einstein, 1915). These equations specify how the geometry of space and time is influenced by whatever matter and radiation are present, and form the core of Einstein's general theory of relativity. According to the general theory of relativity, a massive object (e.g., an asteroid) causes curvature in the geometry of the space-time due to the gravitational field. This is felt as gravity (Einstein, 1916). Einstein's theory of general relativity predicted that the space-time around the Earth would be not only warped but also twisted by the planet's rotation. For example, a visual illustration of the curvature of the space-time is shown in Figure 10.3. Gravity Probe B[*] showed this to be correct (Everitt et al., 2000). A good account of Einstein's theory of general relativity and astrophysics can be found in Davies (1980), Harwit (1998), and Kenyon (1990).

There are numerous asteroids, planets, and stars of massive masses in the universe. These huge masses create space-time curvature due to the gravitational field. An asteroid will be able to accelerate toward the heavy mass around it by the variations in geometry of space-time. According to Newton's law of gravity, the force acting between two masses M and m at a distance r apart from each other is defined by

$$F = \frac{G \cdot M \cdot m}{r^2} = mg \tag{10.48}$$

where G is the gravitational constant and g is the gravitational acceleration. If $M \gg m$ and the absolute position of M is constant, then the gravitational acceleration of m can be described by

$$g_m = \frac{G \cdot M}{r^2} \tag{10.49}$$

The gravitational acceleration g_m of the asteroid m will vary depending on the mass M and its distance r from the asteroid. On the basis of the variation of g_m, the direction of gravitational acceleration of the asteroid can be determined. The possibility that the asteroid will enter the orbiting trajectory of the heavy mass in the universe is very small. This leaves two possibilities for the asteroid: (i) the gravitational force of the heavy mass is too strong, so the asteroid will not be able to escape the gravitational force and crashes on the heavy mass and (ii) gravitational force of the heavy mass is not strong, so the asteroid will circle around the heavy mass for a number of times; as the asteroid moves closer to the heavy mass with increasing kinetic energy, the asteroid gains huge speed and then sling-shot out (thrown out) of the gravitational field of the heavy mass. The thrown-out asteroid then enters into the gravitational field of another heavy mass within the universe. It is to be noted that the gravitational effects between asteroids are ignored in this case.

---

[*] Gravity Probe B is a space experiment testing two fundamental predictions of Einstein's theory of general relativity, the geodetic and frame-dragging effects, by means of cryogenic gyroscopes in Earth orbit (Wang, 2011).

The asteroid continues to search for heavier masses, which are considered as an optimal solution for an optimization problem.

Hsiao et al. (2005) proposed an optimal searching approach, called SGO algorithm using the notion of space gravitational curvature. SGO is a general-purpose search technique for multidimensional and multiobjective optimization problems. SGO is inspired by the concept of Einstein equivalence principle mentioned earlier. In SGO, the search agents are asteroids moving around within the universe (i.e., search space). The agents travel through the search space by Newton's law of gravity, and the solution space is formulated as a curvature of the space-time according to the concept of Einstein's theory of general relativity. The search agents of SGO aggressively search undiscovered regions for heavier masses (i.e., a better solution) based on the geometrical variance around them. The performance of SGO is quite promising for optimization problems.

In SGO, a set of asteroids are randomly initialized. The positions, velocities, and accelerations of the asteroids are denoted by $\{x,y\}$, $\{v_x,v_y\}$, and $\{a_x,a_y\}$, respectively. The accelerations $\{a_x,a_y\}$ are set to 0. The global optimum X_{best} is set to $-\infty$. The variation in geometry of space-time will cause gravitational acceleration or deceleration of asteroids (i.e., the searching agents). The summation of the variations of geometry of space-time in the directions of (x,y)-axes is the acceleration rate. The gravitational acceleration $\{a_x(n),a_y(n)\}$ of the n-th asteroid is computed using the following equations:

$$a_x(n) = G \cdot \{[f(x(n),y(n)) - f(x(n)+r_d,y(n))] + [f(x(n)-r_d,y(n)) - f(x(n),y(n))]\} \quad (10.50)$$

$$\begin{aligned} a_y(n) = G \cdot \{&[f(x(n),y(n)) - f(x(n),y(n)+r_d)] \\ &+ [f(x(n),y(n)) - f(x(n),y(n)-r_d)]\} \end{aligned} \quad (10.51)$$

where r_d is the range of detection, G is the strength of gravity, and $f(x(n),y(n))$ is the evaluation function used for the quality of the solution of the n-th asteroid. The evaluation function $f(x(n),y(n))$ is problem dependent, which can be defined using any quality measure or criteria of optimality.

Velocities and positions of the asteroids are updated after some time instant using the following equations:

$$v_x(n) = v_x(n) + a_x(n) \cdot \Delta t \quad (10.52)$$

$$v_y(n) = v_y(n) + a_y(n) \cdot \Delta t \quad (10.53)$$

$$x(n) = x(n) + v_x(n) \cdot \Delta t \quad (10.54)$$

$$y(n) = y(n) + v_y(n) \cdot \Delta t \quad (10.55)$$

where Δt is the time taken from an initial position to the new position and it is considered unity here for simplicity of computation.

Once the position $\{x(n),y(n)\}$ of the n-th asteroid is obtained, the evaluation function value $f(x(n),y(n))$ is calculated and it is compared with the current global optimum X_{best}.

If $f(x(n),y(n)) < X_{best}$
{
    $X_{best} = f(x(n),y(n))$
}

The solutions $X_{best\_x} = x(n)$ and $X_{best\_y} = y(n)$ are returned. If the asteroid, that is, the search agent, does not reach the global optimum, then it looks for further heavier mass. The main steps of SGO algorithm are stated as follows:

Step 1: Initialize a set of asteroids randomly
Step 2: Compute the gravitational acceleration $\{a_x(n), a_y(n)\}$ using Equations 10.50 and 10.51
Step 3: Compute the velocities $\{v_x(n), v_y(n)\}$ and positions $\{x(n), y(n)\}$ using Equations 10.52 through 10.55
Step 4: Compute evaluation function $f[x(n), y(n)]$
  If $f[x(n), y(n)] < X_{best}$
  $X_{best} = f[x(n), y(n)]$
Step 5: If (termination condition not reached), Goto Step 2
Step 6: Return solutions $X_{best\_x} = x(n)$ and $X_{best\_y} = y(n)$

SGO has been tested through several optimization problems and provided promising results that overcome the difficulties encountered in classical and modern optimization approaches, such as ACO and PSO.

10.6.2 Modified SGO

According to Newton's laws of gravity, the strength of gravity between two masses at a distance is the same all over the universe at all times. In the earlier SGO, the gravitational acceleration between the asteroid and the heavier mass is considered. In the modified SGO, gravitational acceleration between asteroids is added to the calculation of acceleration.

The gravitational acceleration $\{a_x(n), a_y(n)\}$ of the n-th asteroid is thus computed using the following equations:

$$a_x(n) = G \cdot \{[f(x(n), y(n)) - f(x(n) + r_\mathrm{d}, y(n))] + [f(x(n) - r_\mathrm{d}, y(n)) - f(x(n), y(n))]\} + \frac{(\alpha \cdot x_\mathrm{c})}{r_n^2}$$

(10.56)

$$a_y(n) = G \cdot \{[f(x(n), y(n)) - f(x(n), y(n) + r_\mathrm{d})] + [f(x(n), y(n)) - f(x(n), y(n) - r_\mathrm{d})]\} + \frac{(\alpha \cdot y_\mathrm{c})}{r_n^2}$$

(10.57)

where $\{x_\mathrm{c}, y_\mathrm{c}\}$ is the center of mass of all asteroids and α is a parameter that determines the influence of the gravitational effect between all asteroids. The value of α is chosen reasonably small to increase the searching efficiency of the SGO algorithm. A large value of α may start gathering all asteroids together and decrease the search efficiency. r_n is the distance between asteroid n and $\{x_\mathrm{c}, y_\mathrm{c}\}$. The center of all asteroids $\{x_\mathrm{c}, y_\mathrm{c}\}$ can be calculated using the following equations:

$$x_\mathrm{c} = \frac{1}{n}\left(\sum_n x(n)\right)$$

(10.58)

$$y_\mathrm{c} = \frac{1}{n}\left(\sum_n y(n)\right)$$

(10.59)

10.6.3 CONSIDERATION OF SHAPE OF UNIVERSE IN SGO

There is a continuing debate on the ultimate fate of the universe. There is a growing consensus among cosmologists that the fate of the universe depends on its shape. The dark energy will also play a role as the universe ages. The density of the universe also determines its geometry of the space-time. If the density of the universe exceeds the critical density, then the geometry of space is closed and positively curved like the surface of a sphere. If the density of the universe is less than the critical density, then the geometry of space is open (infinite), and negatively curved like the surface of a saddle. If the density of the universe exactly equals the critical density, then the geometry of the universe is flat and infinite in extent. Closed, open, and flat actually refer to the shape (or curvature) of the space-time itself as shown in Figure 10.4. The shape or curvature of the space-time is determined by the parameter Ω_0. The value of Ω_0 defines the shape (or curvature) of the space-time.

If $\Omega_0 > 1$, then the geometry of space is closed like the surface of a sphere (Ryden, 2003) as shown in Figure 10.4a. It is called a closed universe. In a closed universe, it lacks the repulsive effect of dark energy. The gravity eventually stops the expansion of the universe. After this point of time, the universe starts contracting until all matter in the universe collapses to a point. Estimation says that all the matter and space-time in the universe will collapse into a dimensionless singularity. This final singularity is termed as the Big Crunch. The opposite is the Big Bang.

If $\Omega_0 < 1$, the geometry of space is open like the surface of a saddle or hyperbolic (Ryden, 2003), that is, negatively curved, as shown in Figure 10.4b. It is called an open universe. In an open universe, without dark energy, the universe expands forever. The gravity slows down the rate of expansion. With dark energy, the expansion continues. The gravity accelerates the expansion. An open universe will end up in universal heat death. The Big Freeze (or the Big Rip) will occur when the acceleration caused by dark energy eventually becomes so strong that it completely overwhelms the effects of the gravitational, electromagnetic, and strong binding forces.

If $\Omega_0 = 1$, then the geometry of the universe is flat (Ryden, 2003) as shown in Figure 10.4c. It is called a flat universe. There is a growing consensus among cosmologists that the universe is flat. In a flat universe, without dark energy, the universe expands forever at a continually decelerating rate. That is the expansion approaches asymptotically to zero. With dark energy, the expansion rate of the universe initially slows down, due to the effect of gravity, but eventually increases. The ultimate fate of the universe is the same as an open universe.

According to the theory discussed above, the universe might be one of three shapes: closed universe, open universe, or flat universe. For open universe, the universe will keep expanding until all energies of the universe are vaporized. Due to the expanding effect, the distances between stars are increasing exponentially. For flat universe, the size of the universe is constant, as well as the distances between stars. Moreover, the concept of the closed universe is that the size of

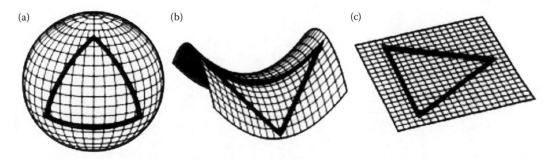

FIGURE 10.4 Shape of the universe: (a) closed universe ($\Omega_0 > 1$), and (b) open universe ($\Omega_0 < 1$), and (c) flat universe ($\Omega_0 = 1$). (From Internet.)

the universe will keep compressing until all matters are combined in a singular point; therefore, the distances between stars are also exponentially decreasing. Therefore, while calculating the velocity of the asteroids, the effects of the shape of the universe should be taken into account. According to the notion of the different types of the universe, calculations of the velocity of the asteroids using Equations 10.52 and 10.53 are modified by incorporating the curvature parameter Ω_0 as follows:

$$v_x(n) = \Omega_0 v_x(n) + a_x(n) \cdot \Delta t \tag{10.60}$$

$$v_y(n) = \Omega_0 v_y(n) + a_y(n) \cdot \Delta t \tag{10.61}$$

As discussed earlier, Ω_0 determines the shape of the universe. If $\Omega_0 = 1$, then it determines the solution space to be flat universe. If $\Omega_0 > 1$, then it turns the solution space into a closing universe, because everything seems to be approaching to the asteroid. Finally, if $\Omega_0 < 1$, then the solution space is turned into an open universe, because everything seems leaving from the asteroid. Hsiao et al. (2005) found that if Ω_0 is slightly greater than 1, it helps SGO algorithm in escaping the local optima. Setting Ω_0 to a big value generally makes the algorithm ineffective due to the high speed of the asteroid.

10.6.4 APPLICATIONS

To verify the performance and efficiency of the proposed SGO algorithm, Hsiao et al. (2005) applied the algorithm to the function optimization problem and the design optimization of a PID controller. SGO algorithm has been applied to a two-dimensional polynomial function defined by

$$\min f(x_1, x_2) = \sum_{i=1}^{2} \left(x_i^4 - 16 x_i^2 + 0.5 x_i \right), \quad \text{subject to} \quad 50 \leq (x_1, x_2) \leq 50 \tag{10.62}$$

SGO algorithm was able to find the global minimal values of the function $f(x_1, x_2)$ using 1000 asteroids (i.e., search agents) with the parameter values of $G = 1.0$, $\alpha = 0.05$, $\Omega_0 = 1.01$, and $r_d = 0.1$ within 10,000 iterations.

SGO algorithm has also been applied to the design optimization of a PID controller defined by

$$u(t) = K_p e(t) + K_i \sum_{t=0}^{T} e(t) + K_d \Delta e(t) \tag{10.63}$$

where $e(t)$ is the error and K_p, K_i, and K_d are the PID controller's gain to be optimized for three different plants. The objective function value depends on the max overshoot (f_{mo}), rise time (f_{rt}), settling time (f_{st}), and integral of absolute error (f_{iae}). Therefore, the objective function is defined as

$$\min f = f_{mo} + f_{rt} + f_{st} + f_{iae} \tag{10.64}$$

SGO algorithm was able to achieve the global optimal performance within 10 iterations using 100 asteroids (i.e., search agents). The parameters values chosen for SGO algorithm are $G = 0.5$, $\alpha = 0.05$, $\Omega_0 = 1.01$, and $r_d = 0.05$. It has been found that the SGO algorithm outperforms other approaches such as Ziegler–Nichols, fuzzy, and ACO.

10.7 INTEGRATED RADIATION OPTIMIZATION

10.7.1 INSPIRATION AND ALGORITHM

Einstein's general theory of relativity is the geometrical theory published between 1915 and 1916 (Einstein, 1915, 1916), which also explains the distortion of space-time by mass, energy, and momentum. According to the general theory of relativity (also called Einstein's equivalence principle), the curved geometry of space-time can be interpreted as the geometrical distribution of gravity. That is, the gravitational force observed locally by two reference frames, one in a space free from gravitational fields and one with uniform acceleration, are physically equivalent. According to Einstein's equivalence principle, the result of a local nongravitational experiment in an inertial frame of reference is independent of its own velocity or location. This means, in other words, the effects of gravity are exactly equivalent to the effects of acceleration.

In physics, gravitational waves are ripples in the curvature of space-time, which propagate as a wave outward from a moving object or system of objects transporting energy as gravitational radiation. Einstein (1915, 1918) first predicted the gravitational waves on the basis of his theory of general relativity (Einstein, 1916). Binary star[*] systems composed of white dwarfs,[†] neutron stars,[‡] or black holes[§] are sources of gravitational waves. The existence of gravitational radiation has been proved indirectly by Hulse and Taylor. The measurement of the Hulse–Taylor binary system suggests that gravitational waves are more than mathematical anomalies. They detected pulsed radio emissions from a pulsar in a binary star system and also discovered that the changing orbit of the binary star system is matched with the loss of energy due to giving off gravitational radiation from the binary star system (Weisberg and Taylor, 2004). It is to be noted that gravitational radiation can occur only for an accelerating object and cannot occur for a static object or for a nonspinning object at a constant velocity. This was a significant discovery in the support of the proof of gravitational radiation for which Hulse–Taylor won the 1993 Nobel Prize in Physics.

On the basis of the theoretical analysis of gravitational radiation, two general cases are considered: (i) two or more stars orbiting to each other and (ii) an isolated nonaxis-symmetrical supernova[¶] expanding the space.

Stars orbiting to each other: Binary star system (also called pulsar such as Hulse–Taylor binary star system) generates strong gravitational radiation. The strength of gravitational radiation depends on the mass of the star. Since the orbiting star and its companion star are both massive, the binary star system generates strong gravitational radiation.

Supernova expanding the space: After stellar explosion of the supernova, massive stellar materials are ejected into space. Depending on the nonsymmetric shape of the ejected material, supernova generates gravitational radiations.

Search space can be imagined as a hyperspace with a number of search agents distributed randomly within the space. A solution provided by the search agent is evaluated against its objective function. The solution having a better objective value is assumed to have a supernova with imperfect symmetrical shape expanding in the space. If a search agent stands alone in the hyperspace, the direction that detects a large quantity of gravitational radiation can directly imply to have one or a

---

[*] A binary star is a star system consisting of two stars orbiting around their common center of mass.

[†] A white dwarf is a stellar remnant composed mostly of electron-degenerate matter and very dense. A white dwarf's mass is comparable to that of the Sun with a volume comparable to that of the Earth.

[‡] A neutron star is a stellar remnant that results from the gravitational collapse of a massive star. Neutron stars are the densest and the smallest stars known to exist in the universe. They can have radius of only about 12–13 km and mass of about two times that of the Sun.

[§] A black hole is a mathematically defined region of space-time exhibiting such a strong gravitational pull that no particle or electromagnetic radiation can escape from it.

[¶] A supernova is a stellar explosion that briefly outshines an entire galaxy. Supernovae can be triggered by a sudden re-ignition of nuclear fusion in a degenerate star or by the gravitational collapse of the core of a massive star. A supernova radiates as much energy as the Sun or any ordinary star is expected to emit over its entire life span. It then fades away from view within several weeks or months (Giacobbe, 2005).

group of massive stars locating at that direction. The search agent will move toward the location of the massive stars and explore the undiscovered regions on the path. Finally, a global optimal solution can be produced. In an optimization process, the search agents in search space are thought of as massive binary star systems moving in the universe. The massive binary star emits gravitational radiation along the curvature of the space-time. The quality of solution is thought of as the mass of the star in binary star systems. The heavier mass represents higher quality solution. The other thing is that the gravitational radiation is proportional to the mass of binary star. Therefore, the search is directed to the direction of the highest amount of gravitational radiation. This phenomenon also agrees with Einstein's general theory of relativity. The power of gravitational radiation (GR) emitted by the massive binary star system is defined by

$$\text{GR} = -\frac{32}{\pi}\frac{G^4}{c^5}\frac{(m_1 m_2)^2 (m_1 + m_2)}{R^5} \tag{10.65}$$

where c is the speed of light, R is the distance between the binary star systems, G is the gravitational constant, and m_1 and m_2 are the total masses of the two binary star systems, respectively. GR is inversely proportional to R^5 meaning GR has exceptionally increasing rate with decreasing distance. The amount of power P_{GR} that can be detected from the source of gravitational radiation at an average distance of R_a is defined by

$$P_{GR} = \frac{GR}{4\pi R_a^{\,2}} = -\frac{8}{\pi^2 R_a^{\,2}}\frac{G^4}{c^5}\frac{(m_1 m_2)^2 (m_1 + m_2)}{R^5} \tag{10.66}$$

When the solution quality of an agent is significantly higher than that of all other agents, then its mass is significantly higher than the rest of the masses. That means, the agent with the best solution becomes isolated from the other agents, meaning that the agent has no or less influence from the rest. The acceleration of the mass m_1 can be computed using Newton's laws of motion as follows:

$$a_1 = \frac{(G(m_1 m_2 / r^2))}{m_1} = G\frac{m_2}{r^2} \tag{10.67}$$

The movement of a search agent is directed to the accumulated gravitational radiation from one or a set of agents fighting against other directions and also depending on the quality of solutions.

On the basis of the above theories, Chuang and Jiang (2007) proposed an algorithm called IRO inspired by the gravitational radiation in the curvature of space-time. IRO algorithm comprises three stages: (i) initialization, (ii) preprocessing, and (iii) ranking and movement.

Initialization: Suppose that there are m decision variables of a system to be optimized within the upper and lower limit defined by $[\text{Min}_m, \text{Max}_m]$. The search space (SS) consists of m-grid points in m-dimensional space. In other words, SS is the convolution of m variables. The quality indices of a solution can be computed from the convolution of the m variables. The size of the convolute SS comprising m-dimensional Gaussian distribution function is specified by

$$M_m = \text{size}(\text{SS}) = \prod_{i=1}^{m} (s_m) \tag{10.68}$$

where M_m is the total number of points (i.e., number of memory elements[*]) required for the construction of search space SS and s_m is the number of grids in the m-th dimension of SS.

---

[*] In the original paper, the authors referred it as the total amount of memory.

A population of search agents X_i, $i = 1,2, ..., N$, is randomly initialized within the m-dimensional hyperspace for the optimization problem. The search agents are evaluated using a fitness function $f(\cdot)$ (also called cost function defined by the mass of the search agent similar to fitness function used in SGO). The fitness of agent $f(X_i)$ corresponds to a position in the m-dimensional hyperspace. The position of the agent X_i in the SS is denoted as $idx(X_i)$. The corresponding position of the search agent X_i in the SS, denoted as $idx(X_i)$, is considered as a special parameter. Therefore, corresponding position $idx(X_i)$ in SS is specified by the fitness of agent $f(X_i)$ as follows:

$$SS[idx(X_i)] - f(X_i) \tag{10.69}$$

where SS[·] denotes the corresponding position in search space.

Preprocessing: The geometry and the strength of the gravitational radiation are estimated using the equations of astrophysics. The equations of astrophysics are of higher order and very complex, which lead to increased computational complexity with increasing number of iterations. An alternative is to approximate the astrophysics equations. Chuang and Jiang (2007) used Gaussian distribution function for the fast approximation of geometry of gravity field and strength of gravitational radiation. The Gaussian distribution function for m-dimensional hyperspace is defined as

$$\text{Gauss}(x_1, ..., x_m) = A \cdot \exp\left[-\sum_{i=1}^{m}\left(\frac{x_i - \mu}{\sigma}\right)^2\right] \tag{10.70}$$

where A is the amplitude, μ is the mean, and σ is the spread (standard deviation) of the distribution.

The power of gravitational radiation GR is approximated by calculating the convolution of Gaussian distribution $\text{Gauss}(x_1, ..., x_m)$ and the size of $\text{SS}(k_1, ..., k_m)$ as follows:

$$GR = -\sum_{k_1=-\infty}^{\infty} \cdots \sum_{k_m=-\infty}^{\infty} \text{Gauss}[(x_1 - k_1), ..., (x_m - k_m)] \cdot SS(k_1, ..., k_m) \tag{10.71}$$

Ranking and movement: The search agents are ranked using percentile ranking. A percentile rank is the percentage of scores that fall below a given score. The percentile of the i-th data point in a sorted data of N samples is defined by

$$\text{pct}(X_i) = 100 \times \frac{i - 0.5}{N} \tag{10.72}$$

where pct(.) denotes the percentile function.

The percentiles of search agents X_i, $i = 1, 2, ..., N$, in search space and gravitational radiation are calculated as follows:

$$\begin{cases} X_{\text{ipct}}^{\text{SS}} = \text{pct}[\text{SS}(idx(X_i))] \\ X_{\text{ipct}}^{\text{GR}} = \text{pct}[\text{GR}(idx(X_i))] \end{cases} \tag{10.73}$$

where $X_{\text{ipct}}^{\text{SS}}$ and $X_{\text{ipct}}^{\text{GR}}$ are the percentiles of X_i in $\text{SS}(idx(X_i))$ and $\text{GR}(idx(X_i))$, respectively.

The displacement Δ of the search agent X_i is denoted as $\Delta(X_i)$. $\Delta(X_i)$ depends on the gravitational radiation from other agents and is defined as

$$\Delta(X_i) = \sum_{\forall j, j \neq i}^{m} \frac{X_{\text{ipct}}^{\text{SS}}}{\|u_{ij}\|^2} \frac{X_{\text{ipct}}^{\text{GR}}}{\|u_{ij}\|^2} (\text{rand}_j \cdot \bar{u}_{ij}) \tag{10.74}$$

where $\bar{u}_{ij}$ is the unit vector for the geometric vector of $u_{ij} = X_j - X_i$ and rand$_j$ is a $1 \times j$ random vector uniformly distributed over [0,1]. This term rand$_j$ allows IRO to incorporate stochastic exploration ability and helps in escaping local minima. The term $X_{\text{ipct}}^{\text{GR}} / \| u_{ij} \|^2$ simulates the amount of power that is detected from the source of gravitational radiation given in Equation 10.66. The term $X_{\text{ipct}}^{\text{SS}} / \| u_{ij} \|^2$ represents the acceleration of search agent X_i caused by the mass of agent X_j given in Equation 10.67. The position of the search agent X_i is updated at every iteration using the following equation:

$$X_i = X_i + \Delta(X_i) \tag{10.75}$$

When updating the positions of all search agents is done, an iteration of the optimization process is then completed. IRO algorithm runs for max iterations or terminates when the termination condition is satisfied. The main steps of the IRO algorithm can be stated as follows:

Step 1: Generate search agents X_i, $i = 1, 2, \ldots, N$, randomly within the m-dimensional hyper space
Step 2: Estimate corresponding position in SS($idx(X_i)$) and in GR($idx(X_i)$) using Equations 10.69 and 10.71
Step 3: Calculate percentile $X_{\text{ipct}}^{\text{SS}}$ and $X_{\text{ipct}}^{\text{GR}}$ of agent X_i in SS($idx(X_i)$) and GR($idx(X_i)$) using Equation 10.73
Step 4: Calculate displacement of agent $\Delta(X_i)$ using Equation 10.74
Step 5: Update position of agent X_i using Equation 10.75
Step 6: If (termination condition not met), Goto Step 2
Step 7: Return solution

10.7.2 APPLICATIONS

In order to verify the performance and efficiency of the proposed IRO algorithm, Chuang and Jiang (2007) applied the algorithm to the function optimization problem and the design optimization of a PID controller. IRO algorithm has been applied to a two-dimensional polynomial function defined by

$$\min f(x_1, x_2) = \sum_{i=1}^{2} \left(x_i^4 - 16x_i^2 + 0.5x_i \right), \quad \text{subject to} \quad 50 \leq (x_1, x_2) \leq 50 \tag{10.76}$$

IRO algorithm was able to find the global minimal values of the function $f(x_1,x_2)$ within 100 iterations using 20 search agents and having a resolution of search space in each dimension of 1000. Chuang and Jiang (2007) also applied the IRO algorithm to the design optimization of a PID controller defined by

$$u(t) = K_p e(t) + K_i \sum_{t=0}^{T} e(t) + K_d \Delta e(t) \tag{10.77}$$

where $e(t)$ is the error and K_p, K_i, and K_d are the PID controller's gain to be manipulated for an optimal performance of the PID controller. The objective function value depends on the max overshoot (f_{mo}), rise time (f_{rt}), settling time (f_{st}), and integral of absolute error (f_{iae}). Therefore, the objective function is defined as

$$\min f = f_{\text{mo}} + f_{\text{rt}} + f_{\text{st}} + f_{\text{iae}} \tag{10.78}$$

IRO algorithm was able to achieve the global optimal performance within 100 iterations using 20 search agents and having a resolution of search space in each dimension of 1000. It has been found that the IRO algorithm outperforms other approaches such as Ziegler–Nichols, fuzzy, and ACO. The IRO algorithm is relatively a new heuristic algorithm. There have not been many applications reported in the literature yet.

10.8   GRAVITATIONAL INTERACTIONS OPTIMIZATION

10.8.1   INSPIRATION AND ALGORITHM

Flores et al. (2010) proposed GIO inspired from Newton's law. According to Newton's laws of motion, the attraction force F between two masses $\{m_1, m_2\}$ is proportional to the masses and inversely proportional to the distance r between them as follows:

$$F = G\frac{m_1 \cdot m_2}{r^2} \qquad (10.79)$$

where G is an empirical physical constant called gravitational constant in Newtonian mechanics.

GIO is a population-based meta-heuristic optimization algorithm. A population of N objects is created randomly with a uniform distribution. The objects are points in an n-dimensional search space. The position of an object $x_i = \{x_i^1, x_i^2, \ldots, x_i^n\}$ represents a point in the n-dimensional search space where $\{x_i^1, x_i^2, \ldots, x_i^n\}$ are the decision variables of an optimization problem. The fitness of object, denoted as $f(x_i)$, is mapped to a mass in a gravitational field. The fitness function is a mapping that transforms the vector $x_i = \{x_i^1, x_i^2, \ldots, x_i^n\}$ to a scalar value $f(x_i)$. Each object represented by x_i is assigned a mass representing its value of the fitness function. Each object stores its current position x_i and its best position x_i^{best} so far according to the fitness function. Objects are allowed to interact with the rest of the objects in the population in a synchronous discrete manner. The objects follow Newton's gravitational law, attracted by the force F according to Equation 10.79 and move due to interaction forces to new locations in such a way so that the whole population moves toward the global optimum (or multiple local optima for multimodal problems) within a number of iterations. The force F_{ij} acting between two objects can be computed according to

$$F_{ij} = \frac{M[f(x_i)] \cdot M[f(x_j)]}{|x_i - x_j|^2} \bar{x}_{ij} \qquad (10.80)$$

where x_i and x_j are the positions of the i-th and j-th objects (or individuals), $|x_i - x_j|$ is the Euclidean distance between x_i and x_j, $\bar{x}_{ij}$, the unit vector between the positions x_i and x_j, provides the direction of the force, $f(x_i)$ is the fitness of object at x_i, and $M[f(x_i)]$ is the corresponding mass of the object.

The mass of an object, denoted as $M[f(x_i)]$, is computed as follows:

$$M[f(x_i)] = \left(\frac{f(x_i) - \min f(x)}{\max f(x) - \min f(x)} \times (1 - \lambda) + \lambda \right)^2 \qquad (10.81)$$

where $\min f(x)$ and $\max f(x)$ are the minimum and maximum of the fitness values of the population, respectively, and λ is a parameter chosen very small close to 0. The term $(1 - \lambda)$ is used to rescale the fitness value $f(x_i)$ within the interval of $[\lambda, 1]$. The whole term on the right-hand side is squared to emphasize the best and worst fitnesses.

In GIO, each object interacts with every other object in the population. The total force F_i is the sum of all component forces between the masses $M[f(x_i)]$ and $M[f(x_j^{\text{best}})]$, which is calculated according to

$$F_i = \sum_{j=1}^{n} \frac{M[f(x_i)] \cdot M[f(x_j^{\text{best}})]}{\left| x_i - x_j^{\text{best}} \right|^2} \overline{x_i x_j^{\text{best}}} \qquad (10.82)$$

where the term $\left| x_i - x_j^{best} \right|$ is the Euclidean distance between the current position x_i and the best position x_j^{best} so far. Flores et al. (2011) calculated the total force when the Euclidean distance $\left| x_i - x_j^{best} \right| \geq 10^{-5}$; otherwise, objects at x_i and x_j^{best} are thought to be located at the same position. In other words, they are collided together to become one object. The unit vector $\overline{x_i x_j^{best}}$ provides the direction of the force.

In order to compute the displacement of x_i in the direction of force F_{ij}, it is required to solve Equation 10.80 for x_j. That is to find the position of an object with unit mass, which means $M[f(x_j)] = 1$. The displacement of $x = |x_i - x_j|$ is defined by

$$x = \sqrt{\frac{M[f(x_i)]}{|F_i|}} \, \bar{F}_i \tag{10.83}$$

Due to the force F_i acting on each object, it will change its velocity from $v_i(t)$ to $v_i(t+1)$. For unimodal optimization problems, the current velocity v_i is updated according to the function below:

$$v_i(t+1) = \chi \cdot [v_i(t) + \rho \cdot c \cdot x] \tag{10.84}$$

where χ is an inertia constriction factor, $\rho \in [0,1]$ is a random number, c is the gravitational interaction coefficient and usually chosen as $c = 2.01$ for unimodal optimization problem, and t is the current iteration (or time). The inertia constriction factor χ helps in avoiding objects to explore out of the search space. χ is defined as

$$\chi = \frac{2k}{\left| 2 - \phi - \sqrt{\phi^2 - 4\phi} \right|} \tag{10.85}$$

where k is an arbitrary value within the interval of $[0,1]$ and $\phi = c > 4$.

The velocity update equation (10.84) works well for unimodal optimization problems. For multimodal optimization problems, Flores et al. (2010) used a cognitive component similar to PSO added to the velocity update equation and the velocity update is performed according to

$$v_i(t+1) = \chi \cdot \left[v_i(t) + \rho_1 \cdot c_1 \left(x^{best} - x_i \right) + \rho_2 \cdot c_2 \cdot x \right] \tag{10.86}$$

where $\rho_1, \rho_2 \in [0,1]$ are random numbers, $\{c_1, c_2\}$ are the gravitational interaction coefficients, and $\phi = c_1 + c_2 > 4$. Flores et al. (2011) used $c_1 = c_2 = 2.01$ in their experiments.

Once the velocity is updated, the positions of the objects are updated according to

$$x_i(t+1) = x_i(t) + v_i(t+1) \cdot \Delta t \tag{10.87}$$

where Δt is the time unit taken to reach the velocity at time $(t+1)$ and Δt is considered unity.

The main steps of the GIO algorithm can now be stated as the following:

Step 1: Initialize objects x_i, $i = 1, \ldots, N$ in n-dimensional space
        Set parameter values
Step 2: Compute fitness $f(x_i)$, update x_i^{best} and mass $M[f(x_i)]$ using Equation 10.81
Step 3: Compute total force F_i using Equation 10.82
Step 4: Update velocity v_i using Equation 10.84 or 10.86 according to the problem

Step 5: Update x_i using Equation 10.87
Step 6: If (termination condition not met), Goto Step 2
Step 7: Return solution

GIO has some similarities with GSA presented in Chapter 2. GIO algorithm was introduced around the same time independently of GSA. The gravitational constant G in GSA decreases linearly with time, whereas GIO uses a hypothetical gravitational constant G as constant. GSA uses K_{best} individuals (or agents) to reduce computation time while GIO allows all masses to interact with each other.

10.8.2 Applications

Flores et al. (2010, 2011) verified the performance of GIO algorithm on a number of unimodal test functions such as Goldstein, Price function, Booth function, and variable Colville function and a number of multimodal test functions such as Branin's function with three global optima, Deb's function with six global maximum, Himmelblau's function with four global optima, and Six-Hump Camelback function with two global optima and four local optima. The experimental results were compared with the results of PSO algorithm and showed more reliable results than did PSO. Flores et al. (2011) did claim that their method was similar to GSA with the addition of some parameters but there was no comparison with GSA provided in their investigation. The GIO algorithm is relatively new. There is still a need to verify its effectiveness and performance by applying to real-world optimization problems.

10.9 CHARGED SYSTEM SEARCH

10.9.1 Inspiration and Algorithm

In physics, Coulomb's law describes the electrostatic interaction between electrically charged particles (CPs). According to Coulomb's law, each particle has a charge in an electromagnetic field and there is electromagnetic force acting between two particles. The theory was first published by French physicist Charles Augustin de Coulomb in 1785 (Coulomb, 1785a,b). He determined that the magnitude of the electric force between two point charges is directly proportional to the product of the charges and inversely proportional to the square of the distance between them. Mathematically, Coulomb's law can be expressed in scalar form as

$$|F| = k_e \frac{|q_i \cdot q_j|}{r^2} \tag{10.88}$$

where k_e is the Coulomb's constant, q_i and q_j are the signed magnitudes of the charges, and r is the distance between the charges.

If the two charges have the same sign, the electrostatic force between them is repulsive. If they have different signs, the force between them is attractive. The vector form of Equation 10.88 can be expressed as

$$F_{ij} = k_e \frac{q_i \cdot q_j}{|r_{ij}|^2} \hat{r}_{ij} \tag{10.89}$$

where the vector r_{ij} is the distance between the charges and $\hat{r}_{ij} = (r_j - r_i)/\|r_j - r_i\|$ is the unit vector acting from q_j to q_i. The vector form of Equation 10.89 calculates the force F_{ij} applied on q_i by q_j.

The charge determines the magnitude of attraction or repulsion. The total force acting on a particle by other particles defines the direction of movement in order to reach equilibrium in the space

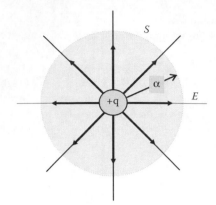

FIGURE 10.5 Electric field around a point charge.

from where the particle will not move any further. Coulomb's law was an essential component to the development of the theory of electromagnetism (Coulomb, 1785a,b). Coulomb's law can be used to derive Gauss's law and vice versa. The Gauss's law was formulated by Carl Friedrich Gauss in 1835 and was first published in 1867. Gauss's law (also known as Gauss's flux theorem) states the distribution of electric charge to the resulting electric field. The magnitude of the electric field at a point inside a charged sphere can be obtained using Gauss's law that it is proportional to the separation distance between the particles. It is one of Maxwell's four equations, which form the basis of classical electrodynamics. Interested readers are directed to the book by Inan and Inan (1998) for the detailed mathematical treatment of the laws of Coulomb, Gauss, Ampere, and Faraday.

The Coulomb and Gauss laws provide the electric field at a point inside and outside a charged insulating solid sphere S with radius α, as shown in Figure 10.5. The magnitude of the electric field for the CP located inside the sphere is proportional to the separation distance between the CPs and inversely proportional to the square of the separation distance between the particles for CPs located outside the sphere, which is expressed by Inan and Inan (1998)

$$
F_{ij} = \begin{cases} k_e \dfrac{q_i}{\alpha^3} r_{ij} \dfrac{r_j - r_i}{\| r_j - r_i \|} & \text{if } r_{ij} < \alpha \\[2em] k_e \dfrac{q_i}{r_{ij}^2} \dfrac{r_j - r_i}{\| r_j - r_i \|} & \text{if } r_{ij} \geq \alpha \end{cases}
\tag{10.90}
$$

where k_e is the Coulomb constant, q_i is the charge of the sphere, r_{ij} is the distance between the two centres of spheres, and r_i and r_j are the radii of the CPs i and j, respectively.

The resulting electric force due to N CPs (considered as charged sphere) having uniform volume charge density, which can exert an electric force to the other CPs in the electric field, is given by

$$
F_i = q_i F_{ij} = q_i k_e \sum_{j=1}^{N} \left(\frac{q_j}{\alpha^3} r_{ij} \cdot w_1 + \frac{q_j}{r_{ij}^2} \cdot w_2 \right) \frac{r_j - r_i}{\| r_j - r_i \|}, \quad \begin{cases} w_1 = 1, w_2 = 0 & \text{if } r_{ij} < \alpha \\ w_1 = 0, w_2 = 1 & \text{if } r_{ij} \geq \alpha \end{cases}
\tag{10.91}
$$

The force F_{ij} due to the electric field is the resultant force acting on a CP i ($i = 1, 2, ..., N$). As the particles continue to attract each other, they come closer decreasing the distance between the particles. When the distance between particles becomes less than α (radius of the sphere), the resultant force becomes proportional to the distance between the particles. α is an arbitrary parameter used in the calculation for the resultant force. According to Newton's second law, the acceleration of a mass (i.e., of a particle) is directly proportional to the total force acting on that particle. Thus, the resultant electrical force acting on a CP results in its acceleration, which causes the particle to change its position and velocity under the electric field. Using Newton's laws of motion, the position of a particle considered as a point-mass having infinitesimal size can be determined at any time when its position, velocity, and acceleration at previous time are known. The changes in position, velocity, and acceleration are defined as follows:

$$\Delta x = x_{\text{new}} - x_{\text{old}} \tag{10.92}$$

$$v = \frac{x_{\text{new}} - x_{\text{old}}}{\Delta t} \tag{10.93}$$

$$a = \frac{v_{\text{new}} - v_{\text{old}}}{\Delta t} \tag{10.94}$$

where x_{old} and x_{new} are the initial and final positions of a particle, respectively, v_{old} and v_{new} are the initial and final velocities of a particle, respectively, v is the velocity of the particle, a is the acceleration of the particle, and Δt is the time taken to reach the new position. Using Equations 10.91 through 10.94 and Newton's second law, the new position of any particle can be obtained. The new position of the particle is calculated as

$$x_{\text{new}} = x_{\text{old}} + v_{\text{old}} \cdot \Delta t + \frac{1}{2} \frac{F}{m} \cdot \Delta t^2 \tag{10.95}$$

where F is the force and m is the point-mass of the particle.

As the particles move to each other under the electric field, the separation distances between particles diminish and acting forces gradually reach an equilibrium state when the particles no longer move. The successive movements of the CPs drive the algorithm toward optimal solutions. From an optimization point of view, this state of equilibrium is a representative of an optimal solution. CSS is inspired from electrostatics in physics, Coulomb's law, Gauss law, and Newtonian mechanics and was first proposed by Kaveh and Talatahari (2010b–d). In CSS, a population of candidate solutions is generated within the variable space of the optimization problem at hand. The candidate solutions x_i, $i = 1, 2, \ldots, N$ within the search space are positions of CPs under the influence of the electric field. By applying electrostatics, Coulomb's law, Gauss law, and Newtonian mechanics, an optimal solution is to be found. CSS algorithm is carried out in three stages, namely, initialization stage, search stage, and termination stage.

In the initialization stage of CSS, a population of CPs with random positions x_i, $i = 1, 2, \ldots, N$ within an n-dimensional search space is generated. In CSS, the positions x_i represent the solutions of the problem in question. The initial velocities of these particles are set to zero. Each particle has a charge. The magnitude of charge q is defined in terms of the fitness of its solution. The charge of a particle is computed according to

$$q_i = \frac{f(x_i) - f(x_{\text{worst}})}{f(x_{\text{best}}) - f(x_{\text{worst}})}, \quad i = 1, 2, \ldots, N \tag{10.96}$$

where $f(x_{\text{best}})$ and $f(x_{\text{worst}})$ are the best and worst fitness of all the particles and $f(x_i)$ is the fitness of the i-th particle.

The separation distance r_{ij} between two CPs i and j is defined as the Euclidean distance between them, which is defined as

$$r_{ij} = \frac{\| x_i - x_j \|}{\| (x_i + x_j)/2 - x_{\text{best}} \| + \varepsilon} \tag{10.97}$$

where x_i and x_j are the positions of the i-th and j-th CPs, respectively, x_{best} is the position of the best current CP, and ε is a small positive number to avoid singularities, that is, division by zero.

CPs are ranked by evaluating and comparing the fitness function values of each CP and sorted in increasing order. A charged memory (CM) is created to store the fitness function values of the CPs. The size of the CM is set equal to the number of the first CPs, that is, size of the population of particles. The CM size (CMS) is a parameter of the CSS algorithm.

In search stage of CSS, attracting force between CPs is determined. A probability value p_{ij} is calculated in determining the moving CP toward the other, considering the following probability function:

$$p_{ij} = \begin{cases} 1 & \dfrac{f(x_j) - f(x_{\text{best}})}{f(x_i) - f(x_j)} > \text{rand or } f(x_i) > f(x_j) \\ 0 & \text{otherwise} \end{cases} \tag{10.98}$$

where rand is a random value uniformly distributed over the interval [0,1].

Handling the attractiveness and repulsiveness of resulting force of CPs with the clever idea of parameter p_{ij} is very effective. The attracting force vector for each particle is determined according to Equation 10.91 and the probability value p_{ij}.

$$F_i = k_c q_i \sum_{\substack{j=1 \\ j \neq i}}^{N} \left(\frac{q_j}{\alpha^3} r_{ij} \cdot w_1 + \frac{q_j}{r_{ij}^2} \cdot w_2 \right) p_{ij}(x_j - x_i), \quad \begin{cases} w_1 = 1, w_2 = 0 & \text{if } r_{ij} < \alpha \\ w_1 = 0, w_2 = 1 & \text{if } r_{ij} \geq \alpha \end{cases} \tag{10.99}$$

where F_i is the resultant electric force acting on the i-th CP, $i = 1, 2, \ldots, N$, and α is an arbitrary parameter of CSS algorithm, which decides the move from the global search (exploration phase) to the local search (exploitation phase).

The force F_i calculated in Equation 10.99 is considered attractive. But the force F_i can be attractive or repulsive. By introducing a force (attractiveness or repulsiveness) parameter ar_{ij}, attractiveness or repulsiveness of force in Equation 10.99 can be determined. The force parameter ar_{ij} is defined as

$$ar_{ij} = \begin{cases} +1 & k_{ar} < \text{rand for attractive force} \\ -1 & k_{ar} > \text{rand for repulsive force} \end{cases} \tag{10.100}$$

where k_{ar} is a parameter to control the effect of force.

Incorporating the force parameter in Equation 10.99, the force becomes

$$F_i = k_c q_i \sum_{\substack{j=1 \\ j \neq i}}^{N} \left(\frac{q_j}{\alpha^3} r_{ij} \cdot w_1 + \frac{q_j}{r_{ij}^2} \cdot w_2 \right) ar_{ij} \cdot p_{ij}(x_j - x_i), \quad \begin{cases} w_1 = 1, w_2 = 0 & \text{if } r_{ij} < \alpha \\ w_1 = 0, w_2 = 1 & \text{if } r_{ij} \geq \alpha \end{cases} \tag{10.101}$$

During the early iterations of the CSS algorithm, the exploration takes place in the global search phase. As the search continues, the particles come closer to each other and the distance between the particles decreases. The resultant force becomes proportional to the distance between the particles. Search procedure starts exploiting the local search space. For an efficient searching algorithm, it is essential to increase the exploitation and to decrease the exploration over the iterations. The appropriate choice of the parameter α can strike a balance between exploration and exploitation. Particles move to new positions depending on the resultant forces acting on CPs. The velocity and position CPs are updated according to Equations 10.93 and 10.95, respectively, as follows:

$$v_i^{\text{new}} = \frac{x_i^{\text{new}} - x_i^{\text{old}}}{\Delta t} \tag{10.102}$$

$$x_i^{\text{new}} = x_i^{\text{old}} + \rho_1 \cdot k_v \cdot v_i^{\text{old}} \cdot \Delta t + \rho_2 \cdot k_a \cdot \frac{F_i}{m_i} \cdot \Delta t^2 \tag{10.103}$$

where ρ_1 and ρ_2 are two random numbers uniformly distributed over the interval [0,1]; (F_i/m_i) is the acceleration term for the i-th particle according to Newton's law of motion; m_i is the mass of the i-th particle, which is considered equal to q_i (calculated in Equation 10.96) according to the algorithm proposed by Kaveh and Talatahari (2010b); Δt is the time step, which is considered unity; k_a is the acceleration coefficient; and k_v is the velocity coefficient to control the influence of the velocity of previous iteration. These parameters can be set to fixed values or can be adaptive during the search process (Kaveh and Talatahari, 2010b–d). However, whether CSS is going to explore or exploit the search space depends on the parameters k_a and k_v. Generally, k_v works as a control parameter of the exploration property (Kaveh and Talatahari, 2010b); therefore, choosing a linear decreasing function defined by Equation 10.104 can improve the performance. But k_a works as a control parameter of the exploitation property (Kaveh and Talatahari, 2010b); therefore, choosing a linear increasing function defined by Equation 10.105 can help in improving the performance of the algorithm.

$$k_v = 0.5\left(1 - \frac{t}{t_{\text{max}}}\right) \tag{10.104}$$

$$k_a = 0.5\left(1 + \frac{t}{t_{\text{max}}}\right) \tag{10.105}$$

where t is the current iteration and t_{max} is the maximum number of iterations.

Initially, values of k_a and k_v are almost same. Gradually, k_a increases and k_v decreases over the iterations. With increasing k_a, the effect of attraction of good solutions also increases. Thus, the algorithm ensures convergence toward better solutions. The CPs can go out of bound of the search space over the iterations. A position correction is necessary to bring the CPs back into the search space. Simple mechanism can be applied to correct the position of the CPs that go out of allowable search space such as one used in CFO (central force optimization) algorithm in Chapter 3. Kaveh and Talatahari (2009b) also suggested such a correction method, which they applied in a hybrid approach in HSA discussed in Chapter 5.

$$\begin{cases} x_i^k = x_L^k + \delta_{\text{rep}} \cdot \left(x_i^k - x_L^k\right) & \text{if } x_i^k < x_L^k \\ x_i^k = x_U^k - \delta_{\text{rep}} \cdot \left(x_U^k - x_i^k\right) & \text{if } x_i^k > x_U^k \end{cases} \tag{10.106}$$

where x_L^k and x_U^k are the lower and upper bounds of the k-th variable of the i-th CP, respectively, and the factor δ_{rep} is an arbitrary reposition factor specified by the user within the range of $0 \leq \delta_{\text{rep}} \leq 1$.

The CPs are evaluated and compared against the new CPs. The CPs are then sorted in increasing order of the fitness function values. CM created in Step 3 is updated with new CP vectors. If the new CPs are better in terms of their objective function values than the CPs already stored in the CM, the worst ones in the CM are replaced by the new vectors.

In termination stage of CSS, it decides to stop the algorithm and return the final solution to the optimization problem. The termination condition can be the maximum number of iterations or a desired level of solution measured by the fitness value.

The main steps of CSS algorithm are as follows:

Initialization stage:
Step 1: Initialize a population of CPs with random positions x_i, $i = 1, 2, \ldots, N$ in n-dimensional space, initialize velocities and set parameters
Step 2: Evaluate $f(x_i)$, determine $f(x_{best})$ and $f(x_{worst})$
      Compute charge of particles q_i using Equation 10.96
      Evaluate fitness of CPs and rank them
Step 3: Store ranked CPs into CM

Search stage:
Step 4: Determine p_{ij} using Equation 10.98
      Calculate force F_i using Equation 10.101
Step 5: Update velocity v_i and position x_i using Equations 10.102 and 10.103
Step 6: Correct positions x_i of CPs using Equation 10.106
Step 7: Evaluate CPs and rank them in increasing order of fitness
Step 8: If (new CPs > old CPs)
      Update CM replacing worst ones

Termination stage:
Step 9: If (termination condition not met), Goto Step 4
Step 10: Return solution

The CSS algorithm has found a number of applications in various domains since its inception. Due to the nature of applications, the algorithm has gone under various modifications. Moreover, in order to address the convergence issue, computational complexity, and handling local optima, researchers have also suggested some variants of CSS algorithm and hybrid approaches. Some well-known variants are discussed in brief.

10.9.2 Discrete CSS

CSS algorithm is essentially a continuous one. Kaveh and Talatahari (2010e, 2012) proposed a discrete version of CSS algorithm where the solution construction step of search stage (i.e., Step 5 of CSS algorithm in Section 10.9.1) is modified to yield a discrete value by applying a function Fix(·) as follows:

$$x_i^{new} = \text{Fix}\left(x_i^{old} + \rho_1 \cdot k_v \cdot v_i^{old} \cdot \Delta t + \rho_2 \cdot k_a \cdot \frac{F_i}{m_i} \cdot \Delta t^2 \right) \qquad (10.107)$$

Fix(z) is a function which rounds off each element of the argument z to the nearest permissible discrete value. Thus, the discrete version of CSS algorithm can be applied to problems where a discrete value is preferred to any real value.

10.9.3 Chaotic CSS

In order to enhance the global search ability for optimization algorithms, chaotic maps are utilized instead of random process due to the nonrepetition of chaotic maps. The overall search speed using chaotic search is higher than stochastic search based on the probabilistic distribution of numbers over a predefined interval. In addition, chaotic optimization approaches can escape from local minima more easily than other stochastic optimization approaches. Therefore, some researchers incorporated chaotic maps into the CSS to improve the convergence speed. Another key factor in optimization algorithms is the striking balance between exploration and exploitation in the search process. The parameters k_a, k_v, k_{ar}, ar_{ij}, and p_{ij} (in Equations 10.100 through 10.103) are random values and are the key factors in CSS to control the balance between exploration and exploitation. The notion of using chaotic systems instead of random processes has been identified as a potential tool in several fields. By simply introducing chaotic systems into CSS-based optimization, the performance of CSS algorithm can be greatly improved. Talatahari et al. (2011) first proposed chaotic CSS (CCSS) using chaotic maps where the role of randomness is replaced by some chaotic dynamics.

Chaos is a mathematical property of dynamical systems. Even though described by simple, deterministic, and first-order difference equation, it exhibits surprising dynamic, pseudo-random, ergodic, and nonperiodic behavior from stable points to bifurcating hierarchy of stable cycles (May, 1976). The behavior is sensitive to the initial value and can be controlled using a set of parameters (Elaydi, 1999). A chaotic map cm representing a discrete-time dynamical system has the generic form as follows:

$$cm_{k+1} = f(cm_k), \quad k = 1, 2, \ldots \tag{10.108}$$

There are many chaotic maps used in chaotic search such as logistic map, tent map, sinusoidal iterator, circle map, sinus map, Chebyshev map, Gauss map, Henon map, Ikeda map, Liebovtech map, and Zaslavskii map (Peitgen et al., 1992). Some of these chaotic maps are provided in Appendix C.

Talatahari et al. (2011) used 10 different chaotic maps such as logistic map, tent map, sinusoidal iterator, circle map, sinus map, Gauss map, Henon map, Ikeda map, Liebovtech map, and Zaslavskii map and incorporated those chaotic maps into the standard CSS in a number of ways in the proposed CCSS algorithms. For example, generating initial positions of CPs using chaotic maps, the attractive or repulsive force is determined using chaotic ar_{ij} (in Equation 10.101), the probability p_{ij} (in Equation 10.101) is determined using cm, and both ar_{ij} and p_{ij} (in Equation 10.101) are determined using cm. In Equation 10.103, they replaced the acceleration parameter k_a and random value ρ_2 with cm yielding Equation 10.109, replaced velocity parameter k_v and random value ρ_1 with cm yielding Equation 10.110, and replaced $\{k_a, \rho_2\}$ and $\{k_v, \rho_1\}$ with selected cm yielding Equation 10.111 for position updates.

$$x_i^{\text{new}} = x_i^{\text{old}} + \rho_1 \cdot k_v \cdot v_i^{\text{old}} \cdot \Delta t + \underbrace{\text{cm}}_{\rho_2 \cdot k_a} \cdot \frac{F_i}{m_i} \cdot \Delta t^2 \tag{10.109}$$

$$x_i^{\text{new}} = x_i^{\text{old}} + \underbrace{\text{cm}}_{\rho_1 \cdot k_v} \cdot v_i^{\text{old}} \cdot \Delta t + \rho_2 \cdot k_a \cdot \frac{F_i}{m_i} \cdot \Delta t^2 \tag{10.110}$$

$$x_i^{\text{new}} = x_i^{\text{old}} + \underbrace{\text{cm}}_{\rho_1 \cdot k_v} \cdot v_i^{\text{old}} \cdot \Delta t + \underbrace{\text{cm}}_{\rho_2 \cdot k_a} \cdot \frac{F_i}{m_i} \cdot \Delta t^2 \tag{10.111}$$

They also investigated the combination of chaotic maps for $\{ar_{ij}, p_{ij}\}$ (in Equation 10.101) together with chaotic maps for $\{k_a, \rho_2\}$ and $\{k_v, \rho_1\}$ in Equation 10.102. Finally, they proposed a combination of chaotic initial CPs with chaotic values for $\{ar_{ij}, p_{ij}, \rho_1, \rho_2\}$. The steps of CCSS algorithm are the same as the standard CSS except the positions are calculated using Equations 10.109 through 10.111 in Step 5. The proposed CCSS approach was tested on unimodal and multimodal optimization benchmark problems such as Griewank, Rosenbrock, Rastring, and Ackley functions by applying all chaotic maps mentioned earlier. Experimental results show that Tent, Sinus, Liebovtech, Gauss, Ikeda, Sinusoidal, Zaslavski, and Circle maps are comparatively better maps for the CCSS methods.

In general, CSS algorithm uses arbitrary, fixed, or random values for the parameters k_a, k_v, k_{ar}, ar_{ij}, and p_{ij} in Equations 10.100 through 10.103. These are the key factors that control the performance of CSS algorithm. There is no definite and reliable approach to determine the suitable values for these parameters. In order to improve the performance, Talatahari et al. (2012b) proposed the use of chaotic maps for the parameters ar_{ij}, k_{ar}, p_{ij}, k_a, k_v, ρ_1, and ρ_2. The proposed CCSS approach has been tested on benchmark engineering optimization problems, namely, piston lever, welded beam, four-story, two-bay frame, design of car side impact, and design of a 4-step-cone pulley. Experimental results show that logistic, Sinusoidal, Tent, Liebovtech, and Gauss maps are comparatively better maps for the CCSS methods. Nouhi et al. (2013) investigated the performance of different combinations of CCSS algorithms and 10 chaotic maps resulting in 90 different methods. These methods are tested on benchmark mathematical functions and constraint design examples. The simulation results are compared to recognize the most efficient and powerful methods.

10.9.4 ADAPTIVE CSS

Niknam et al. (2013) proposed two modifications to the computation of force in Equation 10.101 in the pursuit of improving the performance of CSS, which leads to the self-adaptive CSS (SCSS) method based on the CSS algorithm and some kinds of modification termed as a reformation technique. In SCSS, a certain number of CPs with the best fitness values are allowed to influence other CPs. That is, the N_{best} particles (solutions) are stored in CM and are used in the computation of force. The force equation in (10.101) is modified with two new features that include N_{best} particles and sign of force:

$$F_i = k_c q_i \sum_{\substack{j=1, \\ j \neq i}}^{N_{best}} \left(\frac{q_j}{\alpha^3} r_{ij} \cdot w_1 + \frac{q_j}{r_{ij}^2} \cdot w_2 \right) p_{ij} \cdot \mathrm{sgn} \cdot (x_j - x_i), \quad \begin{cases} w_1 = 1, w_2 = 0 \ \text{ if } r_{ij} < \alpha \\ w_1 = 0, w_2 = 1 \ \text{ if } r_{ij} \geq \alpha \end{cases} \quad (10.112)$$

The value of N_{best} starts with the population size N and terminates with a fraction of N. N_{best} is defined as

$$N_{best} = \mathrm{round}\left[(0.2 * N - N) \cdot \frac{t}{t_{max}} + N \right] \quad (10.113)$$

where round[.] is a function that rounds off values to the nearest integer. ar_{ij} in Equation 10.101 is replaced with sgn that determines the type of force (attractive or repulsive), which is defined by

$$\mathrm{sgn} = \begin{cases} +1 & \text{if } (\mathrm{rand} \leq pr) \\ -1 & \text{else} \end{cases} \quad (10.114)$$

where +1 enforces attractive force and −1 enforces repulsive force. pr is a constant value arbitrarily chosen by the user that controls the type of force to be applied. Niknam et al. (2013) have chosen a value around 0.8 for pr in their experiments.

Using the force calculated in Equation 10.112, solutions are computed by updating positions of CPs. In SCSS, two new updating methods are proposed as reformation. The first method saves the best solutions (x_{best}) found by the population in the past iterations. The other method improves the diversity of the solutions by incorporating mutation technique that helps in escaping local optima. The first method computes the reform of x_i defined as

$$x_i^{reform} = x_i^{new} + \mathrm{rand} \cdot (x_{best} - \lceil (1 + \mathrm{rand}) \rceil \cdot \mathrm{Mean}) \quad (10.115)$$

where $\lceil \cdot \rceil$ denotes the rounding off the value to ceiling, rand is a random value, and Mean[*] is the mean of the solutions.

The second method computes a trial solution for x_j by mutating five randomly selected CPs as follows:

$$x_i^{\text{trial}} = x_{n1} + \text{rand}_1 \cdot (x_{n2} - x_{n3}) + \text{rand}_2 \cdot (x_{n4} - x_{n5}) \tag{10.116}$$

where rand_1 and rand_2 are two random numbers.

The performance of the methods is evaluated by a probability model, which is based on the ability of the methods to provide more optimal solutions. A reformed solution is obtained based on the values of rand_1 and rand_2 as follows:

$$x_{i,\theta}^{\text{reform}} = \begin{cases} x_{i,\theta}^{\text{trial}} & \text{if } (\text{rand}_1 \leq \text{rand}_2) \\ x_{i,\theta}^{\text{new}} & \text{else} \end{cases}, \quad \theta = 1, 2, \ldots, n \tag{10.117}$$

Finally, from the solutions x_i^{reform} and x_i^{new}, the solution with higher fitness value is selected. Using the proposed self-adaptive reformation, in each stage of the optimization process, the algorithm self-adaptively recognizes which updating method is more beneficial to focus on. The effectiveness of the proposed CCSS approach is verified on a multiple-distributed generation–microgrids (DG–MG) in the grid-connected mode.

Talatahari et al. (2012a,c) proposed an adaptive CSS method utilizing the notion of suboptimization mechanism (SOM) based on the principles of finite element method, where the search space is divided into many subdomains and perform the optimization process. The low fit subdomains are removed and the remaining domains are divided again. The process continues until an acceptable solution is reached as per the criterion of optimality.

10.9.5   MAGNETIC CSS

The standard CSS algorithm is based on the Coulomb and Gauss laws, which describe the electric field at a point inside and outside a charged insulating solid sphere shown in Figure 10.5. The forces expressed by Equation 10.90 do not take into account of the magnetic force. It was H.C. Oersted who discovered by accident a new phenomenon (i.e., magnetic field) caused by steady moving charges (i.e., electric current) in wires in 1820 (Oersted, 1820). Shortly after Oersted's finding, Andre-Marie Ampere published his findings on current flowing in wires exerts forces on one another (Ampere, 1820). In the same year of 1820, J.-B. Biot and F. Savart repeated Oersted's experiments and formulated a compact law of static magnetic fields generated by current in a circuit, which later became known as Biot–Savart law (Biot and Savart, 1820). The Biot–Savart law is the most basic law of magnetostatics. It describes how the magnetic field **B** at a given point P is produced by the moving charges (i.e., current) in the neighborhood of that point as shown in Figure 10.6. Biot–Savart law defines the magnitude of magnetic field at any point of the space in terms of the electric current that produces the field.

$$d\mathbf{B} = \frac{\mu_0}{4\pi} \frac{Ids \times \hat{r}}{r^2} \tag{10.118}$$

where μ_0 is the permeability constant of free space, I is the current, Ids is the differential current element, $\hat{r}$ is the unit vector pointing from the location of the current element to the field point, and r is the distance between point P and ds. The cross product ($\times$) indicates that $d\mathbf{B}$ is perpendicular to both ds and $\hat{r}$.

---

[*] The authors of the adaptive CSS in Niknam et al. (2013) did not explain the meaning of this term properly. Also probabilistic model that decides the final solution remains ambiguous.

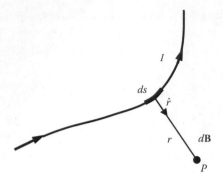

FIGURE 10.6   Magnetic field $d\mathbf{B}$ at point P.

The magnetic field at a point outside the wire with radius α and flowing current I, as shown in Figure 10.7, can be defined using the Biot–Savart law as

$$\mathbf{B}_i = \frac{\mu_0}{2\pi}\frac{I}{r}, \quad \text{when } r \geq \alpha \tag{10.119}$$

where $\mathbf{B}_i$ is the magnetic field of wire i. The magnetic field at a point inside the wire can be defined using the Ampere's law as

$$\mathbf{B}_i = \left(\frac{\mu_0}{2\pi}\frac{I}{\alpha^2}\right)\times r, \quad \text{when } r < \alpha \tag{10.120}$$

For a number of wires ($i = 1, 2, \ldots, n$), the sum of magnetic field is calculated as

$$\mathbf{B} = \sum_{i=1}^{n} \mathbf{B}_i \tag{10.121}$$

The differential current element Ids is equivalent to charge q at a velocity v. The presence of a magnetic field $\mathbf{B}$ would experience a magnetic force F_B on a CP given by

$$F_B = q \cdot v \times \mathbf{B} \tag{10.122}$$

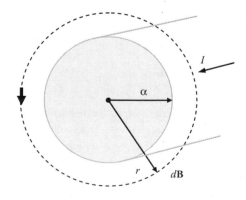

FIGURE 10.7   Magnetic field of wire with flowing current.

The magnitude and direction of the magnetic force F_B depend on the velocity of the particle and magnitude and direction of the magnetic field $\mathbf{B}$. The force F_E considering the electric field computed in Equation 10.91 and the magnetic force F_B are now combined to contribute to the total force, also known as Lorentz force (Inan and Inan, 1998). The Lorentz force acting on a particle with charge q moving at a velocity v in the presence of magnetic field $\mathbf{B}$ and electric field $\mathbf{E}$ is the vector sum of the magnetic and electric forces given by

$$F = F_B + F_E = qv \times \mathbf{B} + q\mathbf{E} \tag{10.123}$$

By considering the population of the CPs, the resultant magnetic force acting on each CP in CSS is calculated as follows:

$$F_{Bi} = k_c q_i \sum_{\substack{j=1, \\ j \neq i}}^{N} \left(\frac{I_j}{\alpha^2} r_{ij} \cdot w_1 + \frac{I_j}{r_{ij}} \cdot w_2 \right) pm_{ij} (x_j - x_i), \quad \begin{cases} w_1 = 1, w_2 = 0 & \text{if } r_{ij} < \alpha \\ w_1 = 0, w_2 = 1 & \text{if } r_{ij} \geq \alpha \end{cases} \tag{10.124}$$

where I_j is the average current for the charge q_i passing through the cross section per unit time, pm_{ij} is the probability for the moving CP under magnetic field, α is the radius of the virtual wire as shown in Figure 10.7, and the separation distance r_{ij} means the distance between the wire (virtual CP) and the CP defined in the same way as Equation 10.97. Two definitions of the electric current I_i are proposed for magnetic CSS.

$$I_i(k) = \frac{q_i(k) - q_i(k-1)}{q_i(k) + \varepsilon} \tag{10.125}$$

$$I_i(k) = \text{sgn}[df_i(k)] \times \frac{|df_i(k)| - df_{\min}(k)}{df_{\max}(k) - df_{\min}(k)} \tag{10.126}$$

$$df_i(k) = f_i(k) - f_i(k-1) \tag{10.127}$$

where $df_i(k)$ is the variation of the objective function value of the i-th CP and $f_i(k)$ is the fitness function value at k-th iteration.

pm_{ij} is the probability of the magnetic influence (attracting or repelling) of the i-th wire (virtual CP) on the j-th CP, which is slightly different from the probability of electric influence defined in Equation 10.98. pm_{ij} is defined as

$$pm_{ij} = \begin{cases} 1 & \text{if } f_i > f_j \\ 0 & \text{else} \end{cases} \tag{10.128}$$

where f_i and f_j are the fitness values of the i-th and j-th CP.

The total force F is now accommodated into the standard CSS algorithm and a new variant of CSS is proposed by Kaveh et al. (2013b) called magnetic CSS (MCSS). The Lorentz force acting on j-th CP is now modified in MCSS.

$$\begin{aligned} F &= F_{Bi} + F_{Ei} \\ &= k_c q_i \sum_{\substack{j=1 \\ j \neq i}}^{N} \left[\left\{ \left(\frac{I_j}{\alpha^2} r_{ij} \cdot w_1 + \frac{I_j}{r_{ij}} \cdot w_2 \right) pm_{ij} + \left(\frac{q_j}{\alpha^3} r_{ij} \cdot w_1 + \frac{q_j}{r_{ij}^2} \cdot w_2 \right) p_{ij} \right\} (x_j - x_i) \right], \end{aligned} \tag{10.129}$$

$$\begin{cases} w_1 = 1, w_2 = 0 & \text{if } r_{ij} < \alpha \\ w_1 = 0, w_2 = 1 & \text{if } r_{ij} \geq \alpha \end{cases}, \quad i = 1, 2, \ldots, N$$

In the early iterations of the MCSS, the agents are far from each other. This results in small magnetic force, which causes the CPs to move slowly exploring the search space. This is an ideal condition for optimization problems for global search ability of the CPs. Over the iterations, CPs move close to each other and gather in small space when a local search is expected to start. Due to decreasing distances, the magnetic force becomes large, causing fast movement of CPs and resulting in poor exploitation of the search space. One of the solutions that can be proposed is that when the distances are relatively small, the magnetic force should be computed using the linear formulation of magnetic field in Equation 10.120. This means that the formulation of the magnetic force for global and local search phases inherent in Equation 10.124 should be separated. A suitable value for α in Equation 10.124 can be unity. The other possible solution is to drop the first term in Equation 10.124 considering the virtual wire as a point (not setting the value of α to 0 in the equation). All of the magnetic fields produced by virtual wires can be calculated based on Equation 10.119. This means when the CPs come closer together in a small search space, the movements become slow, which is the idea for local search for exploitation. Thus, both x_i-x_j and I_i become small. When considering Equation 10.124 for the calculation of the magnetic force, it is to be noted that a large value is multiplied by two small values, which leads the final value of the magnetic force to be a normal value. This brings some advantages to the MCSS algorithm. Due to the ease of implementation and better convergence rate, the second solution is suggested by Kaveh et al. (2013b). The Lorentz force in Equation 10.129 now becomes

$$F = k_c q_i \sum_{\substack{j=1 \\ j \neq i}}^{N} \left\{ \frac{I_j}{r_{ij}} \cdot pm_{ij} + \left(\frac{q_j}{\alpha^3} r_{ij} \cdot w_1 + \frac{q_j}{r_{ij}^2} \cdot w_2 \right) p_{ij} \right\} (x_j - x_i),$$

$$\begin{cases} w_1 = 1, w_2 = 0 & \text{if } r_{ij} < \alpha \\ w_1 = 0, w_2 = 1 & \text{if } r_{ij} \geq \alpha \end{cases}, \quad i = 1, 2, \ldots, N$$

(10.130)

The steps of MCSS algorithm are the same as the standard CSS except the force equation (10.130) along with the necessary computation of I_j and pm_{ij} (in Step 4) is used instead of Equation 10.101. The effectiveness and efficiency of the MCSS are verified by applying the algorithm to 18 well-known benchmark functions. MCSS algorithm has been applied to structural engineering design optimization problems by Kaveh and Zolghadr (2014). The experimental results are compared to those of the standard CSS and other meta-heuristic algorithms. In all cases, MCSS showed competitive performance and demonstrated improvements over the standard CSS.

A number of improved versions of MCSS have also been reported in the literature, which incorporated some new features to apply the MCSS algorithm to optimal design of double-layer barrel vaults, shape–size optimization of single-layer barrel vaults, and optimization of truss structures (Kaveh et al., 2014a,b, 2015).

10.9.6 Hybrid CSS

A hybridization of the CSS and the BB–BC algorithm (discussed earlier in this chapter) with trap recognition capability has been proposed by Kaveh and Zolghadr (2012). A diversity index is computed at every iteration. When trapped in local minima, a disturbance is produced using the Big Bang operator of the BB–BC algorithm to help particles moving out of the trap. The steps of the hybrid CSS-BB–BC algorithm are the same as the standard CSS. The performance and effectiveness of the hybrid approach have been verified on truss optimization problem with natural frequency constraints where the objective is to minimize the weight of the structure while satisfying multiple constraints on natural frequencies.

10.9.7   APPLICATIONS

Since its inception in 2010, the CSS algorithm has found diverse applications in various domains. The diverse applications can be categorized as benchmark functions, engineering design problem, structural design optimization, optimization of composite floor system, economic power dispatch problem, data clustering, fuzzy logic control design, and multiobjective optimization problems. The majority of applications of CSS algorithm are found in the structural design optimization such as truss structures, frame structures, skeletal structures, reinforced concrete 3-D structures, geometry and topology optimization, grillage systems design, seismic structural design, and seismic design of steel frames to damage detection of structures.

Firstly, the effectiveness and performance of the algorithm have been verified on a number of benchmark optimization functions by Kaveh and Talatahari (2010b) followed by other researchers (Nouhi et al., 2013).

Kaveh and Talatahari (2010b) investigated the application of CSS algorithm to engineering design problems, which have been previously solved using a variety of other techniques to demonstrate the validity and effectiveness of CSS algorithm. Experimental results based on tension/compression spring design problem, welded beam design problem, and pressure vessel design problem show competitive performance of the CSS algorithm compared to other methods.

Structural optimization for solving shape and sizing optimization of trusses with multiple frequency constraints is a highly nonlinear problem, which has been considered as a benchmark problem in structural design optimization and used for testing the effectiveness and performance of many optimization algorithms. Mathematical description of truss structures optimization problem is presented in Chapter 5. Possibly, it is the first engineering application of CSS algorithm to optimal design of skeletal structures by Kaveh and Talatahari (2010c). Three truss and two frame structures are chosen as example problems. A population of 20 CPs for the first truss examples and 50 CPs for the second truss examples are selected for the simulation. The experimental results are then compared to the solutions of other advanced heuristic methods to demonstrate the efficiency of this approach. Kaveh and Talatahari (2010e) applied a discrete version of CSS to optimal design of truss examples with discrete variables, namely, 52-bar planar truss, 72-bar spatial truss, and 354-bar dome-shaped truss, which again demonstrate the effectiveness and robustness of the proposed CSS algorithm compared to other meta-heuristics. The simulation results reveal that the CSS performs better than the others. Also, the convergence capability of the CSS method outperformed those of the other approaches.

Grillage systems are widely used in structures to cover large areas in bridge decks, ship hulls, and floors. The optimal design requires the determination of the cross-sectional properties of the longitudinal and transverse beams so that the weight of the system is minimal satisfying the behavioral and performance limitations. Kaveh and Talatahari (2010d) applied CSS algorithm to grillage system with a population of 20 CPs and obtained the optimal design for the grillage system within 250 iterations.

Kaveh and Talatahari (2012) used a discrete version of the CSS for the optimal design of 3-bay 15-story, 24-story planar frame and 290-member 10-story space frame structures. The CSS algorithm required approximately 5500 frame analyses to converge to a solution, which is significantly less than the 15,500 analyses required by the standard ACO, 13,924 analyses required by HS, and 7500 analyses required by ICA. In plastic analysis of frame structures, it is to find the collapse load factor of the frame structure by checking every possible combination of its elementary mechanisms. Eventually, this leads to a search procedure. As the structure becomes more complex, the search procedure becomes extensive consuming huge computation time. This situation requires the use of suitable meta-heuristic approach capable of finding solutions as fast and accurate as possible. Kaveh et al. (2013a) employed CSS algorithm in combination with ACO to optimize the process of finding the collapse load factor of planar frames. The hybrid approach has been applied to three-bay three-story frame, four-bay four-story frame, pitched roof frame, and four-story frame to verify the effectiveness and performance of the algorithm. Simulation results show that CSS converges to the solution in a more monotonic manner resulting in the exact load factor of the studied frame structures. Different methods have been used for the optimal design of reinforced concrete 3-D structures over the past. Kaveh and Behnam (2013) proposed the use of CSS for the design optimization of reinforced concrete 3-D structures considering frequency constraints. Analysis of the structures is performed by the standard stiffness method. The objective function comprises the weight of the structure and constraints consist of the slenderness of compression members, the maximum allowable drift, and the natural frequency of the structure. The effectiveness and performance of CSS-based optimization were verified on a seven-story frame with three spans. The sensitivity analysis is carried out on the optimal design of nine frames having three stories and two spans by varying span lengths and loading conditions. The simulation results demonstrate the effectiveness and simplified process of optimization of 3-D structures compared to other methods, which is proved to be useful for future industrial buildings.

An improved CSS algorithm proposed by Kaveh and Zolghadr (2015) has been applied to damage detection of truss structures using changes in natural frequencies and mode shapes. Kaveh et al. (2015) applied an improved MCSS to the optimization of truss structures with continuous and discrete variables where harmony search was utilized to improve the most effective parameters in the convergence rate of the algorithm.

Designing optimal dome structures is very challenging as they enclose large spanned structures without intermediate supports. The optimal design requires finding the optimal sections for elements (i.e., optimal size), optimal height for the crown (i.e., geometry optimization), and the optimal number of elements (i.e., topology optimization) under determined loading conditions. Kaveh and Talatahari (2011a) applied CSS algorithm to find the optimal design of geodesic domes. In order to investigate the efficiency of the CSS algorithm, two different loading conditions are considered: uniform loading and multiple loading conditions. Kaveh and Talatahari (2011b) applied CSS algorithm to the optimization of configuration design of truss structures, that is, finding the optimal shape of the structure for a topology. The decision variables are the coordinates of certain nodes of the truss and the size of variables for different members. To verify the performance of the CSS algorithm, three example problems from the configuration optimization are considered: an 18-bar planar truss, a 25-bar dome-shaped truss, and a 120-bar dome-shaped truss. To enhance the performance of CSS algorithm, the concept of "fields of forces" is utilized in this application. Kaveh and Ahmadi (2013) proposed to apply the CSS algorithm for simultaneous analysis, design, and optimization of large-scale structures by introducing supervisor agents to enhance the exploration ability of the CSS algorithm leading to a new algorithm called supervised CSS. The objective function is defined using minimum energy methods and members with predefined stress ratios instead of the direct solution of classic equations, which requires the inversion of large matrices. The proposed approach is applied to the design and analysis of some planar and space structures. Simulation results demonstrate the effectiveness and competitive performance of the supervised CSS algorithm over other methods.

In seismic structural design of buildings, it is desirable to reach a proposed service-ability level with the least usage of the material. Performance level is the required behavior of a structure in different situations. Kaveh and Nasrollahi (2014) applied CSS algorithm to performance-based seismic design of steel frames. Two numerical examples have been investigated and the results illustrate significant improvement in structural weight compared to the conventional design methods. Gravity walls derive their capacity to resist lateral loads through dead weight of the wall. The optimal design of a concrete cantilever retaining wall is to be determined by the minimum of the costs of concrete, steel reinforcement, and form working. An enhanced version of CSS has been applied to optimum cost for designing gravity and reinforced cantilever retaining walls by Talatahari and Sheikholeslami (2014). The results are compared to the solution of the PSO, BB–BC algorithm, and heuristic BB–BC methods to demonstrate the efficiency of the present approach.

Apart from the structural design optimization problems, there are many other applications of CSS algorithm reported in the literature. Kaveh and Behnam (2012) applied CSS to cost optimization of a composite floor system. Özyön et al. (2012) used CSS algorithm for the optimization of the emission constrained economic power dispatch problem. Kumar and Sahool (2014, 2015) applied CSS and MCSS and PSO algorithm for data clustering problems. Precup et al. (2014) applied adaptive CSS for tuning of fuzzy controllers. Kaveh and Laknejadi (2011) applied a hybrid CSS and PSO algorithm for multiobjective optimization problems. In all cases, effectiveness and efficiency of the algorithm are evident and the performance of CSS algorithm is competitive.

10.10 HYSTERETIC OPTIMIZATION

10.10.1 INSPIRATION AND ALGORITHM

Magnetization is the process of ordering ferromagnetic material with its unbalanced magnetic dipole moments that are inherent in the material under the external magnetic field. Demagnetization is the reverse process, which reduces or eliminates the magnetization. Demagnetization of magnets can be done in two ways. One way is to heat the magnet above the Curie temperature.[*] The thermal energy can overcome the exchange interaction and destroy the ferromagnetic order. The other way is to pull it out of an electric coil with alternating current running through it, which gives rise to the magnetic fields, and oppose the magnetization. When applied alternating magnetic field to the material, its magnetization will trace out a loop called a hysteretic loop, as illustrated in Figure 10.8. It is traditional to plot the magnetization M of the sample as a function of the magnetic field intensity H. H is a measure of the externally applied field, which drives the magnetization process. Figure 10.8 shows the hysteresis as a plot of magnetization M over the intensity of magnetic field H. The magnetization curve is the property called hysteresis and it is related to the existence of magnetic domains in the material (Lin et al., 2013). It clearly shows that it takes some energy for reoriented magnetic materials to demagnetize.

When an oscillating magnetic field with slowly decreasing amplitude is applied, the magnetic sample comes to a very stable low-energy state called ground state in the process of demagnetization. To improve low-energy state further, a technique called shake-up is used repeatedly after demagnetization. Zaránd et al. (2002) proposed a method of optimization inspired from these two processes of demagnetization. This is a process similar to SA where the material achieves a stable state by slowly decreasing the temperature, that is, finding the ground states of magnetic samples, which is to be compared to finding the optimal point in the search process. The new optimization method is called HO. Zaránd et al. (2002) showed the similarity between the magnetic hysteresis and optimization process using the Ising model.

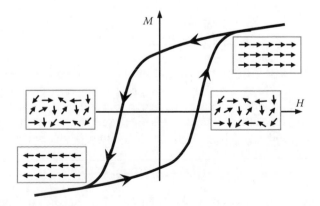

FIGURE 10.8 Hysteresis of magnetization.

---

[*] Curie temperature (T_c) is the temperature at which the permanent magnetism in material changes to induced magnetism.

The Ising model is a mathematical model of ferromagnetism in statistical mechanics (Ising, 1925). It is named after German physicist Ernst Ising. The model consists of discrete variables that represent magnetic dipole moments of atomic spins. Spins can be in one of the two states +1 or −1, that is, $\sigma \in \{+1, -1\}$. Each spin σ_i is assumed to have a random bond J_{ij} with other spin σ_j, allowing each spin to interact with its neighbors. Taking the summation of all pairs of spins yields

$$H = -\sum_{\{i,j\}}^{N} J_{ij} \sigma_i \sigma_j \tag{10.131}$$

where $\{i,j\}$ denotes the pairs of nearest neighbors.

Each spin σ_i is coupled with an external field of amplitude H with a random sign $\xi_i \in \{\pm 1\}$, which can be adjusted during a single demagnetization run. The energy of a configuration σ is given by the Hamiltonian function:

$$\Phi(\sigma) = -\frac{1}{2} \sum_{\{i,j\}}^{N} J_{ij} \sigma_i \sigma_j - H \sum_{i=1}^{N} \xi_i \sigma_i \tag{10.132}$$

The second term in Equation 10.132 is the external demagnetizing field applied to minimize the energy. H is the appropriately oscillating field strength. N is the number of spins, that is, the size of the system. $\xi_i \in \{\pm 1\}$ defines the direction of the field, which is chosen randomly for each i, and it is fixed during each demagnetizing cycle (either a full cycle or a shake-up).

The sign term ξ_i in Equation 10.132 should actually be positive because the electron's magnetic moment is antiparallel to its spin, but the negative term is used conventionally. The minus sign on each term of the Hamiltonian function $\Phi(\sigma)$ is conventional. Using this sign convention, the Ising models can be classified according to the sign of the interaction, that is, for all pairs $\{i,j\}$ the interaction is

$$\begin{cases} \text{Ferromagnetic} & \text{if } J_{ij} > 0 \\ \text{Antiferromagnetic} & \text{if } J_{ij} < 0 \\ \text{Noninteracting} & \text{if } J_{ij} = 0 \end{cases} \tag{10.133}$$

otherwise, the system is called nonferromagnetic.

The demagnetization process starts with $\sigma_i = \xi_i$ aligned with the external field and high value of H. The value of H is decreased slowly until a value of H_s, when a spin becomes unstable. The spin is flipped at $H = H_s$, that is, if $\xi_i = \pm 1$, then change it to $\xi_i = \mp 1$. Flipping this spin makes further spins that were interacting with it within the same unstable field. If there are multiple spins to be flipped at the same time, a random selection is made. Again flipping those spins will cause a whole avalanche. All the unstable spins are flipped one by one until all spins are stable. The value of H is then decreased again slowly until the next spin becomes unstable. Flipping the spin may cause another avalanche. The value of H at which an avalanche occurred can directly be calculated. The process of demagnetization continues; the field value of H is decreased at every avalanche starting from the saturation field value of H_s and gradually is decreased to a value of $-\gamma H_0$. H is decreased according to

$$H_{n+1} = -\gamma_n H_n \tag{10.134}$$

where $0 < \gamma_n < 1$ is the amplitude reduction factor and γ is a parameter of the algorithm.

The average quality of results depends on the reduction parameter γ. The choice of appropriate value of γ is critical. A value closer to 1 seems to provide better results. Empirically found that $\gamma \in [0.9, 0.95]$ seems to provide good results (Pal, 2003, 2006a). H_0 is the maximum amplitude and it is a predetermined value. A value $H_0 \sim H_s$ is appropriate. Once it reaches the minimum, the field is then increased again up to $+\gamma^2 H_0$, then decreased again to $-\gamma^3 H_0$. The process is repeated a number of times until amplitude becomes so small that no more spin flip occurs, that is, the state does not change any more. Every time it goes through the zero fields, the energy is compared and configuration is saved if it is better than the best one found before. It is very often the case that the final configuration of the cycle is not always as good as the previous ones. Another demagnetization cycle called shake-up is used. A good balance between exploration and exploitation is required by any optimization algorithm. The demagnetization process is an explorative measure. Better solutions are searched using an exploitative measure by performing a number of shake-up operations. During the shake-up, it starts from the zero field and goes to the maximum amplitude of $H_{\text{shake-up}} \ll H_s$. Hence, $H_{\text{shake-up}}$ is considered as another parameter of the algorithm. The main steps of HO algorithm are as follows:

Step 1: Set $H = H_0$ large enough such that $\sigma_i = \xi_i$, $i = 1, 2, \ldots, N$
       Set $E_{\min} = H$
Step 2: Decrease H until spin becomes unstable
       If $(\Phi(\sigma) < E_{\min})$
          $E_{\min} = \Phi(\sigma)$
Step 3: Flip all unstable spins (making an avalanche)
Step 4: Reverse direction of H at each $H = H_{n+1} = -\gamma_n H_n$ for $0 < \gamma_n < 1$
Step 5: Stop when amplitude $|H_n| < H_{\min}$
Step 6: Restart algorithm (from Step 1) with new random set of ξ_i for N_{run} times
Step 7: Return solution $E_{\min}$

10.10.2 APPLICATIONS

Though HO is a recent addition to meta-heuristic algorithms, the HO algorithm has found a significant number of applications in various domains. HO algorithm has been applied to TSP (Pal, 2003; Zaránd et al., 2002), spin glass problem (Goncalves and Boettcher, 2008; Pal, 2006b), vehicle routing problem (Yan and Wu, 2012), and protein folding (Zha et al., 2010).

10.11 COLLIDING BODIES OPTIMIZATION

10.11.1 INSPIRATION AND ALGORITHM

According to Newtonian mechanics, an impulse of force causes a change in momentum of a mass. Momentum is defined as the product of mass and its velocity, that is, mv, where m is the mass and v is its velocity. The impulse–momentum equation is useful in dealing with interaction or collision between masses. When two moving bodies collide, the resulting changes in the momentum depend on their masses, elastic properties, and initial velocities. When two moving masses m_1 and m_2 having velocity v_1 and v_2, respectively, collide in an isolated system, as shown in Figure 10.9, it will result in the change of velocities v'_1 and v'_2, respectively.

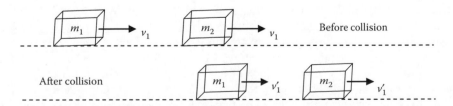

FIGURE 10.9 Collision between two bodies.

The total momentum of the two masses before the collision will be equal to the total momentum of the two masses after the collision. According to the law of conservation of momentum, the momentum of the two masses can be expressed as

$$m_1 v_1 + m_2 v_2 = m_1 v_1' + m_2 v_2' \tag{10.135}$$

This law holds no matter how complicated the force is between masses. Similarly, if there are several masses, the momentum exchanged between each pair of masses adds up to zero, so the total change in momentum is zero.

At the moment of collision, all energy is kinetic. The total energy before the collision should be equal to the total energy after the collision. Some energy is lost through heat, sound, and vibration waves during the collision. The conservation of the energy is thus expressed as

$$\frac{1}{2} m_1 v_1^2 + \frac{1}{2} m_2 v_2^2 = \frac{1}{2} m_1 v_1'^2 + \frac{1}{2} m_2 v_2'^2 + Q \tag{10.136}$$

where Q is the loss of kinetic energy. If there is no loss of energy, then it can be shown that the velocity difference $(v_1 - v_2)$ before collision is equal to velocity difference $(v_2' - v_1')$ after collision. Isaac Newton found that there is always some loss of energy and showed by experiment that $(v_2' - v_1')$ is always less than $(v_1 - v_2)$. That is, the relation between the velocity differences can be established by

$$(v_2' - v_1') = \varepsilon(v_1 - v_2) \tag{10.137}$$

where ε is the coefficient of restitution (COR) of the two colliding bodies. COR is defined as the ratio of relative velocity of separation to relative velocity as follows:

$$\varepsilon = \frac{|v_2' - v_1'|}{|v_1 - v_2|} = \frac{v'}{v} \tag{10.138}$$

The value of ε is between [0,1] and depends on the material properties of the body (i.e., mass). In other words, the momentum lost by Body 1 is equal to the momentum gained by Body 2.

When the velocities v_1 and v_2 and value of ε are known, the final velocities v'_1 and v'_2 can be calculated using Equations 10.135 and 10.138:

$$v_1' = \frac{(m_1 - \varepsilon m_2)v_1 + (m_2 + \varepsilon m_2)v_2}{m_1 + m_2} \tag{10.139}$$

$$v_2' = \frac{(m_2 - \varepsilon m_1)v_2 + (m_1 + \varepsilon m_1)v_1}{m_1 + m_2} \tag{10.140}$$

Two kinds of collision can happen depending on the material properties of the bodies: elastic collision and inelastic collision. In an elastic collision, there is no loss of kinetic energy, which means $Q = 0$ and $\varepsilon = 1$. But macroscopic collision between two masses will convert some kinetic energy into internal energy and other forms of energy. In this case, the velocity of separation is high after collision. In an inelastic collision, kinetic energy is converted into some other forms of energy, for example, heat, sound, and vibration, which means $Q \neq 0$ and $\varepsilon \leq 1$. In this case, the velocity of separation is low after collision. For most of the collisions of real masses, the COR is $\varepsilon \in [0,1]$.

Collision can occur between two bodies in motion or one body in motion with another stationary body. In all cases, law of conservation of energy is preserved resulting in the change of velocity or

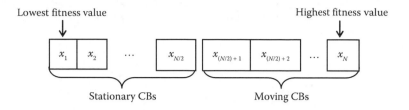

FIGURE 10.10 CBs arranged in ascending order.

change position of bodies. Collision between bodies results in some kind of changes: (i) improve or change the position of the moving body and (ii) improve or change the position of the stationary body.

Through a process of continuous collision between bodies, some bodies will achieve better positions according to an optimality criterion defined by an objective function. The process continues for some time (i.e., for number of iterations) until the solutions meet the desired accuracy or condition. On the basis of this simple principle, Kaveh and Mahdavi (2014a) proposed CBO. In CBO, candidate solutions consist of problem variables and are considered as colliding body (CB). The population of CBs is divided into two equal groups: stationary and moving. Collision occurs between pairs of bodies; that is, moving bodies collide with stationary bodies. This changes the position of the moving bodies and moves the stationary bodies toward better positions. Positions are updated after every collision using the new velocities calculated by Equations 10.139 and 10.140. In this process of optimization, the masses of the bodies are defined in terms of fitness function values of the solutions.

In CBO, a position x_i is a representation of a candidate solution comprising problem variables to be optimized. A population of initial positions of CBs is generated randomly, which are distributed uniformly over the whole search space.

$$x_i = x_i^{\min} + \text{rand}\left(x_i^{\max} - x_i^{\min}\right) \quad i = 1, 2, \ldots, N \tag{10.141}$$

where $x_i^{\min}$ and $x_i^{\max}$ are the minimum and maximum allowable values of variables, respectively, rand is a random number uniformly distributed over the interval [0,1], and N is the number of CBs, that is, size of the population.

A fitness or objective function is defined to assess the quality or fitness of the solution. The fitness of the candidate solution x_i is denoted as $f(x_i)$. In order to run the optimization process, a suitable fitness function $f(\cdot)$ is to be defined, which is very much problem dependent that incorporates different performance indices of the problem at hand. The CBs then undergo evaluation using the fitness function and arranged in ascending order of their fitness values, as shown in Figure 10.10.

The CBs are divided into two equal groups. The lower half of the CBs, that is, $\{x_1, \ldots, x_{N/2}\}$, is the stationary CBs. These CBs have lower fitness values. The velocity of the stationary CBs is set to 0.

$$v_i = 0, \quad i = 1, 2, \ldots, N/2 \tag{10.142}$$

The upper half of the CBs, that is, $\{x_{(N/2)+1}, \ldots, x_N\}$, is the moving CBs. These CBs have higher fitness values and they move toward the lower half CBs and collide with them. The collision is illustrated in Figure 10.11. The velocity v_i of the i-th CB can be defined as the change of position x_i over time where the time taken is considered unity.

$$v_i = \frac{x_i - x_{i-(N/2)}}{\Delta t}, \quad i = \frac{N}{2} + 1, \ldots, N \tag{10.143}$$

where $x_{i-N/2}$ is the i-th CB pair position of x_i in the previous group and Δt is unity.

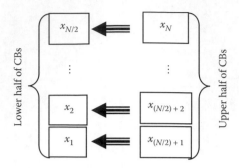

FIGURE 10.11 Collision between upper CBs with lower CBs.

After the collision, the velocity of the moving CBs will change from v_i to v'_i. The new velocities are calculated using Equations 10.139 and 10.140. The velocities of the moving CBs and the stationary CBs are calculated as follows:

$$v'_i = \frac{(m_i - \varepsilon m_{i-N/2})v_i}{m_i + m_{i-N/2}}, \quad i = \frac{N}{2} + 1, \ldots, N \tag{10.144}$$

$$v'_i = \frac{(m_{i+(N/2)} + \varepsilon m_{i+(N/2)})v_{i+(N/2)}}{m_i + m_{i+(N/2)}}, \quad i = 1, \ldots, \frac{N}{2} \tag{10.145}$$

where m_i is the mass of i-th CB, $m_{i-(N/2)}$ is the mass of i-th stationary CB pair, $m_{i+(N/2)}$ is the mass of i-th moving CB pair, and $v_{i+(N/2)}$ is the velocity of the i-th moving CB pair. The COR ε plays an important role in CBO algorithm and controls the balance between exploration and exploitation, that is, controlling the global and local search. In elastic collision, $\varepsilon = 1$. In inelastic collision, COR decreases linearly to 0 and defined as $\varepsilon = 1 - (t/t_{max})$, where t is the iteration number and t_{max} is the maximum iteration.

In CBO, like many other meta-heuristic algorithms, the mass is defined in terms of fitness values. For a minimization problem, the mass is defined as

$$m_i = \frac{1/f(x_i)}{\sum\limits_{i=1}^{N} f(x_i)}, \quad i = 1, 2, \ldots, N \tag{10.146}$$

For a maximization problem, $1/f(x_i)$ is replaced by $f(x_i)$. It appears from the definition that a CB representing a good solution has a larger mass and the larger masses also make fewer moves.

The positions of stationary CBs and moving CBs are updated using the velocity after collision:

$$x_i^{new} = x_{i-(N/2)} + \text{rand} \cdot v'_i \cdot \Delta t, \quad i = \tfrac{N}{2} + 1, \ldots, N \tag{10.147}$$

$$x_i^{new} = x_i + \text{rand} \cdot v'_i \cdot \Delta t \quad i = 1, \ldots, \tfrac{N}{2} \tag{10.148}$$

where x_i^{new} is the new position, $x_{(i-N/2)}$ is the old position of the i-th stationary CB pair, x_i is the old position of the i-th stationary CB, and rand is a random number uniformly distributed over the interval $[-1,1]$. Δt is the time taken to reach from old position to new position, which is considered unity.

The process continues until the position of the CBs, that is, solutions, satisfies a user-defined optimality criterion or reaches the maximum number of iterations. The main steps of the CBO algorithm can be described as follows:

Step 1: Generate population of positions x_i, $i = 1, 2, ..., N$ randomly
Step 2: Calculate mass m_i, $i = 1, 2, ..., N$ of CBs using Equation 10.146
Step 3: Order CBs in ascending values of fitness
      Divide population into equal halves: stationary CBs $\{x_1, ..., x_{N/2}\}$ and moving CBs $\{x_{(N/2)+1}, ..., x_N\}$
      Set initial velocity for stationary CBs and moving CBs
Step 4: Compute velocity of moving CBs using Equations 10.144 and 10.145 after collision
Step 5: Compute new positions of CBs using Equations 10.147 and 10.148
Step 6: If (termination condition not met), Goto Step 2
Step 7: Return solution

The CBO is based on simple formulation and does not involve parameter tuning. The CBO algorithm has been found very effective and good performance has been achieved while applied to benchmark functions and engineering problems. Some modifications to the standard CBO algorithm have also been made to suit applications. A few variants based on applications have been reported in the literature since its inception. Kaveh and Mahdavi (2014b) proposed a discrete version of CBO algorithm to solve discrete problems by introducing a rounding function $F(\cdot)$, which changes the continuous value of solutions to the nearest discrete value as follows:

$$x_{i(\text{discrete})}^{\text{new}} = F\left(x_{i(\text{continuous})}^{\text{new}}\right) \tag{10.149}$$

In general, the standard CBO algorithm does not use any memory to save the position of the CBs from previous iteration. Kaveh and Ghazaan (2014) proposed an enhanced CBO (ECBO) algorithm by introducing a memory element, called colliding memory (CM), to store a number of best solutions (i.e., CB vectors), which helps in improving the convergence and performance of the CBO algorithm without any further computational cost. The solutions saved onto CM replace the same number of worst CBs (i.e., CBs with poor fitness function values) and the CBs are sorted according to their masses in descending order. The steps of the ECBO algorithm are the same as the standard algorithm except the addition of CM in Step 3 of the algorithm. Another modification is made to the ECBO to escape local minima by introducing a parameter pro $\in [0,1]$, which decides whether a component of CB vector should be changed or not. If the mass m_i is less than a threshold value pro, then a variable of x_i is randomly selected and changed as follows:

$$\text{If } (m_i < \text{pro})$$
$$x_i^j = x_{\min}^j + \text{rand}\left(x_{\max}^j - x_{\min}^j\right), \quad i = 1, 2, ..., N$$

where x_i^j is the j-th variable of the i-th CB. The positions are mutated using the simple technique in Step 5 of the CBO algorithm.

Kaveh and Mahdavi (2015a) proposed a two-dimensional CBO algorithm where the direction of the velocities of the CBs after collision is considered. Three features are added to the standard CBO algorithm: (i) memories are added to increase the exploitation ability, (ii) due to the elastic collision formulation, COR ε is set to 1, and (iii) an initial population size of $2N$ is considered. The velocity after collision is calculated as

$$v_i' = \begin{cases} \dfrac{2m_i}{m_i + m_{i+N}} \sin\left(\dfrac{\theta_i}{2}\right) v_i & \text{for} \quad i = 1, \ldots, N \\[4mm] \dfrac{\sqrt{m_{i-N}^2 + m_i^2 + 2m_i m_{i-N}\cos(\theta_{i-N})}}{m_i + m_{i-N}} v_{i-N} & \text{for} \quad i = N+1, \ldots, 2N \end{cases} \tag{10.150}$$

where $\theta_i = \pi - 2\upsilon_i$, $i = 1, \ldots, N$. υ_i is the velocity direction vector of the i-th CB and υ_i is defined as

$$\upsilon_i = \arctan\left(\frac{f(x^{\text{best}}) - f(x_i)}{x^{\text{best}} - x_i + \delta}\right), \quad i = 1, \ldots, N \tag{10.151}$$

where x^{best} is the saved best position of CBs and $f(x^{\text{best}})$ is its corresponding fitness value, respectively, δ is a small positive number to avoid singularities. The steps of the two-dimensional CBO algorithm are the same as the standard CBO except the calculation of the velocity after collision.

As mentioned earlier, standard CBO does not use any memory to preserve the best CBs, but the PSO uses memory to protect the global and local best particles (i.e., CBs in CBO). Kaveh and Mahdavi (2015b) proposed a hybrid CBO–PSO algorithm where the features of preserving the global and local best CBs from PSO are added to CBO, which eventually increases the exploitation power of the CBO algorithm. Three new features are added to the hybrid CBO–PSO approach: (i) the stationary CBs move toward the better positions, (ii) the CBs move toward local best, and (iii) the old velocities are added to the new velocities of CBs before collision.

10.11.2 Applications

CBO algorithm is a very recent addition to the family of meta-heuristic algorithms. It has found a good number of applications to benchmark functions and engineering optimization problems. Apart from the efficiency and robustness of the CBO algorithm, CBO algorithm is advantageous due to the ease of implementation and parameter independence. The application domains can be categorized as benchmark functions, engineering design such as welded beam, cylindrical pressure vessel, tension/compression spring, truss structures, and so on.

Kaveh and Ghazaan (2014) applied ECBO to benchmark unimodal and multimodal functions. Kaveh and Mahdavi (2015a) applied the two-dimensional CBO algorithm to constrained benchmark functions. Simulation results show that CBO algorithm is competitive compared to many other meta-heuristic algorithms.

Hitherto CBO algorithm has been applied mostly to the design optimization of engineering problems and optimization of truss structures by researchers (Kaveh and Mahdavi, 2014a–c, 2015a; Kaveh and Ghazaan, 2014). Kaveh and Mahdavi (2014a) investigated the performance of CBO algorithm on the welded beam, cylindrical pressure vessel capped at both ends by hemispherical heads, minimization of the weight of a tension/compression spring subject to constraints on shear stress, surge frequency, and minimum deflection, weight minimization problem of the 120-bar truss dome, and layout optimization of the forth truss bridge. Kaveh and Mahdavi (2014b) applied the discrete version of CBO algorithm to four classical test problems such as planar 52-bar truss, spatial 72-bar truss, spatial 582-bar tower, and planar 47-bar power line. Kaveh and Mahdavi (2014c) applied CBO algorithm to the optimal design of truss structures. Kaveh and Ghazaan (2014) applied ECBO to planar 17-bar truss problem, spatial 72-bar truss problem, and spatial 582-bar tower problem. Kaveh and Mahdavi (2015a) applied the two-dimensional CBO algorithm to optimal design of truss structures such as 72-bar spatial truss structure, 120-bar dome truss, and 200-bar planar truss. Kaveh and Ghazaan (2015) have also reported a comparative study of standard CBO and ECBO for the optimal design of skeletal structures. Simulation results clearly show that the two-dimensional

CBO algorithm is competitive and outperforms most of the standard CBO algorithm. In all cases, CBO demonstrates its efficiency in terms of quality of solution and the required number of structural analyses as well as the simplicity of implementation of the algorithm.

10.12   RAY OPTIMIZATION (RO) ALGORITHM

10.12.1   INSPIRATION AND ALGORITHM

The law of refraction is also known as Snell's law.* Snell formulated the law in 1621 (Pledge, 1939), which was unknown until Christian Huygens reported Snell law in his book *Dioptrica* in 1703. Rene Descartes also discovered this law independently and published in a treatise *Discours de la Methode* in 1637. When contacted by Descartes' students, Pierre de Fermat found Descartes derivation was based on unsatisfactory description of refraction of light. Fermat then derived the law based on more fundamental principle—light travels in the shortest time path in any medium. In optics, it is known as Fermat's principle (also known as the principle of least time), which states that light ray traverses the path between two points in the least time. This principle is also taken as the definition of a ray of light. Fermat's principle is also used to describe the properties of light rays reflected off mirrors, refracted through different media, and undergoing total internal reflection.

Light travels in a straight line in a transparent dielectric material but when travels from one medium (i.e., dielectric material) to another, it generally refracts. The angle of refraction changes from the angle of incident depending on the medium. Snell's law of refraction is illustrated in Figure 10.12. Refraction involves the angles that the incident ray x and the refracted ray x' make with the normal to the surface at the point of refraction. The angle of refraction depends on the medium through which the light rays travel. Each transparent material (i.e., dielectric material) has an index of refraction. The dependence is made explicit in Snell's law via refractive indices.

When the light ray travels from Medium 1 (e.g., air) into Medium 2 (e.g., water), Snell's law is defined by the following expression:

$$n_1 \sin \theta_1 = n_2 \sin \theta_2 \tag{10.152}$$

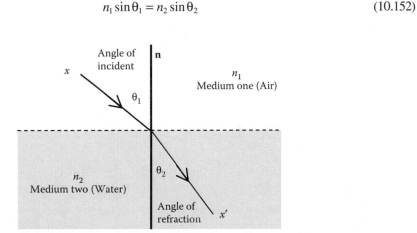

FIGURE 10.12   Snell's law of refraction.

---

* It is named after Dutch astronomer Willebrord Snellius (1580–1626). The law was first described by the scientist Ibn Sahl at Baghdad in 984 AD. In the manuscript *On Burning Mirrors and Lenses*, Sahl used the law to derive lens shapes that focus light with no geometric aberrations.

TABLE 10.1

Refractive Indices of Some Media

Medium	Refractive Index, n
Air (1 atmosphere pressure, 0°C)	1.00029
Water (20°C)	1.33
Crown glass	1.52
Flint glass	1.66

where n_1 and n_2 are constants, also called refractive indices, for Medium 1 and Medium 2, respectively, θ_1 is the angle of incident between the incoming ray and the normal **n**, and θ_2 is the angle of refraction between the refracted ray and the normal.

The refractive indices depend on the wavelength of light and density of the medium. Snell calculated refractive indices for air, water, and glass for light of wavelength of 600 nm, which are shown in Table 10.1.

The law of refraction (or Snell's law) can be used for determining the angle of refraction or incidence. The law is also used in experimental optics to determine the refractive index of dielectric material. It is also applicable to meta-material in which light ray bends at a negative angle of refraction, which has negative refractive index. Rearranging Snell's equation in Equation 10.152 with θ_1 and θ_2 being the incident and refracted angles, the angle of refraction θ_2 can be calculated by

$$\theta_2 = \sin^{-1}\left[\frac{n_1}{n_2}\sin\theta_1\right] \tag{10.153}$$

A qualitative description of refraction becomes clear from the equation. When the light ray is travelling from a medium of higher index to a medium of lower index, the ratio is $(n_1/n_2) > 1$, so the angle θ_2 will be greater than the angle θ_1, that is, the refracted ray is moved away from the normal **n**. When light travels from a medium of lower index to a medium of higher index, the ratio is $(n_1/n_2) < 1$, so the angle θ_2 will be less than the angle θ_1, that is, the refracted ray will move closer toward the normal **n**. This rule of refraction can be utilized for convergence of ray vector. In other words, if light ray passes through successive media with higher refraction indices, the ray vector will converge to the normal. This basic idea of Snell's law is utilized for RO algorithm developed by Kaveh and Khayatazad (2012, 2013). In RO algorithm, a solution comprises a vector of variables, which is simulated by a ray of light passing through space consists of media with different refractive indices.

The light ray can refract in two-dimensional space and in three-dimensional space. For the case of two-dimensional ray tracing, the normalized refracted ray vector x', as shown in Figure 10.12, is defined as

$$x' = -\mathbf{n}\cdot\sqrt{1 - \frac{n_1}{n_2}\cdot\sin^2(\theta_1)} + \frac{n_1}{n_2}[x - (x\cdot\mathbf{n})\cdot\mathbf{n}] \tag{10.154}$$

where x is the ray vector of incident, n_1 and n_2 are the refractive indices and **n** is the normal, and θ_1 is the angle of incident.

For the case of three-dimensional ray tracing, it is considered as a special state of ray tracing in two-dimensional spaces that occurs in a plane with an arbitrary orientation. **n** and x can be rewritten in terms of two normalized vectors i^* and j^*. Considering i^* same as **n**, j^* can be defined by Equation 10.155 such that the dot product of i^* and j^* is zero.

$$j^* = \begin{cases} \dfrac{\left(\mathbf{n} + \dfrac{x}{-\mathbf{n}\cdot x}\right)}{\mathrm{norm}\left(\mathbf{n} + \dfrac{x}{-\mathbf{n}\cdot x}\right)} & \text{if } -1 \le \mathbf{n}\cdot x \le 0 \\[4ex] \dfrac{\left(\mathbf{n} - \dfrac{x}{\mathbf{n}\cdot x}\right)}{\mathrm{norm}\left(\mathbf{n} - \dfrac{x}{\mathbf{n}\cdot x}\right)} & \text{if } 0 \le \mathbf{n}\cdot x \le 1 \end{cases} \tag{10.155}$$

where norm() is a function that computes the magnitude of a vector.

Once the i^* and j^* are obtained, x is transformed to a coordinate system as $x^* = [x \cdot i^*, x \cdot j^*]$. Using Equation 10.154, x'^* is calculated as $x'^* = [x_1'^*, x_2'^*]$ and x'^* is transformed back to x'. Finally, x' is calculated for three-dimensional space as

$$x' = x_1'^* \cdot i^* + x_2'^* \cdot j^* \tag{10.156}$$

Once the refracted ray x' in two-dimension or three-dimension is computed, RO algorithm can be formulated. In RO algorithm, a light ray is thought of a vector consisting of variables of the problem at hand. The light ray passes from medium to medium with higher and higher refractive indices. The light ray changes its course of direction toward the normal obeying the Snell's law. This phenomenon of refraction of light ray is considered as exploration of search space in terms of optimization algorithm. It is clearly visible that the light ray will converge. It is anticipated that the solution vector will lead to an optimal or near-optimal solution in the final stages.

The RO algorithm is implemented in three main levels such as scattering and evaluation, movement vector generation and refinement, origin making, and convergence.

A suitable objective function is defined for the problem at hand. A solution vector $x_i = \{x_{ij}\}$, $i = 1, \ldots, N$, consists of n variables, that is, $j = 1, \ldots, n$, of the problem to be optimized. The solution vectors should be well distributed over the search space, which can be defined by

$$x_{ij} = x_j^{\min} + \mathrm{rand}\cdot\left(x_j^{\max} - x_j^{\min}\right), \quad \forall i = 1,\ldots,N \tag{10.157}$$

where $x_j^{\max}$ and $x_j^{\min}$ are the upper and lower limits of the j-th variable, respectively, and rand is a random number uniformly generated over the interval [0,1].

Since the ray tracing can be carried out in two-dimensional or three-dimensional space, variables are divided into groups of two variables and three variables. If the solution vector consists of even number of variables, it is divided into groups of two variables. If the solution vector consists of odd number of variables, then it is divided into groups of two variables and one group with three variables. For example, a solution vector with even number of variables $x_i = \{x^1, x^2, x^3, x^4\}$ is divided into two groups $x_{i1} = \{x^1, x^2\}$ and $x_{i2} = \{x^3, x^4\}$. A solution vector with odd number of variables $x_i = \{x^1, x^2, x^3, x^4, x^5\}$ is divided into two groups $x_{i1} = \{x^1, x^2\}$ and $x_{i2} = \{x^3, x^4, x^5\}$. After the movement operation, the variables are joined together to form the complete solution vector. For example, $x_i = \{x_{i1}, x_{i2}\}$.

Ray tracing is applied to compute the movement. A movement vector is assigned to each of the solution vector. It is performed on each component of the groups of the solution vectors.

$$v_{ij} = -1 + 2\cdot\mathrm{rand} \tag{10.158}$$

where v_{ij} is the j-th component of the i-th solution vector.

On obtaining the components, they are converted to normalized vectors. By adding the movement vector of each solution vector, they move to new position, which might fall outside the boundary. These are brought back within the boundary by applying refinement. Vectors are evaluated against the objective function and the so-far best vector is selected as global best at this stage and so-far best position is selected as its local best. Each of the solution is moved to new position. The new position is termed as origin and specified by

$$O_i(t) = \frac{(t_{max} + t) \cdot x^{gbest} + (t_{max} - t) \cdot x_i^{best}}{2 \cdot t_{max}} \qquad (10.159)$$

where $O_i(t)$ is the origin of the i-th solution at t-th iteration, t is the current iteration, t_{max} is the maximum number of iterations, x^{gbest} is the global best, and x_i^{best} is the local best solution.

The origin is at the middle of local best and global best at the beginning. This ensures the exploration of the search space. As the iteration goes on, the origin moves closer toward the global best. This ensures the exploitation of the search space.

Each ray is a normalized vector and goes through the medium. In case of optimization, the movement vector is a positive multiple of normalized vector. When ray goes into a new medium with different refraction index, its course of direction is changed according to Snell's law described by Equation 10.152. If the refraction index of the new medium is higher, the angle of refraction will be less than the angle of incident, that is, the ray will move closer to the normal. In other words, passing the light ray through successive media with higher refraction indices, the ray vector will converge to the normal. In the process of optimization, if the normal is selected as a vector with the origin at O and the end at the current position, it is anticipated that the solution vector will converge to O.

The direction of the new movement vector is determined and normalized using Equation 10.154. The refinement of the movement vector is given by

$$v_{il}' = v_{il} \times \text{norm}\left(x_{il} - O_{il}\right) \qquad (10.160)$$

where v'_{il} is the normalized movement vector, x_{il} is the current position of the ray, O_{il} is the origin, and v_{il} is the refined movement vector of the i-th solution vector of the l-th group. In case where the origin O_{il} and the current position x_{il} are same, the direction of the normal cannot be obtained. In such case, the direction is obtained according to

$$v_{il}(t+1) = \frac{v_{il}(t)}{\text{norm}\left(v_{il}(t)\right)} \times \text{rand} \times (0.001) \qquad (10.161)$$

where $v_{il}(t+1)$ is the movement of the vector at $(t+1)$-th iteration, $v_{il}(t)$ is the movement vector at t-th iteration, and rand is a random number within the range [0,1].

A stochastic feature is brought into RO algorithm by introducing a parameter stoch, which specifies whether a movement vector must be changed or not. If ($r >$ stoch), then the new movement vector is computed by

$$v_{ijl}'(t+1) = -1 + 2 \times \text{rand}, \quad \text{if} \quad (r > \text{stoch}) \qquad (10.162)$$

where $v'_{ijl}(t+1)$ is the j-th component of the l-th group of i-th solution at $(t+1)$-th iteration. The length of the vector is refined by

$$v_{il}(t+1) = \frac{v_{il}'(t+1)}{\text{norm}\left|v_{il}'(t+1)\right|} \times \frac{a}{d} \times \text{rand} \qquad (10.163)$$

where d is a number that divides a into smaller length for effective search. The value of d and stoch are chosen arbitrarily. Kaveh and Khayatazad (2013) found that $d = 7.5$ and stoch $= 0.35$ provide best result in their experiments. a is defined by

$$a = \sqrt{\sum_{i=1}^{n} \left(x_i^{max} - x_i^{min}\right)^2} \quad \text{with} \quad n = \begin{cases} 2 & \text{for 2 variable groups} \\ 3 & \text{for 3 variable groups} \end{cases} \quad (10.164)$$

where x_i^{max} and x_i^{min} are the upper and lower bounds of the i-th component of the movement vector, respectively.

The RO algorithm is implemented in three levels. The main steps are as follows:

Level 1: Scattering and evaluation
    Step 1: Initialize parameters of RO algorithm
        Initialize an array of positions x_i, $i = 1, \ldots, N$ in n-dimensional space randomly
        Create array of velocity vector
    Step 2: Evaluate the solutions $f(x_i)$ and determine x_i^{best}
        Set $x^{gbest} = x_i^{best}$ and $x_i^{best} = x_i$
Level 2: Movement vector and motion refinement
    Step 3: Create movement vectors v_{ij} using Equation 10.158
        Calculate new solutions by adding position vector to corresponding movement
            vector
    Step 4: Refine movement vectors if solution vectors go out of bound
        Evaluate the solutions $f(x_i)$ and determine x_i^{best}
        Set $x^{gbest} = x_i^{best}$ and $x_i^{best} = x_i$
Level 3: Origin making and convergence
    Step 5: Calculate origin of solution vector O_i using Equation 10.159
    Step 6: Calculate new movement vector $v_{il}(t)$ for each solution using Equation 10.160
        or using any of Equations 10.161 through 10.163 accordingly
    Step 7: If (termination condition not met), Goto Step 2
    Step 8: Return solution

The termination condition can be the maximum iteration t_{max}. But if after executing a predefined number of iterations, there is no improvement in the fitness function value, the algorithm can be terminated. Sometimes an error goal function is used. The algorithm is terminated when a minimum of the error goal function is achieved.

10.12.2 Applications

The RO algorithm is a very recent addition to meta-heuristic family. The effectiveness and performance of RO algorithm have been verified on a number of benchmark functions (Kaveh and Khayatazad, 2012). The RO algorithm has been applied to engineering design problems such as tension/compression spring design problem, welded beam design problem, and truss structure design problem (Kaveh and Khayatazad, 2012, 2013). The simulation results show that RO algorithm has good efficiency and is comparable to other meta-heuristic algorithms.

10.13 EXTREMAL OPTIMIZATION (EO) ALGORITHM

10.13.1 Inspiration and Algorithm

Self-organized criticality (SOC) is a statistical physics concept to describe a class of dynamical systems that have a critical point as an attractor. The theory of SOC states that large interactive systems and nonequilibrium systems evolve naturally through avalanches of change and dissipations that reach up to the highest scales of the system (Onody and de Castro, 2002). It is believed that SOC governs the dynamics of some natural systems that have these burst-like phenomena such as landscape formation, earthquakes, evolution, and the granular dynamics of rice and sand piles. It was Bak and Sneppen who developed the evolutionary SOC model by observing these critical points in combinatorial optimization problems (Bak et al., 1987). The Bak–Sneppen model of SOC is of special interest due to the fact that it is able to describe evolution via punctuated equilibrium (Bak and Sneppen, 1993). The probability distribution of the sizes s of these avalanches is described by a power-law in the form $p(s) \sim s^{-\tau}$, where τ is a positive parameter. That means, smaller avalanches occur more frequently than big ones, but even avalanches as big as the whole system may occur with a nonnegligible probability. The Bak–Sneppen model based on SOC led to the development of EO proposed by Boettcher (1999, 2000) with inspiration from the recent progress in understanding far-from-equilibrium phenomena found in physical systems. EO is a meta-heuristic optimization algorithm. Rather than breeding the better variables, EO successively updates extremely undesirable variables of a suboptimal solution by assigning new and random values to them (Boettcher and Percus, 1999, 2002). EO improves on a single candidate solution $x \in S$ by treating each of its components as species coevolving according to Darwinian principles where S is the whole search space. A solution x consists of a large number of variables $x = \{x^1, x^2, ..., x^n\} = \{x^k\}$ with $k = 1, ..., n$. In the Bak–Sneppen model, $\{x^1, x^2, ..., x^n\}$ are considered as species. The species $x = \{x^k\}, k = 1, ..., n$, have an associated fitness degree $\lambda^k = \text{fit}(x^k) \in [0,1]$. In EO algorithm, a selection process against the extremely bad components is applied. At each iteration, the species (i.e., the components x_j in x) having the smallest fitness degree, that is, x_j with $\lambda^j < \min\{\lambda^k\}$, are identified and selected for a random modification, which impacts the fitness of interconnected species. A new solution $x' \in N(x)$ in the neighborhood of x is generated. In generating the new solution, typically only a small number of components change state such that only a few components need re-evaluation and re-ranking. The cost function value $C(x)$ of a solution x is computed. Ideally, the cost $C(x)$ is assumed to be a linear function of the fitness degrees λ^k of each component x^k defined as

$$C(x) = -\sum_{k=1}^{n} \lambda^k \tag{10.165}$$

If the cost of new solution x' is less than the cost of old solution x, that is, $C(x') < C(x)$, then x is replaced with x'. After a sufficient number of iterations, the system reaches a highly correlated state known as SOC (Bak et al., 1987) in which all species have reached a fitness of optimal adaptation. The main steps of EO algorithm are described as follows:

Step 1: Initialize a random configuration $x = \{x^k\}, k = 1, ..., n$
Step 2: Evaluate fitness degree λ^k for each component x^k, that is, $\lambda^k = \text{fit}(x^k)$
        Calculate $C(x)$ using Equation 10.165
Step 3: Rank all components according to fitness
        Find the lowest fitness degree $\lambda^{\min} = \min\{\lambda^1, ..., \lambda^n\}$
        Identify components x^j with $\lambda^j \leq \lambda^{\min}$
Step 4: Select a new solution $x' \in N(x)$ in the neighborhood of x by changing x^j
Step 5: Calculate cost $C(x')$
        If $C(x') < C(x)$
            $x = x'$
Step 6: If (termination condition not satisfied), Goto Step 2
Step 7: Return solution

The termination condition is generally set to the maximum number of iterations or to a desired level of solution quality.

10.13.2 Variants of EO

Boettcher and Percus (2001a) modified basic EO to τ-EO algorithm. Its mutation is based on probability distribution over the rank order. In the τ-EO implementation, the ecosystem of N variables evolves, that is, the search of space for the global optimum, as one variable is mutated with probability $P(k) \propto k^{-\tau}$, where k is the rank of the variable. It ranks all variables according to fitness λ^k where rank order $k = 1$ is the worst fitness and rank $k = n$ is the best fitness. For a given value of τ, a power-law probability distribution over the rank order k is considered:

$$P(k) \propto k^{-\tau}, \quad 1 \le k \le n \tag{10.166}$$

At each update, select a rank k according to $P(k)$ and update the state of the variable x^k. The worst variable (with Rank 1) will be chosen most frequently, while the best ones (with higher ranks) will sometimes be updated. In this way, a bias against the worst variables is maintained and no rank gets completely excluded. The search process performance depends on the value of the parameter τ. For $\tau = 0$, the algorithm becomes a random walk through the search space. While for a large values of τ, only a small number of variables with low fitness would be chosen at each iteration and, in this way, the process tends to a deterministic local search. Boettcher and Percus (2001a) have established a relation between τ, run time t, and n, the number of variables of the system to estimate the optimal value of τ. Let $t = A \cdot n$, where A is a constant $(1 \ll A \ll n)$, then

$$\tau \sim 1 + \frac{\ln(A / \ln(n))}{\ln(n)}, \quad n \to \infty \tag{10.167}$$

At this optimal value of τ, the best fitness variables are not completely excluded from the selection process and hence, more space configurations can be reached so that the greatest performance can be obtained. The asymptotic choice of $\tau - 1 \sim [\ln(n)]^{-1}$ optimizes the performance of τ-EO algorithm (Boettcher and Grigni, 2002), which has been verified on a number of problems (Boettcher and Percus, 2001a,b, 2002).

Menai and Batouche (2003) introduced the Bose–Einstein distribution (see Appendix E for details) used in quantum physics to EO, which provides a new stochastic initialization scheme to the EO procedure. The approach is called Bose–Einstein EO (BE-EO). The performance of BE-EO has been verified on the weighted maximum satisfiability problem. The experimental results show that BE-EO outperforms EO and SA.

De Sousa et al. (2004) demonstrated the efficacy of the generalized EO (GEO) algorithm on dealing with complex design spaces, which is illustrated through the application of the method to the design of a heat pipe for satellite thermal control. Experimentation carried out on the GEO algorithm, and it can be said that it seems to be a good candidate to be incorporated to the designer's tools suitcase. Further work is under way on the implementation of hybrids, parallelization of the algorithm, and other practical applications.

An adaptive EO has been proposed by Hamacher (2007) based on the study and analysis of the dynamics of EO where an online measure of the performance of EO is used to estimate the optimal values of the internal parameters. The effectiveness and performance of the adaptive EO were verified on a particular difficult optimization problem such as spin glass.

Xie et al. (2009d) proposed a hybrid GEO (HGEO) approach by combining GA and GEO. To verify the performance of the proposed HGEO, it has been applied to optimal power consumption in semi-track air-cushion vehicle, which confirms the effectiveness and applicability of the approach.

It is widely believed that the power-law is the proper probability distribution for τ-EO. Zeng et al. (2010a) found that those distributions, for example, exponential distributions or hybrid ones (e.g., power-laws with exponential cutoff), can be replaced with a modified τ-EO method called self-organized algorithm (SOA). They proposed the SOA. Experimental results based on a number of test problems such as random Euclidean TSP and nonuniform instances show that SOA provides better performances than other statistical physics oriented methods.

A binary coded EO (BCEO) is introduced by Zeng et al. (2014) where the decision variables are encoded into binary strings. The proposed BCEO has been applied to the optimization of the design parameters of PID controllers and the performance was verified on benchmark plants.

Zeng et al. (2015a) proposed a real-coded population-based EO (RPEO) where the decision variables are encoded into real values and a population of decision variables is employed in the optimization process. The generation of new population is based on multi-non-uniform mutation, and the updating of population is done by accepting the new population unconditionally. The approach is applied to design multivariable PID controllers. The effectiveness and performance are verified by extensive simulation on multivariable benchmark problems, which showed that the proposed RPEO algorithm can obtain better performance than those reported in the literature.

A number of hybrid approaches of EO combining with other meta-heuristic algorithms have been reported in the literature. Chen et al. (2010) proposed a novel hybrid approach, called PSO-EO, by combining the exploration ability of PSO with exploitation ability of EO. The effectiveness and performance of the hybrid approach have been verified on a suite of six unimodal and multimodal benchmark functions, for example, Michalewicz, Schwefel, Griewank, Rastrigin, Ackley, and Rosenbrock, and compared with other meta-heuristics such as standard PSO, GA, and EO. In all cases, it showed competitive performance.

Jahan and Akbarzadeh-T (2012) proposed a hybrid algorithm that combines the long range effect in spin glasses with EO-SA. The strategy aims to choose the next spin and selectively exploiting the optimization landscape, which leads to faster rate of convergence and improved performance. The effectiveness and performance have been verified on five of the world's major stock markets. The experimental results were compared with other heuristic methods such as neural network, Tabu search, and GA to demonstrate the effectiveness of the proposed method.

Li et al. (2012) proposed a hybrid approach combining the global search features of the SFLA and local exploration features of EO. The effectiveness and performance have been verified on six widely used benchmark continuous functions and the performances are compared with those of standard SFLA, modified SFLA, and PSO. The hybrid approach exhibits strong robustness and fast convergence for high-dimensional continuous function optimization problems. Luo and Chen (2014) also proposed a multiphase SFLA with EO. The experimental results show that the algorithm can achieve a high-quality solution within a short runtime for the multidepot vehicle routing problem.

10.13.3 Applications

EO algorithm has been applied to solve hard discrete optimization problems such as the phase transition in the three-coloring problem. Boettcher and Percus (2001a,b) provided independent confirmation of previously reported extrapolations for the ground-state energy of $\pm J$ spin glasses (Nobre, 2003). Ground states of Ising spin glasses on fully connected graphs are studied for a broadly distributed bond family by Boettcher (2014) and determined approximate ground states for a large number of fully connected graphs for sizes from 31 to 255 with bond matrices filled with random bonds drawn from the power-law distribution. De Falco et al. (2015) applied EO with guided state changes to processor load balancing during the execution of distributed applications. The experimental results are compared against a greedy fully deterministic approach, GA, and an EO-based algorithm with random placement of migrated tasks. In all cases, it shows competitive performance.

EO and τ-EO have been successfully applied to a number of NP-hard problems, such as graph bipartitioning (Boettcher and Percus, 1999, 2000), TSP (Boettcher and Percus, 2000; Chen et al., 2007; Zeng et al., 2010a), MAX-SAT (Menai and Batouche, 2003), graph coloring (Boettcher and Percus, 2004), heat pipe design (De Sousa et al., 2004), spin glasses (Boettcher, 2005, 2014), community detection in complex networks (Duch and Arenas, 2005), dynamic combinatorial problems (Moser and Hendtlass, 2006), image segmentation (Melkemi et al., 2006), protein folding simulations on the lattice (Lu and Yang, 2009), controllability of directed networks (Ding et al., 2013), Lennard–Jones cluster (Zhou et al., 2005) and quadratic unconstrained binary optimization (QUBO) problems (Boettcher, 2015; Chen et al., 2006). In general, τ-EO showed superior performance over the EO.

EO has been applied to a number of multiobjective optimization problems. Chen and Lu (2008) investigated the effectiveness of EO in a numerical multiobjective optimization and proposed a new novel elitist $(1 + \lambda)$ multiobjective EO (MOEO) algorithm and verified on five popular benchmark functions. Zeng et al. (2015b) applied an improved MOEO to design a fractional-order PID controller with three objective functions, namely, integral of absolute error, absolute steady-state error, and settling time for automatic regulator voltage system. Randall et al. (2014) proposed a generic local search strategy for inseparable problems. To supplement the local search strategy, a diversification strategy that draws from the external archive is incorporated into the local search strategy. This strategy has been employed for multiobjective optimization based on EO. Using benchmark problems and a real-world airfoil design problem, it is shown that this combination leads to improved solutions.

10.14   PARTICLE COLLISION ALGORITHM

10.14.1   INSPIRATION AND ALGORITHM

PCA is loosely inspired by the physics of nuclear particle collision (Duderstadt and Hamilton, 1976), particularly absorption and scattering of particles. In absorption, the incident particle from collision is absorbed by the target nucleus. In scattering, the incident particle is scattered by a target nucleus. This simple mechanism is applied to an optimization problem where a particle that reaches a promising or high-fitness area of search space is absorbed and will explore the region, while a particle that hits a low-fitness region will be scattered to other regions. PCA was proposed by Sacco and de Oliveira (2005) based on the particle collision phenomena in nuclear physics and further developed by Sacco et al. (2006). The scattering mechanism allows the algorithm to simulate the exploration of search space and exploitation of fitness landscape through successive scattering and absorption mechanism.

PCA is a meta-heuristic algorithm conceptually similar to SA (Kirkpatrick et al., 1983) and can also be considered as a Metropolis algorithm (Metropolis et al., 1953) where a trial solution is accepted with a probability. The PCA has two main operations: absorption operation and scatter operation. In absorption operation, a new solution is created by adding small perturbation to old solution. If the new solution is better, then it is absorbed. Absorption operation on x is performed using the following pseudocodes:

For $t = 0$ to max_iteration
    {
    Create a stochastic perturbation $\Delta x = \text{rand}(0,1)$
    $x' = x + \Delta x$
    If fit$(x') >$ fit(x)
        $x = x'$
    }

where rand(0,1) is a random number distributed over [0,1].

In scatter operation, a probability p_{Scatter} is used to accept a random new solution. p_{Scatter} is defined as $p_{\text{Scatter}} = 1 - (\text{fit}(x')/\text{fit}(x^{\text{best}}))$. The scatter operation on x is performed using the following pseudocodes:

$$p_{\text{Scatter}} = 1 - \frac{\text{fit}(x')}{\text{fit}(x^{\text{best}})}$$

If $P_{Scatter} > \mathrm{rand}(0,1)$
    $x = $ random solution
Else
    Perform absorption operation on x

The main steps of PCA are as follows:

Step 1: Generate an initial solution x in n-dimensional space
Step 2: Create a random perturbation $\Delta x = \mathrm{rand}(0, 1)$
        $x' = x + \Delta x$
Step 3: If $\mathrm{fit}(x') > \mathrm{fit}(x)$
        $x = x'$
        Perform absorption operation on x
    Else
        Perform scatter operation on x
Step 4: If (max_iteration not reached), Goto Step 2
Step 5: Return solution

A number of variants of PCA have been reported in the literature. Sacco et al. (2007) proposed a population-based PCA (called PopPCA) by hybridizing PCA and GA where a population of initial solutions is used. At the end of a generation, particles reproduce and the fittest individuals survive to next generation. The performance of the PopPCA has been verified by applying it to the optimization of nuclear reactor cell parameters such as dimensions, enrichment, and materials. The comparative study based on simulation results shows that PopPCA outperforms GA, SA, and PCA.

Multiparticle PCA, called MPCA, has been proposed by Luz et al. (2008, 2011) based on the canonical PCA where several particles are used instead of one used in standard PCA. The best fitness information is shared among all particles through a blackboard-type mechanism. The effectiveness and performance of MPCA have been verified on a group of inverse problems such as radiative transference and localization of polluting sources. The experimental results showed that MPCA is a viable alternative to the solutions of inverse problems. Torres et al. (2015) proposed four new variants of the MPCA. These are the opposition-based MPCA (O-MPCA), quasi-opposition-based MPCA (QO-MPCA), quasi-reactive MPCA (QR-MPCA), and the center-based sampling MPCA (CB-MPCA). The effectiveness and performance of all four variants are verified on well-known benchmark functions such as Ackley, Griewank, Rastrigin, and Sphere. The algorithms with opposition and reflection mixed with randomness achieved better results than MPCA and MPCA with simple opposition over all four functions of simple and median complexity.

Abuhamdah and Ayob (2009b) proposed a multineighborhood PCA to enhance the performance of PCA approach by introducing two separate stages of neighborhood structure to the solution mechanism. The first stage is employed in the solution construction phase and the second stage is employed in the exploration phase to improve the solution. In both stages, the neighborhoods are randomly selected. The effectiveness and performance of the proposed approach were verified on standard benchmark university course timetabling problem and compared with other meta-heuristic algorithms. Results show that the proposed approach outperforms other approaches. A hybrid PCA (HPCA) is proposed by Abuhamdah and Ayob (2009c) by combining multineighborhood PCA with great deluge using composite neighborhood structures. In order to enhance the quality of solution, the HPCA accepts the worst solution in the scattering phase using hybridization of PCA and great deluge acceptance criteria. Firstly, the solution is accepted by the PCA acceptance criterion. If it is not accepted, it then checks whether it can be accepted by the great deluge acceptance criterion, which thus increases the acceptance probability of the worst solution to produce better quality results. HPCA also enhances further the trial solution by exploring different neighborhood structures. The HPCA approach has been verified on university course timetabling problems, which performed well obtaining good quality solutions for small, medium, and large data sets. Abuhamdah and Ayob (2011) proposed hybridization between MPCA (Abuhamdah and Ayob, 2009b) and adaptive randomized descent algorithm (ARDA) (Abuhamdah and Ayob, 2010) acceptance criterion to solve university course timetabling problems. This hybrid MPCA-ARDA is an adaptive approach, which enhances the performance of both HPCA and MPCA.

10.14.2 APPLICATIONS

The PCA has found many applications since its inception. The effectiveness and performance of PCA have been tested on some benchmark functions such as Easom function, Rosenbrock function, and De Jong function (Sacco and de Oliveira, 2005). Sacco and de Oliveira (2006, 2007) also applied PCA to a nuclear reactor design optimization problem. PCA and some of its variants have been applied to university course timetabling problems by Abuhamdah and Ayob (2009a–c, 2011). The results obtained by PCA are very promising compared to other meta-heuristic approaches. One of the unique characteristics of PCA lies in that it does not require the parameters to be set by the user. Neural networks have been used for prediction, identification, and control where finding the optimal network topology is mostly done by trial-and-error method. MPCA has been applied to automatic configuration of neural network architecture (i.e., topology) by formulating the topological parameters as an optimization problem. A number of researchers applied MPCA to topology learning of the neural network that has been used for mesoscale climate prediction (Anochi and Velho, 2014; Sambatti et al., 2012).

Parameter identification associated with the aerodynamic performance of the aircraft can be formulated as an optimization problem. Sumida et al. (2014) applied MPCA for determining of flight dynamics parameters, specifically helicopter parameter identification. Simulation results show that the performance of MPCA is comparable to that of GA.

10.15   RIVER FORMATION DYNAMICS

10.15.1   INSPIRATION AND ALGORITHM

Water drops flow toward the sea over the land. As they flow, they change the landscape by eroding land at high altitudes and deposit sediments in flatter or low altitude areas. By removing or depositing sediments by the flowing water drops, the altitudes in the landscape change. By increasing or decreasing the altitude at some nodes, the gradients of those nodes do change. This change in turn affects the flow of water. Eventually, the decreasing gradients will depict the paths of flowing water from the points of rains to the sea, and thus form rivers. The river formation phenomena can be used for finding the solution of an optimization problem. Considering this natural phenomenon of river formation by eroding the land and depositing sediments along the paths from the origin to destination, Rabanal et al. (2007) proposed a meta-heuristic optimization algorithm called RFD.

In many meta-heuristic optimization procedures, problems are represented as graphs with the objective of minimizing the total distance between the starting node and the destination node. The distance between nodes is considered as cost of edge. The cost of edge can incur any other costs such as computation time, complexity or material costs, and so on depending on the problem at hand and the algorithm itself. For example, pheromone values in ACO algorithm are used for edges as cost. Unlike many other meta-heuristic algorithms, altitude values are included to nodes toward the edge of cost in RFD. That is, the cost of edge depends on the difference of altitudes between nodes and the distance in RFD. Water drops erode the land resulting in the reduction of the altitude of nodes or deposit the sediment in flat areas resulting in increase in the altitudes of nodes. The probability of the water drop having selected an edge (or path) over others is proportional to the gradient of the slope in the edge, which in turn depends on the difference of altitudes between the nodes and the distance (i.e., the cost of the edge).

The RFD algorithm is provided with a flat environment where all the nodes of the graph are initialized with the same altitude values except the destination node. The altitude of the destination node is set to 0. In terms of the RFD, the destination node represents the sea, which is the final goal of all water drops. The exception is the destination node, which is a hole. Initially, all water drops are initialized at random and put in the initial node. Water drops flow across the nodes of the graph, which spread around the flat environment until some of them fall in the destination node, in partially random way. This erodes adjacent nodes, which creates new downward slopes, and in this way, the erosion process is propagated. The probability that a water drop k at node i will move to the node j is defined by

$$P_k(i,j) = \begin{cases} \dfrac{\nabla(i,j)}{\displaystyle\sum_{l \in V_k(i)} \nabla(i,l)} & \text{if} \quad j \in V_k(i) \\ 0 & \text{if} \quad j \notin V_k(i) \end{cases} \tag{10.168}$$

where $V_k(i)$ is the set of neighboring nodes of node i that can be visited by the water drop k, $\nabla(i,j)$ represents the negative gradient between nodes (i,j). $\nabla(i,j)$ is defined by

$$\nabla(i,j) = \frac{\text{altitude}(j) - \text{altitude}(i)}{D(i,j)} \tag{10.169}$$

where altitude($\cdot$) defines the altitude of a node and $D(i,j)$ defines the length of the edge between nodes i and j.

Since the altitudes of all nodes are set to equal values at the beginning, the sum of negative gradients of nodes $\sum_{l \in V_k(i)} \nabla(i,l)$ will be 0, which will result in a probability value of zero for $P_k(i,j)$. This means that the water drop will not move to anywhere. In order to get the water drop moving, $\nabla(i,j)$ must be nonzero at the beginning of the RFD algorithm. Therefore, to initiate the move at the beginning, $\nabla(i,j)$ is set to a small value, that is, $\nabla(i,j) = \varepsilon$. The decreasing gradient is expressed as follows:

$$\text{Decreasing } \nabla(i,j) = \begin{cases} 90 - \gamma & \text{if} \quad \nabla(i,j) < 0 \\ 90 - \eta & \text{if} \quad \nabla(i,j) = 0 \\ 180 - \gamma & \text{if} \quad \nabla(i,j) > 0 \end{cases} \tag{10.170}$$

where $\gamma = |\text{arc} \tan(|\nabla(i,j)|)| \cdot (180/\pi)$, $\eta = \text{edgeCost}/\mu$, and μ is an arbitrary parameter usually chosen as $\mu > 0$. Rabanal et al. (2010) used $\mu = 5$ in their experiments. The decreasing $\nabla(i,j)$ should be within the range of $[0,180°]$. That means, if decreasing $\nabla(i,j) = 0$, water drops move vertically up, if decreasing $\nabla(i,j) = 90$, water drops move in a flat environment, and if decreasing $\nabla(i,j) = 180$, water drops move vertically down. The graphical illustration of the parameter is shown in Figure 10.13.

The gradient enables the water drop to spread around the environment and the erosion process to propagate. The erosion is defined as

$$\text{Erosion }(i,j) = \frac{\varphi \cdot M \cdot \nabla(i,j)}{S_{\text{cost}}} \tag{10.171}$$

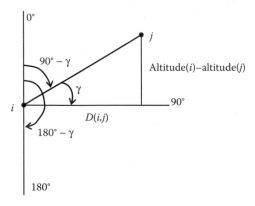

FIGURE 10.13   Graphical illustration of parameters of $\nabla(i,j)$.

where φ is an arbitrary parameter to be chosen by the user, M is the number of water drops, and S_{cost} is the cost of solution in the previous phase. Rabanal et al. (2010) found that $\varphi = 0.1$ is useful in their experiments.

The erosion causes the altitude to change. The erosion is high, if the negative $\nabla(i,j)$ is high. If the edge (i,j) is flat and $\nabla(i,j)$ is positive, small erosion occurs. That is, if water drop moves from node i to node j, the altitude of node i is reduced depending on the current gradient between nodes i and j. The altitude change is defined as

$$\text{altitude}(i) = \text{altitude}(i) - \text{erosion}(i,j) \tag{10.172}$$

As the drops move from node to node of the graph, they deposit sediments in nodes. Consequently, this implies the gradients between nodes to go positive, drops cannot move up an edge with positive gradient, and so the water drops deposit sediments they carry. The deposition of sediment causes the altitudes of the nodes to increase slightly and uniformly. The increase in altitudes is proportional to the amount of accumulated sediment. Two different sedimentation processes are used: uniform sedimentation process applied at each iteration and one sedimentation applied to particular node when drops get blocked. The altitude of the final destination, that is, sea, is fixed to 0 and is never modified. New drops are added to the origin node to transform paths and reinforce the erosion of promising paths. After some iteration, good paths from the origin to the destination are created. These paths are given in the form of sequences of decreasing edges from the origin to the destination, which eventually results in an optimal solution. The main steps of the RFD algorithm are described as follows:

Step 1: Generate M water drops randomly
        Put all the drops in the initial nodes
        Set the initial parameters of RFD
Step 2: Initialize all nodes of the graph
        Set the altitude of the destination node to a fixed value 0
        Set some equal value to the rest of the nodes
Step 3: Determine the probability $P_k(i,j)$ using Equation 10.168
        Move drop k at node i to node j of the graph
Step 4: Erode paths according to the probability of movements of drops
        Modify altitudes of nodes according to Equation 10.172
Step 5: Apply sedimentation process
Step 6: Analyze all paths found by drops
Step 7: If (termination criterion not met), Goto Step 3
Step 8: Return solution

RFD optimization can be seen as a gradient-based version of ACO (Dorigo and Stützle, 2004). RFD uses altitudes of edges of a graph and ACO uses pheromone values of the edges. In RFD, the probability of a water drop to move to a destination is proportional to the difference of altitudes. In ACO, the probability of an ant to move to a destination is proportional to the amount of pheromone trail. The gradient of an edge from node i to node i will never be decreasing. This is the advantage of using the gradient information in RFD that helps drops in avoiding local cycles. Shortcuts are reinforced in RFD by connecting the same origin and destination at lower cost. The overall decreasing gradient is immediately higher. Therefore, edges in the shortcut are preferable over others. The sediment process also provides a localized way of punishing wrong paths. That is, when drops get blocked, they increase the node altitude by depositing sediments, which eventually prevents other drops from taking the same path.

10.15.2 APPLICATIONS

Though RFD is a very recent addition to the literature of meta-heuristic algorithms, RFD has been applied to solve several NP-complete problems (Kalayci and Surendra, 2013; Rabanal et al., 2007, 2009a; Rabanal and Rodríguez, 2009). RFD algorithm actually fits particularly well for problems optimizing kind of covering tree (Rabanal et al., 2008a,b, 2009b) and therefore found many interesting applications such as Steiner tree problem (STP), minimum spanning tree, minimum distance tree, TSP, and mobile robot navigation. Moreover, the performances of RFD has been tested and verified on a number of benchmark functions.

Rabanal et al. (2008a) applied RFD to two NP-complete problems such as minimum spanning tree and minimum distance tree in a variable cost graph. A variable cost graph is tuple $G = (N,O,d,V,A,E)$, where N denotes the set of nodes, $O \subseteq N$ is the set of origin nodes, d is the destination node, $V = \{v_1, v_2, ..., v_n\}$ is a finite set of values, $A: O \rightarrow V$ is the value function that assigns an initial value to each origin node, and E is the set of edges. Each edge $e \in E$ is tuple of (n_1, n_2, C, T), where $n_1, n_2 \in N$ denotes origin and destination node, respectively, C is the cost function for e, and T is the transformation function of e. A spanning tree of a variable cost graph G is a subgraph (or tree) that connects all the nodes together. A single graph can have many different spanning trees. A minimum spanning tree $G' \subset G$ is then a spanning tree with cost(G') less than or equal to the cost of every other spanning trees.

Rabanal et al. (2008b) applied RFD to TSP, which is considered as an NP-complete problem. A detailed description of TSP is provided in Chapter 2. A graph with 100 nodes is randomly generated for TSP. RFD has to find a route starting from a node of origin, traversing through all the nodes, and coming back to the node of origin forming a cycle. In general, RFD always avoids cycle, whereas TSP solution requires a cycle. In order to form a cycle, a dummy node is created by cloning the node of origin where the drop should reach its destination. The solution obtained by RFD algorithm is compared with that of ACO. It is found that ACO provides better solution in short term but RFD provides better solution in long term. RFD algorithm works faster in the case of dynamic TSP. RFD algorithm always finds a solution after a modification to graphs, whereas ACO cannot adapt the solution to modifications. This reveals that the exploration of graphs for solution is deeper in RFD than in ACO.

Rabanal et al. (2010) applied RFD to STP. Finding the cheapest tree containing all terminal nodes from a given undirected graph where the subset of terminal nodes and cost of edges are provided. In order to apply RFD to STP, they computed a matrix denoting the shortest available path between any pair of nodes called shortest path matrix (SPM). SPM is defined in a way such that the position (i,j) of SPM contains the length of the shortest path between change and j. The original graph is modified by adding or substituting some edges. For example, if edge (i,j) does not exist in the graph, then an edge (i,j) with cost SPM(i,j) is added to the graph. If edge (i,j) exists in the matrix but the cost is higher than SPM(i,j), it is then substituted by a new edge with the cost of SPM(i,j). After doing the preprocessing, an equivalent graph is obtained and manipulated using RFD.

Redlarski et al. (2013) applied RFD to a mobile robot navigation problem. The original algorithm was modified, which increased the convergence for the application. The probability at which the drop moved from one node to another was defined as follows:

$$P_k(i,j) = \begin{cases} \dfrac{\nabla(i,j)}{\displaystyle\sum_{l\in V_k(i)} (d_j)^\alpha} & \text{if} \quad j \in V_k(i) \\[4mm] \dfrac{\omega/|\nabla(i,j)|}{\displaystyle\sum_{l\in U_k(i)} (d_j)^\alpha} & \text{if} \quad j \in U_k(i) \\[4mm] \dfrac{\delta}{\displaystyle\sum_{l\in F_k(i)} (d_j)^\alpha} & \text{if} \quad j \in F_k(i) \end{cases} \tag{10.173}$$

where $V_k(i)$ is the set of neighbors with a positive gradient, that is, altitude of node i is greater than that of node j, $U_k(i)$ is the set of neighbors with negative gradient, and $F_k(i)$ is the set of neighbors with zero gradient, that is, i and j have the same altitude. Neighbors are chosen from up to eight surrounding nodes except an obstacle. The gradient $\nabla(i,j)$ is defined as the difference of altitudes between nodes. ω and δ are constants to be chosen as small values. d_j is the length of the path from node j to the goal and α is a tuning parameter for convergence of the algorithm. Erosion is defined as

$$\text{Erosion}(i,j) = \begin{cases} \dfrac{\varepsilon_V \nabla(i,j)}{(N-1)\cdot M \cdot L_k} & \text{if} \quad j \in V_k(i) \\[4mm] \dfrac{\varepsilon_U}{\nabla(i,j)(N-1)\cdot M \cdot L_k} & \text{if} \quad j \in U_k(i) \\[4mm] \dfrac{\varepsilon_F}{(N-1)\cdot M \cdot L_k} & \text{if} \quad j \in F_k(i) \end{cases} \tag{10.174}$$

where ε_V, ε_U, and ε_F are the parameters associated with the respective group of neighbors, N is the number of nodes, M is the number of drops, and L_k is the length of path traversed by drop k. Redlarski et al. (2013) used small amount of sediment to be added to all nodes when the altitude of a node goes close to 0. The modified RFD algorithm was applied to a real robot navigation problem. The algorithm was able to identify a sequence of points for the decomposed terrain through which the robot should traverse. The robot was able to find the shortest route to the goal correctly to sufficient accuracy and avoiding obstacles. The tests were carried out on different landscape layouts and RFD algorithm outperforms Dijkstra's algorithm. Experimental results presented in Rabanal et al. (2010) show that RFD meets a competitive trade-off between execution time and quality of constructed solutions for STP.

10.16 WATER CYCLE ALGORITHM

10.16.1 INSPIRATION AND ALGORITHM

The water cycle is an essential process that helps in sustaining life on Earth, which is basically driven by the solar energy and the gravity of the Earth. In this process, water is constantly being cycled between the ocean, the atmosphere, and the land. The surface waters are evaporated during the day by the solar energy and raised into the atmosphere. The water vapors in the air are

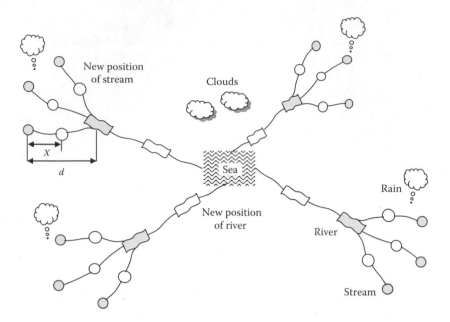

FIGURE 10.14 Water cycle process. A stream is shown as a circle and a river is shown as a rectangle.

condensed, form clouds, and are transported to the different places over the land by wind. The droplets merge together and become cool and large enough to fall over to the land by the gravity. The water drops reach to the land in the form of rain, hail, or snow. Most of the water from the rain stays on the surface of the earth and flows from the higher levels to the lower levels due to gravity. Mostly, the flowing surface waters form small streams. The streams join each other to form larger streams or rivers. The rivers flow toward to the big natural water storages, such as the lakes, seas, and oceans. Some of the streams also flow directly to the ocean. All rivers and streams end up in the ocean, which is to be considered as the optimal stage or point in the water cycling process. There is always the evaporation of the surface waters in the open area, but the main water evaporation occurs in the oceans (Gat, 2010), which starts the cycling process again. Applying this natural phenomenon, Eskandar et al. (2012) proposed a new meta-heuristic algorithm called WCA for optimizing constrained functions and engineering problems.

The WCA begins with an initial population of streams consisting of rain water. The best individual (i.e., best stream) is chosen as a sea. A number of good streams (evaluated as per the cost function value) are chosen as a river and the rest of the streams flow to the rivers and sea directly. Depending on the magnitude of flow, each river takes in water from the streams. The amount of water in a stream joining rivers and/or sea varies from other streams. All rivers flow to the sea, which is the ultimate destination and optimal solution in terms of optimization. The water cycle process is illustrated in Figure 10.14.

In an n-dimensional optimization problem, a raindrop is an array of n variables to be optimized, which is defined as

$$\text{raindrop} = \{x^1, x^2, \ldots, x^n\} \tag{10.175}$$

where the variables x^1, x^2, …, x^n can be of real values in continuous optimization problems or they can be selected from a predefined set of discrete values.

A population of raindrops is defined as

$$
\begin{bmatrix} \text{raindrop}_1 \\ \text{raindrop}_2 \\ \vdots \\ \text{raindrop}_N \end{bmatrix} = \begin{bmatrix} x_1 \\ x_2 \\ \vdots \\ x_N \end{bmatrix} = \begin{bmatrix} x_1^1 & x_1^2 & \cdots & x_1^n \\ x_2^1 & x_2^2 & \cdots & x_2^n \\ \vdots & \vdots & \ddots & \vdots \\ x_N^1 & x_N^2 & \cdots & x_N^n \end{bmatrix} \tag{10.176}
$$

where N is the size of population meaning the total number of raindrops and n is the number of variables.

The fitness of a raindrop is computed by evaluating a fitness function f, which is very much problem dependent, as follows:

$$
f_i = f(x_i^1, x_i^2, \ldots, x_i^n), \quad i = 1, 2, \ldots, N \tag{10.177}
$$

On the basis of the fitness function value $f_i(\cdot)$, the raindrop that has the minimum value among others is considered as the sea. A number of $N_{sr} \subseteq N$ from the best individuals (those have lower fitness values) are selected as sea and rivers. In other words, N_{sr} is the total of number of rivers and number of seas, which is a user-defined parameter. N_{sr} is defined as

$$
N_{sr} = N_{\text{rivers}} + \overset{\text{Sea}}{\overset{\frown}{1}} \tag{10.178}
$$

where N_{rivers} is the number of rivers and $\overset{\text{Sea}}{\overset{\frown}{1}}$ denotes a single sea in Equation 10.178. The rest of the population of raindrops, denoted as N_{raindrop} (which flow into the rivers or may directly flow to the sea), is given by

$$
N_{\text{raindrop}} = N - N_{sr} \tag{10.179}
$$

The remaining raindrops N_{raindrop} form NS_k streams that flow to the specific rivers or sea. The number of streams NS_k is defined by the following (Eskandar et al., 2012):

$$
NS_k = \text{round}\left[\left| \frac{f_k}{\sum_{i=1}^{N_{sr}} f_i} \right| \times N_{\text{raindrop}} \right], \quad k = 1, 2, \ldots, N_{sr} \tag{10.180}
$$

Streams are formed from rain water and they join each other to generate new rivers. Let us assume that there are N streams out of which $N_{sr} - 1$ are selected as rivers and one is selected as sea.

The distance between streams is denoted as X and distance between stream and river or sea is denoted as d. X and d are shown in Figure 10.14. X is chosen as a random number uniformly distributed between 0 and $(C \times d)$ defined as

$$
X \in [0, C \times d], \quad 1 \leq C \leq 2 \tag{10.181}
$$

where C is an arbitrary parameter within the range of $1 \leq C \leq 2$.

The best value for C can be chosen as 2. Choosing $C > 1$ allows the streams flowing in different directions toward rivers. Similarly, choosing appropriate values for $C \in \{[1,2]|1 \leq C \leq 2\}$ will help describing rivers flowing to the sea or streams flowing to rivers. Streams and rivers flow toward their destination and reach new positions. Let the positions of i-th stream, river, and sea at time t be denoted as $X_i^{\text{Stream}}(t)$, $X_i^{\text{River}}(t)$, and $X_i^{\text{Sea}}(t)$, respectively. The new positions of the streams $X_i^{\text{Stream}}(t+1)$ and rivers $X_i^{\text{River}}(t+1)$ at time $(t+1)$ can be estimated using C as a parameter (Eskandar et al., 2012) as follows:

$$X_i^{\text{Stream}}(t+1) = X_i^{\text{Stream}}(t) + \text{rand} \times C \times \left(X_i^{\text{River}}(t) - X_i^{\text{Stream}}(t) \right) \tag{10.182}$$

$$X_i^{\text{River}}(t+1) = X_i^{\text{River}}(t) + \text{rand} \times C \times \left(X_i^{\text{Sea}}(t) - X_i^{\text{River}}(t) \right) \tag{10.183}$$

where rand is a random number uniformly distributed over [0,1].

If the position of stream $X_i^{\text{Stream}}(t+1)$ is better than the position of river $X_i^{\text{River}}(t+1)$, the positions of river and stream are exchanged, that is, stream becomes river and river becomes stream. The swapping of streams and rivers is illustrated in Figure 10.15, which is the best solution among other streams and rivers. Similarly, if the solution $X_i^{\text{River}}(t+1)$ is better than $X_i^{\text{Sea}}(t+1)$, then the positions of sea and river are exchanged, that is, river becomes sea and sea becomes river.

To balance between exploitation and exploration properties of the algorithm, an evaporation mechanism is added to WCA. Evaporation is an important process that is directly related to the rain. The evaporation process causes the sea water to evaporate, transform into clouds, and come back to the earth as rain. The rain creates the new streams and the new streams flow into the rivers, which flow to the sea again (Gat, 2010). The key idea of the evaporation is that the rivers that are very near the sea die off and then regenerate again by the raining process at another random position in the search space. That is, evaporation and raining processes are required for the rivers to flow into the sea. Otherwise, WCA is assumed to be trapped into local minima. Therefore, the evaporation is considered as an important operator that helps in preventing the WCA from premature convergence. The necessary condition for the evaporation and raining processes can be determined from the following rule:

If $\left| X_i^{\text{Sea}} - X_i^{\text{River}} \right| < d_{\max}$

    Perform evaporation and raining processes

where $d_{\max}$ is a small value.

If the distance between river and sea is less than $d_{\max}$, it indicates that the river has reached the sea and an evaporation process should start for raining to happen and to add sufficient water to the

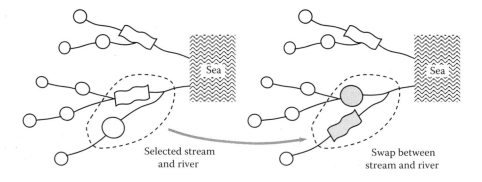

Selected stream and river   Swap between stream and river

FIGURE 10.15 Swapping of the position of streams and rivers.

rivers to flow. d_{max} is an arbitrary parameter chosen by the user. A large value for d_{max} reduces the search, while a small value intensifies the search near the sea. In other words, d_{max} controls the exploitation of the search space near the sea for an optimal solution. For a better exploration and exploitation of the search space, d_{max} should be decreased over iterations as follows:

$$d_{max}(t+1) = d_{max}(t) - \frac{d_{max}(t)}{t_{max}}$$ (10.184)

where t_{max} is the maximum iteration number.

Raining process adds some fresh water to streams in different locations. Thus, it helps in improving the solution quality. This mechanism is called mutation, which is applied to solutions at random. New locations of newly formed streams are computed as follows:

$$X_{new}^{Stream} = LB + rand \times (UB - LB)$$ (10.185)

where LB and UB are lower and upper bounds defined by the given problem, respectively. The best newly formed raindrop is considered as a river flowing to the sea. The rest of the raindrops are assumed to form new streams, which flow to rivers or directly flow to the sea. In order to improve the solution quality and enhance the convergence, the streams that flow directly to the sea are updated according to the following equation:

$$X_{new}^{Stream} = X_{Sea} + \sqrt{\mu} \times rand(1, N)$$ (10.186)

where rand is a random number with normal distribution and μ is a parameter that decides between exploration or exploitation of the search region near the sea. Mathematically, μ can be considered as variance and the term $\sqrt{\mu}$ represents standard deviation. Using this notion, the individuals with variance μ are distributed around the optimal solution, which is the sea. A larger value of μ increases the possibility of moving away from the feasible region. A smaller value of μ exploits the search region near the sea. A suitable value for μ is 0.1.

The main steps of the WCA are described as follows:

Step 1: Set the initial parameters of the WCA
        Generate an initial random population of streams, rivers, and sea
Step 2: Calculate the fitness function value of each raindrops using Equation 10.177
Step 3: Determine the intensity of flow for rivers and sea using Equation 10.180
Step 4: Streams flow to the rivers according to Equation 10.182
Step 5: Rivers flow to the sea according to Equation 10.183
Step 6: Swap positions of river with a stream that gives the best solution
Step 7: If a river finds better solution than the sea, swap the position of river with the sea
Step 8: Check the evaporation condition
Step 9: If (evaporation condition not satisfied)
        Apply raining process using Equations 10.185 and 10.186
Step 10: Update the value of d_{max} using Equation 10.184
Step 11: If (stopping criterion not met), Goto Step 4
Step 12: Return solution

The widely used criteria for stopping the WCA are the maximum number of iterations, computation time, and a small nonnegative tolerance value between two solutions. It is to be noted that rivers are not fixed points and generally move toward the sea as the optimal solution. The procedure of the WCA is illustrated as a flow diagram in Figure 10.16.

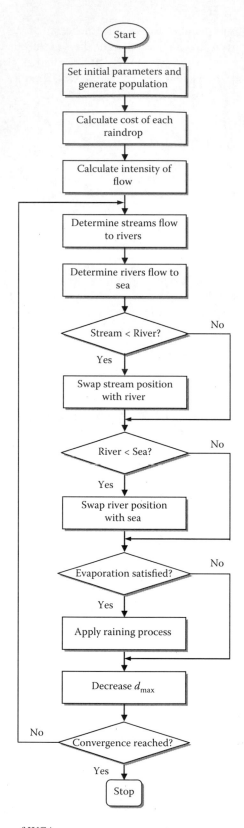

FIGURE 10.16 Flow diagram of WCA.

10.16.2 VARIANTS OF WCA

WCA is a very recent algorithm based on the water cycle process. So far, only a very few variants are reported in the literature such as quantum-based WCA. A quantized WCA (QWCA) is proposed by Guney and Basbug (2015) for the antenna array pattern synthesis with low side-lobe levels (SLLs) and null depth levels (NDLs) at desired directions by using four-bit digital phase shifters.

Two modifications have been made to the standard WCA. The first modification is the quantization of the phase results as only discrete phase values can be assigned to the digital phase shifter applications. In addition to the standard features of a meta-heuristic algorithm, QWCA has an internal quantization mechanism and a precalculated array factor method (PAFM). The internal quantization mechanism of QWCA is utilized to achieve digital values matching to the discrete values of the phase shifter instead of the simple rounding up/down routines after optimization. It is rather applied as an inner process of the method before the candidate solutions are used for the function evaluation. The second modification is a speed boosting, which compensates the disadvantage of the quantization for the optimization process. This acceleration is based on the reality that the accessing data in the memory need less time than calculating the mathematical functions throughout the optimization process. Instead of calculating the subfunctions of the array factor per each evaluation, the precalculated values are stored in a three-dimensional array and used when required. PAFM provides a speedup for the optimization in case where accessing memory is faster than calculating cost function elements as in the synthesis examples in this article.

In QWCA, a quantization process has been carried out for each candidate solution before they are used for the cost function evaluation. Before the algorithm utilizes the candidate solutions to evaluate the cost function, $X_{j,g}^i$ are converted to quantized value $Q_{j,g}^i$ by using the following formula:

$$Q_{j,g}^i = \left\lfloor X_{j,g}^i \times 2^B \right\rfloor \qquad (10.187)$$

where B is the number of bits for the digital phase shifter and $\lfloor \cdot \rfloor$ is the floor function that rounds off the value to the closest integer. In this way, the real quantized values are used to calculate the cost function value instead of the continuous values. While the quantized values $Q_{j,g}^i$ are used only to evaluate the cost function, the continuous values $X_{j,g}^i$ are used at every point in the subsequent recurring steps of the algorithm.

The experimental results show that QWCA can obtain very good SLLs and NDLs on the synthesized pattern. Moreover, the optimization results are achieved within a remarkably short time. SLL and NDL values obtained by QWCA are also compared with the results using other available methods in the literature, which reveal that QWCA is able to produce better results than the other methods.

10.16.3 APPLICATIONS

In order to validate WCA approach and verify the performance of the algorithm, it has been applied to a number of constrained benchmark optimization problems and engineering design problems that are widely used in the literature. These benchmarks problems are examined and the optimization results are compared with other approaches to demonstrate the effectiveness of the approach.

The benchmark problems considered by different researchers include functions of various types such as quadratic, cubic, polynomial, and nonlinear with various numbers of the design variables, different types, and a number of inequality and equality constraints. Among engineering design problems include truss design, speed reducer design, pressure vessel design, tension/compression spring design, welded beam design, rolling element bearing design, disk brake design, gear train design, antenna array pattern synthesis, attribute reduction in rough sets, and water distribution system (WDS) design problems. The experimental results and analysis carried out based on the comparisons of the WCA against other optimization methods illustrate the efficiency of the WCA for handling constraints. In general, the WCA offers competitive solutions compared with other meta-heuristic optimizers based on the reported and experimental results carried out in different research reported in the literature.

Eskandar et al. (2012) demonstrated the applicability and effectiveness of WCA by applying the algorithm to four constrained benchmark problems and to seven constrained engineering and mechanical design problems. The benchmark problems include various types of quadratic, cubic, polynomial, and nonlinear functions with various numbers of the design variables, different types, and a number of inequality and equality constraints. The engineering design problems include three-bar truss design, speed reducer design, pressure vessel design, tension/compression spring design, welded beam design, rolling element bearing design, and multiple disk clutch brake design problem. They showed the efficiency and performance of the WCA by extensive experimental results, which proves that the algorithm offers better solutions than other meta-heuristic algorithms considered in the literature. Sadollah et al. (2015a) also applied WCA to four truss structures, for example, planar 52-bar truss, spatial 72-bar truss, planar 200-bar truss, and spatial 582-bar tower problems, and compared the performance with that of mine blast algorithm (MBA) (Sadollah et al., 2013) and improved MBA (Sadollah et al., 2015a), which shows that the performances are competitive.

Jabbaer and Zainudin (2014) applied WCA for attribute reduction problems in rough set theory as a mathematical tool for assessing the quality of solutions. Experimental results from regular benchmark data sets and comparison with other approaches demonstrate the viability of the application of WCA to such problem domain. It is also evident that the WCA is capable of producing good results compared to other meta-heuristic methods. WDS design problem is considered as a class of large combinatorial nonlinear optimization problem. Due to the complexity and large feasible solution, traditional optimization techniques are difficult to apply to WDS. Sadollah et al. (2014) first showed the application of WCA to WDS problem using a well-known Balerma benchmark problem widely used in the literature. The WCA achieved about 11% cost reduction for Balerma network problem compared to other meta-heuristic algorithms such as genetic algorithm, SA, and harmony search. It also shows better performance of the WCA in terms of convergence rate and solution quality over other methods.

Guney and Basbug (2015) applied QWCA to linear antenna arrays synthesis. Experimental results show that QWCA was able to achieve desired performance for the antenna arrays meeting the constraints and shows competitive performance compared to other algorithms.

Sadollah et al. (2015b) proposed the multiobjective WCA (MOWCA) for constrained multiobjective optimization problems (CMOPs). The MOWCA has been applied to 12 benchmark CMOPs and engineering design problems to verify the efficiency and performance of the proposed algorithm for handling multifaceted mathematical problems. The benchmark problems include constrained optimization problems and engineering design problems such as four-bar truss design, speed reducer design, disk brake design, welded beam design, spring design, and gear train design problems. Extensive simulations were carried out to show the efficiency and performance of the MOWCA. Three popular criteria, for example, generational distance, metric of spacing, and spread metric, are used to compare the performances, which again revealed that the proposed MOWCA is able to approach a full optimal Pareto front. The MOWCA also provides a superior quality of solutions in comparison to a variety of state-of-the art algorithms.

10.17   ARTIFICIAL CHEMICAL PROCESS ALGORITHM

10.17.1   INSPIRATION AND ALGORITHM

Artificial chemistry is a metaphor based on the abstraction of natural chemical molecular process and the dynamics of these complex systems (Dittrich et al., 2001). Artificial chemistry deals with combinatorial elements that change or maintain themselves and are able to produce new components.

These mechanisms can be utilized for modeling, information processing, or optimization. Artificial chemistry is defined by a triple (S, R, A), where $S = \{s_1, s_2,\ldots, s_n\}$ is the set of n valid molecules that participate in chemical reaction R determined by the reactor algorithm A. The molecules s_i can be abstract symbols, binary strings, numbers, character sequences, lambda expressions, or any other hierarchical data structures. The chemical reaction $R = \{r_1, r_2,\ldots, r_m\}$ is the set of reaction rules that describe the interaction between molecules $s_i \in S$. A rule $r_j \in R$ can be written in the form of chemical reaction where s_i reactants participate in a reaction and results in s'_j products as follows:

$$r_j: s_1 + s_2 + \cdots + s_n \rightarrow s'_1 + s'_2 + \cdots + s'_m \tag{10.188}$$

A is a reactor algorithm that determines how the set of reaction rules R is applied to the collection of molecules $P \subset S$ such that $P \subseteq S$ and $P \neq S$, that is, P cannot be identical to S since some molecules might be present in many exemplars. P is called a reactor or population. The algorithm A depends on the representation of P. The population P can be represented explicitly as a multiset or implicitly as a concentration vector. A number of reaction algorithms are formulated:

1. *Stochastic molecular collisions*: Population is represented by a multiset P. A sample of molecules is randomly drawn from P and checks whether a rule $r_j \in R$ can be applied. If so, molecules s_i are replaced by right-hand side molecules s'_j given by r_j. A reaction r_j can be written as

$$s_1, s_2 = \mathrm{draw}(P) \tag{10.189}$$

$$r_j: s_1 + s_2 \rightarrow s'_1 + s'_2 + \cdots + s'_m \tag{10.190}$$

where function $\mathrm{draw}(\cdot)$ selects randomly two molecules from the multiset P.

2. *Continuous differential or discrete difference equations*: Commonly, chemical systems are described by differential rate equations that reflect the development of the concentrations of molecular species. A reaction r can be written as

$$r_j: a_1 s_1 + a_2 s_2 + \cdots + a_n s_n \rightarrow b_1 s_1 + b_2 s_2 + \cdots + b_n s_n \tag{10.191}$$

where $\{a_i, b_i\}$ are the coefficients of stoichiometric factors of the reaction. If $\{a_i, b_i\}$ are zero, then s_i does not participate in the reaction, that is, if $a_i = 0$, then s_i is not a reactant and if $b_i = 0$, then s_i is not a product. s_i is also used to describe concentration of the species. The change of the overall concentration can be expressed by the following system of differential equations:

$$\frac{\mathrm{d}s_i}{\mathrm{d}t} = (b_i - a_i) \prod_{j=1}^{n} s_j^{a_j}, \quad i = 1,\ldots,n \tag{10.192}$$

Taking every reaction into account, the equation is written as

$$\frac{\mathrm{d}s_i}{\mathrm{d}t} = \sum_{r \in R} \left[(b_i^r - a_i^r) \prod_{j=1}^{n} s_j^{a_j^r} \right], \quad i = 1,\ldots,n \tag{10.193}$$

3. *Metadynamics*: The equations at a given time represent the dominant species. A dominant species is a population of molecules with concentration above a certain threshold. As the concentration changes, the dominant species also change, which results in a modification to the differential equations by adding or removing equations.
4. *Mixed approach*: Single macromolecules are simulated explicitly and a small number of molecules are represented by their concentrations.
5. *Symbolic analysis of equations*: Symbolic analysis is possible if the differential equation systems appear to be simple enough. By solving the equations symbolically, the steady-state behavior can be derived.

On the basis of the artificial chemistry model (S, R, A) and theoretical underpinning thereof described by Equations 10.188 through 10.193 in the preceding section, a number of algorithms can be developed. Irizarry (2004) proposed ACPA based on the principles of artificial chemical process. The notion of artificial chemical process is that low-level materials (i.e., reactants) evolved into high-level products by a series of chemical reactions. The quality of product is measured using a set of criteria called an objective function $f(\cdot)$. In ACPA, solutions are represented as vectors of decision variables $\theta = \{\theta^1, \theta^2, \ldots, \theta^n\}$ encoded into a set of independent discrete variables $x_j, j = 1, \ldots, V$, called molecules. In other words, the vector of decision variables is denoted as $\theta \in \Omega$, where Ω is the feasible solution space and decision variables are the transformation of molecules defined by $\theta = H(x)$, $x = \{x_j\}$, with $j = 1, \ldots, V$. The search procedure of ACPA operates on these molecules. At each iteration, a set of selected molecules undergo different chosen chemical reactions and create new molecules. The encoding scheme, for example, binary, integer, real, logical, or combinatorial, allows the algorithm to search over a large functional space with a reduced number of decision variables.

ACPA is an iterative improvement methodology, which considers one solution at a time. Given the decision variables θ encoded into molecules, a perturbation is applied to a randomly selected set of molecules to move between four disjunctive set of processes (also called compartment) named as L (load tank), AR (activation reactor), E (extraction unit), and S (separation unit) as shown in Figure 10.17. The algorithm operates on the molecules to create new trial states through the processing compartments L, AR, E, and S and the algorithm is called LARES after these four compartments.

At each iteration, a subset of molecules moves from one compartment to another as illustrated in Figure 10.17. At the beginning, a random initial set of molecules $x^g = \left\{x_1^g, x_2^g, \ldots, x_V^g\right\}$ is generated (as shown on the top-right box in Figure 10.17). x^g is considered the initial best vector. x^g is then placed in the load tank (L). The algorithm starts with all the molecules from load tank L. A subset of molecules from L is transferred to activation reactor (AR). The molecules undergo change to a random new state, that is, $x_j^g \xrightarrow{AR} x_j^a$, and $x_j^a \neq x_j^g$, $\forall j \in AR$, through a chemical process. This event generates a new trial vector $\theta^t = \left\{x_j^t\right\}$. In a different iteration, some reacted molecules are sent to extraction unit (E) where they are deactivated back to previous state upon entering the AR, that is, $x_j^g \xleftarrow{E} x_j^t$, $\forall j \in E$. These extracted molecules could be sent to separation unit (S) or recycled back to AR unit where the molecules change to random new states.

At each iteration, a trial vector $x_j^t = \left\{x_1^t, x_2^t, \cdots, x_V^t\right\}$ is generated comprising activated molecules in AR unit and deactivated molecules in other three sets $\left\{x_j^t, x_j^g, x_j^a\right\}$ with $x_j^a \neq x_j^g$.

$$x_j^t = \begin{cases} x_j^g & \text{if} \quad x_j \notin AR \\ x_j^a \neq x_j^g & \text{if} \quad x_j \in AR \end{cases} \tag{10.194}$$

If in the current iteration, a better objective function is found, that is, $f(\theta^t) \leq f(\theta^b)$, the AR can be emptied into separation unit S. In this case, all molecules retain the new state $x_j^g = x_j^a$, $\forall j \in AR$, and a new batch is started. ACPA is implemented mainly in two loops. The outer loop performs a

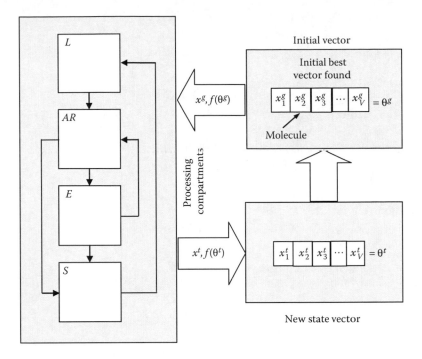

FIGURE 10.17   Process in ACPA.

perturbation to form AR. The inner loop performs improvement to AR in an iterative way. The main steps of the ACPA are as follows:

Step 1: The algorithm starts by initializing x^g randomly and placing all molecules in L and setting $L = \left\{ x_j^g \right\}$, $AR = \emptyset$, $E = \emptyset$, and $S = \emptyset$

Formation of AR *by perturbation in the outer loop:*

Step 2: Select a subset of molecules $T \subseteq L$

Move the elements to AR: $AR = AR \cup T$; $L = L \backslash T$

Step 3: Select a random new value $x_j^a \neq x_j^g$, $\forall j \in T$ for each molecule in T

Form new trial vector using Equation 10.194

$$x_j^t = \begin{cases} x_j^g & \text{if } x_j \notin AR \\ x_j^a \neq x_j^g & \text{if } x_j \in AR \end{cases}$$

Step 4: If $f(x^t) < f(x^g)$, accept trial vector as a new best solution $x^g \leftarrow x^t$

Move all molecules in AR to S: $S = S \cup AR$; $AR = \emptyset$

If (termination criterion met),

Return x^g as solution and exit algorithm

If (better solution found),

Goto Step 2 to perform another outer loop

Step 5: Set parameters: $RP = f(x^t)$ for goodness test in inner loop

Set the initial number of molecules in AR: $|AR|_0 = |AR|$ before starting next inner loop

Where $|AR|_0$ is the initial number of molecules in AR

Iterative improvement of AR *in the inner loop:*

Step 6: Select a random number $|E| < |AR|$ using a probability density function

Select a subset $E \subseteq AR$; $AR = AR \backslash E$

Where $|E|$ denotes number of elements in E

Step 7: Generate new trial vector x^t using

$$x_j^t = \begin{cases} x_j^g & \text{if } x_j \notin AR \\ x_j^a \neq x_j^g & \text{if } x_j \in AR \end{cases}$$

Step 8: If $f(x^t) < f(x^g)$,

    Accept trial vector as a new best solution $x^g \leftarrow x^t$

    Move all elements in AR to S: $S \cup AR$; $AR = \emptyset$

    If (termination criterion met),

        Return x^g as solution and exit algorithm

Step 9: Improvement criterion for AR

    If $f(x^t) \leq RP$,

        Move all molecules in E to S: $S = S \cup E$; $E = \emptyset$

    Update parameter RP: $RP = f(x^t)$

Step 10: If $f(x^t) > RP$,

    Generate new activated state for all elements in $E\left(x_j = x_j^a \neq x_j^g, \forall j \in E\right)$

    Transfer all elements in E back to AR: $AR = AR \cup E$; $E = \emptyset$

Step 11: If one of the following conditions is satisfied, exit inner loop

    {

    If better solution found in Step 8

    If $|AR| \leq 1$ or $|AR| \leq c$

    If $RR = (rec/|AR|_0) \geq RRT$

    Where c is a threshold, RR is the recycle-ratio, rec is the counter number of times E
        is sent back to AR, and RRT is an adjustable parameter.

    }

    Otherwise Goto Step 6 to perform a new inner loop

Step 12: Check the number of elements in L

    If $|L| < LT$,

    Move all elements in S to L: $L = L \cup S$; $S = \emptyset$

    If $|L| \ll LT$,

    Move all elements in AR to L: $L = L \cup AR$; $AR = \emptyset$

    Where $|L|$ denotes the number of elements in L and LT is a threshold value

Step 13: Goto Step 2 to perform an outer loop

The choice of parameters c, RRT, and LT in APCA is arbitrary. Irizarry (2005a) used $c = 0.3$, $RRT = 1.0$, and $LT = V/2$. A simplified flowchart explaining the mechanism of the algorithm is given in Figure 10.18. An extended version of the APCA, called LARES-PR, is proposed by Irizarry (2006) for a class of hybrid dynamic systems consisting of a finite set of dynamic subsystems where the sequence and duration are not established *a priori*. The efficiency of LARES-PR has been verified by applying it to a multicomponent nanoparticle alloy synthesis.

10.17.2   APPLICATIONS

ACPA has found applications in many domains (Irizarry, 2005a,b, 2006, 2008, 2011). Irizarry (2004) verified the performance of ACPA (also known as LARES) on test beds of random multimodal random problem generators, random LSAT problem generators with various degrees of epistasis, and real-valued functions with different degrees of multimodality, discontinuity, and flatness. The simulation results showed that LARES performed very well in terms of robustness and efficiency in all cases. Irizarry (2005a) applied the LARES algorithm to optimal control problem of a set of benchmark problems such as Van der Pol oscillator, plug-flow reactor, production of secreted protein in a fed-batch reactor, and Nishida problem. The efficiency of LARES algorithm has also been verified on large-scale discrete-time dynamic models with multiple control laws such

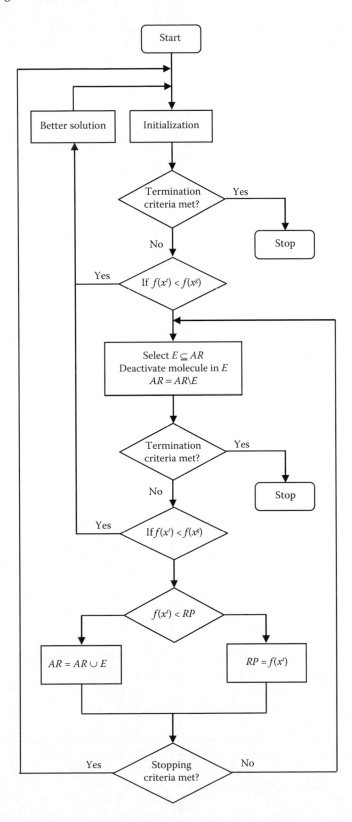

FIGURE 10.18 Flow diagram of ACPA.

as climate-economy model for global warming and population balance equations. The performance study with these benchmark problems showed that LARES is capable of finding the global maximum in all cases with a competitive number of function evaluations. Irizarry (2005b) proposed an ACPA-based fuzzy classification algorithm, which generates a fuzzy inference system for automatic pattern classification from data using LARES as the learning tool. The simulation results demonstrated the efficiency of ACPA. Irizarry (2008) applied LARES to global optimization of noisy black-box functions where evolutionary algorithm and ACPA are modified using concepts of optimal comparison and stochastic rule during the search. The modified algorithms accelerated the optimization process significantly. Irizarry (2011) applied ACPA and fast Monte Carlo methods to global and dynamic optimization for the solution of population balance models. The algorithms are compared in terms of robustness and speed in finding the global optimum.

10.18   ARTIFICIAL CHEMICAL REACTION OPTIMIZATION ALGORITHM

10.18.1   INSPIRATION AND ALGORITHM

Artificial chemistry demonstrated how a chemical process leads to a transformation of a set of chemical substances to another, which takes place in various rates leading to new products. The chemical reaction metaphor can be applied to other domains by encoding appropriate information into molecule-like objects and performing a set of chemical reaction-like operations onto them to obtain a certain kind of derivative information suitable for the problem in question. ACROA was proposed by Alatas (2011b, 2012) based on the concept of chemical reactions that occur between reactants and change to products by performing a set of chemical reactions. Molecules are encoded in ACROA using an appropriate scheme for the optimization problems. Compared to other chemistry-based algorithms, ACROA introduces seven types of chemical reactions: synthesis, decomposition, single-displacement, double-displacement, combustion, redox, and reversible reactions.

Synthesis reaction: A synthesis reaction or direct combination reaction is one of the most common types of chemical reactions. In a synthesis reaction, two or more chemical species combine to form a more complex product. An example of synthesis reaction can be expressed symbolically as $A + B \rightarrow AB$. The combination of hydrogen and oxygen to form water is an example of a synthesis reaction: $2H_2 + O_2 \rightarrow 2H_2O$. More description on synthesis reaction is provided in Chapter 9.

Decomposition reaction: A decomposition reaction is a type of chemical reaction where one reactant yields two or more products. In general, a decomposition reaction requires auxiliary energy source such as heat or light. The general form of a decomposition reaction can be expressed as $AB \rightarrow A + B$. Water can be separated by electrolysis into hydrogen gas and oxygen gas through the decomposition reaction: $2H_2O \rightarrow 2H_2 + O_2$. More description on decomposition reaction is provided in Chapter 9.

A substitution or displacement reaction is a type of reaction where a part of one reactant is replaced by another reactant. Single and double replacement reactions are the most common reactions in a chemical process.

Single-displacement reaction: A substitution or single-displacement reaction is characterized by one element being displaced from a compound by another element. In this reaction, two or more reactants participate in chemical reaction and an element of a compound is replaced by another element to produce two or more products. An example of a single-displacement reaction can be expressed symbolically as $A + BC \rightarrow AC + B$.

Double-displacement reaction: In this reaction, two or more reactants participate in a chemical reaction and elements of compounds are interchanged by one another to produce two new products. An example of a double-displacement reaction can be expressed symbolically as $AB + CD \rightarrow AD + CB$.

Combustion reaction: Any combustion involves a reaction between any combustible material and an oxidizer to form an oxidized product. A combustion reaction is a major class of chemical reactions, which occurs when a hydrocarbon reacts with oxygen to produce carbon dioxide and water. Combustion is an exothermic reaction and it releases energy in the form of heat and light. Sometimes, the temperature change is not noticeable when the reaction process is very slow. The general form of a combustion reaction can be expressed as hydrocarbon + oxygen → carbon dioxide + water. An example of a combustion reaction is the combustion of methane gas with oxygen that produces carbon dioxide and water: $CH_4 + 2O_2 \rightarrow CO_2 + 2H_2O$.

Redox reaction: Any chemical reaction in which the oxidation numbers (oxidation states) of the atoms are changed is an oxidation–reduction reaction. Such reactions transfer electrons from one reactant to another. The chemical agent that gains electrons is called an oxidizing agent. The agent that loses electrons is called an oxidized agent. Such reactions are also known as redox reactions. Redox is short for *red*uction *ox*idation (REDOX) reactions. An example is the reaction between hydrogen and fluorine where the oxidation–reduction is taking place: $H_2 + F_2 \rightarrow 2HF$.

Reversible reaction: In a reversible reaction, the chemical reactants form products that, in turn, react together to give the reactants back. Such reactions take place in both forward and backward directions under certain suitable conditions. An example of a reversible reaction is the calcium carbonate: $CaCO_3 \leftrightarrow CaO + CO_2$. Reversible reactions will reach an equilibrium point where the concentrations of reactants and products will no longer change.

Let there are R_i, $i = 1, \ldots, N$, chemical reactants that participate in M possible chemical reactions. The chemical system tends toward the highest entropy and lowest enthalpy. Enthalpy can be used as a measure or objective function for a minimization problem and entropy for a maximizing problem. The observable properties of all participating reactants become stable as the chemical system reaches an equilibrium state when the process terminates. The main steps of CROA are as follows:

Step 1: Parameters of CROA are initialized
Step 2: Population of initial reactants are generated randomly
      Evaluate reactants
Step 3: Decide on monomolecular or bimolecular reaction
      If (bimolecular reaction to be performed), Goto Step 7
Step 4: Select one molecule to perform monomolecular reaction operation
Step 5: On select chemical reaction: Perform decomposition or redox 1 operation
      Goto Step 8 for enthalpy computation
Step 6: Select two molecules
      Perform bimolecular reaction operation
Step 7: On select chemical reaction: Perform synthesis, displacement, or redox 2 operation
Step 8: Compute enthalpy
      If (enthalpy reduced), Goto Termination Step 10
Step 9: Perform reversible reaction operation
Step 10: If (termination criterion not met), Goto Step 3
Step 11: Return solution

The flow diagram of CROA is shown in Figure 10.19.

Yang et al. (2011) proposed an improved ACROA, called global-best-guided ACROA (GACR for short) by incorporating global best solution information (an approach borrowed from PSO) into the solution of ACROA. GACR method has been verified on a number of benchmark functions such as Griwangk, Rastrigin, Rosenbrock, Schaffer, and Ackley functions, which show that GACR with appropriate parameters outperforms ACROA.

10.18.2 APPLICATIONS

ACROA has been found to possess efficient objects, states, process, and events that can be designed as a computational method. ACROA found applications in many domains (Alatas, 2011b, 2012; Yang et al., 2011). Alatas (2011b) applied ACROA to multiple sequence alignment, and association rules mining problems and benchmark unimodal and multimodal functions such as Griwangk,

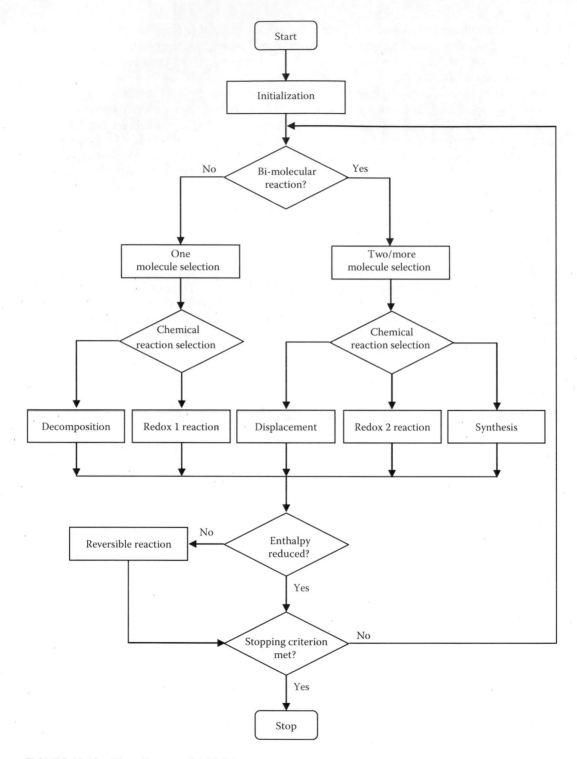

FIGURE 10.19   Flow diagram of ACROA.

Rastrigin, and Rosenbrock. Simulation results show promising performance of ACROA in all cases. Alatas (2012) performed one of the first applications of ACROA in classification rule discovery in the field of data mining and demonstrated the efficiency. ACROA has the advantage that the algorithm does not need to apply any local search method as it has the ability to perform both global and local search. The quality of reactants is measured by computing enthalpy.

The ACROA is chemistry-inspired computational method of optimization, which is different from CRO proposed by Lam and Li (2010a,b) and presented in Chapter 9 where the operators introduced to ACROA are based on different set of bimolecular and unimolecular operators in chemical reactions. Among bimolecular operators are bimolecular reaction, synthesis reaction, displacement reaction, and redox 2 reaction. Among monomolecular operators are decomposition reaction and redox 1 reaction. These chemical reaction operators are different from that of CRO operators. This method does not use an extra function for the determination of quality of reactants. The effectiveness and performance of ACRO were tested on three benchmark function optimization problems, namely, Griewangk, Rastrigin, and Rosenbrock functions with a population (reactant) size of 20. The algorithm reached a minimum error within 100 iterations and provided near-optimal solutions, which are competitive compared to ABC and BBO algorithms.

10.19 CHEMICAL REACTION ALGORITHM

10.19.1 INSPIRATION AND ALGORITHM

CRA is a population-based meta-heuristic algorithm proposed by Melin et al. (2013) based on the principles of artificial chemistry discussed in the earlier sections. A solution is represented as an element or compound in CRA. The elements and the compounds are selected in a pool, called reactor, where they participate in a series of chemical reactions until a state of equilibrium. The state of equilibrium is the optimal state for all participating elements or compounds, which is considered as the optimal solution. The CRA is described by a quintuple expressed as

$$CRA = f\left(S, \sigma, CR, O, RI\right) \tag{10.195}$$

where $S = \{s_1, s_2, ..., s_n\}$ is the set of n valid elements or compounds that may participate in the chemical reaction, σ is the selection strategy, R is the set of reactions that can be applied, O is the objective function against which the solution is to be evaluated, and RI is the reinsertion strategy. The main components of CRA are discussed in the sequel.

Elements/compounds (S): CRA uses elements or compounds for the parametric representation of the candidate solutions. Each element or compound comprising the parameters of the problem to be optimized represents a solution. A population of compounds is generated randomly, which is uniformly distributed over the entire search space. The compounds participate in chemical interaction with each other implicitly to produce new compounds. The interaction is independent of molecular structure, potential, or kinetic energy. The initial set of compounds S is randomly generated using the expression

$$x_{ij} = x_{LB} + (x_{UB} - x_{LB}) \times \rho \tag{10.196}$$

where $i = 1, ..., n, j = 1, ..., d$, d is the dimension of the compound, and $\rho \in [0,1]$ is a random number. The initial population of compounds is represented by

$$S = \begin{bmatrix} x_{11} & x_{12} & \cdots & x_{1d} \\ x_{21} & x_{22} & \cdots & x_{2d} \\ \vdots & \vdots & \ddots & \vdots \\ x_{n1} & x_{n2} & \cdots & x_{nd} \end{bmatrix} \tag{10.197}$$

Selection strategy (σ): The compounds are selected based on stochastic universal sampling. Other types of sampling can also be applied to CRA. A selection rate is used to select the number of compounds that will participate in the chemical reaction. The number of selected compounds is given by $SC = CRR \times n$, where $CRR \in [0,1]$ is the chemical reaction rate for each of the chemical reactions used in CRA such as synthesis, decomposition, single-substitution, and double-substitution reactions and n is the number of compounds available for the particular reaction. CRR decides on the number (i.e., percentage) of compounds of the population that should undergo the particular chemical reaction.

Chemical reaction (CR): A chemical reaction is a process in which at least one participating compound (i.e., reactant) changes its chemical composition and properties to yield products. Four types of chemical reactions are considered in CRA such as synthesis, decomposition, single-substitution, and double-substitution reactions. The synthesis and decomposition reactions are applied as diversifying mechanisms for exploring the search space. These reactions are highly effective and lead to faster convergence of the CRA. The single-substitution and double-substitution reactions are applied as intensification mechanisms for exploiting the search space. They are used as local search technique and applied in the neighborhood of a good solution found using a global search technique such as synthesis and decomposition. CRA can trigger single reaction or multiple reactions at a time depending on the nature of the problem.

In synthesis reaction, two reactants participate in the reaction to produce one product, symbolically expressed as $B + C \rightarrow BC$, where B and C are reactants and BC is a product. The synthesis vector SV is obtained by

For $i = 1$ to SCS/sf
$$SV = x_{2i-1} + x_{2i}$$

where $i = 1,\ldots,$ SCS and SCS is the number of selected compounds for synthesis defined by $SCS = SR \times n$. SR is the reaction rate for synthesis and *sf* is a synthesis factor generally set to 2.

In the decomposition reaction, one compound participates in the reaction and decomposes into two products, symbolically expressed as $BC \rightarrow B + C$. The decomposition vector DV is obtained by

For $i = 1$ to SCD
{
$$DV_j = x_i \times \rho$$
$$DV_{j+1} = x_i \times (1-\rho)$$
}

where $i = 1,\ldots,$ SCD and SCD is the number of selected compounds for decomposition defined by $SCD = DR \times n$. DR is the reaction rate for decomposition. ρ is a random number within the range of $0 < \rho < 1$. A decomposition factor df can be used to decide on the number of elements the decomposition should yield. In general, df is set to 2.

In the single-substitution reaction, a free element reacts with a compound, replaces an element of the compound when it is more reactive than the other element and produces a new compound and a free element. It is symbolically expressed as $AB + C \rightarrow AC + B$. The single-substitution reaction requires a synthesis and a decomposition operation. The single-substitution vector SSV is obtained by

For $i = 1$ to SCSS
{
$$DV_j = x_{2i} \times \rho$$
$$DV_{j+1} = x_{2i} \times (1-\rho)$$

$$SSV_j = SP(DV_j, x_i)$$
$$SSV_{j+1} = DV_{j+1}$$
$$\}$$

where $i = 1,\ldots,$ SCSS and SCSS is the number of selected compounds for single substitution defined by SCSS $=$ SSR $\times n$. SSR is the reaction rate for single substitution. SP($\cdot$) is the synthesis procedure.

In the double-substitution reaction, two compounds react, swap the more reactive elements between them, and produce two new compounds. It is symbolically expressed as $AB + CD \to AC + BD$. The double substitution reaction requires a synthesis and a decomposition operation. The double-substitution vector DSV is obtained by

For $i = 1$ to SCDS
$$\{$$
$$DV_j = x_i \times \rho$$
$$DV_{j+1} = x_i \times (1-\rho)$$
$$DSV_j = SP(DV_j, DV_{i+2})$$
$$DSV_{j+1} = SP(DV_{j+1}, DV_{j+3})$$
$$\}$$

where $i = 1,\ldots,$ SCDS and SCDS is the number of selected compounds for double substitution defined by SCDS $=$ DSR $\times n$. DSR is the reaction rate for double substitution.

Objection function (O): A feasible solution obtained from the chemical reaction is evaluated against an objective function defined for the specific problem. The objective function is to be minimized or maximized by searching the solution space.

Reinsertion strategy (RI): An elitist reinsertion strategy is applied to a new set of compounds generated by the chemical reaction. This ensures the inclusion of the best compounds in the population for the next iteration. An elitist reinsertion strategy is adopted so as to ensure the selection of the best elements in the pool. Thus, the average fitness of the entire element pool increases with iterations.

Each candidate solution is evaluated using the objective function to yield its potential energy (PE), which is a measure of fitness of the solution. Reactants are selected based on the values of PE and undergo CRA operations to create new products. The population converges to an optimal solution over a finite number of iterations. The main steps of the CRA are described as follows:

Step 1: Set initialial values to parameters: size of pool, number of trials, upper and lower boundary, synthesis and decomposition rate, and single- and double-substitution rates
Step 2: Generate an initial pool randomly
Step 3: Perform evaluation of the initial pool
Step 4: Identify the best solution
Step 5: Perform synthesis operation on agents to get synthesis vector
Step 6: Perform decomposition operation on agents to get decomposition vector
Step 7: Perform single-substitution operation on agents to get single-substitution vector
Step 8: Perform double-substitution operation on agnts to get double-substitution vector
Step 9: Perform evaluation of synthesis vector, decomposition vector, single-substitution vector, and double-substitution vector
Step 10: Apply elitist reinsertion to improve average fitness of the pool
Step 11: Update best solution
Step 12: If (termination condition not met), Goto Step 5
Step 13: Return solution

The general flow diagram of the CRA is shown in Figure 10.20.

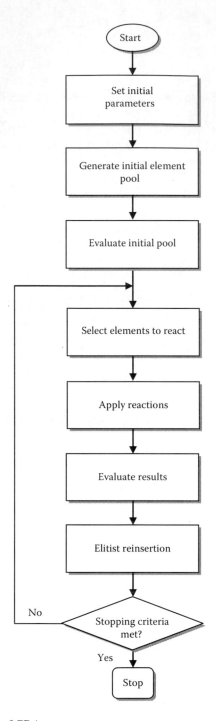

FIGURE 10.20   Flow diagram of CRA.

10.19.2 APPLICATIONS

CRA has found applications in controller designs (Astudillo et al., 2015; Melin et al., 2013). A type-1 and a type-2 fuzzy logic controller were designed. The goal was to find the gain constants involved in the tracking controller for the dynamic model of a unicycle mobile robot (Melin et al., 2013). Simulation results show that the proposed optimization method is able to outperform the results previously obtained applying a GA optimization technique. The optimal type-1 fuzzy logic controller obtained with CRA paradigm has been able to reach smaller error faster than GA. Also, the type-2 fuzzy controller was able to perform better under the presence of disturbance despite a large initial error. Astudillo et al. (2015) verified the performance of CRA on 10 complex numerical benchmark functions and compared with that of other meta-heuristic optimization algorithms such as GA, PSO, ACO, GSA, HS, and hybrid algorithms. Simulation results showed that CRA was able to reach near the optimal values for some functions and outperformed many of the paradigms.

10.20 GASES BROWNIAN MOTION OPTIMIZATION (GBMO) ALGORITHM

10.20.1 INSPIRATION AND ALGORITHM

Molecules of ideal gas are in a constant irregular motion. On the basis of Maxwell–Boltzmann distribution of molecular speed (Hirschfelder, 1966), the velocity v of a molecule of mass m at temperature T can be defined by

$$v_i = \sqrt{\frac{3K_B T}{m_i}} \tag{10.198}$$

where K_B is Boltzmann's constant that relates energy and temperature in environment, T is the absolute temperature, and v_i and m_i are the velocity and mass of i-th molecule in the gass, respectively.

The velocity is proportional to the square root of temperature. This means as temperature T approaches to 0, the velocity v will tend to 0. This will lead to a stable system at $T = 0$. Therefore, the velocity of each molecule can be computed iteratively applying Equation 10.198 as follows:

$$v_i(t+1) = v_i(t) + \sqrt{\frac{3K_B T}{m_i}} \tag{10.199}$$

where t is the current iteration number.

The change in position of a molecule Δx is caused by the velocity. Δx is defined by

$$\Delta x = x_i(t+1) - x_i(t) = v_i \times \Delta t \tag{10.200}$$

where $x_i(t + 1)$ is the new position after the time interval Δt. Δt is is considered unity and is used to convert the velocity into position. The new position $x_i(t + 1)$ of the molecule can be computed from Equation 10.200 as follows:

$$x_i(t+1) = x_i(t) + v_i(t) \tag{10.201}$$

GBMO algorithm was proposed by Abdechiri et al. (2011, 2013) based on the laws of Brownian motion (Gallavotti, 1999; Ornstein, 1917) and turbulent rotation motion of gas molecules. A population of molecules is randomly generated. Each molecule has a position, mass, velocity, and radius of turbulent that represents part of the solution. The population represents an array of molecules with random positions. Molecules move in the solution space toward equilibrium state, which is to be considered as an optimal solution. Higher temperature causes the molecules to move at higher velocity, which are gases Brownian motion (Uhlenbeck and Ornstein, 1930). Lower temperature causes the molecules to move at lower velocity, which are gases turbulent rotation motion. As the iterative process continues, velocity and position are updated leading to an optimal solution. The algorithm starts at a high temperature with gases Brownian random motion. The higher velocities of the molecules help the algorithm to explore the solution space. The velocity decreases with decreasing temperature. A temperature T is assigned to the system. In the early stage of the search, high temperature in the gas causes to increase the kinetic energy and the velocity of the molecules. Brownian motion causes global exploration of the full range of the search space. Turbulent rotational motion causes local exploitation of the search space and produces optimal solution with higher accuracy. The turbulent rotation motion for each molecule is generated using the chaotic sequence generator called circle map defined by

$$x_i(t+1) = x_i(t) + b - \left(\frac{a}{2\pi}\right)\sin(2\pi \cdot x_i(t)) \bmod (1) \tag{10.202}$$

where $a = 0.5$ and $b = 0.2$ are two parameters chosen for the turbulent rotation motion.

At a later stage of the search, low temperature in the gas causes to decrease the kinetic energy and the velocity of the molecules. The decrease of temperature is calculated according to the rule

$$T = T - \left(\frac{1}{\text{mean}\left[\text{fit}(x_i)\right]}\right) \tag{10.203}$$

where $\text{fit}(x_i)$ is the fitness value of the i-th molecule.

In GBMO, the mass of the i-th molecule is defined by

$$m_i(t) = \frac{\text{fit}(x_i) - x^{\text{worst}}(t)}{x^{\text{best}}(t) - x^{\text{worst}}(t)} \tag{10.204}$$

where $x^{\text{best}}(t)$ and $x^{\text{worst}}(t)$ are the best fitness and worst fitness of molecues in the population at iteration t and are defined (for minimization problem) as follows:

$$\begin{cases} x^{\text{best}} = \min\{\text{fit}(x_i), \quad i = 1,\ldots,N\} \\ x^{\text{worst}} = \max\{\text{fit}(x_i), \quad i = 1,\ldots,N\} \end{cases} \tag{10.205}$$

When temperature is decreased, the velocity is also decreased at turbulent rotation motion, which helps the algorithm to exploit the solution space.

The temperature plays a decisive role in controlling the exploration and exploitation. The convergence of the GBMO algorithm is ensured by decreasing the temperature to $T = 0$. Mass and temperatures are updated. The main steps of GBMO algorithm are described in the following:

Step 1: Initialize a population of molecules x_i, $i = 1,\ldots, N$, in n-dimensional space randomly
       Initialize velocities v_i randomly
       Set random radius of turbulence for each molecule
Step 2: Assign temperature T to the system
Step 3: Update velocity v_i using Equation 10.199
       Update position x_i using Equation 10.201
Step 4: Evaluate fitness of molecules
Step 5: Compute turbulent rotation motion using Equation 10.202
Step 6: Evaluate fitness of new positions of molecules
       Determine x^{best} and x^{worst}
       If $(\text{fit}[x_i(t + 1)] < \text{fit}[x_i(t)])$
       $x_i(t) = x_i(t + 1)$

Step 7: Decrease temperature T using Equation 10.203
  Update mass m_i using Equation 10.204
Step 8: If $(T \neq 0)$, Goto Step 2
Step 9: Return solution

The flowchart of the GBMO algorithm is shown in Figure 10.21.

Rathore and Roy (2014) proposed a modified GBMO to speed up the convergence by incorporating a sign function into the velocity, position, and turbulent rotation motion equations in Equations 10.199, 10.201, and 10.202 as follows:

$$v_i(t+1) = v_i(t) + \text{sign}(\rho) \times \sqrt{\frac{3K_B T}{m_i}} \qquad (10.206)$$

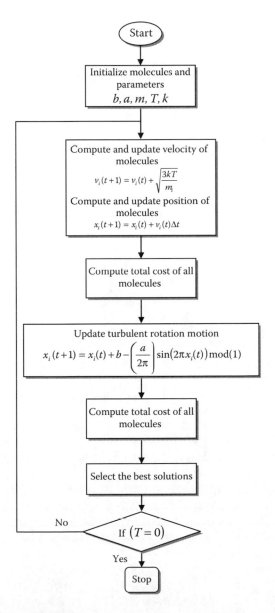

FIGURE 10.21 Flow diagram of GBMO.

$$x_i(t+1) = x_i(t) + \text{sign}(\rho) \times v_i(t) \tag{10.207}$$

The turbulent rotation motion for each molecule is generated using the chaotic sequence generator called circle map defined by

$$x_i(t+1) = x_i(t) + b \times \text{sign}(\rho) - \left(\frac{a}{2\pi}\right)\sin(2\pi \cdot x_i(t)) \times \text{sign}(\rho) \tag{10.208}$$

where $\rho \in [0,1]$ is a random number and $\text{sign}(\rho)$ is defined as

$$\text{sign}(\rho) = \begin{cases} -1 & \text{if} \quad (\rho) \le 0.5 \\ 1 & \text{if} \quad (\rho) > 0.5 \end{cases} \tag{10.209}$$

The temperature decrease is a linear function of time (i.e., iteration) defined by

$$T(t+1) = T(t) - \rho \times T(t) \tag{10.210}$$

where $\rho \in [0,1]$ is a random number.

The modified GBMO was verified on static TNEP problem. A detailed description of TNEP problem is provided in Chapter 5.

In order to improve the local search and quality of initial population, Elyas et al. (2014) proposed a hybrid CSA and GBMO. They verified the performance of the hybrid CSA–GBMO algorithm on ELD problem. Abdechiri and Faez (2015) proposed a hybrid approach combining the global search feature of GBMO algorithm with the fast convergence feature of structured nonlinear parameter optimization method (SNPOM). The performance of GBMO–SNPOM approach was verified on training of RBF networks. Xing and Gao (2014) have reported a survey on chemistry-based algorithms, which briefly describes the methods with useful references therein.

10.20.2 APPLICATIONS

GBMO algorithm has been applied to satisfiability problem, benchmark functions such as sphere, Rastrigrin, Rosenbrock, Griewank, Ackley, Booth, and Zakharoz functions, and real-world problems such as Lennard–Jones potential problem and Tersoff potential function. The efficiency of the GBMO was compared with that of well-known methods such as PSO, ICA, GSA, and GA. The experimental results show that GBMO is tractable for medium-size problems and more successful for high-dimensional problems (Abdechiri et al., 2011, 2013). Rathore and Roy (2014) applied modified GBMO to static TNEP problem. The effectiveness was proved by testing the algorithm on Garver's 6-bus, IEEE 24-bus, and IEEE 25-bus test systems. Elyas et al. (2014) applied hybrid CSA–GBMO algorithm to ELD problem. A detailed description of ELD problem is provided in Chapter 2. Abdechiri and Faez (2015) applied GBMO–SNPOM approach to training of radial basis function (RBF) networks for solving multi-instance multi-label learning framework. The widths and centres of RBF networks were obtained by GBMO, and gradient-based fast converging parameter estimation (i.e., weights) was performed by SNPOM. The effectiveness was verified on scene classification, text categorization, and Corel data set. Zamani et al. (2016) applied GBMO algorithm to the design of a fractional-order PID controller for load frequency control. Load frequency control is a challenging issue in power system operation. GBMO is used to optimize the mitigation of frequency and exchanged power deviation.

10.21 CONCLUSION

In general, all the algorithms discussed in the above sections can be classified into two groups: physics-based algorithms and chemistry-based algorithms. This chapter introduced and briefly discussed the most recent physics- and chemistry-based algorithms. Physics-based algorithms can be further classified as gravitational force-based algorithms, astrophysics-based algorithms, electromagnetism-based algorithms, optics-based algorithms, and Newton's laws of motion-based algorithms. There are also a few hydro-dynamics-based algorithms. The chemistry-based algorithms can be further classified as ACPA, ACROA, CRA, and GBMO.

Among gravitational force-based algorithms, SGO, GIO, and APO algorithms are reported in the literature. There are also a few algorithms based on astrophysics reported in the literature. Among them, BB–BC algorithm, black hole search, galaxy-based search, and integrated radiation search algorithms are reported. These algorithms are getting popular. Among electromagnetism-based algorithms are electromagnetism optimization, CSS, and HO. A good number of applications of electromagnetism optimization and CSS algorithms have been reported. It appears that these two algorithms are getting popular with the scientific community. Among optics-based algorithm is the RO algorithm. It is a very recent addition to the meta-heuristic family. CBO is based on Newton's laws of motion. The implementation of the algorithm is very simple. There are a few hydro-dynamics-based algorithms such as RFD and WCA. There are a good number of variants and applications that have been reported in the literature. It appears that these algorithms are slowly getting popular.

There are also a few other algorithms reported recently. These are less-known algorithms such as atmospheric clouds model optimization (Yan and Hao, 2012, 2013), cloud model-based optimization (Wang et al., 2012), gravitational field algorithm (Zheng et al., 2010), gravitational clustering algorithm (Kundu, 1999), gravitational emulation local search algorithm (Webster, 2004), light ray optimization (Shen and Li, 2008; 2009), magnetic optimization (Tayarani-N and Akbarzadeh-T, 2008), water flow algorithm (Brodic and Milivojevic, 2010, 2011), water flow-like algorithm (Yang and Wang, 2007), interior search algorithm (Gandomi, 2014), and MBA (Sadollah et al., 2012, 2013).

Sadollah et al. (2012) recently developed MBA, which mimics the explosion of landmines. MBA is inspired by the process of landmines explosion; shrapnel pieces fly away and collide with other mines in the vicinity of the explosion area causing further explosions. Consider a landmine field where the goal is to clear landmines. To clear all the mines, the position of the most explosive mine must be located. This position corresponds to the optimal design. Landmines of different sizes and explosive power are planted under the ground. Landmine explosions cause many pieces of shrapnel to be propelled in the air. The casualties of each piece of shrapnel are evaluated using a cost function (fitness function) and, then, related to the presence of other landmines with different explosive power (Sadollah et al., 2012, 2013). The MBA was successfully applied to discrete sizing optimization of truss structures and constrained engineering optimization problems (Sadollah et al., 2012, 2013). An improved MBA with enhanced operators is reported in Sadollah et al. (2015a).

REFERENCES

Aarts, E. H. K. and Lenstra, J. K. 1997. *Local Search in Combinatorial Optimisation*, London, UK: John Wiley.

Abbott, B. P., Abbott, R., Abbott, T. D., Abernathy, M. R., Acernese, F., Ackley, K., Adams, C., Adams, T., Addesso, P., Adhikari, R. X. et al. 2016. Observation of gravitational waves from a binary black hole merger, *Physics Review Letters*, 116(6), 061102.

Abdechiri, M. and Faez, K. 2015. Efficacy of utilising a hybrid algorithmic method in enhancing the functionality of multi-instance multi-label radial basis function network, *Applied Soft Computing*, 34, 788–798.

Abdechiri, M., Meybodi, M. R., and Bahrami, H. 2011. A new algorithm: Inspired of gases Brownian motion, *National Conference of Computer Society of Iran CSI2011*, 2011, Tehran, Iran, pp. 171–176.

Abdechiri, M., Meybodi, M. R., and Bahrami, H. 2013. Gases Brownian motion optimization: An algorithm for optimization (GBMO), *Applied Soft Computing*, 13, 2932–2946.

Abuhamdah, A. and Ayob, M. 2009a. Experimental result of particle collision algorithm for solving course timetabling problems, *International Journal of Computer Science and Network Security*, 9(9), 134–142.

Abuhamdah, A. and Ayob, M. 2009b. Multi-neighbourhood particle collision algorithm for solving course timetabling problems, *Proceedings of the IEEE 2nd Conference on Data Mining and Optimization*, October 27–28, Selangor, Malaysia, pp. 21–27.

Abuhamdah, A. and Ayob, M. 2009c. Hybridization multi-neighbourhood particle collision algorithm and great deluge for solving course timetabling problems, *Proceedings of the IEEE 2nd Conference on Data Mining and Optimization*, October 27–28, Selangor, Malaysia, pp. 108–114.

Abuhamdah, A. and Ayob, M. 2010. Adaptive randomized descent algorithm for solving course timetabling problems, *International Journal of the Physical Sciences*, 5(16), 2516–2522.

Abuhamdah, A. and Ayob, M. 2011. MPCA-ARDA for solving course timetabling problems, *Proceedings of the IEEE 3rd Conference on Data Mining and Optimization (DMO)*, June 28–29, Selangor, Malaysia, pp. 171–177.

Alatas, B. 2011a. Uniform Big Bang–Chaotic Big Crunch optimization, *Communications in Nonlinear Science and Numerical Simulation*, 16(9), 3696–3703.

Alatas, B. 2011b. ACROA: Artificial chemical reaction optimization algorithm for global optimization, *Expert Systems with Applications*, 38, 13170–13180.

Alatas, B. 2012. A novel chemistry based meta-heuristic optimization method for mining of classification rules, *Expert Systems with Applications*, 39, 11080–11088.

Aliasghary, M., Eksin, I., and Guzelkaya, M. 2011. Fuzzy-sliding model reference learning control of inverted pendulum with Big Bang–Big Crunch optimization method, *Proceedings of the11th International Conference on Intelligent Systems Design and Applications (ISDA'11)*, November 22–24, 2011, Cordoba, Spain, pp. 380–384.

Ampere, A.-M. 1820. Memoir on the mutual action of two electric currents, *Annales de Chimie et Physique*, 15, 59.

Anochi, J. A. and Velho, H. F. de C. 2014. Optimization of feedforward neural network by multiple particle collision algorithm, *IEEE Symposium on Foundations of Computational Intelligence (FOCI)*, December 9–12, Orlando, Florida, pp. 128–134.

Astudillo, L., Melin, P., and Castillo, O. 2015. Introduction to an optimization algorithm based on the chemical reactions, *Information Sciences*, 291, 85–95.

Bak, P. and Sneppen, K. 1993. Punctuated equilibrium and criticality in a simple model of evolution, *Physical Review letters*, 59, 381–384.

Bak, P., Tang, C., Wiesenfeld, K. 1987. Self-organized criticality: An explanation of 1/f-noise, *Physical Review Letters*, 86(23), 5211–5214.

Bars, I., Chen, S.-H., Steinhardt, P. J., and Turok, N. 2012. Antigravity and the Big Crunch/Big Bang transition, *Physics Letters B*, 715, 278–281.

Biot, J.-B. and Savart, F. 1820. Note sur le magnétisme de la pile de Volta, *Annales de Chimie et Physique*, 15, 222–223.

Biswas, A., Mishra, K. K., Tiwari, S., and Misra, A. K. 2013. Physics-inspired optimization algorithms: A survey, *Journal of Optimization*, 2013, 438152, 16 pp.

Boettcher, S. 1999. Extremal optimization and graph partitioning at the percolation threshold, *Journal of Physics A: Mathematical and General*, 32, 5201–5211.

Boettcher, S. 2000. Extremal optimization: Heuristics via co-evolutionary avalanches, *Computing in Science and Engineering*, 2(6), 75.

Boettcher, S. 2005. Extremal optimization for the Sherrington–Kirkpatrick spin glass, *The European Physical Journal B*, 46, 501–505.

Boettcher, S. 2014. Extremal optimization for ground states of the Sherrington–Kirkpatrick spin glass with Levy bonds, *Physics Procedia*, 53, 24–27.

Boettcher, S. 2015. Extremal optimization for quadratic unconstrained binary problems, *Physics Procedia*, 64, 16–19.

Boettcher, S. and Grigni, M. 2002. Jamming model for the extremal optimization heuristic, *Journal of Physics A: Mathematical and General*, 35, 1109–1123.

Boettcher, S. and Percus, A. G. 1999. Extremal optimization: Methods derived from co-evolution, in: *GECCO-99: Proceedings of the Genetic and Evolutionary Computation Conference*, July 13–17, 1999, Orlando, Florida, pp. 825–832.

Boettcher, S. and Percus, A. G. 2000. Nature's way of optimizing, *Artificial Intelligence*, 119, 275–286.

Boettcher, S. and Percus, A. G. 2001a. Optimisation with extremal dynamics, *Physical Review Letters*, 86, 5211–5214.

Boettcher, S. and Percus, A. G. 2001b. Extremal optimization for graph partitioning, *Physical Review E*, 64, 026114.

Boettcher, S. and Percus, A. G. 2002. Extremal optimisation: An evolutionary local-search algorithm, *Physical Review Letters*, 86, 5211–5214.

Boettcher, S. and Percus, A. G. 2004. Extremal optimization at the phase transition of the three-coloring problem, *Physical Review E*, 69, 066703.

Brodic, D. and Milivojevic, Z. 2010. An approach to modification of water flow algorithm for segmentation and text parameters extraction, in: L. M. Camarinha-Matos, P. Pereira, and L. Ribeiro, eds., *Emerging Trends in Technological Innovation, IFIP Advances in Information and Communication Technology*, Berlin, Germany: Springer, Vol. 314, pp. 324–331.

Brodic, D. and Milivojevic, Z. 2011. A new approach to water flow algorithm for text line segmentation, *Journal of Universal Computer Science*, 17, 30–47.

Camp, C.-V. 2007. Design of space trusses using Big Bang–Big Crunch optimization, *Journal of Structural Engineering*, 133, 999–1008.

Chandrasekar, P. and Krishnamoorthi, M. 2014. BHOHS: A two stage novel algorithm for data clustering, *2014 International Conference on Intelligent Computing Applications*, March 6–7, 2014, pp. 138–142.

Chandrasekhar, S. 1931a. The Highly collapsed configurations of a stellar mass, *Monthly Notices of the Royal Astronomical Society*, 91, 456–466.

Chandrasekhar, S. 1931b. The maximum mass of ideal white dwarfs, *Astrophysical Journal*, 74, 81–82.

Chandrasekhar, S. 1998. *The Mathematical Theory of Black Holes*, Oxford: Oxford University Press.

Chen, M.-R., Li, X., Zhang, X., and Lu, Y.-Z. 2010. A novel particle swarm optimizer hybridized with extremal optimization, *Applied Soft Computing*, 10, 367–373.

Chen, M.-R. and Lu, Y.-Z. 2008. A novel elitist multiobjective optimization algorithm: Multiobjective extremal optimization, *European Journal of Operation Research*, 188, 637–651.

Chen, M.-R., Lu, Y.-Z., and Yang, G. 2006. Population-based extremal optimization with adaptive lévy mutation for constrained optimization, in: Y. Wang, Y.-M. Cheung, and H. Liu, eds., *Proceedings of the 2006 International Conference on Computational Intelligence and Security (CIS'06)*, vol. 4456 of Lecture Notes in Computer Science, Springer, Berlin, Heidelberg, pp. 144–155.

Chen, Y.-W., Lu, Y.-Z., and Chen, P. 2007. Optimization with extremal dynamics for the travelling salesman problem, *Physica A*, 385, 115–123.

Chuang, C. and Jiang, J. 2007. Integrated radiation optimization: Inspired by the gravitational radiation in the curvature of space–time, *IEEE Congress on Evolutionary Computation (CEC)*, September 25–28, Singapore, pp. 3157–3164.

Coulomb, C. A. 1785a. Premier mémoire sur l'électricité et le magnétisme, Histoire de l'Académie Royale des Sciences, *Imprimerie Royale*, Paris, 569–577.

Coulomb, C. A. 1785b. Second mémoire sur l'électricité et le magnétisme, Histoire de l'Académie Royale des Sciences, *Imprimerie Royale*, Paris, 578–611.

Davies, P. C. W. 1978. Thermodynamics of Black Holes, *Reports on Progress in Physics*, 41(8), 1313–1355.

Davies, P. C. W. 1980. *The Search for Gravity Waves*, Cambridge: Cambridge University Press.

De Castro, L. N. 2007. Fundamentals of natural computing: An overview, *Physics of Life Reviews*, 4, 1–36.

De Falco, I., Laskowski, E., Olejnik, R., Scafuri, U., Tarantino, E., Tudruj, M. 2015. Extremal optimization applied to load balancing in execution of distributed programs, *Applied Soft Computing*, 30, 501–513.

De Sousa, F. L., Vlassov, V., and Ramos, F. M. 2004. Generalized extremal optimization: An application in heat pipe design, *Applied Mathematical Modelling*, 28, 911–931.

Ding, J., Lu, Y.-Z., and Chu, J. 2013. Studies on controllability of directed networks with extremal optimization, *Physica A*, 392, 6603–6615.

Dittrich, P., Ziegler, J., and Banzhaf, W. 2001. Artificial chemistries—A review, *Artificial Life*, 7, 225–275.

Dorigo, M. and Stützle, T. 2004. *Ant Colony Optimization*, Cambridge, MA: MIT Press.

Duch, J. and Arenas, A. 2005. Community detection in complex networks using extremal optimization, *Physical Review E: Statistical, Nonlinear, and Soft Matter Physics*, 72, 027104.

Duderstadt, J. J. and Hamilton, L. J. 1976. *Nuclear Reactor Analysis*, New York: John Wiley and Sons.

Einstein, A. 1905. Über einen die Erzeugung und Verwandlung des Lichtes betreffenden heuristischen Gesichtspunkt, *Annalen der Physik*, 17(6), 132–148 (in German).

Einstein, A. 1915. Die Feldgleichungun der Gravitation, *Sitzungsberichte der Preussischen Akademie der Wissenschaften zu Berlin*, 844–847.

Einstein, A. 1916. The foundation of the general theory of relativity, *Annalen der Physik*, 354(7), 769–822.

Einstein, A. 1918. Über gravitationswellen, *Sitzungsberichte der Königlich Preussischen Akademie der Wissenschaften Berlin*, Part 1, 154–167.

Elaydi, S. N. 1999. *Discrete Chaos*, 2nd Edt. Chapman & Hall/CRC Press, Boca Ratan, London, New York.

Elyas, S. H., Mandal, P., Haque, A. U., Giani, A., and Tseng, T.-L. 2014. A new hybrid optimisation algorithm for solving economic load dispatch problem with valve-point effect, *North American Power Symposium (NAPS)*, September 7–9, 2014, Pullman, WA, pp. 1–6.

Erol, O. K. and Eksin, I. 2006. A new optimization method: Big Bang–Big Crunch, *Advances in Engineering Software*, 37(2), 106–111.

Eskandar, H., Sadollah, A., Bahreininejad, A., and Hamdi, M. 2012. Water cycle algorithm—A novel meta-heuristic optimization method for solving constrained engineering optimization problems, computers & structures, *Journal of Computers and Structures*, 110–111, 151–166.

Everitt, C. W. F., DeBra, D. B., Parkinson, B. W., Turneaure, J. P., Conklin, J. W., Heifetz, M. I., Keiser, G. M., Silbergleit, A. S., Holmes, T., Kolodziejczak, J. et al. 2000. *Cosmology: The Science of the Universe*, Cambridge: Cambridge University Press.

Flores, J. J., Lopez, R., and Barrera, J. 2010. Particle swarm optimization with gravitational interactions for multimodal and unimodal problems, in: G. Sidorov, A. H. Aguirre, R. Garcia, and C. Alberto, eds., *Proceedings of the 9th Mexican International Conference on Artificial Intelligence (MICAI 2010)*, Berlin, Germany: Springer, pp. 361–370.

Flores, J. J., Lopez, R., and Barrera, J. 2011. Gravitational interactions optimization, in: Carlos A. Coello, ed., *Learning and Intelligent Optimization*, Lecture Notes in Computer Science, Berlin, Germany: Springer, pp. 226–237.

Gallavotti, G. 1999. *Statistical Mechnics: Short Treatise*, Rome, Italy: University of Rome.

Gandomi, A. H. 2014. Interior search algorithm (ISA): A novel approach for global optimization, *ISA Transactions*, 53, 1168–1183.

Gat, J. 2010. *Isotope Hydrology: A Study of the Water Cycle*, Singapore: World Scientific Publisher.

Genc, H. M., Eksin, I., and Erol, O. K. 2010. Big Bang–Big Crunch optimization algorithm hybridized with local directional moves and application to target motion analysis problem, in: *Proceedings of the IEEE International Conference on Systems, Man and Cybernetics (SMC'10)*, October 10–13, 2010, Istanbul, Turkey, pp. 881–887.

Genc, H. M. and Hocaoglu, A. K. 2008. Bearing-only target tracking based on Big Bang–Big Crunch algorithm, in: *Proceedings of the 3rd International Multi-Conference on Computing in the Global Information Technology (ICCGI'08)*, July 27–August 1, 2008, IEEE Computer Society, Washington DC, pp. 229–233.

Gendreau, M. and Potvin, J.-Y. 2010. *Handbook of Metaheuristics*, Berlin, Germany: Springer.

Giacobbe, F. W. 2005. How a type II supernova explodes, *Electronic Journal of Theoretical Physics*, 2(6), 30–38.

Gibbons, G. W. and Hawking, S. W. 1993. *Euclidean Quantum Gravity*, Singapore: World Scientific Publisher.

Gillispie, C. C. 2000. *Pierre-Simon Laplace, 1749–1827: A Life in Exact Science*, Princeton: Princeton University Press.

Goncalves, B. and Boettcher, S. 2008. Hysteretic optimization for spin glasses, *Journal of Statistical Mechanics*, 2008(1), P01003.

Guney, K. and Basbug, S. 2015. A quantized water cycle optimization algorithm for antenna array synthesis by using digital phase shifters, *International Journal of RF and Microwave Computer-Aided Engineering*, 25(1), 21–29.

Hamacher, K. 2007. Adaptive extremal optimization by detrended fluctuation analysis, *Journal of Computational Physics*, 227, 1500–1509.

Harwit, M. 1998. *Astrophysical Concepts*, New York: Springer.

Hasan, Z. and El-Hawary, M. E. 2014. Optimal power flow by black hole optimization algorithm, in: *2014 IEEE Electrical Power and Energy Conference*, November 12–14, Calgary, Canada, pp. 134–141.

Hatamlou, A. 2013. Black hole: A new heuristic optimization approach for data clustering. *Information Sciences*, 222, 175–184.

Hatamlou, A., Abdullah, S., and Hatamlou, M. 2011. Data clustering using Big Bang–Big Crunch algorithm, *Communications in Computer and Information Science*, December 13–15, 2011, Tehran, Iran, pp. 383–388.

Hawking, S. 1976. Black holes and thermodynamics, *Physical Review D*, 13(2), 191–197.

Hawking, S. 1998. *A Brief History of Time—From the Big Bang to Black Hole*, London: Bantam Press.

Hawking, S. W. 1992. Evaporation of two-dimensional black holes, *Physical Review Letters*, 69(3), 406–409.

Hawking, S. W. 1993. *The Big Bang and Black Holes, Advanced Series in Astrophysics and Cosmology*, Vol. 8, Singapore: World Scientific Publisher.

Hawking, S. W. and Penrose, R. 1970. The singularities of gravitational collapse and cosmology, *Proceedings of the Royal Society A: Mathematical, Physical and Engineering Sciences*, 314(1519), 529–548.

Hirschfelder, J. 1966. *Molecular Theory of Gases and Liquids*, Wisconsin: John Wiley and Sons Inc., pp. 250–262.

Hosseini, H. S. 2011a. Principal component analysis by galaxy-based search algorithm: A novel meta-heuristic for continuous optimisation, *International Journal of Computational Science and Engineering*, 6(1–2), 132–140.

Hosseini, H. S. 2011b. Otsu's criterion-based multilevel thresholding by a nature-inspired metaheuristic called galaxy-based search algorithm, *2011 Third World Congress on Nature and Biologically Inspired Computing*, (NaBIC), October 19–21, 2011, Salamanca, Spain, pp. 383–388.

Hsiao, Y. T., Chuang, C. L., Jiang, J. A., and Chien, C. C. 2005. A novel optimization algorithm: Space gravitational optimization, *Proceeding of 2005 IEEE International Conference on Systems, Man and Cybernetics*, October 10–12, Waikoloa, HI, Vol. 3, pp. 2323–2328.

Hubble, E. 1929. A relation between distance and radial velocity among extra-galactic nebulae, *Proceedings of the National Academy of Sciences of the United States of America*, 15(3), 168–173.

Hubble, E. P. 1936. *The Realm of the Nebulae*, New Haven: Yale University Press.

Inan, U. S. and Inan, A. S. 1998. *Engineering Electromagnetics*, California: Addison-Wesley.

Irizarry, R. 2004. LARES: An artificial chemical process approach for optimization, *Evolutionary Computation Journal*, 12(4), 435–459.

Irizarry, R. 2005a. A generalized framework for solving dynamic optimization problems using the artificial chemical process paradigm: Applications to particulate processes and discrete dynamic systems, *Chemical Engineering Science*, 60, 5663–5681.

Irizarry, R. 2005b. Fuzzy classification with an artificial chemical process, *Chemical Engineering Science*, 60(2), 399–412.

Irizarry, R. 2006. Hybrid dynamic optimization using artificial chemical process: Extended LARES-PR, *Industrial & Engineering Chemistry Research*, 45, 8400–8412.

Irizarry, R. 2008. Global optimization of noisy black box functions using the artificial chemical process and evolutionary algorithms, *Proceedings of the AIChE*, November 16–21, 2008, Philadelphia, PA, Paper 577.

Irizarry, R. 2011. Global and dynamic optimization using the artificial chemical process paradigm and fast Monte Carlo methods for the solution of population balance models, in: I.Dritsasm, ed., *Stochastic Optimization—Seeing the Optimal for the Uncertain*, Rijeka: InTech, Chapter 16.

Ising, E. 1925. Beitrag zur Theorie des Ferromagnetismus, *Zeitschrift für Physik*, 31(1), 253–258.

Jabbaer, A. and Zainudin, S. 2014. Water cycle algorithm for attribute reduction problems in Rough Set Theory, *Journal of Theoretical and Applied Information Technology*, 61(1), 107–117.

Jackson, K. A., Horoi, M., Chaudhuri, I., Frauenheim, T., and Shvartsburg, A. A. 2006. Statistical evaluation of the big bang search algorithm, *Computational Materials Science*, 35, 232–237.

Jahan, M. V. and Akbarzadeh-T, M.-R. 2012. Extremal optimization vs. learning automata: Strategies for spin selection in portfolio selection problems, *Applied Soft Computing*, 12, 3276–3284.

Jaradat, G. M. and Ayob, M. 2010. Big Bang–Big Crunch optimization algorithm to solve the course time-tabling problem, *Proceedings of the 10th International Conference on Intelligent Systems Design and Applications (ISDA'10)*, November 29–December 1, 2010, Cairo, Egypt, pp. 1448–1452.

Kalayci, C. B. and Surendra, M. G. 2013. River formation dynamics approach for sequence-dependent disassembly line balancing problem, Book Chapter 12, in: Surendra, M. G., ed., *Reverse-Supply Chain—Issues and Analysis*, CRC Press/Taylor & Francis, Boca Ratan, London, New York, pp. 289–312.

Kant, I. 1781. *Universal Natural History and Theory of the Heavens, translated by Stephen Palmquist in Kant's Critical Religion, 2000*, Ashgate: Aldershot.

Karci, A. 2007. Theory of saplings growing up algorithm, *Lecture Notes on Computer Science*, 4431, 450–460.

Kaveh, A. and Ahmadi, B. 2013. Simultaneous analysis, design and optimization of structures using the force method and supervised charged system search algorithm, *Scientia Iranica A*, 20(1), 65–76.

Kaveh, A., Bakhshpoori, T., and Kalateh-Ahani, M. 2013a. Optimum plastic analysis of planar frames using ant colony system and charged system search algorithms, *Scientia Iranica A*, 20(3), 414–421.

Kaveh, A. and Behnam, A. F. 2012. Cost optimization of a composite floor system, one-way waffle slab, and concrete slab formwork using a charged system search algorithm, *Scientia Iranica A*, 19(3), 410–416.

Kaveh, A. and Behnam, A. F. 2013. Design optimization of reinforced concrete 3D structures considering frequency constraints via a charged system search, *Scientia Iranica A*, 20(3), 387–396.

Kaveh, A. and Ghazaan, M. I. 2014. Enhanced colliding bodies optimization for design problems with continuous and discrete variables, *Advances in Engineering Software*, 77, 66–75.

Kaveh, A. and Ghazaan, M. I. 2015. A comparative study of CBO and ECBO for optimal design of skeletal structures, *Computers and Structures*, 153, 137–147.

Kaveh, A. and Khayatazad, M. 2012. A new meta-heuristic method, *Ray Optimization, Computers and Structures*, 112–113, 283–294.

Kaveh, A. and Khayatazad, M. 2013. Ray optimization for size and shape optimization of truss structures, *Computers and Structures*, 117, 82–94.

Kaveh, A. and Laknejadi, K. 2011. A novel hybrid charge system search and particle swarm optimization method for multi-objective optimization, *Expert Systems with Applications*, 38, 15475–15488.

Kaveh, A. and Mahdavi, V. R. 2014a. Colliding bodies optimization: A novel meta-heuristic method, *Computers and Structures*, 139, 18–27.

Kaveh, A. and Mahdavi, V. R. 2014b. Colliding bodies optimization method for optimum discrete design of truss structures, *Computers and Structures*, 139, 43–53.

Kaveh, A. and Mahdavi, V. R. 2014c. Colliding bodies optimization method for optimum design of truss structures with continuous variables, *Advances in Engineering Software*, 70, 1–12.

Kaveh, A. and Mahdavi, V. R. 2015a. Two-dimensional colliding bodies algorithm for optimal design of truss structures, *Advances in Engineering Software*, 83, 70–79.

Kaveh, A. and Mahdavi, V. R. 2015b. A hybrid CBO–PSO algorithm for optimal design of truss structures with dynamic constraints, *Applied Soft Computing*, 34, 260–273.

Kaveh, A., Mirzaei, B., and Jafarvand, A. 2014a. Optimal design of double layer barrel vaults using improved magnetic charged system search, *Asian Journal of Civil Engineering (BHRC)*, 15(1), 135–154.

Kaveh, A., Mirzaei, B., and Jafarvand, A. 2014b. Shape-size optimization of single-layer barrel vaults using improved magnetic charged system search, *International Journal of Civil Engineering*, 12(4), 447–465.

Kaveh, A., Mirzaei, B., and Jafarvand, A. 2015. An improved magnetic charged system search for optimization of truss structures with continuous and discrete variables, *Applied Soft Computing*, 28, 400–410.

Kaveh, A., Motie Share, M. A., and Moslehi, M. 2013b. Magnetic charged system search: A new meta-heuristic algorithm for optimization, *Acta Mechanica*, 224(1), 85–107.

Kaveh, A. and Nasrollahi, A. 2014. Performance-based seismic design of steel frames utilizing charged system search optimization, *Applied Soft Computing*, 22, 213–221.

Kaveh, A. and Talatahari, S. 2009a. Size optimization of space trusses using Big Bang–Big Crunch algorithm, *Computers & Structures*, 87(17–18), 1129–1140.

Kaveh, A. and Talatahari, S. 2009b. Particle swarm optimizer, ant colony strategy and harmony search scheme hybridized for optimization of truss structures, *Computers and Structures*, 87(5-6), 267–283.

Kaveh, A. and Talatahari, S. 2010a. Optimal design of Schwedler and ribbed domes via hybrid Big Bang–Big Crunch algorithm, *Journal of Constructional Steel Research*, 66(3), 412–429.

Kaveh, A. and Talatahari, S. 2010b. A novel heuristic optimization method: Charged system search, *Acta Mechanica*, 213(3–4), 267–289.

Kaveh, A. and Talatahari, S. 2010c. Optimal design of skeletal structures via the charged system search algorithm, *Structural and Multidisciplinary Optimisation*, 41, 893–911.

Kaveh, A. and Talatahari, S. 2010d. Charged system search for optimum grillage systems design using the LRFD-AISC code, *Journal of Constructional Steel Research*, 66, 767–771.

Kaveh, A. and Talatahari, S. 2010e. A charged system search with a fly to boundary method for discrete optimum design of truss structures, *Asian Journal of Civil Engineering*, 11(3), 277–293.

Kaveh, A. and Talatahari, S. 2011a. Geometry and topology optimization of geodesic domes using charged system search, *Structural and Multidisciplinary Optimisation*, 43(2), 215–229.

Kaveh, A. and Talatahari, S. 2011b. An enhanced charged system search for configuration optimization using the concept of fields of forces, *Structural and Multidisciplinary Optimisation*, 43, 339–351.

Kaveh, A. and Talatahari, S. 2012. Charged system search for optimal design of frame structures, *Applied Soft Computing*, 12, 382–393.

Kaveh, A. and Zolghadr, A. 2012. Truss optimization with natural frequency constraints using a hybridized CSS-BBBC algorithm with trap recognition capability, *Computers and Structures*, 102–103, 14–27.

Kaveh, A. and Zolghadr, A. 2014. Magnetic charged system search for structural optimization, *Periodica Polytechnica, Civil Engineering*, 58(3), 203–216.

Kaveh, A. and Zolghadr, A. 2015. An improved CSS for damage detection of truss structures using changes in natural frequencies and mode shapes, *Advances in Engineering Software*, 80, 93–100.

Kenyon, I. R. 1990. *General Relativity*, Oxford: Oxford University Press.

Kirkpatrick, S., Gelatt, C. D., and Vecchi, M. P. 1983. Optimisation by simulated annealing, *Science*, 220, 671–680.

Kumar, Y. and Sahool, G. 2014. A charged system search for approach for data clustering, *Progress in Artificial Intelligence*, 2, 153–166.

Kumar, Y. and Sahool, G. 2015. Hybridization of magnetic charge system search and particle swarm optimization for efficient data clustering using neighborhood search strategy, *Soft Computing*, 19, 3621–3645, doi: 10.1007/s00500-015-1719-0.

Kumbasar, T., Eksin, I., Guzelkaya, M., and Yesil, E. 2008a. Big Bang Big Crunch optimization method based fuzzy model inversion, in: A. Gelbukh and E. F. Morales, eds., *MICAI 2008: Advances in Artificial Intelligence*, Berlin, Germany: Springer, 2008, pp. 732–740.

Kumbasar, T., Yesil, E., Eksin, I., and Guzelkaya, M. 2008b. Inverse fuzzy model control with online adaptation via Big Bang–Big Crunch optimization, *2008 3rd International Symposium on Communications, Control, and Signal Processing (ISCCSP'08)*, St. Julians, Malta, pp. 697–702.

Kumbasar, T., Yesil, E., Eksin, I., and Guzelkaya, M. 2011. Adaptive fuzzy model based inverse controller design using BB–BC optimization algorithm, *Expert Systems with Applications*, 38, 12356–12364.

Kundu, S. 1999. Gravitational clustering: A new approach based on the spatial distribution of the points, *Pattern Recognition*, 32, 1149–1160.

Lam, A. Y. S. and Li, V. O. K. 2010a. Chemical reaction optimization for cognitive radio spectrum allocation, *Proceedings of the IEEE Global Telecommunications Conference* (GLOBECOM 2010), December 6–10, 2010, Miami, Florida, USA, pp. 1–5.

Lam, A. Y. S. and Li, V. O. K. 2010b. Chemical-reaction-inspired metaheuristic for optimization, *IEEE Transaction on Evolutionary Computation*, 14(3), 381–399.

Lenin, K., Reddy, B. R., and Kalavathi, M. S. 2014. Black hole algorithm for solving optimal reactive power dispatch problem, *International Journal of Research in Management, Science & Technology*, 2(1), 10–15.

Li, X., Luo, J., Chen, M.-R., and Wang, N. 2012. An improved shuffled frog-leaping algorithm with extremal optimisation for continuous optimisation, *Information Sciences*, 192, 143–151.

Lin, C.-J., Yau, H.-T., Lin, C.-R., and Hsu, C.-R. 2013. Simulation and experimental analysis for hysteresis behaviour of a piezoelectric actuated micro stage using modified charge system search, *Microsystems Technology*, 19, 1807–1815.

Lu, H. and Yang, G. 2009. Extremal optimization for protein folding simulations on the lattice, *Computers and Mathematics with Applications*, 57, 1855–1861.

Luo, J. and Chen, M.-R. 2014. Multi-phase modified shuffled frog leaping algorithm with extremal optimization for the MDVRP and the MDVRPTW, *Computers & Industrial Engineering*, 72, 84–97.

Luz, E. F. P. D., Becceneri, J. C., and Velho, H. F. D. C. 2008. A new multi-particle collision algorithm for optimization in a high performance environment, *Journal of Computational Interdisciplinary Sciences*, 1, 3–10.

Luz, E. F. P. D., Becceneri, J. C., and Velho, H. F. D. C. 2011. Multiple particle collision algorithm applied to radiative transference and pollutant localization inverse problems, *IEEE International Symposium on Parallel and Distributed Processing Workshops and PhD Forum (IPDPSW)*, pp. 347–351.

Ma, J., Zhao, J.-L., Zhang, F.-P., and Peng, Q.-H. 2000. Some statistical properties of spiral galaxies along the Hubble sequence, *Chinese Astronomy and Astrophysics*, 24, 435–443.

May, R. M. 1976. Simple mathematical models with very complicated dynamics, *Nature*, 261, 459–467.

Melin, P., Astudillo, L., Castillo, O., Valdez, F., and Valdez, F. 2013. Optimal design of type-2 and type-1 fuzzy tracking controllers for autonomous mobile robots under perturbed torques using a new chemical optimization paradigm, *Expert Systems with Applications*, 40, 3185–3195.

Melkemi, K. E., Batouche, M., and Foufou, S. 2006. A multiagent system approach for image segmentation using genetic algorithms and extremal optimization heuristics, *Pattern Recognition Letters*, 27, 1230–1238.

Menai, M. E. and Batouche, M. 2003. Efficient initial solution to extremal optimization algorithm for weighted MAXSAT problem, in: P. W. H. Chung, C. J. Hinde, and M. Ali, eds., *Lecture Notes in AI, IEA/AIE 2003*, Berlin, Germany: Springer, Vol. 2718, pp. 592–603.

Metropolis, N., Rosenbluth, A., Rosenbluth, M., Teller, A., and Teller, E., J. 1953. Equation of state calculation by fast computing machines, *Journal of Chemical Physics*, 21, 1087–1091.

Michalewicz, Z. and Dcholenauer, M. 1996. Evolutionary algorithms for constrained parameter optimization problems, *Evolutionary Computation*, 4, 1–32.

Michell, J. 1784. On the means of discovering the distance, magnitude, and of the fixed stars, *Philosophical Transactions of the Royal Society*, 74(0), 35–57.

Mo, S. and Zeng, J. 2009. Performance analysis of the artificial physics optimization algorithm with simple neighborhood topologies, in: *2009 International Conference on Computational Intelligence and Security*, December 11–14, Beijing, China, IEEE Computer Society, pp. 155–160.

Moser, I. and Hendtlass, T. 2006. Solving problems with hidden dynamics-comparison of extremal optimization and ant colony system, *Proceedings of the 2006 IEEE Congress on Evolutionary Computation (CEC'2006)*, July 16–21, 2006, Vancouver, Canada, pp. 1248–1255.

Nemati, M. and Momeni, H. 2014. Black holes algorithm with fuzzy hawking radiation, *International Journal of Science and Technology Research*, 3(6), 85–88.

Nemati, M., Salami, R., and Bazrkar, N. 2013. Black holes algorithm: A swarm algorithm inspired of black holes for optimization problems, *IAES International Journal of Artificial Intelligence (IJ-AI)*, 2(3), 143–150.

Niknam, T., Golestaneh, F., and Shafiei, M. 2013. Probabilistic energy management of a renewable micro-grid with hydrogen storage using self-adaptive charge search algorithm, *Energy*, 49, 252–267.

Nobre, F. D. 2003. The two-dimensional ±J Ising spin glass: A model at its lower critical dimension, *Physica A: Statistical Mechanics and Its Applications*, 319, 362–370.

Nouhi, B., Talatahari, S., Kheiri, H., and Cattani, C. 2013. Chaotic charged system search with a feasible-based method for constraint optimization problems, *Mathematical Problems in Engineering*, 2013, 391765, 8 pp.

Oersted, H. C. 1820. Experiments on the effect of current of electricity on the magnetic needle (privately distributed pamphlet dated July 21, 1820), *English Translation in: Annals of Philosophy*, 16, 273.

Onody, R. N. and de Castro, P. A. 2002. Self-organized criticality, optimization and biodiversity, *International Journal of Modern Physics C*, 14, 911–916.

Oppenheimer, R. and Snyder, H. 1939. On continued gravitational contraction, *Physical Review*, 56, 455.

Ornstein, L. S. 1917. On the Brownian motion, *Proceedings of Royal Academy of Amsterdam*, 21, 96–108.

Otsu, N. 1979. A threshold selection method from gray-level histogram, *IEEE Transaction on Systems, Man, and Cybernetics*, 9(1), 62–66.

Özyön, S., Temurtas, H., Durmus, B., and Kuvat, G. 2012. Charged system search algorithm for emission constrained economic power dispatch problem, *Energy*, 46, 420–430.

Pal, K. F. 2003. Hysteretic optimization for the travelling salesman problem, *Physica A: Statistical Mechanics and Its Applications*, 329(1–2), 287–297.

Pal, K. F. 2006a. Hysteretic optimization, faster and simpler, *Physica A: Statistical Mechanics and Its Applications*, 360(2), 525–533.

Pal, K. F. 2006b. Hysteretic optimization for the Sherrington–Kirkpatrick spin glass, *Physica A: Statistical Mechanics and Its Applications*, 367, 261–268.

Peebles, P. J. E., Schramm, D. N., Turner, E. L., and Kron, R. G. 1994. The evolution of the universe, *Scientific America*, 271(4), 52–57.

Peitgen, H., Jurgens, H., and Saupe, D. 1992. *Chaos and Fractals*, Berlin, Germany: Springer.

Pledge, H. T. 1939. *Science Since 1500: A Short History of Mathematics, Physics, Chemistry and Biology*, London: Ten Shillings Net.

Precup, R.-E., David, R.-C., Petriu, E. M., Preitl, S., and Radac, M.-B. 2014. Novel adaptive charged system search algorithm for optimal tuning of fuzzy controllers, *Expert Systems with Applications*, 41, 1168–1175.

Prudhvi, P. 2011. A complete copper optimization technique using BB–BC in a smart home for a smarter grid and a comparison with GA, *Proceedings of the 24th Canadian Conference on Electrical and Computer Engineering (CCECE'11)*, May 2011, Ontario, Canada, pp. 69–72.

Rabanal, P. and Rodríguez, I. 2009. Testing restorable systems by using RFD, in: *10th International Work Conference on Artificial Neural Networks, IWANN'09*, June 10–12, 2009, Salamanca, Spain, pp. 351–358.

Rabanal, P., Rodríguez, I., and Rubio, F. 2007. Using river formation dynamics to design heuristic algorithms, in: S. G. Akl et al. eds., *Unconventional Computation, UC 2007, LNCS 4618*, Springer-Verlag, Berlin, Heidelberg, 2007, pp. 163–177.

Rabanal, P., Rodríguez, I., and Rubio, F. 2008a. Finding minimum spanning/distances trees by using river formation dynamics, in: M. Dorigo, M. Birattari, C. Blum, M. Clerc, Th. Stützle, and A. Winfield, eds., *Ant Colony Optimization and Swarm Intelligence, ANTS'08*, Theoretical Computer Science and General Issues: *LNCS 5217*. Berlin, Germany: Springer, pp. 60–71.

Rabanal, P., Rodríguez, I., and Rubio, F. 2008b. Solving dynamic TSP by using river formation dynamics, *Fourth IEEE International Conference on Natural Computation*, October 18–20, 2008, Jinan, China, pp. 246–250.

Rabanal, P., Rodríguez, I., and Rubio, F. 2009a. Applying river formation dynamics to solve NP-complete problems, in: R.Chiong, ed., *Nature-Inspired Algorithms for Optimisation, Studies in Computational Intelligence*, Berlin, Germany: Springer, Vol. 193, pp. 333–368.

Rabanal, P., Rodríguez, I., and Rubio, F. 2009b. A formal approach to heuristically test restorable systems, in: M. Leucker and C. Morgan, eds., *6th International Colloquium on Theoretical Aspects of Computing—ICTAC 2009, LNCS 5684*, Kuala Lumpur, Malaysia, August 16–20, 2009. Springer-Verlag, Berlin, Heidelberg, pp. 292–306.

Rabanal, P., Rodríguez, I., and Rubio, F. 2010. Applying river formation dynamics to the Steiner tree problem, in: F. Sun, Y. Wang, J. Lu, B. Zhang, W. Kinsner, and L. A. Zadeh, eds., *Proceedings of the 9th IEEE International Conference on Cognitive Informatics (ICCI'10)*, pp. 704–711.

Randall, M., Lewis, A., Hettenhausen, J., and Kipouros, T. 2014. Local search enabled extremal optimisation for continuous inseparable multi-objective benchmark and real-world problems, *Procedia Computer Science*, 29, 1904–1914.

Rathore, C. and Roy, R. 2014. A novel modified GBMO algorithm based static transmission network expansion planning, *Electrical Power and Energy System*, 62, 519–531.

Redlarski, G., Palkowski, A., and Dabkowski, M. 2013. Using river formation dynamics algorithm in mobile robot navigation, *Solid State Phenomena*, 198, 138–143.

Ruffini, R. and Wheeler, J. A. 1971. Introducing the black hole, *Physics Today*, 24, 30–41.

Ryden, B. 2003. *Introduction to Cosmology*, San Francisco: Addison-Wesley.

Sacco, W. F. and de Oliveira, C. R. E. 2005. A new stochastic optimization algorithm based on particle collisions, *Transactions of the American Nuclear Society*, 92, 657–659.

Sacco, W. F., de Oliveira, C. R. E., and Pereira, C. M. N. A. 2006. Two stochastic optimization algorithms applied to a nuclear reactor core design, *Progress in Nuclear Energy*, 48, 525–539.

Sacco, W. F., Filho, H. A., and de Oliveira, C. R. E. 2007. A populational particle collision algorithm applied to a nuclear reactor core design optimisation, *Joint International Topical Meeting on Mathematics & Computation and Supercomputing in Nuclear Applications (M&C + SNA 2007)*, April 15–19, Monterey, CA, pp. 1–10.

Sadollah, A., Bahreininejad, A., Eskandar, H., and Hamdi, M. 2012. Mine blast algorithm for optimization of truss structures with discrete variables, *Computers and Structures*, 102–103, 49–63.

Sadollah, A., Bahreininejad, A., Eskandar, H., and Hamdi, M. 2013. Mine blast algorithm: A new population based algorithm for solving constrained engineering optimization problems, *Applied Soft Computing*, 13, 2592–612.

Sadollah, A., Eskandar, H., Bahreininejad, A., and Kim, J. H. (2015a) Water cycle, mine blast and improved mine blast algorithms for discrete sizing optimization of truss structures, *Computers and Structures*, 149, pp. 1–16.

Sadollah, A., Eskandar, H., and Kim, J. H. (2015b) Water cycle algorithm for solving constrained multi-objective optimization problems, *Applied Soft Computing*, 27, 279–298.

Sadollah, A., Yoo, D. G., Yazdi, J., and Kim, J. H. 2014. Application of water cycle algorithm for optimal cost design of water distribution system, *11th International Conference on Hydro-informatics (HIC 2014)*, August 17–21, New York, NY.

Sambatti, S. B. M., Anochi, J. A., Luz, E. F. P., Shiguemori, E. H., Carvalho, A. R., and Velho, H. F. D. C. 2012. Automatic configuration for neural network applied to atmospheric temperature profile identification, *International Conference on Engineering Optimization (EngOpt 2012)*, July 1–5, Rio de Janeiro, Brazil, pp. 1–9.

Sandage, A. 1975. *Galaxies and Universe-Stars and Stellar Systems*, Chicago, IL: University of Chicago Press.

Schwarzschild, K. 1916. Über das Gravitationsfeld eines Massenpunktes nach der Einsteinschen Theorie, Sitzungsberichte der Königlich-Preussischen Akademie der Wissenschaften, Sitzung 3, Februar 1916, Deutsche Akademie der Wissenschaften zu Berlin, pp. 189–196.

Shen, J. and Li, Y. 2008. An optimisation algorithm based on optical principles, *Adavances in Systems Science and Application*, 5, 1–8.

Shen, J. and Li, Y. 2009. Light ray optimisation and its parameter analysis, *IEEE International Joint Conference on Computational Science and Optimisation*, Sanya, China, pp. 918–922.

Siddique, N. and Adeli, H. 2013. Computational intelligence: Synergies of fuzzy logic, in: *Neural Networks and Evolutionary Computing*, Chichester, West Sussex, UK: John Wiley.

Siddique, N. and Adeli, H. 2016. Physics-based Search and Optimisation: Inspiration from Nature, *Expert Systems*, 33, 607–623.

Siddique, N. and Adeli, H. 2015a. Harmony search algorithm and its variants, *International Journal of Pattern Recognition and Artificial Intelligence*, 29(8), 1539001 (22 pages).

Siddique, N. and Adeli, H. 2015b. Hybrid harmony search algorithms, *International Journal on Artificial Intelligence Tools*, 24(6), 1530001 (16 pages).

Siddique, N. H. and Adeli, H. 2015c. Applications of harmony search algorithms in engineering, *International Journal on Artificial Intelligence Tools*, 24(6), 1530002 (15 pages).

Spears, D. F., Kerr, W., Kerr, W., and Hettiarachchi, S. 2005a. An overview of physicomimetics, *Lecture Notes in Computer Science-State of the Art Series*, 3324, 84–97.

Spears, D. F., Kerr, W., and Spears, W. F. 2006. Physics-based robots swarms for coverage problems, *International Journal on Intelligent Control and Systems*, 11(3), 11–23.

Spears, D. F. and Spears, W. M. 2003. Analysis of a phase transition in a physics-based multiagent system, *Lecture Notes in Computer Science*, 2699, 193–207.

Spears, W. M., Heil, R., Spears, D. F., and Zarzhitsky, D. 2004. Physicomimetics for mobile robot formations, *Proceedings of the Third International Joint Conference on Autonomous Agents and Multi Agent Systems (AAMAS-04)*, July 19–23, New York, NY, Vol. 3, pp. 1528–1529.

Spears, W. M., Heil, R., and Zarzhitsky, D. 2005b. Artificial physics for mobile robot formations, *2005 IEEE International Conference on Systems, Man and Cybernetics*, October 10–12, 2005, Waikoloa, HI, USA, pp. 2287–2292.

Sumida, I. Y., Velho, H. F. D. C., Luz, E. F. P., Cruz, R. V., and Góes, L. C. S. 2014. MPCA for flight dynamics parameters determination, *Computer Assisted Methods in Engineering and Science*, 21, 257–265.

Sun, H., Feng, D., Liu, Y., and Feng, M. Q. 2015. Statistical regularization for identification of structural parameters and external loadings using state space models, *Computer-Aided Civil and Infrastructure Engineering*, 30(11), 843–858.

Tang, H., Zhou, J., Xue, S., and Xie, L. 2010. Big Bang–Big Crunch optimization for parameter estimation in structural systems, *Mechanical Systems and Signal Processing*, 24(8), 2888–2897.

Talatahari, S., Kaveh, A., and Rahbari, N. M. 2012a. Parameter identification of Bouc-Wen model for MR fluid dampers using adaptive charged system search optimization, *Journal of Mechanical Science and Technology*, 26(8), 2523–2534.

Talatahari, S., Kaveh, A., and Sheikholeslami, R. 2011. An efficient charged system search using chaos for optimization problems, *International Journal of Optimization in Civil Engineering*, 1(2), 305–325.

Talatahari, S., Kaveh, A., and Sheikholeslami, R. 2012b Engineering design optimization using chaotic enhanced charged system search algorithms, *Acta Mechanica*, 223(10), 2269–2285.

Talatahari, S. and Sheikholeslami, R. 2014. Optimum design of gravity and reinforced retaining walls using enhanced charged system search algorithm, *KSCE Journal of Civil Engineering*, 18(5), 1464–1469.

Talatahari, S., Sheikholeslami, R., Shadfaran, M., and Porbaba, M. 2012c. Charged system search algorithm for optimum design of gravity retaining walls subject to seismic loading, *Mathematical Problems in Engineering*, 2012, 301628, 10 pp.

Tayarani-N, M. H. and Akbarzadeh-T, M. R. 2008. Magnetic optimisation algorithms: A new synthesis, *IEEE Congress on Evolutionary Computation (CEC 2008)*, June 1–6, 2008, Hong Kong, pp. 2659–2664.

Torres, R. H., Luz, E. F. P. D., and Velho, H. F. D. C. 2015. Multi-particle collision algorithm with reflected points, *Proceeding Series of the Brazilian Society of Applied and Computational Mathematics*, 3(1), 010433-1-6.

Uhlenbeck, G. E. and Ornstein, L. S. 1930. On the theory of Brownian motion, *Physical Review*, 36, 823–841.

Wang, L., Li, W., Fei, R., and Hei, X. 2012. Cloud droplets evolutionary algorithm on reciprocity mechanism for function optimisation, in: T. Tan, Y. Shi, and Z. Li, eds., *ICSI 2012, Part I, LNCS 7331*, Berlin, Germany: Springer, pp. 268–275.

Wang, S. 2011. Gravity Probe B: Final results of a space experiment to test general relativity, *Physical Review Letters*, 106(22), 221101.

Wang, N. and Adeli, H. 2012. Algorithms for chattering reduction in system control, *The Journal of the Franklin Institute*, 349(8), 2687–2703.

Webster, B. L. 2004. Solving combinatorial optimisation problems using a new algorithm based on gravitational attraction, PhD thesis, Florida Institute of Technology.

Weisberg, J. M. and Taylor, J. H. 2004. Relativistic binary pulsar B1913+16: Thirty years of observations and analysis, in: F. A. Rasio and I. H. Stairs, eds., *Binary Radio Pulsars, Proceeding of the Aspen Conference, vol. 328 of ASP Conference Series*, Astronomical Society of the Pacific, San Francisco, U.S.A., pp. 25–32.

White, N. E. and Diaz, A. V. 2004. Beyond Einstein: From the Big Bang to black holes, *Advances in Space Research*, 34, 651–658.

Wood, A. J. and Wollenberg, B. F. 1994. *Power Generation, Operation and Control*, New York: John Wiley.

Woolfson, M. M. 1993. Solar system—Its origin and evolution, *The Quarterly Journal of the Royal Astronomical Society*, 34, 1–20.

Xie, D., Luo, Z., and Yu, F. 2009d. The computing of the optimal power consumption for semi-track air-cushion vehicle using hybrid generalized extremal optimization, *Applied Mathematical Modelling*, 33, 2831–2844.

Xie, L., Tan, Y., and Zeng, J. 2012. Artificial physics optimization algorithm for global optimization, in: W. M. Spears and D. F. Spears, eds., *Physicomimetics*, Springer, Berlin, Heidelberg, pp. 565–589, Chapter 18.

Xie, L. and Zeng, J. 2011. A hybrid vector artificial physics optimization for constrained optimization problems, in: *2011 First International Conference on Robot, Vision and Signal Processing*, November 21–23, 2011, Kaohsiung City, Taiwan, 145–148.

Xie, L., Zeng, J., and Cai, X. 2011a. A hybrid vector artificial physics optimization with multi-dimensional search method, *2011 Second International Conference on Innovations in Bio-inspired Computing and Applications*, December 16–18, 2011, Shenzhen, China, pp. 116–119.

Xie, L., Zeng, J., and Cui, Z. 2009a. General framework of artificial physics optimization algorithm, *2009 World Congress on Nature & Biologically Inspired Computing (NaBIC 2009)*, Coimbatore, India, December 2009, pp. 1321–1326.

Xie, L., Zeng, J., and Cui, Z. 2009b. Using artificial physics to solve global optimization problems, in: G. Baclu, Y. Wang, Y. Y. Yae, W. Kinsner, K. Chan, and, L. A. Zadeh, eds., *Proceedings of the 8th IEEE International Conference on Cognitive Informatics (ICCI'09)*, pp. 502–508.

Xie, L., Zeng, J., and Cui, Z. 2009c. The vector model of artificial physics optimization algorithm for global optimization problems, *The 10th International Conference on Intelligent Data Engineering and Automated Learning(IDEAL)*, September 2009, Spain, pp. 610–617.

Xie, L., Zeng, J., and Formato, R. A. 2011b. Convergence analysis and performance of the extended artificial physics optimization algorithm, *Applied Mathematics and Computing*, 218, 4000–4011.

Xie, L. P. and Zeng, J. C. 2009a. Physicomimetics for Swarm Robots search, *Patter Recognition and Artificial Intelligence*, 22(4), 647–652.

Xie, L. P. and Zeng, J. C. 2009b. A global optimization based on physicomimetics framework, *The 2009 World Summit on Genetic and Evolutionary Computation (GEC'09)*, June 12–14, Shanghai, China.

Xie, L. P. and Zeng, J. C. 2009c. An extended artificial physics optimization algorithm for global optimization problems, *Fourth International Conference on Innovative Computing, Information and Control (ICICIC 2009)*, December 7–9, Kaohsiung, Taiwan, pp. 881–884.

Xie, L. P. and Zeng, J. C. 2010. The performance analysis of artificial physics optimization algorithm driven by different virtual forces, *ICIC Express Letters (ICIC-EL)*, 4(1), 239–244.

Xie, L. P., Tan, Y., and Zeng, J. C. 2010a. The convergence analysis and parameter selection of artificial physics optimization algorithm, *Proceedings of the 2010 International Conference on Modelling, Identification and Control*, Okayama, Japan, July, 2010, pp. 562–567.

Xie, L. P., Zeng, J. C., and Cui, Z. H. 2010b. On mass effects to artificial physics optimization algorithm for global optimization problems, *International Journal of Innovative Computing and Applications*, 2(2), 69–76.

Xie, L. P., Tan, Y., and Zeng, J. C. 2011c. A study on the effect of Vmax in artificial physics optimization algorithm with high dimension, *2011 International Conference of Soft Computing and Pattern Recognition (SoCPaR)*, October 14–16, Dalian, China, pp. 550–555.

Xing, B. and Gao, W.-J. 2014. Emerging chemistry-based CI algorithms, in: B. Xing and W.-J. Gao, eds., *Innovative Computational Intelligence: A Rough Guide to 134 Clever Algorithms, Intelligent Systems Reference Library*, Switzerland: Springer International Publishing.

Yan, G.-W. and Hao, Z. 2012. A novel atmospheric clouds model optimisation algorithm, *IEEE International Conference on Computing, Measurement, Control and Sensor Network (CMCSN)*, Taiyuan, China, pp. 217–220.

Yan, G.-W. and Hao, Z. 2013. A novel optimisation algorithm based on atmospheric clouds model, *International Journal of Computational Intelligence and Applications*, 12, 1–16.

Yan, W. and Lian-chao, Z. 2010. A constraint multi-objective artificial physics optimization algorithm, *2010 Second International Conference on Computational Intelligence and Natural Computing (CINC)*, September 13–14, 2010, Wuhan, China, pp. 107–112.

Yan, X. and Wu, W. 2012. Hysteretic optimization for the capacitated vehicle routing problem, *Proceedings of the 9th IEEE International Conference on Networking, Sensing and Control (ICNSC'12)*, April 11–14, 2012, Beijing, China, pp. 12–15.

Yang, F.-C. and Wang, Y.-P. 2007. Water flow-like algorithm for object grouping problems, *Journal of the Chinese Institute of Industrial Engineers*, 24, 475–488.

Yang, G. J., Xie, L. P., Tan, Y., and Zeng, J. C. 2010. A hybrid vector artificial physics optimization with one-dimensional search method, *2010 International Conference on Computational Aspects of Social Networks (CASoN)*, September 26–28, 2010, Taiyuan, China, pp. 19–22.

Yang, S.-D., YI, Y.-L., and Shan, Z.-Y. 2011. Gbest-guided artificial chemical reaction algorithm for global numerical optimization, *Procedia Engineering*, 24, 197–201.

Yin, J., Xie, L. P., Zeng, J., and Tan, Y. 2010. Artificial physics optimization algorithm with a feasibility-based rule for constrained optimization problems, *2010 IEEE International Conference on Intelligent Computing and Intelligent Systems (ICIS)*, October 29–31, Xiamen, China, pp. 488–492.

Zamani, A., Barakati, S. M., and Yousofi-Darmian, S. 2016. Design of a fractional order PID controller using GBMO algorithm for load-frequency control with governor saturation consideration, *ISA Transactions*, 64, 56–66, doi: 10.1016/j.isatra.2016.04.021.

Zandi, Z., Afjei, E., and Sedighizadeh, M. 2012. Reactive power dispatch using Big Bang–Big Crunch optimization algorithm for voltage stability enhancement, *Proceeding of 2012 IEEE International Conference on Power and Energy (PECon)*, December 2–5, 2012, Kota Kinabalu, Malaysia, pp. 239–244.

Zaránd, G., Pázmándi, F., Pál, K. F., and Zimányi, G. T. 2002. Hysteretic optimization, *Physical Review Letters*, 89(15), 1502011–1502014.

Zeng, G.-Q., Chen, J., Chen, M.-R., Dai, Y.-X., Li, L.-M., Lu, K.-D., and Zheng, C.-W. 2015a. Design of multi-variable PID controllers using real-coded population-based extremal optimization, *Neurocomputing*, 151, 1343–1353.

Zeng, G.-Q., Chen, J., Dai, Y.-X., Li, L.-M., Zheng, C.-W., and Chen, M.-R. 2015b. Design of fractional order PID controller for automatic regulator voltage system based on multi-objective extremal optimization, *Neurocomputing*, 160, 173–184.

Zeng, G.-Q., Lu, K.-D., Dai, Y.-X., Zhang, Z.-J., Chen, M.-R., Zheng, C.-W., Wu, D., and Peng, W.-W. 2014. Binary-coded extremal optimization for the design of PID controllers, *Neurocomputing*, 138, 180–188.

Zeng, G.-Q., Lu, Y.-Z., and Mao, W.-J. 2010a. Multi-stage extremal optimization for hard travelling salesman problem, *Physica A*, 389, 5037–5044.

Zeng, G.-Q., Lu, Y.-Z., Mao, W.-J., and Chu, J. 2010b. Study on probability distributions for evolution in modified extremal optimization, *Physica A*, 389, 1922–1930.

Zerigat, D. H., Benasla, L., Belmadani, A., and Rahli, M. 2013. Solution of combined economic and emission dispatch problems using galaxy-based search algorithm, *Journal of Electrical Systems*, 9-4, 468–480.

Zerigat, D. H., Benasla, L., Belmadani, A., and Rahli, M. 2014. Galaxy-based search algorithm to solve combined economic and emission dispatch, *U.P.B. Science Bulletin Series C*, 76(1), 209–220.

Zha, J., Zeng, G., and Lu, Y. 2010. Hysteretic optimization for protein folding on the lattice, *Proceedings of the International Conference on Computational Intelligence and Software Engineering (CiSE'10)*, December 10–12, 2010, Wuhan, China, pp. 1–4.

Zhang, J., Liu, K., Tan, Y., and He, X. 2008. Random black hole particle swarm optimization and its application, *2008 IEEE International Conference Neural Networks and Signal Processing, ICNNSP*, June 7–11, Zhenjiang, China, pp. 359–365.

Zheng, M., Liu, G. X., Zhou, C.-G., Liang, Y.-C., and Wang, Y. 2010. Gravitational field algorithm and its application in gene cluster, *Algorithms for Molecular Biology*, 5, 1–11.

Zhou, T., Bai, W.-J., Cheng, L.-J., and Wang, B.-H. 2005. Continuous extremal optimization for Lennard–Jones Clusters, *Physical Review E*, 72, 016702.

Appendix A: Vector and Matrix

A.1   VECTOR AND VECTOR SPACES

A column vector is defined by an array of n elements in a column

$$\mathbf{a} = \begin{bmatrix} a_1 \\ a_2 \\ a_3 \\ \vdots \\ a_n \end{bmatrix} \tag{A.1}$$

The element $a_i \in R$, $i = 1, 2, \ldots, n$, is called the i-th element of the vector where R denotes the set of all real numbers. R^n denotes an n-dimensional real vector space.

A row vector is defined as an array of n elements in a row

$$\mathbf{a} = [a_1 \quad a_2 \quad \ldots \quad a_n] \tag{A.2}$$

The transpose of a column vector in Equation A.1 is a row vector written as $\mathbf{a}^T = [a_1 \; a_2 \; \ldots \; a_n]$. The vector $\mathbf{a}$ can be written as $\mathbf{a} = [a_1 \quad a_2 \quad \ldots \quad a_n]^T$. The superscript T denotes transpose of a vector.

A.2   MATRIX AND MATRIX SPACES

A matrix is a two-dimensional array. An $n \times m$ dimensional matrix is defined by

$$\mathbf{A} = \begin{bmatrix} a_{11} & a_{12} & \cdots & a_{1m} \\ a_{21} & a_{22} & \cdots & a_{2m} \\ \vdots & \vdots & \ddots & \vdots \\ a_{n1} & a_{n2} & \cdots & a_{nm} \end{bmatrix} \tag{A.3}$$

The element $a_{ij} \in R$, $i = 1, 2, \ldots, n$ and $j = 1, 2, \ldots, m$, is the element of the i-th row and j-th column of the matrix $\mathbf{A}$. $R^{n \times m}$ denotes an $n \times m$-dimensional real matrix space.

The transpose of an $n \times m$-dimensional matrix $\mathbf{A}$ is an $m \times n$-dimensional matrix $\mathbf{A}^T$ denoted as

$$\mathbf{A}^T = \begin{bmatrix} a_{11} & a_{21} & \cdots & a_{n1} \\ a_{12} & a_{22} & \cdots & a_{n2} \\ \vdots & \vdots & \ddots & \vdots \\ a_{1m} & a_{2m} & \cdots & a_{nm} \end{bmatrix} \tag{A.4}$$

An $n \times m$-dimensional matrix $\mathbf{A}$ is a square matrix when the numbers of rows and columns are equal, that is, $m = n$.

$$\mathbf{A} = \begin{bmatrix} a_{11} & a_{12} & \cdots & a_{1n} \\ a_{21} & a_{22} & \cdots & a_{2n} \\ \vdots & \vdots & \ddots & \vdots \\ a_{n1} & a_{n2} & \cdots & a_{nn} \end{bmatrix} \tag{A.5}$$

An $n \times n$-dimensional square matrix $\mathbf{A}$ is a symmetric matrix when $a_{ij} = a_{ji}$ with $\{i, j\} = 1, 2, \ldots,$ n. That means matrix $\mathbf{A}$ is equal to its transpose

$$\mathbf{A}^T = \mathbf{A} \tag{A.6}$$

The superscript T denotes transpose of a matrix.

A matrix $\mathbf{A}$ is called singular if its determinant is zero, that is, $\det(\mathbf{A}) = 0$. A matrix $\mathbf{A}$ is called nonsingular if its determinant is nonzero, that is, $\det(\mathbf{A}) \neq 0$. A matrix $\mathbf{A}$ is invertible if $\det(\mathbf{A}) \neq 0$. That is

$$\mathbf{C} = \mathbf{A}^{-1} \tag{A.7}$$

The superscript -1 denotes inverse of a matrix.

If $\mathbf{A}$ and $\mathbf{B}$ are two compatible matrices, then the following property holds for them

$$(\mathbf{AB})^T = \mathbf{B}^T \mathbf{A}^T \tag{A.8}$$

If $\mathbf{A}$ and $\mathbf{B}$ are two compatible and nonsingular matrices, then the following property holds for them:

$$(\mathbf{AB})^{-1} = \mathbf{B}^{-1} \mathbf{A}^{-1} \tag{A.9}$$

An $n \times n$-dimensional square matrix $\mathbf{A}$ is a diagonal matrix when $a_{ij} \neq 0$ for all $i = j$ and $a_{ij} = 0$ for all $i \neq j$ with $\{i, j\} = 1, 2, \ldots, n$.

$$\mathbf{A} = \begin{bmatrix} a_{11} & 0 & \cdots & 0 \\ 0 & a_{22} & \cdots & 0 \\ \vdots & \vdots & \ddots & \vdots \\ 0 & 0 & \cdots & a_{nn} \end{bmatrix} \tag{A.10}$$

An identity matrix, denoted as $\mathbf{I}$, is a square diagonal matrix when $a_{ij} = 1$ for all $i = j$ and $a_{ij} = 0$ for all $i \neq j$ with $\{i, j\} = 1, 2, \ldots, n$.

$$\mathbf{I} = \begin{bmatrix} 1 & 0 & \cdots & 0 \\ 0 & 1 & \cdots & 0 \\ \vdots & \vdots & \ddots & \vdots \\ 0 & 0 & \cdots & 1 \end{bmatrix} \tag{A.11}$$

A matrix $\mathbf{A}$ can be written in block forms as a matrix containing blocks of smaller matrices where the blocks are dictated by the application. For example, the matrix $\mathbf{A}$ defined as

$$\mathbf{A} = \begin{bmatrix} 1 & 2 & 3 \\ 4 & 5 & 6 \\ 7 & 8 & 9 \end{bmatrix} \tag{A.12}$$

The matrix $\mathbf{A}$ in Equation A.12 can be written in block form as

$$\mathbf{A} = \begin{bmatrix} \mathbf{A}_1 & \mathbf{A}_2 \\ \mathbf{A}_3 & \mathbf{A}_4 \end{bmatrix} \tag{A.13}$$

where

$$\mathbf{A}_1 = \begin{bmatrix} 1 & 2 \\ 4 & 5 \end{bmatrix}, \quad \mathbf{A}_2 = \begin{bmatrix} 3 \\ 6 \end{bmatrix}, \quad \mathbf{A}_3 = [7 \quad 8], \quad \text{and} \quad \mathbf{A}_4 = [9]$$

Transpose and product of block matrices can be performed easily by applying the rules on the blocks. Let $\mathbf{A}$ and $\mathbf{B}$ are two block matrices defined by

$$\begin{cases} \mathbf{A} = \begin{bmatrix} \mathbf{A}_1 & \mathbf{A}_2 \\ \mathbf{A}_3 & \mathbf{A}_4 \end{bmatrix} \\ \mathbf{B} = \begin{bmatrix} \mathbf{B}_1 & \mathbf{B}_2 \\ \mathbf{B}_3 & \mathbf{B}_4 \end{bmatrix} \end{cases} \tag{A.14}$$

The transpose of $\mathbf{A}$ can be obtained as

$$\mathbf{A}^T = \begin{bmatrix} \mathbf{A}_1^T & \mathbf{A}_2^T \\ \mathbf{A}_3^T & \mathbf{A}_4^T \end{bmatrix} \tag{A.15}$$

The product of $\mathbf{A}$ and $\mathbf{B}$ can be obtained as

$$\mathbf{AB} = \begin{bmatrix} \mathbf{A}_1\mathbf{B}_1 + \mathbf{A}_2\mathbf{B}_3 & \mathbf{A}_1\mathbf{B}_2 + \mathbf{A}_2\mathbf{B}_4 \\ \mathbf{A}_3\mathbf{B}_1 + \mathbf{A}_4\mathbf{B}_3 & \mathbf{A}_3\mathbf{B}_2 + \mathbf{A}_4\mathbf{B}_4 \end{bmatrix} \tag{A.16}$$

The maximum number of linearly independent columns of a matrix $\mathbf{A}$ is called the rank of the matrix denoted by rank($\mathbf{A}$). The k-th column of the matrix $\mathbf{A}$ in Equation A.3 is $\mathbf{a}_k$ given by

$$\mathbf{a}_k = \begin{bmatrix} a_{1k} \\ a_{2k} \\ a_{3k} \\ \vdots \\ a_{nk} \end{bmatrix} \tag{A.17}$$

A rank-one matrix is a matrix which can be written as $r = ab^T$ where a and b are vectors, which has at most one nonzero eigenvalue, and this eigenvalue can be calculated by $b^T a$.

A.3   NORMS, PRODUCTS, AND ORTHOGONALITY OF VECTORS

The Euclidean norm of an n-tuple vector $\mathbf{a} = [a_1 \quad a_2 \quad \ldots \quad a_n]^T$, denoted as $|\mathbf{a}|$, is defined by

$$|\mathbf{a}| = \sqrt{\mathbf{a}^T \mathbf{a}} \tag{A.18}$$

$|\mathbf{a}|$ is computed by using vector multiplication as follows:

$$|\mathbf{a}| = \left([a_1 \quad a_2 \quad \ldots \quad a_n] \begin{bmatrix} a_1 \\ a_2 \\ \vdots \\ a_n \end{bmatrix} \right)^{\frac{1}{2}} \tag{A.19}$$

The scalar product of two n-component vectors $\mathbf{a} = [a_1 \quad a_2 \quad \ldots \quad a_n]^T$ and $\mathbf{b} = [b_1 \quad b_2 \quad \ldots \quad b_n]^T$ is a scalar s defined by

$$s = \mathbf{a}^T \mathbf{b} = [a_1 \quad a_2 \quad \ldots \quad a_n] \begin{bmatrix} b_1 \\ b_2 \\ \vdots \\ b_n \end{bmatrix} = a_1 b_1 + a_2 b_2 + \cdots + a_n b_n = \sum_{i=1}^{n} a_i b_i \tag{A.20}$$

The scalar product of two vectors is commutative, that is,

$$\mathbf{a}^T \mathbf{b} = \mathbf{b}^T \mathbf{a} \tag{A.21}$$

Two n-component vectors $\mathbf{a} = [a_1 \quad a_2 \quad \ldots \quad a_n]^T$ and $\mathbf{b} = [b_1 \quad b_2 \quad \ldots \quad b_n]^T$ are said to be orthogonal if and only if their scalar product is zero, that is,

$$\mathbf{a}^T \mathbf{b} = \mathbf{b}^T \mathbf{a} = 0 \tag{A.22}$$

The angle between two n-component vectors $\mathbf{a} = [a_1 \quad a_2 \quad \ldots \quad a_n]^T$ and $\mathbf{b} = [b_1 \quad b_2 \quad \ldots \quad b_n]^T$ can be obtained by

$$\cos \psi = \frac{\mathbf{a}^T \mathbf{b}}{|\mathbf{a}| \cdot |\mathbf{b}|} \tag{A.23}$$

A.4   AFFINE FUNCTIONS

A function $\mathbf{F}$ is called affine if there exist a linear function $\Gamma: R^n \to R^m$ and a vector constant $\mathbf{c}$ such that

$$\mathbf{F} = \Gamma(\mathbf{x}) + \mathbf{c} \tag{A.24}$$

A function $\Gamma(\mathbf{x})$ is called linear if for any vectors $\mathbf{x}$ and $\mathbf{y}$, and any scalar $\alpha \in R$, it satisfies

$$\begin{cases} \Gamma(\mathbf{x} + \mathbf{y}) = \Gamma(\mathbf{x}) + \Gamma(\mathbf{y}) \\ \quad \Gamma(\alpha \mathbf{x}) = \alpha \Gamma() \end{cases} \tag{A.25}$$

In general, an affine function is a linear function with a translation, which can be written as a matrix form

$$\mathbf{F} = \mathbf{A}\mathbf{x} + \mathbf{c} \tag{A.26}$$

where $\mathbf{A}$ is an $m \times n$-dimensional matrix and $\mathbf{c}$ is a column vector in R^n.

A.5   QUADRATIC FORMS

Let $\mathbf{A} = [a_{ij}]$ be an $n \times n$-dimensional square matrix and $\mathbf{x} = [x_1 \quad x_2 \quad \ldots \quad x_n]^T$ be a column vector. A quadratic form $f: R^n \to R$ is a function in $\mathbf{x}$ with matrix $\mathbf{A}$ defined as

$$f(\mathbf{x}) = \mathbf{x}^T \mathbf{A} \mathbf{x} \tag{A.27}$$

A can be assumed symmetric, that is, $A = A^T$, without any loss of generality. If $\mathbf{A}$ is not symmetric, it can always be replaced with the symmetric matrix as follows:

$$\mathbf{A}_0 = \mathbf{A}_0^T = \frac{1}{2}(\mathbf{A} + \mathbf{A}^T) \tag{A.28}$$

which leads to

$$\mathbf{x}^T \mathbf{A} \mathbf{x} = \mathbf{x}^T \mathbf{A}_0 \mathbf{x} = \mathbf{x}^T \left(\frac{1}{2}\mathbf{A} + \frac{1}{2}\mathbf{A}^T \right) \mathbf{x} \tag{A.29}$$

Quadratic form $f(\mathbf{x})$ of a vector variable $\mathbf{x}$ is of fundamental importance for studying multi-input and multi-output systems. In short, the quadratic form can be written as

$$\mathbf{x}^T \mathbf{A} \mathbf{x} = \sum_{i=1}^{n} \sum_{j=1}^{n} a_{ij} x_i x_j \tag{A.30}$$

A.6   LINEAR SYSTEMS

Consider a linear system of n equations with m unknowns $x_1, x_2, \ldots, x_m$ defined by

$$\begin{cases} a_{11}x_1 + a_{12}x_2 + \cdots + a_{1m}x_m & = & b_1 \\ a_{21}x_1 + a_{22}x_2 + \cdots + a_{2m}x_m & = & b_2 \\ \quad\vdots & \ddots & \vdots \\ a_{n1}x_1 + a_{n2}x_2 + \cdots + a_{nm}x_m & = & b_n \end{cases} \tag{A.31}$$

where $a_{ij} \in R$, $i = 1, 2, \ldots, n$ and $j = 1, 2, \ldots, m$ are the coefficients of the unknowns $x_1, x_2, \ldots, x_m$.

The system of linear equations can be written in the vector–matrix form as follows:

$$\mathbf{A}\mathbf{x} = \mathbf{b} \tag{A.32}$$

where $\mathbf{A}$ is the known coefficient matrix (also called system matrix), $\mathbf{b}$ is the known vector, and $\mathbf{x}$ is the vector of unknowns or parameter defined as

$$\mathbf{A} = \begin{bmatrix} a_{11} & a_{12} & \cdots & a_{1m} \\ a_{21} & a_{22} & \cdots & a_{2m} \\ \vdots & \vdots & \ddots & \vdots \\ a_{n1} & a_{n2} & \cdots & a_{nm} \end{bmatrix} \tag{A.33}$$

$$\mathbf{b} = \begin{bmatrix} b_1 \\ b_2 \\ \vdots \\ b_n \end{bmatrix} \tag{A.34}$$

$$\mathbf{x} = \begin{bmatrix} x_1 \\ x_2 \\ \vdots \\ x_n \end{bmatrix} \tag{A.35}$$

The system described by the linear equations of the form $\mathbf{Ax} = \mathbf{b}$ is called a linear system. The linear system has a solution if and only if

$$\text{rank}(\mathbf{A}) = \text{rank}[\mathbf{A}, \mathbf{b}] \tag{A.36}$$

To identify uniquely the unknown parameter vector $\mathbf{x} = [x_1 \quad x_2 \quad \ldots \quad x_n]^T$, it is a necessary condition that $n \geq m$, that is, the number of equations is greater than or equal to the number of unknowns. If $\mathbf{A}$ is square (i.e., $n = m$) and nonsingular matrix (i.e., $\det(\mathbf{A}) \neq 0$), then the solution of the linear system $\mathbf{Ax} = \mathbf{b}$ is given by

$$\mathbf{x} = \mathbf{A}^{-1}\mathbf{b} \tag{A.37}$$

If $\mathbf{A}$ is full column rank ($\text{rank}(\mathbf{A}) = m \leq n$), that means ($\mathbf{A}^T\mathbf{A}$) is nonsingular, then the (Moore–Penrose) pseudo-inverse of $\mathbf{A}$ (denoted as $\mathbf{A}^+$) is defined by $A^+ = (\mathbf{A}^T\mathbf{A})^{-1}\mathbf{A}^T$ and the solution to Equation A.32 is given by an estimate of $\hat{\mathbf{x}}$:

$$\hat{\mathbf{x}} = (\mathbf{A}^T\mathbf{A})^{-1}\mathbf{A}^T\mathbf{b} = \mathbf{A}^+\mathbf{b} \tag{A.38}$$

If $\mathbf{A}$ is full row rank ($\text{rank}(\mathbf{A}) = n \leq m$), that means ($\mathbf{AA}^T$) is nonsingular, then the pseudo-inverse $\mathbf{A}^+$ is defined by $A^+ = \mathbf{A}^T(\mathbf{AA}^T)^{-1}$ and the solution to Equation A.32 is given by an estimate of $\hat{\mathbf{x}}$:

$$\hat{\mathbf{x}} = \mathbf{A}^T(\mathbf{AA}^T)^{-1}\mathbf{b} = \mathbf{A}^+\mathbf{b} \tag{A.39}$$

Practical problem of the estimate of $\hat{\mathbf{x}}$ is that a solution is not always guaranteed as data used for estimation may not be representative of the whole input space or may be data collected using noisy measurements, and data cannot be representative of all the features of the data that will be presented to the model. Also, the collected data may contain leaking values or some very extreme values that are unlikely to occur during the execution/usage of the model. Removal of the wrong values will increase the capabilities of the model to be better behaved and to better capture the underlying regularities. A very common and practical problem in observed sample or measured data is missing data. Another very frequent situation in data distribution is that the data can be concentrated in a very small portion of the input range, that input to the system may have very little effect on the resulting model. It means that the singularity of the term $(\mathbf{A}^T\mathbf{A})$ or $(\mathbf{A}\mathbf{A}^T)$ cannot be guaranteed always. A weighting is therefore essential for a robust estimation. In Equations A.38 and A.39, the different observations are given equal weight. The observation could be of varying reliability. Some observation could, for example, be subject to more disturbances and should therefore be penalized or down weighted. The observation could be of varying relevance. It is perhaps not believed that a linear model holds over all ranges of $\mathbf{A}$. An observation, corresponding to $\mathbf{A}$ in such a questionable region, even if accurate, should therefore carry less weight. That means the term $(\mathbf{A}^T\mathbf{A})$ or $(\mathbf{A}\mathbf{A}^T)$ may suffer from singularity. In this case, a weighting matrix $\mathbf{Q}$ is introduced as

$$\hat{\mathbf{x}}_Q = (\mathbf{A}^T\mathbf{Q}\mathbf{A})^{-1}\mathbf{A}^T\mathbf{Q}\mathbf{b} \tag{A.40}$$

$$\hat{\mathbf{x}}_Q = \mathbf{A}^T(\mathbf{A}\mathbf{Q}\mathbf{A}^T)^{-1}\mathbf{Q}\mathbf{b} \tag{A.41}$$

The weighting matrix $\mathbf{Q}$ is symmetric and positive definite. The elements of $\mathbf{Q}$ are chosen arbitrarily so that a robust solution can be obtained by ensuring the nonsingularity of the system in Equations A.40 and A.41.

Usually, $n > m$ indicates that there are more data pairs than the number of parameters. In this case, an exact solution satisfying all the n equations is not always possible because there might be noise in the data or the model is not representative of the target system. In order to address such issues, Equation A.32 is modified by incorporating an error vector $\mathbf{e}$ that accounts for model error or random noise as follows:

$$\mathbf{A}\mathbf{x} + \mathbf{e} = \mathbf{b} \tag{A.42}$$

Instead of looking for an exact solution to Equation A.32, an estimation of $\mathbf{x} = \hat{\mathbf{x}}$ is searched, which minimizes the sum of squared error defined by

$$E(\mathbf{x}) = \sum_{i=1}^{n}\left(b_i - a_i^T\mathbf{x}\right)^2 = \mathbf{e}^T\mathbf{e} = (\mathbf{b} - \mathbf{A}\mathbf{x})^T(\mathbf{b} - \mathbf{A}\mathbf{x}) \tag{A.43}$$

where $\mathbf{e} = (\mathbf{b} - \mathbf{A}\mathbf{x})$ is the error vector depending on specific choice of $\mathbf{x}$. $E(\mathbf{x})$ in Equation A.43 is in quadratic form and has a unique minimum at $\mathbf{x} = \hat{\mathbf{x}}$.

A.7   GRADIENT OF FUNCTIONS

Let $f(\mathbf{x})$ be a scalar function where $\mathbf{x}$ is defined as $\mathbf{x} = [x_1, x_2, \ldots, x_n]^T$. The derivative of $f(\mathbf{x})$ with respect to $\mathbf{x}$, called the gradient vector of gradient of $f(\mathbf{x})$, is defined as

$$\nabla f(\mathbf{x}) = \begin{bmatrix} \partial f(\mathbf{x})/\partial x_1 \\ \partial f(\mathbf{x})/\partial x_2 \\ \vdots \\ \partial f(\mathbf{x})/\partial x_n \end{bmatrix} \tag{A.44}$$

Let $f(\mathbf{x})$ be a vector function of $\mathbf{x} = [x_1, x_2, \ldots, x_n]^T$ defined by $f(\mathbf{x}) = [f_1(\mathbf{x}) \quad f_2(\mathbf{x}) \quad \ldots \quad f_m(\mathbf{x})]$. The derivative of $f(\mathbf{x})$ with respect to $\mathbf{x}$, called the Jacobian matrix or Jacobian of $f(\mathbf{x})$, is defined as

$$\mathbf{J}_f = \begin{bmatrix} \dfrac{\partial f_1}{\partial x_1} & \dfrac{\partial f_1}{\partial x_2} & \cdots & \dfrac{\partial f_1}{\partial x_n} \\[2mm] \dfrac{\partial f_2}{\partial x_1} & \dfrac{\partial f_2}{\partial x_2} & \cdots & \dfrac{\partial f_2}{\partial x_n} \\[2mm] \vdots & \vdots & \ddots & \vdots \\[2mm] \dfrac{\partial f_m}{\partial x_1} & \dfrac{\partial f_m}{\partial x_2} & \cdots & \dfrac{\partial f_m}{\partial x_n} \end{bmatrix} = \begin{bmatrix} \nabla^T f_1(\mathbf{x}) \\[2mm] \nabla^T f_2(\mathbf{x}) \\[2mm] \vdots \\[2mm] \nabla^T f_m(\mathbf{x}) \end{bmatrix} \tag{A.45}$$

Let $f(\mathbf{x})$ be a scalar function where $\mathbf{x}$ is defined as $\mathbf{x} = [x_1, x_2, \ldots, x_n]^T$. The second derivative of $f(\mathbf{x})$ with respect to $\mathbf{x}$ is called the Hessian of $f(\mathbf{x})$. The Hessian of $f(\mathbf{x})$ is an $n \times n$ matrix denoted as

$$H_f = \begin{bmatrix} \dfrac{\partial^2 f}{\partial x_1^2} & \dfrac{\partial^2 f}{\partial x_1 \partial x_2} & \cdots & \dfrac{\partial^2 f}{\partial x_1 \partial x_n} \\[2mm] \dfrac{\partial^2 f}{\partial x_2 \partial x_1} & \dfrac{\partial^2 f}{\partial x_2^2} & \cdots & \dfrac{\partial^2 f}{\partial x_2 \partial x_n} \\[2mm] \vdots & \vdots & \ddots & \vdots \\[2mm] \dfrac{\partial^2 f}{\partial x_n \partial x_1} & \dfrac{\partial^2 f}{\partial x_n \partial x_2} & \cdots & \dfrac{\partial^2 f}{\partial x_n^2} \end{bmatrix} \tag{A.46}$$

The Hessian can be expressed in other way as follows:

$$H_f = \begin{bmatrix} \dfrac{\partial}{\partial x_1}\left(\dfrac{\partial f}{\partial x_1}\right) & \dfrac{\partial}{\partial x_2}\left(\dfrac{\partial f}{\partial x_1}\right) & \cdots & \dfrac{\partial}{\partial x_n}\left(\dfrac{\partial f}{\partial x_1}\right) \\[2mm] \dfrac{\partial}{\partial x_1}\left(\dfrac{\partial f}{\partial x_2}\right) & \dfrac{\partial}{\partial x_2}\left(\dfrac{\partial f}{\partial x_2}\right) & \cdots & \dfrac{\partial}{\partial x_n}\left(\dfrac{\partial f}{\partial x_2}\right) \\[2mm] \vdots & \vdots & \ddots & \vdots \\[2mm] \dfrac{\partial}{\partial x_1}\left(\dfrac{\partial f}{\partial x_n}\right) & \dfrac{\partial}{\partial x_2}\left(\dfrac{\partial f}{\partial x_n}\right) & \cdots & \dfrac{\partial}{\partial x_n}\left(\dfrac{\partial f}{\partial x_n}\right) \end{bmatrix} \tag{A.47}$$

The other common notation used for Hessian is $\nabla^2 f(\mathbf{x})$ or $((\partial^2 f(\mathbf{x}))/(\partial \mathbf{x} \partial \mathbf{x}^T))$. The Hessian H_f is symmetric if the function $f(\mathbf{x})$ is continuous.

Appendix B: Random Numbers

A random number is a number chosen from some specified distribution such as Gaussian and Cauchy. The selection of a large set of these numbers should also be able to reproduce the distribution again. Such random numbers are required to be independent so that there are no correlations between successive numbers. Random numbers are generated using a computational or physical device that is capable of generating a sequence of numbers that do not have any pattern. All random numbers generated by computers are sometimes called pseudo-random numbers, while the term random is reserved for the output of unpredictable physical processes.

A random number generator requires an initial value, also called a seed, as the starting point. The goodness of random numbers generated by a given algorithm can be analyzed by examining its noise sphere. Random numbers have huge applications in scientific computing, simulation, sampling, engineering, business, and gambling. These applications are sensitive to random numbers. Therefore, generating random numbers is challenging and demands use of clever methods. Methods for generating a sequence of random real numbers $U_n \in [0,1]$ with a uniform distribution will be considered. By generating integer numbers $X_n \in [1,m]$, the random real number U_n can be generated as follows:

$$U_n = \frac{X_n}{m} \qquad (B.1)$$

where m is some integer number.

There are several classes of random number generators such as congruential generators, Fibonacci generators, inverse transform generators, and twisters. The simplest of all these generators is the linear congruence generator. It was formulated by D.H. Lehmer using an iterative procedure. The desired sequence of random numbers $\langle X_n \rangle$ can be obtained by

$$X_{n+1} = (aX_n + c)\%m, \quad \text{with } n \geq 0 \qquad (B.2)$$

where a, c, and m are integers; $m > 0$ is the modulus; $0 \leq a < m$ is a multiplier; $0 \leq c < m$ is the increment; $0 \leq X_0 < m$ is the initial value; and $\langle X_n \rangle$ is called the linear congruent sequence. The starting or initial value X_0 is called the seed of the generator. The problem of congruent sequence $\langle X_n \rangle$ is that there is a possible maximum period of m (i.e., a cycle of numbers repeated in the sequence), which is attainable only if $c > 0$, m are relatively prime and $(a - 1)$ is a multiple of 4. This is a common problem of random sequence generators of the form $X_{n+1} = f(X_n)$ as $f(\cdot)$ transforms a finite set into itself.

An improvement to the linear congruent sequence in Equation B.2 is the quadratic congruent method defined by

$$X_{n+1} = (b[X_n]^2 + aX_n + c)\%m \qquad (B.3)$$

where $b > 0$ and the necessary and sufficient conditions for a, b, and c are chosen such that Equation B.3 has a period of the maximum length m.

An interesting quadratic congruent sequence is defined by using m as a power of 2 as follows:

$$X_{n+1} = X_n(X_n + 1)\%2^e \qquad (B.4)$$

The generators in Equations B.1 through B.4 use extensively multiplications and divisions, which are computationally expensive. If additions and subtractions are used in the formulations, the generators will be computationally efficient. Lagged Fibonacci pseudo-random generator uses the following formulation:

$$X_{n+1} = (X_n + X_{n-1})\%m \tag{B.5}$$

where the sequence X_{n+1} depends on more than one of the preceding values. The sequence in Equation B.5 gives a period length greater than m.

MATLAB provides a number of functions to generate random numbers, for example, functions rand(), randn(), and randi() can be used to create sequences of pseudo-random numbers. The function rng() can be used to control the repeatability of results. The function RandStream class is used when more advanced control over random number generation is required.

Appendix C: Chaotic Maps

Chaotic map is an evolution function that exhibits chaotic behavior. Such maps can be parameterized using a discrete-time or continuous-time parameter. Chaotic maps are very useful in the study of dynamical systems and simulation of algorithms. There are many chaotic maps used in chaotic search. Some of the chaotic maps are discussed in brief.

C.1 HENON MAP

An important example of an algebraically defined two-dimensional mapping is the Henon map defined by

$$\begin{cases} x_{n+1} = 1 - a \cdot x_n^2 + y_n \\ y_{n+1} = b \cdot x_n \end{cases} \tag{C.1}$$

where $x_n \in [0,1]$ is the n-th chaotic number, n is the iteration number, and a and b are arbitrary parameters. The suggested parameter values are $a = 1.4$ and $b = 0.3$. The Henon map is a nonlinear two-dimensional map. It is employed for testing purposes most frequently. When $b = 0$, all points in the two-dimensional space (i.e., plane) are mapped onto the x-axis. Thus, the two-dimensional Henon map is converted into one-dimensional map defined by

$$x_{n+1} = 1 - a \cdot x_n^2 \tag{C.2}$$

The map in Equation C.2 can be written as

$$x_{n+1} = c - a \cdot x_n^2 \tag{C.3}$$

where c is a scaling parameter.

The one-dimensional maps in Equations C.2 and C.3 are also known as quadratic maps. The quadratic maps have strikingly rich chaotic behavior though they are defined very simply.

C.2 LOGISTIC MAP

Another variant of the quadratic map is the logistic map. Logistic map appears in nonlinear dynamics of biological population. The chaotic behavior logistic map is defined by

$$x_{n+1} = a \cdot x_n (1 - x_n) \tag{C.4}$$

where $x_n \in [0,1]$ is the n-th chaotic number, n is the iteration number, $x_0 \in [0,1]$ is the initial number, and a is an arbitrary parameter.

The quadratic map and the logistic map are two examples of a wider class of maps that are topologically equivalent unimodal one-dimensional maps. Both of them are defined on finite interval and have a single smooth maximum.

C.3   TENT MAP

Tent map is very much similar to logistic map that generates chaotic sequences within the interval of [0,1]. The chaotic behavior of tent map is defined by

$$x_{n+1} = \begin{cases} x_n/0.7 & x_n < 0.7 \\ 10/3x_n(1-x_n) & \text{otherwise} \end{cases} \tag{C.5}$$

The tent map and the logistic map are topologically conjugate, and thus, the behaviors of the two maps are in this sense identical under iteration.

C.4   SINUSOIDAL ITERATOR

The chaotic behavior of sinusoidal iterator is defined by

$$x_{n+1} = a \cdot x_n^2 \sin(\pi \cdot x_n) \tag{C.6}$$

where a is an arbitrary parameter. For $a = 2.3$ and the initial number $x_0 = 0.7$, the simplified version of sinusoidal iterator is given by

$$x_{n+1} = \sin(\pi \cdot x_n) \tag{C.7}$$

Sinusoidal iterator generates chaotic sequence within the interval of [0,1].

C.5   GAUSS MAP

The chaotic behavior of the Gauss map is defined by

$$x_{n+1} = \begin{cases} 0 & x_n = 0 \\ 1/x_n \bmod(1) & \text{otherwise} \end{cases} \tag{C.8}$$

The term $1/x_n \bmod(1)$ is further defined by

$$1/x_n \bmod(1) = \frac{1}{x_n} - \left\lfloor \frac{1}{x_n} \right\rfloor \tag{C.9}$$

The Gauss map generates chaotic sequence within the interval of [0,1].

C.6   CIRCLE MAP

The chaotic behavior of circle map is defined by

$$x_{n+1} = x_n + b - (a/2\pi)\sin(\pi \cdot x_n) \bmod(1) \tag{C.10}$$

where a and b are arbitrary parameters. For $a = 0.5$ and $b = 0.2$, circle map generates a chaotic sequence within the range of [0,1].

C.7 SINUS MAP

The chaotic behavior of sinus map is defined by

$$x_{n+1} = 2.3(x_n)^{2\sin(\pi \cdot x_n)} \tag{C.11}$$

C.8 IKEDA MAP

Ikeda map is a discrete-time dynamical system defined by

$$\begin{cases} x_{n+1} = 1 + 0.7(x_n \cos(\theta_n) - y_n \sin(\theta_n)) \\ y_{n+1} = 0.7(x_n \sin(\theta_n) + y_n \cos(\theta_n)) \end{cases} \tag{C.12}$$

where θ_n is defined as

$$\theta_n = 0.4 - \frac{6}{1 + x_n^2 + y_n^2} \tag{C.13}$$

C.9 LIEBOVITCH AND TOTH MAP

The Liebovitch and Toth map (known as Liebovitch map) consists of three piecewise linear segments on nonoverlapping subintervals within the range of [0,1]. The chaotic behavior of the Liebovitch map is defined by

$$x_{n+1} = \begin{cases} \alpha_1 x_n & 0 < x_n \le d_1 \\ \dfrac{d_2 - x_n}{d_2 - d_1} & d_1 < x_n \le d_2 \\ 1 - \alpha_2(1 - x_n) & d_1 < x_n \le 1 \end{cases} \tag{C.14}$$

where $d_1, d_2 \in (0,1)$ with $d_1 < d_2$. α_1 and α_2 are defined as

$$\alpha_1 = \frac{d_2}{d_1}[1 - (d_2 - d_1)] \tag{C.15}$$

$$\alpha_2 = \frac{1}{d_2 - 1}[(d_2 - 1) - d_1(d_2 - d_1)] \tag{C.16}$$

C.10 ZASLAVSKII MAP

The Zaslavskii map is a discrete-time dynamical system introduced by George M. Zaslavsky. It is an example of a dynamical system that exhibits very interesting chaotic behavior defined by

$$\begin{cases} x_{n+1} = [x_n + v + a y_{n+1}](\bmod 1) \\ y_{n+1} = \cos(2\pi x_n) + e^{-r} y_n \end{cases} \tag{C.17}$$

where mod is the modulus after division and $v = 400$, $r = 3$, and $a = 12.6695$.

C.11 CAT MAPPING

Cat mapping function has good ergodic uniformity in the interval of [0,1] and does not easily fall into a minor cycle. Two-dimensional Cat mapping function is described by

$$\begin{cases} x_{n+1} = (x_n + y_n) \bmod 1 \\ y_{n+1} = (x_n + 2y_n) \bmod 1 \end{cases} \tag{C.18}$$

where $(x \bmod 1) = x - [x]$.

Appendix D: Optimization

In general, an optimization problem requires finding a set of parameters $\mathbf{x} \in M$ of a system under consideration such that a certain quality criterion $f\colon M \to R$ (also called objective function) is maximized (or equivalently minimized), that is,

$$f(\mathbf{x}) \to \max \tag{D.1}$$

The global optimization problem in Equation D.1 requires finding a vector $\mathbf{x}$ such that

$$\forall \mathbf{x} \in M : f(\mathbf{x}) \le f(\mathbf{x}^*) = f^* \tag{D.2}$$

Multimodality is a common characteristic in optimization problems, that is, there can be several local maxima $\mathbf{x}'$ such that

$$\exists \varepsilon > 0 : \forall \mathbf{x} \in M : \rho(\mathbf{x}, \mathbf{x}') < \varepsilon \Rightarrow f(\mathbf{x}) \le f(\mathbf{x}') \tag{D.3}$$

where $\rho(\cdot)$ denotes a distance measure on M.

Constraints (i.e., restrictions) on the set M by functions $g_j\colon M \to R$ such that the set of feasible solutions $F \subseteq M$ yields only a subset of the domain of the variables

$$F = \{\mathbf{x} \in M \mid g_j(\mathbf{x}) \ge 0 \ \forall j\} \tag{D.4}$$

D.1 MULTIOBJECTIVE OPTIMIZATION

Real-world problems are complex, and the definition of optimality is not simple as they need to satisfy multiple competing objective functions at the same time. Moreover, some of these objectives may have conflicting relations with others, which in fact makes the optimization a difficult task. Problems requiring simultaneous optimization of more than one objective functions are known as MOOP. They can be defined as the problems consisting of multiple objectives, which are to be minimized or maximized while maintaining some constraints. Formally, it can be defined as follows:

$$\text{Minimize/maximize } f(x) \tag{D.5}$$

$$\text{Subject to } g_j(x) \ge 0, \quad j = 1, 2, 3, \ldots, J \tag{D.6}$$

$$h_k(x) = 0, \quad k = 1, 2, 3, \ldots, K \tag{D.7}$$

where $f(x) = \{f_1(x), f_2(x), \ldots, f_n(x)\}$ is a vector of objective functions, $x = \{x_1, x_2, \ldots, x_p\}$ is a vector of decision variables, n is the number of objectives, and p is the number of decision variables. Here, the problem optimizes n objectives and satisfies J inequality and K equality constraints. This type of problem has no unique perfect solution. In traditional multiobjective optimization, it is very common to simply aggregate all the objectives together to form a single (scalar) fitness function. However, the obtained solution using a single scalar is sensitive to the weight vector used in the scaling process. This requires knowledge about the underlying problem which is not known *a priori* in most of the cases. Moreover, the objectives can interact or conflict with each other. Therefore, trade-offs are sought when dealing with such MOOPs rather than a single solution. Most MOOPs do not provide a single solution; rather, they offer a set of solutions. Such solutions are the "trade-offs" or good compromises among the objectives. In order to generate these trade-off solutions, an old notion of optimality called Pareto-optimum set is normally adopted.

In multiobjective optimization, the definition of quality of solution is substantially more complex than for single-objective optimization problems. The main challenges in a multiobjective optimization environment are as follows: converge as close as possible to the Pareto-optimal front, and maintain as diverse a set of solutions as possible. The first task ensures that the obtained set of solutions is near optimal, while the second task ensures that a wide range of trade-off solutions is obtained.

Due to the advantageous features of derivative-freeness and population-based approach to solutions of optimization problems, EAs are applied in MOOPs and the combination became known as multiobjective evolutionary algorithm (MOEA). An MOEA will be considered good only if both of the goals of convergence and diversity are satisfied simultaneously. MOEA's population-based approach helps preserving and emphasizing nondominated diverse set of solutions in a population. MOEA converges to the Pareto-optimal front with a good spread of solutions in some reasonable number of generations. Most MOEAs use the concept of domination to attain the set of Pareto-optimum solutions. In the total absence of information for preferences of the objectives, solutions to multiobjective problems are compared using the notion of Pareto dominance. For the problems having more than one objective function, any two solutions $x^{(1)}$ and $x^{(2)}$ can have one of the two possibilities: one dominates the other, or none dominates the other. A particular solution $x^{(1)}$ with performance vector u is said to be dominated or better than the solution $x^{(2)}$ with performance vector v, if both the following conditions hold: (i) the solution $x^{(1)}$ is no worse than $x^{(2)}$ in all objectives and (ii) the solution $x^{(1)}$ is strictly better than $x^{(2)}$ is at least one objective. This notion can be generalized in the following equation:

$$u \prec v \quad \text{iff} \quad [\forall i \in \{1, 2, \ldots, n\}, u_i \leq v_i] \cap [\exists i \subset \{1, 2, \ldots, n\} \mid u_i < v_i] \tag{D.8}$$

where it holds $u \prec v \Leftrightarrow x^{(1)} \prec x^{(2)}$. For a given finite set of solutions, we need to perform pairwise comparisons to find out which solutions dominate and which are dominated by each other. From these comparisons, we can find a subset of the finite set of solutions such that any two solutions of which do not dominate each other and all the other solutions of the finite set are dominated by one or more members of this subset. This subset is called the nondominated set for the given set of solutions. A solution is said to be the Pareto optimal if it is not dominated by any other possible solution. This is described as follows:

$$x^{(1)} \in x_{PO} \quad \text{iff} \quad \nexists x^{(2)} \in \Psi \mid x^{(2)} \prec x^{(1)} \tag{D.9}$$

where x_{PO} is the set of the Pareto-optimal solutions and Ψ is the set of all feasible solutions. The Pareto front is the set of points in the criterion space that corresponds to the Pareto-optimal solutions.

In MOEA, a randomly selected population is generated within a specific range. Each individual of the population is evaluated with the objective functions. Figure D.1 shows many solutions trading-off differently between the objectives for a two-objective minimization problem. Any two solutions from the feasible objective space can be compared. For a pair of solutions, it can be seen that one solution is better than the other in the first objective but worse in the second objective. The individuals that fall close to either axes or origin of two-dimensional objective space are better than those away from axes or origin. In the objective space, some individuals may be found, such as the individuals denoted by E, A, G, and F in Figure D.1, falling on outer edge and close to axes or origin and having one objective better than another. For clarity, these individuals are joined by a dotted

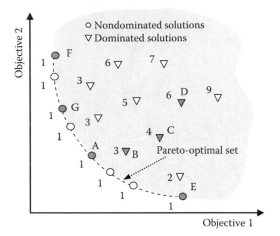

FIGURE D.1 Dominated and nondominated solutions with ranking.

line in Figure D.1. All these individuals lying on this curve form a set called nondominated solution set or the Pareto-optimal set. The curve formed by joining these solutions is called Pareto-optimal front.

Individuals A, E, F, and G are called nondominated because no other individuals provide better performance in the objective space. On the other hand, individuals, falling away from edges, such as B, C, and D, are called dominated solutions since many individuals provide better performance than these in terms of both objectives. The dominated and nondominated solutions are shown in Figure D.1. For example, individual A dominates individual B, similarly B dominates C, and C dominates D in the objective space in terms of both objectives. In the process, each individual is ranked according to their degree of dominance. An individual's ranking equals the number of individuals better than that in terms of both objectives plus one. Individuals on the Pareto-optimal front (denoted with a small circle) are nondominated and have a ranking of one. The individuals inside the Pareto-optimal front (denoted with a small triangle) have higher ranking than one. The numbers shown in Figure D.1 correspond to their ranking. The main goal of an ideal multiobjective optimization is to find as many Pareto-optimal solutions as possible.

Appendix E: Probability Distribution Function

Probability distribution is typically defined in terms of PDF. In general, PDF has two parameters: location and scale parameter. There is a number of distribution functions used in modeling and statistical applications. Some widely used distributions are discussed in the following.

E.1 UNIFORM DISTRIBUTION

A random variable is said to be uniformly distributed over [a,b] if its probability density is constant within the interval [a,b] and 0 outside. The PDF of uniform distribution is given by

$$p(x) = \frac{1}{(b-a)} \quad \text{for } a \leq x \leq b \tag{E.1}$$

For the case $a = 0$ and $b = 1$, the distribution is called standard uniform distribution with $p(x) = 1$ for $0 \leq x \leq 1$.

E.2 NORMAL OR GAUSSIAN DISTRIBUTION

The Gaussian distribution is a continuous function that approximates the exact binomial distribution of events. A Gaussian distribution with mean μ, standard deviation σ, and variance σ^2 is a statistical distribution over the interval $x \in [-\infty, +\infty]$ with PDF defined by

$$p(x) = \frac{1}{\sqrt{2\pi}\sigma} \exp\left[-\frac{(x-\mu)^2}{2\sigma^2} \right] \tag{E.2}$$

It is one of the most commonly used probability distribution function and widely known as normal distribution. De Moivre developed the normal distribution as an approximation to the binomial distribution. It was later used by Gauss in 1809 for analysis of astronomical data.

E.3 LOG NORMAL DISTRIBUTION

Log normal distribution is a continuous distribution in which the logarithm of a variable, for example, $\ln(x)$, has a normal distribution. The PDF of log normal distribution is given by

$$p(x) = \frac{1}{\sigma\sqrt{2\pi}x} \exp\left[-\frac{(\ln x - \mu)^2}{2\sigma^2} \right] \tag{E.3}$$

E.4 CAUCHY DISTRIBUTION

The Cauchy distribution is a continuous distribution, which describes resonance behavior. It is also called the Lorentz distribution. The PDF of the Cauchy distribution is given by

$$p(x) = \frac{1}{\pi\beta} \frac{1}{1 + \left(\dfrac{x-\mu}{\beta}\right)^2} \tag{E.4}$$

where μ is the location parameter (or mean) and β is the scale parameter.

The Cauchy distribution looks similar to normal distribution, but the Cauchy distribution has fatter tail.

E.5 BOLTZMANN DISTRIBUTION

The Boltzmann distribution describes the statistical distribution of energy E among particles in a system. The PDF of the Boltzmann distribution is given by

$$p(E) = A \exp\left[-\frac{E}{k_B T}\right] \tag{E.5}$$

where A is the amplitude of the distribution, k_B is the Boltzmann constant, and T is the temperature.

E.6 MAXWELL–BOLTZMANN DISTRIBUTION

The Maxwell–Boltzmann distribution is the classical distribution, which describes the amount of energy E among identical and distinguishable particles in a system. The PDF of the Maxwell–Boltzmann distribution is given by

$$p(E) = \frac{1}{A \exp\left[\dfrac{E}{k_B T}\right]} \tag{E.6}$$

E.7 BOSE–EINSTEIN DISTRIBUTION

The Bose–Einstein distribution describes the quantum behavior of identical and indistinguishable particles with integer spin (Bosons). Bosons behave very differently at low temperature. The PDF of the Bose–Einstein distribution is given by

$$p(E) = \frac{1}{A \exp\left[\dfrac{E}{k_B T}\right] - 1} \tag{E.7}$$

E.8 FERMI–DIRAC DISTRIBUTION

The Fermi–Dirac distribution describes the quantum behavior of identical and indistinguishable particles with half integer spin (Fermions). The PDF of the Fermi–Dirac distribution is given by

$$p(E) = \frac{1}{A \exp\left[\dfrac{E}{k_B T}\right] + 1} \tag{E.8}$$

E.9 EXPONENTIAL DISTRIBUTION

A random variable is said to be exponentially distributed if its PDF is given by

$$p(x) = \frac{1}{\beta} e^{-(x-\mu)/\beta} \quad \text{for } x \geq 0 \tag{E.9}$$

where μ is the mean or location parameter and β is a scale parameter of the distribution. The scale parameter is often referred to as shape parameter λ with positive value defined as $\lambda = 1/\beta$. A good example of exponential distribution is the lifespan of an electric bulb.

E.10 BINOMIAL DISTRIBUTION

A random variable is said to be binomially distributed with parameters n and p when it takes the possible values 0, 1, ... , n with probabilities $p(x;p,n)$. Binomial distribution is used to obtain the probability of observing x successes in n trials when there are exactly two mutually exclusive outcomes of a trial. The parameters n and p determine the binomial distribution where n is the total number of trials and p is the outcome of each trial, that is, probability of success on a single trial. The PDF of binomial distribution is given by

$$p(x; p,n) = \binom{n}{x} p^x (1-p)^{n-x} \quad \text{for } x = 0,1,\dots,n \tag{E.10}$$

where

$$\binom{n}{x} = \frac{n!}{x!(n-x)!}$$

E.11 POISSON DISTRIBUTION

A random quantity is said to have the Poisson distribution if it takes denumerably many values 0, 1, 2, ... The PDF of the Poisson distribution is given by

$$p(x) = \frac{\lambda^x}{x!} e^{-\lambda} \quad \text{for } x = 0,1,2,\dots \tag{E.11}$$

where λ is a shape parameter of the distribution, which indicates the average number of events in the given time interval. The Poisson distribution is used to model the number of events occurring within a given time interval.

E.12 WEIBULL DISTRIBUTION

The PDF of the Weibull distribution is given by

$$p(x) = \frac{\lambda}{\beta} \left(\frac{x-\mu}{\beta} \right)^{(\lambda-1)} \exp\left[-\frac{(x-\mu)}{\beta} \right]^{\lambda} \tag{E.12}$$

where λ is a shape parameter, μ is the location parameter (i.e., mean), and β is scale parameter of the distribution. The Weibull distribution is widely used in reliability applications to model failure times of a system. When $\lambda = 1$, the Weibull distribution becomes exponential distribution.

Index

Note: Page numbers followed by "*fn*" indicate footnotes.